Stochastic Differential Equations
and
Diffusion Processes

Second Edition

North-Holland Mathematical Library

VOLUME 24

NORTH-HOLLAND PUBLISHING COMPANY
AMSTERDAM • OXFORD • NEW YORK

Stochastic Differential Equations and Diffusion Processes

Second Edition

by

Nobuyuki IKEDA
Department of Mathematics
Osaka University

Shinzo WATANABE
Department of Mathematics
Kyoto University

NORTH-HOLLAND PUBLISHING COMPANY
AMSTERDAM • OXFORD • NEW YORK

1989

KODANSHA LTD.
TOKYO

KODANSHA SCIENTIFIC BOOKS

ISBN 978-0-444-86172-6

Published in co-edition with North-Holland Publishing Company

Distributors for Japan:
KODANSHA LTD.
12-21 Otowa 2-chome, Bunkyo-ku, Tokyo 112, Japan

Distributors outside Japan, U.S.A. and Canada:
ELSEVIER SCIENCE PUBLISHERS B.V.
P. O. Box 103
1000 AC Amsterdam
The Netherlands

Sole distributors for the U.S.A. and Canada:
ELSEVIER SCIENCE PUBLISHING COMPANY, INC.
655 Avenue of the Americas
New York, NY 10010
U.S.A.

Transferred To Digital Printing 2011

It is with our deepest gratitude and warmest affection
that we dedicate this book
to our teacher

Kiyosi Itô

who has been a constant source of
knowledge and inspiration.

Preface to the Second Edition

A considerable number of corrections and improvements have been made in this second edition. In particular, major and substantial changes are in Chapter III and Chapter V where the sections treating on excursions of Brownian motion and the Malliavin calculus have been much more expanded and refined. Also, sections discussing complex (conformal) martingales and Kähler diffusions have been inserted.

We would express our sincere thanks to all those who kindly communicated us mistakes in and gave us helpful comments on the first edition of this book.

November 1988

N. IKEDA
S. WATANABE

Preface

The theory of stochastic integrals and stochastic differential equations was initiated and developed by K. Itô. In 1942 ([57], cf. also [62]) this theory was first applied to Kolmogorov's problem of determining Markov processes [85]. Let y_t be a Markov process on the real line $\boldsymbol{R}^1$ and, for each instant t_0, let $F_{t_0,t} = F_{t_0,t}(y_{t_0})$ be the conditional probability distribution of y_t given y_{t_0}. In almost all interesting cases we may assume that $F_{t_0,t}^{*[(t-t_0)^{-1}]}$ converges as $t \downarrow t_0$ to a probability distribution on $\boldsymbol{R}^1$, which we shall denote by Dy_{t_0}. (Here $[a]$ is the integer part of a and $*k$ denotes the k-fold convolution.) Dy_{t_0} is thus an infinitely divisible distribution. Kolmogorov's problem is now formulated as follows: given a system $L(t,y)$ of infinitely divisible distributions, find the Markov process y_t with given initial distribution such that

$$(0.1) \qquad Dy_t = L(t,y_t).$$

Kolmogorov [85] and Feller [26] both succeeded in obtaining Markov processes by solving Kolmogorov's differential equations (equations equivalent to (0.1) for the transition probabilities), thus establishing an analytical method in probability theory. This method has been further developed in connection with Hille-Yosida's theory of semigroups.

In contrast with these analytical methods, a probabilistic approach suggested by Lévy and established by Itô enables one to construct sample functions of the process y_t directly. Consider the case when $L(t,y) = G(a(t,y),b(t,y))$, where $G(\alpha,\beta)$ is the Gaussian distribution with mean α and standard deviation β. The intuitive meaning of (0.1) then is that the infinitesimal change of the conditional distribution given $y_t = y$ is $G(a(t,y)dt, b(t,y)\sqrt{dt})$. On the other hand, if x_t is a Brownian motion* (Wiener

* The Brownian motion is a random movement of microscopic particles which was discovered by the English botanist R.Brown. Its physical theory was investigated by Einstein ([24]). The mathematical theory was initiated and developed by Wiener [177] and Lévy [99].

process), the "stochastic differential"* $dx_t = x_{t+dt} - x_t$ conditioned on the past $\{x_s, s \leq t\}$ satisfies the law $G(0, \sqrt{dt})$. This suggests that we write

$$(0.2) \qquad dy_t = a(t,y_t)dt + b(t,y_t)dx_t;$$

thus y_t should be determined as a solution of the integral equation

$$(0.3) \qquad y_t = y_0 + \int_0^t a(s,y_s)\,ds + \int_0^t b(s,y_s)\,dx_s.$$

However, Wiener and Lévy had already shown that x_t is nowhere differentiable for almost all sample paths and so the integral with respect to dx_t cannot be defined in the ordinary sense. In order to get around this difficulty, Itô introduced the notion of "stochastic integrals". Using this idea he was then able to obtain y_t as the unique solution of (0.3) for a given initial value y_0 under a Lipschitz condition on $a(t,y)$ and $b(t,y)$; furthermore, he showed that this y_t actually satisfies the original equation (0.1). Independently in the Soviet Union, S. Bernstein ([4] and [5]) introduced a stochastic difference equation and showed that the random variable determined by this equation has the fundamental solution of Kolmogorov's equation as its limiting distribution. Gihman ([35], [36] and [37]) carried out Bernstein's program independently of Itô and succeeded in constructing a theory of stochastic differential equations.

Today Itô's theory is applied not only to Markov processes (diffusion processes) but also to a large class of stochastic processes. This framework provides us with a powerful tool for describing and analyzing stochastic processes. Since Itô's theory may be considered as an integral-differential calculus for stochastic processes, it is often called Itô's stochastic analysis or stochastic calculus.

The main aim of the present book is to give a systematic treatment of the modern theory of stochastic integrals and stochastic differential equations. As is customary nowadays, we will develop this theory within the martingale framework. The theory of martingales, which was initiated and developed by J.L. Doob, plays an indispensable role in the modern theory of stochastic analysis. The class of stochastic processes to which Itô's theory can be applied (usually called Itô processes or locally infinitely divisible processes) is now extended to a class of stochastic processes called *semimartingales*. Such processes appear to be the most gen-

---

* The notion of stochastic differentials was considered by Lévy ([99], [100] and [101]); he used a suggestive notation $\xi\sqrt{dt}$ to denote dx_t where ξ is a random variable with distribution $G(0,1)$.

eral for which a unified theory of stochastic calculus can be developed.* A somewhat different type of stochastic calculus has been introduced by Stroock and Varadhan under the name of *martingale problems* (cf. [160] for this elegant and powerful approach). In this book, however, we prefer to follow Itô's original approach although an influence of Stroock and Varadhan will be seen in many places.

Chapter I contains some preliminary materials that are necessary for the development and understanding of our book. In particular, we review the theory of martingales.

The notion of stochastic integrals and Itô's formula are discussed in Chapter II. These constitute the core of stochastic analysis.

In Chapter III, the results of Chapter II are reformulated in a more convenient form for applications. In particular, a type of stochastic integral introduced by Stratonovich [153] and Fisk [27] will be formulated as an operation (called *symmetric multiplication*) in the framework of the theory developed there. It will play an important role in Chapters V and VI.

The general theory of stochastic differential equations is discussed in Chapter IV. The solutions of these equations will not necessarily be non-anticipative functionals of the accompanying Brownian paths; we distinguish those solutions having this property as *strong solutions*. Stochastic differential equations are then used to construct diffusion processes for given differential operators and, in the case of a state space with boundary, for given boundary conditions. In discussing stochastic differential equations with boundary conditions, we are naturally led to stochastic differential equations which are based on semimartingales more general than Brownian motions and on Poisson random measures over general state spaces.

The main objects to be studied in Chapter V are the *flows of diffeomorphisms* defined on a differentiable manifold which are associated with a given system of vector fields. Stochastic calculus enables us to transfer many important operations in analysis and differential geometry which are defined on smooth curves through ordinary differential-integral calculus to a class of stochastic curves. Flows of diffeomorphisms are one such example. By making use of affine connections or Riemannian connections we can construct flows of diffeomorphisms over the space of frame bundles. The most general non-singular diffusion can then be obtained by projection. We can also realize Itô's stochastic parallel displacement of tensor fields by this method. These ideas are due to Eells, Elworthy and Malliavin. In discussing a similar problem in the case of a manifold with boundary, we

* The modern theory of semimartingales and the stochastic calculus based on them have been extensively developed in France by Meyer, Dellacherie, Jacod etc. (cf. Jacod [77]).

exhibit a probabilistic condition which characterizes the reflecting diffusion processes in the normal direction among all reflecting diffusion processes in oblique directions.

Many of the mathematical objects obtained through stochastic analysis,e.g.,strong solutions of stochastic differential equations, are functionals of the accompanying Brownian paths, i.e., *Brownian functionals* or *Wiener functionals.* In Sections 7 and 8 of Chapter V we introduce a recent work of Malliavin which analyzes solutions of stochastic differential equations as Wiener functionals. There he obtains the remarkable result that the smoothness problem for the heat equation can be treated by this probabilistic approach.

Some miscellaneous materials are presented in Chapter VI. In the first half we discuss some topics related to the *comparison theorem.* Although the results obtained by this method are usually a little weaker than those obtained via partial differential equations, the technique is simpler and it sometimes produces sharp results. The topics discussed in the second half are more or less related to the notion of *stochastic area* as introduced by Lévy. The approximation theorems considered in Section 7 are concerned with the transfer of concepts defined on smooth curves to stochastic curves.

There are many topics in the area of stochastic differential equations and their applications which are not considered in this book: e.g., the theory of stochastic control, filtering and stability, applications to limit theorems, applications to partial differential equations including non-linear equations, etc. We especially regret that we could not include the important works by N.V. Krylov on the estimates for the distributions of stochastic integrals and their applications (cf. e.g. [91] and [92]).

We wish to express our gratitude to G. Maruyama for his constant interest and encouragement during the writing of this book. We are also indebted to H. Kunita for his critical comments and constructive suggestions and to S. Nakao who greatly assisted us in the proofs of Theorems VI-7.1 and VI-7.2. Among the many others who have made contributions, we would particularly thank H. Asano, S. Kotani, S. Kusuoka, S. Manabe, Y. Okabe, H. Ôkura and I. Shigekawa for their useful comments, suggestions and corrections. Finally, we also express our appreciation to T.H. Savits who read the manuscript and suggested grammatical improvements, and to the editorial staff of Kodansha Scientific who provided much help in the publication of this book.

Osaka and Kyoto, February 1980

Nobuyuki IKEDA
Shinzo WATANABE

Contents

General Notation

Theorem IV-3.2, for example, means Theorem 3.2 (the second theorem in Section 3) in Chapter IV. If this theorem is quoted in Chapter IV, it is written as Theorem 3.2 only.

The following notations are frequently used.

$A := B$ means that A is defined by B or A is denoted by B.

$A(x) \equiv B(x)$	: $A(x)$ and $B(x)$ are identically equal, i.e. $A(x) = B(x)$ for all x.
$f\|_A$	: the restriction (of a function f) to a subset A of the domain of definition of f.
$A \triangle B$	: the symmetric difference of A and B, i.e., $(A \backslash B) \cup (B \backslash A)$.
I_A	: the indicator function of A, i.e., $I_A(x) = 1$ or 0 accordingly as $x \in A$ or $x \notin A$.
$a \vee b$	: the maximum of a and b.
$a \wedge b$	: the minimum of a and b.
$E(X:B)$	: the expectation of a random variable X on an event B, i.e., $\int_B X(\omega)P(d\omega) = E(XI_B)$.
$\delta_{ij}, \delta^{ij}, \delta^i_j$	: the Kronecker's δ.
$\delta_{(a)}$	: the unit (Dirac) measure at a.
l.i.p.	: limit in probability.
a.s.	: almost surely.
$t \longmapsto X(t)$, $[t \longmapsto X(t)]$	: the mapping $t \longrightarrow X(t)$.
$\boldsymbol{R} = \boldsymbol{R}^1$	: the real line.
$\boldsymbol{C}$	: the complex plane.
$\boldsymbol{R}^d$	: the d-dimensional Euclidean space.
$\boldsymbol{N}$	: the set of all positive integers.
$\boldsymbol{Z}$	: the set of all integers.
$\boldsymbol{Z}^+$	: the set of all non-negative integers.

$\boldsymbol{Q}$	:	the set of all rational numbers.
$[x]$	:	the largest integer not exceeding x.

"smooth" usually means C^∞(i.e., infinitely differentiable); "sufficiently smooth" is the same as "sufficiently differentiable".

Other notations will be explained where they first appear.

CHAPTER I

Preliminaries

1. Basic notions and notations

The reader is assumed to be familiar with some of the basic notions in measure theory (cf. [45] and [129]), especially the notions of a *σ-field* (called also a *σ-algebra* or a *Borel field*), a *measurable space* (a pair consisting of an abstract space and a σ-field on it) and a *measurable mapping*. When S is a topological space, the smallest σ-field $\mathscr{B}(S)$ on S which contains all open sets is called the *topological σ-field* and an element $B \in \mathscr{B}(S)$ is called a *Borel set in S*. A mapping f from a topological space into another topological space S' which is $\mathscr{B}(S)/\mathscr{B}(S')$-measurable (i.e., $f^{-1}(B) = \{x; f(x) \in B\} \in \mathscr{B}(S)$ for all $B \in \mathscr{B}(S')$) is called *Borel measurable*. Any σ-additive non-negative measure P on a measurable space $(\Omega, \mathscr{F})$ such that $P(\Omega) = 1$ is called a *probability* on $(\Omega, \mathscr{F})$ and the triple $(\Omega, \mathscr{F}, P)$ is called a *probability space*. If P is a probability on $(\Omega, \mathscr{F})$, then $\overline{\mathscr{F}}^P = \{A \subset \Omega; {}^{\exists}B_1, B_2 \in \mathscr{F}$ such that $B_1 \subset A \subset B_2$ and $P(B_1) = P(B_2)\}$ is a σ-field on Ω containing $\mathscr{F}$. The probability P can be naturally extended to $\overline{\mathscr{F}}^P$ and the space $(\Omega, \overline{\mathscr{F}}^P, P)$ is called the *completion* of $(\Omega, \mathscr{F}, P)$. A probability P on $(\Omega, \mathscr{F})$ such that $\mathscr{F} = \overline{\mathscr{F}}^P$ is called *complete* and in this case $(\Omega, \mathscr{F}, P)$ is called a *complete probability space*. If S is a topological space, we set

$$(1.1) \qquad \mathscr{E}(S) = \bigcap_{\mu} \overline{\mathscr{B}(S)}^{\mu},$$

where μ runs over all probabilities on $(S, \mathscr{B}(S))$. An element in $\mathscr{E}(S)$ is called a *universally measurable set* in S and a mapping f from S into another topological space S' which is $\mathscr{E}(S)/\mathscr{B}(S')$-measurable is called *universally measurable*. Let $\{\mathscr{F}_\alpha\}$ be a family of σ-fields on Ω. We denote by $\bigvee_\alpha \mathscr{F}_\alpha$ the smallest σ-field on Ω which contains all $\mathscr{F}_\alpha$. If $\mathscr{C}$ is a class of subsets of Ω, we denote by $\sigma[\mathscr{C}]$ the smallest σ-field on Ω which contains $\mathscr{C}$. Also, if $\{X_\alpha\}_{\alpha \in A}$ is a family of mappings from Ω into a measurable space $(\Omega', \mathscr{F}')$, then the smallest σ-field $\mathscr{F}$ on Ω such that each X_α is $\mathscr{F}/\mathscr{F}'$-measurable is denoted by $\sigma[X_\alpha; \alpha \in A]$. In particular, $\sigma[X] = X^{-1}(\mathscr{F}')$ for every mapping $X: \Omega \longrightarrow \Omega'$.

Let $(\Omega, \mathscr{F}, P)$ be a probability space and $(S, \mathscr{B}(S))$ be a topological space S with the topological σ-field $\mathscr{B}(S)$. A mapping X from Ω into S which is $\mathscr{F}/\mathscr{B}(S)$-measurable is called an *S-valued random variable.*[*1] If, in particular, $S = \boldsymbol{R}$, $S = \boldsymbol{C}$, or $S = \boldsymbol{R}^d$, then X is called a *real random variable*, a *complex random variable*, or a *d-dimensional random variable*, respectively. If X is an S-valued random variable, then

$$(1.2) \qquad P^X(B) = P[X^{-1}(B)] = P[\omega; X(\omega) \in B] = P[X \in B],^{*2} \quad B \in \mathscr{B}(S),$$

defines a probability on $(S, \mathscr{B}(S))$. P^X is called the *probability law* (or *probability distribution*) of the random variable X. P^X is nothing but the *induced measure*, or the *image measure* of the measurable mapping X.

2. Probability measures on a metric space

Let S be a separable metric space with the metric ρ and $\mathscr{B}(S)$ be the topological σ-field.

Proposition 2.1. Let P be a probability on $(S, \mathscr{B}(S))$. Then, for every $B \in \mathscr{B}(S)$,

$$(2.1) \qquad P(B) = \sup_{F \subset B,\, F:\, \text{closed}} P(F) = \inf_{B \subset G,\, G:\, \text{open}} P(G).$$

Proof. Set $\mathscr{C} = \{B \in \mathscr{B}(S);\ (2.1) \text{ holds}\}$. If $B \in \mathscr{C}$, then clearly B^C (the complement of B) $\in \mathscr{C}$. If $B_n \in \mathscr{C}$, $n = 1, 2, \ldots,$ then $\bigcup B_n \in \mathscr{C}$. Indeed, for given $\varepsilon > 0$, we can choose an open set G_n and a closed set F_n such that $F_n \subset B_n \subset G_n$ and

$$P(G_n \setminus F_n) \leq \varepsilon/2^{n+1} \qquad n = 1, 2, \ldots .$$

Set $G = \bigcup_{n=1}^{\infty} G_n$ and $F = \bigcup_{n=1}^{n_0} F_n$ where n_0 is chosen so that

$$P\left(\bigcup_{n=1}^{\infty} F_n \setminus \bigcup_{n=1}^{n_0} F_n\right) \leq \varepsilon/2.$$

---

*1 More generally, a random variable is any measurable mapping from $(\Omega, \mathscr{F})$ into a measurable space.

*2 We often suppress the argument ω in this way.

Then G is open, F is closed, $F \subset \bigcup_{n=1}^{\infty} B_n \subset G$ and

$$P(G \setminus F) \leq \sum_{n=1}^{\infty} P(G_n \setminus F_n) + P\left(\bigcup_{n=1}^{\infty} F_n \setminus F\right) \leq \varepsilon/2 + \varepsilon/2 = \varepsilon.$$

Thus, $\bigcup_{n=1}^{\infty} B_n \in \mathscr{C}$. Therefore, $\mathscr{C}$ is a σ-field. If G is open, then $F_n = \{x; \rho(x, G^C) \geq 1/n\}$ * is closed, $F_n \subset F_{n+1}$ and $\bigcup_n F_n = G$. Since $P(G) = \lim_{n\to\infty} P(F_n)$, it follows that $G \in \mathscr{C}$. Therefore, $\mathscr{C} = \mathscr{B}(S)$.

We denote, by $C_b(S)$, the set of all bounded, continuous real-valued functions on S. $C_b(S)$ is a Banach space with the usual norm $\|f\| = \sup_{x\in S} |f(x)|$, $f \in C_b(S)$.

Proposition 2.2. Let P and Q be probabilities on $(S, \mathscr{B}(S))$. If $\int_S f(x)P(dx) = \int_S f(x)Q(dx)$ for every $f \in C_b(S)$, then $P = Q$.

Proof. By Proposition 2.1, it is sufficient to show that $P(F) = Q(F)$ for every closed set F. If we set $f_n(x) = \phi(n\rho(x, F))$ where $\phi(t)$ is the function defined by

$$\phi(t) = \begin{cases} 1 \quad , & t \leq 0, \\ 1 - t \quad , & 0 \leq t \leq 1, \\ 0 \quad , & t \geq 1, \end{cases}$$

then $\lim_{n\to\infty} f_n(x) = I_F(x)$ for every $x \in S$. Consequently, by the dominated convergence theorem,

$$P(F) = \lim_{n\to\infty} \int_S f_n(x)P(dx) = \lim_{n\to\infty} \int_S f_n(x)Q(dx) = Q(F).$$

Proposition 2.3. Suppose that S is complete with respect to the metric ρ, that is, every ρ-Cauchy sequence is a convergent sequence. Then any probability P on $(S, \mathscr{B}(S))$ is *inner regular* in the sense that for every $B \in \mathscr{B}(S)$

(2.2) $$P(B) = \sup_{K \subset B,\, K:\, \text{compact}} P(K).$$

Proof. We first prove that for every $\varepsilon > 0$ there exists a compact set

---

* $\rho(x, A) = \inf_{y\in A} \rho(x, y)$.

$K \subset S$ such that $P(K) > 1-\varepsilon$. Since S is separable, S can be covered by a countable number of balls of radius δ for any given $\delta > 0$. Let $\delta_n \downarrow 0$, and, for each n, let $\sigma_k^{(n)}$, $k = 1, 2, \ldots$ be a sequence of closed balls of radius δ_n covering S. Then

$$1 = P(S) = \lim_{l \to \infty} P\left(\bigcup_{k=1}^{l} \sigma_k^{(n)}\right)$$

and consequently we can find l_n such that

$$P\left(\bigcup_{k=1}^{l_n} \sigma_k^{(n)}\right) > 1 - \varepsilon/2^n. \tag{2.3}$$

Set $K = \bigcap_{n=1}^{\infty} \bigcup_{k=1}^{l_n} \sigma_k^{(n)}$. Clearly, for every $\delta > 0$, K can be covered by a finite number of balls with radius δ and hence K is totally bounded. Since S is complete, this implies that K is compact. From (2.3), we conclude that $P(K) > 1 - \varepsilon$.

Next, let $B \in \mathscr{B}(S)$. By Proposition 2.1, we can choose a closed set $F \subset B$ such that $P(B) \leq P(F) + \varepsilon$. Now $F' = F \cap K$ is compact and $P(F) - P(F') \leq P(K^c) \leq \varepsilon$. Hence $P(B) \leq P(F') + 2\varepsilon$ which implies that (2.2) holds.

Definition 2.1. A sequence $\{P_n\}$ of probabilities on $(S, \mathscr{B}(S))$ is said to be *weakly convergent* to a probability P on $(S, \mathscr{B}(S))$ if for every $f \in C_b(S)$

$$\lim_{n \to \infty} \int_S f(x) P_n(dx) = \int_S f(x) P(dx).$$

P is uniquely determined from $\{P_n\}$ by Proposition 2.2, and we write $P = w\text{-}\lim_{n \to \infty} P_n$, or $P_n \xrightarrow{w} P$ as $n \longrightarrow \infty$.

Proposition 2.4. The following five conditions are equivalent:

(i) $P_n \xrightarrow{w} P$.

(ii) $\lim_{n \to \infty} \int_S f(x) P_n(dx) = \int_S f(x) P(dx)$ for every *uniformly continuous* $f \in C_b(S)$.

(iii) $\overline{\lim_{n \to \infty}}\, P_n(F) \leq P(F)$ for every closed set F.

(iv) $\underline{\lim_{n \to \infty}}\, P_n(G) \geq P(G)$ for every open set G.

(v) $\lim_{n\to\infty} P_n(A) = P(A)$ for every $A \in \mathscr{B}(S)$ such that $P(\partial A) = 0$.*

Proof. "(i) $\Rightarrow$ (ii)" is obvious. To show "(ii) $\Rightarrow$ (iii)", note that the functions $f_k(x)$ used in the proof of Proposition 2.2 are uniformly continuous. Then, (ii) implies that

$$\overline{\lim_{n\to\infty}} P_n(F) \leq \lim_{n\to\infty} \int_S f_k(x) P_n(dx) = \int_S f_k(x) P(dx)$$

and, letting $k \longrightarrow \infty$, we have (iii). "(iii) $\Leftrightarrow$ (iv)" follows by taking complements. Next, we show "(iii) $\Rightarrow$ (i)". Let $f \in C_b(S)$. By a linear transformation, we may assume, without loss of generality that $0<f<1$. Then

$$\begin{aligned} &\sum_{i=1}^{k} (i-1)/k\, P\{x; (i-1)/k \leq f(x) < i/k\} \\ (2.4)\qquad &\leq \int_S f(x)P(dx) \leq \sum_{i=1}^{k} i/k\, P\{x; (i-1)/k \leq f(x) < i/k\}. \end{aligned}$$

If we set $F_i = \{x;\ i/k \leq f(x)\}$, the right-hand side of (2.4) is equal to $\sum_{i=0}^{k-1} P(F_i)/k$ and the left-hand side is equal to $\sum_{i=0}^{k-1} P(F_i)/k - 1/k$. From (iii), it follows that

$$\begin{aligned} \overline{\lim_{n\to\infty}} \int_S f(x)P_n(dx) &\leq \overline{\lim_{n\to\infty}} \sum_{i=0}^{k-1} P_n(F_i)/k \leq \sum_{i=0}^{k-1} P(F_i)/k \\ &\leq 1/k + \int_S f(x)P(dx). \end{aligned}$$

Since k was arbitrary, we deduce that

$$\overline{\lim_{n\to\infty}} \int_S f(x)P_n(dx) \leq \int_S f(x)P(dx).$$

By replacing f with $1-f$ in the above argument, we obtain

$$\underline{\lim_{n\to\infty}} \int_S f(x)P_n(dx) \geq \int_S f(x)P(dx).$$

Consequently, $\lim_{n\to\infty}\int_S f(x)P_n(dx) = \int_S f(x)P(dx)$.

---

* $\partial A = \bar{A} \setminus \mathring{A}$ is the boundary of the set A, $\bar{A}$ is the closure of A and $\mathring{A}$ is the interior of A.

Finally, we show "(iii) $\Leftrightarrow$ (v)". If $P(\partial A) = 0$, then, assuming (iii) ($\Leftrightarrow$ (iv)), we have

$$P(A) = P(\mathring{A}) \leq \varliminf_{n\to\infty} P_n(\mathring{A}) \leq \varlimsup_{n\to\infty} P_n(\bar{A}) \leq P(\bar{A}) = P(A)$$

showing that $\lim_{n\to\infty} P_n(A) = P(A)$. Conversely, assume (v). Let F be a closed set and set $F_\delta = \{x;\ \rho(x,F) \leq \delta\}$. Then $\partial F_\delta \subset \{x;\ \rho(x,F) = \delta\} := A_\delta$ and, since the A_δ's are disjoint for different δ, the set of δ such that $P(A_\delta) > 0$ is at most countable. Therefore, we can choose $\delta_l \downarrow 0$ such that $P(A_{\delta_l}) = 0$ and hence $P(\partial F_{\delta_l}) = 0$. Consequently

$$P(F) = \lim_{l\to\infty} P(F_{\delta_l}) = \lim_{l\to\infty} \lim_{n\to\infty} P_n(F_{\delta_l}) \geq \varlimsup_{n\to\infty} P_n(F).$$

Example 2.1. If $S = \boldsymbol{R}$, there is a one-to-one correspondence between a probability P on $(S,\mathscr{B}(S))$ and its distribution function $F(x) = P((-\infty, x])$. Then "$P_n \xrightarrow{w} P$" is equivalent to "$F_n(x) \longrightarrow F(x)$ for every continuity point x of F". The former implies the latter by (v) of the above proposition and the converse implication is proved easily by approximating the integrals $\int_{-\infty}^{\infty} f(x)dF_n(x)$ and $\int_{-\infty}^{\infty} f(x)dF(x)$ with the Riemann sums.

Proposition 2.5. Weak convergence of probabilities is a metric concept. To be precise, we can define a metric d on the totality $\mathscr{P}(S)$ of probabilities on $(S, \mathscr{B}(S))$ such that

$$P_n \xrightarrow{w} P \quad \text{is equivalent to} \quad d(P_n, P) \longrightarrow 0 \quad \text{as } n \longrightarrow \infty.$$

Proof. The well known Prohorov metric, which is a generalization of that of Lévy in the case $S = \boldsymbol{R}$, is such a metric (cf. [142]). Here we give an equivalent metric in the following way (cf. [165]). If S is a separable metric space, we can choose an equivalent metric so that S is totally bounded under this metric.* Then, the set of all uniformly continuous functions has a countable dense subset $\{f_n\}$ with respect to the uniform norm and we set

$$d(P, Q) = \sum_{j=1}^{\infty} 2^{-j}\left\{1 \wedge \left(\left| \int_S f_j(x)P(dx) - \int_S f_j(x)Q(dx) \right|\right)\right\}.$$

---

* Indeed, it is well known that a separable metric space is homeomorphic to a subset of the Hilbert cube $[0, 1]^N$, [98].

It is easy to see, by using Proposition 2.4 (ii), that d is a metric satisfying the condition of the theorem.

Thus, the totality $\mathscr{P}(S)$ of probabilities on $(S, \mathscr{B}(S))$ is a metric space under weak convergence. We now want to characterize a relatively compact*[1] set in $\mathscr{P}(S)$. For this, we give the following definition:

Definition 2.2. A family $\Lambda \subset \mathscr{P}(S)$ is called *tight* if for every $\varepsilon > 0$ there exists a compact subset $K \subset S$ such that $P(K) \geq 1 - \varepsilon$ for every $P \in \Lambda$; i.e., $\inf_{P \in \Lambda} P(K) \geq 1 - \varepsilon$.

Example 2.2. If S is complete under the metric ρ, then every finite set Λ is tight by Proposition 2.3. Generally, if Λ_1 and Λ_2 are tight, then so is $\Lambda_1 \cup \Lambda_2$.

Theorem 2.6. Let $\Lambda \subset \mathscr{P}(S)$.

(1) If Λ is tight, then Λ is relatively compact in $\mathscr{P}(S)$.

(2) When S is complete under the metric ρ we have the converse of (1): namely, if Λ is relatively compact in $\mathscr{P}(S)$, then Λ is tight.

Proof. For the proof of (1), we first note that, if S is a compact metric space then $\mathscr{P}(S)$ is compact and hence every $\Lambda \subset \mathscr{P}(S)$ is relatively compact. Indeed, by Riesz's theorem, $\mathscr{P}(S) = \{\mu \in \boldsymbol{C}^*(S); \mu(f) \geq 0 \text{ for } f \geq 0 \text{ and } \mu(\mathbf{1}) = 1\}$ *[2] and, since $\boldsymbol{C}(S) = \boldsymbol{C}_b(S)$, weak convergence is equivalent to convergence in the weak $*$-topology on $\boldsymbol{C}^*(S)$. Thus $\mathscr{P}(S)$ is compact since it is a weak $*$-closed subset of the unit ball in $\boldsymbol{C}^*(S)$ and, as is well known, the unit ball is weak $*$-compact.

In the general case, we note that S is homeomorphic to a subset of a compact metric space (actually, a subset of $[0,1]^N$) and hence we may assume that S is a subset of a compact metric space $\tilde{S}$. We want to show that for every sequence $\{\mu_n\}$ from a tight family Λ we can always choose a convergent subsequence. For a probability μ on $(S,\mathscr{B}(S))$, we define a probability on $(\tilde{S},\mathscr{B}(\tilde{S}))$ by $\tilde{\mu}(\tilde{A}) = \mu(\tilde{A} \cap S)$, $\tilde{A} \in \mathscr{B}(\tilde{S})$. Note that $A \subset S$ is in $\mathscr{B}(S)$ if and only if it is expressed as $A = \tilde{A} \cap S$ for some $\tilde{A} \in \mathscr{B}(\tilde{S})$. Now, $\{\tilde{\mu}_n\}$ is a sequence in $\mathscr{P}(\tilde{S})$ and hence, by the above remark, we can choose a subsequence, which we denote again by $\{\tilde{\mu}_n\}$, converging weakly to a probability ν on $(\tilde{S},\mathscr{B}(\tilde{S}))$. We now show that there is a probability μ on $(S,\mathscr{B}(S))$ such that $\tilde{\mu} = \nu$ and μ_n converges

*[1] i.e., its closure is compact, or equivalently, from every infinite sequence contained in this set we can choose a convergent subsequence.

*[2] $\boldsymbol{C}(S)$ is the Banach lattice under the natural order of all real continuous functions on S and $\boldsymbol{C}^*(S)$ is the dual space. $\mathbf{1}$ denotes the function $f(x) \equiv 1$.

weakly to μ. Indeed, for every $r = 1, 2, \dots$, there exists a compact subset K_r of S (and hence a compact subset in $\tilde{S}$) such that $\mu_n(K_r) > 1 - 1/r$ for all n. Clearly $K_r \in \mathscr{B}(S) \cap \mathscr{B}(\tilde{S})$ and $\tilde{\mu}_n(K_r) = \mu_n(K_r)$. Since $\tilde{\mu}_n \to \nu$ weakly, we have

$$\nu(K_r) \geq \overline{\lim_{n\to\infty}}\ \mu_n(K_r) \geq 1 - 1/r.$$

Therefore $\cup K_r \equiv E \subset S$ is both in $\mathscr{B}(S)$ and $\mathscr{B}(\tilde{S})$ and $\nu(E) = 1$. If $A \in \mathscr{B}(S)$, then $A \cap E \in \mathscr{B}(\tilde{S})$ since $A \cap E = \tilde{A} \cap S \cap E = \tilde{A} \cap E$ for some $\tilde{A} \in \mathscr{B}(\tilde{S})$. We set $\mu(A) = \nu(A \cap E)$ for every $A \in \mathscr{B}(S)$. Now it is easy to see that μ is a probability on $(S, \mathscr{B}(S))$ and $\tilde{\mu} = \nu$. Finally we show that $\mu_n \longrightarrow \mu$ weakly in $\mathscr{P}(S)$. Let A be closed in S. Then $A = \tilde{A} \cap S$ for some closed set in $\tilde{S}$ and $\tilde{\mu}_n(\tilde{A}) = \mu_n(A)$. Consequently

$$\overline{\lim_{n\to\infty}}\ \mu_n(A) = \overline{\lim_{n\to\infty}}\ \tilde{\mu}_n(\tilde{A}) \leq \tilde{\mu}(\tilde{A}) = \mu(A)$$

and the assertion follows from Proposition 2.4, (iii). We omit the proof of (2) (cf. [6], [138]).

Consider an S-valued random variable X; i.e., an $\mathscr{F}/\mathscr{B}(S)$-measurable mapping from a probability space $(\Omega, \mathscr{F}, P)$ into S. The probability law P^X of X is the image measure of the mapping $X\colon \Omega \longrightarrow S$.

Definition 2.3. Let X_n, $n = 1, 2, \dots$, and X be S-valued random variables.* We say X_n *converges to* X *in law* if $P^{X_n} \longrightarrow P^X$ weakly.

Suppose X_n, $n = 1, 2, \dots$ and X are defined on the same probability space $(\Omega, \mathscr{F}, P)$. Then X_n is said to *converge to* X *almost everywhere* (or *almost surely*) if

$$P\{\omega;\ \rho(X_n(\omega), X(\omega)) \longrightarrow 0 \quad \text{as} \quad n \longrightarrow \infty\} = 1$$

and X_n is said to *converge to* X *in probability*, if, for every $\varepsilon > 0$,

$$P\{\omega;\ \rho(X_n(\omega), X(\omega)) > \varepsilon\} \longrightarrow 0 \qquad \text{as} \quad n \longrightarrow \infty.$$

It is well known that almost everywhere convergence implies convergence in probability and convergence in probability implies convergence in law. The following theorem due to Skorohod asserts that, if S

* They may be defined on different probability spaces.

is complete under the metric ρ, a converse of the above implications holds in a certain sense.

Theorem 2.7. Let (S, ρ) be a complete separable metric space and $P_n, n = 1, 2, \ldots$, and P be probabilities on $(S, \mathscr{B}(S))$ such that $P_n \xrightarrow{w} P$ as $n \longrightarrow \infty$. Then, on a probability space $(\tilde{\Omega}, \tilde{\mathscr{B}}, \tilde{P})$, we can construct S-valued random variables X_n, $n = 1, 2, \ldots$, and X such that

(i) $P_n = \tilde{P}^{X_n}$, $n = 1, 2, \ldots$ and $P = \tilde{P}^X$.

(ii) X_n converges to X almost everywhere.

Thus, in particular, convergence in law of random variables X_n can be realized by an almost everywhere convergence without changing the law of any X_n.

Proof. We prove this theorem by taking $\tilde{\Omega} = [0, 1)$, $\tilde{\mathscr{B}} = \mathscr{B}([0, 1))$ and $\tilde{P}(d\omega) = d\omega$: the Lebesgue measure. To every finite sequence $(i_1, i_2, \ldots, i_k)$ $(k = 1, 2, \ldots)$ of natural numbers, we associate a set $S_{(i_1, i_2, \ldots, i_k)} \in \mathscr{B}(S)$ as follows:

(1) If $(i_1, i_2, \ldots, i_k) \neq (j_1, j_2, \ldots, j_k)$, then
$S_{(i_1, i_2, \ldots, i_k)} \cap S_{(j_1, j_2, \ldots, j_k)} = \phi$;

(2) $\bigcup_{j=1}^{\infty} S_j = S$ and $\bigcup_{j=1}^{\infty} S_{(i_1, i_2, \ldots, i_k, j)} = S_{(i_1, i_2, \ldots, i_k)}$;

(3) diam $S_{(i_1, i_2, \ldots, i_k)} \leq 2^{-k}$ *[1];

(4) $P_n(\partial S_{(i_1, i_2, \ldots, i_k)}) = 0$, $n = 1, 2, \ldots$ and $P(\partial S_{(i_1, i_2, \ldots, i_k)}) = 0$.

Thus, by (1) and (2), $\{S_{(i_1, i_2, \ldots, i_k)}\}$ forms for each fixed k a disjoint covering of S which is a refinement of that corresponding to $k' < k$. We can construct such a system of subsets as follows. For each k, let $\sigma_m^{(k)}$, $m = 1, 2, \ldots$ be balls with radius $\leq 2^{-(k+1)}$ covering the whole space S and satisfying $P_n(\partial\sigma_m^{(k)}) = 0$, $P(\partial\sigma_m^{(k)}) = 0$ for every n, k, m. Set, for each k, $D_1^{(k)} = \sigma_1^{(k)}$, $D_2^{(k)} = \sigma_2^{(k)} \setminus \sigma_1^{(k)}, \ldots, D_n^{(k)} = \sigma_n^{(k)} \setminus (\sigma_1^{(k)} \cup \cdots \cup \sigma_{n-1}^{(k)}), \ldots$ and $S_{(i_1, i_2, \ldots, i_k)} = D_{i_1}^{(1)} \cap D_{i_2}^{(2)} \cap \cdots \cap D_{i_k}^{(k)}$. It is easy to verify that the system of sets so defined possesses the above properties. Fixing k, we order all $(i_1, i_2, \ldots, i_k)$ lexicographically. Define intervals*[2] $\Delta_{(i_1, i_2, \ldots, i_k)}$, $\Delta_{(i_1, i_2, \ldots, i_k)}^{(n)}$ in $[0, 1)$ with the following properties:

(i) $|\Delta_{(i_1, i_2, \ldots, i_k)}| = P(S_{(i_1, i_2, \ldots, i_k)})$, $|\Delta_{(i_1, i_2, \ldots, i_k)}^{(n)}| = P_n(S_{(i_1, i_2, \ldots, i_k)})$;

(ii) if $(i_1, i_2, \ldots, i_k) < (j_1, j_2, \ldots, j_k)$, then the interval $\Delta_{(i_1, i_2, \ldots, i_k)}$ $(\Delta_{(i_1, i_2, \ldots, i_k)}^{(n)})$ is located to the left of the interval $\Delta_{(j_1, j_2, \ldots, j_k)}$ (resp. $\Delta_{(j_1, j_2, \ldots, j_k)}^{(n)}$);

*[1] diam $A = \sup_{x, y \in A} \rho(x, y)$.

*[2] Here, by intervals in $[0, 1)$, we mean intervals of the form $[a, b)$ $(a \leq b)$ only. $|\Delta|$ stands for the length of the interval Δ.

(iii) $\bigcup_{(i_1,i_2,\dots,i_k)} \Delta_{(i_1,i_2,\dots,i_k)} = [0, 1), \quad \bigcup_{(i_1,i_2,\dots,i_k)} \Delta^{(n)}_{(i_1,i_2,\dots,i_k)} = [0, 1).$

Clearly these intervals are uniquely determined by these properties. For each $(i_1, i_2, \dots, i_k)$ such that $\mathring{S}_{(i_1,i_2,\dots,i_k)} \neq \phi$, we choose a point $x_{i_1,i_2,\dots,i_k} \in \mathring{S}_{(i_1,i_2,\dots,i_k)}$. For $\omega \in [0, 1)$, we set

$$X^k_n(\omega) = x_{i_1,i_2,\dots,i_k} \quad \text{if} \quad \omega \in \Delta^{(n)}_{(i_1,i_2,\dots,i_k)} \quad \text{and}$$
$$X^k(\omega) = x_{i_1,i_2,\dots,i_k} \quad \text{if} \quad \omega \in \Delta_{(i_1,i_2,\dots,i_k)},$$

for $k = 1, 2, \dots,\ \ n = 1, 2, \dots$.* Clearly,

$$\rho(X^k_n(\omega), X^{k+p}_n(\omega)) \leq 1/2^k, \quad \rho(X^k(\omega), X^{k+p}(\omega)) \leq 1/2^k$$

and hence $X_n(\omega) = \lim_{k\to\infty} X^k_n(\omega)$, $X(\omega) = \lim_{k\to\infty} X^k(\omega)$ exist by the completeness of (S, ρ). Since $P_n(S_{(i_1,i_2,\dots,i_k)}) = |\Delta^{(n)}_{(i_1,i_2,\dots,i_k)}| \longrightarrow |\Delta_{(i_1,i_2,\dots,i_k)}| = P(S_{(i_1,i_2,\dots,i_k)})$, if $\omega \in \mathring{\Delta}_{(i_1,i_2,\dots,i_k)}$ then there exists n_k such that $\omega \in \mathring{\Delta}^{(n)}_{(i_1,i_2,\dots,i_k)}$ for all $n \geq n_k$. Then $X^k_n(\omega) = X^k(\omega)$ and hence

$$\rho(X_n(\omega), X(\omega)) \leq \rho(X_n(\omega), X^k_n(\omega)) + \rho(X^k_n(\omega), X^k(\omega))$$
$$+ \rho(X^k(\omega), X(\omega)) \leq 2/2^k, \quad \text{if} \quad n \geq n_k.$$

Therefore, if we set $\Omega_0 = \bigcap_{k=1}^{\infty} (\bigcup_{(i_1,i_2,\dots,i_k)} \mathring{\Delta}_{(i_1,i_2,\dots,i_k)})$, $X_n(\omega) \longrightarrow X(\omega)$ for $\omega \in \Omega_0$ as $n \longrightarrow \infty$ and clearly $\tilde{P}(\Omega_0) = 1$.

Finally, we show that $\tilde{P}^{X_n} = P_n$ and $\tilde{P}^X = P$. Since

$$\tilde{P}\{\omega;\ X^{k+p}_n(\omega) \in \bar{S}_{(i_1,i_2,\dots,i_k)}\} = \tilde{P}\{\omega;\ X^{k+p}_n(\omega) \in \mathring{S}_{(i_1,i_2,\dots,i_k)}\}$$
$$= P_n(S_{(i_1,i_2,\dots,i_k)})$$

and since every open set in S is expressed as a disjoint countable union of $S_{(i_1,i_2,\dots,i_k)}$'s, we have, by Fatou's lemma,

$$\varliminf_{p\to\infty} \tilde{P}^{X^p_n}(O) \geq P_n(O)$$

for every open set O in S. Then, by Proposition 2.4, $\tilde{P}^{X^p_n}$ converges weakly to P_n as $p \longrightarrow \infty$ showing that $\tilde{P}^{X_n} = P_n$. Similarly, we have $\tilde{P}^X = P$.

---

* If $\Delta_{(i_1,i_2,\dots,i_k)} \neq \phi$ or $\Delta^{(n)}_{(i_1,i_2,\dots,i_k)} \neq \phi$, then $\mathring{S}_{(i_1,i_2,\dots,i_k)} \neq \phi$ and hence this is well defined.

3. Expectations, conditional expectations and regular conditional probabilities

Let X be a real (or complex) random variable defined on a probability space $(\Omega,\mathscr{F},P)$. Two random variables X and Y are identified if $P[\omega;X(\omega) \neq Y(\omega)] = 0$. X is called *integrable* if

$$\int_\Omega |X(\omega)| P(d\omega) < \infty.$$

More generally, if

$$\int_\Omega |X(\omega)|^p P(d\omega) < \infty, \qquad (p > 0),$$

it is called p-th integrable.*[1] Let $p \geq 1$. The totality of p-th integrable random variables, denoted by $\mathscr{L}_p(\Omega,\mathscr{F},P)$ or simply by $\mathscr{L}_p(\Omega)$ or $\mathscr{L}_p(P)$, forms a Banach space with the norm

$$\|X\|_p = \left(\int_\Omega |X(\omega)|^p P(d\omega)\right)^{1/p}.$$

$\mathscr{L}_\infty(\Omega,\mathscr{F},P)$ is the Banach space of essentially bounded random variables with the norm $\|X\|_\infty = \text{ess.sup } |X(\omega)|$.

For an integrable random variable X, $E(X) = \int_\Omega X(\omega)P(d\omega)$ is called the *expectation* of X. For a square-integrable random variable X, $V(X) = E(X^2) - E(X)^2$ $(= E((X - E(X))^2))$ is called the *variance* of X.

A family $\{\mathscr{F}_\alpha\}_{\alpha\in\Lambda}$ of sub σ-fields of $\mathscr{F}$ is called *mutually independent* if for every distinct choice of $\alpha_1, \alpha_2, \ldots, \alpha_k \in \Lambda$ and $A_i \in \mathscr{F}_{\alpha_i}$, $i = 1, 2, \ldots, k$,

$$P(A_1 \cap A_2 \cap \cdots \cap A_k) = P(A_1)P(A_2) \cdots P(A_k).$$

A family of random variables $\{X_\alpha\}_{\alpha\in\Lambda}$ is called *mutually independent**[2] if $\{\sigma[X_\alpha]\}_{\alpha\in\Lambda}$ is mutually independent. A family $\{X_\alpha, \alpha \in \Lambda\}$ of random variables is called *independent of a σ-field $\mathscr{G} \subset \mathscr{F}$* if $\sigma[X_\alpha; \alpha \in \Lambda]$ and $\mathscr{G}$ are mutually independent.

Let X be an integrable random variable and $\mathscr{G} \subset \mathscr{F}$ be a sub σ-field of $\mathscr{F}$. Then $\mu(B) = E(X:B) := \int_B X(\omega)P(d\omega)$, $B \in \mathscr{G}$, defines a

*[1] If $p = 2$, it is called square-integrable.

*[2] This definition applies for general S-valued random variables.

σ-additive set function on $\mathscr{G}$ with finite total variation and is clearly absolutely continuous with respect to $\nu = P|_{\mathscr{G}}$. The Radon-Nikodym derivative $d\mu/d\nu(\omega)$ is denoted by $E(X|\mathscr{G})(\omega)$; thus $E(X|\mathscr{G})$ is the unique (up to identification) $\mathscr{G}$-measurable integrable random variable Y such that $E(Y{:}B) = E(X{:}B)$ for every $B \in \mathscr{G}$.

Definition 3.1. $E(X|\mathscr{G})(\omega)$ is called *the conditional expectation of X given $\mathscr{G}$.*

The following properties of conditional expectations are easily proved. (X, Y, X_n, below are integrable real random variables and a, b are real numbers.)

(E.1) $E(aX + bY|\mathscr{G}) = aE(X|\mathscr{G}) + bE(Y|\mathscr{G})$ a.s.

(E.2) If $X \geq 0$ a.s., then $E(X|\mathscr{G}) \geq 0$ a.s.

(E.3) $E(\mathbf{1}|\mathscr{G}) = \mathbf{1}$ a.s.

(E.4) If X is $\mathscr{G}$-measurable, then $E(X|\mathscr{G}) = X$, a.s., more generally, if XY is integrable and X is $\mathscr{G}$-measurable

$$E(XY|\mathscr{G}) = XE(Y|\mathscr{G}) \quad \text{a.s.}$$

(E.5) If $\mathscr{H}$ is a sub σ-field of $\mathscr{G}$, then

$$E(E(X|\mathscr{G})|\mathscr{H}) = E(X|\mathscr{H}) \quad \text{a.s.}$$

(E.6) If $X_n \longrightarrow X$ in $\mathscr{L}_1(\Omega)$, then $E(X_n|\mathscr{G}) \longrightarrow E(X|\mathscr{G})$ in $\mathscr{L}_1(\Omega)$.

(E.7) (Jensen's inequality) If ψ: $\boldsymbol{R}^1 \longrightarrow \boldsymbol{R}^1$ is convex and $\psi(X)$ integrable then

$$\psi(E(X|\mathscr{G})) \leq E(\psi(X)|\mathscr{G}) \quad \text{a.s.}$$

In particular, $|E(X|\mathscr{G})| \leq E(|X|\,|\mathscr{G})$ and, if X is square-integrable, $|E(X|\mathscr{G})|^2 \leq E(|X|^2|\mathscr{G})$.

(E.8) X is independent of $\mathscr{G}$ if and only if for every Borel measurable function f such that $f(X)$ is integrable, $E(f(X)|\mathscr{G}) = E(f(X))$ a.s.

Let ξ be a mapping from Ω into a measurable space $(S, \mathscr{B})$ such that it is $\mathscr{F}/\mathscr{B}$-measurable. Then $\mu(B) = E(X{:}\{\omega;\ \xi(\omega) \in B\})$ is a σ-additive set function on $\mathscr{B}$ which is absolutely continuous with respect to the image measure $\nu = P^{\xi}$. The Radon-Nikodym density $d\mu/d\nu(x)$ is denoted by $E(X|\xi = x)$ and is called *the conditional expectation* of X *given* $\xi = x$. It possesses similar properties as above.

Definition 3.2. Let $(\Omega,\mathscr{F},P)$ be a probability space and $\mathscr{G}$ be a sub σ-field of $\mathscr{F}$. A system $\{p(\omega, A)\}_{\omega\in\Omega, A\in\mathscr{F}}$ is called a *regular conditional probability given* $\mathscr{G}$ if it satisfies the following conditions:

(i) for fixed ω, $A \longmapsto p(\omega, A)$ is a probability on $(\Omega, \mathscr{F})$;

(ii) for fixed $A \in \mathscr{F}$, $\omega \longmapsto p(\omega, A)$ is $\mathscr{G}$-measurable;

(iii) for every $A \in \mathscr{F}$ and $B \in \mathscr{G}$,

$$P(A \cap B) = \int_B p(\omega, A)P(d\omega).$$

Clearly property (iii) is equivalent to

(iii)′ for every non-negative random variable X and $B \in \mathscr{G}$

$$E(X\colon B) = \int_\Omega \{I_B(\omega) \int_\Omega X(\omega')p(\omega, d\omega')\}\, P(d\omega),$$

that is, $\int_\Omega X(\omega')p(\omega, d\omega')$ coincides with $E(X|\mathscr{G})(\omega)$ a.s.

We say that *the regular conditional probability is unique* if whenever $\{p(\omega, A)\}$ and $\{p'(\omega, A)\}$ possess the above properties, then there exists a set $N \in \mathscr{G}$ of P-measure 0 such that, if $\omega \notin N$ then $p(\omega, A) = p'(\omega, A)$ for all $A \in \mathscr{F}$.

Definition 3.3. A measurable space $(\Omega, \mathscr{F})$ is called a *standard measurable space* if it is Borel isomorphic* to one of the following measurable spaces: $(\langle 1, n\rangle, \mathscr{B}(\langle 1, n\rangle))$, $(\boldsymbol{N}, \mathscr{B}(\boldsymbol{N}))$ or $(\boldsymbol{M}, \mathscr{B}(\boldsymbol{M}))$, where $\langle 1, n\rangle = \{1, 2, \ldots, n\}$ with the discrete topology, $\boldsymbol{N} = \{1, 2, \ldots\}$ with the discrete topology and $\boldsymbol{M} = \{0,1\}^{\boldsymbol{N}} = \{\omega = (\omega_1, \omega_2, \ldots), \omega_i = 0$ or $1\}$ with the product topology.

It is well known that a Polish space (a complete separable metric space) with the topological σ-field is a standard measurable space and every measurable subset of a standard measurable space with the induced σ-field is a standard measurable space (cf. [98], [138]).

Theorem 3.1. Let $(\Omega, \mathscr{F})$ be a standard measurable space and P be a probability on $(\Omega, \mathscr{F})$. Let $\mathscr{G}$ be a sub σ-field of $\mathscr{F}$. Then a regular conditional probability $\{p(\omega, A)\}$ given $\mathscr{G}$ exists uniquely.

Proof. We consider the case where $(\Omega, \mathscr{F})$ is isomorphic to $(\boldsymbol{M}, \mathscr{B}(\boldsymbol{M}))$ and hence we may assume that $\Omega = \boldsymbol{M}$ and $\mathscr{F} = \mathscr{B}(\boldsymbol{M})$. Let $\pi_n\colon \Omega \ni \omega \longmapsto (\omega_1, \omega_2, \ldots, \omega_n) \in \{0,1\}^n$ be the projection and

* $(\Omega, \mathscr{F})$ and $(\Omega', \mathscr{F}')$ are Borel isomorphic if there exstis a bijection $f\colon \Omega \longrightarrow \Omega'$ such that f is $\mathscr{F}/\mathscr{F}'$-measurable and f^{-1} is $\mathscr{F}'/\mathscr{F}$-measurable, i.e. $f(\mathscr{F}) = \mathscr{F}'$.

$\mathscr{F}_n = \pi_n^{-1}\mathscr{B}[\{0,1\}^n]$. Clearly $\{\mathscr{F}_n\}$ is an increasing family of *finite* σ-fields and $\bigvee_n \mathscr{F}_n = \mathscr{F}$. If P_n, $n = 1, 2, \ldots$ are probabilities on $(\Omega, \mathscr{F}_n)$ and $\{P_n\}$ is consistent in the sense that $P_{n+1}|_{\mathscr{F}_n} = P_n$, $n = 1, 2, \ldots$, then there exists a unique probability P on $(\Omega, \mathscr{F})$ such that $P|_{\mathscr{F}_n} = P_n$. Indeed, by consistency, P is well defined on $\bigcup_{n=1}^{\infty} \mathscr{F}_n$ and if $B_n \in \bigcup_{k=1}^{\infty} \mathscr{F}_k$, $n = 1, 2, \ldots$, is such that $B_n \supset B_{n+1}$ and $\lim_{n\to\infty} P(B_n) > 0$ then $\bigcap_n B_n \neq \phi$, since $\{B_n\}$ is a system of closed sets in a compact space Ω having the finite intersection property. P is then extended to $\sigma[\bigcup_n \mathscr{F}_n] = \bigvee_n \mathscr{F}_n = \mathscr{F}$ by Hopf's extension theorem.

We set $p_n(\omega, A) = E(I_A | \mathscr{G})(\omega)$, $A \in \mathscr{F}_n$. Clearly, there exists a set $N_n \in \mathscr{G}$ of P-measure 0 such that if $\omega \notin N_n$, then $p_n(\omega, A)$ is a probability on $\mathscr{F}_n$ and $p_n(\omega, A) = p_{n-1}(\omega, A)$ for $A \in \mathscr{F}_{n-1}$, $n = 1, 2, \ldots$. If we set $N = \bigcup_n N_n$, then for each $\omega \notin N$, $\{p_n(\omega, \cdot)\}$ is a consistent family and hence determines a unique probability $p(\omega, \cdot)$ on $(\Omega, \mathscr{F})$. Take a probability ν on $(\Omega, \mathscr{F})$ and define $p(\omega, \cdot) = \nu$ if $\omega \in N$. Then the system $\{p(\omega, \cdot)\}$ is a regular conditional probability given $\mathscr{G}$. Indeed, properties (ii) and (iii) are obvious for $A \in \bigcup_n \mathscr{F}_n$ and extend to $\mathscr{F}$ by a standard monotone lemma. If $\{p(\omega, \cdot)\}$ and $\{p'(\omega, \cdot)\}$ are two regular conditional probabilities, then the set $N = \{\omega;\ p(\omega, A) \neq p'(\omega, A)$ for some $A \in \bigcup_{n=1}^{\infty} \mathscr{F}_n\}$ has P-measure 0 and if $\omega \notin N$, then $p(\omega, A) = p'(\omega, A)$ for all $A \in \mathscr{F}$ again by the monotone lemma. This proves the uniqueness of the regular conditional probability.

Definition 3.4. Let $(\Omega, \mathscr{F})$ be a measurable space. We say that $\mathscr{F}$ is *countably determined* if there exists a countable subset $\mathscr{F}_0 \subset \mathscr{F}$ such that whenever any two probabilities agree on $\mathscr{F}_0$ they must coincide.

Clearly, if $(\Omega, \mathscr{F})$ is a standard measurable space then $\mathscr{F}$ is countably determined.

Theorem 3.2. Let $(\Omega, \mathscr{F})$ be a standard measurable space and P be a probability on $(\Omega, \mathscr{F})$. Let $\mathscr{G}$ be a sub σ-field of $\mathscr{F}$ and $p(\omega, d\omega')$ be a regular conditional probability given $\mathscr{G}$. If $\mathscr{H}$ is a countably determined sub σ-field of $\mathscr{G}$, then there exists a set $N \in \mathscr{G}$ of P-measure 0 such that $\omega \notin N$ implies $p(\omega, A) = I_A(\omega)$ for every $A \in \mathscr{H}$.

Proof. Let $\mathscr{H}_0 \subset \mathscr{H}$ be a countable set in Definition 3.4 for $\mathscr{H}$. Clearly if $A \in \mathscr{H}_0$, then there exists $N_A \in \mathscr{G}$ of P-measure 0 such that $p(\omega, A) = I_A(\omega)$ if $\omega \notin N_A$. Set $N = \bigcup_{A \in \mathscr{H}_0} N_A$, then $p(\omega, A) = I_A(\omega)$ holds for all $A \in \mathscr{H}$ if $\omega \notin N$.

Corollary. Let $(\Omega, \mathscr{F})$ be a standard measurable space and P be a probability on $(\Omega, \mathscr{F})$. Let $\mathscr{G}$ be a sub σ-field of $\mathscr{F}$ and $p(\omega, \cdot)$ be a regular conditional probability given $\mathscr{G}$. Let $\xi(\omega)$ be a mapping from Ω into a measurable space $(S, \mathscr{B})$ such that it is $\mathscr{G}/\mathscr{B}$-measurable. Suppose further that $\mathscr{B}$ is countably determined and $\{x\} \in \mathscr{B}$ for every $x \in S$ (this is true, for example, if $(S, \mathscr{B})$ is a standard measurable space). Then

$$(3.1) \qquad p(\omega, \{\omega'; \xi(\omega') = \xi(\omega)\}) = 1, \quad \text{a.a. } \omega.$$

Proof. Since $\mathscr{B}$ is countably determined, there exists a countable subset $\mathscr{B}_0 \subset \mathscr{B}$ with the property of Definition 3.4. Hence, if we set $\mathscr{H} = \{\xi^{-1}(B); B \in \mathscr{B}\}$, $\mathscr{H}$ is a countably determined sub σ-field of $\mathscr{G}$ with $\mathscr{H}_0 = \{\xi^{-1}(B); B \in \mathscr{B}_0\}$. By Theorem 3.2, there exists a set $N \in \mathscr{G}$ of P-measure 0 such that if $\omega \notin N$, $p(\omega, A) = I_A(\omega)$ for every $A \in \mathscr{H}$. By taking $A_\omega = \{\omega'; \xi(\omega') = \xi(\omega)\} \in \mathscr{H}$, we have (3.1) for $\omega \notin N$.

In the same way as Theorem 3.1, we can prove the following result.

Theorem 3.3. Let $(\Omega, \mathscr{F})$ be a standard measurable space and P be a probability on $(\Omega, \mathscr{F})$. Let $\xi(\omega)$ be a mapping from Ω into a measurable space $(S, \mathscr{B})$ such that it is $\mathscr{F}/\mathscr{B}$-measurable and let P^ξ be the induced measure on $(S, \mathscr{B})$ by ξ. Then there exists a system $\{p(x, A)\}_{x\in S, A\in\mathscr{F}}$ such that

(i) for fixed $x \in S$, $A \longmapsto p(x, A)$ is a probability on $(\Omega, \mathscr{F})$;
(ii) for fixed $A \in \mathscr{F}$, $x \longmapsto p(x, A)$ is $\mathscr{B}$-measurable;
(iii) for every $A \in \mathscr{F}$ and $B \in \mathscr{B}$, we have

$$P(A \cap \{\omega; \xi(\omega) \in B\}) = \int_B p(x, A) P^\xi(dx).$$

Furthermore, if $\{p'(x, A)\}$ is any other such system, then there exists a set $N \in \mathscr{B}$ of P^ξ-measure 0 such that $x \notin N$ implies

$$p(x, A) = p'(x, A) \qquad \text{for all} \quad A \in \mathscr{F}.$$

Thus, for every integrable random variable X, $\int_\Omega X(\omega) p(x, d\omega)$ coincides with $E(X \mid \xi = x)$, P^ξ-a.e. x. $p(x, A)$ is called the *regular conditional probability given* $\xi = x$. In the same way as Theorem 3.2 and its corollary we can prove the following:

Corollary. We assume further in Theorem 3.3 that $\mathscr{B}$ is countably determined and $\{x\} \in \mathscr{B}$ for every $x \in S$. Then there exists a set $N \in \mathscr{B}$ of P^ξ-measure 0 such that if $x \notin N$, $p(x, \{\omega; \xi(\omega) \in B\}) = I_B(x)$ for every $B \in \mathscr{B}$. In particular, if $x \notin N$ then

(3.2) $\quad p(x, \{\omega; \xi(\omega) = x\}) = 1.$

4. Continuous stochastic processes

Let $\boldsymbol{W}^d = \boldsymbol{C}([0, \infty) \longrightarrow \boldsymbol{R}^d)$ be the set of all continuous functions $w: [0, \infty) \ni t \longmapsto w(t) \in \boldsymbol{R}^d$. We define a metric ρ on $\boldsymbol{W}^d$ by

$$(4.1) \qquad \rho(w_1, w_2) = \sum_{n=1}^{\infty} 2^{-n}[(\max_{0 \leq t \leq n} |w_1(t) - w_2(t)|) \wedge 1], \quad w_1, w_2 \in \boldsymbol{W}^d.$$

It is easy to see that $\boldsymbol{W}^d$ is complete and separable under this metric. Clearly, $w_n \longrightarrow w$ with respect to the metric ρ if and only if $w_n(t)$ converges to $w(t)$ uniformly in t on each bounded interval. Let $\mathscr{B}(\boldsymbol{W}^d)$ be the topological σ-field. By a *Borel cylinder set* we mean a set $B \subset \boldsymbol{W}^d$ of the following form

$$B = \{w; (w(t_1), w(t_2), \ldots, w(t_n)) \in E\}$$

for some sequence $0 \leq t_1 < t_2 < \cdots < t_n$ and $E \in \mathscr{B}(\boldsymbol{R}^{nd})$. The totality of Borel cylinder sets is denoted by $\mathscr{C}$. Since the mapping $w \in \boldsymbol{W}^d \longmapsto (w(t_1), w(t_2), \ldots, w(t_n)) \in \boldsymbol{R}^{nd}$ is continuous, it is clear that $\mathscr{C} \subset \mathscr{B}(\boldsymbol{W}^d)$.

Proposition 4.1. $\sigma[\mathscr{C}] = \mathscr{B}(\boldsymbol{W}^d)$.

Proof. It is only necessary to show that $\sigma[\mathscr{C}] \supset \mathscr{B}(\boldsymbol{W}^d)$. The totality of sets of the form $\{w; \max_{0 \leq t \leq n} |w(t) - w_0(t)| \leq \varepsilon\}$, $w_0 \in \boldsymbol{W}^d$, $\varepsilon > 0$, $n = 1, 2, \ldots$, forms a basis of neighborhoods in $\boldsymbol{W}^d$ and we have

$$\{w; \max_{0 \leq t \leq n} |w(t) - w_0(t)| \leq \varepsilon\} = \bigcap_{r \in \boldsymbol{Q}, 0 \leq r \leq n} \{w; w(r) \in U(w_0(r), \varepsilon)\}$$

where $U(a, \varepsilon) = \{x \in \boldsymbol{R}^d; |x - a| \leq \varepsilon\}$. Thus, such a set is representable as a countable intersection of sets in $\mathscr{C}$. Now $\mathscr{B}(\boldsymbol{W}^d) \subset \sigma[\mathscr{C}]$ is obvious.

Corollary. Every probability on $(\boldsymbol{W}^d, \mathscr{B}(\boldsymbol{W}^d))$ is uniquely determined by its values on $\mathscr{C}$.

Definition 4.1. By a *d-dimensional continuous process* X we mean a $\boldsymbol{W}^d$-valued random variable defined on a probability space $(\Omega, \mathscr{F}, P)$, i.e., a mapping $X: \Omega \longrightarrow \boldsymbol{W}^d$ which is $\mathscr{F}/\mathscr{B}(\boldsymbol{W}^d)$-measurable.

Thus, if X is a d-dimensional continuous process, then for each ω,

$X(\omega) \in W^d$. The value of $X(\omega)$ at $t \in [0, \infty)$ * is denoted by $X_t(\omega)$ or $X(t, \omega)$. For fixed t, $\omega \longmapsto X_t(\omega)$ is a d-dimensional random variable. Conversely, a collection $\{X_t(\omega)\}_{t\in[0,\infty)}$ of d-dimensional random variables determines a d-dimensional continuous process if $t \longmapsto X(t)$ is continuous with probability one.

We say that two d-dimensional processes X and X' have the same law, denoted by $X \overset{\mathscr{L}}{\approx} X'$, if their probability laws P^X and $P^{X'}$ coincide. Since P^X and $P^{X'}$ are determined by their values on $\mathscr{C}$, $X \overset{\mathscr{L}}{\approx} X'$ if and only if all their finite dimensional distributions coincide : a finite dimensional distribution of X for a given sequence of times $0 \leq t_1 < t_2 < \cdots < t_n$, is the probability law of nd-dimensional random variable $(X_{t_1}, X_{t_2}, \ldots, X_{t_n})$.

Theorem 4.2. Let $X_n = \{X_n(t)\}$, $n = 1,2, \ldots$, be a sequence of d-dimensional continuous processes satisfying the following two conditions:

(4.2) $$\lim_{N\to\infty} \sup_n P\{|X_n(0)| > N\} = 0;$$

(4.3) for every $T > 0$ and $\varepsilon > 0$,
$$\lim_{h\downarrow 0} \sup_n P\{\max_{\substack{t,s\in[0,T]\\|t-s|\leq h}} |X_n(t) - X_n(s)| > \varepsilon\} = 0.$$

Then there exists a subsequence $n_1 < n_2 < \cdots < n_k \cdots \longrightarrow \infty$, a probability space $(\hat{\Omega}, \hat{\mathscr{F}}, \hat{P})$ and d-dimensional continuous processes $\hat{X}_{n_k} = (\hat{X}_{n_k}(t))$, $k = 1, 2, \ldots$ and $\hat{X} = (\hat{X}(t))$ defind on it such that

(4.4) $$\hat{X}_{n_k} \overset{\mathscr{L}}{\approx} X_{n_k}, \quad k = 1,2, \ldots,$$

(4.5) $\hat{X}_{n_k}$ converges to $\hat{X}$ almost everywhere as $k \longrightarrow \infty$, i.e.,
$$\hat{P}\{\hat{\omega};\ \rho(\hat{X}_{n_k}(\hat{\omega}), \hat{X}(\hat{\omega})) \longrightarrow 0 \text{ as } k \longrightarrow \infty\} = 1.$$

Furthermore, if every finite dimensional distribution of P^{X_n} converges as $n \longrightarrow \infty$, then we need not take a subsequence: we can construct $\hat{X}_n$, $n = 1, 2, \ldots$ and $\hat{X}$ such that (4.4) and (4.5) hold for $n = 1, 2, \ldots$ and as $n \longrightarrow \infty$ respectively.

Proof. First we show that $\{P^{X_n}\}$ is tight (cf. Definition 2.2) if (4.2) and (4.3) are satisfied. By the Ascoli-Arzelà theorem, a subset $A \subset W^d$ is relatively compact in W^d if and only if it is both

(i) uniformly bounded, i.e., for every $T > 0$, $\sup_{w\in A} \max_{t\in[0,T]} |w(t)| < \infty$,

* $t \in [0, \infty)$ is considered as *time*.

and

(ii) equi-continuous, i.e., for every $T > 0$, $\lim_{h \downarrow 0} \sup_{w \in A} V_h^T(w) = 0$, where $V_h^T(w) = \max_{\substack{t,s \in [0,T] \\ |t-s| \leq h}} |w(s) - w(t)|$.

We see by (4.2) that for every $\varepsilon > 0$ there exists $a > 0$ such that $P^{X_n}\{w; |w(0)| \leq a\} > 1 - \varepsilon/2$ for all n. Also, by (4.3), for every $\varepsilon > 0$ and $k = 1, 2, \ldots$ there exists $h_k > 0$ such that $h_k \downarrow 0$ and $P^{X_n}\{w;\ V_{h_k}^k(w) > 1/k\} \leq \varepsilon/2^{k+1}$ for all n. Thus $P^{X_n}[\bigcap_{k=1}^{\infty} \{w;\ V_{h_k}^k(w) \leq 1/k\}] > 1 - \varepsilon/2$. Set $K_\varepsilon = \{w; |w(0)| \leq a\} \cap (\bigcap_{k=1}^{\infty} \{w; V_{h_k}^k(w) \leq 1/k\})$. Then clearly K_ε satisfies (i) and (ii) and hence it is compact. Now $P^{X_n}(K_\varepsilon) > 1 - \varepsilon$ showing that $\{P^{X_n}\}$ is tight. So, by Theorem 2.6 (1), a subsequence $\{n_k\}$ exists such that $P^{X_{n_k}} \xrightarrow{w} P$ for some probability P on $(W^d, \mathscr{B}(W^d))$. We now apply Theorem 2.7 to construct $\hat{X}_{n_k}$ and $\hat{X}$ with properties as above. If every finite dimensional distribution of P^{X_n} converges, then clearly a limit point of $\{P^{X_n}\}$ is unique and hence P^{X_n} itself converges weakly to P as $n \longrightarrow \infty$.

Theorem 4.3. Let $X_n = (X_n(t))$, $n = 1, 2, \ldots$, be a sequence of d-dimensional continuous processes satisfying the following two conditions:

(4.6) there exist positive constants M and γ such that $E\{|X_n(0)|^\gamma\} \leq M$ for every $n = 1, 2, \ldots$,

(4.7) there exist positive constants α, β, M_k, $k = 1,2, \ldots$, such that $E\{|X_n(t) - X_n(s)|^\alpha\} \leq M_k|t - s|^{1+\beta}$ for every n and $t,s \in [0,k]$, $(k = 1, 2, \ldots)$.

Then $\{X_n\}$ satisfies the conditions (4.2) and (4.3) of Theorem 4.2.

Proof. By Chebyshev's inequality,

$$P\{|X_n(0)| > N\} \leq M/N^\gamma, \qquad n = 1, 2, \ldots,$$

and (4.2) is clearly satisfied. We now prove (4.3). We may take T to be an integer. By (4.7), $Y(t) = X_n(t)$, $n = 1, 2, \ldots$, satisfies

$$E\{|Y(t) - Y(s)|^\alpha\} \leq M_T|t - s|^{1+\beta}, \qquad t, s \in [0, T].$$

By Chebyshev's inequality,

$$(4.8) \qquad \begin{aligned} P\{|Y((i+1)/2^m) - Y(i/2^m)| > 1/2^{ma}\} &\leq M_T 2^{-m(1+\beta)} 2^{ma\alpha} \\ &= M_T 2^{-m(1+\beta-a\alpha)}, \qquad i = 0,1,2, \ldots, 2^m T - 1. \end{aligned}$$

We choose a such that $0 < a < \beta/\alpha$. By (4.8), we have

$$P\{\max_{0\le i\le 2^m T-1} |Y((i+1)/2^m) - Y(i/2^m)| > 1/2^{ma}\} \le M_T T 2^{-m(\beta-a\alpha)}. \tag{4.9}$$

Let $\varepsilon > 0$ and $\delta > 0$ be given. Choose $\nu = \nu(\delta, \varepsilon)$ such that $(1 + 2/(2^a - 1))/2^{\nu a} \le \varepsilon$ and

$$\begin{aligned} &P[\bigcup_{m=\nu}^{\infty} \{\max_{0\le i\le 2^m T-1} |Y((i+1)/2^m) - Y(i/2^m)| > 1/2^{ma}\}] \\ &\qquad \le M_T T \sum_{m=\nu}^{\infty} 2^{-m(\beta-a\alpha)} < \delta. \end{aligned} \tag{4.10}$$

Set $\Omega_\nu = \bigcup_{m=\nu}^{\infty} \{\max_{0\le i\le 2^m T-1} |Y((i+1)/2^m) - Y(i/2^m)| > 1/2^{ma}\}$. Then $P(\Omega_\nu) < \delta$ and if $\omega \notin \Omega_\nu$

$$|Y((i+1)/2^m) - Y(i/2^m)| \le 1/2^{ma}$$

for all $m \ge \nu$ and i such that $(i+1)/2^m \le T$. Let $\boldsymbol{D}_T$ be the set of all binary rationals in the interval $[0, T]$. If $s \in \boldsymbol{D}_T$ is in the interval $[i/2^\nu, (i+1)/2^\nu)$, then s is written as $s = i/2^\nu + \sum_{l=1}^{j} \alpha_l/2^{\nu+l}$, where α_l is 0 or 1, and hence, if $\omega \notin \Omega_\nu$,

$$\begin{aligned} |Y(s) - Y(i/2^\nu)| &\le \sum_{k=1}^{j} |Y(i/2^\nu + \sum_{l=1}^{k} \alpha_l/2^{\nu+l}) - Y(i/2^\nu + \sum_{l=1}^{k-1} \alpha_l/2^{\nu+l})| \\ &\le \sum_{k=1}^{j} 1/2^{(\nu+k)a} \le \sum_{k=1}^{\infty} 1/2^{(\nu+k)a} = 1/(2^a - 1)2^{\nu a}. \end{aligned}$$

Therefore, if $s, t \in \boldsymbol{D}_T$, $|s - t| \le 1/2^\nu$ and $\omega \notin \Omega_\nu$, then

$$|Y(s) - Y(t)| \le (1 + 2/(2^a - 1))/2^{\nu a} \le \varepsilon. \tag{4.11}$$

Indeed, if $t \in [(i-1)/2^\nu, i/2^\nu)$ and $s \in [i/2^\nu, (i+1)/2^\nu)$, then

$$\begin{aligned} |Y(t) - Y(s)| &\le |Y(s) - Y(i/2^\nu)| + |Y(t) - Y((i-1)/2^\nu)| \\ &\quad + |Y(i/2^\nu) - Y((i-1)/2^\nu)| \le (1 + 2/(2^a - 1))/2^{\nu a} \end{aligned}$$

and if $t, s \in [i/2^\nu, (i+1)/2^\nu)$, then

$$\begin{aligned} |Y(t) - Y(s)| &\le |Y(s) - Y(i/2^\nu)| + |Y(t) - Y(i/2^\nu)| \\ &\le 2/(2^a - 1)2^{\nu a}. \end{aligned}$$

Since $\boldsymbol{D}_T$ is dense in $[0, T]$, (4.11) holds for every $s, t \in [0, T]$ such that $|s - t| \le 1/2^\nu$. Therefore,

$$P\{\max_{\substack{t,s\in[0,T]\\|t-s|\leq 1/2^\nu}} |Y(t)-Y(s)| > \varepsilon\} \leq P(\Omega_\nu) < \delta.$$

Since $\nu = \nu(\varepsilon, \delta)$ is independent of n, we have proved (4.3).

Corollary. Let $\{X(t)\}_{t\in[0,\infty)}$ be a system of d-dimensional random variables such that for some positive constants α, β and M_k, $k = 1, 2, \ldots,$ the following condition (4.12) holds:

$$(4.12) \qquad E(|X(t)-X(s)|^\alpha) \leq M_k|t-s|^{1+\beta}$$

for every $t, s \in [0, k]$, $k = 1, 2, \ldots$.

Then there exists a d-dimensional continuous process $\hat{X} = (\hat{X}(t))$ such that for every $t \in [0, \infty)$, $P[X(t) = \hat{X}(t)] = 1$.

Proof. Just as in the above proof, (4.9) holds for $Y(t) = X(t)$ and so, by Borel-Cantelli's lemma ([129]), we have for almost all ω

$$|X((i+1)/2^m) - X(i/2^m)| \leq 1/2^{ma}, \quad i = 0, 1, 2, \ldots, 2^mT-1,$$
$$\text{for all} \quad m \geq \nu = \nu(\omega).$$

This implies, as in the above proof, that $|X(t)-X(s)| \leq (1+2/(2^a-1))/2^{ma}$ if $t,s\in\boldsymbol{D}_T$, $|t-s| \leq 1/2^m$ and $m \geq \nu$. Consequently, $t \in \boldsymbol{D}_T \longmapsto X(t)$ is uniformly continuous almost surely. Let $\hat{X}(t)$ be the continuous extension of $X(t)|_{\bigcup_{k=1}^\infty \boldsymbol{D}_k}$. Then $\hat{X} = (\hat{X}(t))$ is a continuous process and it is easy to prove that $P[X(t) = \hat{X}(t)] = 1$ for every $t \in [0, \infty)$.

5. Stochastic processes adapted to an increasing family of sub σ-fields

Let $(\Omega, \mathscr{F}, P)$ be a probability space and $(\mathscr{F}_t)_{t\geq 0}$ be an increasing family of sub σ-fields of $\mathscr{F}$: i.e.,

$$(5.1) \qquad \mathscr{F}_t \subset \mathscr{F}_s \qquad \text{if} \quad 0 \leq t \leq s.$$

$(\mathscr{F}_t)$ is called *right-continuous* if $\mathscr{F}_{t+0} := \bigcap_{\varepsilon>0} \mathscr{F}_{t+\varepsilon} = \mathscr{F}_t$ for every $t \in [0, \infty)$. *In the following, unless otherwise stated,* $(\mathscr{F}_t)$ *is assumed to be right continuous.* Such a family $(\mathscr{F}_t)$ is called a *reference family*. Suppose we are given $(\Omega, \mathscr{F}, P)$ and $(\mathscr{F}_t)$. By a d-dimensional stochastic process*

* We consider here the d-dimensional case (i.e., the state space is $\boldsymbol{R}^d$) but the generalization to an arbitrary state space is obvious.

we mean a family of d-dimensional random variables $X = (X_t)$. In this section, a d-dimensional stochastic process is simply called a process.

Definition 5.1. A process $X = (X_t)_{t\geq 0}$ is called *adapted to* $(\mathscr{F}_t)$ *[1] if X_t is $\mathscr{F}_t$-measurable for every t.

Generally, a process $X = (X_t)_{t\geq 0}$ is called *measurable* if the mapping

$$(t, \omega) \in [0, \infty) \times \Omega \longmapsto X_t(\omega) \in \boldsymbol{R}^d$$

is $\mathscr{B}([0, \infty)) \times \mathscr{F}/\mathscr{B}(\boldsymbol{R}^d)$-measurable.

Let $\mathscr{S}$ be the smallest σ-field on $[0, \infty) \times \Omega$ such that all left-continuous $(\mathscr{F}_t)$-adapted processes Y: $[0,\infty) \times \Omega \ni (t,\omega) \longmapsto Y_t(\omega) \in \boldsymbol{R}^d$ are measurable. If we replace left-continuous Y by right-continuous Y in the above, the corresponding σ-field is denoted by $\mathscr{T}$.

Definition 5.2. A process $X = (X_t)$ is called *predictable**[2] if the mapping $(t,\omega) \longmapsto X_t(\omega)$ is $\mathscr{S}/\mathscr{B}(\boldsymbol{R}^d)$-measurable; it is called *well measurable**[3] if the corresponding mapping is $\mathscr{T}/\mathscr{B}(\boldsymbol{R}^d)$-measurable.

Clearly, both predictable and well measurable processes are measurable and adapted to $(\mathscr{F}_t)$.

Remark 5.1. A predictable process is a well measurable process. Indeed, if $X = (X_t)$ is left-continuous, i.e., $t \longmapsto X_t$ is left-continuous for every ω, then $X_n = (X^{(n)}_t)$ defined by $X^{(n)}_t = X_{k/2^n}$ for $t \in [k/2^n, (k+1)/2^n)$ is right continuous and $X^{(n)}_t(\omega) \longrightarrow X_t(\omega)$ as $n \longrightarrow \infty$. Thus X is $\mathscr{T}$-measurable. This implies that $\mathscr{S} \subset \mathscr{T}$.

Proposition 5.1.*[4] Let $\boldsymbol{\Phi}$ be a linear space of real and bounded*[5] measurable processes satisfying the following two conditions:

(i) $\boldsymbol{\Phi}$ contains all bounded, left (resp. right)-continuous $(\mathscr{F}_t)$-adapted processes;

(ii) if $\{\Phi_n\}$ is a monotone increasing sequence of processes in $\boldsymbol{\Phi}$ such that $\Phi = \sup_n \Phi_n$ is bounded, then $\Phi \in \boldsymbol{\Phi}$.

Then $\boldsymbol{\Phi}$ contains all bounded predictable (resp. well measurable) processes.

Proof. A bounded predictable process, i.e., a bounded $\mathscr{S}$-measurable

*[1] We also say $(\mathscr{F}_t)$-adapted.
*[2] To be precise, *predictable with respect to* $(\mathscr{F}_t)$. We also say $(\mathscr{F}_t)$-*predictable*.
*[3] To be precise, $(\mathscr{F}_t)$-*well measurable*. It is also called *optional*.
*[4] In this proposition, we assume $d = 1$.
*[5] That is, it is bounded as a function from $[0, \infty) \times \Omega$ into $\boldsymbol{R}$.

function, is a monotone increasing limit of $\mathscr{S}$-measurable simple functions. Hence, it is enough to prove $I_B \in \Phi$ for $B \in \mathscr{S}$. Set

$$\mathscr{S}' = \{B; B \subset [0, \infty) \times \Omega \text{ such that } I_B \in \Phi\}.$$

Then, since Φ is a linear space and $\mathbf{1} \in \Phi$, $\mathscr{S}'$ has the following properties:

(i) $[0, \infty) \times \Omega \in \mathscr{S}'$;

(ii) $A, B \in \mathscr{S}', A \subset B \Rightarrow B \setminus A \in \mathscr{S}'$;

(iii) $A_n \in \mathscr{S}', A_n \subset A_{n+1}, n = 1, 2, \ldots \Rightarrow \bigcup_n A_n \in \mathscr{S}'$.

Let Y_i $(i = 1, 2, \ldots, k)$ be a left-continuous $(\mathscr{F}_t)$-adapted process and E_i $(i = 1, 2, \ldots, k)$ be an open set in $\mathbf{R}^1$. Then $\bigcap_{i=1}^k Y_i^{-1}(E_i) \in \mathscr{S}'$. Indeed, $I_{\bigcap_{i=1}^k Y_i^{-1}(E_i)}(t, \omega) = \prod_{i=1}^k I_{E_i}(Y_i(t, \omega))$ and since there exists a sequence of bounded continuous functions ϕ_n^i on $\mathbf{R}$ such that $\phi_n^i(x) \uparrow I_{E_i}(x)$ as $n \longrightarrow \infty$,

$$\prod_{i=1}^k \phi_n^i(Y_i(t, \omega)) \uparrow \prod_{i=1}^k I_{E_i}(Y_i(t, \omega)).$$

The left-hand side being a bounded left-continuous $\mathscr{F}_t$-adapted process is in Φ and hence so is the right-hand side.

But the totality $\mathscr{C}$ of sets of the form $\bigcap_{i=1}^k Y_i^{-1}(E_i)$ is closed under the finite intersection and $\sigma[\mathscr{C}] = \mathscr{S}$ by the definition. Then $\mathscr{S} \subset \mathscr{S}'$ follows from the lemma below.

Lemma 5.1.*[1] Generally, if a family $\mathscr{C}$ of subsets of $[0, \infty) \times \Omega$ satisfies (i), (ii) and (iii) above, it is called a *d-system.* If $\mathscr{C}$ is closed under finite intersections, it is called a *π-system.* Let the smallest d-system containing $\mathscr{C}$ be denoted by $d[\mathscr{C}]$. Then for every π-system $\mathscr{C}$, $d[\mathscr{C}] = \sigma[\mathscr{C}]$.

The proof is standard and is left to the reader.

Definition 5.3. Let $(\Omega, \mathscr{F}, P)$ and $(\mathscr{F}_t)_{t \geq 0}$ be given. A mapping $\sigma: \Omega \longrightarrow [0, \infty]$ is called a *stopping time**[2] (with respect to $(\mathscr{F}_t)$), if, for every $t \geq 0$, $\{\omega; \sigma(\omega) \leq t\} \in \mathscr{F}_t$. For a stopping time σ, we set

$$(5.2) \qquad \mathscr{F}_\sigma = \{A \in \mathscr{F}; \forall t \in [0, \infty), A \cap \{\sigma(\omega) \leq t\} \in \mathscr{F}_t\}.$$

Clearly, $\mathscr{F}_\sigma$ is a sub σ-field of $\mathscr{F}$ and if $\sigma(\omega) \equiv t$, then $\mathscr{F}_\sigma = \mathscr{F}_t$.

---

*[1] Due to Dynkin (cf. [21]).

*[2] Sometimes, it is called a *Markov time.*

Example 5.1. If $X = (X_t)$ is a right-continuous, $(\mathscr{F}_t)$-adapted process and $E \subset \boldsymbol{R}^d$ is an open set in $\boldsymbol{R}^d$, then the *first hitting time* σ_E to the set E, defined by

(5.3) $$\sigma_E(\omega) = \inf\{t > 0;\ X_t(\omega) \in E\}\ ^{*1}$$

is a stopping time. Indeed,

$$\begin{aligned}\{\sigma_E(\omega) \leq t\} &= \bigcap_n \{\sigma_E(\omega) < t + 1/n\} \\ &= \bigcap_n \bigcup_{\substack{r \in \boldsymbol{Q} \\ r < t + 1/n}} \{X_r(\omega) \in E\} \in \mathscr{F}_{t+0} = \mathscr{F}_t.\end{aligned}$$

It is known that if $(\mathscr{F}_t)$ satisfies $\overline{\mathscr{F}}_t^P = \mathscr{F}_t$ for every t or, more generally, if for some system $\{P_\alpha\}$ of probabilities on $(\Omega, \mathscr{F})$, $\bigcap_\alpha \overline{\mathscr{F}}_t^{P_\alpha} = \mathscr{F}_t$ for every t, then σ_E is a stopping time for every Borel subset $E \subset \boldsymbol{R}^d$ for every well measurable X (cf. [15] and [117]).

Proposition 5.2. Let $\sigma, \tau, \sigma_n, n = 1, 2, \ldots$ be stopping times. Then
(i) $\sigma \vee \tau, \quad \sigma \wedge \tau$,
(ii) $\sigma = \lim_n \sigma_n$, when $\sigma_n \uparrow$ or $\sigma_n \downarrow$,

are all stopping times.

Proof. Since $\{\sigma \vee \tau \leq t\}^{*2} = \{\sigma \leq t\} \cap \{\tau \leq t\}$, $\sigma \vee \tau$ is a stopping time. Similarly, $\sigma \wedge \tau$ is also a stopping time. If $\sigma_n \uparrow \sigma$ then $\{\sigma \leq t\} = \bigcap_n \{\sigma_n \leq t\}$ and hence σ is a stopping time. If $\sigma_n \downarrow \sigma$, then $\{\sigma < t\} = \bigcup_n \{\sigma_n < t\}$ and by the following lemma, σ is a stopping time.

Lemma 5.2. σ is a stopping time if and only if $\{\sigma < t\} \in \mathscr{F}_t$ for every t.

Proof. If σ is a stopping time then $\{\sigma < t\} = \bigcup_n \{\sigma(\omega) \leq t - 1/n\} \in \mathscr{F}_t$. Conversely, if $\{\sigma < t\} \in \mathscr{F}_t$ for every t, then $\{\sigma(\omega) \leq t\} = \bigcap_n \{\sigma(\omega) < t + 1/n\} \in \mathscr{F}_{t+0} = \mathscr{F}_t$.

Proposition 5.3. Let $\sigma, \tau, \sigma_n, n = 1, 2, \ldots,$ be stopping times.
(1) If $\sigma(\omega) \leq \tau(\omega)$ for all ω, then $\mathscr{F}_\sigma \subset \mathscr{F}_\tau$.
(2) If $\sigma_n(\omega) \downarrow \sigma(\omega)$ for all ω, then $\bigcap_n \mathscr{F}_{\sigma_n} = \mathscr{F}_\sigma$.

---

*1 We define always $\inf \phi = \infty$ unless otherwise stated.
*2 We often write $\{X \in A\}$ for $\{\omega;\ X(\omega) \in A\}$.

Proof. (1) follows from the definition. If $\sigma_n \downarrow \sigma$, then $\bigcap_n \mathscr{F}_{\sigma_n} \supset \mathscr{F}_\sigma$ by (1). Let $A \in \bigcap_n \mathscr{F}_{\sigma_n}$. Then, as in Lemma 5.2, $A \cap \{\sigma_n < t\} \in \mathscr{F}_t$, for every $(t, n) \in [0, \infty) \times \{1, 2, \dots\}$. Hence $A \cap \{\sigma < t\} = \bigcup_n [A \cap \{\sigma_n < t\}] \in \mathscr{F}_t$ and so $A \in \mathscr{F}_\sigma$.

Proposition 5.4. Let σ be a stopping time. Then the following hold.

(1) $\sigma: \Omega \ni \omega \longmapsto \sigma(\omega) \in [0, \infty]$ is $\mathscr{F}_\sigma/\mathscr{B}([0, \infty])$-measurable.

(2) If $X = (X_t)_{t \geq 0}$ is well-measurable, the mapping
$$X_\sigma: \Omega_\sigma = \{\omega; \sigma(\omega) < \infty\} \ni \omega \longmapsto X_{\sigma(\omega)}(\omega) \in \boldsymbol{R}^d$$
is $\mathscr{F}_\sigma|_{\Omega_\sigma}/\mathscr{B}(\boldsymbol{R}^d)$-measurable.

Proof. (1) is easy and hence omitted. To prove (2), we may assume X is right-continuous by Proposition 5.1. Let $\sigma_n(\omega) = k/2^n$ if $\sigma(\omega) \in [(k-1)/2^n, k/2^n)$. Then σ_n is a stopping time and $\sigma_n \downarrow \sigma$. Hence $X_\sigma = \lim_n X_{\sigma_n}$ on Ω_σ. On the other hand $\{X_{\sigma_n} \in E\} \cap \{\sigma_n \leq t\} = \bigcup_k [\{X_{k/2^n} \in E\} \cap \{\sigma_n = k/2^n\} \cap \{\sigma_n \leq t\}] \in \mathscr{F}_t$. This implies that X_{σ_n} is $\mathscr{F}_{\sigma_n}|_{\Omega_\sigma}$-measurable and hence X_σ is $\bigcap_n \mathscr{F}_{\sigma_n}|_{\Omega_\sigma} = \mathscr{F}_\sigma|_{\Omega_\sigma}$-measurable.

Intuitive reasonings on stopping times are often justified by the following proposition.

Proposition 5.5.* Let σ and τ be stopping times and X be an integrable random variable. Then the following hold:

$$E(I_{\{\sigma>\tau\}} X | \mathscr{F}_\tau) = I_{\{\sigma>\tau\}} E(X | \mathscr{F}_{\sigma\wedge\tau}) \tag{5.4}$$

$$E(I_{\{\sigma\geq\tau\}} X | \mathscr{F}_\tau) = I_{\{\sigma\geq\tau\}} E(X | \mathscr{F}_{\sigma\wedge\tau}) \tag{5.5}$$

$$E\{E(X | \mathscr{F}_\tau) | \mathscr{F}_\sigma\} = E(X | \mathscr{F}_{\tau\wedge\sigma}). \tag{5.6}$$

Proof. First we note that $I_{\{\sigma>\tau\}}$ and $I_{\{\sigma\geq\tau\}}$ are $\mathscr{F}_{\sigma\wedge\tau}$-measurable. Indeed $\{\sigma > \tau\} \cap \{\sigma\wedge\tau \leq t\} = \{\sigma > \tau, \ \tau \leq t\} = \{\sigma > t, \ \tau \leq t\} \cup \{\tau < \sigma \leq t\} \in \mathscr{F}_t$ for every $t \geqslant 0$ and hence $\{\sigma > \tau\} \in \mathscr{F}_{\sigma\wedge\tau}$. Now, $\{\sigma \geq \tau\} = \{\sigma < \tau\}^c \in \mathscr{F}_{\sigma\wedge\tau}$ is obvious. In order to prove (5.4), it is sufficient to show that $E(I_{\{\sigma>\tau\}} X | \mathscr{F}_\tau) = I_{\{\sigma>\tau\}} E(X | \mathscr{F}_\tau)$ is $\mathscr{F}_{\sigma\wedge\tau}$-measurable. This is true because

$$I_{\{\sigma>\tau\}} E(X | \mathscr{F}_\tau) I_{\{\sigma\wedge\tau\leq t\}} = E(X | \mathscr{F}_\tau) I_{\{\tau\leq t\}} I_{\{\sigma>\tau, \tau\leq t\}}$$

is $\mathscr{F}_t$-measurable for every $t \geq 0$. The proof of (5.5) is similar and (5.6) is easily obtained from (5.4) and (5.5).

* Communicated by H. Asano.

So far, we have only considered the continuous time parameter case (i.e., $t \in [0, \infty)$). The discrete time parameter case (i.e., $t = 0, 1, 2, \ldots$) can be handled in an obvious way. The notion of adaptedness and stopping times are defined as before. In this case, however, the notion of measurable processes is trivial. The notion of predictable processess is defined as follows. A process $X = (X_n)_{n=1,2,\ldots}$ is *predictable* with respect to a reference family $(\mathscr{F}_n)$ if X_n is $\mathscr{F}_{n-1}$-measurable for each $n = 1, 2, \ldots$.

6. Martingales

Let $\boldsymbol{T}$ be the time parameter set: $\boldsymbol{T} = \{0, 1, 2, \ldots\}$ in the discrete time case or $\boldsymbol{T} = [0, \infty)$ in the continuous time case. Let $\bar{\boldsymbol{T}} = \boldsymbol{T} \cup \{\infty\}$ be the one-point compactification of $\boldsymbol{T}$. Let $(\Omega, \mathscr{F}, P)$ be a probability space and $(\mathscr{F}_t)_{t \in \boldsymbol{T}}$ be a reference family.*[1]

Definition 6.1. A real stochastic process $X = (X_t)_{t \in \boldsymbol{T}}$ is called a *martingale**[2] (*supermartingale, submartingale*) with respect to $(\mathscr{F}_t)$ if

(i) X_t is integrable for each $t \in \boldsymbol{T}$,

(ii) $X = (X_t)$ is $(\mathscr{F}_t)$-adapted,

(iii) $E(X_t \mid \mathscr{F}_s) = X_s$ (resp. $E(X_t \mid \mathscr{F}_s) \leq X_s$, $E(X_t \mid \mathscr{F}_s) \geq X_s$) a.s. for every $t, s \in \boldsymbol{T}$ such that $s < t$.

First, we consider the discrete time case. Let $f = (f_n)_{n=1,2,\ldots}$ be a bounded and non-negative predictable process; i.e., there exists a constant $M > 0$ such that $0 \leq f_n(\omega) \leq M$ for all n and f_n is $\mathscr{F}_{n-1}$-measurable, $n = 1, 2, \ldots$. Let $X = (X_n)$ be a martingale (supermartingale, submartingale) relative to $(\mathscr{F}_n)$. We define a stochastic process $Y = (Y_n)_{n \in \boldsymbol{T}}$ by

$$
(6.1) \qquad \begin{cases} Y_0 = X_0 \\ Y_n = Y_{n-1} + f_n \cdot (X_n - X_{n-1}), \qquad n = 1, 2, \ldots . \end{cases}
$$

Then it is easy to see that Y is a martingale (resp., supermartingale, submartingale) relative to $(\mathscr{F}_n)$. We denote $Y = f \cdot X$ and call it the *martingale transform* of X by f.

Example 6.1. (Optional stopping). Let $\sigma: \Omega \longrightarrow \bar{\boldsymbol{T}}$ be a stopping time and set $f_n = I_{\{n \leq \sigma\}}$, $n = 1, 2, \ldots$. $f = (f_n)$ is predictable because $\{n \leq \sigma\} = [\bigcup_{k=0}^{n-1} \{\sigma = k\}]^c$ and clearly $X^\sigma := f \cdot X$ is given by $X_n^\sigma = X_{n \wedge \sigma}$, $n =$

*[1] In the continuous time case $(\mathscr{F}_t)$ is assumed to be right-continuous.

*[2] Sometimes, X is said to be an $(\mathscr{F}_t)$-martingale or a $(P, \mathscr{F}_t)$-martingale.

0, 1, 2, ... Thus the process X^σ obtained by stopping a process X at time σ is a martingale (supermartingale or submartingale) if X is one.

Theorem 6.1. (Optional sampling theorem).*[1] Let $X = (X_n)_{n\in T}$ be a martingale (supermartingale, submartingale) relative to $(\mathscr{F}_n)$ and let σ and τ be bounded*[2] stopping times such that $\sigma(\omega) \leq \tau(\omega)$ for all ω. Then

$$(6.2) \qquad E(X_\tau | \mathscr{F}_\sigma) = X_\sigma \qquad (\text{resp.} \leq, \geq) \quad \text{a.s.}$$

In particular,

$$(6.3) \qquad E(X_\tau) = E(X_\sigma) \qquad (\text{resp.} \leq, \geq).$$

Proof. First we show (6.3). Let $m \in T$ be such that $\tau(\omega) \leq m$ for all ω. Set $f_n = I_{\{\sigma<n\leq\tau\}} = I_{\{n\leq\tau\}} - I_{\{n\leq\sigma\}}$, for $n = 1, 2, \ldots$. Then $f = (f_n)$ is predictable as Example 6.1 and $(f\cdot X)_n - X_0 = X_{\tau\wedge n} - X_{\sigma\wedge n}$. In particular, $(f\cdot X)_m - X_0 = X_\tau - X_\sigma$ and hence (6.3) is obvious.

Next we show (6.2). Clearly it is sufficient to prove that if $B \in \mathscr{F}_\sigma$ then

$$(6.4) \qquad E(X_\tau: B) = E(X_\sigma: B) \qquad (\text{resp.} \leq, \geq).$$

Set

$$\sigma_B(\omega) = \begin{cases} \sigma(\omega) , & \omega \in B, \\ m , & \omega \in B^c, \end{cases}$$

and

$$\tau_B(\omega) = \begin{cases} \tau(\omega) , & \omega \in B, \\ m , & \omega \in B^c. \end{cases}$$

Then σ_B and τ_B are stopping times. Indeed,

$$\{\sigma_B \leq j\} = \begin{cases} \Omega , & \text{if } j \geq m, \\ \{\sigma \leq j\} \cap B \in \mathscr{F}_j , & \text{if } j < m, \end{cases}$$

and similarly for τ_B. Consequently by (6.3)

---

*[1] Due to Doob [18].

*[2] σ is bounded if a constant $m \in T$ exists such that $\sigma(\omega) \leq m$, for all ω.

$$E(X_{\tau_B}) = E(X_{\sigma_B}), \qquad \text{(resp. } \leq, \geq \text{)}.$$

That is

$$E(X_\tau: B) + E(X_m: B^c) = E(X_\sigma: B) + E(X_m: B^c) \quad \text{(resp. } \leq, \geq\text{)}$$

proving (6.4).

Corollary. Let $X = (X_t)_{t \in \boldsymbol{T}}$ satisfy the conditions (i) and (ii) of Definition 6.1. Then it is a martingale (supermartingale, submartingale) if and only if (6.3) holds for every bounded stopping times σ and τ such that $\sigma \leq \tau$.

Remark 6.1. In view of Proposition 5.5, (6.2) is equivalent to

$$E(X_\tau | \mathscr{F}_\sigma) = X_{\tau \wedge \sigma} \qquad \text{(resp. } \leq, \geq \text{)}$$

for any bounded stopping times σ and τ.

In the following, we call simply that $X = (X_n)$ is a martingale (supermartingale or submartingale) if X is a martingale (resp. supermartingale or submartingale) with respect to some reference family. Clearly this is equivalent to saying that X is a martingale (supermartingale or submartingale) with respect to the *proper reference family* $(\mathscr{F}_n^X)$ *of* X, where $\mathscr{F}_n^X = \sigma[X_0, X_1, \ldots, X_n]$, $n = 1, 2, \ldots$.

Theorem 6.2.* Let $X = (X_n)_{n \in \boldsymbol{T}}$ be a submartingale. Then for every $\lambda > 0$ and $N \in \boldsymbol{T}$,

$$\lambda P(\max_{0 \leq n \leq N} X_n \geq \lambda) \leq E(X_N: \max_{0 \leq n \leq N} X_n \geq \lambda) \leq E(X_N^+) \leq E(|X_N|). \tag{6.5}$$

and

$$\begin{aligned} \lambda P(\min_{0 \leq n \leq N} X_n \leq -\lambda) &\leq -E(X_0) + E(X_N: \min_{0 \leq n \leq N} X_n > -\lambda) \\ &\leq E(|X_0|) + E(|X_N|). \end{aligned} \tag{6.5$'$}$$

Proof. Set

$$\sigma = \begin{cases} \min\{n \leq N;\ X_n \geq \lambda\}, & \\ N & , \text{ if } \{\ \} = \phi. \end{cases}$$

* Due to Doob (cf. [18]).

Clearly σ is a bounded stopping time such that $\sigma \leq N$. By (6.2), we have

$$\begin{aligned} E(X_N) \geq E(X_\sigma) &= E(X_\sigma\colon \max_{0\leq n\leq N} X_n \geq \lambda) + E(X_N\colon \max_{0\leq n\leq N} X_n < \lambda) \\ &\geq \lambda P(\max_{0\leq n\leq N} X_n \geq \lambda) + E(X_N\colon \max_{0\leq n\leq N} X_n < \lambda). \end{aligned}$$

Thus

$$\begin{aligned} \lambda P(\max_{0\leq n\leq N} X_n \geq \lambda) &\leq E(X_N) - E(X_N\colon \max_{0\leq n\leq N} X_n < \lambda) \\ &= E(X_N\colon \max_{0\leq n\leq N} X_n \geq \lambda). \end{aligned}$$

The inequality (6.5)′ is obtained from $E(X_0) \leq E(X_\tau)$ where

$$\tau = \begin{cases} \min\{n \leq N;\, X_n \leq -\lambda\} \\ N, \qquad \text{if } \{\quad\} = \phi. \end{cases}$$

Corollary. Let $X = (X_n)$ be a martingale such that $E(|X_n|^p) < \infty$, $n = 1, 2, \ldots$ for some $p \geq 1$. Then, for every N,

$$P(\max_{0\leq n\leq N} |X_n| \geq \lambda) \leq E(|X_N|^p)/\lambda^p \tag{6.6}$$

and if $p > 1$,

$$E\{\max_{0\leq n\leq N} |X_n|^p\} \leq (p/(p-1))^p\, E(|X_N|^p). \tag{6.7}$$

The case of $p = 2$ in (6.6) is known as *Doob-Kolmogorov's inequality.*

Proof. By Jensen's inequality (Section 3, (E.7)), $n \longmapsto |X_n|^p$ is a submartingale and so (6.6) follows from Theorem 6.2. As for (6.7), setting $Y = \max_{0\leq n\leq N} |X_n|$ we have by Theorem 6.2 that

$$\lambda P(Y \geq \lambda) \leq \int_\Omega I_{(Y\geq\lambda)}\, |X_N|\, dP.$$

Hence

$$\begin{aligned} E(Y^p) &= \int_\Omega dP \int_0^Y p\lambda^{p-1} d\lambda = \int_\Omega dP \int_0^\infty I_{(Y\geq\lambda)}\, p\lambda^{p-1} d\lambda \\ &= p\int_0^\infty \lambda^{p-1}\, P(Y \geq \lambda) d\lambda \leq p\int_0^\infty\!\!\int_\Omega \lambda^{p-2}\, I_{(Y\geq\lambda)}\, |X_N|\, dP d\lambda \\ &= (p/(p-1)) \int_\Omega Y^{p-1}\, |X_N|\, dP \end{aligned}$$

and, by Hölder's inequality,

$$\begin{aligned} E(Y^p) &\le (p/(p-1))\, E(Y^{p-1}|X_N|) \\ &\le (p/(p-1))\, E(|X_N|^p)^{1/p} E(Y^p)^{(p-1)/p} \end{aligned}$$

from which (6.7) follows.

For a real $(\mathscr{F}_n)$-adapted process $(X_n)_{n\in \boldsymbol{T}}$ and an interval $[a, b]$, we set*

$$\begin{aligned} \tau_1 &= \min\{n;\, X_n \le a\}, \\ \tau_2 &= \min\{n \ge \tau_1;\, X_n \ge b\}, \\ &\vdots \\ \tau_{2k+1} &= \min\{n \ge \tau_{2k};\, X_n \le a\}, \\ \tau_{2k+2} &= \min\{n \ge \tau_{2k+1};\, X_n \ge b\}, \\ &\vdots \end{aligned} \tag{6.8}$$

$\{\tau_n\}$ is clearly an increasing sequence of stopping times. We set

$$U_N^X(a, b)(\omega) = \max\{k;\, \tau_{2k}(\omega) \le N\}. \tag{6.9}$$

Clearly, $U_N^X(a, b)$ is the number of upcrossings of $(X_n)_{n=0}^N$ for the interval $[a, b]$.

Theorem 6.3. Let $X = (X_n)_{n\in \boldsymbol{T}}$ be a submartingale. Then for every $N \in \boldsymbol{T}$ and real numbers a and b such that $a < b$,

$$E(U_N^X(a, b)) \le \frac{1}{b-a}(E\{(X_N - a)^+ - (X_0 - a)^+\}). \tag{6.10}$$

Proof. By Jensen's inequality, $Y = (Y_n)$ where $Y_n = (X_n - a)^+$, $n = 0, 1, 2, \ldots$, is also a submartingale and $U_N^X(a, b) = U_N^Y(0, b-a)$. Let $\tau_1, \tau_2, \ldots$ be defined as in (6.8) with X, a and b replaced by Y, 0 and $b - a$ respectively. Set $\tau'_n = \tau_n \wedge N$. Then if $2k > N$,

$$\begin{aligned} Y_N - Y_0 &= \sum_{n=1}^{2k} (Y_{\tau'_n} - Y_{\tau'_{n-1}}) \\ &= \sum_{n=1}^{k} (Y_{\tau'_{2n}} - Y_{\tau'_{2n-1}}) + \sum_{n=0}^{k-1} (Y_{\tau'_{2n+1}} - Y_{\tau'_{2n}}) \end{aligned} \tag{6.11}$$

* We always set $\min \phi = +\infty$ unless otherwise stated.

and it is not difficult to see that the first term in the right-hand side is greater than or equal to $(b-a)U_N^Y(0, b-a)$. Also $E(Y_{\tau'_{2n+1}}) \geq E(Y_{\tau'_{2n}})$ since Y is a submartingale. Thus (6.10) is obtained by taking expectations in (6.11).

Theorem 6.4. If $X = (X_n)_{n \in T}$ is a submartingale such that

$$(6.12) \qquad \sup_n E(X_n^+) < \infty,$$

then $X_\infty = \lim_{n\to\infty} X_n$ exists almost surely and X_∞ is integrable.*[1]

Proof. Since $E(|X_n|) = 2E(X_n^+) - E(X_n) \leq 2E(X_n^+) - E(X_0)$, we have $\sup_n E(|X_n|) < \infty$ by (6.12). Thus, if $X_\infty = \lim_{n\to\infty} X_n$ exists, then X_∞ is integrable by Fatou's lemma. Clearly, if we set $U_\infty^X(a, b) = \lim_{N\to\infty} U_N^X(a, b)$,

$$\{\omega;\ \underline{\lim}_{n\to\infty} X_n(\omega) < \overline{\lim}_{n\to\infty} X_n(\omega)\} = \bigcup_{r, r' \in Q, r<r'} \{\omega;\ U_\infty^X(r, r') = \infty\}.$$

But, by Theorem 6.3,

$$\begin{aligned} E(U_\infty^X(r, r')) &= \lim_{N\to\infty} E(U_N^X(r, r')) \\ &\leq \frac{1}{r' - r} \lim_{N\to\infty} E\{(X_N - r)^+ - (X_0 - r)^+\} \end{aligned}$$

and this is finite by (6.12). Consequently $P[\underline{\lim}_{n\to\infty} X_n < \overline{\lim}_{n\to\infty} X_n] = 0$, which proves that $\lim_{n\to\infty} X_n$ exists a.s.

Theorem 6.5. Let $X = (X_n)_{n \in T}$ be a submartingale satisfying the condition (6.12) and let $X_\infty = \lim_{n\to\infty} X_n$. In order that $\bar{X} = (X_n)_{n \in \bar{T}}$ be a submartingale, i.e., $X_n \leq E(X_\infty | \mathscr{F}_n)$, $n = 0, 1, 2, \dots$, it is necessary and sufficient that $\{X_n^+\}_{n \in T}$ be equi-integrable.*[2]

*[1] In particular, non-positive submartingales (X_n) or non-negative supermartingales (X_n) possess finite limits $\lim_{n\to\infty} X_n$ a.s.

*[2] The reader is assumed to be familiar with the notion of equi-integrability: a family $\Lambda \subset \mathscr{L}_1(\Omega, \mathscr{F}, P)$ is *equi-integrable* if $\lim_{\lambda\to\infty} \sup_{X \in \Lambda} E(|X| : |X| > \lambda) = 0$. Λ is equi-integrable if and only if Λ is relatively compact in the weak topology $\sigma(\mathscr{L}_1, \mathscr{L}_\infty)$, (cf. [129]).

Proof. If $X_n \leq E(X_\infty | \mathscr{F}_n)$, $n = 0, 1, 2, \ldots$, then, by Jensen's inequality, $X_n^+ \leq E(X_\infty^+ | \mathscr{F}_n)$ and hence $E(X_n^+ : X_n^+ > \lambda) \leq E(X_\infty^+ : X_n^+ > \lambda)$. Also, $P(X_n^+ > \lambda) \leq E(X_n^+)/\lambda \leq E(X_\infty^+)/\lambda$ and thus it is clear that $\{X_n^+\}_{n \in T}$ is equi-integrable.

Conversely, if $\{X_n^+\}_{n \in T}$ is equi-integrable then the almost sure convergence $\lim_{n\to\infty} X_n^+ = X_\infty^+$ is clearly a convergence in $\mathscr{L}_1(\Omega)$. In the same way, $\lim_{n\to\infty} X_n \vee (-a) = X_\infty \vee (-a)$ in the sense of $\mathscr{L}_1(\Omega)$. Since $(X_n \vee (-a))_{n \in T}$ is a submartingale,

$$E(X_\infty \vee (-a) | \mathscr{F}_n) = \lim_{m\to\infty} E(X_m \vee (-a) | \mathscr{F}_n) \geq X_n \vee (-a),$$

and so, letting $a \uparrow \infty$, $E(X_\infty | \mathscr{F}_n) \geq X_n$.

Theorem 6.6. For $Y \in \mathscr{L}_1(\Omega, \mathscr{F}, P)$ set $X_n = E(Y | \mathscr{F}_n)$, $n \in T$. Then $X = (X_n)$ is an equi-integrable martingale and $\lim_{n\to\infty} X_n = X_\infty$ exists almost surely and in $\mathscr{L}_1(\Omega)$. Furthermore,

$$(6.13) \qquad X_\infty = E(Y | \mathscr{F}_\infty),$$

where $\mathscr{F}_\infty = \bigvee_n \mathscr{F}_n$.

Proof. Since $|X_n| \leq E(|Y| | \mathscr{F}_n)$, $n \in T$, $\{X_n\}$ is equi-integrable as in the proof of Theorem 6.5. Also (X_n) is a martingale since $E(X_m | \mathscr{F}_n) = E(E(Y | \mathscr{F}_m) | \mathscr{F}_n) = E(Y | \mathscr{F}_n) = X_n$ for $m > n$. Thus $X_\infty = \lim_{n\to\infty} X_n$ in $\mathscr{L}_1(\Omega)$ and $\bar{X} = (X_n)_{n \in T}$ is a martingale. In order to prove (6.13), consider the totality $\mathscr{C}$ of sets $B \in \mathscr{F}$ such that $E(X_\infty : B) = E(Y : B)$. If $B \in \mathscr{F}_n$, then $E(Y: B) = E(X_n : B) = E(X_\infty : B)$ and hence $\mathscr{C} \supset \bigcup_n \mathscr{F}_n$. Clearly $\mathscr{C}$ is a d-system and hence, by applying Lemma 5.1, we see $\mathscr{C} \supset \mathscr{F}_\infty = \sigma[\bigcup_n \mathscr{F}_n]$. This proves (6.13).

We consider, for a moment, martingales with "reversed" time. Let $(\Omega, \mathscr{F}, P)$ be a probability space and $(\mathscr{F}_n)_{n=0,-1,-2,\ldots}$ be a family of sub σ-fields of $\mathscr{F}$ such that $\mathscr{F}_0 \supset \mathscr{F}_{-1} \supset \mathscr{F}_{-2} \supset \ldots$. $X = (X_n)_{n=0,-1,-2,\ldots}$ is called a martingale (supermartingale, submartingale) if X_n is integrable, $\mathscr{F}_n$-measurable random variable such that $E(X_n | \mathscr{F}_m) = X_m$ (resp. $\leq$, $\geq$) for every $n, m \in \{0, -1, -2, \ldots\}$ satisfying $n > m$.

Theorem 6.7. Let $X = (X_n)_{n=0,-1,-2,\ldots}$ be a submartingale such that

$$(6.14) \qquad \inf_n E(X_n) > -\infty.$$

Then X is equi-integrable and $\lim_{n\to-\infty} X_n = X_{-\infty}$ exists almost surely and in $\mathscr{L}_1(\Omega)$.

Proof. Since $E(X_n)$ is decreasing as $n \downarrow -\infty$, (6.14) implies that $\lim_{n\to-\infty} E(X_n)$ exists finitely as $n \downarrow -\infty$. Let $\varepsilon > 0$ and take k such that $E(X_k) - \lim_{n\to-\infty} E(X_n) < \varepsilon$. Then if $n \leq k$,

$$\begin{aligned} E(|X_n| : |X_n| > \lambda) &= E(X_n : X_n > \lambda) + E(X_n : X_n \geq -\lambda) - E(X_n) \\ &\leq E(X_k : X_n > \lambda) + E(X_k : X_n \geq -\lambda) - E(X_k) + \varepsilon \\ &\leq E(|X_k| : |X_n| > \lambda) + \varepsilon. \end{aligned}$$

Also

$$\begin{aligned} P(|X_n| > \lambda) &\leq \frac{1}{\lambda} E(|X_n|) = \frac{1}{\lambda}(2E(X_n^+) - E(X_n)) \\ &\leq \frac{1}{\lambda}(2E(X_0^+) - \lim_{n\to-\infty} E(X_n)). \end{aligned}$$

Now, we can easily conclude the equi-integrability of (X_n). The almost sure existence of $\lim_{n\to-\infty} X_n$ is proved similarly as in Theorem 6.4.

Now, we consider the continuous time case, i.e., $\boldsymbol{T} = [0, \infty)$. Let $X = (X_t)_{t\in\boldsymbol{T}}$ be a submartingale.

Theorem 6.8. With probability one, $t \in \boldsymbol{T} \cap \boldsymbol{Q} \longmapsto X_t$ is bounded and possesses

$$\lim_{\boldsymbol{Q}\cap\boldsymbol{T}\ni s\downarrow t} X_s \text{ and } \lim_{\boldsymbol{Q}\cap\boldsymbol{T}\ni s\uparrow t} X_s$$

for every $t \geq 0$.

Proof. Let $T > 0$ be given and $\{r_1, r_2, \ldots\}$ be an enumeration of the set $Q \cap [0,T]$. For every n, if $[s_1, s_2, \ldots, s_n]$ is the set $[r_1, r_2, \ldots, r_n]$ arranged according to the natural order, then $Y_i = X_{s_i}$, $i = 1, 2, \ldots n$, defines a submartingale $Y = (Y_i)_{i=1}^n$. Also $\tilde{Y} = (Y_i)_{i=0}^{n+1}$ where $Y_0 = X_0$ and $Y_{n+1} = X_T$ is a submartingale. Therefore by Theorem 6.2 and Theorem 6.3, we have

$$P(\max_{1\leq i\leq n} |Y_i| > \lambda) \leq \frac{1}{\lambda}\{E(|X_0|) + E(|X_T|)\}$$

and

$$E(U_n^Y(a, b)) \leq \frac{1}{b-a} E\{(Y_n - a)^+\} \leq \frac{1}{b-a} E\{(X_T - a)^+\}.$$

Since this holds for every n, we have

$$P(\sup_{t \in Q \cap [0,T]} |X_t| > \lambda) \leq \frac{1}{\lambda}\{E(|X_0|) + E(|X_T|)\}$$

and

$$E(U_\infty^{X|Q \cap [0,T]}(a, b)) \leq \frac{1}{b-a} E\{(X_T - a)^+\}.$$

By letting λ and $a < b$ run over positive integers and pairs of rationals respectively, the assertion of the theorem follows.

By Theorem 6.8, if $X = (X_t)_{t \in \boldsymbol{T}}$ is a submartingale with respect to a reference family $(\mathscr{F}_t)$, then $\hat{X}_t = \lim_{r \downarrow t,\, r \in Q} X_r$, $t \in \boldsymbol{T}$, exists a.s. and it is easy to see that $t \longmapsto \hat{X}_t$ is right-continuous with left-hand limits. $\hat{X} = (\hat{X}_t)$ is also a submartingale with respect to $(\mathscr{F}_t)$. Indeed, $\hat{X}_t$ is $\mathscr{F}_{t+0} = \mathscr{F}_t$-adapted and since for every sequence $\varepsilon_n \downarrow 0$, $\{X_{t+\varepsilon_n}\}$ is equi-integrable by Theorem 6.7, we have

$$E(\hat{X}_t: B) = \lim_{n \to \infty} E(X_{t+\varepsilon_n}: B) \leq \lim_{n \to \infty} E(X_{s+\varepsilon_n}: B) = E(\hat{X}_s: B)$$

for $s > t$ and $B \in \mathscr{F}_t$. Similarly, $E(X_t: B) \leq E(\hat{X}_t: B)$ for every $B \in \mathscr{F}_t$ and hence $X_t \leq \hat{X}_t$ a.s. We summarize these results in the next theorem.

Theorem 6.9. Let $X = (X_t)_{t \in \boldsymbol{T}}$ be a submartingale. Then $\hat{X}_t = \lim_{r \downarrow t,\, r \in Q} X_r$ exists a.s. and $\hat{X} = (\hat{X}_t)_{t \in \boldsymbol{T}}$ is a submartingale such that $t \longmapsto \hat{X}_t$ is right-continuous with left-hand limits a.s. Furthermore, $X_t \leq \hat{X}_t$ a.s. for every $t \in \boldsymbol{T}$.

$\hat{X} = (\hat{X}_t)$ in this theorem is called the *right-continuous modification* of $X = (X_t)$. Clearly $P[X_t = \hat{X}_t] = 1$ for every $t \in \boldsymbol{T}$ if and only if $t \longmapsto E(X_t)$ is right-continuous.

In the remainder of this section, we consider right-continuous martingale only. The first theorem is an easy consequence of Theorem 6.2 and Corollary.

Theorem 6.10. If $X = (X_t)_{t \in \boldsymbol{T}}$ is a right-continuous martingale such that $E(|X_t|^p) < \infty$, then for every $T > 0$,

(6.15) $$P[\sup_{t\in[0,T]} |X_t| > \lambda] \leq E[|X_T|^p]/\lambda^p, \qquad (p \geq 1),$$

(6.16) $$E[\sup_{t\in[0,T]} |X_t|^p] \leq (p/(p-1))^p\, E[|X_T|^p], \qquad (p > 1).$$

The next theorem is a consequence of Theorem 6.1: we approximate a stopping time σ by stopping times σ_n as in the proof of Proposition 5.4, apply Theorem 6.1 for σ_n and then take limits as $n \longrightarrow \infty$ by using Theorem 6.7.

Theorem 6.11. (Optional sampling theorem). Let $X = (X_t)_{t\in \boldsymbol{T}}$ be a right-continuous submartingale with respect to $(\mathscr{F}_t)$ and $(\sigma_t)_{t\in[0,\infty)}$ be a family of bounded stopping times such that $P[\sigma_t \leq \sigma_s] = 1$ if $t < s$. Set $\tilde{X}_t = X_{\sigma_t}$ and $\tilde{\mathscr{F}}_t = \mathscr{F}_{\sigma_t}$ for $t \in \boldsymbol{T}$. Then $\tilde{X} = (\tilde{X}_t)$ is a submartingale with respect to $(\tilde{\mathscr{F}}_t)$.

Finally, we will discuss *Doob-Meyer's decomposition theorem* for submartingales. In the discrete time case, any submartingale $X = (X_n)$ with respect to $(\mathscr{F}_n)$ is decomposed as

(6.17) $$X_n = M_n + A_n, \qquad n = 0, 1, 2, \ldots,$$

where $M = (M_n)$ is a martingale with respect to $(\mathscr{F}_n)$ and $A = (A_n)$ is an increasing process adapted to $(\mathscr{F}_n)$, i.e., $A_n \leq A_{n+1}$ a.s., $n = 0, 1, 2, \ldots$. $A = (A_n)$ is always chosen to be a predictable process (i.e., A_n is $\mathscr{F}_{n-1}$-measurable, $n = 1, 2, \ldots$) such that $A_0 = 0$ a.s. and, under these conditions, the decomposition (6.17) is unique. Indeed, the process defined by

(6.18) $$\begin{cases} A_0 = 0, \\ A_n = A_{n-1} + E(X_n - X_{n-1}|\mathscr{F}_{n-1}), \qquad n = 1, 2, \ldots, \end{cases}$$

is a predictable increasing process and clearly $M = (M_n)$, where $M_n = X_n - A_n$, is a martingale. If we have two decompositions $X_n = M_n + A_n = M'_n + A'_n$, then $M_n - M'_n = A'_n - A_n$ is $\mathscr{F}_{n-1}$-measurable and hence $M_n - M'_n = E[M_n - M'_n|\mathscr{F}_{n-1}] = M_{n-1} - M'_{n-1}$. Since $M_0 - M'_0 = X_0 - X_0 = 0$, we have $M_n - M'_n = 0$ for all n thus proving the uniqueness of such a decomposition.

In the continuous time case, the situation is more difficult.*

---

* In the following, we fix a probability space $(\Omega, \mathscr{F}, P)$ with a reference family $(\mathscr{F}_t)$. Martingales, submartingales, adaptedness, stopping times etc., are all relative to this reference family $(\mathscr{F}_t)$.

Definition 6.2. By an *integrable increasing process*, we mean a process $A = (A_t)$ with the following properties:

(i) A is $(\mathscr{F}_t)$-adapted,

(ii) $A_0 = 0$, $t \longmapsto A_t$ is right continuous and increasing* a.s. (hence $A_t \geq 0$ a.s.),

(iii) $E(A_t) < \infty$ for every $t \in [0, \infty)$.

Definition 6.3. An integrable increasing process $A = (A_t)$ is called *natural* if for every bounded martingale $M = (M_t)$,

$$(6.19) \qquad E\left[\int_0^t M_s dA_s\right] = E\left[\int_0^t M_{s-} dA_s\right]$$

holds for every $t \in [0, \infty)$.

It is known that an integrable increasing process is natural if and only if it is predictable, cf. [15]. If an integrable increasing process is continuous (i.e., $t \longmapsto A_t$ is continuous a.s.), then it is natural. Indeed, $\int_0^t I_{\{M_s \neq M_{s-}\}} dA_s = 0$ a.s. and hence $\int_0^t M_s dA_s = \int_0^t M_{s-} dA_s$ a.s. Note also that (6.19) is equivalent to

$$(6.20) \qquad E(M_t A_t) = E\left(\int_0^t M_{s-} dA_s\right)$$

for every bounded martingale $M = (M_t)$. Indeed, for a partition Δ: $0 = t_0 < t_1 < \cdots < t_n = t$, we define $M^\Delta = (M_t^\Delta)$ by $M_t^\Delta = M_{t_{k+1}}$, $t \in (t_k, t_{k+1}]$. Then $E[M_t A_t] = \sum_{k=1}^{n} E[M_t(A_{t_k} - A_{t_{k-1}})] = \sum_{k=1}^{n} E[M_{t_k}(A_{t_k} - A_{t_{k-1}})] = E[\int_0^t M_s^\Delta dA_s]$, and letting $|\Delta| (:= \max_k (t_k - t_{k-1})) \longrightarrow 0$ we have $E[M_t A_t] = E[\int_0^t M_s dA_s]$.

For $a > 0$, let $\boldsymbol{S}_a$ be the set of all stopping times σ such that $\sigma \leq a$ a.s.

Definition 6.4. A submartingale $X = (X_t)$ is said to be of *class* (DL) if for every $a > 0$ the family of random variables $\{X_\sigma; \sigma \in \boldsymbol{S}_a\}$ is equi-integrable.

Every martingale $M = (M_t)$ is of class (DL) because, by optional sampling theorem (Theorem 6.11),

$$\int_{\{|M_\sigma| > c\}} |M_\sigma| dP \leq \int_{\{|M_\sigma| > c\}} |M_a| dP, \qquad \sigma \in \boldsymbol{S}_a,$$

* When we say "*increasing*", we mean increasing in the wide sense, i.e., "*non-decreasing*". Otherwise, we shall write "*strictly increasing*".

and

$$\sup_{\sigma \in S_a} P(|M_\sigma| > c) \leq \sup_{\sigma \in S_a} E(|M_\sigma|)/c \leq E(|M_a|)/c.$$

Therefore, if a submartingale $X = (X_t)$ is represented as

$$(6.21) \qquad X_t = M_t + A_t$$

where $M = (M_t)$ is a martingale and $A = (A_t)$ is an integrable increasing process, then X is of class (DL). Conversely, we have the following result.

Theorem 6.12. If $X = (X_t)$ is a submartingale of class (DL), then it is expressible as the sum of a martingale $M = (M_t)$ and an integrable increasing process $A = (A_t)$. Furthermore, A can be chosen to be natural and, under this condition, the decomposition is unique.

This theorem is known as Doob-Meyer's decomposition theorem for submartingales.

*Proof.** First we prove that the decomposition (6.21) with A natural is unique. Indeed, if $X_t = M_t + A_t = M'_t + A'_t$ are two such decompositions, then since $A_t - A'_t = M'_t - M_t$ is a martingale, we have, for any bounded martingale m_t,

$$E\left[\int_0^t m_{s-} d(A_s - A'_s)\right] = \lim_{|\Delta| \to 0} E\left[\sum_{k=0}^{n-1} m_{t_k} \{(A_{t_{k+1}} - A'_{t_{k+1}}) - (A_{t_k} - A'_{t_k})\}\right] = 0,$$

where Δ is a partition $0 = t_0 < t_1 < t_2 < \cdots < t_n = t$. Consequently, by (6.20), $E[m_t A_t] = E[m_t A'_t]$. If ξ is a bounded random variable, $m_t = E[\xi \,|\, \mathscr{F}_t]$ is a bounded martingale and hence $E[\xi A_t] = E[m_t A_t] = E[m_t A'_t] = E[\xi A'_t]$. Therefore $A_t = A'_t$ a.s. and so, by the right-continuity of A and A', $A_t = A'_t$ for all t, a.s.

Next we prove that a submartingale $X = (X_t)$ of class (DL) has the decomposition (6.21) with A natural. Because of the uniqueness result, it is sufficient to prove the existence of the decomposition (6.21) on the interval $[0, a]$ for every $a > 0$. Set $Y_t = X_t - E[X_a \,|\, \mathscr{F}_t]$, $t \in [0, a]$. Then Y_t is a non-positive submartingale on $[0, a]$ such that $Y_a = 0$ a.s. For each $n = 1, 2, \ldots,$ let Δ_n be the partition $0 = t_0^{(n)} < t_1^{(n)} < \cdots < t_{2^n}^{(n)} = a$ of $[0, a]$ given by $t_j^{(n)} = ja/2^n$. Define a discrete time increasing process $A_t^{(n)}$, $t \in \Delta_n$, by

* We follow here the proof of Kunita [94] which was originally given by Rao [143].

$$A^{(n)}_{t^{(n)}_k} = \sum_{j=0}^{k-1} \{E(Y_{t^{(n)}_{j+1}} | \mathscr{F}_{t^{(n)}_j}) - Y_{t^{(n)}_j}\}, \qquad t^{(n)}_k \in \Delta_n.$$

Lemma 6.1. $\{A^{(n)}_a, n = 1, 2, \dots\}$ is equi-integrable.

Proof. Let $c > 0$ be fixed and set

$$\sigma^{(n)}_c = \begin{cases} \inf \{t^{(n)}_{k-1};\ A^{(n)}_{t^{(n)}_k} > c\}, \\ a \quad , \quad \text{if } \{\ \ \} = \phi. \end{cases}$$

Then $\sigma^{(n)}_c \in \boldsymbol{S}_a$. Since $Y_t = -E[A^{(n)}_a | \mathscr{F}_t] + A^{(n)}_t$, $t \in \Delta_n$, we have by the optional sampling theorem that $Y_{\sigma^{(n)}_c} = -E[A^{(n)}_a \mid \mathscr{F}_{\sigma^{(n)}_c}] + A^{(n)}_{\sigma^{(n)}_c}$. Hence, noting that $A^{(n)}_{\sigma^{(n)}_c} \le c$,

$$\begin{aligned} E[A^{(n)}_a : A^{(n)}_a > c] &= -E[Y_{\sigma^{(n)}_c} : \sigma^{(n)}_c < a] + E[A^{(n)}_{\sigma^{(n)}_c} : \sigma^{(n)}_c < a] \\ &\le -E[Y_{\sigma^{(n)}_c} : \sigma^{(n)}_c < a] + cP[\sigma^{(n)}_c < a]. \end{aligned}$$

On the other hand

$$\begin{aligned} -E[Y_{\sigma^{(n)}_{c/2}} : \sigma^{(n)}_{c/2} < a] &= E[A^{(n)}_a - A^{(n)}_{\sigma^{(n)}_{c/2}} : \sigma^{(n)}_{c/2} < a] \\ &\ge E[A^{(n)}_a - A^{(n)}_{\sigma^{(n)}_{c/2}} : \sigma^{(n)}_c < a] \ge (c/2)P[\sigma^{(n)}_c < a]. \end{aligned}$$

Consequently

$$E[A^{(n)}_a : A^{(n)}_a > c] \le -E[Y_{\sigma^{(n)}_c} : \sigma^{(n)}_c < a] - 2E[Y_{\sigma^{(n)}_{c/2}} : \sigma^{(n)}_{c/2} < a].$$

By the assumption, $\{X_\sigma, \sigma \in \boldsymbol{S}_a\}$ is equi-integrable and hence $\{Y_\sigma, \sigma \in \boldsymbol{S}_a\}$ is also equi-integrable. Since $P(\sigma^{(n)}_c < a) = P(A^{(n)}_a > c) \le E(A^{(n)}_a)/c \le -E(Y_0)/c \longrightarrow 0$ as $c \uparrow \infty$, we can now conclude that $\{A^{(n)}_a;\ n = 1, 2, \dots\}$ is equi-integrable.

Now we return to the original proof. The equi-integrability of $\{A^{(n)}_a;\ n = 1, 2, \dots\}$ implies that this set is relatively compact in the weak topology $\sigma(\mathscr{L}_1, \mathscr{L}_\infty)$ of the space $\mathscr{L}_1(\Omega)$. Thus, there exists a subsequence n_l, $l = 1, 2, \dots$, and $A_a \in \mathscr{L}_1(\Omega)$ such that $A^{(n_l)}_a \longrightarrow A_a$ in $\sigma(\mathscr{L}_1, \mathscr{L}_\infty)$. Define A_t by $A_t = Y_t + E[A_a | \mathscr{F}_t]$, $t \in [0, a]$.* Since also $A^{(n_l)}_t \longrightarrow A_t$

* $E[A_a | \mathscr{F}_t]$ is taken to be right-continuous.

in $\sigma(\mathscr{L}_1, \mathscr{L}_\infty)$ for each $t \in \bigcup_n \Delta_n$, it is clear that $t \longmapsto A_t$ is increasing. Since $X_t = E[X_a - A_a | \mathscr{F}_t] + A_t$, what remains to be proven is that $A = (A_t)$ is natural. If m_t is a bounded martingale, then

$$\begin{aligned}
E[m_a A_a^{(n)}] &= \sum_{k=0}^{2^n-1} E[m_a(A^{(n)}_{t^{(n)}_{k+1}} - A^{(n)}_{t^{(n)}_k})] \\
&= \sum_{k=0}^{2^n-1} E[m_{t^{(n)}_k}(A^{(n)}_{t^{(n)}_{k+1}} - A^{(n)}_{t^{(n)}_k})] \\
&= \sum_{k=0}^{2^n-1} E[m_{t^{(n)}_k}(Y_{t^{(n)}_{k+1}} - Y_{t^{(n)}_k})] \\
&= \sum_{k=0}^{2^n-1} E[m_{t^{(n)}_k}(A_{t^{(n)}_{k+1}} - A_{t^{(n)}_k})].
\end{aligned}$$

Letting $n \longrightarrow \infty$, we have

$$E[m_a A_a] = E\left[\int_0^a m_{s-} dA_s\right].$$

Replacing $m = (m_s)$ by $m = (m_{t\wedge s})$ for each $t \in [0, a]$ it is easy to conclude that

$$E[m_t A_t] = E\left[\int_0^t m_{s-} dA_s\right].$$

Thus $A = (A_t)$ is natural.

Definition 6.5. A submartingale $X = (X_t)$ is called *regular*, if for every $a > 0$ and $\sigma_n \in \boldsymbol{S}_a$ such that $\sigma_n \uparrow \sigma$, we have

$$E(X_{\sigma_n}) \longrightarrow E(X_\sigma).$$

Theorem 6.13. Let $X = (X_t)$ be a submartingale of class (DL) and $A = (A_t)$ be the natural integrable increasing process in the decomposition (6.21). Then A is continuous if and only if X is regular.

Proof. If A is continuous, then clearly X is regular. Indeed, every martingale X is regular because $E(X_\sigma) = E(X_a)$ for every $\sigma \in \boldsymbol{S}_a$. Conversely, suppose that $X = (X_t)$ is regular. Then $A = (A_t)$ is regular and it is easy to see that if $\sigma_n \uparrow \sigma, \sigma_n \in \boldsymbol{S}_a$, then $A_{\sigma_n} \uparrow A_\sigma$. Define the sequence Δ_n of partitions of $[0, a]$ as in the proof of Theorem 6.12. Set, for $c > 0$,

$$A_t^n = E[A_{t^{(n)}_{k+1}} \wedge c | \mathscr{F}_t], \qquad t \in (t_k^{(n)}, t_{k+1}^{(n)}].$$

Since A_t^n is a martingale on the interval $(t_k^{(n)}, t_{k+1}^{(n)}]$, it is easy to see that

$$(6.22)\qquad E\left[\int_0^t A_s^n dA_s\right] = E\left[\int_0^t A_{s-}^n dA_s\right], \qquad \text{for every } t \in [0, a].$$

We prove now that there exists a subsequence n_l such that $A_t^{n_l}$ converges to $A_t \wedge c$ uniformly in $t \in [0, a]$ as $l \longrightarrow \infty$. For $\varepsilon > 0$ we define

$$\sigma_{n,\varepsilon} = \begin{cases} \inf\{t \in [0, a];\ A_t^n - A_t \wedge c > \varepsilon\}, \\ a \qquad\qquad ,\ \text{if}\ \{\ \ \} = \phi. \end{cases}$$

Since $A_a^n = A_a \wedge c$ for every n, $\sigma_{n,\varepsilon} = a$ implies that $A_t^n - A_t \wedge c \leq \varepsilon$ for all $t \in [0, a]$. Let $\phi_n(t)$ be defined by $\phi_n(t) = t_{k+1}^{(n)}$ for $t \in (t_k^{(n)}, t_{k+1}^{(n)}]$. Then $\sigma_{n,\varepsilon}$ and $\phi_n(\sigma_{n,\varepsilon})$ belong to $\boldsymbol{S}_a$. Since A_t^n is decreasing in n, $\sigma_{n,\varepsilon}$ is increasing in n. Let $\sigma_\varepsilon = \lim_{n\to\infty} \sigma_{n,\varepsilon}$. Then $\sigma_\varepsilon \in \boldsymbol{S}_a$ and $\lim_{n\to\infty} \phi_n(\sigma_{n,\varepsilon}) = \sigma_\varepsilon$. By the optional sampling theorem,* $E[A_{\sigma_{n,\varepsilon}}^n] = E[A_{\phi_n(\sigma_{n,\varepsilon})} \wedge c]$ and hence

$$\begin{aligned} E[A_{\phi_n(\sigma_{n,\varepsilon})} \wedge c - A_{\sigma_{n,\varepsilon}} \wedge c] &= E[A_{\sigma_{n,\varepsilon}}^n - A_{\sigma_{n,\varepsilon}} \wedge c] \\ &\geq \varepsilon P(\sigma_{n,\varepsilon} < a). \end{aligned}$$

It is easy to see by the regularity of A_t, that the left-hand side tends to zero as $n \longrightarrow \infty$. Hence $\lim_{n\to\infty} P(\sigma_{n,\varepsilon} < a) = 0$, which implies that $\lim_{n\to\infty} P(\sup_{t\in[0,a]} |A_t^n - A_t \wedge c| > \varepsilon) = 0$. From this it follows that there exists a subsequence n_l such that

$$\sup_{t\in[0,a]} |A_t^{n_l} - A_t \wedge c| \longrightarrow 0 \qquad \text{as} \quad l \longrightarrow \infty \quad \text{a.s.}$$

Thus, by (6.22), we have

$$E\left[\int_0^t (A_s \wedge c) dA_s\right] = E\left[\int_0^t (A_{s-} \wedge c) dA_s\right]$$

and so

$$0 = E\left[\int_0^t (A_s \wedge c - A_{s-} \wedge c) dA_s\right] \geq E[\sum_{s\leq t} \{(A_s \wedge c) - (A_{s-} \wedge c)\}^2].$$

This clearly implies that $t \longmapsto A_t \wedge c$ is continuous a.s. and hence, since c is arbitrary, $t \longmapsto A_t$ is continuous a.s.

* If we consider the expectation on the set where $\sigma_{n,\varepsilon} \in (t_k^{(n)}, t_{k+1}^{(n)}]$, we may then apply the optional sampling theorem since A_t^n is a martingale on $(t_k^{(n)}, t_{k+1}^{(n)}]$. Now sum over k.

7. Brownian motions

Let $p(t, x)$, $t > 0$, $x \in \boldsymbol{R}^d$, be defined by

$$p(t, x) = (2\pi t)^{-d/2}\exp[-|x|^2/2t]. \tag{7.1}$$

Let $X = (X_t)_{t\in[0,\infty)}$ be a d-dimensional continuous process such that for every $0 < t_1 < \cdots < t_m$ and $E_i \in \mathscr{B}(\boldsymbol{R}^d)$, $i = 1, 2, \ldots, m$,

$$\begin{aligned} &P\{X_{t_1} \in E_1, X_{t_2} \in E_2, \ldots, X_{t_m} \in E_m\} \\ &\quad = \int_{\boldsymbol{R}^d}\mu(dx)\int_{E_1}p(t_1, x_1 - x)dx_1\int_{E_2}p(t_2 - t_1, x_2 - x_1)dx_2 \\ &\qquad \cdots\int_{E_m}p(t_m - t_{m-1}, x_m - x_{m-1})dx_m. \end{aligned} \tag{7.2}$$

Here, μ is a probability on $(\boldsymbol{R}^d, \mathscr{B}(\boldsymbol{R}^d))$. This is equivalent to saying that $X = (X_t)$ is a d-dimensional continuous process such that for every $0 = t_0 < t_1 < \cdots < t_m$, $X_{t_0}, X_{t_1} - X_{t_0}, X_{t_2} - X_{t_1}, \ldots, X_{t_m} - X_{t_{m-1}}$ are mutually independent, $P^{X_{t_0}} = \mu$ and $P^{X_{t_i} - X_{t_{i-1}}}$ = the Gaussian distribution $p(t_i - t_{i-1}, x)dx$, $i = 1, 2, \ldots, m$.

Definition 7.1. A process $X = (X_t)$ with the above property is called a *d-dimensional Brownian motion (*or *Wiener process) with the initial distribution* (or *law*) μ. The probability law P^X on $(\boldsymbol{W}^d, \mathscr{B}(\boldsymbol{W}^d))$ is called the *d-dimensional Wiener measure with the initial distribution* (or *law*) μ.

Thus, the d-dimensional Wiener measure P with the initial law μ is a probability on $(\boldsymbol{W}^d, \mathscr{B}(\boldsymbol{W}^d))$ characterized by the property that $P\{w;\ w(t_1) \in E_1, w(t_2) \in E_2, \ldots, w(t_m) \in E_m\}$ is given by the right-hand side of (7.2).

Theorem 7.1. For any probability μ on $(\boldsymbol{R}^d, \mathscr{B}(\boldsymbol{R}^d))$ the d-dimensional Wiener measure P_μ with the initial distribution μ exists uniquely.

Proof. The uniqueness is obvious. We will show its existence. First, we consider the case $d = 1$ and $\mu = \delta_0$. On a probability space, we construct a family of real random variables $\{X(t),\ t \in [0, \infty)\}$ such that $X(0) = 0$ and for every $0 < t_1 < t_2 < \cdots < t_m$ and $E_i \in \mathscr{B}(\boldsymbol{R}^1)$, $i = 1, 2, \ldots, m$,

$$\begin{aligned} &P\{X(t_1) \in E_1, X(t_2) \in E_2, \ldots, X(t_m) \in E_m\} \\ &\quad = \int_{E_1}p(t_1, x_1)dx_1\int_{E_2}p(t_2 - t_1, x_2 - x_1)dx_2 \end{aligned}$$

$$\cdots\int_{E_m} p(t_m - t_{m-1}, x_m - x_{m-1})dx_m.$$

By Kolmogorov's extension theorem ([45]), such a system exists. Then it is easy to see that

$$(7.3) \qquad E\{|X(t) - X(s)|^4\} = 3|t - s|^2, \qquad t > s \geq 0.$$

By the corollary of Theorem 4.3, there exists a continuous process $\hat{X} = (\hat{X}(t))$ such that for every $t \in [0, \infty)$, $X(t) = \hat{X}(t)$ a.s. The probability law P_0 of $\hat{X}$ on $(\boldsymbol{W}^1, \mathscr{B}(\boldsymbol{W}^1))$ is the Wiener measure with the initial law δ_0. More generally, for $x \in \boldsymbol{R}^1$, the probability law P_x of the process $Y_x = (Y(t))$ with $Y(t) = x + \hat{X}(t)$ is the Wiener measure with the initial law δ_x. Now let $x = (x^1, x^2, \ldots, x^d) \in \boldsymbol{R}^d$ be given and consider the one dimensional Wiener measures P_{x^i}, $i = 1, 2, \ldots, d$. The product measure $P_{x^1} \otimes P_{x^2} \otimes \cdots \otimes P_{x^d}$ on $\boldsymbol{W}^1 \times \boldsymbol{W}^1 \times \cdots \times \boldsymbol{W}^1 = \boldsymbol{W}^d$ is denoted by P_x. It is easy to verify that P_x is the d-dimensional Wiener measure with the initial law δ_x. For any probability μ on $(\boldsymbol{R}^d, \mathscr{B}(\boldsymbol{R}^d))$, $P_\mu(B) = \int_{\boldsymbol{R}^d} P_x(B)\mu(dx)$ defines the d-dimensional Wiener measure with the initial law μ.

On the probability space $(\boldsymbol{W}^d, \mathscr{B}(\boldsymbol{W}^d), P_\mu)$, where P_μ is the d-dimensional Wiener measure with the initial law μ, the coordinate process $X(t, w) = w(t)$, $w \in \boldsymbol{W}^d$, defines a d-dimensional Brownian motion with the initial law μ. This process is called the *canonical realization* of a d-dimensional Brownian motion.

Suppose we are given a probability space $(\Omega, \mathscr{F}, P)$ with a reference family $(\mathscr{F}_t)_{t\in[0,\infty)}$.

Definition 7.2. A d-dimensional continuous process $X = (X(t))_{t\in[0,\infty)}$ is called a *d-dimensional $(\mathscr{F}_t)$-Brownian motion* if it is $(\mathscr{F}_t)$-adapted and satisfies

$$(7.4) \qquad E[\exp[i\langle \xi, X_t - X_s\rangle]|\mathscr{F}_s] = \exp[-(t-s)|\xi|^2/2], \quad \text{a.s.,}$$
$$\text{for every} \quad \xi \in \boldsymbol{R}^d \text{ and } 0 \leq s < t.$$

(7.4) implies that $X_t - X_s$ is independent of $\mathscr{F}_s$ (and hence independent of $\sigma[X_u; u \leq s]$) and that the probability law of $X_t - X_s$ is the Gaussian distribution $p(t - s, x)dx$. Thus X satisfies (7.2) and hence it is a d-dimensional Brownian motion in the sense of Definition 7.1. Conversely, any d-dimensional Brownian motion $X = (X_t)$ is a d-dimensional $(\mathscr{F}_t)$-Brownian motion if $(\mathscr{F}_t)$ is the proper reference family of X; i.e., $\mathscr{F}_t = \bigcap_{\varepsilon>0} \sigma[X_u; u \leq t + \varepsilon]$.

It is easy to verify the following result.

Theorem 7.2. If $X = \{X_t = (X_t^1, X_t^2, \ldots, X_t^d)\}$ is a d-dimensional $(\mathscr{F}_t)$-Brownian motion, then for every $t > s \geq 0$,

(7.5) $\quad E(X_t^i - X_s^i | \mathscr{F}_s) = 0, \quad$ a.s., $\qquad$ and

(7.6) $\quad E((X_t^i - X_s^i)(X_t^j - X_s^j) | \mathscr{F}_s) = (t-s)\delta_{ij} \qquad$ a.s.

Thus, if $E(|X_0|^2) < \infty$, then each (X_t^i), $i = 1, 2, \ldots, d$, is a square-integrable martingale relative to $(\mathscr{F}_t)$ such that $X_t^i X_t^j - \delta_{ij}t$ is a martingale relative to $(\mathscr{F}_t)$, $i, j = 1, 2, \ldots d$. We shall see in Theorem II-6.1 that these properties characterize a d-dimensional $(\mathscr{F}_t)$-Brownian motion.

8. Poisson random measure

Let $(\boldsymbol{X}, \mathscr{B}_{\boldsymbol{X}})$ be a measurable space. Let $\boldsymbol{M}$ be the totality of non-negative (possibly infinite) integral-valued measures on $(\boldsymbol{X}, \mathscr{B}_{\boldsymbol{X}})$ and $\mathscr{B}_{\boldsymbol{M}}$ be the smallest σ-field on $\boldsymbol{M}$ with respect to which all $\mu \in \boldsymbol{M} \longmapsto \mu(B) \in \boldsymbol{Z}^+ \cup \{\infty\}$, $B \in \mathscr{B}_{\boldsymbol{X}}$, are measurable.

Definition 8.1. An $(\boldsymbol{M}, \mathscr{B}_{\boldsymbol{M}})$-valued random variable μ (i.e., a mapping μ: $\Omega \longrightarrow \boldsymbol{M}$ defined on a probability space $(\Omega, \mathscr{F}, P)$ which is $\mathscr{F}/\mathscr{B}_{\boldsymbol{M}}$-measurable) is called a *Poisson random measure* if

(i) for each $B \in \mathscr{B}_{\boldsymbol{X}}$, $\mu(B)$ is Poisson distributed; i.e., $P(\mu(B) = n) = \lambda(B)^n \exp[-\lambda(B)]/n!$, $n = 0, 1, 2, \ldots$, where $\lambda(B) = E(\mu(B))$, $B \in \mathscr{B}_{\boldsymbol{X}}$;*

(ii) if $B_1, B_2, \ldots, B_n \in \mathscr{B}_{\boldsymbol{X}}$ are disjoint, then $\mu(B_1), \mu(B_2), \ldots, \mu(B_n)$ are mutually independent.

Theorem 8.1. Given a σ-finite measure λ on $(\boldsymbol{X}, \mathscr{B}_{\boldsymbol{X}})$, there exists a Poisson random measure μ with $E(\mu(B)) = \lambda(B)$ for every $B \in \mathscr{B}_{\boldsymbol{X}}$.

Proof. Let $U_n \in \mathscr{B}_{\boldsymbol{X}}$ be disjoint, $0 < \lambda(U_n) < \infty$ and $\bigcup_n U_n = \boldsymbol{X}$. On a probability space we construct the following:

(i) for each $n = 1, 2, \ldots$ and $i = 1, 2, \ldots$, $\xi_i^{(n)}$ is a U_n-valued random variable with $P(\xi_i^{(n)} \in du) = \lambda(du)/\lambda(U_n)$;

(ii) p_n, $n = 1, 2, \ldots$, is an integral-valued random variable such that $P(p_n = k) = \lambda(U_n)^k \exp[-\lambda(U_n)]/k!$, $k = 0, 1, \ldots$;

(iii) $\xi_i^{(n)}, p_n$, $n = 1, 2, \ldots$, $i = 1, 2, \ldots$ are mutually independent. Set

---

* If $\lambda(B) = \infty$, then we understand that $\mu(B) = \infty$ a.s.

$$\mu(B) = \sum_{n=1}^{\infty} \sum_{i=1}^{p_n} I_{B \cap U_n}(\xi_i^{(n)}) I_{\{p_n \geq 1\}}, \quad B \in \mathscr{B}_X.$$

It is a simple exercise to prove that $\mu = \{\mu(B)\}$ is a Poisson random measure with $E(\mu(B)) = \lambda(B)$, $B \in \mathscr{B}_X$: just verify for every disjoint $B_1, B_2, \ldots, B_m \in \mathscr{B}_X$ and $\alpha_i > 0$, $i = 1, 2, \ldots, m$, that

$$E(\exp[-\sum_{i=1}^{m} \alpha_i \mu(B_i)]) = \exp[\sum_{i=1}^{m} (e^{-\alpha_i} - 1)\lambda(B_i)].$$

Clearly the probability law of μ is uniquely determined by the measure λ. λ is called the *mean measure* or the *intensity measure* of the Poisson random measure μ.

9. Point processes and Poisson point processes

Let $(\boldsymbol{X}, \mathscr{B}_X)$ be a measurable space. By a *point function p* on $\boldsymbol{X}$ we mean a mapping $p: \boldsymbol{D}_p \subset (0, \infty) \longrightarrow \boldsymbol{X}$, where the domain $\boldsymbol{D}_p$ is a countable subset of $(0, \infty)$. p defines a counting measure $N_p(dtdx)$ on $(0, \infty) \times \boldsymbol{X}$* by

(9.1) $$N_p((0, t] \times U) = \# \{s \in \boldsymbol{D}_p; s \leq t, p(s) \in U\}, \quad t > 0, U \in \mathscr{B}_X.$$

A point process is obtained by randomizing the notion of point functions. Let Π_X be the totality of point functions on $\boldsymbol{X}$ and $\mathscr{B}(\Pi_X)$ be the smallest σ-field on Π_X with respect to which all $p \longmapsto N_p((0, t] \times U)$, $t > 0$, $U \in \mathscr{B}_X$, are measurable.

Definition 9.1. A *point process p* on $\boldsymbol{X}$ is a $(\Pi_X, \mathscr{B}(\Pi_X))$-valued random variable, that is, a mapping $p: \Omega \longrightarrow \Pi_X$ defined on a probability space $(\Omega, \mathscr{F}, P)$ which is $\mathscr{F}/\mathscr{B}(\Pi_X)$-measurable.

A point process p is called *stationary* if for every $t > 0$, p and $\theta_t p$ have the same probability law, where $\theta_t p$ is defined by $\boldsymbol{D}_{\theta_t p} = \{s \in (0, \infty); s + t \in \boldsymbol{D}_p\}$ and $(\theta_t p)(s) = p(s + t)$. A point process p is called *Poisson* if $N_p(dtdx)$ is a Poisson random measure on $(0, \infty) \times \boldsymbol{X}$. A Poisson point process is stationary if and only if its intensity measure $n_p(dt\,dx) = E(N_p(dtdx))$ is of the form

(9.2) $$n_p(dtdx) = dtn(dx)$$

for some measure $n(dx)$ on $(\boldsymbol{X}, \mathscr{B}_X)$. $n(dx)$ is called the *characteristic*

* We endow $(0, \infty) \times \boldsymbol{X}$ with the product σ-field $\mathscr{B}((0, \infty)) \times \mathscr{B}_X$.

measure of p. Given a measure $n(dx)$ on $(\boldsymbol{X}, \mathscr{B}_{\boldsymbol{X}})$, p is a stationary Poisson point process with the characteristic measure n if and only if for every $t > s \geq 0$, disjoint $U_1, U_2, \ldots, U_m \in \mathscr{B}_{\boldsymbol{X}}$ and $\lambda_i > 0$,

$$E(\exp[-\sum_{i=1}^{m} \lambda_i N_p((s, t]\times U_i)] \,|\, \sigma[N_p((0, s']\times U); s' \leq s, U \in \mathscr{B}_{\boldsymbol{X}}])$$

$$= \exp\left[(t-s)\sum_{i=1}^{m}(e^{-\lambda_i}-1)n(U_i)\right] \qquad \text{a.s.}$$

Theorem 9.1. Given a σ-finite measure n on $(\boldsymbol{X},\mathscr{B}_{\boldsymbol{X}})$ there exists a stationary Poisson point process on $\boldsymbol{X}$ with the characteristic measure n.

The following construction is essentially the same as in Theorem 8.1; indeed, p may be identified with a Poisson random measure on $(0, \infty)\times \boldsymbol{X}$ having the intensity measure $dtn(dx)$.

Let $U_k \in \mathscr{B}_{\boldsymbol{X}}$, $k = 1, 2, \ldots$, be such that they are disjoint, $n(U_k) < \infty$ and $\boldsymbol{X} = \bigcup_k U_k$. Let $\xi_i^{(k)}$, $k,i = 1, 2, \ldots$, be U_k-valued random variables with $P(\xi_i^{(k)} \in dx) = n(dx)/n(U_k)$ and $\tau_i^{(k)}$, $k,i = 1, 2, \ldots$, be non-negative random variables such that $P(\tau_i^{(k)} > t) = \exp[-tn(U_k)]$ for $t \geq 0$. We also want all $\xi_i^{(k)}$, $\tau_i^{(k)}$ to be mutually independent. After constructing such random variables on a probability space $(\Omega, \mathscr{F}, P)$, we set

$$\boldsymbol{D}_p = \bigcup_{k=1} \{\tau_1^{(k)}, \tau_1^{(k)} + \tau_2^{(k)}, \ldots, \tau_1^{(k)} + \tau_2^{(k)} + \ldots + \tau_m^{(k)}, \ldots\}^*$$

and

$$p(\tau_1^{(k)} + \tau_2^{(k)} + \ldots + \tau_m^{(k)}) = \xi_m^{(k)}, \qquad k, m = 1, 2, \ldots.$$

It is easy to see that the point process so defined is what we want.

* It is easy to see that this union is disjoint a.s.

CHAPTER II

Stochastic Integrals and Itô's Formula

1. Itô's definition of stochastic integrals

Let $(\Omega, \mathscr{F}, P)$ be a complete probability space with a right-continuous increasing family $(\mathscr{F}_t)_{t\geq 0}$ of sub σ-fields of $\mathscr{F}$ each containing all P-null sets. Let $B = (B(t))_{t\geq 0}$ be a one-dimensional $(\mathscr{F}_t)$-Brownian motion (cf. Definition I–7.2). Since with probability one the function $t \longmapsto B(t)$ is nowhere differentiable, the integral $\int_0^t f(s)dB(s)$ can not be defined in the ordinary way. However, we can define the integral for a large class of functions by making use of the stochastic nature of Brownian motion. This was first defined by K. Itô [59] and is now known as *Itô's stochastic integral.*

Definition 1.1. Let $\mathscr{L}_2$ be the space of all real measurable processes $\Phi = \{\Phi(t, \omega)\}_{t\geq 0}$ on Ω adapted to $(\mathscr{F}_t)$ *1 such that for every $T > 0$,

$$(1.1) \qquad \|\Phi\|_{2,T}^2 = E\left[\int_0^T \Phi^2(s, \omega)\, ds\right] < \infty.$$

We identify Φ and Φ' in $\mathscr{L}_2$ if $\|\Phi - \Phi'\|_{2,T} = 0$ for every $T > 0$, and in this case write $\Phi = \Phi'$. For $\Phi \in \mathscr{L}_2$, we set

$$(1.2) \qquad \|\Phi\|_2 = \sum_{n=1}^{\infty} 2^{-n}(\|\Phi\|_{2,n} \wedge 1).$$

Clearly $\|\Phi - \Phi'\|_2$ defines a metric on $\mathscr{L}_2$; furthermore, $\mathscr{L}_2$ is complete in this metric.

Remark 1.1. For every $\Phi \in \mathscr{L}_2$ there exists a predictable*2 $\Phi' \in \mathscr{L}_2$

---

*1 Cf. Chapter I, §5.

*2 Cf. Definition I-5.2.

such that $\Phi = \Phi'$: for example, we may take

$$\Phi'(t, \omega) = \overline{\lim_{h\downarrow 0}} \frac{1}{h} \int_{t-h}^{t} \Phi(s, \omega)ds.*$$

Thus without loss of generality we may assume that $\Phi \in \mathscr{L}_2$ is predictable.

Definition 1.2. Let $\mathscr{L}_0$ be the subcollection of those real processes $\Phi = \{\Phi(t,\omega)\} \in \mathscr{L}_2$ having the property that there exists a sequence of real numbers $0 = t_0 < t_1 < \cdots < t_n < \cdots \longrightarrow \infty$ and a sequence of random variables $\{f_i(\omega)\}_{i=0}^{\infty}$ such that f_i is $\mathscr{F}_{t_i}$-measurable, $\sup_i \|f_i\|_\infty < \infty$ and

$$\Phi(t, \omega) = \begin{cases} f_0(\omega) \quad , & \text{if} \quad t = 0, \\ f_i(\omega) \quad , & \text{if} \quad t \in (t_i, t_{i+1}], \qquad i = 0, 1, \ldots . \end{cases}$$

Clearly such Φ is expressed as

$$\Phi(t, \omega) = f_0(\omega)I_{\{t=0\}}(t) + \sum_{i=0}^{\infty} f_i(\omega)I_{(t_i, t_{i+1}]}(t).$$

Lemma 1.1. $\mathscr{L}_0$ is dense in $\mathscr{L}_2$ with respect to the metric $\|\cdot\|_2$.

Proof. Take $\Phi \in \mathscr{L}_2$ and set $\Phi^M(t,\omega) = \Phi(t,\omega)I_{[-M,M]}(\Phi(t,\omega))$. Then $\Phi^M \in \mathscr{L}_2$ and $\|\Phi - \Phi^M\|_2 \longrightarrow 0$ as $M \longrightarrow \infty$. Thus it is enough to show that for any *bounded* $\Phi \in \mathscr{L}_2$ we can find $\Phi_n \in \mathscr{L}_0$, $n = 1, 2, \ldots,$ such that $\|\Phi - \Phi_n\|_2 \longrightarrow 0$ as $n \longrightarrow \infty$. Let $\boldsymbol{\Phi} = \{\Phi \in \mathscr{L}_2;$ Φ is bounded and there exists $\Phi_n \in \mathscr{L}_0$ such that $\|\Phi - \Phi_n\|_2 \longrightarrow 0$ as $n \longrightarrow \infty\}$. $\boldsymbol{\Phi}$ is a linear space and it is easy to see that if $\Phi_n \in \boldsymbol{\Phi}$, $|\Phi_n| \leq M$ for some constant $M > 0$ and $\Phi_n \uparrow \Phi$ then $\Phi \in \boldsymbol{\Phi}$. Suppose Φ is a left-continuous bounded $(\mathscr{F}_t)$-adapted process. Then if we set

$$\Phi_n(t, \omega) = \begin{cases} \Phi(0, \omega), & t = 0, \\ \Phi\left(\dfrac{k}{2^n}, \omega\right), & t \in \left(\dfrac{k}{2^n}, \dfrac{k+1}{2^n}\right], \quad k = 0, 1, \ldots, \end{cases}$$

it is clear that $\Phi_n \in \mathscr{L}_0$ and $\|\Phi_n - \Phi\|_2 \longrightarrow 0$ as $n \longrightarrow \infty$ by the bounded convergence theorem. Now, by Proposition I-5.1, we can conclude that

* To prove this rigorously, we have to appeal to Theorem 4.6 in [117], p. 68 to guarantee that Φ has a modification $\tilde{\Phi}$ progressively measurable with respect to $(\mathscr{F}_t)$ (*i. e.*, for every $T > 0$, the mapping $(t, w) \longrightarrow \tilde{\Phi}(t \wedge T, W)$ is $\mathscr{B}([0, T]) \times \mathscr{F}_T$-measurable).

Φ contains all bounded $(\mathscr{F}_t)$-predictable processes. By Remark 1.1, Φ contains all bounded $\Phi \in \mathscr{L}_2$.

Remark 1.2. A more direct proof may be found in [18] p. 440–441.

Definition 1.3. $\mathscr{M}_2 = \{X = (X_t)_{t \geq 0};\ X$ is a square integrable martingale*[1] on $(\Omega, \mathscr{F}, P)$ with respect to $(\mathscr{F}_t)_{t \geq 0}$ and $X_0 = 0$ a.s.$\}$. $\mathscr{M}_2^c = \{X \in \mathscr{M}_2;\ t \longmapsto X_t$ is continuous a.s.$\}$.

We identify two $X, X' \in \mathscr{M}_2$ if $t \longmapsto X_t$ and $t \longmapsto X'_t$ coincide a.s.

Definition 1.4. For $X \in \mathscr{M}_2$, we set

$$\|X\|_T = E[X_T^2]^{1/2}, \quad T > 0, \tag{1.3}$$

and

$$\|X\| = \sum_{n=1}^{\infty} 2^{-n}(\|X\|_n \wedge 1). \tag{1.4}$$

We note that since X is a martingale, $\|\cdot\|_t$ is non-decreasing in t.

Lemma 1.2. $\mathscr{M}_2$ is a complete metric space in the metric $\|X - Y\|$, $X, Y \in \mathscr{M}_2$, and $\mathscr{M}_2^c$ is a closed subspace of $\mathscr{M}_2$.

Proof. First we note that if $\|X - Y\| = 0$, then $X = Y$; indeed, $\|X - Y\| = 0$ implies $X_n = Y_n$ a.s., $n = 1, 2, \ldots$, and so $X_t = E[X_n | \mathscr{F}_t] = E[Y_n | \mathscr{F}_t] = Y_t$ for $t \leq n$. By the right-continuity of $t \longmapsto X_t$ and $t \longmapsto Y_t$ we conclude that $X = Y$.

Next, let $X^{(n)}$, $n = 1, 2, \ldots$, be a Cauchy sequence; i.e., $\lim_{n,m \to \infty} \|X^{(n)} - X^{(m)}\| = 0$.

Then by the Kolmogorov-Doob inequality for martingales*[2] we have for every $T > 0$ and $C > 0$

$$P[\sup_{0 \leq t \leq T} |X_t^{(n)} - X_t^{(m)}| \geq C] \leq \frac{1}{C^2} \|X^{(n)} - X^{(m)}\|_T^2.$$

Therefore, there exists $X = (X_t)$ such that

$$\sup_{0 \leq t \leq T} |X_t^{(n)} - X_t| \longrightarrow 0 \qquad \text{in probability}$$

---

*[1] It is always assumed that $t \longmapsto X_t$ is right continuous a.s.

*[2] Theorem I-6.10.

as $n \longrightarrow \infty$ for every $T > 0$.

Clearly, for every $t > 0$, $E[|X_t^{(n)} - X_t|^2] \longrightarrow 0$ as $n \longrightarrow \infty$ and we can conclude from this that $X \in \mathcal{M}_2$ and $\|X^{(n)} - X\| \longrightarrow 0$ as $n \longrightarrow \infty$. Finally, it is also clear from this proof that if $X^{(n)} \in \mathcal{M}_2^c$, then $X \in \mathcal{M}_2^c$.

We will now define the stochastic integral with respect to an $(\mathscr{F}_t)$-Brownian motion as a mapping

$$\Phi \in \mathscr{L}_2 \longmapsto I(\Phi) \in \mathcal{M}_2^c.$$

Suppose we are given an $(\mathscr{F}_t)$-Brownian motion $B(t)$ on $(\Omega, \mathscr{F}, P)$. If $\Phi \in \mathscr{L}_0$ and

$$\Phi(t, \omega) = f_0(\omega)I_{\{t=0\}}(t) + \sum_{i=0}^{\infty} f_i(\omega)I_{(t_i, t_{i+1}]}(t)$$

then we set

$$\begin{aligned} I(\Phi)(t, \omega) = \sum_{i=0}^{n-1} f_i(\omega)(B(t_{i+1}, \omega) - B(t_i, \omega)) \\ + f_n(\omega)(B(t, \omega) - B(t_n, \omega)) \end{aligned} \tag{1.5}$$

for $t_n \leq t \leq t_{n+1}$, $n = 0, 1, 2, \ldots$. Clearly $I(\Phi)$ may be expressed as

$$I(\Phi)(t) = \sum_{i=0}^{\infty} f_i(B(t \wedge t_{i+1}) - B(t \wedge t_i)) \tag{1.6}$$

(This sum is, in fact, a finite sum.) We can easily verify that for $s \leq t$

$$E[f_i(B(t \wedge t_{i+1}) - B(t \wedge t_i))|\mathscr{F}_s] = f_i(B(s \wedge t_{i+1}) - B(s \wedge t_i)),$$

and hence $I(\Phi)(t)$ is a continuous $\mathscr{F}_t$-martingale, i.e., $I(\Phi) \in \mathcal{M}_2^c$. Also it is easy to verify that

$$E(I(\Phi)(t)^2) = \sum_{i=0}^{\infty} E[f_i^2(t \wedge t_{i+1} - t \wedge t_i)] = E[\int_0^t \Phi^2(s, \omega)ds].^* \tag{1.7}$$

Thus,

$$\|I(\Phi)\|_T = \|\Phi\|_{2,T} \tag{1.8}$$

* From this formula we see that $I(\Phi)$ is uniquely determined by Φ and is independent of the particular choice of $\{t_i\}$.

and

$$(1.9) \qquad \| I(\Phi) \| = \|\Phi\|_2.$$

Next, let $\Phi \in \mathscr{L}_2$. Then by Lemma 1.1 we can find $\Phi_n \in \mathscr{L}_0$ such that $\|\Phi - \Phi_n\|_2 \longrightarrow 0$ as $n \longrightarrow \infty$. Since $\| I(\Phi_n) - I(\Phi_m) \| = \|\Phi_n - \Phi_m\|_2$, $I(\Phi_n)$ is a Cauchy sequence in $\mathscr{M}_2$ and consequently by Lemma 1.2, it converges to a unique element $X = (X_t) \in \mathscr{M}_2^c$. Clearly X is determined uniquely from Φ and is independent of the particular choice of Φ_n. We denote X by $I(\Phi)$.

Definition 1.5. $I(\Phi) \in \mathscr{M}_2^c$ defined above is called the *stochastic integral of $\Phi \in \mathscr{L}_2$ with respect to the Brownian motion $B(t)$.* We shall often denote $I(\Phi)(t)$ by $\int_0^t \Phi(s, \omega)dB(s, \omega)$ or more simply $\int_0^t \Phi(s)dB(s)$.

Clearly if $\Phi, \Psi \in \mathscr{L}_2$ and $\alpha, \beta \in \boldsymbol{R}$, then

$$(1.10) \qquad I(\alpha\Phi + \beta\Psi)(t) = \alpha I(\Phi)(t) + \beta I(\Psi)(t) \qquad \text{for each } t \geq 0 \text{ a.s.}$$

Remark 1.3. Thus a stochastic integral $I(\Phi)$ is defined as a stochastic process (in fact, a martingale). But for each fixed t the random variable $I(\Phi)(t)$ itself is also called a stochastic integral.

Proposition 1.1. The stochastic integral with respect to an $(\mathscr{F}_t)$-Brownian motion has the following properties:

(i) $I(\Phi)(0) = 0$ a.s.

(ii) For each $t > s \geqq 0$,

$$(1.11) \qquad E[(I(\Phi)(t) - I(\Phi)(s)) | \mathscr{F}_s] = 0 \qquad \text{a.s.}$$

and

$$(1.12) \qquad E[(I(\Phi)(t) - I(\Phi)(s))^2 | \mathscr{F}_s] = E[\int_s^t \Phi^2(u, \omega)du | \mathscr{F}_s] \qquad \text{a.s.}$$

More generally, if σ, τ are $(\mathscr{F}_t)$-stopping times such that $\tau \geqq \sigma$ a.s., then for every constant $t > 0$

$$(1.13) \qquad E[(I(\Phi)(t \wedge \tau) - I(\Phi)(t \wedge \sigma)) | \mathscr{F}_\sigma] = 0 \qquad \text{a.s.}$$

and

$$(1.14) \qquad E[(I(\Phi)(t \wedge \tau) - I(\Phi)(t \wedge \sigma))^2 | \mathscr{F}_\sigma] = E[\int_{t\wedge\sigma}^{t\wedge\tau} \Phi^2(u, \omega)du | \mathscr{F}_\sigma] \text{ a.s.}$$

(iii) (1.12) and (1.14) have the following generalization: if $\Phi, \Psi \in \mathscr{L}_2$, then

(1.15)
$$E[(I(\Phi)(t) - I(\Phi)(s))(I(\Psi)(t) - I(\Psi)(s)) \mid \mathscr{F}_s] = E[\int_s^t (\Phi \cdot \Psi)(u, \omega)du \mid \mathscr{F}_s] \quad \text{a.s.}$$

and

(1.16)
$$E[(I(\Phi)(t\wedge\tau) - I(\Phi)(t\wedge\sigma))(I(\Psi)(t\wedge\tau) - I(\Psi)(t\wedge\sigma)) \mid \mathscr{F}_\sigma] = E[\int_{t\wedge\sigma}^{t\wedge\tau} (\Phi \cdot \Psi)(u, \omega)du \mid \mathscr{F}_\sigma] \quad \text{a.s.}$$

(iv) If σ is an $(\mathscr{F}_t)$-stopping time, then

(1.17) $\quad I(\Phi)(t\wedge\sigma) = I(\Phi')(t) \qquad$ for every $t \geq 0$,

where $\Phi'(t, \omega) = I_{\{\sigma(\omega)\geq t\}} \cdot \Phi(t, \omega)$.*[1]

Proof. (i) is obvious. (1.11) is clear since $I(\Phi)$ is a martingale. (1.12) is easily proved first for $\Phi \in \mathscr{L}_0$ and then by taking limits. (1.13) and (1.14) are consequences of Doob's optional sampling theorem. Therefore, we need only prove (iv).*[2] First consider the case when $\Phi \in \mathscr{L}_0$.

Let $\{s_i^{(n)}\}_{i=0}^{\infty}$ $(n = 1, 2, \ldots)$ be a refinement of subdivisions $\{t_i\}_{i=0}^{\infty}$ and $\{i2^{-n}\}_{i=0}^{\infty}$. Suppose Φ has the expression $\Phi(t, \omega) = f_0(\omega)I_{\{t=0\}}(t) + \sum_i f_i^{(n)}(\omega)I_{(s_i^{(n)}, s_{i+1}^{(n)}]}(t)$ for each $n = 1, 2, \ldots$. Define

$$\sigma^n(\omega) = s_{i+1}^{(n)} \quad \text{if} \quad \sigma(\omega) \in (s_i^{(n)}, s_{i+1}^{(n)}].$$

It is easy to see that σ^n is an $(\mathscr{F}_t)$-stopping time for each $n = 1, 2, \ldots$ and $\sigma^n \downarrow \sigma$ as $n \longrightarrow \infty$. If $s \in (s_i^{(n)}, s_{i+1}^{(n)}]$, then $I_{\{\sigma^n \geq s\}} = I_{\{\sigma > s_i^{(n)}\}}$ and therefore, if we set $\Phi_n'(s, \omega) = \Phi(s, \omega)\, I_{\{\sigma^n(\omega)\geq s\}}$, $\Phi_n' \in \mathscr{L}_0$. Clearly, for every $t > 0$,

$$\|\Phi_n' - \Phi'\|_{2,t}^2 = E[\int_0^t \Phi^2(s, \omega)I_{\{\sigma^n \geq s > \sigma\}}ds] \longrightarrow 0$$

as $n \longrightarrow \infty$. Hence $I(\Phi_n') \longrightarrow I(\Phi')$ in $\mathscr{M}_2$. Also,

---

*[1] Clearly $\Phi' \in \mathscr{L}_2$.

*[2] A simpler proof is given in Remark 2.2 below.

$$I(\Phi_n')(t) = \sum_{k=0}^{\infty} f_k^{(n)}(\omega) I_{\{\sigma > s_k^{(n)}\}} (B(t \wedge s_{k+1}^{(n)}) - B(t \wedge s_k^{(n)}))$$
$$= \sum_{k=0}^{\infty} f_k^{(n)}(\omega) I_{\{\sigma > s_k^{(n)}\}} (B(t \wedge \sigma^n \wedge s_{k+1}^{(n)}) - B(t \wedge \sigma^n \wedge s_k^{(n)}))$$
$$= \sum_{k=0}^{\infty} f_k^{(n)}(\omega) (B(t \wedge \sigma^n \wedge s_{k+1}^{(n)}) - B(t \wedge \sigma^n \wedge s_k^{(n)}))$$
$$= \int_0^{t \wedge \sigma^n} \Phi(s, \omega) dB(s)$$

since $\sigma^n \leq s_k^{(n)}$ if $\sigma \leq s_k^{(n)}$. Consequently

$$I(\Phi')(t) = \lim_{n \to \infty} I(\Phi_n')(t) = \int_0^{t \wedge \sigma} \Phi(s, \omega) dB(s) = I(\Phi)(t \wedge \sigma).$$

The general case can be proved by approximating Φ with $\Phi_n \in \mathscr{L}_0$.

Let $B(t) = (B^1(t), B^2(t), \ldots, B^r(t))$ be an r-dimensional $(\mathscr{F}_t)$-Brownian motion and let $\Phi_1(t,\omega), \ldots, \Phi_r(t, \omega) \in \mathscr{L}_2$. Then the stochastic integrals $\int_0^t \Phi_i(s) dB^i(s)$ are defined for $i = 1, 2, \ldots, r$.

Proposition 1.2. For $t > s \geq 0$,

$$E[\int_s^t \Phi_i(u) dB^i(u) \int_s^t \Phi_j(u) dB^j(u) \mid \mathscr{F}_s] = \delta_{ij} E[\int_s^t \Phi_i(u)\Phi_j(u) du \mid \mathscr{F}_s], \quad i,j = 1,2, \ldots, r. \tag{1.18}$$

Proof. It is easy to prove this when $\Phi_i \in \mathscr{L}_0$, $i = 1, 2, \ldots, r$. The general case follows by taking limits.

So far we have defined the stochastic integral for elements in $\mathscr{L}_2$. This can be extended to a more general class of integrands as follows.

Definition 1.6. $\mathscr{L}_2^{loc} = \{\Phi = \{\Phi(t)\}_{t \geq 0}$; Φ is a real measurable process on Ω adapted to $(\mathscr{F}_t)$ such that for every $T > 0$

$$\int_0^T \Phi^2(t, \omega) dt < \infty \qquad \text{a.s.}\}. \tag{1.19}$$

We identify Φ and Φ' in $\mathscr{L}_2^{loc}$ if, for every $T > 0$,

$$\int_0^T |\Phi(t, \omega) - \Phi'(t, \omega)|^2 dt = 0 \qquad \text{a.s.}$$

In this case we write $\Phi = \Phi'$.

Similarly as in Remark 1.1, we may always assume that $\Phi \in \mathscr{L}_2^{loc}$ is a predictable process.

Definition 1.7. A real stochastic process $X = (X_t)_{t\geq 0}$ on $(\Omega, \mathscr{F}, P)$ is called a *local* $(\mathscr{F}_t)$*-martingale* if it is adapted to $(\mathscr{F}_t)$ and if there exists a sequence of $(\mathscr{F}_t)$-stopping times σ_n such that $\sigma_n < \infty$, $\sigma_n \uparrow \infty$ and $X_n = (X_n(t))$ is an $(\mathscr{F}_t)$-martingale for each $n = 1, 2, \ldots$ where $X_n(t) = X(t \wedge \sigma_n)$. If, in addition, X_n is a square integrable martingale for each n, then we call X a *locally square integrable* $(\mathscr{F}_t)$*-martingale.*

By taking an appropriate modification if necessary, we may and shall always assume for such X that $t \longmapsto X_t$ is right continuous a.s.

Definition 1.8. $\mathscr{M}_2^{loc} = \{X = (X_t)_{t\geq 0};\ X$ is a locally square integrable $(\mathscr{F}_t)$-martingale and $X_0 = 0$ a.s.$\}$. $\mathscr{M}_2^{c,loc} = \{X \in \mathscr{M}_2^{loc};\ t \longmapsto X_t$ is continuous a.s.$\}$.

Let $B = (B(t))_{t\geq 0}$ be an $(\mathscr{F}_t)$-Brownian motion and $\Phi \in \mathscr{L}_2^{loc}$. Let $\sigma_n(\omega) = \inf\{t;\ \int_0^t \Phi^2(s,\omega)ds \geq n\} \wedge n$, $n = 1,2,\ldots$. Then σ_n is a sequence of $(\mathscr{F}_t)$-stopping times such that $\sigma_n \uparrow \infty$ a.s. Set $\Phi_n(s,\omega) = I_{\{\sigma_n(\omega)\geq s\}}\Phi(s, \omega)$. Clearly

$$\int_0^\infty \Phi_n^2(s, \omega)ds = \int_0^{\sigma_n} \Phi_n^2(s, \omega)ds \leq n$$

and hence $\Phi_n \in \mathscr{L}_2$, $n = 1,2,\ldots$. By Proposition 1.1 (iv), we have for $m < n$ that

$$I(\Phi_n)(t \wedge \sigma_m) = I(\Phi_m)(t).$$

Consequently if we define $I(\Phi)$ by

$$I(\Phi)(t) = I(\Phi_n)(t) \qquad \text{for} \quad t \leq \sigma_n,$$

then this definition is well-defined and determines a continuous process $I(\Phi)$ such that

$$I(\Phi)(t \wedge \sigma_n) = I(\Phi_n)(t), \qquad n = 1, 2, \ldots.$$

Therefore $I(\Phi) \in \mathscr{M}_2^{c,loc}$.

Definition 1.9. $I(\Phi) \in \mathscr{M}_2^{c,loc}$ defined above is called the *stochastic integral* of $\Phi \in \mathscr{L}_2^{loc}$ with respect to the Brownian motion $B(t)$. We shall often denote $I(\Phi)(t)$ by $\int_0^t \Phi(s,\omega)dB(s,\omega)$ or more simply $\int_0^t \Phi(s)dB(s)$. As in Remark 1.3, we shall also call the random variable $I(\Phi)(t)$, for each fixed t, a stochastic integral.

2. Stochastic integrals with respect to martingales

Let $(\Omega, \mathscr{F}, P)$ and $(\mathscr{F}_t)_{t\geq 0}$ be given as in Section 1. Suppose we are given $M \in \mathscr{M}_2$. We shall now define a stochastic integral $\int_0^t \Phi(s)dM(s)$ which coincides with that in Section 1 when M is an $(\mathscr{F}_t)$-Brownian motion (Kunita-Watanabe [97]).

Proposition 2.1. (i) Let $M = (M_t) \in \mathscr{M}_2$. Then there exists a natural integrable increasing process*[1] $A = (A_t)$ such that $M_t^2 - A_t$ is an $(\mathscr{F}_t)$-martingale. Furthermore, A is uniquely determined.*[2]

(ii) Let $M = (M_t)$ and $N = (N_t)$ be in $\mathscr{M}_2$. Then there exists $A = (A_t)$ which is expressible as the difference of two natural integrable increasing processes such that $M_tN_t - A_t$ is an $(\mathscr{F}_t)$-martingale. Furthermore, A is uniquely determined.

Proof. Let $M = (M_t) \in \mathscr{M}_2$. Then $t \longmapsto M_t^2$ is a non-negative submartingale of the class (DL) and hence by Doob-Meyer's decomposition theorem*[3], there exists a unique natural integrable increasing process $A = (A_t)$ such that $t \longmapsto M_t^2 - A_t$ is an $(\mathscr{F}_t)$-martingale. If $M, N \in \mathscr{M}_2$, then $M_1 = (M + N)/2 \in \mathscr{M}_2$ and $M_2 = (M - N)/2 \in \mathscr{M}_2$. Let A^1 and A^2 be the natural integrable increasing processes corresponding as above to M_1 and M_2 respectively. Then $A = A^1 - A^2$ satisfies that $t \longmapsto M_tN_t - A_t$ is an $(\mathscr{F}_t)$-martingale. The uniqueness of A is a consequence of the uniqueness of Doob-Meyer's decomposition.

Definition 2.1. (i) $A = (A_t)$ in Proposition 2.1 (i) is denoted by $\langle M\rangle = (\langle M\rangle_t)_{t\geq 0}$.

(ii) $A = (A_t)$ in Proposition 2.1 (ii) is denoted by $\langle M, N\rangle = (\langle M, N\rangle_t)_{t\geq 0}$.

Note that $\langle M\rangle = \langle M, M\rangle$. We call $\langle M, N\rangle$ the *quadratic variational process* corresponding to M and N.

By Theorem I-6.13, we see that $\langle M, N\rangle$ is continuous if at least one of the following conditions is satisfied:

(i) $(\mathscr{F}_t)_{t\geq 0}$ has no time of discontinuity, that is, if σ_n is an increasing

*[1] Cf. Definition I-6.2 and Definition I-6.3.
*[2] i.e., if A' is another such process, then $t \longmapsto A_t$ and $t \longmapsto A'_t$ coincide a.s.
*[3] Theorem I-6.12.

sequence of $(\mathscr{F}_t)$-stopping times and $\sigma = \lim_n \sigma_n$, then $\mathscr{F}_\sigma = \bigvee_n \mathscr{F}_{\sigma_n}$;

(ii) $M, N \in \mathscr{M}_2^c$.

Indeed, if (i) is satisfied, then $M_{t\wedge\sigma_n} = E[M_t | \mathscr{F}_{t\wedge\sigma_n}] \longmapsto E[M_t | \mathscr{F}_{t\wedge\sigma}] = M_{t\wedge\sigma}$ as $\sigma_n \uparrow \sigma$ and hence $M(t)^2$ is regular.

Example 2.1. Let $B(t) = (B^1(t), B^2(t), \dots, B^d(t))$ be a d-dimensional $(\mathscr{F}_t)$-Brownian motion such that $B(0) = 0$ a.s. Then, for each i, $B^i \in \mathscr{M}_2^c$ and $\langle B^i, B^j \rangle(t) = \delta_{ij} t$, $i, j = 1, 2, \dots, d$. A famous theorem of Lévy asserts that a d-dimensional $(\mathscr{F}_t)$-Brownian motion is characterized by this property (cf. Theorem 6.1).

Let $M \in \mathscr{M}_2$ and $\langle M \rangle$ be the corresponding quadratic variational process.

Definition 2.2. $\mathscr{L}_2(M) = \{\Phi = (\Phi(t,\omega))_{t\geq 0};\ \Phi$ is a real $(\mathscr{F}_t)$-predictable process and, for every $T > 0$,

$$(2.1) \qquad (\|\Phi\|_{2,T}^M)^2 = E\left[\int_0^T \Phi^2(s,\omega) d\langle M \rangle(s)\right] < \infty\}.$$

For $\Phi \in \mathscr{L}_2(M)$, we set

$$(2.2) \qquad \|\Phi\|_2^M = \sum_{n=1}^\infty 2^{-n}(\|\Phi\|_{2,n}^M \wedge 1).$$

We identify Φ and Φ' in $\mathscr{L}_2(M)$ if $\|\Phi - \Phi'\|_{2,T}^M = 0$ for every T and write $\Phi = \Phi'$. If M is an $(\mathscr{F}_t)$-Brownian motion, $\mathscr{L}_2(M)$ coincides with $\mathscr{L}_2$ of Definition 1.1. Note that $\mathscr{L}_2(M) \supset \mathscr{L}_0$,* where $\mathscr{L}_0$ is defined by Definition 1.2. In exactly the same way as in Lemma 1.1, we have the following result.

Lemma 2.1. $\mathscr{L}_0$ is dense in $\mathscr{L}_2(M)$ with repect to the metric $\|\cdot\|_2^M$.

Using this lemma, we can now define stochastic integrals in the same way as in the case of a Brownian motion. First let $\Phi \in \mathscr{L}_0$. Then $\Phi(t) = f_0(\omega) I_{\{t=0\}}(t) + \sum_{i=0}^\infty f_i(\omega) I_{(t_i, t_{i+1}]}(t)$, and we set

$$I^M(\Phi)(t) = \sum_{i=0}^{n-1} f_i(\omega)(M(t_{i+1},\omega) - M(t_i,\omega)) + f_n(\omega)(M(t,\omega) - M(t_n,\omega)) \quad \text{for} \quad t_n \leq t \leq t_{n+1}, \quad n = 1, 2, \dots .$$

* Every $\Phi \in \mathscr{L}_0$ is left-continuous and so it is predictable.

As before $I^M(\Phi) \in \mathscr{M}_2$ and $\|I^M(\Phi)\| = \|\Phi\|_2^M$. Using this isometry, $\Phi \in \mathscr{L}_0 \longmapsto I^M(\Phi) \in \mathscr{M}_2$ is extended to $\Phi \in \mathscr{L}_2(M) \longmapsto I^M(\Phi) \in \mathscr{M}_2$ as in Section 1.

Definition 2.3. $I^M(\Phi)$ is called the *stochastic integral* of $\Phi \in \mathscr{L}_2(M)$ with respect to $M \in \mathscr{M}_2$. We shall also denote $I^M(\Phi)(t)$ by $\int_0^t \Phi(s)dM(s)$.

Thus we have defined the stochastic integral with respect to a martingale $M \in \mathscr{M}_2$. It is clear that if $M \in \mathscr{M}_2^c$ then $I^M(\Phi) \in \mathscr{M}_2^c$ and if M is an $(\mathscr{F}_t)$-Brownian motion, then $I^M(\Phi)$ coincides with $I(\Phi)$ defined in Section 1.

Proposition 2.2. The stochastic integral $I^M(\Phi)$, $\Phi \in \mathscr{L}_2(M)$, $M \in \mathscr{M}_2$, has the following properties:

(i) $I^M(\Phi)(0) = 0$, a.s.

(ii) For each $t > s \geq 0$,

$$E[(I^M(\Phi)(t) - I^M(\Phi)(s)) \,|\, \mathscr{F}_s] = 0, \quad \text{a.s.} \tag{2.3}$$

and

$$E[(I^M(\Phi)(t) - I^M(\Phi)(s))^2 \,|\, \mathscr{F}_s] = E[\int_s^t \Phi^2(u)d\langle M\rangle(u) \,|\, \mathscr{F}_s], \quad \text{a.s.} \tag{2.4}$$

More generally, if σ, τ are $\mathscr{F}_t$-stopping times such that $\tau \geq \sigma$ a.s., then for every $t > 0$,

$$E[(I^M(\Phi)(t\wedge\tau) - I^M(\Phi)(t\wedge\sigma)) \,|\, \mathscr{F}_\sigma] = 0, \quad \text{a.s.} \tag{2.5}$$

and

$$\begin{aligned} &E[(I^M(\Phi)(t\wedge\tau) - I^M(\Phi)(t\wedge\sigma))^2 \,|\, \mathscr{F}_\sigma] \\ &\qquad = E[\int_{t\wedge\sigma}^{t\wedge\tau} \Phi^2(u)d\langle M\rangle(u) \,|\, \mathscr{F}_\sigma], \quad \text{a.s.} \end{aligned} \tag{2.6}$$

(iii) (2.4) and (2.6) can be generalized as follows: if Φ, $\Psi \in \mathscr{L}_2(M)$, then

$$\begin{aligned} &E[(I^M(\Phi)(t) - I^M(\Phi)(s))(I^M(\Psi)(t) - I^M(\Psi)(s)) \,|\, \mathscr{F}_s] \\ &\qquad = E[\int_s^t (\Phi\Psi)(u)d\langle M\rangle(u) \,|\, \mathscr{F}_s], \quad \text{a.s.} \end{aligned} \tag{2.7}$$

and

$$(2.8)\qquad \begin{aligned} &E[(I^M(\Phi)(t\wedge\tau) - I^M(\Phi)(t\wedge\sigma))(I^M(\Psi)(t\wedge\tau) - I^M(\Psi)(t\wedge\sigma))\,|\,\mathscr{F}_\sigma] \\ &\quad = E[\int_{t\wedge\sigma}^{t\wedge\tau}(\Phi\Psi)(u)d\langle M\rangle(u)\,|\,\mathscr{F}_\sigma], \qquad \text{a.s.} \end{aligned}$$

(iv) If σ is an $(\mathscr{F}_t)$-stopping time, then

$$(2.9)\qquad I^M(\Phi)(t\wedge\sigma) = I^M(\Phi')(t) \qquad \text{for} \qquad t \geq 0,$$

where $\Phi'(t) = I_{\{\sigma\geq t\}}\Phi(t)$.*

The proof is same as in Proposition 1.1.

If $M,N \in \mathscr{M}_2$, $\Phi \in \mathscr{L}_2(M)$ and $\Psi \in \mathscr{L}_2(N)$, then $I^M(\Phi)$ and $I^N(\Psi)$ are defined as elements in $\mathscr{M}_2$.

Proposition 2.3. For $t > s \geq 0$,

$$(2.10)\qquad \begin{aligned} &E[(I^M(\Phi)(t) - I^M(\Phi)(s))(I^N(\Psi)(t) - I^N(\Psi)(s))\,|\,\mathscr{F}_s] \\ &\quad = E[\int_s^t (\Phi\Psi)(u)d\langle M,N\rangle(u)\,|\,\mathscr{F}_s]. \end{aligned}$$

The proof is easily obtained first for $\Phi,\Psi \in \mathscr{L}_0$ and then by a limiting procedure. (2.10) is another way of saying that

$$(2.11)\qquad \langle I^M(\Phi), I^N(\Psi)\rangle(t) = \int_0^t (\Phi\Psi)(u)d\langle M,N\rangle(u),$$

or more symbolically,

$$(2.11)'\qquad d\langle I^M(\Phi), I^N(\Psi)\rangle_t = \Phi(t)\Psi(t)d\langle M,N\rangle_t.$$

Remark 2.1. (2.10) implicitly implies the following result: if $M, N \in \mathscr{M}_2$, $\Phi \in \mathscr{L}_2(M)$ and $\Psi \in \mathscr{L}_2(N)$ then

$$E[\int_0^t |\Phi\Psi|(u)d\|\langle M,N\rangle\|(u)] < \infty$$

where $\|\langle M,N\rangle\|(t)$ denotes the total variation of $s \in [0, t] \longmapsto \langle M, N\rangle(s)$. In fact, we have the inequality

$$(2.12)\qquad \begin{aligned} \int_0^t |\Phi\Psi|(u)d\|\langle M,N\rangle\|(u) &\leq \left\{\int_0^t \Phi(u)^2 d\langle M\rangle(u)\right\}^{1/2} \\ &\quad \times \left\{\int_0^t \Psi(u)^2 d\langle N\rangle(u)\right\}^{1/2}. \end{aligned}$$

---

* $I_{\{\sigma\geq t\}}$ is left continuous in t and so it is predictable. Hence $\Phi' \in \mathscr{L}_2(M)$.

(2.12) is easily proved when $\Phi, \Psi \in \mathscr{L}_0$ and the general case follows by a limiting procedure. Note that if $t \longmapsto A_t$ is of bounded variation on finite intervals and is continuous, then there exists a predictable process Φ_t such that $|\Phi_t| = 1$ a.s. and $\|A\|_t = \int_0^t \Phi(u)dA_u$.

The above definition of stochastic integrals is extended in an obvious manner to the case of local martingales. Let M, $N \in \mathscr{M}_2^{loc}$. Then there exists a sequence of $(\mathscr{F}_t)$-stopping times σ_n such that $\sigma_n \uparrow \infty$ a.s. and $M^{\sigma_n} = (M_{t\wedge\sigma_n})$ and $N^{\sigma_n} = (N_{t\wedge\sigma_n})$ are in $\mathscr{M}_2$. It follows from the uniqueness of the quadratic variational process that if $m < n$ then

$$\langle M^{\sigma_m}, N^{\sigma_m}\rangle(t) = \langle M^{\sigma_n}, N^{\sigma_n}\rangle(t \wedge \sigma_m).$$

Hence there exists a unique predictable process $\langle M,N\rangle$ such that $\langle M, N\rangle (t \wedge \sigma_n) = \langle M^{\sigma_n}, N^{\sigma_n}\rangle(t)$ for all n and $t > 0$. We write $\langle M, M\rangle$ as $\langle M\rangle$.

Definition 2.4. Let $M \in \mathscr{M}_2^{loc}$. $\mathscr{L}_2^{loc}(M) = \{\Phi = (\Phi(t))$; Φ is a real $(\mathscr{F}_t)$-predictable process on Ω such that there exists a sequence of $(\mathscr{F}_t)$-stopping times σ_n such that $\sigma_n \uparrow \infty$ a.s. and

$$(2.13) \qquad E[\int_0^{T\wedge\sigma_n} \Phi^2(t, \omega)d\langle M\rangle(t)] < \infty$$

for every $T > 0$ and $n = 1, 2, \dots\}$.*

Let $M \in \mathscr{M}_2^{loc}$ and $\Phi \in \mathscr{L}_2^{loc}(M)$. Then clearly we may choose a sequence of $(\mathscr{F}_t)$-stopping times σ_n such that $\sigma_n \uparrow \infty$ a.s., $M^{\sigma_n} = (M(t \wedge \sigma_n)) \in \mathscr{M}_2$ and (2.13) is satisfied. Hence for $\Phi_n(t, \omega) = I_{\{\sigma_n(\omega) \geqq t\}} \Phi(t, \omega)$ and $M_n = M^{\sigma_n}$, we can define $I^{M_n}(\Phi_n)$ and it is easy to see that $I^{M_m}(\Phi_m)(t) = I^{M_n}(\Phi_n)(t \wedge \sigma_m)$ for $m < n$. Thus there exists a unique process $I^M(\Phi)(t)$ such that $I^{M_n}(\Phi_n)(t) = I^M(\Phi)(t \wedge \sigma_n)$, $n = 1, 2, \dots$. It is clear that $I^M(\Phi) \in \mathscr{M}_2^{loc}$.

Definition 2.5. $I^M(\Phi)$ is called the *stochastic integral* of $\Phi \in \mathscr{L}_2^{loc}(M)$ *with respect to* $M \in \mathscr{M}_2^{loc}$. $I^M(\Phi)(t)$ is also denoted by $\int_0^t \Phi(s)dM(s)$.

It is clear that Propositions 2.2 and 2.3 easily extend to this general case. In particular, we note that (2.11) holds.

As a special case of (2.11), we obtain

$$(2.14) \qquad \langle I^M(\Phi), N\rangle(t) = \int_0^t \Phi(u)d\langle M,N\rangle(u).$$

This property completely characterizes the stochastic integral; more precisely, we state the following proposition.

* If $\langle M\rangle_t$ is continuous, (2.13) is equivalent to that $\int_0^T \Phi^2(t,\omega)d\langle M\rangle(t) < \infty$ for all $T > 0$, a.s.

Proposition 2.4. Let $M \in \mathscr{M}_2$ and $\Phi \in \mathscr{L}_2(M)$ (or $M \in \mathscr{M}_2^{loc}$, $\Phi \in \mathscr{L}_2^{loc}(M)$). Then $X = I^M(\Phi)$ is characterized as the unique $X \in \mathscr{M}_2$ ($X \in \mathscr{M}_2^{loc}$) such that

$$\langle X, N\rangle(t) = \int_0^t \Phi(u)d\langle M, N\rangle(u) \tag{2.15}$$

for every $N \in \mathscr{M}_2$ ($N \in \mathscr{M}_2^{loc}$) and all $t \geq 0$.

Proof. We need only show uniqueness. If $X' \in \mathscr{M}_2$ also satisfies (2.15), then $\langle X - X', N\rangle = 0$ for every $N \in \mathscr{M}_2$ and so, by taking $N = X - X'$, $\langle X - X'\rangle = 0$. Hence $X = X'$.

Remark 2.2. For example, we can prove (2.9) using this characterization. Denoting generally by X^σ the stopped process $\{X(\sigma \wedge t)\}$, by Proposition I-5.5 and Doob's optional sampling theorem,

$$\begin{aligned}\langle (I^M)(\Phi)^\sigma, N\rangle(t) &= \langle (I^M)(\Phi)^\sigma, N^\sigma\rangle(t) = \langle I^M(\Phi), N\rangle^\sigma(t) \\ &= \int_0^{t\wedge\sigma} \Phi(u)d\langle M, N\rangle(u) = \int_0^t \Phi'(u)d\langle M, N\rangle(u),\end{aligned}$$

where $\Phi'(t) = I_{\{\sigma \geq t\}}\Phi(t)$. Consequently $I^M(\Phi') = I^M(\Phi)^\sigma$ by Proposition 2.4.

Proposition 2.5. (i) Let $M, N \in \mathscr{M}_2^{loc}$ and $\Phi \in \mathscr{L}_2^{loc}(M) \cap \mathscr{L}_2^{loc}(N)$. Then $\Phi \in \mathscr{L}_2^{loc}(M + N)$ and

$$\int_0^t \Phi(u)d(M+N)(u) = \int_0^t \Phi(u)dM(u) + \int_0^t \Phi(u)dN(u). \tag{2.16}$$

(ii) Let $M \in \mathscr{M}_2^{loc}$ and $\Phi, \Psi \in \mathscr{L}_2^{loc}(M)$. Then

$$\int_0^t (\Phi+\Psi)(u)dM(u) = \int_0^t \Phi(u)dM(u) + \int_0^t \Psi(u)dM(u). \tag{2.17}$$

(iii) Let $M \in \mathscr{M}_2^{loc}$ and $\Phi \in \mathscr{L}_2^{loc}(M)$. Set $N = I^M(\Phi)$ and let $\Psi \in \mathscr{L}_2^{loc}(N)$. Then $\Phi\Psi \in \mathscr{L}_2^{loc}(M)$ and

$$\int_0^t (\Phi\Psi)(u)dM(u) = \int_0^t \Psi(u)dN(u).$$

(iv) Let $M \in \mathscr{M}_2^{loc}$. Let $\Phi \in \mathscr{L}_2^{loc}(M)$ be a *stochastic step function* in the sense that there exists a sequence of increasing $(\mathscr{F}_t)$-stopping times σ_n such that

$$\Phi(t, \omega) = \sum_n f_n(\omega) I_{(\sigma_n, \sigma_{n+1}]}(t),$$

where f_n is $\mathscr{F}_{\sigma_n}$-measurable. Then

$$I^M(\Phi)(t) = \sum_n f_n(\omega)(M_{t\wedge\sigma_{n+1}} - M_{t\wedge\sigma_n}).$$

The proof is easy and so we omit it. It is proved most simply by using Proposition 2.4.

3. Stochastic integrals with respect to point processes

The notion of point processes was given in Section 9 of Chapter I. Here we discuss point processes adapted to an increasing family of σ-fields and their stochastic calculus. Let $(\Omega, \mathscr{F}, P)$ and $(\mathscr{F}_t)_{t\geq 0}$ be given as above. A point process $p = (p(t))$ on $\boldsymbol{X}$ defined on Ω is called $(\mathscr{F}_t)$-*adapted* if for every $t > 0$ and $U \in \mathscr{B}(\boldsymbol{X})$, $N_p(t, U) = \sum_{s\in D_p, s\leq t} I_U(p(s))$ is $\mathscr{F}_t$-measurable. p is called σ-*finite* if there exist $U_n \in \mathscr{B}(\boldsymbol{X})$, $n = 1, 2, \ldots$ such that $U_n \uparrow \boldsymbol{X}$ (i.e. $\bigcup_n U_n = \boldsymbol{X}$) and $E[N_p(t, U_n)] < \infty$ for all $t > 0$ and $n = 1, 2, \ldots$. For a given $(\mathscr{F}_t)$-adapted, σ-finite point process p, let $\boldsymbol{\Gamma}_p = \{U \in \mathscr{B}(\boldsymbol{X});\ E[N_p(t,U)] < \infty \text{ for all } t > 0\}$. If $U \in \boldsymbol{\Gamma}_p$, then $t \longmapsto N_p(t, U)$ is an adapted, integrable increasing process and hence there exists a natural integrable increasing process $\hat{N}_p(t,U)$ such that $\tilde{N}_p$: $t \longmapsto \tilde{N}_p(t, U) = N_p(t, U) - \hat{N}_p(t, U)$ is a martingale. In general, $t \longmapsto \hat{N}_p(t,U)$ is not continuous but it seems reasonable (at least in applications discussed in this book) to assume that $t \longmapsto \hat{N}_p(t,U)$ is continuous for every U. $U \longmapsto \hat{N}_p(t,U)$ may not be σ-additive, but if the space $\boldsymbol{X}$ is a nice space e.g., a standard measurable space (Definition I-3.3), it is well known that a modification of $\hat{N}_p$ exists such that it is a measure with respect to U. Keeping these considerations in mind, we give the following definition:

Definition 3.1. An $(\mathscr{F}_t)$-adapted point process p on $(\Omega, \mathscr{F}, P)$ is said to be *of the class* (QL)* (with respect to $(\mathscr{F}_t)$) if it is σ-finite and there exists $\hat{N}_p = (\hat{N}_p(t,U))$ such that

(i) for $U \in \boldsymbol{\Gamma}_p$, $t \longmapsto \hat{N}_p(t,U)$ is a continuous $(\mathscr{F}_t)$-adapted increasing process,

* Quasi left-continuous.

(ii) for each t and a.a. $\omega \in \Omega$, $U \longmapsto \hat{N}_p(t,U)$ is a σ-finite measure on $(\boldsymbol{X},\mathscr{B}(\boldsymbol{X}))$,

(iii) for $U \in \boldsymbol{\Gamma}_p$, $t \longmapsto \tilde{N}_p(t,U) = N_p(t,U) - \hat{N}_p(t,U)$ is an $(\mathscr{F}_t)$-martingale.

The random measure $\{\hat{N}_p(t,U)\}$ is called the *compensator* of the point process p (or $\{N_p(t,U)\}$).

Definition 3.2. A point process p is called an $(\mathscr{F}_t)$-*Poisson point process* if it is an $(\mathscr{F}_t)$-adapted, σ-finite Poisson point process such that $\{N_p(t+h,U) - N_p(t,U)\}_{h>0,\, U\in\mathscr{B}(X)}$, is independent of $\mathscr{F}_t$.

An $(\mathscr{F}_t)$-Poisson point process is of class (QL) if and only if $t \longmapsto E(N_p(t,U))$ is continuous for $U \in \boldsymbol{\Gamma}_p$; in this case, the compensator $\hat{N}_p$ is given by $\hat{N}_p(t,U) = E[N_p(t,U)]$. In particular, a stationary $(\mathscr{F}_t)$-Poisson point process is of the class (QL) with compensator $\hat{N}_p(t,U) = tn(U)$ where $n(dx)$ is the characteristic measure of p (Chapter I, Section 9). This property characterizes a stationary $(\mathscr{F}_t)$-Poisson point process (cf. Section 6).

Theorem 3.1. Let p be a point process of the class (QL). Then for $U \in \boldsymbol{\Gamma}_p$, $\tilde{N}_p(\cdot,U) \in \mathscr{M}_2$ and we have

$$(3.1)\qquad \langle \tilde{N}_p(\cdot,\, U_1),\, \tilde{N}_p(\cdot,\, U_2)\rangle(t) = \hat{N}_p(t,\, U_1 \cap U_2).$$

For the proof, we need the following lemma.

Lemma 3.1. If $U \in \boldsymbol{\Gamma}_p$ and $f(s) = f(s,\omega)$ is a bounded $(\mathscr{F}_t)$-predictable process, then

$$X(t) = \int_0^t f(s)d\tilde{N}_p(s,U)^* \ (= \sum_{\substack{s\le t\\ s\in \boldsymbol{D}_p}} f(s)I_U(p(s)) - \int_0^t f(s)d\hat{N}_p(s,U))$$

is an $(\mathscr{F}_t)$-martingale.

Proof. By Proposition I-5.1, it is sufficient to assume that $s \longmapsto f(s)$ is a bounded, left continuous adapted process. Then, for every $s \in [0,\infty)$,

$$f_n(s) = f(0)I_{\{s=0\}}(s) + \sum_{k=0}^{\infty} f\left(\frac{k}{2^n}\right) I_{(k/2^n,\,(k+1)/2^n]}(s) \longrightarrow f(s).$$

Thus it is sufficient to prove the result for $f_n(s)$. But

* In the sense of the Stieltjes integral.

$$\int_0^t f_n(s)d\tilde{N}_p(s,U) = \sum_{k=0}^{\infty} f\left(\frac{k}{2^n}\right)\left[\tilde{N}_p\left(\frac{k+1}{2^n}\wedge t,U\right) - \tilde{N}_p\left(\frac{k}{2^n}\wedge t,U\right)\right]$$

is obviously an $(\mathscr{F}_t)$-martingale.

Proof of the theorem. It is clear that there exists a sequence of $(\mathscr{F}_t)$-stopping times σ_n such that $\sigma_n \uparrow \infty$ a.s. and both $\tilde{N}_p^{(n)}(t, U_1) = \tilde{N}_p(t\wedge\sigma_n, U_1)$ and $\tilde{N}_p^{(n)}(t, U_2) = \tilde{N}_p(t\wedge\sigma_n, U_2)$ are bounded in t. Consequently it suffices to show for each n

$$\tilde{N}_p^{(n)}(t, U_1)\tilde{N}_p^{(n)}(t, U_2) = \text{a martingale} + \hat{N}_p(t\wedge\sigma_n, U_1\cap U_2).$$

By an integration by parts,

$$\begin{aligned}
&\tilde{N}_p^{(n)}(t, U_1)\tilde{N}_p^{(n)}(t, U_2)\\
&\quad= \int_0^t \tilde{N}_p^{(n)}(s-, U_1)\tilde{N}_p^{(n)}(ds, U_2) + \int_0^t \tilde{N}_p^{(n)}(s, U_2)\tilde{N}_p^{(n)}(ds, U_1)\\
&\quad= \int_0^t \tilde{N}_p^{(n)}(s-, U_1)\tilde{N}_p^{(n)}(ds, U_2) + \int_0^t \tilde{N}_p^{(n)}(s-, U_2)\tilde{N}_p^{(n)}(ds, U_1)\\
&\qquad\qquad + \int_0^t [\tilde{N}_p^{(n)}(s, U_2) - \tilde{N}_p^{(n)}(s-, U_2)]\tilde{N}_p^{(n)}(ds, U_1).
\end{aligned}$$

But the first two terms are martingales by Lemma 3.1 and the last term is equal to $\sum_{\substack{s\leq t\wedge\sigma_n\\ s\in D_p}} I_{U_1\cap U_2}(p(s)) = N_p(t\wedge\sigma_n, U_1\cap U_2) = \hat{N}_p(t\wedge\sigma_n, U_1\cap U_2) + \tilde{N}(t\wedge\sigma_n, U_1\cap U_2)$. This proves the theorem.

We are now going to discuss stochastic integrals with respect to a given point process of the class (QL). For this it is convenient to generalize the notion of predictable processes.

Definition 3.3. A real function $f(t, x, \omega)$ defined on $[0, \infty)\times \boldsymbol{X}\times\Omega$ is called $(\mathscr{F}_t)$-predictable if the mapping $(t,x,\omega) \longmapsto f(t,x,\omega)$ is $\mathscr{S}/\mathscr{B}(\boldsymbol{R}^1)$-measurable where $\mathscr{S}$ is the smallest σ-field on $[0, \infty)\times \boldsymbol{X}\times\Omega$ with respect to which all g having the following properties are measurable:
(i) for each $t > 0$, $(x, \omega) \longmapsto g(t, x, \omega)$ is $\mathscr{B}(\boldsymbol{X})\times\mathscr{F}_t$-measurable;
(ii) for each (x, ω), $t \longmapsto g(t, x, \omega)$ is left continuous.

We introduce the following classes:

$$F_p = \{f(t, x, \omega); f \text{ is } (\mathscr{F}_t)\text{-predictable and for each } t > 0,$$

$$\int_0^{t+}\int_X |f(s, x, \omega)|\,N_p(dsdx) < \infty \quad \text{a.s.}\}, \tag{3.2}$$

(3.3) $$F_p^1 = \{f(t, x, \omega); f \text{ is } (\mathscr{F}_t)\text{-predictable and for every } t > 0,\ E[\int_0^t \int_X |f(s, x, \cdot)| \hat{N}_p(dsdx)] < \infty\},$$

(3.4) $$F_p^2 = \{f(t, x, \omega); f \text{ is } (\mathscr{F}_t)\text{-predictable and for every } t > 0,\ E[\int_0^t \int_X |f(s, x, \cdot)|^2 \hat{N}_p(dsdx)] < \infty\},$$

(3.5) $F_p^{2,loc} = \{f(t, x, \omega);\ f$ is $(\mathscr{F}_t)$-predictable and there exists a sequence of $(\mathscr{F}_t)$-stopping times σ_n such that $\sigma_n \uparrow \infty$ a.s. and $I_{[0,\sigma_n]}(t)f(t, x, \omega) \in F_p^2,\ n = 1, 2, \ldots\}$.

For $f \in F_p$,

(3.6) $$\int_0^{t+} \int_X f(s, x, \cdot) N_p(dsdx)$$

is well defined a.s. as the Lebesgue-Stieltjes integral and equals the absolutely convergent sum

(3.7) $$\sum_{\substack{s \le t \\ s \in D_p}} f(s, p(s), \cdot).$$

Next let $f \in F_p^1$. By the same argument as in the proof of Lemma 3.1, it is easy to see that

$$E[\int_0^{t+} \int_X |f(s, x, \cdot)| N_p(dsdx)] = E[\int_0^t \int_X |f(s, x, \cdot)| \hat{N}_p(dsdx)].$$

This implies, in particular, that $F_p^1 \subset F_p$. Set

(3.8) $$\int_0^{t+} \int_X f(s, x, \cdot) \tilde{N}_p(dsdx) = \int_0^{t+} \int_X f(s, x, \cdot) N_p(dsdx) - \int_0^t \int_X f(s, x, \cdot) \hat{N}_p(dsdx).$$

Then $t \longmapsto \int_0^{t+} \int_X f(s, x, \cdot) \tilde{N}_p(dsdx)$ is an $(\mathscr{F}_t)$-martingale by the same proof as in Lemma 3.1.

If we assume $f \in F_p^1 \cap F_p^2$, then

$$t \longmapsto \int_0^{t+} \int_X f(s, x, \cdot) \tilde{N}_p(dsdx) \in \mathscr{M}_2$$

and

(3.9) $$\langle \int_0^{t+} \int_X f(s, x, \cdot) \tilde{N}_p(dsdx) \rangle = \int_0^t \int_X f^2(s, x, \cdot) \hat{N}_p(dsdx).$$

(3.9) is proved by the same argument as in the proof of Theorem 3.1 and Lemma 3.1.

Let $f \in \boldsymbol{F}_p^2$. If we set

$$f_n(s, x, \omega) = I_{(-n,n)}(f(s, x, \omega))I_{U_n}(x)f(s, x, \omega)^{*1}$$

then $f_n \in \boldsymbol{F}_p^1 \cap \boldsymbol{F}_p^2$ and so $\int_0^{t+}\int_X f_n(s, x, \cdot)\tilde{N}_p(dsdx)$ is defined for each n. By (3.9),

$$E\left[\left\{\int_0^{t+}\int_X f_n(s, x, \cdot)\tilde{N}_p(dsdx) - \int_0^{t+}\int_X f_m(s, x, \cdot)\tilde{N}_p(dsdx)\right\}^2\right]$$
$$= E\left[\int_0^t\int_X [f_n(s, x, \cdot) - f_m(s, x, \cdot)]^2\hat{N}_p(dsdx)\right]$$

and thus $\{\int_0^{t+}\int_X f_n(s, x, \cdot)\tilde{N}_p(dsdx)\}_{n=1}^{\infty}$ is a Cauchy sequence in $\mathscr{M}_2$. We denote its limit by $\int_0^{t+}\int_X f(s, x, \cdot)\tilde{N}_p(dsdx)$. Note that (3.8) no longer holds: each term in the right-hand side has no meaning in general. The integral $\int_0^{t+}\int_X f(s, x, \cdot)\tilde{N}_p(dsdx)$ may be called a "*compensated sum*".

Finally, if $f \in \boldsymbol{F}_p^{2,loc}$, the stochastic integral $\int_0^{t+}\int_X f(s, x, \cdot)\tilde{N}_p(dsdx)$ is defined as the unique element $X \in \mathscr{M}_2^{loc}$ having the property

$$X(t \wedge \sigma_n) = \int_0^{t+}\int_X I_{[0,\sigma_n]}(s)f(s, x, \cdot)\tilde{N}_p(dsdx), \qquad n = 1, 2, \ldots$$

where $\{\sigma_n\}$ is a sequence of stopping times in (3.5).

4. Semi-martingales

The time evolution of a physical system is usually described by a differential equation. Besides such a deterministic motion, it is sometimes necessary to consider a random motion and its mathematical model is a stochastic process. Many important stochastic processes have the following common feature: they are expressed as the sum of a *mean motion* and a *fluctuation* from the mean motion.*2 A typical example is a stochastic process $X(t)$ of the form

$$X(t) = X(0) + \int_0^t f(s)ds + \int_0^t g(s)dB(s)$$

*1 $\{U_n\}$ is chosen so that $U_n \uparrow X$ and $E[N_p(t, U_n)] < \infty$ for every $t > 0$ and all n.
*2 It may be considered as a *noise*.

where $f(s)$ and $g(s)$ are suitable adapted processes and $\int \cdot dB$ is the stochastic integral with respect to a Brownian motion $B(t)$. Here the process $\int_0^t f(s)ds$ is the mean motion and $\int_0^t g(s)dB(s)$ is the fluctuation. Such a stochastic process is called an *Itô process* and it constitutes a very important class of stochastic processes. The essential structure of such a process is that it is the sum of a process with sample functions of bounded variation and a martingale. Generally, a stochastic process $X(t)$ defined on $(\Omega, \mathscr{F}, P)$ with a reference family $(\mathscr{F}_t)$ is called a *semi-martingale* if

$$X(t) = X(0) + M(t) + A(t)$$

where $M(t)$ ($M(0) = 0$, a.s.) is a local $(\mathscr{F}_t)$-martingale and $A(t)$ ($A(0) = 0$, a.s.) is a right-continuous $(\mathscr{F}_t)$-adapted process whose sample functions $t \longmapsto A(t)$ are of bounded variation on any finite interval a.s. We will restrict ourselves to a sub-class of semi-martingales which is easier to handle and yet is sufficient for applications discussed in this book.* For simplicity, we will call an element of this sub-class a semi-martingale. Namely we give the following

Definition 4.1. Let $(\Omega, \mathscr{F}, P)$ and $(\mathscr{F}_t)_{t \geq 0}$ be given as usual. A stochastic process $X = (X(t))_{t \geq 0}$ defined on this probability space is called a semi-martingale if it is expressed as

$$X(t) = X(0) + M(t) + A(t) + \int_0^{t+} \int_X f_1(s, x, \cdot) N_p(dsdx) + \int_0^{t+} \int_X f_2(s, x, \cdot) \tilde{N}_p(dsdx) \tag{4.1}$$

where
(i) $X(0)$ is an $\mathscr{F}_0$-measurable random variable,
(ii) $M \in \mathscr{M}_2^{c, loc}$ (so, in particular, $M(0) = 0$ a.s.),
(iii) $A = (A(t))$ is a continuous $(\mathscr{F}_t)$-adapted process such that a.s. $A(0) = 0$ and $t \longmapsto A(t)$ is of bounded variation on each finite interval,
(iv) p is an $(\mathscr{F}_t)$-adapted point process of class (QL) on some state space $(\boldsymbol{X}, \mathscr{B}(\boldsymbol{X}))$, $f_1 \in \boldsymbol{F}_p$ and $f_2 \in \boldsymbol{F}_p^{2, loc}$ such that

$$f_1 f_2 = 0. \tag{4.2}$$

It is not difficult to see that $M(t)$ in the expression (4.1) is uniquely determined from $X(t)$; it is called a *continuous martingale part* of $X(t)$. The

* For a detailed treatment of the general theory of semi-martingales, we refer the reader to Meyer [121] and Jacod [77].

discontinuities of $X(t)$ come from the sum of the last two terms and, by (4.2), these two terms do not have any common discontinuities.

Example 4.1. (Lévy processes).

Let $X(t)$ be a d-dimensional time homogeneous Lévy process (i.e. a right continuous process with stationary independent increments) and $(\mathscr{F}_t)_{t\geq 0}$ be generated by the sample paths $X(t)$. Let $\boldsymbol{D}_p = \{t > 0; X(t) \neq X(t-)\}$ and, for $t \in \boldsymbol{D}_p$, let $p(t) = X(t) - X(t-)$. Then p defines a stationary $(\mathscr{F}_t)$-Poisson point process on $\boldsymbol{X} = \boldsymbol{R}^d \setminus \{0\}$ (cf. Definition 3.1). The famous Lévy-Itô theorem*[1] states that there exist a d'-dimensional $(\mathscr{F}_t)$-Brownian motion $B(t) = (B^k(t))_{k=1}^{d'}$, $0 \leq d' \leq d$, a $d \times d'$ matrix $A = (a_k^i)$ of the rank d' and a d-dimensional vector $B = (b^i)$ such that $X(t) = (X^1(t), X^2(t), \ldots, X^d(t))$ can be represented as

$$
(4.3)\quad \begin{aligned} X^i(t) = X^i(0) + \sum_{k=1}^{d'} a_k^i B^k(t) + b^i t + \int_0^{t+}\int_{\boldsymbol{R}^d\setminus\{0\}} x^i I_{\{|x|\geq 1\}} N_p(dsdx) \\ + \int_0^{t+}\int_{\boldsymbol{R}^d\setminus\{0\}} x^i I_{\{|x|<1\}} \tilde{N}_p(dsdx), \qquad i = 1, 2, \ldots, d. \end{aligned}
$$

In this case, the compensator $\hat{N}_p(dsdx)$ of p is of the form $\hat{N}(dsdx) = dsn(dx)$, where $n(dx)$ is the characteristic measure of p. $n(dx)$ is also called the *Lévy measure* of the process X. It is a σ-finite measure on $\boldsymbol{R}^d \setminus \{0\}$ such that $\int_{\boldsymbol{R}^d\setminus\{0\}} \{|x|^2/(1 + |x|^2)\} n(dx) < \infty$.

In the above expression, the Brownian motion $\{B^k(t)\}$ and p are automatically independent (cf. Section 6). By (4.3), we get the following Lévy-Khinchin formula:

$$
(4.4)\quad E[e^{i\langle \xi, X(t)-X(s)\rangle} \mid \mathscr{F}_s] = e^{(t-s)\psi(\xi)} \quad \text{a.s.,} \qquad t \geq s \geq 0, \;\; \xi \in \boldsymbol{R}^d,
$$

where

$$
(4.5)\quad \begin{aligned} \psi(\xi) = -\frac{1}{2}\langle AA^*\xi, \xi\rangle + i\langle B, \xi\rangle \\ + \int_{\boldsymbol{R}^d\setminus\{0\}} (e^{i\langle \xi, x\rangle} - 1 - iI_{\{|x|<1\}} \langle \xi, x\rangle) n(dx).^{*2} \end{aligned}
$$

Let $X(t)$ be a given semi-martingale with respect to $(\mathscr{F}_t)$ and let $(\mathscr{F}_t^X)$ be generated by the sample paths $X(t)$. Clearly $\mathscr{F}_t^X \subset \mathscr{F}_t$ for each $t \geq 0$. It is not trivial that $X(t)$ is also a semi-martingale with respect to $(\mathscr{F}_t^X)$.*[3]

*[1] Cf. Lévy [99] and Itô [58], [69].
*[2] A^* stands for the transposed matrix of A.
*[3] Cf. Stricker [155].

The continuous martingale part of $X(t)$ with respect to $(\mathscr{F}_t^X)$ differs from that with respect to $(\mathscr{F}_t)$ and generally it is an important problem to discuss how the martingale part differs under a change of a reference family.*[1]

5. Itô's formula

Itô's formula is one of the most important tools in the study of semi-martingales. It provides us with the differential-integral calculus for sample functions of stochastic processes.

Let $(\Omega,\mathscr{F},P)$ with $(\mathscr{F}_t)_{t\geq 0}$ be given as above. Suppose on this probability space the following are given:

(i) $M^i(t) \in \mathscr{M}_2^{c,loc}$, $(i = 1, 2, \ldots, d)$;

(ii) $A^i(t)$ $(i = 1, 2, \ldots, d)$: a continuous $(\mathscr{F}_t)$-adapted process whose almost all sample functions are of bounded variation on each finite interval and $A^i(0) = 0$;

(iii) p: a point process of the class (QL) with respect to $(\mathscr{F}_t)$ on some state such that $f^i(t, x, \omega)g^j(t, x, \omega) = 0$, $i, j = 1, 2, \cdots, d$; furthermore, we assume that $g(t, x, \omega)$ is bounded, i.e. a constant $M > 0$ exists such that

$$|g^i(t, x, \omega)| \leqq M \qquad \text{for all } i, t, x, \omega.$$

(iv) $X^i(0)$ $(i = 1, 2, \cdots, d)$: an $\mathscr{F}_0$-measurable random variable.

Define a d-dimensional semi-martingale $X(t) = (X^1(t), X^2(t), \cdots, X^d(t))$ by

$$\begin{aligned} X^i(t) = {} & X^i(0) + M^i(t) + A^i(t) \\ & + \int_0^{t+}\int_X f^i(s, x, \cdot)N_p(dsdx) + \int_0^{t+}\int_X g^i(s, x, \cdot)\tilde{N}_p(dsdx), \\ & \qquad i = 1, 2, \ldots d. \end{aligned} \tag{5.1}$$

Denote also $f = (f^1, f^2, \ldots, f^d)$ and $g = (g^1, g^2, \ldots, g^d)$.

Theorem 5.1. (Itô's formula).*[2] Let F be a function of class C^2 on $\boldsymbol{R}^d$ and $X(t)$ a d-dimensional semi-martingale given above. Then the stochastic process $F(X(t))$ is also a semi-martingale (with respect to $(\mathscr{F}_t)_{t\geq 0}$) and the following formula holds:*[3]

$$F(X(t)) - F(X(0))$$

*[1] Cf. e.g. Jeulin [78] for a general theory and applications. In filtering theory,it is related to the notion of "innovation", Fujisaki-Kallianpur-Kunita [29].

*[2] Cf. Itô [63], Kunita-Watanabe [97] and Doléans-Dade-Meyer [17].

*[3] $F_i' = \dfrac{\partial F}{\partial x_i}$, $F_{ij}'' = \dfrac{\partial^2 F}{\partial x_i \partial x_j}$, $i, j = 1, 2, \ldots.$

$$= \sum_{i=1}^{d} \int_0^t F_i'(X(s))dM^i(s) + \sum_{i=1}^{d} \int_0^t F_i'(X(s))dA^i(s)$$

$$+ \frac{1}{2} \sum_{i,j=1}^{d} \int_0^t F_{ij}''(X(s))d\langle M^i, M^j\rangle(s)$$

$$(5.2) \qquad + \int_0^{t+} \int_X \{F(X(s-) + f(s, x, \cdot)) - F(X(s-))\} N_p(dsdx)$$

$$+ \int_0^{t+} \int_X \{F(X(s-) + g(s, x, \cdot)) - F(X(s-))\} \tilde{N}_p(dsdx)$$

$$+ \int_0^t \int_X \{F(X(s) + g(s, x, \cdot)) - F(X(s))$$

$$- \sum_{i=1}^{d} g^i(s, x, \cdot) F_i'(X(s))\} \hat{N}_p(dsdx).$$

Proof. To avoid notational complexity, we shall assume $d = 1$; there is no essential change in the multi-dimensional case.

First, we will prove the result in the case of continuous semi-martingale:

$$(5.3) \qquad X(t) = X(0) + M(t) + A(t).$$

Formula (5.2) then reduces to

$$(5.4) \qquad F(X(t)) - F(X(0)) = \int_0^t F'(X(s))dM(s) + \int_0^t F'(X(s))dA(s) + \frac{1}{2} \int_0^t F''(X(s))d\langle M\rangle(s).$$

Let

$$\tau_n = \begin{cases} 0 \quad , & \text{if } \ |X(0)| > n, \\ \inf\{t;\ |M(t)| > n \text{ or } \|A\|(t) > n \text{ or } |\langle M\rangle(t)| > n\}, & \text{if } \ |X(0)| \leq n. \end{cases}$$

Clearly $\tau_n \uparrow \infty$ a.s. Consequently, if we can prove (5.4) for $X(t \wedge \tau_n)$ on the set $\{\tau_n > 0\}$, then by letting $n \uparrow \infty$ we see immediately that (5.4) holds. Therefore, we may assume that $X(0)$, $M(t)$, $\|A\|(t)$, $\langle M\rangle(t)$ are bounded in (t, ω) and also that $F(x)$ is a C^2-function with compact support.

Fix $t > 0$ and let Δ be a division of $[0, t]$ given by $0 = t_0 < t_1 < \cdots < t_n = t$. By the mean value theorem,

$$F(X(t)) - F(X(0)) = \sum_{k=1}^{n} \{F(X(t_k)) - F(X(t_{k-1}))\}$$

$$= \sum_{k=1}^{n} F'(X(t_{k-1}))\{X(t_k) - X(t_{k-1})\}$$

$$+\frac{1}{2}\sum_{k=1}^{n} F''(\xi_k)\{X(t_k)-X(t_{k-1})\}^2,$$

where ξ_k satisfies $X(t_k)\wedge X(t_{k-1}) \leq \xi_k \leq X(t_k)\vee X(t_{k-1})$. Firstly

$$\sum_{k=1}^{n} F'(X(t_{k-1}))\{X(t_k)-X(t_{k-1})\}$$

$$=\sum_{k=1}^{n} F'(X(t_{k-1}))\{M(t_k)-M(t_{k-1})\}$$

$$+\sum_{k=1}^{n} F'(X(t_{k-1}))\{A(t_k)-A(t_{k-1})\}$$

$$= I_1^{\Delta} + I_2^{\Delta}, \quad \text{say.}$$

It is clear that, as $|\Delta| = \max_k |t_k - t_{k-1}| \longrightarrow 0, I_2^{\Delta} \longrightarrow \int_0^t F'(X(s))dA(s)$ a.s. Also, if we set

$$\Phi^{\Delta}(s,\omega) = I_{\{s=0\}}(s)F'(X(0)) + \sum_{k=1}^{n} I_{(t_{k-1},t_k]}(s)F'(X(t_{k-1}))$$

and

$$\Phi(s,\omega) = F'(X(s)),$$

then $\Phi^{\Delta} \in \mathscr{L}_0$ and

$$\|\Phi^{\Delta}-\Phi\|_{2,t}^{M} = E[\int_0^t |\Phi^{\Delta}(s,\omega)-\Phi(s,\omega)|^2 d\langle M\rangle(s)]^{1/2} \longrightarrow 0$$

as $|\Delta| \longrightarrow 0$. Hence $I_1^{\Delta} = \int_0^t \Phi^{\Delta}(s,\omega)dM(s) \longrightarrow \int_0^t F'(X(s))dM(s)$ in $\mathscr{L}_2(\Omega)$ as $|\Delta| \longrightarrow 0$.

Secondly,

$$\frac{1}{2}\sum_{k=1}^{n} F''(\xi_k)\{X(t_k)-X(t_{k-1})\}^2$$

$$=\frac{1}{2}\sum_{k=1}^{n} F''(\xi_k)\{A(t_k)-A(t_{k-1})\}^2$$

$$+\sum_{k=1}^{n} F''(\xi_k)\{M(t_k)-M(t_{k-1})\}\{A(t_k)-A(t_{k-1})\}$$

$$+\frac{1}{2}\sum_{k=1}^{n} F''(\xi_k)\{M(t_k)-M(t_{k-1})\}^2$$

$$= I_3^{\Delta} + I_4^{\Delta} + I_5^{\Delta}, \quad \text{say.}$$

It is not hard to show that I_3^Δ and I_4^Δ tend to 0 a.s. as $|\Delta| \longrightarrow 0$; for example,

$$|I_4^\Delta| \leq \sup_{x\in R} |F''(x)| \max_{1\leq k\leq n} |M(t_k) - M(t_{k-1})| \| A \|(t) \longrightarrow 0 \quad \text{a.s.} \quad \text{as}$$

$|\Delta| \longrightarrow 0$. We now show that $I_5^\Delta \longrightarrow \frac{1}{2}\int_0^t F''(X(s))d\langle M\rangle(s)$ in $\mathscr{L}_1(\Omega)$. For this we need the following lemma.

Lemma 5.1. Let $C > 0$ be a constant such that $|M(s)| \leq C$, $s \in [0, t]$. Set $V_l^\Delta = \sum_{k=1}^{l} \{M(t_k) - M(t_{k-1})\}^2$, $l = 1, 2, \ldots, n$. Then

$$E[(V_n^\Delta)^2] \leq 12\, C^4.$$

Proof. It is easy to see that

$$(V_n^\Delta)^2 = \sum_{k=1}^{n} \{M(t_k) - M(t_{k-1})\}^4 + 2\sum_{k=1}^{n} (V_n^\Delta - V_k^\Delta)(M(t_k) - M(t_{k-1}))^2$$

and

$$E[(V_n^\Delta - V_k^\Delta)|\mathscr{F}_{t_k}] = E[\sum_{l=k+1}^{n} \{M(t_l) - M(t_{l-1})\}^2|\mathscr{F}_{t_k}]$$
$$= E[(M(t) - M(t_k))^2|\mathscr{F}_{t_k}] \leq (2C)^2.$$

Hence,

$$E[\sum_{k=1}^{n} \{(V_n^\Delta - V_k^\Delta)(M(t_k) - M(t_{k-1}))^2\}]$$
$$\leq (2C)^2 E(V_n^\Delta) = (2C)^2 E(M(t)^2) \leq 4C^4.$$

Finally,

$$E[\sum_{k=1}^{n} \{M(t_k) - M(t_{k-1})\}^4] \leq (2C)^2 E(V_n^\Delta) \leq 4C^4,$$

and the proof is complete.

Now, returning to the proof, we set

$$I_6^\Delta = \frac{1}{2}\sum_{k=1}^{n} F''(X(t_{k-1}))\{M(t_k) - M(t_{k-1})\}^2.$$

Then

$$
(5.5)\qquad \begin{aligned} E(|I_5^\Delta - I_6^\Delta|) &\le \frac{1}{2}(E\{\max_{1\le k\le n}|F''(\xi_k)-F''(X(t_{k-1}))|^2\})^{1/2}(E\{(V_n^\Delta)^2\})^{1/2} \\ &\le (\sqrt{12C^4}/2)(E\{\max_{1\le k\le n}|F''(\xi_k) - F''(X(t_{k-1}))|^2\})^{1/2} \longrightarrow 0 \end{aligned}
$$

as $|\Delta| \longrightarrow 0$ by the dominated convergence theorem. If we set

$$
I_7^\Delta = \frac{1}{2}\sum_{k=1}^{n} F''(X(t_{k-1}))\{\langle M\rangle(t_k) - \langle M\rangle(t_{k-1})\},
$$

then clearly

$$
(5.6)\qquad E[|I_7^\Delta - \frac{1}{2}\int_0^t F''(X(s))d\langle M\rangle(s)|] \longrightarrow 0 \qquad \text{as} \qquad |\Delta| \longrightarrow 0
$$

by the dominated convergence theorem. Finally,

$$
\begin{aligned} &E\{|I_6^\Delta - I_7^\Delta|^2\} \\ &= \frac{1}{4}E\{[\sum_{k=1}^{n} F''(X(t_{k-1}))\{(M(t_k) - M(t_{k-1}))^2 \\ &\qquad - (\langle M\rangle(t_k) - \langle M\rangle(t_{k-1}))\}]^2\}. \end{aligned}
$$

Noting that

$$
E\{[(M(t_k) - M(t_{k-1}))^2 - (\langle M\rangle(t_k) - \langle M\rangle(t_{k-1}))]|\mathscr{F}_{t_{k-1}}\} = 0 \qquad \text{for all } k,
$$

the above equals

$$
\begin{aligned} &\frac{1}{4}E\{\sum_{k=1}^{n}[F''(X(t_{k-1}))^2\{(M(t_k) - M(t_{k-1}))^2 \\ &\qquad\qquad - (\langle M\rangle(t_k) - \langle M\rangle(t_{k-1}))\}^2]\} \\ &\le \frac{1}{2}\max_{x\in R^1}|F''(x)|^2 E\{\sum_{k=1}^{n}(M(t_k) - M(t_{k-1}))^4\} \\ &\quad + \frac{1}{2}\max_{x\in R^1}|F''(x)|^2 E\{\sum_{k=1}^{n}(\langle M\rangle(t_k) - \langle M\rangle(t_{k-1}))^2\} \end{aligned}
$$

$$
(5.7)\qquad \begin{aligned} &\le \frac{1}{2}\max_{x\in R^1}|F''(x)|^2 E[\max_{1\le k\le n}((M(t_k) - M(t_{k-1}))^2 V_n^\Delta] \\ &\quad + \frac{1}{2}\max_{x\in R^1}|F''(x)|^2 E[\max_{1\le k\le n}(\langle M\rangle(t_k) - \langle M\rangle(t_{k-1}))\langle M\rangle(t)] \\ &\le \frac{1}{2}\max_{x\in R^1}|F''(x)|^2(E[(V_n^\Delta)^2])^{1/2}(E[\max_{1\le k\le n}|M(t_k) - M(t_{k-1})|^4])^{1/2} \\ &\quad + \frac{1}{2}\max_{x\in R^1}|F''(x)|^2 E[\max_{1\le k\le n}(\langle M\rangle(t_k) - \langle M\rangle(t_{k-1}))\langle M\rangle(t)]. \end{aligned}
$$

This last expression tends to zero as $|\Delta| \longrightarrow 0$ by Lemma 5.1 and the dominated convergence theorem.* By (5.5), (5.6) and (5.7), we have

* Note that $\langle M\rangle(t)$ is bounded.

proved that $I_3^4 \longrightarrow \frac{1}{2}\int_0^t F''(X(s))d\langle M\rangle(s)$ in $\mathscr{L}_1(\Omega)$. Thus (5.4) is true for a fixed time t but, since both sides are continuous in t a.s., (5.4) holds for all $t \geq 0$ a.s.

Next we will prove (5.2) for the general $X(t)$ in the case $d = 1$:

$$X(t) = X(0) + M(t) + A(t) + \int_0^{t+}\int_X f(s, x, \cdot)N_p(dsdx) + \int_0^{t+}\int_X g(s, x, \cdot)\tilde{N}_p(dsdx).$$

First we note that the proof is easily reduced to the case that $|f(s, x, \cdot)|$ is bounded; i.e. there exists $M > 0$ such that

$$|f(s, x, \omega)| \leqq M \qquad \text{for all } (s, x, \omega);$$

because the set $\{s;\ |f(s, p(s), \omega)| > M\}$ is discrete in $(0, \infty)$ almost surely. Then, by the usual method of truncation, the proof can be reduced to the case where $g \in F_p^2$ and F is such that F, F' and F'' are bounded. For the point process p, let $U_n \in \mathscr{B}(X)$, $n = 1, 2, \cdots$ be such that $U_n \subset U_{n+1}$, $\bigcup_n U_n = X$ and $E(N_p((0, t] \times U_n)) < \infty$ for all $t \geq 0$. For each n set

$$X_n(t) = X(0) + M(t) + A(t) + \int_0^{t+}\int_X f^{(n)}(s, x, \cdot)N_p(dsdx) + \int_0^{t+}\int_X g^{(n)}(s, x, \cdot)\tilde{N}_p(dsdx), \tag{5.8}$$

where $f^{(n)}(s, x, \omega) = f(s, x, \omega)I_{U_n}(x)$ and $g^{(n)}(s, x, \omega) = g(s, x, \omega)I_{U_n}(x)$. First we prove (5.2) for the semi-martingale $X_n(t)$. The point process p_n defined by $D_{p_n} = \{s \in D_p;\ p(s) \in U_n\}$ and $p_n(s) = p(s)$ for $s \in D_{p_n}$ is discrete in the sense that $\#\{s; s \leq t, s \in D_{p_n}\}$ is finite a.s. for each $t > 0$. If we order the set D_{p_n} according to magnitude, say $0 < \sigma_1 < \sigma_2 < \cdots < \sigma_m < \cdots$ it is easy to see that σ_m is an $(\mathscr{F}_t)$-stopping time. Now $X_n(t)$ is represented as

$$X_n(t) = X(0) + M(t) + A(t) + \sum_{\sigma_m \leq t} f(\sigma_m, p(\sigma_m), \cdot) + \sum_{\sigma_m \leq t} g(\sigma_m, p(\sigma_m), \cdot) - \int_0^t\int_X g^{(n)}(s, x, \cdot)\hat{N}_p(dsdx).$$

Then, setting $\sigma_0 \equiv 0$,

$$F(X_n(t)) - F(X(0)) = \sum_m \{F(X_n(\sigma_m \wedge t)) - F(X_n(\sigma_m \wedge t-))\}$$

$$+\sum_m \{F(X_n(\sigma_m \wedge t-)) - F(X_n(\sigma_{m-1} \wedge t))\}^*$$

$$= I_1(t) + I_2(t) \quad \text{say.}$$

We can apply (5.4) to obtain

$$\begin{aligned} F(X_n(\sigma_m \wedge t-)) &- F(X_n(\sigma_{m-1} \wedge t)) \\ &= \int_{\sigma_{m-1}\wedge t}^{\sigma_m\wedge t} F'(X_n(s))\, dM(s) + \int_{\sigma_{m-1}\wedge t}^{\sigma_m\wedge t} F'(X_n(s))dA(s) \\ &+ \frac{1}{2}\int_{\sigma_{m-1}\wedge t}^{\sigma_m\wedge t} F''(X_n(s))d\langle M\rangle(s) - \int_{\sigma_{m-1}\wedge t}^{\sigma_m\wedge t} F'(X_n(s))dA^{g_n}(s), \end{aligned}$$

where

$$A^{g_n}(t) = \int_0^t\int_X g^{(n)}(s, y, \cdot)\hat{N}_p(dsdy).$$

Thus

$$\begin{aligned} I_2(t) &= \int_0^t F'(X_n(s))dM(s) + \int_0^t F'(X_n(s))dA(s) \\ &+ \frac{1}{2}\int_0^t F''(X_n(s))d\langle M\rangle(s) - \int_0^t F'(X_n(s))dA^{g_n}(s). \end{aligned} \tag{5.9}$$

Noting the assumption $f(s,x,\omega)g(s,x,\omega) = 0$, we have

$$\begin{aligned} I_1(t) &= \sum_m \{F(X_n((\sigma_m)) - F(X_n(\sigma_m-))\} I_{\{\sigma_m\le t, f(\sigma_m, p(\sigma_m), \cdot)\neq 0\}} \\ &+ \sum_m \{F(X_n(\sigma_m)) - F(X_n(\sigma_m-))\} I_{\{\sigma_m\le t, g(\sigma_m, p(\sigma_m), \cdot)\neq 0\}} \\ &= \int_0^{t+}\int_X \{F(X_n(s-) + f^{(n)}(s, x, \cdot)) - F(X_n(s-))\} N_p(dsdx) \\ &+ \int_0^{t+}\int_X \{F(X_n(s-) + g^{(n)}(s, x, \cdot)) - F(X_n(s-))\} N_p(dsdx) \\ &= \int_0^{t+}\int_X \{F(X_n(s-) + f^{(n)}(s, x, \cdot)) - F(X_n(s-))\} N_p(dsdx) \\ &+ \int_0^{t+}\int_X \{F(X_n(s-) + g^{(n)}(s, x, \cdot)) - F(X_n(s-))\} \tilde{N}_p(dsdx) \\ &+ \int_0^t\int_X \{F(X_n(s) + g^{(n)}(s, x, \cdot)) - F(X_n(s))\} \hat{N}_p(dsdx). \end{aligned} \tag{5.10}$$

By (5.9) and (5.10), we see that (5.2) holds for the process $X_n(t)$. The for-

---

* To be precise,

$$F(X_n(\sigma_m \wedge t-)) = \begin{cases} F(X_n(\sigma_m-)) & \text{if } \sigma_m \le t \\ F(X_n(t)) & \text{if } t < \sigma_m. \end{cases}$$

mula (5.2) for the process $X(t)$ is obtained by letting $n \longrightarrow \infty$. First, $\int_0^{t+}\int_X g^{(n)}(s, x, \cdot)\tilde{N}_p(dsdx)$ converges to $\int_0^{t+}\int_X g(s, x, \cdot)\tilde{N}_p(dsdx)$ in $\mathscr{M}_2$ as $n \longrightarrow \infty$ and hence, by taking a subsequence if necessary, we may assume that this convergence is uniform on every finite interval a.s. Also $\int_0^{t+}\int_X f^{(n)}(s, x, \cdot)N_p(dsdx)$ converges to $\int_0^{t+}\int_X f(s, x, \cdot)N_p(dsdx)$ as $n \longrightarrow \infty$ uniformly in t on each finite interval a.s. Consequently $X_n(t)$ converges to $X(t)$ as $n \longrightarrow \infty$ uniformly in t on each finite interval a.s. and hence $F(X_n(t)) - F(X(0)) \longrightarrow F(X(t)) - F(X(0))$ a.s. Also, it is easy to see, by the dominated convergence theorem, that

$$\int_0^t F'(X_n(s))dM(s) \longrightarrow \int_0^t F'(X(s))dM(s) \quad \text{in } \mathscr{M}_2,$$

$$\int_0^t F'(X_n(s))dA(s) \longrightarrow \int_0^t F'(X(s))dA(s) \quad \text{a.s.},$$

$$\int_0^t F''(X_n(s))d\langle M\rangle(s) \longrightarrow \int_0^t F''(X(s))d\langle M\rangle(s) \quad \text{a.s.},$$

$$\begin{aligned}\int_0^t\int_X \{F(X_n(s) + g^{(n)}(s, x, \cdot)) - F(X_n(s)) \\ - g^{(n)}(s, x, \cdot)F'(X_n(s))\}\hat{N}_p(dsdx) \\ \longrightarrow \int_0^t\int_X \{F(X(s) + g(s, x, \cdot)) - F(X(s)) \\ - g(s, x, \cdot)F'(X(s))\}\hat{N}_p(dsdx) \quad \text{a.s.},\end{aligned}$$

$$\begin{aligned}\int_0^{t+}\int_X \{F(X_n(s-) + f^{(n)}(s,x, \cdot)) - F(X_n(s-))\}N_p(dsdx) \\ \longrightarrow \int_0^{t+}\int_X \{F(X(s-) + f(s,x, \cdot)) - F(X(s-))\}N_p(dsdx) \quad \text{a.s.},\end{aligned}$$

and

$$\begin{aligned}\int_0^{t+}\int_X \{F(X_n(s-) + g^{(n)}(s, x, \cdot)) - F(X_n(s-))\}\tilde{N}_p(dsdx) \\ \longrightarrow \int_0^{t+}\int_X \{F(X(s-) + g(s,x, \cdot)) - F(X(s-))\}\tilde{N}_p(dsdx) \quad \text{in } \mathscr{M}_2.\end{aligned}$$

Thus the proof of (5.2) for $X(t)$ is now complete.

6. Martingale characterization of Brownian motions and Poisson point processes

It is a remarkable fact that many interesting stochastic processes are characterized as semi-martingales whose characteristics (e.g., the quadratic variational process of the continuous martingale part, the compensator of the point process describing discontinuities of the sample path) are given functionals of sample paths.*[1] *Martingale problems* (first introduced by Stroock and Varadhan [157]) are just such ways of determining stochastic processes. They are based on the fact that the basic stochastic processes such as Brownian motions and Poisson point processes are characterized in terms of the characteristics of semi-martingales. This fact itself may be considered as a typical martingale problem.

Theorem 6.1. Let $X(t) = (X^1(t), X^2(t), \ldots, X^d(t))$ be a d-dimensional $(\mathscr{F}_t)$-semi-martingale such that

$$M^i(t) = X^i(t) - X^i(0) \in \mathscr{M}_2^{c, loc} \tag{6.1}$$

and

$$\langle M^i, M^j\rangle(t) = \delta_{ij}t, \qquad i,j = 1,2, \ldots, d. \tag{6.2}$$

Then $X(t)$ is a d-dimensional $(\mathscr{F}_t)$-Brownian motion*[2].

Proof. It is enough to prove that

$$E[e^{i\langle \xi, X(t)-X(s)\rangle} | \mathscr{F}_s] = e^{-\frac{1}{2}|\xi|^2(t-s)} \qquad \text{a.s.} \tag{6.3}$$

for every $\xi \in \boldsymbol{R}^d$ and $t > s \geq 0$. Let $F(x) = e^{i\langle \xi, x\rangle}$ and apply Itô's formula. Then we have

$$\begin{aligned} &e^{i\langle \xi, X(t)\rangle} - e^{i\langle \xi, X(s)\rangle} \\ &\quad = \sum_{k=1}^d \int_s^t i\xi_k e^{i\langle \xi, X(u)\rangle} dM^k(u) + \frac{1}{2}\sum_{k=1}^d \int_s^t (-\xi_k^2) e^{i\langle \xi, X(u)\rangle} du. \end{aligned} \tag{6.4}$$

Clearly (6.2) implies that $M^k \in \mathscr{M}_2^c$ and hence

$$E[\int_s^t e^{i\langle \xi, X(u)\rangle} dM^k(u) | \mathscr{F}_s] = 0 \qquad \text{a.s.} \tag{6.5}$$

Take any $A \in \mathscr{F}_s$. Then, multiplying both sides of (6.4) by $e^{-i\langle \xi, X(s)\rangle} I_A$ and taking the expectation,

---

*[1] Grigelionis [44].

*[2] Cf. Definition I-7.2.

$$E[e^{i\langle\xi, X(t)-X(s)\rangle}I_A] - P(A) = -\frac{|\xi|^2}{2}\int_s^t E[e^{i\langle\xi, X(u)-X(s)\rangle}I_A]du.$$

From this integral equation we see at once

$$E[e^{i\langle\xi, X(t)-X(s)\rangle}: A] = P(A)e^{-\frac{1}{2}|\xi|^2(t-s)}.$$

This proves (6.3).

In particular, this theorem implies that a continuous process $X(t)$ is a one-dimensional Brownian motion if and only if both $t \longmapsto X(t)$ and $t \longmapsto X(t)^2 - t$ are martingales. This result is known as a theorem of P. Lévy (cf. Doob [18], Chapter VII, Theorem 11.9).

Example 6.1. Let $X(t) = (X^1(t), X^2(t), \ldots, X^d(t))$ be a d-dimensional $(\mathscr{F}_t)$-Brownian motion and $p = (p_l^k(t, \omega))$ be a process with values in orthogonal $d\times d$-matrices such that each component $p_l^k(t,\omega)$ is an $(\mathscr{F}_t)$-predictable process. Set

$$M^i(t) = X^i(t) - X^i(0)$$

and

$$\tilde{X}^k(t) = \tilde{X}^k(0) + \sum_{l=1}^d \int_0^t p_l^k(t, \omega)dM^l(s), \quad k = 1, 2, \ldots, d$$

where $\tilde{X}^k(0)$ is an $\mathscr{F}_0$-measurable random variable. Then $\tilde{X}(t) = (\tilde{X}^1(t), \tilde{X}^2(t), \ldots, \tilde{X}^d(t))$ is a d-dimensional $(\mathscr{F}_t)$-Brownian motion. Indeed, setting $\tilde{M}^k(t) = \tilde{X}^k(t) - \tilde{X}^k(0)$,

$$\begin{aligned}\langle \tilde{M}^k, \tilde{M}^l\rangle(t) &= \int_0^t \sum_{m,n=1}^d p_m^k(s, \omega)p_n^l(s, \omega)d\langle M^m, M^n\rangle(s) \\ &= \int_0^t \sum_{m=1}^d p_m^k(s, \omega)p_m^l(s, \omega)ds \\ &= \delta_{kl}t.\end{aligned}$$

Theorem 6.2. Let p be a point process of class (QL) with respect to $(\mathscr{F}_t)$ on some state space $(\boldsymbol{X}, \mathscr{B}(\boldsymbol{X}))$ such that its compensator $\hat{N}_p(dtdx)$ is a *non-random* σ-finite measure on $[0, \infty)\times \boldsymbol{X}$. Then p is an $(\mathscr{F}_t)$-Poisson point process. If, in particular, $\hat{N}_p(dtdx) = dtn(dx)$ where $n(dx)$ is a non-random σ-finite measure on $\boldsymbol{X}$, p is a stationary $(\mathscr{F}_t)$-Poisson point process with n as its characteristic measure.

Proof. The idea of the proof is essentially same as in Theorem 6.1. Let $t > s \geq 0$ and let $U_1, U_2, \ldots, U_m \in \mathscr{B}(X)$ be disjoint sets such that $\hat{N}_p((0, t] \times U_k) < \infty$, $k = 1, 2, \ldots, m$. It is enough to prove for $\lambda_1, \lambda_2, \ldots, \lambda_m > 0$

$$
\begin{aligned}
&E[\exp(-\sum_{i=1}^{m} \lambda_i N_p((s, t] \times U_i)) | \mathscr{F}_s] \\
&\qquad = \exp[-\sum_{k=1}^{m} (e^{-\lambda_k} - 1)\hat{N}_p((s, t] \times U_k)].
\end{aligned} \tag{6.6}
$$

In fact, one can easily deduce from (6.6) that $N_p(E_1), N_p(E_2), \ldots$ are independent if $E_1, E_2, \ldots \in \mathscr{B}((0, \infty)) \times \mathscr{B}(X)$ are disjoint. Let $F(x^1, x^2, \ldots, x^m) = \exp[-\sum_{k=1}^{m} \lambda_k x^k]$ and $f^k(t,x,\omega) = I_{U_k}(x)$, $f = (f^1, f^2, \ldots, f^m)$. Then

$$\int_0^{t+} \int_X f^k(s, x, \cdot) N_p(dsdx) = N_p((0,t] \times U_k)$$

and by Itô's formula (setting $N(t) = (N_p((0,t] \times U_1), \ldots, N_p((0,t] \times U_m))$)

$$
\begin{aligned}
F(N(t)) - F(N(s)) &= \int_{s+}^{t+} \int_X [F(N(u-) + f(u, x, \cdot)) \\
&\qquad\qquad - F(N(u-))] N_p(dudx) \\
&= \text{a martingale} + \int_s^t \int_X [F(N(u) + f(u, x, \cdot)) - F(N(u))] \hat{N}_p(dudx).
\end{aligned}
$$

But

$$F(N(u) + f(u, x, \cdot)) - F(N(u)) = e^{-\sum_{k=1}^{m} \lambda_k N((0,u] \times U_k)} (e^{-\sum_{k=1}^{m} \lambda_k I_{U_k}(x)} - 1).$$

Consequently, as in the proof of Theorem 6.1 we have from this

$$
\begin{aligned}
&E[e^{-\sum_{k=1}^{m} \lambda_k N((s,t] \times U_k)} : A] - P(A) \\
&\quad = \int_s^t \int_X E[\exp(-\sum_{k=1}^{m} \lambda_k N((s, u] \times U_k)) : A](e^{-\sum_{l=1}^{m} \lambda_l I_{U_l}(x)} - 1) \hat{N}_p(dudx)
\end{aligned}
$$

for any $A \in \mathscr{F}_s$. Hence,

$$E[e^{-\sum_{k=1}^{m} \lambda_k N_p((s,t] \times U_k)} : A]$$

$$= P(A) \exp \left[\sum_{k=1}^{m} (e^{-\lambda_k} - 1)\hat{N}_p((s, t] \times U_k)\right].$$

This proves (6.6). If, in particular, $\hat{N}_p(dtdx) = dtn(dx)$, we have that

$$E[e^{-\sum_{k=1}^{m} \lambda_k N_p((s,t] \times U_k)} \mid \mathscr{F}_s] = \exp[(t-s) \sum_{k=1}^{m} (e^{-\lambda_k} - 1)n(U_k)] \tag{6.7}$$

and so p is a stationary Poisson point process with $n(dx)$ as its characteristic measure.

Theorem 6.1 and Theorem 6.2 can be combined as in Theorem 6.3: an interesting point is that we have *automatically* the independence of Brownian motion and Poisson point process.

Theorem 6.3. Let $X(t) = (X^1(t), X^2(t), \ldots, X^d(t))$ be a d-dimensional $(\mathscr{F}_t)$ semi-martingale and $p_1, p_2, \ldots, p_n$ be point processes of class (QL) with respect to $(\mathscr{F}_t)$ on state spaces $\boldsymbol{X}_1, \boldsymbol{X}_2, \ldots, \boldsymbol{X}_n$ respectively. Suppose that

$$M^i(t) = X^i(t) - X^i(0) \in \mathscr{M}_2^{c,loc}, \tag{6.8}$$

$$\langle M^i, M^j \rangle(t) = \delta_{ij} t, \qquad i,j = 1,2, \ldots, d, \tag{6.9}$$

(6.10) the compensator $\hat{N}_{p_i}(dtdx)$ of p_i is a *non-random* σ-finite measure on $[0, \infty) \times \boldsymbol{X}_i$, $i = 1,2, \ldots, n$ and

(6.11) with probability one, the domains $\boldsymbol{D}_{p_i}$ are mutually disjoint. Then $X(t)$ is a d-dimensional $(\mathscr{F}_t)$-Brownian motion and p_i $(i = 1, 2, \ldots, n)$ is an $(\mathscr{F}_t)$-Poisson point process such that they are *mutually independent*.

Proof. Let $\boldsymbol{D}_p = \bigcup_{i=1}^{n} \boldsymbol{D}_{p_i}$ and set, $p(t) = p_i(t)$ if $t \in \boldsymbol{D}_{p_i}$. Then we have a point process p on the sum $\bigcup_{i=1}^{n} \boldsymbol{X}_i$ * which is clearly a point process of the class (QL) with compensator $\hat{N}_p(dtdx) = \sum_{i=1}^{n} I_{\boldsymbol{X}_i}(x)\hat{N}_{p_i}(dtdx)$. Therefore, p is an $(\mathscr{F}_t)$-Poisson point process. This clearly implies that p_i, $i = 1, 2, \ldots, n$, are mutually independent Poisson point processes since the $\boldsymbol{X}_i$

* The sum of $\boldsymbol{X}_1, \boldsymbol{X}_2, \ldots, \boldsymbol{X}_n$ is a set $\boldsymbol{H}$ such that there exists a family of subsets $\boldsymbol{H}_1, \boldsymbol{H}_2, \ldots, \boldsymbol{H}_n$ with the following property: $\boldsymbol{H}_i$ are mutually disjoint, $\bigcup_{i=1}^{n} \boldsymbol{H}_i = \boldsymbol{H}$ and there exists a bijection between $\boldsymbol{X}_i$ and $\boldsymbol{H}_i$ for each i. Identifying $\boldsymbol{H}_i$ with $\boldsymbol{X}_i$, we often denote the sum as $\bigcup_{i=1}^{n} \boldsymbol{X}_i$.

are supposed to be disjoint (by the definition of sum $\bigcup_{i=1}^{n} \boldsymbol{X}_i$). Thus it is sufficient to prove the independence of $X(t)$ and p and this will be accomplished if we show that for every $t > s \geq 0$,

$$
\begin{aligned}
&E[e^{i\langle\xi, X(t)-X(s)\rangle} e^{-\sum_{k=1}^{m} \lambda_k N_p((s,t]\times U_k)} \mid \mathscr{F}_s] \\
(6.12) \qquad &= e^{-\frac{1}{2}|\xi|^2(t-s)} \exp\left[\sum_{k=1}^{m} (e^{-\lambda_k} - 1)\hat{N}_p((s,t]\times U_k)\right],
\end{aligned}
$$

where $\xi, \lambda = (\lambda_k), \{U_i\}$ have the same meaning as in the proof of Theorems 6.1 and 6.2. Let $F(x^1, \ldots, x^d, y^1, \ldots, y^m) = e^{i\langle\xi, x\rangle} e^{-\sum_{k=1}^{m} \lambda_k y^k}$ and apply Itô's formula to the $(d+m)$-dimensional semi-martingale $(X_i(t), N_p((0,t]\times U_k))_{i=1,2,\ldots,d, k=1,2,\ldots,m}$. (6.12) then follows as in the proof of previous theorems.

Finally we note that the strong Markov property of Brownian motions and Poisson point processes are simple consequences of Theorems 6.1 and 6.2.

Theorem 6.4. Let $X(t) = (X^1(t), X^2(t), \ldots, X^d(t))$ be a d-dimensional $(\mathscr{F}_t)$-Brownian motion and σ be an $(\mathscr{F}_t)$-stopping time such that $\sigma < \infty$ a.s.[*1] Let $X^*(t) = X(t+\sigma)$ and $\mathscr{F}_t^* = \mathscr{F}_{t+\sigma}$, $t \in [0, \infty)$. Then $X^* = \{X^*(t)\}$ is a d-dimensional $(\mathscr{F}_t^*)$-Brownian motion. In particular, $B^*(t) = X(t+\sigma) - X(\sigma)$ is a d-dimensional Brownian motion which is independent of $\mathscr{F}_0^* = \mathscr{F}_\sigma$.

Proof. By Doob's optional sampling theorem, $M^{*i}(t) = X^i(t+\sigma) - X^i(\sigma)$ is a local martingale with respect to $(\mathscr{F}_t^*)$ and also $\langle M^{*i}, M^{*j}\rangle(t) = \delta_{ij}(t+\sigma-\sigma) = \delta_{ij}t$, $i,j = 1, 2, \ldots, d$. Then the assertion of the theorem follows from Theorem 6.1.

In the same way, we have

Theorem 6.5.[*2] Let p be a stationary $(\mathscr{F}_t)$-Poisson point process on some space $\boldsymbol{X}$ with the characteristic measure $n(dx)$ and σ be an $(\mathscr{F}_t)$-stopping time such that $\sigma < \infty$ a.s. Let a point process p^* on $\boldsymbol{X}$ be defined by

$$
\boldsymbol{D}_{p^*} = \{t; t+\sigma \in \boldsymbol{D}_p\}
$$

[*1] If we only assume that $P(\sigma < \infty) > 0$, we have the same conclusion by restricting Ω to $\Omega \cap \{\sigma < \infty\}$ and by substituting P by $P(\cdot) = P(\cdot \cap \{\sigma < \infty\})/P(\sigma < \infty)$.

[*2] Cf. Itô [70], Theorem 5.1 where it is called the *strong renewal property*.

and $p^*(t) = p(t + \sigma)$, $t \in \boldsymbol{D}_{p^*}$. Let $\mathscr{F}_t^* = \mathscr{F}_{t+\sigma}$. Then p^* is a stationary $(\mathscr{F}_t^*)$-Poisson point process with the characteristic measure n.

Let $X(t) = (X^1(t), X^2(t), \dots, X^d(t))$ be a d-dimensional Brownian motion on a complete probability space and let $(\mathscr{F}_t^X)$ be the family of σ-fields generated by the sample paths $X(t)$: $\mathscr{F}_t^X = \sigma\{X(s); s \le t\} \vee \mathscr{N}$. Here $\mathscr{N}$ stands for the totality of P-null sets.

Lemma 6.1. $\mathscr{F}_{t+0}^X = \mathscr{F}_t^X$.

Proof. Let $p(t, x)$ be given by (I-7.1) and set

$$(H_t f)(x) = \int_{R^d} p(t, x - y) f(y) dy, \qquad f \in \boldsymbol{C}_0(\boldsymbol{R}^d).^*$$

$\{H_t\}$ constitutes a strongly continuous semigroup of operators on $\boldsymbol{C}_0(\boldsymbol{R}^d)$. We can rewrite (I-7.2) in the following form

$$\begin{aligned} &E[f_1(X(t_1))f_2(X(t_2)) \cdots f_n(X(t_n))] \\ &\quad = \int_{R^d} \mu(dx) H_n(t_1, t_2, \dots, t_n; f_1, f_2, \dots, f_n)(x), \end{aligned}$$

where $0 \le t_1 < t_2 < \cdots < t_n, f_1, f_2, \dots, f_n \in \boldsymbol{C}_0(\boldsymbol{R}^d)$ and $H_n(t_1, t_2, \dots, t_n; f_1, f_2, \dots, f_n) \in \boldsymbol{C}_0(\boldsymbol{R}^d)$ is defined inductively by

$$\begin{cases} H_n(t_1, t_2, \dots, t_n; f_1, f_2, \dots, f_n) \\ \qquad = H_{n-1}(t_1, t_2, \dots, t_{n-1}; f_1, f_2, \dots, f_{n-2}, f_{n-1}H_{t_n - t_{n-1}} f_n) \\ H_1(t; f) = H_t f. \end{cases}$$

Consequently, if $t_{k-1} \le t < t_k$ we have

$$\begin{aligned} &E[f_1(X(t_1))f_2(X(t_2)) \cdots f_n(X(t_n)) | \mathscr{F}_t^X] \\ &\quad = \prod_{i=1}^{k-1} f_i(X(t_i)) H_{n-k+1}(t_k - t, t_{k+1} - t, \dots, t_n - t; f_k, f_{k+1}, \dots, f_n)(X(t)) \end{aligned}$$

and hence

$$E[f_1(X(t_1))f_2(X(t_2)) \cdots f_n(X(t_n)) | \mathscr{F}_{t+0}^X]$$

---

* $\boldsymbol{C}_0(\boldsymbol{R}^d)$ is the Banach space of all continuous functions on $\boldsymbol{R}^d$ such that $\lim_{|x| \to \infty} |f(x)| = 0$ with the maximum norm.

$$= \lim_{h \downarrow 0} E[f_1(X(t_1))f_2(X(t_2)) \cdots f_n(X(t_n)) \mid \mathscr{F}_{t+h}^X]$$

$$= E[f_1(X(t_1))f_2(X(t_2)) \cdots f_n(X(t_n)) \mid \mathscr{F}_t^X].$$

This proves $\mathscr{F}_{t+0}^X = \mathscr{F}_t^X$.

Lemma 6.2. For any increasing sequence σ_n of $(\mathscr{F}_t^X)$-stopping times,

$$\bigvee_n \mathscr{F}_{\sigma_n}^X = \mathscr{F}_\sigma^X$$

where $\sigma = \lim_{n \to \infty} \sigma_n$.

Proof. By the strong Markov property,

$$E[f_1(X(t_1))f_2(X(t_2)) \cdots f_n(X(t_n)) \mid \mathscr{F}_\tau^X]$$

$$= \sum_{k=1}^n I_{\{t_{k-1} \leq \tau < t_k\}} \prod_{i=1}^{k-1} f_i(X(t_i)) H_{n-k+1}(t_k - \tau, t_{k+1} - \tau, \ldots, t_n - \tau;$$

$$f_k, f_{k+1}, \ldots, f_n)(X(\tau)) + \prod_{i=1}^n f_i(X(t_i)) I_{\{t_n \leq \tau\}}$$

for any $(\mathscr{F}_t^X)$-stopping time τ. Using this, the lemma can be proved as in Lemma 6.1.

Let $B^i(t) = X^i(t) - X^i(0)$, $i = 1, 2, \ldots, d$. Then $B^i \in \mathscr{M}_2^c(\mathscr{F}_t^X)$.* The following theorem, first proved by Itô as an application of the multiple Wiener-Itô integrals (c.f. [64]), is very useful and will be used often in this book. The proof which we give is based on Theorem 6.1 and is due to Dellacherie [16].

Theorem 6.6. Let $M = (M_t) \in \mathscr{M}_2(\mathscr{F}_t^X)$ $(\mathscr{M}_2^{loc}(\mathscr{F}_t^X))$. Then there exist $\Phi_i \in \mathscr{L}_2(\mathscr{F}_t^X)$ $(\mathscr{L}_2^{loc}(\mathscr{F}_t^X))$, $i = 1, 2, \ldots, d$, such that

$$M(t) = \sum_{i=1}^d \int_0^t \Phi_i(s) dB^i(s). \tag{6.13}$$

That is, every martingale with respect to the proper reference family of $X(t)$ can be represented as a sum of stochastic integrals with respect to the *basic* martingales B^i, $i = 1, 2, \ldots, d$.

Proof. We assume that $X(0)$ is constant. The proof is easily reduced

* $\mathscr{M}_2^c(\mathscr{F}_t^X)$ is the space $\mathscr{M}_2^c$ with respect to $(\mathscr{F}_t^X)$.

to this case by a standard argument. For simplicity of notations, we assume $d = 1$; i.e., $X(t)$ is a one-dimensional Brownian motion and $B(t) = X(t) - X(0)$. It sufficies to prove (6.13) on each finite interval $[0, T]$; indeed, it is easy to see that $\Phi(s)$ is determined consistently on different intervals and thus defines the expression (6.13) on $[0, \infty)$. The spaces $\mathcal{M}_2$, $\mathcal{M}_2^c$, $\mathcal{L}_2$, etc. all refer to the proper reference family $(\mathcal{F}_t^X)$ and time is restricted on $[0, T]$. Let $\mathcal{M}_2^* \subset \mathcal{M}_2^c$ be defined by

$$\mathcal{M}_2^* = \{M(t) = \int_0^t \Phi(s)dB(s);\ \Phi \in \mathcal{L}_2\}.$$

The theorem then asserts that $\mathcal{M}_2 = \mathcal{M}_2^*$. To prove this, we first show that every $M \in \mathcal{M}_2$ can be expressed as

$$M(t) = M_1(t) + M_2(t) \tag{6.14}$$

where $M_1 \in \mathcal{M}_2^*$ and $M_2 \in \mathcal{M}_2$ satisfies $\langle M_2, N\rangle = 0$ for all $N \in \mathcal{M}_2^*$. Clearly, such a decomposition is unique if it exists.

Let $\mathcal{H} = \{M_1(T);\ M_1 \in \mathcal{M}_2^*\}$. It is easy to see that $\mathcal{H}$ is a closed subspace of $\mathcal{L}_2(\Omega,P)$. Let $\mathcal{H}^\perp$ be the orthogonal complement of $\mathcal{H}$. Now, let $M \in \mathcal{M}_2$ be given. Then, since $M(T) \in \mathcal{L}_2(\Omega,P)$, we have the orthogonal decomposition

$$M(T) = H_1 + H_2,$$

where $H_1 \in \mathcal{H}$ and $H_2 \in \mathcal{H}^\perp$. By definition, H_1 is of the form $H_1(\omega) = \int_0^T \Phi(s)dB(s)$ for some $\Phi \in \mathcal{L}_2$. Let $M_2(t)$ be the right-continuous modification of $E[H_2 | \mathcal{F}_t^X]$. Then clearly

$$M(t) = M_1(t) + M_2(t), \quad t \in [0, T],$$

where $M_1(t) = \int_0^t \Phi(s)dB(s)$. It remains to show that $\langle M_2, N\rangle(t) = 0$ on $[0, T]$ for every $N \in \mathcal{M}_2^*$: that is, $t \longmapsto M_2(t)N(t)$ is an $(\mathcal{F}_t^X)$-martingale on $[0, T]$. For this it is sufficient to show that, for every $(\mathcal{F}_t^X)$-stopping time σ such that $\sigma \leq T$,

$$E[M_2(\sigma)N(\sigma)] = 0.\text{*1}$$

But if $N(t) = \int_0^t \Psi(s)dB(s)$, then $N^\sigma(t) = \int_0^t \Psi(s)I_{\{s\leq\sigma\}}dB(s) \in \mathcal{M}_2^*$ *2 and hence

*1 Cf. Corollary of Theorem I-6.1 and its continuous time version.

*2 For $X = (X(t))$, $X^\sigma = (X^\sigma(t))$ is defined by $X^\sigma(t) = X(t \wedge \sigma)$.

$$E[N(\sigma)M_2(\sigma)] = E[N(\sigma)E[M_2(T)\,|\,\mathscr{F}_\sigma]]$$
$$= E[N(\sigma)M_2(T)] = E[N^\sigma(T)H_2] = 0.$$

Thus, we have now completed the proof of the decomposition (6.14).

To prove the theorem, it is sufficient to show that, for a dense subspace $\mathscr{N} \subset \mathscr{M}_2$, the M_2-part in the decompositions (6.14) of $M \in \mathscr{N}$ vanishes: indeed, we then have that $\mathscr{N} \subset \mathscr{M}_2^*$ and, since $\mathscr{M}_2^*$ is closed, we have $\mathscr{M}_2 = \mathscr{M}_2^*$.

Let $\tilde{\mathscr{N}} = \{M \in \mathscr{M}_2;\ M \text{ is bounded}\}$. It is easy to see that $\tilde{\mathscr{N}}$ is dense in $\mathscr{M}_2$ because, in the space $\mathscr{H} = \{M(T);\ M \in \mathscr{M}_2\} = \mathscr{L}_2(\Omega, \mathscr{F}_T^X, P)$,* $\mathscr{H}_0 = \{F \in \mathscr{H};\ \text{bounded}\}$ is dense and the norm $\|\cdot\|_T$ of $\mathscr{M}_2$ (restricted to the interval $[0, T]$) was defined by the $\mathscr{L}_2$-norm of $\mathscr{H}$. Let $M \in \tilde{\mathscr{N}}$ and $M = M_1 + M_2$ be the decomposition of (6.14). Since M_1 is a continuous martingale, there exists a sequence σ_n $(= \sigma_n(M_1))$ of $(\mathscr{F}_t^X)$-stopping times such that $\sigma_n \in [0,T]$, $\sigma_n \uparrow T$ and $M_1^{\sigma_n} = (M_1(t \wedge \sigma_n))$ is a bounded martingale, $n = 1, 2, \ldots$. As we know, $M_1^{\sigma_n} \in \mathscr{M}_2^*$ and $M^{\sigma_n} = M_1^{\sigma_n} + M_2^{\sigma_n}$ is the decomposition of (6.14) for M^{σ_n} since $\langle N, M_2^{\sigma_n}\rangle = \langle N^{\sigma_n}, M_2^{\sigma_n}\rangle = \langle N, M_2\rangle^{\sigma_n} = 0$ for every $N \in \mathscr{M}_2^*$. Set

$$\mathscr{N} = \{M^{\sigma_n};\ n = 1, 2, \ldots, M \in \tilde{\mathscr{N}}\}.$$

Then by Lemma 6.2 it is easy to see that $\mathscr{N}$ is dense in $\mathscr{M}_2$ and if $M = M_1 + M_2$ is the decomposition of (6.14) for $M \in \mathscr{N}$ then both M_1 and M_2 are bounded. It is sufficient to show $M_2 = 0$. This follows from the next lemma.

Lemma 6.3. Let $M \in \mathscr{M}_2$ be bounded and suppose that $\langle M, N\rangle = 0$ for every $N \in \mathscr{M}_2^*$. Then $M = 0$.

Remark 6.1. The condition $\langle M, N\rangle = 0$ for every $N \in \mathscr{M}_2^*$ is equivalent to the condition $\langle M, B\rangle = 0$ since $\langle M, N\rangle(t) = \int_0^t \Phi(s)d\langle M, B\rangle(s)$ if $N(t) = \int_0^t \Phi(s)dB(s)$.

Proof. Assume $|M(t)| \le \alpha$ where α is a positive constant and set $D(\omega) = 1 + M(T, \omega)/2\alpha$. Then $D(\omega) \ge 1/2$ and $E[D(\omega)] = 1$. Define a new probability measure $\tilde{P}$ on $\mathscr{F}_T^X$ by

$$\tilde{P}(B) = E[D(\omega)I_B(\omega)], \quad B \in \mathscr{F}_T^X.$$

* $\mathscr{L}_2(\Omega, \mathscr{F}_T^X, P) = \{F \in \mathscr{L}_2(\Omega, P);\ F \text{ is } \mathscr{F}_T^X\text{-measurable}\}$.

Then for every $\mathscr{F}_t^X$-stopping time $\sigma \in [0, T]$,

$$\tilde{E}[B(\sigma)] = E[D(\omega)B(\sigma)] = E[E[D(\omega)\,|\,\mathscr{F}_\sigma]B(\sigma)]$$
$$= E[B(\sigma)] + \frac{1}{2\alpha} E[M(\sigma)B(\sigma)] = E[B(\sigma)] = 0$$

because $\langle M, B\rangle = 0$.

Similarly, $\tilde{E}[B(\sigma)^2 - \sigma] = 0$ because $B(t)^2 - t = 2\int_0^t B(s)dB(s) \in \mathscr{M}_2^*$ and hence $\langle B(t)^2 - t, M(t)\rangle = 0$. That is, both $t \longmapsto B(t)$ and $t \longmapsto B(t)^2 - t$ are continuous $\mathscr{F}_t^X$-martingales with respect to the probability $\tilde{P}$. By Theorem 6.1, $t \longmapsto B(t)$ is an $(\mathscr{F}_t^X)$-Brownian motion with respect to $\tilde{P}$. This clearly implies that $P = \tilde{P}$ on $\mathscr{F}_T^X$ and hence we must have $D = 1$ a.s., i.e., $M = 0$ a.s.

Corollary 1. $\mathscr{M}_2(\mathscr{F}_t^X) = \mathscr{M}_2^c(\mathscr{F}_t^X)$.

Corollary 2. Let $F \in \mathscr{L}_2(\Omega, \mathscr{F}_T^X, P)$ for a positive constant $T > 0$. Then there exists an $(\mathscr{F}_t^X)$-predictable process $f(s)$ $(0 \leq s \leq T)$ such that

$$E[\int_0^T f^2(s)ds] < \infty$$

and

$$F = E[F\,|\,\mathscr{F}_0^X] + \int_0^T f(s)dB(s).^* \tag{6.15}$$

Similar proof (based on Theorem 6.2) applies for Poisson point processes.

Theorem 6.7. Let p be a Poisson point process on some state space $(X, \mathscr{B}(X))$ and $\mathscr{F}_t^p = \bigcap_{\varepsilon>0} \sigma[N_p(s,E); s \leq t + \varepsilon,\ E \in \mathscr{B}(X)]$. Then every $M \in \mathscr{M}_2(\mathscr{F}_t^p)$ $(\mathscr{M}_2^{loc}(\mathscr{F}_t^p))$ can be expressed in the form

$$M(t) = \int_0^{t+}\int f(s, x, \cdot)\tilde{N}_p(dsdx) \tag{6.16}$$

for some $f \in \mathbf{F}_p^2(\mathscr{F}_t^p)$ $\quad (f \in \mathbf{F}_p^{2,loc}(\mathscr{F}_t^p))$.

As we saw above, the proof of representation theorems for martingales (like Theorem 6.6 and Theorem 6.7) is based on the martingale characterization of the basic processes. Generally, we can say that a martingale representation theorem holds if the process is determined as a *unique*

* This corollary is also true for $T = \infty$ if we set $\mathscr{F}_\infty^X = \bigvee_t \mathscr{F}_t^X$.

solution of a martingale problem. More generally Jacod [77] showed that the validity of such a representation theorem is equivalent to the extremality of the basic probability in the convex set of all solutions of a martingale problem.

7. Representation theorem for semi-martingales

In this section, we will see how semi-martingales are represented in terms of Brownian motions and Poisson point processes. The results of this section will play an important role in the study of stochastic differential equations.

Let $(\Omega, \mathscr{F}, P)$ be a given probability space with $(\mathscr{F}_t)_{t\geq 0}$ as usual.

Theorem 7.1. Let $M^i \in \mathscr{M}_2^{c,loc}$, $i = 1, 2, \ldots, d$. Suppose that $\Phi_{ij}(s) \in \mathscr{L}_1^{loc}$ * and $\Psi_{ik}(s) \in \mathscr{L}_2^{loc}$, $i,j,k = 1, 2, \ldots, d$, exist such that

$$(7.1) \qquad \langle M^i, M^j\rangle(t) = \int_0^t \Phi_{ij}(s, \omega)ds,$$

$$(7.2) \qquad \Phi_{ij}(s) = \sum_{k=1}^d \Psi_{ik}(s)\Psi_{jk}(s)$$

and

$$(7.3) \qquad \det(\Psi_{ik}(s)) \neq 0 \qquad \text{a.s.} \qquad \text{for every } s.$$

Then there exists a d-dimensional $(\mathscr{F}_t)$-Brownian motion $B(t) = (B^1(t), B^2(t), \ldots, B^d(t))$ with $B(0) = 0$ a.s. such that

$$(7.4) \qquad M^i(t) = \sum_{k=1}^d \int_0^t \Psi_{ik}(s)dB^k(s), \quad i = 1, 2, \ldots, d.$$

Proof. We will consider the case where $M^i \in \mathscr{M}_2^c$, $\Phi_{ij} \in \mathscr{L}_1$ and $\Psi_{ik} \in \mathscr{L}_2$; the general case is easily reduced to this case. For $N > 0$ we set

$$(7.5) \qquad \theta_{ik}^{(N)}(s, \omega) = \begin{cases} (\Psi^{-1})_{ik}(s, \omega), & \text{if } |(\Psi^{-1})_{ik}(s, \omega)| \leq N \quad \text{for all } i, k, \\ 0, & \text{otherwise,} \end{cases}$$

where Ψ^{-1} denotes the inverse of $\Psi = (\Psi_{ik}(s))$. Then clearly $\theta_{ik}^{(N)} \in \mathscr{L}_2$ and for every i and j

* $\mathscr{L}_1^{loc} = \{\Phi = (\Phi(t))_{t\geq 0};\ \Phi$ is a real $(\mathscr{F}_t)$-predictable process and $\int_0^t |\Phi(s)|\, ds < \infty$ a.s. for every $t > 0\}$. $\mathscr{L}_1 = \{\Phi \in \mathscr{L}_1^{loc};\ E[\int_0^t |\Phi(s)|\, ds] < \infty$ for every $t > 0\}$.

$$(7.6) \qquad \int_0^t E[|\sum_{k,k'=1}^d \theta_{ik}^{(N)}(s,\omega)\theta_{jk'}^{(N)}(s,\omega)\Phi_{kk'}(s,\omega) - \delta_{ij}|^2]ds \longrightarrow 0 \quad \text{as} \quad N \longrightarrow \infty.$$

If we set

$$B_{(N)}^i(t) = \sum_{k=1}^d \int_0^t \theta_{ik}^{(N)}(s,\omega)dM^k(s,\omega)$$

then $B_{(N)}^i \in \mathscr{M}_2^c$ and

$$(7.7) \qquad \langle B_{(N)}^i, B_{(N)}^j\rangle(t) = \int_0^t \sum_{k,k'=1}^d \theta_{ik}^{(N)}(s,\omega)\theta_{jk'}^{(N)}(s,\omega)\Phi_{kk'}(s,\omega)ds.$$

From (7.6) and (7.7) we see that $B_{(N)}^i$ converges to some B^i in $\mathscr{M}_2^c$ as $N \longrightarrow \infty$ and $\langle B^i, B^j\rangle(t) = \delta_{ij}t$. By Theorem 6.1, $B(t) = (B^1(t), B^2(t), \ldots, B^d(t))$ is a d-dimensional $(\mathscr{F}_t)$-Brownian motion. Since

$$\sum_{k=1}^d \int_0^t \Psi_{ik}(s)dB_{(N)}^k(s) = \int_0^t I_N(s)dM^i(s),$$

where

$$I_N(s,\omega) = \begin{cases} 1, & \text{if } \ |(\Psi^{-1})_{ik}(s,\omega)| \le N \quad \text{for all } \ i, k, \\ 0, & \text{otherwise}, \end{cases}$$

we have, by letting $N \longrightarrow \infty$,

$$M^i(t) = \sum_{k=1}^d \int_0^t \Psi_{ik}(s)dB^k(s).$$

Theorem 7.2. Let $M \in \mathscr{M}_2^{c,loc}$ such that $\lim_{t\uparrow\infty}\langle M\rangle(t) = \infty$ a.s. Then, if we set

$$(7.8) \qquad \tau_t = \inf\{u; \langle M\rangle(u) > t\},$$

and $\hat{\mathscr{F}}_t = \mathscr{F}_{\tau_t}$, the time changed process $B(t) = M(\tau_t)$ is an $(\hat{\mathscr{F}}_t)$-Brownian motion. Consequently, we can represent M by an $(\hat{\mathscr{F}}_t)$-Brownian motion $B(t)$ and an $(\hat{\mathscr{F}}_t)$-stopping time $\langle M\rangle(t)$:*

$$(7.9) \qquad M(t) = B(\langle M\rangle(t)).$$

* Note that $\langle M\rangle(t)$ is an $(\hat{\mathscr{F}}_t)$-stopping time for each $t \ge 0$. Indeed, $\{\langle M\rangle(t) \le u\} = \{\tau_u \ge t\} \in \mathscr{F}_{\tau_u} = \hat{\mathscr{F}}_u$.

Proof. First, we remark that, with probability one, $t \longmapsto B(t) = M(\tau_t)$ is continuous. It is sufficient to show that, for any fixed $r < r'$, we have, except a set of probability zero,

$$\{\langle M\rangle(r') = \langle M\rangle(r)\} \subset \{M(u) \equiv M(r),\ {}^\forall u \in [r, r']\}. \tag{7.10}$$

Indeed, if (7.10) holds, we can conclude by a standard argument that the following is true with probability one: for any $r < r'$, $\langle M\rangle(r') = \langle M\rangle(r)$ implies that $M(u) = M(r)$ on the interval $[r, r']$. This clearly implies that, with probability one, $t \longmapsto M(\tau_t)$ is continuous.

To prove (7.10), set $\sigma = \inf\{s > r;\ \langle M\rangle(s) > \langle M\rangle(r)\}$. Then σ is an $(\mathscr{F}_t)$-stopping time and hence $N(s) = M(\sigma \wedge (r+s)) - M(r)$ is a local-martingale with respect to $(\hat{\mathscr{F}}_t)$ where $\hat{\mathscr{F}}_s = \mathscr{F}_{\sigma\wedge(r+s)}$. Since $\langle N\rangle(s) = \langle M\rangle(\sigma \wedge (r+s)) - \langle M\rangle(r) = 0$, $N = 0$. This implies that $M(\sigma \wedge (r+s)) = M(r)$ for all $s \geq 0$ a.s., and hence (7.10) holds.

By Doob's optional sampling theorem, $E[M(\tau_t \wedge n)^2] = E[\langle M\rangle(\tau_t \wedge n)] \leq E[\langle M\rangle(\tau_t)] = t$. Letting $n \longrightarrow \infty$, $E[M(\tau_t)^2] = t$. Then we can conclude by the same theorem that $B(t) = M(\tau_t)$ is $\mathscr{M}_2^c$ with respect to $(\hat{\mathscr{F}}_t)$, $\hat{\mathscr{F}}_t = \mathscr{F}_{\tau_t}$ and $\langle B\rangle(t) = \langle M\rangle(\tau_t) = t$. By Theorem 6.1 $B(t)$ is an $(\hat{\mathscr{F}}_t)$-Brownian motion.

This theorem was generalized by Knight [84] (cf. also [119]) as follows:

Theorem 7.3. Let $M^i \in \mathscr{M}_2^{c,loc}$, $i = 1, 2, \ldots, d$, such that $\langle M^i, M^j\rangle = 0$ if $i \neq j$ and $\lim_{t\uparrow\infty} \langle M^i\rangle(t) = \infty$ a.s. Set

$$\tau_t^i = \inf\{u;\ \langle M^i\rangle(u) > t\}, \qquad i = 1,2, \ldots, d. \tag{7.11}$$

Then if we set $B^i(t) = M^i(\tau_t^i)$, $i = 1, 2, \ldots, d$, $B(t) = (B^1(t), B^2(t), \ldots, B^d(t))$ is a d-dimensional Brownian motion.

Proof. By the previous theorem, $B^i(t)$ is a one-dimensional Brownian motion for each $i = 1, 2, \ldots, d$. Consequently, we only need to prove that the processes $B^1(t), B^2(t), \ldots, B^d(t)$ are mutually independent.

We shall show this by induction. Suppose that $B^1(t), B^2(t), \ldots, B^i(t)$ are mutually independent and we will show that $(B^1(t), B^2(t), \ldots, B^i(t))$ and $B^{i+1}(t)$ are mutually independent. Let $\mathscr{G} = \sigma[B^1(t), B^2(t), \ldots, B^i(t),\ t \in [0, \infty)]$ and $\mathscr{G}_t = \bigcap_{\varepsilon>0} \sigma[B^1(s), B^2(s), \ldots, B^i(s);\ s \leq t + \varepsilon]$; let $\mathscr{H} = \sigma[B^{i+1}(t),\ t \in [0, \infty)]$ and $\mathscr{H}_t = \bigcap_{\varepsilon>0} \sigma[B^{i+1}(s);\ s \leq t + \varepsilon]$. We may assume, without loss of generality, that our probability space $(\Omega,$

$\mathscr{F},P)$ is such that $(\Omega, \mathscr{F})$ is a standard measurable space (cf. Chapter I, Section 3). Let $P(\cdot \mid \mathscr{G})$ be the regular conditional probability given $\mathscr{G}$. Clearly $(B^1(t), B^2(t), \ldots, B^i(t))$ and $B^{i+1}(t)$ are mutually independent if and only if $B^{i+1}(t)$ is a one-dimensional Brownian motion with respect to $P(\cdot \mid \mathscr{G})$ a.s. Hence, noting Theorem 6.1, it is sufficient to prove that for every $t > s$ and $\mathscr{H}_s$-measurable bounded function $F_1(\omega)$

$$E[(B^{i+1}(t) - B^{i+1}(s))F_1(\omega) \mid \mathscr{G}] = 0 \qquad \text{a.s.} \tag{7.12}$$

and

$$E\{[(B^{i+1}(t) - B^{i+1}(s))^2 - (t - s)]F_1(\omega) \mid \mathscr{G}\} = 0 \qquad \text{a.s.} \tag{7.13}$$

Thus, it is sufficient to prove the following: for every $t > s$, every bounded $\mathscr{H}_s$-measurable function $F_1(\omega)$ and every bounded $\mathscr{G}$-measurable function $F_2(\omega)$,

$$E[(B^{i+1}(t) - B^{i+1}(s))F_1(\omega)F_2(\omega)] = 0 \tag{7.14}$$

and

$$E\{[(B^{i+1}(t) - B^{i+1}(s))^2 - (t - s)]F_1(\omega)F_2(\omega)\} = 0. \tag{7.15}$$

We will prove (7.14) only: the proof of (7.15) can be given similarly. Let $\mathscr{G}^{(k)} = \sigma[B^k(s); s \in [0, \infty)]$, $k = 1, 2, \ldots, i$. Since $F_2(\omega)$ can be approximated by a linear combination of functions of the form $\prod_{k=1}^{i} G_k(\omega)$ where $G_k(\omega)$ is $\mathscr{G}^{(k)}$-measurable, we may assume from the beginning that $F_2(\omega) = \prod_{k=1}^{i} G_k(\omega)$. By Corollary 2 of Theorem 6.6, $F_1(\omega)$ and $G_k(\omega)$, $k = 1, 2, \ldots, i$, can be expressed as

$$F_1(\omega) = c + \int_0^s \Phi(u)dB^{i+1}(u),$$

$$G_k(\omega) = c_k + \int_0^\infty \Psi_k(u)dB^k(u),$$

where c and c_k are constants and Φ is an $(\mathscr{H}_t)$-predictable process and Ψ_k is a $(\mathscr{G}_t^{(k)})$-predictable process. Here $\mathscr{G}_t^{(k)} = \bigcap_{\varepsilon>0} \sigma[B^k(s); s \leq t + \varepsilon]$, $k = 1, 2, \ldots, i$. By setting $\tau_s = \tau_s^{i+1}$, we can write

$$F_1(\omega) = c + \int_0^{\tau_s} \tilde{\Phi}(u)dM^{i+1}(u)$$

and

$$G_k(\omega) = c_k + \int_0^{\infty} \tilde{\Psi}_k(u)dM^k(u),$$

where $\tilde{\Phi}(u) = \Phi(\langle M^{i+1}\rangle(u))$ and $\tilde{\Psi}_k(u) = \Psi_k(\langle M^k\rangle(u))$ are $(\mathscr{F}_t)$-predictable processes. Since $\langle M^i, M^j\rangle(t) = 0$ for $i \neq j$, we have by Itô's formula

$$\begin{aligned} F_2(\omega) &= \prod_{k=1}^{i} G_k(\omega) \\ &= c_1c_2 \cdots c_i + \sum_{k=1}^{i}\int_0^{\infty}\prod_{\substack{l=1\\ l\neq k}}^{i}(c_l + \int_0^t \tilde{\Psi}_l(u)dM^l(u))\tilde{\Psi}_k(t)dM^k(t) \\ &= c' + \sum_{k=1}^{i}\int_0^{\infty}\theta_k(t)dM^k(t). \end{aligned}$$

Then the left-hand side of (7.14) becomes

$$\begin{aligned} &E\{(B^{i+1}(t) - B^{i+1}(s))F_1(\omega)F_2(\omega)\} \\ &= E\{(M^{i+1}(\tau_t) - M^{i+1}(\tau_s))(c + \int_0^{\tau_s}\tilde{\Phi}(u)dM^{i+1}(u)) \\ &\quad \times (c' + \sum_{k=1}^{i}\int_0^{\infty}\theta_k(u)dM^k(u))\} \\ &= \sum_{k=1}^{i} E\{(M^{i+1}(\tau_t) - M^{i+1}(\tau_s)) \\ &\quad \times (c + \int_0^{\tau_s}\tilde{\Phi}(u)dM^{i+1}(u))\int_0^{\infty}\theta_k(u)dM^k(u)\} \\ &= \sum_{k=1}^{i} E\{(M^{i+1}(\tau_t) \\ &\quad - M^{i+1}(\tau_s))\int_{\tau_s}^{\infty}\theta_k(u)dM^k(u)(c + \int_0^{\tau_s}\tilde{\Phi}(u)dM^{i+1}(u))\} \\ &\quad + \sum_{k=1}^{i} E\{(M^{i+1}(\tau_t) - M^{i+1}(\tau_s))\int_0^{\tau_s}\theta_k(u)dM^k(u) \\ &\quad \times (c + \int_0^{\tau_s}\tilde{\Phi}(u)dM^{i+1}(u))\}. \end{aligned}$$

Now the first term vanishes because $\langle M^{i+1}, M^k\rangle = 0$ and hence

$$E\{(M^{i+1}(\tau_t) - M^{i+1}(\tau_s))\int_{\tau_s}^{\infty}\theta_k(u)dM^k(u)\,|\,\mathscr{F}_{\tau_s}\} = 0$$

and the second term vanishes because

$$E\{(M^{i+1}(\tau_t) - M^{i+1}(\tau_s))\,|\,\mathscr{F}_{\tau_s}\} = 0.$$

Theorems 7.1, 7.2 and 7.3 also hold under weaker assumptions: however, it is generally necessary to extend the given probability space in order to guarantee the existence of Brownian motion. Before stating these results we shall first make precise the notion of extension of a probability space.

Definition 7.1. We say a probability space $(\tilde{\Omega},\tilde{\mathscr{F}},\tilde{P})$ with a reference family $(\tilde{\mathscr{F}}_t)$ an *extension* of a probability space $(\Omega, \mathscr{F}, P)$ with a reference family $(\mathscr{F}_t)$ if there exists a mapping π: $\tilde{\Omega} \longrightarrow \Omega$ which is $\tilde{\mathscr{F}}/\mathscr{F}$-measurable such that

(i) $\tilde{\mathscr{F}}_t \supset \pi^{-1}(\mathscr{F}_t)$,

(ii) $P = \pi(\tilde{P})\ (:= \tilde{P}\circ\pi^{-1})$ and

(iii) for every $X(\omega) \in \mathscr{L}_\infty(\Omega,\mathscr{F},P)$

$$\tilde{E}(\tilde{X}(\tilde{\omega})\,|\,\tilde{\mathscr{F}}_t) = E(X\,|\,\mathscr{F}_t)(\pi\tilde{\omega}), \qquad \tilde{P}\text{-a.s.,}$$

where we set $\tilde{X}(\tilde{\omega}) = X(\pi\tilde{\omega})$ for $\tilde{\omega} \in \tilde{\Omega}$.

Definition 7.2. Let $(\Omega,\mathscr{F},P)$ be a probability space with a reference family $(\mathscr{F}_t)$. Let $(\Omega',\mathscr{F}',P')$ be another probability space and set

$$\tilde{\Omega} = \Omega\times\Omega', \quad \tilde{\mathscr{F}} = \mathscr{F}\times\mathscr{F}', \quad \tilde{P} = P\times P'$$

and

$$\pi\tilde{\omega} = \omega \qquad \text{for} \qquad \tilde{\omega} = (\omega, \omega') \in \tilde{\Omega}.$$

If $(\tilde{\mathscr{F}}_t)$ is a reference family on $(\tilde{\Omega},\tilde{\mathscr{F}},\tilde{P})$ such that $\mathscr{F}_t\times\mathscr{F}' \supset \tilde{\mathscr{F}}_t \supset \mathscr{F}_t\times\{\Omega', \phi\}$, then $(\tilde{\Omega},\tilde{\mathscr{F}},\tilde{P})$ with $(\tilde{\mathscr{F}}_t)$ is called a *standard extension* of $(\Omega,\mathscr{F},P)$ with $(\mathscr{F}_t)$.

It is easy to see that a standard extension is an extension in the sense of Definition 7.1.

Let $(\tilde{\Omega},\tilde{\mathscr{F}},\tilde{P})$ with $(\tilde{\mathscr{F}}_t)$ be an extension of $(\Omega,\mathscr{F},P)$ with $(\mathscr{F}_t)$. If $M \in \mathscr{M}_2$ with respect to $(\Omega,\mathscr{F},P)$ and $(\mathscr{F}_t)$, then $\tilde{M} = (\tilde{M}_t(\tilde{\omega}))$, where $\tilde{M}_t(\tilde{\omega}) = M_t(\pi\tilde{\omega})$, belongs to $\tilde{\mathscr{M}}_2$, i.e., the space $\mathscr{M}_2$ with respect to $(\tilde{\Omega},\tilde{\mathscr{F}},\tilde{P})$ and $(\tilde{\mathscr{F}}_t)$. Also, if $M,N \in \mathscr{M}_2$ then $\langle\tilde{M},\tilde{N}\rangle_t(\tilde{\omega}) = \langle M,N\rangle_t(\pi\tilde{\omega})$ holds. This is an easy consequence of the property (iii) in Definition 7.1. Therefore, the space $\mathscr{M}_2$ ($\mathscr{M}_2^c$, $\mathscr{M}_2^{loc}$ etc.) is naturally imbedded into the space $\tilde{\mathscr{M}}_2$ ($\tilde{\mathscr{M}}_2^c$, $\tilde{\mathscr{M}}_2^{loc}$, etc.) and $M \in \mathscr{M}_2$ may be regarded as a martingale on the extension $\tilde{\Omega}$ by identifying M and $\tilde{M}$. The following three theorems are natural extensions of the above three theorems.

Theorem 7.1′. Let $(\Omega, \mathscr{F}, P)$ be a probability space with a reference family $(\mathscr{F}_t)$ and $M^i \in \mathscr{M}_2^{c,loc}, i = 1, 2, \ldots, d$. Let $\Phi_{ij}, i,j = 1, 2, \ldots, d$ and Ψ_{ik}, $i = 1, 2, \ldots, d$, $k = 1, 2, \ldots r$, be $(\mathscr{F}_t)$-predictable processes such that $\int_0^t |\Phi_{ij}(s)| ds < \infty$ and $\int_0^t |\Psi_{ik}(s)|^2 ds < \infty$ for all $t \geq 0$ a.s.,

$$\langle M^i, M^j\rangle(t) = \int_0^t \Phi_{ij}(s)ds \tag{7.16}$$

and

$$\Phi_{ij}(s) = \sum_{k=1}^{r} \Psi_{ik}(s)\Psi_{jk}(s). \tag{7.17}$$

Then on an extension $(\tilde{\Omega}, \tilde{\mathscr{F}}, \tilde{P})$ and $(\tilde{\mathscr{F}}_t)$ of $(\Omega, \mathscr{F}, P)$ and $(\mathscr{F}_t)$, there exists an r-dimensional $(\tilde{\mathscr{F}}_t)$-Brownian motion $B(t) = (B^1(t), B^2(t), \ldots, B^r(t))$ such that

$$M^i(t) = \sum_{k=1}^{r} \int_0^t \Psi_{ik}(s)dB^k(s), \qquad i = 1,2, \ldots, d. \tag{7.18}$$

Proof. We may assume $d = r$ by setting $M^i(t) \equiv 0$ or $\Psi_{ik}(t) = 0$ if necessary. Since $\Phi = (\Phi_{ij}(s))$ is for each (s, ω) a $d \times d$ symmetric non-negative matrix, $\Phi^{1/2}$ is uniquely determined as a $d \times d$ symmetric non-negative matrix such that $\Phi^{1/2}\Phi^{1/2} = \Phi$; moreover, $s \longmapsto \Phi^{1/2}(s)$ is $(\mathscr{F}_t)$-predictable and $\int_0^t \|\Phi^{1/2}(s)\|^2 ds < \infty$ a.s. for every $t \geq 0$. We may assume without loss of generality that $\Psi = \Phi^{1/2}$. Indeed in the general case there exists a ($d \times d$ orthogonal matrices-valued) predictable process $P = (P_{ij}(s))$ such that $\Phi^{1/2} = \Psi \cdot P$. Consequently, if we have a representation $M^i(t) = \sum_{k=1}^{d} \int_0^t (\Phi^{1/2})_{ik}(s)dB^k(s)$, we can write $M^i(t) = \sum_{k=1}^{d} \int_0^t \Psi_{ik}(s)d\tilde{B}^k(s)$, where $\tilde{B}^k(t) = \sum_{l=1}^{d} \int_0^t P_{kl}(s)dB^l(s)$ is another d-dimensional $(\tilde{\mathscr{F}}_t)$-Brownian motion by Example 6.1. Therefore, we assume $\Psi = \Phi^{1/2}$. Set

$$\tilde{\Psi}(u) = \lim_{\varepsilon \downarrow 0} \Phi^{1/2}(u)(\Phi(u) + \varepsilon I)^{-1}$$

where I is the identity matrix. Let $E_R(u)$ be the matrix corresponding to the orthogonal projection onto range $\Phi(u)\boldsymbol{R}^d$ and set $E_N(u) = I - E_R(u)$. Then clearly $\tilde{\Psi}(u)\Psi(u) = \Psi(u)\tilde{\Psi}(u) = E_R(u)$. We prepare, on a probability space

$(\Omega', \mathscr{F}', P')$ with a reference family $(\mathscr{F}_t')$, a d-dimensional $(\mathscr{F}_t')$-Brownian motion $B'(t) = (B'^1, B'^2(t), \ldots, B'^d(t))$ and construct a standard extension $(\tilde{\Omega}, \tilde{\mathscr{F}}, \tilde{P})$ and $(\tilde{\mathscr{F}}_t)$ of $(\Omega, \mathscr{F}, P)$ and $(\mathscr{F}_t)$ by $\tilde{\Omega} = \Omega \times \Omega'$, $\tilde{\mathscr{F}} = \mathscr{F} \times \mathscr{F}'$, $\tilde{P} = P \times P'$ and $\tilde{\mathscr{F}}_t = \mathscr{F}_t \times \mathscr{F}_t'$. On this extension, $M^i, B'^j \in \mathscr{M}_2^{c,loc}$ such that

$$(7.19) \qquad \begin{cases} \langle M^i, M^j \rangle(t) = \int_0^t \Phi_{ij}(u)du, \\ \langle M^i, B'^j \rangle(t) = 0 \qquad \text{and} \\ \langle B'^i, B'^j \rangle(t) = \delta_{ij} t \end{cases}$$

for $i,j = 1, 2, \ldots, d$. Now set

$$B^i(t) = \sum_{k=1}^d \int_0^t \tilde{\Psi}_{ik}(u)dM^k(u) + \sum_{k=1}^d \int_0^t (E_N)_{ik}(u)dB'^k(u).$$

Then, by (7.19),

$$\begin{aligned} \langle B^i, B^j \rangle(t) &= \int_0^t \sum_{k,l=1}^d \tilde{\Psi}_{ik}(u)\tilde{\Psi}_{jl}(u)\Phi_{kl}(u)du + \int_0^t (E_N)_{ij}(u)du \\ &= \int_0^t (E_R)_{ij}(u)du + \int_0^t (E_N)_{ij}(u)du = \delta_{ij}t, \end{aligned}$$

and hence $\{B^i(t)\}$ is a d-dimensional $(\tilde{\mathscr{F}}_t)$-Brownian motion. Also, noting that $\Psi(u)E_N(u) = E_N(u)\Psi(u) = 0$,

$$\begin{aligned} \sum_{k=1}^d \int_0^t \Psi_{ik}(u)dB^k(u) &= \sum_{k,l=1}^d \int_0^t \Psi_{ik}(u)\tilde{\Psi}(u)_{kl}dM^l(u) \\ &\quad + \sum_{k,l=1}^d \int_0^t \Psi_{ik}(u)(E_N)_{kl}(u)dB'^l(u) \\ &= M^i(t) - \sum_{l=1}^d \int_0^t (E_N)_{il}(u)dM^l(u) + \sum_{l=1}^d \int_0^t (\Psi(u)E_N(u))_{il}dB'^l(u) \\ &= M^i(t). \end{aligned}$$

The second term in the middle line is zero because of

$$\langle \sum_{l=1}^d \int_0^t (E_N)_{il}(u)dM^l(u) \rangle = \int_0^t (E_N(u)\Phi(u)E_N(u))_{ii}du = 0.$$

Theorem 7.2′. Let $(\Omega, \mathscr{F}, P)$ be a probability space with a reference family $(\mathscr{F}_t)$ and $M \in \mathscr{M}_2^{c,loc}$. Set

(7.20) $$\tau_t = \begin{cases} \inf\{u;\ \langle M\rangle(u) > t\}, \\ \infty, \quad \text{if} \quad t \geq \langle M\rangle(\infty) = \lim_{t\uparrow\infty}\langle M\rangle(t), \end{cases}$$

and $\hat{\mathscr{F}}_t = \bigvee_{s>0} \mathscr{F}_{\tau_t\wedge s}$. Then on an extension $(\tilde{\Omega},\tilde{\mathscr{F}},\tilde{P})$ and $(\tilde{\mathscr{F}}_t)$ of $(\Omega,\mathscr{F},P)$ and $(\hat{\mathscr{F}}_t)$ there exists an $(\tilde{\mathscr{F}}_t)$-Brownian motion $B(t)$ such that $B(t) = M(\tau_t)$, $t \in [0, \langle M\rangle(\infty))$. Consequently we can represent M by an $(\tilde{\mathscr{F}}_t)$-Brownian motion $B(t)$ and an $(\tilde{\mathscr{F}}_t)$-stopping time $\langle M\rangle(t)$:

(7.21) $$M(t) = B(\langle M\rangle(t)).^*$$

Proof. By the optional sampling theorem (Theorem I-6. 11), $E(M_{\tau_u\wedge s}|\mathscr{F}_{\tau_v\wedge s'}) = M_{\tau_v\wedge s'}$ and $E((M_{\tau_u\wedge s} - M_{\tau_v\wedge s'})^2|\mathscr{F}_{\tau_v\wedge s'}) = E(\langle M\rangle_{\tau_u\wedge s} - \langle M\rangle_{\tau_v\wedge s'}|\mathscr{F}_{\tau_v\wedge s'})$ for every $s \geq s'$ and $u \geq v$. Hence, $\tilde{B}(u) = \lim_{s\uparrow\infty} M_{\tau_u\wedge s}$ exists a.s. and

(7.22) $$E(\tilde{B}(u)|\hat{\mathscr{F}}_v) = \tilde{B}(v),$$

(7.23) $$E((\tilde{B}(u) - \tilde{B}(v))^2|\hat{\mathscr{F}}_v) = E(\langle M\rangle_\infty\wedge u - \langle M\rangle_\infty\wedge v|\hat{\mathscr{F}}_v)$$

for every $u \geq v$. We prepare, on a probability space $(\Omega',\mathscr{F}',P')$ with a reference family $(\mathscr{F}_t')$, an $(\mathscr{F}_t')$-Brownian motion $B'(t)$. We construct a standard extension $(\tilde{\Omega}, \tilde{\mathscr{F}}, \tilde{P})$ and $(\tilde{\mathscr{F}}_t)$ of $(\Omega, \mathscr{F}, P)$ and $(\hat{\mathscr{F}}_t)$ by setting $\tilde{\Omega} = \Omega\times\Omega'$, $\tilde{\mathscr{F}} = \mathscr{F}\times\mathscr{F}'$, $\tilde{P} = P\times P'$ and $\tilde{\mathscr{F}}_t = \hat{\mathscr{F}}_t \times \mathscr{F}_t'$. On this extension let

$$B(t) = B'(t) - B'(t \wedge \langle M\rangle(\infty)) + \tilde{B}(t).$$

Then $B(t)$ is a continuous $(\tilde{\mathscr{F}}_t)$-martingale such that $\langle B\rangle(t) = t$ and hence it is an $(\tilde{\mathscr{F}}_t)$-Brownian motion. The rest of the proof is obvious.

Theorem 7.3′. Let $(\Omega,\mathscr{F},P)$ be a probability space with a reference family $(\mathscr{F}_t)$. Let $M^i \in \mathscr{M}_2^{c,loc}(\mathscr{F}_t)$, $i = 1, 2, \ldots, d$, such that $\langle M^i,M^j\rangle(t) \equiv 0$, $i \neq j$. Then, on an extension $(\tilde{\Omega},\tilde{\mathscr{F}},\tilde{P})$ of $(\Omega,\mathscr{F},P)$, there exists a d-dimensional Brownian motion $B(t) = (B^1(t), B^2(t), \ldots, B^d(t))$ such that

(7.24) $$B^i(t) = M^i(\tau_t^i) \qquad t \in [0, \langle M^i\rangle(\infty)),$$

where

* As in Theorem 7.2, $\langle M\rangle(t)$ is an $(\tilde{\mathscr{F}}_t)$-stopping time for each $t \geq 0$.

$$\tau_i^t = \begin{cases} \inf\{u;\ \langle M^i\rangle(u) > t\}, \\ \infty \quad , t \geq \langle M^i\rangle(\infty). \end{cases} \tag{7.25}$$

Consequently, $(M^1(t), M^2(t), \ldots, M^d(t))$ can be obtained from a d-dimensional Brownian motion $B(t) = (B^1(t), B^2(t), \ldots, B^d(t))$ as

$$M^i(t) = B^i(\langle M_i\rangle(t)), \qquad i = 1, 2, \ldots, d.$$

The proof is similar to that of Theorem 7.2′ and therefore omitted.

Finally, we will discuss a similar representation theorem for a class of point processes by means of Poisson point processes.

Theorem 7.4.*[1] Let $(\Omega,\mathscr{F},P)$ be a probability space with a reference family $(\mathscr{F}_t)$. Let $(\boldsymbol{X},\mathscr{B}_{\boldsymbol{X}})$ be a measurable space and p be an $(\mathscr{F}_t)$-point process of class (QL) on $\boldsymbol{X}$ with the compensator $\hat{N}_p(dtdx) = q(t, dx, \omega)dt$. Suppose that there exists a σ-finite measure m on a standard measurable space $(\boldsymbol{Z}, \mathscr{B}_{\boldsymbol{Z}})$ and a predictable $\boldsymbol{X}^* = \boldsymbol{X} \cup \{\Delta\}$*[2]-valued process

$$\theta(t, z, \omega) : [0, \infty) \times \boldsymbol{Z} \times \Omega \longrightarrow \boldsymbol{X}^*$$

such that

$$m(\{z;\ \theta(t, z, \omega) \in E\}) = q(t, E, \omega) \qquad \text{for every } E \in \mathscr{B}_{\boldsymbol{X}}. \tag{7.26}$$

Then, on an extension $(\tilde{\Omega},\tilde{\mathscr{F}},\tilde{P})$ and $(\tilde{\mathscr{F}}_t)$ of $(\Omega,\mathscr{F},P)$ and $(\mathscr{F}_t)$, there exists a stationary $(\tilde{\mathscr{F}}_t)$-Poisson point process q on $\boldsymbol{Z}$ with the characteristic measure m such that

$$\begin{aligned} N_p((0, t] \times E) &= \int_0^{t+}\int_{\boldsymbol{Z}} I_E(\theta(s, z, \omega))N_q(dsdz) \\ &= \#\ \{s \in \boldsymbol{D}_q;\ s \leq t,\ \theta(s, q(s), \omega) \in E\} \\ &\qquad\qquad \text{for every } E \in \mathscr{B}_{\boldsymbol{X}}. \end{aligned} \tag{7.27}$$

Proof. First we prove several lemmas.

Lemma 7.1. There exists a predictable probability kernel $Q(t,x,dz,\omega)$ on $[0, \infty)\times\boldsymbol{X}\times\mathscr{B}_{\boldsymbol{Z}}\times\Omega$ (i.e., for a fixed $A \in \mathscr{B}_{\boldsymbol{Z}}$, $(t, x, \omega) \longmapsto Q(t, x, A, \omega)$ is predictable and for fixed $(t, x, \omega) \in [0, \infty)\times\boldsymbol{X}\times\Omega$, $A \in \mathscr{B}_{\boldsymbol{Z}} \longmapsto Q(t, x, A, \omega)$ is a probability on $\mathscr{B}_{\boldsymbol{Z}}$), such that for every non-negative $\mathscr{B}_{\boldsymbol{X}}\times\mathscr{B}_{\boldsymbol{Z}}$-measurable function $f(x,z)$

*[1] Cf. Grigelionis [43], Karoui-Lepeltier [81] and Tanaka [161]. Proof given here was suggested by [161].

*[2] Δ is an extra point attached to $\boldsymbol{X}$ and $\mathscr{B}_{\boldsymbol{X}*}$ is the σ-field generated by $\mathscr{B}_{\boldsymbol{X}}$ and $\{\Delta\}$.

$$\int_Z \{I_{[\theta(t,z,\omega)\neq\Delta]} f(\theta(t, z, \omega), z)\} m(dz) \tag{7.28}$$
$$= \int_X \{\int_Z f(x, z) Q(t, x, dz, \omega)\} q(t, dx, \omega).$$

The proof of this lemma is standard and is left to the reader (cf. Chapter I, Section 3).

Lemma 7.2. On an extension $(\tilde{\Omega}, \mathscr{F}, \tilde{P})$ and $(\mathscr{F}_t)$ of $(\Omega, \mathscr{F}, P)$ and $(\mathscr{F}_t)$, there exists an $(\mathscr{F}_t)$-point process $\tilde{p}$ of the class (QL) on $\boldsymbol{X}\times[0,1]$ such that

(i) $\boldsymbol{D}_{\tilde{p}} = \boldsymbol{D}_p$ and $\pi(\tilde{p}(s)) = p(s)$ for $s \in \boldsymbol{D}_{\tilde{p}}$, where $\pi(x, \alpha) = x$, $(x, \alpha) \in \boldsymbol{X}\times[0, 1]$;

(ii) the compensator $\hat{N}_{\tilde{p}}(dt, dxd\alpha)$ of $\tilde{p}$ is given by

$$\hat{N}_{\tilde{p}}(dt, dxd\alpha) = q(t, dx, \omega)dtd\alpha, \tag{7.29}$$

where $d\alpha$ is the Lebesgue measure on [0, 1].

Proof. We prepare a sequence of independent identically distributed random variables $\xi_{n,k}$, $n,k = 1, 2, \ldots$ on a probability space $(\Omega', \mathscr{F}', P')$ such that $0 \leq \xi_{n,k} \leq 1$ a.s. and are uniformly distributed. Set $\tilde{\Omega} = \Omega\times\Omega'$, $\mathscr{F} = \mathscr{F}\times\mathscr{F}'$, $\tilde{P} = P\times P'$. p may be regarded as a point process defined on this product space. There exist disjoint $U_n \in \mathscr{B}_{\boldsymbol{X}}$, $n = 1, 2, \ldots$ such that $\bigcup_n U_n = \boldsymbol{X}$ and $E(N_p((0, t]\times U_n)) < \infty$ for every $t \in [0, \infty)$ and n. Let $\boldsymbol{D}_{p_n} = \{s \in \boldsymbol{D}_p;\ p(s) \in U_n\}$. We can order each $\boldsymbol{D}_{p_n}$ according to magnitude: i.e., $\boldsymbol{D}_{p_n} = \{s_1^n < s_2^n < \cdots < s_k^n < \cdots\}$. Since $\bigcup_n \boldsymbol{D}_{p_n} = \boldsymbol{D}_p$, there exists for each $s \in \boldsymbol{D}_p$ a unique pair (n, k) such that $s = s_k^n$. We set $\tilde{p}(s) = (p(s), \xi_{n,k})$. Then the point process $\tilde{p}$ with the domain $\boldsymbol{D}_{\tilde{p}} = \boldsymbol{D}_p$ and $\mathscr{F}_t = \bigcap_{\varepsilon>0} \sigma[\tilde{p}(s), s \leq t + \varepsilon]$ is what we want.

Proof of the theorem. Let $Q(t, x, dz, \omega)$ be the probability kernel in Lemma 7.1. Then there exists a predictable process $f(t, x, \alpha, \omega)$: $[0, \infty)\times \boldsymbol{X}\times[0, 1]\times\Omega \longrightarrow \boldsymbol{Z}$ such that the Lebesgue measure of $\{\alpha; f(t, x, \alpha, \omega) \in A\} = Q(t, x, A, \omega)$ for every $A \in \mathscr{B}_{\boldsymbol{Z}}$.* Let $\tilde{p}$ be the point process on $\boldsymbol{X}\times[0, 1]$ given in Lemma 7.2. We write $\tilde{p}(s) = (p_1(s), p_2(s))$ for $s \in \boldsymbol{D}_{\tilde{p}}$, where $p_1(s) = p(s) \in \boldsymbol{X}$ and $p_2(s) \in [0, 1]$. Define a point process q_2 on $\boldsymbol{Z}$ by $\boldsymbol{D}_{q_2} = \boldsymbol{D}_{\tilde{p}} = \boldsymbol{D}_p$ and $q_2(s) = f(s, p_1(s), p_2(s), \omega)$ for $s \in \boldsymbol{D}_{q_2}$. Thus

$$N_{q_2}((0, t] \times A) = \int_0^{t+}\int_{\boldsymbol{X}\times[0,1]} I_A(f(s, x, \alpha, \omega))N_{\tilde{p}}(ds, dxd\alpha)$$

---

* Cf. [81].

and the compensator is given by

$$\begin{aligned}\hat{N}_{q_2}((0, t] \times A) &= \int_0^t \int_{X\times[0,1]} I_A(f(s, x, \alpha, \omega))q(s, dx, \omega)d\alpha ds \\ &= \int_0^t \int_X Q(s, x, A, \omega)q(s, dx, \omega)ds \\ &= \int_0^t (\int_A I_{[\theta(s,z,\omega)\neq\Delta]}m(dz))ds\end{aligned} \tag{7.30}$$

for all $A \in \mathscr{B}_Z$. On an extension $(\tilde{\tilde{\Omega}},\tilde{\tilde{\mathscr{F}}},\tilde{\tilde{P}})$ and $(\tilde{\tilde{\mathscr{F}}}_t)$ of $(\tilde{\Omega},\tilde{\mathscr{F}},\tilde{P})$ and $(\tilde{\mathscr{F}}_t)$ we construct a stationary $(\tilde{\tilde{\mathscr{F}}}_t)$-Poisson point process q_1 on $\boldsymbol{Z}$ with the characteristic measure $m(dz)$ such that q_1 and q_2 are mutually independent. Clearly such a construction is possible by taking a standard extension. Define a point process q_3 on $\boldsymbol{Z}$ by $\boldsymbol{D}_{q_3} = \{s \in \boldsymbol{D}_{q_1};\ \theta(s, q_1(s), \omega) = \Delta\}$ and $q_3(s) = q_1(s)$ for $s \in \boldsymbol{D}_{q_3}$. Then

$$N_{q_3}((0, t]\times A) = \int_0^{t+}\int_A I_{[\theta(s,z,\omega)=\Delta]}\, N_{q_1}(dsdz), \qquad A \in \mathscr{B}_Z$$

and its compensator is

$$\hat{N}_{q_3}((0, t]\times A) = \int_0^t\int_A I_{[\theta(s,z,\omega)=\Delta]}m(dz)ds. \tag{7.31}$$

Finally, a point process q on $\boldsymbol{Z}$ is defined by $\boldsymbol{D}_q = \boldsymbol{D}_{q_2} \cup \boldsymbol{D}_{q_3}$ and

$$q(t) = \begin{cases} q_2(t) , & t \in \boldsymbol{D}_{q_2}, \\ q_3(t) , & t \in \boldsymbol{D}_{q_3}. \end{cases}$$

By the independence of q_1 and q_2, $\boldsymbol{D}_{q_2}$ and $\boldsymbol{D}_{q_3}$ are disjoint a.s., and hence q is well defined; moreover q is a stationary $(\tilde{\tilde{\mathscr{F}}}_t)$-Poisson point process on $\boldsymbol{Z}$ with the characteristic measure $m(dz)$ by (7.30) and (7.31) (c.f. Theorem 6.2).

Thus the only thing remaining to be proven is

$$N_p((0, t]\times E) = \int_0^{t+}\int_Z I_E(\theta(s, z, \omega))N_q(ds\, dz), \quad E \in \mathscr{B}_X.$$

Let $\bar{p}$ be the point process on $\boldsymbol{X}$ whose counting measure $N_{\bar{p}}((0, t] \times E)$ coincides with the right-hand side. Then

$$N_p((0, t]\times E) = \int_0^{t+}\int_{X\times[0,1]} I_{E\times[0,1]}(x, \alpha)N_{\tilde{p}}(ds, dxd\alpha)$$

and

$$N_{\tilde{p}}((0, t]\times E) = \int_0^{t+}\int_Z I_E(\theta(s, z, \omega))N_q(dsdz)$$
$$= \int_0^{t+}\int_Z I_E(\theta(s, z, \omega))N_{q_2}(dsdz)$$
$$= \int_0^{t+}\int_{X\times[0,1]} I_E(\theta(s, f(s, x, \alpha, \omega), \omega))N_{\tilde{p}}(dsdxd\alpha).$$

Hence the compensator $\hat{N}_{\tilde{p}}((0, t] \times E)$ is given as

$$\hat{N}_{\tilde{p}}((0, t]\times E) = \int_0^t ds\int_{X\times[0,1]} I_E(\theta(s, f(s, x, \alpha, \omega),\omega))q(s, dx, \omega)d\alpha$$
$$= \int_0^t ds\int_X\int_Z I_E(\theta(s, z, \omega))Q(s, x, dz, \omega)q(s, dx, \omega)$$
$$= \int_0^t ds\int_Z I_E(\theta(s, z, \omega))m(dz)$$
$$= \int_0^t q(s, E, \omega)ds = \hat{N}_p((0, t]\times E).$$

Finally,

$$E(\{N_p((0, t]\times E) - N_{\tilde{p}}((0, t]\times E)\}^2)$$
$$= E(\{\tilde{N}_p((0, t]\times E) - \tilde{N}_{\tilde{p}}((0, t]\times E)\}^2)$$
$$= E\{[\int_0^{t+}\int_{X\times[0,1]}(I_{E\times[0,1]}(x, \alpha) - I_E(\theta(s, f(s,x,\alpha,\omega),\omega))$$
$$\times \tilde{N}_{\tilde{p}}(ds, dxd\alpha)]^2\}$$
$$= E\{\int_0^t ds\int_{X\times[0,1]}[I_E(x) - I_E(\theta(s, f(s, x, \alpha, \omega), \omega)]^2 q(s, dx, \omega)d\alpha\}$$
$$= E\{\int_0^t ds\int_Z\int_X[I_E(x) - I_E(\theta(s, z, \omega))]^2 Q(s, x, dz, \omega)q(s, dx, \omega)\}$$
$$= E\{\int_0^t ds\int_Z I_{[\theta(s,z,\omega)\neq\Delta]}[I_E(\theta(s, z, \omega)) - I_E(\theta(s, z, \omega))]^2 m(dz)\}$$
$$= 0,$$

and this concludes the proof.

CHAPTER III

Stochastic Calculus

1. The space of stochastic differentials

In Chapter II, we introduced the notion of stochastic integrals and derived Itô's formula. These are fundamental notions in stochastic calculus and its applications. Based on these results, we will now give a systematic treatment of *stochastic differentials* for continuous semi-martingales.*1 One of the important notions to be introduced here is that of *symmetric multiplication* ($\mathscr{S}.\mathscr{M}.$) which corresponds to the so-called *Stratonovich integral* or *Fisk integral* ([153], [27] and [133]). Under this multiplication, the chain rule (Itô's formula) takes the same form as in the ordinary calculus.

Let $(\Omega, \mathscr{F}, P)$ be a probability space and $(\mathscr{F}_t)_{t\geq 0}$ be a reference family (assumed to be right continuous as usual). We introduce the following notations.

$\mathscr{M}$ (previously denoted by $\mathscr{M}_2^{c,loc}$ in Chapter II)
= the family of all continuous locally square integrable martingales $M = (M_t)$ relative to $(\mathscr{F}_t)$ such that $M_0 = 0$ a.s.;

$\mathscr{A}_+$ = the family of all continuous $(\mathscr{F}_t)$-adapted processes $A = (A_t)$ such that $A_0 = 0$ and $t \longmapsto A_t$ is non-decreasing a.s.;

$\mathscr{A}$ = the family of all continuous $(\mathscr{F}_t)$-adapted processes $A = (A_t)$ such that $A_0 = 0$ and $t \longmapsto A_t$ is of bounded variation on every finite interval a.s.;

$\mathscr{B}$ = the family of all $(\mathscr{F}_t)$-predictable processes $\Phi = (\Phi_t)$ such that, with probability one, $t \longmapsto \Phi_t$ is bounded on each bounded interval.

It is easy to show that $A = (A_t) \in \mathscr{A}$ if and only if there exist $A^{(i)} = (A_t^{(i)}) \in \mathscr{A}_+$ $(i = 1, 2)$ such that $A_t = A_t^{(1)} - A_t^{(2)}$, $t \geq 0$.*2

---

*1 The material in this section is adapted from Itô [71].

*2 We simply write $A = A^{(1)} - A^{(2)}$.

Let $X = (X_t)$ be a continuous semimartingale i.e., a process represented in the form

(1.1) $\quad X_t = X_0 + M_t + A_t$

where X_0 is an $\mathscr{F}_0$-measurable random variable, $M = (M_t) \in \mathscr{M}$ and $A \in \mathscr{A}$. Following Itô [71] we also call it a *quasimartingale*.

Definition 1.1. We denote by $\mathscr{Q}$ the totality of quasimartingales. Every $X \in \mathscr{Q}$ is expressed uniquely as (1.1): $M = M_X$ is called the *martingale part* and $A = A_X$ is called the *bounded variation part*, respectively. This decomposition is called the *canonical decomposition* of $X \in \mathscr{Q}$.

The uniqueness of the canonical decomposition is seen as follows. Let

$$X(0) + M_t^{(1)} + A_t^{(1)} = X(0) + M_t^{(2)} + A_t^{(2)}$$

be two such decompositions. Then

$$M_t = M_t^{(1)} - M_t^{(2)} = A_t^{(2)} - A_t^{(1)} := A_t.$$

As we saw in Chapter II, Section 5, $\langle M \rangle_t$ is given as the limit in probability of $\sum_{i=1}^{n} (M_{t_i} - M_{t_{i-1}})^2$ as $|\Delta| \longrightarrow 0$, where Δ is a partition $0 = t_0 < t_1 < \cdots < t_n = t$ and $|\Delta| = \max_{1 \leq i \leq n} |t_i - t_{i-1}|$. But $\sum_{i=1}^{n} (M_{t_i} - M_{t_{i-1}})^2 = \sum_{i=1}^{n} (A_{t_i} - A_{t_{i-1}})^2 \leq V_t(A) \max_{1 \leq i \leq n} |A_{t_i} - A_{t_{i-1}}| \longrightarrow 0$ where $V_t(A) =$ the total variation of $s \in [0, t] \longmapsto A_s$. Therefore, $\langle M_t^{(1)} - M_t^{(2)} \rangle = 0$ which implies that $M_t^{(1)} = M_t^{(2)}$.

The space $\mathscr{Q}$ is closed under addition and multiplication; more generally, if $f(x^1, x^2, \ldots, x^n) \in C^2(\boldsymbol{R}^n \longrightarrow \boldsymbol{R})$ and $X^1, X^2, \ldots, X^n \in \mathscr{Q}$, then $Y = f(X^1, X^2, \ldots, X^n) \in \mathscr{Q}$ (c.f. Theorem II-5.1).

Definition 1.2. For $X, Y \in \mathscr{Q}$, we say that X and Y are equivalent and write $X \sim Y$ if, with probability one,

(1.2) $\quad X(t) - X(s) = Y(t) - Y(s) \qquad \text{for every} \quad 0 \leq s \leq t.$

Clearly this is an equivalence relation. The equivalence class containing X is denoted by dX and is called the *stochastic differential* of X. $\int_s^t dX(u)$ is, by definition, the process $X(t) - X(s)$. Let $d\mathscr{Q} = \{dX;\ X \in \mathscr{Q}\}$, $d\mathscr{M} = \{dM;\ M \in \mathscr{M}\}$ and $d\mathscr{A} = \{dA;\ A \in \mathscr{A}\}$. We introduce the following operations in $d\mathscr{Q}$.

$\mathscr{A}$. *Addition*:

$$dX + dY = d(X + Y), \qquad \text{for} \quad X, Y \in \mathscr{Q}. \tag{1.3}$$

$\mathscr{P}$. *Product*:

$$dX \cdot dY = d\langle M_X, M_Y\rangle, \qquad \text{for} \quad X, Y \in \mathscr{Q}, \tag{1.4}$$

where M_X and M_Y are the martingale parts of X and Y respectively.
Next we define a multiplication between $\mathscr{B}$ and $\mathscr{Q}$.

$\mathscr{M}$. *$\mathscr{B}$-Multiplication*: If $\Phi \in \mathscr{B}$ and $X \in \mathscr{Q}$, then

$$(\Phi \cdot X) = X(0) + \int_0^t \Phi(s, \omega) dM_X(s) + \int_0^t \Phi(s, \omega) dA_X(s), \quad t \geq 0, \tag{1.5}$$

is defined as an element in $\mathscr{Q}$. Hence $d(\Phi \cdot X)$ is uniquely defined from Φ and dX. Now we define an element $\Phi \cdot dX$ of $d\mathscr{Q}$ by

$$\Phi \cdot dX = d(\Phi \cdot X). \tag{1.6}$$

Sometimes we write $\Phi \cdot dX$ simply as ΦdX.

Theorem 1.1. The space $d\mathscr{Q}$ with the operations $\mathscr{A}$, $\mathscr{M}$ and $\mathscr{P}$ is a commutative algebra over $\mathscr{B}$, i.e., a commutative ring with the operations $\mathscr{A}$ and $\mathscr{P}$ satisfying the relations

$$\begin{aligned} &\Phi \cdot (dX + dY) = \Phi \cdot dX + \Phi \cdot dY, \\ &\Phi \cdot (dX \cdot dY) = (\Phi \cdot dX) \cdot dY, \\ &(\Phi + \Psi) \cdot dX = \Phi \cdot dX + \Psi \cdot dX \\ &(\Phi \Psi) \cdot dX = \Phi \cdot (\Psi \cdot dX) \end{aligned} \tag{1.7}$$

for $\Phi, \Psi \in \mathscr{B}$ and $dX, dY \in d\mathscr{Q}$. We also have that

$$d\mathscr{Q} \cdot d\mathscr{Q} \subset d\mathscr{A}, \quad d\mathscr{A} \cdot d\mathscr{Q} = 0 \qquad \text{and} \quad d\mathscr{Q} \cdot d\mathscr{Q} \cdot d\mathscr{Q} = 0. \tag{1.8}$$

Proof. It follows almost immediately from the property of stochastic integrals established in Chapter II that $d\mathscr{Q}$ is a commutative algebra over $\mathscr{B}$. (1.8) follows at once because $\langle M_X, M_Y \rangle \in \mathscr{A}$ for $X, Y \in \mathscr{Q}$.

Now Theorem II-5.1 in continuous case can be rephrased in this context as follows: if $X^1, X^2, \ldots, X^d \in \mathscr{Q}$ and $f \in \boldsymbol{C}^2(\boldsymbol{R}^d \longrightarrow \boldsymbol{R})$, then $Y = f(X^1, X^2, \ldots, X^d) \in \mathscr{Q}$ and

$$dY = \sum_{i=1}^d (\partial_i f) \cdot dX^i + \frac{1}{2} \sum_{i,j=1}^d (\partial_i \partial_j f) \cdot dX^i \cdot dX^j, \tag{1.9}$$

where $\partial_i f$ and $\partial_i \partial_j f$ are elements in $\mathscr{B}$ defined by $\frac{\partial f}{\partial x^i}(X^1, X^2, \ldots, X^d)$ and $\frac{\partial^2 f}{\partial x^i \partial x^j}(X^1, X^2, \ldots, X^d)$ respectively. Also Theorem II-6.1 can be rephrased as follows: if $dX^1, dX^2, \ldots, dX^d \in d\mathscr{M}$ and $dX^i \cdot dX^j = \delta_{ij}\, dt$, $i, j = 1, 2, \ldots, d$, then $(X^1(t), X^2(t), \ldots, X^d(t))$ is a d-dimensional Wiener process. Such a system of martingales $(X^1, X^2, \ldots, X^d)$ is called a d-dimensional *Wiener martingale*.

Now we introduce the fourth operation.

$\mathscr{S}.\mathscr{M}$. Symmetric $\mathscr{Q}$-Multiplication:

$$Y \circ dX = Y \cdot dX + \frac{1}{2} dX \cdot dY \quad \text{for} \quad dX \in d\mathscr{Q} \quad \text{and} \quad Y \in \mathscr{Q}. \tag{1.10}$$

Theorem 1.2. The space $d\mathscr{Q}$ with the operations $\mathscr{A}$, $\mathscr{S}.\mathscr{M}$. and $\mathscr{P}$ is a commutative algebra over $\mathscr{Q}$; we have, for $X, Y, Z \in \mathscr{Q}$,

$$\begin{aligned} &X \circ (dY + dZ) = X \circ dY + X \circ dZ, \\ &(X + Y) \circ dZ = X \circ dZ + Y \circ dZ, \\ &X \circ (dY \cdot dZ) = (X \circ dY) \cdot dZ = X \cdot (dY \cdot dZ), \\ &(XY) \circ dZ = X \circ (Y \circ dZ). \end{aligned} \tag{1.11}$$

Proof. We note that since $d\mathscr{Q} \cdot d\mathscr{A} = 0$ and $d\mathscr{Q} \cdot d\mathscr{Q} \cdot d\mathscr{Q} = 0$, we have that

$$X \circ dY = X \cdot dY \quad \text{if} \quad X \quad \text{or} \quad Y \in \mathscr{A}, \tag{1.12}$$

and

$$(Z \circ dX) \cdot dY = Z \cdot dX \cdot dY. \tag{1.13}$$

Then, for example,

$$\begin{aligned} X \circ (Y \circ dZ) &= X \cdot (Y \circ dZ) + \frac{1}{2} dX \cdot (Y \circ dZ) \\ &= X \cdot (Y \cdot dZ) + \frac{1}{2} X \cdot (dY \cdot dZ) + \frac{1}{2} dX \cdot (Y \cdot dZ) \\ &= (XY) \cdot dZ + \frac{1}{2} d(XY) \cdot dZ \\ &= XY \circ dZ. \end{aligned}$$

The other properties are also easily proved and so we omit the details.

A remarkable fact for the operation $\mathscr{S}.\mathscr{M}.$ is that the chain rule takes the same form as in the ordinary calculus. Namely we have

Theorem 1.3. If $X^1, X^2, \ldots, X^d \in \mathscr{Q}$ and $f \in C^3(\boldsymbol{R}^d \longrightarrow \boldsymbol{R})$, then for $Y = f(X^1, X^2, \ldots, X^d) \in \mathscr{Q}$ we have

$$dY = \sum_{i=1}^{d} \partial_i f \circ dX^i. \tag{1.14}$$

Proof. By Theorem 1.2

$$\begin{aligned}
&\sum_{i=1}^{d} \partial_i f \circ dX^i \\
&\quad = \sum_{i=1}^{d} \left(\partial_i f \cdot dX^i + \frac{1}{2} d(\partial_i f) \cdot dX^i\right) \\
&\quad = \sum_{i=1}^{d} \partial_i f \cdot dX^i + \frac{1}{2} \sum_{i=1}^{d} \Big(\sum_{j=1}^{d} \partial_j \partial_i f \cdot dX^j \\
&\qquad\qquad + \frac{1}{2} \sum_{j,k=1}^{d} \partial_j \partial_k \partial_i f \cdot dX^j \cdot dX^k\Big) \cdot dX^i \\
&\quad = \sum_{i=1}^{d} \partial_i f \cdot dX^i + \frac{1}{2} \sum_{i,j=1}^{d} \partial_i \partial_j f \cdot dX^i \cdot dX^j, \quad \text{(by (1.8))}, \\
&\quad = dY.
\end{aligned}$$

The stochastic integral $\int_0^t Y \circ dX$ is called the *Stratonovich integral* or the *Fisk integral* or sometimes the *Fisk-Stratonovich symmetric integral*. Indeed, we have the following:

Theorem 1.4. For every X and Y in $\mathscr{Q}$,

$$\int_0^t Y \circ dX = \underset{|\Delta| \to 0}{\text{l.i.p.}} \sum_{i=1}^{n} \frac{Y(t_i) + Y(t_{i-1})}{2} (X(t_i) - X(t_{i-1})) \tag{1.15}$$

where Δ denotes a partition $0 = t_0 < t_1 < \cdots < t_n = t$ and $|\Delta| = \max_{1 \le i \le n} (t_i - t_{i-1})$.

Proof.

$$\begin{aligned}
&\sum_{i=1}^{n} \frac{Y(t_i) + Y(t_{i-1})}{2} (X(t_i) - X(t_{i-1})) \\
&\quad = \sum_{i=1}^{n} Y(t_{i-1})(X(t_i) - X(t_{i-1})) + \frac{1}{2} \sum_{i=1}^{n} (Y(t_i) - Y(t_{i-1}))(X(t_i) - X(t_{i-1})).
\end{aligned}$$

Thus the assertion follows by the same arguments as in the proof of Theorem II-5.1.

Finally, we discuss the notion of *stochastic time change*. This is an important operation on quasimartingales.

Definition 1.3. By a *process of time change* ϕ we mean any process $\phi = (\phi_t) \in \mathscr{A}_+$ such that, with probability one, $t \longmapsto \phi_t$ is strictly increasing and $\lim_{t \uparrow \infty} \phi_t = \infty$.

For a given process of time change ϕ, we set

$$\tau_t = \inf\{u;\ \phi_u > t\}. \tag{1.16}$$

Then, with probability one, $\tau_0 = 0$, $t \longmapsto \tau_t$ is strictly increasing and continuous and $\lim_{t \uparrow \infty} \tau_t = \infty$. Furthermore, τ_t is an $(\mathscr{F}_t)$-stopping time because $\{\tau_t \leq u\} = \{t \leq \phi_u\} \in \mathscr{F}_u$. Set

$$\tilde{\mathscr{F}}_t = \mathscr{F}_{\tau_t}. \tag{1.17}$$

Thus $(\tilde{\mathscr{F}}_t)$ is a reference family on $(\Omega, \mathscr{F}, P)$. Let $X = (X_t)$ be an $(\mathscr{F}_t)$-well measurable process and define $T^\phi X = ((T^\phi X_t))$ by $(T^\phi X)_t = X_{\tau_t}$. Then $T^\phi X$ is an $(\tilde{\mathscr{F}}_t)$-well measurable process by Proposition I-5.4. $T^\phi X$ is called the *time change of* X by ϕ. It is easy to see that if $\mathscr{Q}$ and $\tilde{\mathscr{Q}}$ are the space of quasimartingales relative to $(\mathscr{F}_t)$ and $(\tilde{\mathscr{F}}_t)$ respectively, where $\tilde{\mathscr{F}}_t$ is defined by (1.17), then T^ϕ: $\mathscr{Q} \longrightarrow \tilde{\mathscr{Q}}$ is a bijection which preserves all structures on the space of quasimartingales: $T^\phi(M_X) = M_{T^\phi X}$, $T^\phi(A_X) = A_{T^\phi X}$, $X \sim Y$ if and only if $T^\phi X \sim T^\phi Y$. Thus T^ϕ induces a bijection between $d\mathscr{Q}$ and $d\tilde{\mathscr{Q}}$. Also, the space $\tilde{\mathscr{B}}$ with respect to $(\tilde{\mathscr{F}}_t)$ coincides with $T^\phi(\mathscr{B})$ and T^ϕ is an isomorphism between the $\mathscr{B}$-algebra $d\mathscr{Q}$ and the $\tilde{\mathscr{B}}$-algebra $d\tilde{\mathscr{Q}}$. In particular, T^ϕ commutes with the operation $\mathscr{S}.\mathscr{M}.$: $T^\phi(X \circ dY) = (T^\phi X) \circ T^\phi(dY)$.

The proof of these facts proceeds as follows. By Doob's optional sampling theorem (Theorem I-6.11), $M \in \mathscr{M}$ if and only if $T^\phi(M) \in \tilde{\mathscr{M}}$, where $\tilde{\mathscr{M}}$ is defined with respect to $(\tilde{\mathscr{F}}_t)$ and furthermore $T^\phi\langle M,N\rangle = \langle T^\phi M,\ T^\phi N\rangle$. Also, $A \in \mathscr{A}$ if and only if $T^\phi A \in \tilde{\mathscr{A}}$. Consequently $T^\phi(\int \Phi dM) = \int T^\phi(\Phi) d(T^\phi M)$ for $\Phi \in \mathscr{B}$ and $M \in \mathscr{M}$ because $N = \int \Phi dM$ is characterized as the unique $N \in \mathscr{M}$ such that $\langle N,L\rangle = \int \Phi d\langle M,L\rangle$ for all $L \in \mathscr{M}$. Since all operations in $\mathscr{Q}$ are defined in terms of addition (which is clearly preserved by T^ϕ), the operation $\langle M, N\rangle$ and stochastic integration, the assertion is obvious.

2. Stochastic differential equations with respect to quasimartingales

Let $(\Omega, \mathscr{F}, P)$ and $(\mathscr{F}_t)$ be given as in section 1. Suppose that we are given $X^1, X^2, \dots, X^r \in \mathscr{Q}$ and a system $(\sigma_j^i(x))_{i=1,2,\dots,d,\, j=1,2,\dots,r,}$ of real locally bounded Borel measurable functions on $\boldsymbol{R}^d$. We want to find $Y^1, Y^2, \dots, Y^d \in \mathscr{Q}$ such that

$$dY^i(t) = \sum_{j=1}^{r} \sigma_j^i(Y(t)) \cdot dX^j(t), \qquad i = 1, 2, \dots, d, \tag{2.1}$$

where $Y(t) = (Y^1(t), Y^2(t), \dots, Y^d(t))$.*

Theorem 2.1. Suppose that $\sigma_j^i(x)$, $i = 1, 2, \dots, d$, $j = 1, 2, \dots, r$, satisfy a Lipschitz condition; i.e., there exists a constant $K > 0$ such that

$$|\sigma_j^i(x) - \sigma_j^i(y)| \le K|x - y|, \qquad \text{for all} \qquad x, y \in \boldsymbol{R}^d. \tag{2.2}$$

Then for each given $y = (y^1, y^2, \dots, y^d) \in \boldsymbol{R}^d$, there exists a unique $Y = (Y^1, Y^2, \dots, Y^d)$ such that $Y^i \in \mathscr{Q}$, $Y^i(0) = y^i$ and (2.1) is satisfied.

Proof. Since there is no essential change, we assume that $d = 1$ and that (2.1) is of the form

$$dY = a_1(Y(t)) \cdot dM + a_2(Y(t)) \cdot dA \tag{2.3}$$

where $M \in \mathscr{M}$, $A \in \mathscr{A}$ and $a_i(x)$ $(i = 1, 2)$ are given, and the $a_i(x)$ satisfy (2.2). The problem is equivalent to finding $Y \in \mathscr{Q}$ such that

$$Y(t) = y + \int_0^t a_1(Y(s))dM_s + \int_0^t a_2(Y(s))dA_s. \tag{2.4}$$

Set $\phi(t) = t + \langle M \rangle_t + \| A \|_t$ where $\| A \|_t$ denotes the total variation of $[0, t] \ni s \longmapsto A_s$. Then ϕ is a process of time change. By applying the time change T^ϕ to (2.4) we have

$$\tilde{Y}(t) = y + \int_0^t a_1(\tilde{Y}(s))d\tilde{M}_s + \int_0^t a_2(\tilde{Y}(s))d\tilde{A}_s, \tag{2.5}$$

where $\tilde{Y} = T^\phi Y$, $\tilde{M} = T^\phi M$ and $\tilde{A} = T^\phi A$. It is sufficient to show the unique existence of $\tilde{Y}$ satisfying (2.5). Since $\langle \tilde{M} \rangle_t = \langle M \rangle_{\phi^{-1}(t)}$ and $\tilde{A}_t =$

* Clearly each $\sigma_j^i(Y(t)) \in \mathscr{B}$.

$A_{\phi^{-1}(t)}$, it is easy to see that $t \longmapsto t - \langle \tilde{M} \rangle_t - \| \tilde{A} \|_t$ is increasing. In particular we have $d\langle \tilde{M} \rangle_s \leq ds$ and $d\| \tilde{A} \|_s \leq ds$ as Stieltjes measures a.s. Let us construct a solution by successive approximations:

$$Y^{(0)}(t) = y,$$

$$Y^{(n)}(t) = y + \int_0^t a_1(Y^{(n-1)}(s))d\tilde{M}_s + \int_0^t a_2(Y^{(n-1)}(s))d\tilde{A}_s,$$

$$n = 1, 2, \dots .$$

Let $T > 0$ be given and fixed. Set $K_T = 2K^2(1 + T)$ where K is the Lipschitz constant in (2.2). Then, if $t \in [0, T]$,

$$E\{|Y^{(1)}(t) - Y^{(0)}(t)|^2\} = E\{[a_1(y)\tilde{M}_t + a_2(y)\tilde{A}_t]^2\} \leq C_1,$$

where $C_1 = 2(a_1(y)^2T + a_2(y)^2T^2)$. Suppose now that

$$E\{|Y^{(n)}(t) - Y^{(n-1)}(t)|^2\} \leq C_1 K_T^{n-1} \frac{t^{n-1}}{(n-1)!}$$

for some $n \geq 1$. Then

$$E\{|Y^{(n+1)}(t) - Y^{(n)}(t)|^2\} \leq 2E(\{\int_0^t [a_1(Y^{(n)}(s)) - a_1(Y^{(n-1)}(s))]d\tilde{M}_s\}^2)$$

$$+ 2E(\{\int_0^t [a_2(Y^{(n)}(s)) - a_2(Y^{(n-1)}(s))]d\| \tilde{A} \|_s\}^2)$$

$$= I_1 + I_2.$$

But

$$I_1 = 2E\{\int_0^t [a_1(Y^{(n)}(s)) - a_1(Y^{(n-1)}(s))]^2 d\langle \tilde{M} \rangle_s\}$$

$$\leq 2E\{\int_0^t [a_1(Y^{(n)}(s)) - a_1(Y^{(n-1)}(s))]^2 ds\}$$

$$\leq 2K^2E\{\int_0^t |Y^{(n)}(s) - Y^{(n-1)}(s)|^2\} ds$$

$$= 2K^2 \int_0^t E\{|Y^{(n)}(s) - Y^{(n-1)}(s)|^2\} ds$$

and

$$I_2 \leq 2E\{\|\tilde{A}\|_t \int_0^t [a_2(Y^{(n)}(s)) - a_2(Y^{(n-1)}(s))]^2 d\|\tilde{A}\|_s\}$$
$$\leq 2E\{t \int_0^t [a_2(Y^{(n)}(s)) - a_2(Y^{(n-1)}(s))]^2 ds\}$$
$$\leq 2TK^2 \int_0^t E\{|Y^{(n)}(s) - Y^{(n-1)}(s)|^2\} ds.$$

Hence

$$E\{|Y^{(n+1)}(t) - Y^{(n)}(t)|^2\}$$
$$\leq (2K^2 + 2TK^2) \int_0^t E\{|Y^{(n)}(s) - Y^{(n-1)}(s)|^2\} ds$$
$$\leq K_T \int_0^t C_1 K_T^{n-1} \frac{s^{n-1}}{(n-1)!} ds$$
$$= C_1 K_T^n \frac{t^n}{n!}.$$

Therefore the inequality

$$E\{|Y^{(n+1)}(t) - Y^{(n)}(t)|^2\} \leq C_1 K_T^n \frac{t^n}{n!}, \quad t \in [0, T], \tag{2.6}$$

follows by induction. Also, by Theorem I-6.10,

$$E\{\sup_{0\leq t\leq T} |Y^{(n+1)}(t) - Y^{(n)}(t)|^2\}$$
$$\leq 2E\{\sup_{0\leq t\leq T} |\int_0^t [a_1(Y^{(n)}(s)) - a_1(Y^{(n-1)}(s))] d\tilde{M}(s)|^2\}$$
$$+ 2E\{\sup_{0\leq t\leq T} |\int_0^t [a_2(Y^{(n)}(s)) - a_2(Y^{(n-1)}(s))] d\tilde{A}(s)|^2\}$$
$$\leq 8E(\{\int_0^T [a_1(Y^{(n)}(s)) - a_1(Y^{(n-1)}(s))] d\tilde{M}(s)\}^2)$$
$$+ 2E(\{\int_0^T |a_2(Y^{(n)}(s)) - a_2(Y^{(n-1)}(s))| d\|\tilde{A}\|(s)\}^2)$$

and by the same calculations as above, this is dominated by

$$8K^2 \int_0^T E\{|Y^{(n)}(s) - Y^{(n-1)}(s)|^2\} ds$$
$$+ 2TK^2 \int_0^T E\{|Y^{(n)}(s) - Y^{(n-1)}(s)|^2\} ds$$

$$\leq (8K^2+2TK^2)C_1K_T^{n-1}\frac{T^n}{n!}.$$

Consequently

$$P\Big(\sup_{0\leq t\leq T}|Y^{(n+1)}(t)-Y^{(n)}(t)|>\frac{1}{2^n}\Big)\leq \text{const.}\times\frac{(4K_TT)^n}{n!}$$

and, by a standard application of Borel-Cantelli's lemma, we see that $Y_t^{(n)}$ converges uniformly on $[0, T]$, a.s. The limit $Y(t)$ is a continuous $(\mathscr{F}_t)$-adapted process and, by (2.6), $E(|Y_n(t)-Y(t)|^2)\longrightarrow 0$ as $n\longrightarrow\infty$, $t\in[0,T]$. Now it is easy to see that $Y=(Y(t))$ satisfies (2.5). To prove the uniqueness, let Y_1 and Y_2 satisfy (2.5). Then, by the same calculations as above, we have

$$E\{|Y_1(t)-Y_2(t)|^2\}\leq K_T\int_0^t E\{|Y_1(s)-Y_2(s)|^2\}\,ds.$$

By truncating Y_1 and Y_2 with stopping times if necessary, we may assume that $s\longmapsto E\{|Y_1(s)-Y_2(s)|^2\}$ is bounded in $[0, T]$. Now it is easy to conclude that $Y_1=Y_2$.

Corollary. Let $\sigma_j^i(x)$, $i=1,2,\ldots,d$, $j=1,2,\ldots,r$, be real continuous functions on $\mathbf{R}^d$ such that they are twice continuously differentiable with bounded derivatives of the first and second orders. Then, for given $dX^1, dX^2,\ldots, dX^r\in d\mathscr{Q}$ and $y=(y^1,y^2,\ldots,y^d)\in\mathbf{R}^d$, there exist unique $Y^1, Y^2,\ldots,Y^d\in\mathscr{Q}$ such that

$$\text{(2.7)}\quad\begin{cases}Y^i(0)=y^i,\\ dY^i(t)=\sum_{j=1}^r\sigma_j^i(Y(t))\circ dX^j(t)\end{cases},\quad i=1,2,\ldots,d.$$

Proof. This follows at once if we note that (2.7) is equivalent to

$$\text{(2.7)}'\quad\begin{aligned}&Y^i(0)=y^i\\ &dY^i(t)=\sum_{j=1}^r\sigma_j^i(Y(t))\cdot dX^j(t)+\frac{1}{2}\sum_{k=1}^d\sum_{j=1}^r\Big(\frac{\partial}{\partial x^k}\sigma_j^i\Big)(Y(t))\cdot(dY^k\cdot dX^j)(t)\\ &\qquad=\sum_{j=1}^r\sigma_j^i(Y(t))\cdot dX^j(t)+\frac{1}{2}\sum_{j,l=1}^r\sum_{k=1}^d\Big(\frac{\partial}{\partial x^k}\sigma_j^i\cdot\sigma_l^k\Big)(Y(t))\cdot\\ &\qquad(dX^j\cdot dX^l)(t).\end{aligned}$$

This equation is a particular case of (2.1) with respect to dX^i and $dX^j\cdot dX^l$. The coefficients σ_j^i and $\sum_{k=1}^d\frac{\partial}{\partial x^k}\sigma_j^i\cdot\sigma_l^k$ satisfy the Lipschitz condition by the assumption in the corollary.

Thus a general theory of existence and uniqueness of the equations for semimartingales is established. Below there are some examples for which solution can be written down explicitly.

Example 2.1.* Let $d = 1$ and consider, for a given $X \in \mathcal{Q}$ such that $X_0 = 0$ the following equation

$$\begin{cases} dY_t = a(Y_t)\circ dX_t + b(Y_t)\cdot dt \\ Y_0 = y, \end{cases}$$

where $a \in C^2(\mathbf{R}^1 \longrightarrow \mathbf{R})$ with bounded a' and a'' and b is Lipschitz continuous. By the corollary of Theorem 2.1, we know that the solution exists uniquely and it is given in the following way. Let $u(x, z)$ be the solution of

$$\begin{cases} \dfrac{\partial u}{\partial z}(x, z) = a(u(x, z)) \\ u(x, 0) = x. \end{cases}$$

Let D_t be the solution of

$$\begin{cases} \dfrac{dD_t}{dt} = \exp\{-\int_0^{X_t} a'(u(D_t, s))ds\}\, b(u(D_t, X_t)) \\ D_0 = y. \end{cases}$$

Then the solution Y is given by

$$Y_t = u(D_t, X_t).$$

Indeed, by the chain rule (1.14),

$$dY_t = a(u(D_t, X_t))\circ dX_t + \left(\frac{\partial u}{\partial x}\right)(D_t, X_t) \exp\{-\int_0^{X_t} a'(u(D_t, s))ds\} \times b(u(D_t, X_t))\cdot dt.$$

But $\dfrac{\partial}{\partial z}\dfrac{\partial u}{\partial x} = a'(u(x, z))\dfrac{\partial u}{\partial x}$ and $\dfrac{\partial u}{\partial x}(x, 0) = 1$ imply that

$$\frac{\partial u}{\partial x}(x, z) = \exp\{\int_0^z a'(u(x, s))ds\},$$

* Doss [19].

and hence $dY_t = a(Y_t)\circ dX_t + b(Y_t)\cdot dt$.

Example 2.2. Let $A_k = \sum_{j=1}^{d} A_k^j(x)\frac{\partial}{\partial x^j}$ be a C^∞-vector field*[1] on $\boldsymbol{R}^d$, $k = 1, 2, \ldots, r$. We assume that the first and second order derivatives of all coefficients are bounded. For given $X^1, X^2, \ldots, X^r \in \mathcal{Q}$ such that $X_0^i = 0$, $i = 1, 2, \ldots, r$, and $y = (y^1, y^2, \ldots, y^d) \in \boldsymbol{R}^d$, consider the equation

$$\text{(2.8)} \qquad \begin{cases} dY^i(t) = \sum_{k=1}^{r} A_k^i(Y(t))\circ dX^k(t) \\ Y^i(0) = y^i, \end{cases} \quad , \quad i = 1, 2, \ldots, d.$$

[I] *[2] If the vector fields $A_1, A_2, \ldots, A_r$ are commutative, i.e., $[A_p, A_q] = 0$, $p,q = 1, 2, \ldots, r$, then this implies the integrability of

$$\begin{cases} \frac{\partial u^i}{\partial z^j}(x, z) = A_j^i(u(x, z)), \quad i = 1, 2, \ldots, d, \quad j = 1, 2, \ldots, r \\ u(x, 0) = x \in \boldsymbol{R}^d \end{cases}$$

and so we have the solution $u(x, z) = (u^1(x, z), \ldots, u^d(x, z))$. If we set $Y_t^i = u^i(y, X_t)$, $i = 1, 2, \ldots, d$, where $X_t = (X_t^1, X_t^2, \ldots, X_t^r)$, it follows immediately from the chain rule (1.14) that $Y = (Y_t^1, Y_t^2, \ldots, Y_t^d)$ is a solution of (2.8).

[II] *[3] Next we consider the non-commutative case but we assume that the vector fields $A_1, A_2, \ldots, A_r$ satisfy

$$\text{(2.9)} \qquad [A_i, [A_j, A_k]] = 0, \qquad i,j,k = 1,2, \ldots, r,$$

i.e., the Lie algebra $\mathfrak{L}(A_1, A_2, \ldots, A_r)$ generated by $A_1, A_2, \ldots, A_r$ is nilpotent in two steps. Then the solution of (2.8) is given as follows. Let $\tilde{\mathscr{D}} = \{(i, j); 1 \le i < j \le r\}$ and $\mathscr{D} = \{1, 2, \ldots, r\} \cup \tilde{\mathscr{D}}$. For $z = (z^I)_{I\in\mathscr{D}}$, consider the following system of equations

$$\text{(2.10)} \qquad \begin{cases} \frac{\partial u^h}{\partial z^i} = A_i^h(u(z)) - \sum_{p=1}^{i-1} z^p A_{p,i}^h(u(z)) \\ \frac{\partial u^h}{\partial z^{(j,k)}} = A_{j,k}^h(u(z)) \end{cases} \quad , \quad \begin{array}{l} i = 1,2, \ldots, r, \\ 1 \le j < k \le r, \quad h = 1,2, \ldots, d, \end{array}$$

---

*[1] Cf. Chapter V, §1.

*[2] Doss [19].

*[3] Additional information on this subject can be found in Yamato [184].

where $A^h_{j,k}(x)$ is defined by

$$\sum_{h=1}^{d} A^h_{j,k}(x)\frac{\partial}{\partial x^h} = [A_j, A_k] := A_{j,k}.$$

We can prove that (2.9) implies the integrability condition of (2.10) and so, for given $x \in \mathbf{R}^d$, we have the solution $(u^i(x, z))_{i=1,2,\dots,d}$ of (2.10) such that $u^i(x, 0) = x^i$, $i = 1, 2, \dots, d$. Let

$$X^{j,k}_t = \int_0^t X^j_s \circ dX^k_s \quad , \quad 1 \le j < k \le r$$

and $X_t = (X^J_t)_{J \in \mathscr{J}}$. Then $Y^i_t = u^i(y, X_t)$, $i = 1, 2, \dots, d$ is a solution of (2.8). Indeed

$$\begin{aligned} dY^i_t &= \sum_{j=1}^{r} \frac{\partial u^i}{\partial z^j}(y, X_t)\circ dX^j_t + \sum_{1\le j<k\le r} \frac{\partial}{\partial z^{(j,k)}} u^i(y, X_t)\circ dX^{j,k}_t \\ &= \sum_{j=1}^{r} A^i_j(Y_t)\circ dX^j_t - \sum_{j=1}^{r}\sum_{k=1}^{j-1} A^i_{k,j}(Y_t)X^k_t\circ dX^j_t \\ &\quad + \sum_{1\le j<k\le r} A^i_{j,k}(Y_t)X^j_t\circ dX^k_t \\ &= \sum_{j=1}^{r} A^i_j(Y_t)\circ dX^j_t. \end{aligned}$$

For example*, if $d = 3$, $r = 2$ and $A_1 = \dfrac{\partial}{\partial x^1} + 2x^2\dfrac{\partial}{\partial x^3}$, $A_2 = \dfrac{\partial}{\partial x^2} - 2x^1\dfrac{\partial}{\partial x^3}$, the solution $Y_t = (Y^1_t, Y^2_t, Y^3_t)$ is given by $Y^1_t = y^1 + X^1_t$, $Y^2_t = y^2 + X^2_t$ and $Y^3_t = y^3 + 2(y^2X^1_t - y^1X^2_t) + 2(\int_0^t X^2_s\circ dX^1_s - \int_0^t X^1_s\circ dX^2_s)$.

Consider the solution $Y(t) = (Y^1(t), Y^2(t), \dots, Y^d(t))$ of the equation (2.8). If a C^2 function $f(x)$ defined on $\mathbf{R}^d$ satisfies $A_k f = 0$, $k = 1, 2, \dots, r$, then $f(Y(t)) = f(y)$ for all $t > 0$ a.s. Indeed,

$$\begin{aligned} df(Y(t)) &= \sum_{i=1}^{d} \partial_i f(Y(t))\circ dY^i(t) = \sum_{i=1}^{d}\sum_{k=1}^{r} A^i_k(Y(t))\partial_i f(Y(t))\circ dX^k_t \\ &= \sum_{k=1}^{r} (A_k f)(Y(t))\circ dX^k_t = 0. \end{aligned}$$

* Gaveau [33].

For example, if $A_k f = \sum_{i=1}^{d} A_k^i(x) \frac{\partial f}{\partial x^i}(x)$, $k = 1, 2, \ldots, d$, with $A_k^i(x) = \delta_{ik} - x^i x^k / |x|^2$, $x = (x^1, x^2, \ldots, x^d) \in \boldsymbol{R}^d \setminus \{0\}$, $f(x) = \sum_{i=1}^{d} (x^i)^2$ satisfies $A_k f = 0$, $k = 1, 2, \ldots, d$. Thus the solution $Y(t)$ always stays on the sphere with center 0 and radius $|y|$ $(= |Y(0)|)$. If we take $X^k \in \mathscr{M}$, $k = 1, 2, \ldots, d$, such that $dX^k \cdot dX^l = \delta_{kl} \cdot dt$, i.e., $(X^1, X^2, \ldots, X^d)$ is a d-dimensional Wiener process, then the solution

$$\begin{cases} dY^i(t) = \sum_{k=1}^{d} A_k^i(Y(t)) \circ dX_t^k & \qquad i = 1, 2, \ldots, d, \\ y^i(t) = y^i, \end{cases}$$

defines the Brownian motion on the sphere with center 0 and radius $|y|$.*[1]

3. Moment inequalities for martingales

Various inequalities for moments of martingales are discussed, e.g., by Meyer [120] and Garsia [32] in connection with a martingale version of the theory of H^p-spaces. Here we obtain fundamental inequalities for continuous local martingales as an application of stochastic calculus.

Theorem 3.1. There exist universal constants c_p, C_p $(0 < p < \infty)$ such that for every $M \in \mathscr{M} (= \mathscr{M}_2^{c, loc})$ and $t \geq 0$,

$$(3.1) \qquad c_p E(M_t^{*2p}) \leq E(\langle M, M \rangle_t^p) \leq C_p E(M_t^{*2p})$$

where $M_t^* = \max_{0 \leq s \leq t} |M_s|$.

*Proof.**[2] It is sufficient to prove (3.1) for a bounded martingale $M = (M_t)$ since the general case follows easily by a truncation argument. Indeed, setting $T_n = \inf\{t;\ |M_t| \geq n \text{ or } \langle M \rangle_t \geq n\}$, we have $T_n \uparrow \infty$ a.s., and if (3.1) holds for $M^{T_n} = (M_{T_n \wedge t})$ with c_p and C_p independent of n, then (3.1) holds for M by letting $n \to \infty$. In the proof we shall write A for $\langle M, M \rangle$. By the inequality (6.16) of Chapter I,

$$(3.2) \qquad E(M_t^{*p}) \leq \left(\frac{p}{p-1}\right)^p E(|M_t|^p), \qquad p > 1.$$

*[1] This representation of a spherical Brownian motion is due to Stroock [156].
*[2] The following proof is adapted from Getoor-Sharpe [34].

Case 1. If $p = 1$,

$$E(\langle M, M\rangle_t) = E(M_t^2)$$

and hence, noting (3.2), we have (3.1) with $c_p = 1/4$ and $C_p = 1$.

Case 2. If $p > 1$,

$$E(M_t^{*2p}) \leq (2p/(2p-1))^{2p} E(|M_t|^{2p}).$$

Since $|x|^{2p}$ is of class C^2 we can apply Itô's formula to obtain

$$|M_t|^{2p} = \int_0^t 2p|M_s|^{2p-1} \operatorname{sgn}(M_s) dM_s + p(2p-1)\int_0^t |M_s|^{2p-2} dA_s.$$

By taking the expectations, we obtain

$$\begin{aligned} E(|M_t|^{2p}) &\leq p(2p-1)E\Big(\int_0^t |M_s|^{2p-2} dA_s\Big) \\ &\leq p(2p-1)E(M_t^{*2p-2} A_t) \\ &\leq p(2p-1)E(M_t^{*2p})^{1-1/p} E(A_t^p)^{1/p}. \end{aligned}$$

Therefore,

$$E(M_t^{*2p}) \leq (2p/(2p-1))^{2p} p(2p-1) E(M_t^{*2p})^{1-1/p} E(A_t^p)^{1/p},$$

from which the left side of (3.1) follows. To prove the other side of (3.1), set $N_t = \int_0^t A_s^{(p-1)/2} dM_s$. Then $p\langle N,N\rangle = p\int_0^t A_s^{p-1} dA_s = A_t^p$ and so $E(A_t^p) = pE(N_t^2)$. By Itô's formula,

$$\begin{aligned} M_t A_t^{(p-1)/2} &= \int_0^t A_s^{(p-1)/2} dM_s + \int_0^t M_s d(A_s^{(p-1)/2}) \\ &= N_t + \int_0^t M_s d(A_s^{(p-1)/2}) \end{aligned}$$

and hence $|N_t| \leq 2M_t^* A_t^{(p-1)/2}$. Thus

$$\frac{1}{p} E(A_t^p) = E(N_t^2) \leq 4E(M_t^{*2} A_t^{p-1}) \leq 4E(M_t^{*2p})^{1/p} E(A_t^p)^{1-1/p},$$

and so

$$E(A_t^p) \leq (4p)^p E(M_t^{*2p}).$$

Case 3. Let $0 < p < 1$. Set $N_t = \int_0^t A_s^{(p-1)/2} dM_s$. Then, as above, $E(A_t^p) = pE(N_t^2)$ and $M_t = \int_0^t A_s^{(1-p)/2} dN_s$. By Itô's formula,

$$N_t A_t^{(1-p)/2} = \int_0^t A_s^{(1-p)/2} dN_s + \int_0^t N_s d(A_s^{(1-p)/2})$$

$$= M_t + \int_0^t N_s d(A_s^{(1-p)/2}),$$

and hence

$$|M_t| \leq 2N_t^* A_t^{(1-p)/2}.$$

Thus $M_t^* \leq 2N_t^* A_t^{(1-p)/2}$, and by Hölder's inequality,

$$\begin{aligned} E(M_t^{*2p}) &\leq 2^{2p} E(N_t^{*2p} A_t^{p(1-p)}) \leq 2^{2p} E(N_t^{*2})^p E(A_t^p)^{1-p} \\ &\leq 2^{2p} 4^p E(N_t^2)^p E(A_t^p)^{1-p} = (16/p)^p E(A_t^p)^p E(A_t^p)^{1-p} \\ &= (16/p)^p E(A_t^p). \end{aligned}$$

Finally we must show that $E(A_t^p) \leq C_p E(M_t^{*2p})$. Let α be a positive con stant. By applying Hölder's inequality to the identity

$$A_t^p = [A_t^p (\alpha + M_t^*)^{-2p(1-p)}](\alpha + M_t^*)^{2p(1-p)},$$

we obtain

$$E(A_t^p) \leq \{E(A_t(\alpha + M_t^*)^{2(p-1)})\}^p \{E((\alpha + M_t^*)^{2p})\}^{1-p}.$$

Setting $N_t = \int_0^t (\alpha + M_s^*)^{p-1} dM_s$, we have

$$\langle N, N\rangle_t = \int_0^t (\alpha + M_s^*)^{2(p-1)} dA_s \geq A_t(\alpha + M_t^*)^{2(p-1)}.$$

By Itô's formula,

$$M_t(\alpha + M_t^*)^{p-1} = \int_0^t (\alpha + M_s^*)^{p-1} dM_s + \int_0^t M_s d\{(\alpha + M_s^*)^{p-1}\}$$

$$= N_t + (p-1)\int_0^t M_s(\alpha + M_s^*)^{p-2} dM_s^*,$$

and hence

$$|N_t| \leq M_t^{*p} + (1-p)\int_0^t M_s^{*(p-1)}dM_s^* = \frac{1}{p}M_t^{*p}.$$

Therefore $E(N_t^2) \leq \frac{1}{p^2}E(M_t^{*2p})$ for every $\alpha > 0$, so

$$E(A_t^p) \leq p^{-2p}\{E(M_t^{*2p})\}^p\{E((\alpha + M_t^*)^{2p})\}^{1-p}.$$

Letting $\alpha \downarrow 0$, we conclude that

$$E(A_t^p) \leq p^{-2p}\, E(M_t^{*2p}).$$

4. Some applications of stochastic calculus to Brownian motions

4.1. Brownian local time. Let $X = (X_t)$ be a one-dimensional Brownian motion defined on a probability space $(\Omega, \mathscr{F}, P)$.

Definition 4.1. By the *local time* or the *sojourn time density* of X we mean a family of non-negative random variables $\{\phi(t, x, \omega), t \in [0, \infty), x \in \boldsymbol{R}^1\}$ such that, with probability one, the following holds:

(i) $(t, x) \longmapsto \phi(t, x)$ is continuous,

(ii) for every Borel subset A of $\boldsymbol{R}^1$ and $t \geq 0$

$$\int_0^t I_A(X_s)ds = 2\int_A \phi(t, x)dx.$$

It is clear that if such a family $\{\phi(t, x)\}$ exists, then it is unique and is given by

$$\phi(t, x) = \lim_{\varepsilon \downarrow 0} \frac{1}{4\varepsilon}\int_0^t I_{(x-\varepsilon, x+\varepsilon)}(X_s)ds.$$

The notion of the local time of Brownian motion was first introduced by Lévy [101] and the following theorem was first established by H. Trotter [163].

Theorem 4.1. The local time $\{\phi(t, x)\}$ of X exists.

Proof. We will prove this theorem by using stochastic calculus. This idea is due to H. Tanaka (McKean [113] and [114]). Let $(\mathscr{F}_t) = (\mathscr{F}_t^X)$ be the proper reference family of X. Then X is an $(\mathscr{F}_t)$-Brownian motion and $X_t - X_0$ belongs to the space $\mathscr{M}$. Let $g_n(x)$ be a continuous function on $\boldsymbol{R}^1$ such that its support is contained in $(-1/n + a,\ 1/n + a)$, $g_n(x) \geq 0$, $g_n(a + x) = g_n(a - x)$ and

$$\int_{-\infty}^{\infty} g_n(x)dx = 1.$$

Set

$$u_n(x) = \int_{-\infty}^{x} dy \int_{-\infty}^{y} g_n(z)dz.$$

By Itô's formula,

$$u_n(X_t) - u_n(X_0) = \int_0^t u_n'(X_s)dX_s + \frac{1}{2}\int_0^t u_n''(X_s)ds$$

and if the local time $\{\phi(t, x)\}$ does exist, then

$$\frac{1}{2}\int_0^t u_n''(X_s)ds = \frac{1}{2}\int_0^t g_n(X_s)ds = \int_{-\infty}^{\infty} g_n(y)\phi(t, y)dy \longrightarrow \phi(t, a) \quad \text{as} \quad n \longrightarrow \infty.$$

Also, it is clear that

$$u_n(x) \longrightarrow (x - a)^+, \qquad u_n'(x) \longrightarrow \begin{cases} 1, & x > a \\ \frac{1}{2}, & x = a, \\ 0, & x < a \end{cases} \qquad \text{as} \quad n \longrightarrow \infty.$$

Hence, $\phi(t, a)$ should be given as

$$\text{(4.1)} \qquad \phi(t, a) = (X_t - a)^+ - (X_0 - a)^+ - \int_0^t I_{(a,\infty)}(X_s)dX_s.$$

In the sequel we will show that the family of random variables $\phi(t,a)$, $t \geq 0$, $a \in \boldsymbol{R}^1$ defined by (4.1) satisfies the properties (i) and (ii) above. It is obvious that $(X_t - a)^+ - (X_0 - a)^+$ is continuous in (t, a). We will show that there exists a process $\psi(t,a)$ which is continuous in (t, a), a.s., and for each t, a,

$$\psi(t, a) = \int_0^t I_{(a,\infty)}(X_s)dX_s \qquad \text{a.s.}$$

For each $a \in \boldsymbol{R}^1$ and $T > 0$, $[0, T] \ni t \longmapsto Y_a(t) = \int_0^t I_{(a,\infty)}(X_s)dX_s$ is a continuous process, i.e., $\boldsymbol{C}([0, T] \longrightarrow \boldsymbol{R})$-valued random variable. If we set

$$\|Y_a - Y_b\| = \max_{0 \le t \le T} |Y_a(t) - Y_b(t)|,$$

then

$$E\{\|Y_a - Y_b\|^4\} \le K|b - a|^2 \tag{4.2}$$

for some constant $K = K(T) > 0$. Indeed, if $a < b$,

$$(Y_a - Y_b)(t) = \int_0^t I_{(a,b]}(X_s)dX_s \in \mathscr{M}$$

and $\langle Y_a - Y_b\rangle_t = \int_0^t I_{(a,b]}(X_s)ds$. Now applying the inequality (3.1)

$$\begin{aligned}
E(\|Y_a - Y_b\|^4) &\le \frac{1}{c_2}E(\{\int_0^T I_{(a,b]}(X_s)ds\}^2) \\
&\le \frac{2}{c_2}E(\int_0^T I_{(a,b]}(X_s)ds \int_s^T I_{(a,b]}(X_u)du) \\
&= \frac{2}{c_2}\int_0^T ds \int_s^T du E(I_{(a,b]}(X_s)I_{(a,b]}(X_u)) \\
&= \frac{2}{c_2}\int_0^T ds \int_s^T du \int_{\boldsymbol{R}^1} \mu(dx) \int_a^b dy \int_a^b dz \\
&\quad \times \frac{1}{\sqrt{2\pi s}} \exp\left\{-\frac{(x-y)^2}{2s}\right\} \frac{1}{\sqrt{2\pi(u-s)}} \exp\left\{-\frac{(y-z)^2}{2(u-s)}\right\} \\
&\le \frac{2}{c_2}(b-a)^2 \int_0^T ds \int_s^T du \frac{1}{2\pi\sqrt{s(u-s)}} \\
&= K(T)(b-a)^2.
\end{aligned}$$

Here μ is the initial distribution of X, i.e., $\mu = P^{X_0}$. (4.2) is proved. By the corollary of Theorem I-4.3,* there exists a family $\{\psi(a)\}$ of $\boldsymbol{C}([0, T] \longrightarrow \boldsymbol{R})$-valued random variables such that

---

* This corollary applies for any system of random variables taking values in a metric space.

$$a \longmapsto \psi(a) \in C([0, T] \longrightarrow R)$$

is continuous a.s. and for each fixed a, $\psi(a) = Y_a(\cdot)$ a.s. Clearly $\psi(t,a) = \psi(a)(t)$ is what we want. Thus by choosing this modification we see that $\phi(t, a)$ satisfies (i). In order to prove (ii), it is clearly sufficient to show that

$$\int_0^t f(X_s)ds = 2\int_{R^1} \phi(t, a)f(a)da \qquad \text{a.s.} \tag{4.3}$$

for any continuous function $f(x)$ with compact support. Set

$$F(x) = \int_{-\infty}^{\infty} f(a)(x-a)^+ da.$$

Then $F \in C^2(R)$, $F'(x) = \int_{-\infty}^{\infty} f(a)I_{(a,\infty)}(x)da = \int_{-\infty}^{x} f(a)da$, $F''(x) = f(x)$, and so by Itô's formula,

$$F(X_t) - F(X_0) - \int_0^t F'(X_s)dX_s = \frac{1}{2}\int_0^t f(X_s)ds.$$

The left-hand side is equal to

$$\int_{-\infty}^{\infty} f(a)\{(X_t - a)^+ - (X_0 - a)^+\}\, da - \int_0^t \{\int_{-\infty}^{\infty} f(a)I_{(a,\infty)}(X_s)da\}\, dX_s.$$

If we now apply the next lemma to the second term, the above reduces to

$$\int_{-\infty}^{\infty} f(a)\{(X_t - a)^+ - (X_0 - a)^+ - \int_0^t I_{(a,\infty)}(X_s)dX_s\}\, da$$
$$= \int_{-\infty}^{\infty} f(a)\phi(t, a)da$$

and hence (4.3) is established.

Lemma 4.1. (A Fubini-type theorem for stochastic integrals). Let $(\Omega, \mathscr{F}, P)$ be a probability space and $(\mathscr{F}_t)$ be a reference family. Let $M \in \mathscr{M}_2^c$, (i.e., a continuous square-integrable martingale such that $M_0 = 0$ a.s.). Let $\{\Phi(t,a,\omega)\}$, $t \in [0, \infty)$, $a \in R^1$, be a family of real random variables such that

(i) $((t, \omega), a) \in ([0, \infty) \times \Omega) \times R^1 \longrightarrow \Phi(t, a, \omega)$ is $\mathscr{S} \times \mathscr{B}(R^1)$-measurable ($\mathscr{S}$ is defined in Chapter I, p. 21),

(ii) there exists a non-negative Borel measurable function $f(a)$ such that

$$|\Phi(t, a, \omega)| \leq f(a) \qquad \text{for every} \quad t, a, \omega.$$

By (i) and (ii), $\int_0^t \Phi(s, a, \omega)dM_s \in \mathscr{M}_2^c$ is well-defined. We assume further that

(iii) $(a, \omega) \longmapsto \int_0^t \Phi(s, a, \omega)dM_s$ is $\mathscr{B}(\boldsymbol{R}^1)\times\mathscr{F}$-measurable for each $t \geq 0$.

Let $\mu(da)$ be a non-negative Borel measure on $\boldsymbol{R}^1$ such that $\int_{\boldsymbol{R}^1} f(a)\mu(da) < \infty$. Then

$$(4.4) \qquad t \longrightarrow \int_{\boldsymbol{R}^1} \Phi(t, a, \omega)\mu(da) \in \mathscr{L}_2(\langle M\rangle)$$

(i.e., it is predictable and $E[\int_0^t \{\int_{\boldsymbol{R}^1} \Phi(s, a, \cdot)\,\mu(da)\}^2 d\langle M\rangle_s] < \infty$ for every t) and we have

$$(4.5) \qquad \int_0^t \{\int_{\boldsymbol{R}^1} \Phi(s, a, \omega)\mu(da)\}\, dM_s = \int_{\boldsymbol{R}^1} \{\int_0^t \Phi(s, a, \omega)dM_s\}\, \mu(da).$$

Proof. It is clear that $\int_{\boldsymbol{R}^1} \Phi(s, a, \omega)\mu(da)$ is $(\mathscr{F}_t)$-predictable and bounded. Hence it is obvious that

$$E[\int_0^t \{\int_{\boldsymbol{R}^1} \Phi(s, a, \omega)\mu(da)\}^2 d\langle M\rangle_s] < \infty.$$

Thus the left-hand side of (4.5) is well-defined as an element in $\mathscr{M}_2^c$. On the other hand, $a \longmapsto \int_0^t \Phi(s, a, \omega)dM_s$ is Borel measurable by assumption (iii), and for every $T > 0$,

$$\begin{aligned} &E[\int_{\boldsymbol{R}^1} \mu(da) \max_{0\leq t\leq T} |\int_0^t \Phi(s, a, \omega)dM_s|] \\ &\leq \int_{\boldsymbol{R}^1} \mu(da) \{E[\max_{0\leq t\leq T} |\int_0^t \Phi(s, a, \omega)dM_s|^2]\}^{1/2} \\ &\leq 2\int_{\boldsymbol{R}^1} \mu(da)(E[\{\int_0^T \Phi(s, a, \omega)dM_s\}^2])^{1/2} \\ &= 2\int_{\boldsymbol{R}^1} \mu(da)(E[\int_0^T \Phi^2(s, a, \omega)d\langle M\rangle_s])^{1/2} \\ &\leq 2\int_{\boldsymbol{R}^1} f(a)\mu(da)\{E[\langle M\rangle(T)]\}^{1/2} \\ &< \infty. \end{aligned}$$

Hence

$$\int_{R^1} \mu(da) \max_{0 \le t \le T} \left| \int_0^t \Phi(s, a, \omega) dM_s \right| < \infty \qquad \text{a.s.}$$

and this implies that

$$t \longmapsto \int_{R^1} \mu(da) \int_0^t \Phi(s, a, \omega) dM_s$$

is continuous a.s. Thus the right-hand side of (4.5) is well-defined and defines an $(\mathscr{F}_t)$-adapted continuous process. It is square-integrable because

$$\begin{aligned}
&E\{[\int_{R^1} (\int_0^t \Phi(s, a, \omega) dM_s) \mu(da)]^2\} \\
&= \int_{R^1} \mu(da_1) \int_{R^1} \mu(da_2) E[\int_0^t \Phi(s, a_1, \omega) dM_s \int_0^t \Phi(s, a_2, \omega) dM_s] \\
&= \int_{R^1} \mu(da_1) \int_{R^1} \mu(da_2) E[\int_0^t \Phi(s, a_1, \omega) \Phi(s, a_2, \omega) d\langle M \rangle_s] \\
&\le \int_{R^1} f(a_1) \mu(da_1) \int_{R^1} f(a_2) \mu(da_2) E[\langle M \rangle(t)] \\
&= (\int_{R^1} f(a) \mu(da))^2 E[\langle M \rangle(t)] \\
&< \infty.
\end{aligned}$$

It is an $(\mathscr{F}_t)$-martingale because if $t > s > 0$ and $A \in \mathscr{F}_s$,

$$\begin{aligned}
&E[I_A \int_{R^1} \mu(da) \int_s^t \Phi(u, a, \omega) dM_u] \\
&\qquad = \int_{R^1} \mu(da) E[I_A \int_s^t \Phi(u, a, \omega) dM_u] = 0.
\end{aligned}$$

Similarly, if $N \in \mathscr{M}_2$, then

$$\begin{aligned}
&E[I_A \{\int_{R^1} \mu(da) \int_s^t \Phi(u, a, \omega) dM_u\} (N_t - N_s)] \\
&= \int_{R^1} \mu(da) E[I_A \int_s^t \Phi(u, a, \omega) dM_u (N_t - N_s)] \\
&= \int_{R^1} \mu(da) E[I_A \int_s^t \Phi(u, a, \omega) d\langle M, N \rangle_u]
\end{aligned}$$

$$= E[I_A \int_s^t \{\int_{R^1} \Phi(u, a, \omega)\mu(da)\} d\langle M, N\rangle_u].$$

Thus $t \longmapsto \int_{R^1} \mu(da) \int_0^t \Phi(u, a, \omega)dM_u = L_t$ is an element in $\mathcal{M}_2^c$ such that for every $N \in \mathcal{M}_2$,

$$\langle N, L\rangle_t = \int_0^t \{\int_{R^1} \Phi(u, a, \omega)\mu(da)\} d\langle M, N\rangle_u.$$

Now we can conclude that $L_t = \int_0^t \{\int_{R^1} \Phi(u, a, \omega)\mu(da)\} dM_u$ by Proposition II-2.4. This completes the proof of (4.5).

4.2. Reflecting Brownian motion and the Skorohod equation. Let $X = (X_t)$ be a one-dimensional Brownian motion and let $X^+ = (X_t^+)$ be a continuous stochastic process on $[0, \infty)$ defined by

$$X_t^+ = |X_t|. \tag{4.6}$$

It is easy to see that

$$\begin{aligned} &P[X_{t_1}^+ \in A_1, X_{t_2}^+ \in A_2, \ldots, X_{t_n}^+ \in A_n] \\ &= \int_{[0,\infty)} \mu^+(dx)\int_{A_1} p^+(t_1, x, x_1)dx_1 \int_{A_2} p^+(t_2 - t_1, x_1, x_2)dx_2 \\ &\quad \cdots \int_{A_n} p^+(t_n - t_{n-1}, x_{n-1}, x_n)dx_n, \\ &\quad 0 < t_1 < t_2 < \cdots < t_n, \quad A_i \in \mathscr{B}([0, \infty)), \end{aligned} \tag{4.7}$$

where

$$p^+(t, x, y) = \frac{1}{\sqrt{2\pi t}}\left(\exp\left\{-\frac{(x-y)^2}{2t}\right\} + \exp\left\{-\frac{(x+y)^2}{2t}\right\}\right) \tag{4.8}$$

and μ^+ is the probability law of $X_0^+ = |X_0|$. The process X^+ is called the *one-dimensional reflecting Brownian motion.* Reflecting Brownian motion can be characterized in different ways. We shall now present one such characterization due to Skorohod [149].

We set $W_0^1 = \{f \in C([0, \infty) \longrightarrow R^1);\ f(0) = 0\}$ and $C^+ = \{f \in C([0, \infty) \longrightarrow R);\ f(t) \geq 0$ for all $t \geq 0\}$.

Lemma 4.2. Given $f \in W_0^1$ and $x \in R^+$, there exist unique $g \in C^+$ and $h \in C^+$ such that

(i) $g(t) = x + f(t) + h(t)$,

(ii) $h(0) = 0$ and $t \longmapsto h(t)$ is increasing,

(iii) $\int_0^t I_{\{0\}}(g(s))dh(s) = h(t)$,
i.e., $h(t)$ increases only on the set of t when $g(t) = 0$.

Proof. Set

$$g(t) = x + f(t) - \min_{0 \le s \le t} \{(x + f(s)) \wedge 0\}, \tag{4.9}$$

$$h(t) = - \min_{0 \le s \le t} \{(x + f(s)) \wedge 0\}. \tag{4.10}$$

Then it is easy to verify that $g(t)$ and $h(t)$ satisfy the above conditions (i), (ii) and (iii). We shall prove the uniqueness. Suppose $\tilde{g}(t)$ and $\tilde{h}(t) \in C^+$ also satisfy the conditions (i), (ii) and (iii). Then

$$g(t) - \tilde{g}(t) = h(t) - \tilde{h}(t) \qquad \text{for all} \quad t \ge 0.$$

If there exists $t_1 > 0$ such that $g(t_1) - \tilde{g}(t_1) > 0$, we set $t_2 = \max\{t < t_1; g(t) - \tilde{g}(t) = 0\}$. Then $g(t) > \tilde{g}(t) \ge 0$ for all $t \in (t_2, t_1]$ and hence, by (iii), $h(t_1) - h(t_2) = 0$. Since $\tilde{h}(t)$ is increasing, we have

$$0 < g(t_1) - \tilde{g}(t_1) = h(t_1) - \tilde{h}(t_1) \le h(t_2) - \tilde{h}(t_2) = g(t_2) - \tilde{g}(t_2) = 0.$$

This is a contradiction. Therefore $g(t) \le \tilde{g}(t)$ for all $t \ge 0$. By symmetry, $g(t) \ge \tilde{g}(t)$ for all $t \ge 0$. Hence $g(t) \equiv \tilde{g}(t)$ and so $h(t) \equiv \tilde{h}(t)$.

The mappings $(x, f) \longmapsto g$ and $(x, f) \longmapsto h$ given by (4.9) and (4.10) are denoted by $g = \Gamma_1(x, f)$ and $h = \Gamma_2(x, f)$ respectively.

Theorem 4.2. Let $\{X(t), B(t), \phi(t)\}$ be a system of real continuous stochastic processes defined on a probability space such that $B(t)$ is a one-dimensional Brownian motion with $B(0) = 0$, $X(0)$ and the process $\{B(t)\}$ are independent and with probability one the following holds:

(i) $X(t) \ge 0$ for all $t \ge 0$ and $\phi(t)$ is increasing with $\phi(0) = 0$ such that

$$\int_0^t I_{\{0\}}(X(s))d\phi(s) = \phi(t);$$

(ii)

$$X(t) = X(0) + B(t) + \phi(t). \tag{4.11}$$

Then $X = (X(t))$ is a reflecting Brownian motion on $[0, \infty)$.

Equation (4.11) is called the *Skorohod equation*.

Proof. By Lemma 4.2, $X = (X(t))$ and $\phi = (\phi(t))$ are uniquely determined by $X(0)$ and $B = (B(t))$: $X = \Gamma_1(X(0), B)$ and $\phi = \Gamma_2(X(0), B)$. In order to prove the theorem we have only to show that if x_t is a one-dimensional Brownian motion, then $X(t) = |x_t|$ satisfies, with some processes $B(t)$ and $\phi(t)$, the above properties. Let $g_n(x)$ be a non-negative continuous function on $\boldsymbol{R}^1$ with support in $(0,1/n)$ such that $\int_0^\infty g_n(x)dx = 1$. Set

$$u_n(x) = \int_0^{|x|} dy \int_0^y g_n(z)dz.$$

Then it is easy to see that $u_n \in \boldsymbol{C}^2(\boldsymbol{R}^1)$, $|u_n'| \leq 1$, $u_n(x) \uparrow |x|$ and $u_n'(x) \longrightarrow \operatorname{sgn} x$ as $n \longrightarrow \infty$.* By Itô's formula,

$$\begin{aligned} u_n(x_t) - u_n(x_0) &= \int_0^t u_n'(x_s)dx_s + \frac{1}{2}\int_0^t u_n''(x_s)ds \\ &= \int_0^t u_n'(x_s)dx_s + \int_{-\infty}^0 g_n(-y)\phi(t, y)dy + \int_0^\infty g_n(y)\phi(t,y)dy, \end{aligned}$$

where $\phi(t, y)$ is the local time of x_t. Letting $n \longrightarrow \infty$, we have

$$X(t) - X(0) = \int_0^t \operatorname{sgn}(x_s)dx_s + 2\phi(t, 0).$$

Set

$$B(t) = \int_0^t \operatorname{sgn}(x_s)dx_s \quad \text{and} \quad \phi(t) = 2\phi(t, 0).$$

Then, since $\langle B \rangle_t = t$, $B(t)$ is an $(\mathscr{F}_t)$-Brownian motion, where $(\mathscr{F}_t) = (\mathscr{F}_t^x)$ is the proper reference family for x_t. Since $X(0)$ is $\mathscr{F}_0$-measurable, $X(0)$ and $\{B(t)\}$ are independent. Since

$$\phi(t) = \lim_{\varepsilon \downarrow 0} \frac{1}{2\varepsilon}\int_0^t I_{[0,\varepsilon)}(X(s))ds,$$

it is clear that

---

* $$\operatorname{sgn} x = \begin{cases} 1\ , & x > 0, \\ 0\ , & x = 0, \\ -1\ , & x < 0. \end{cases}$$

$$\int_0^t I_{(0)}(X(s))d\phi(s) = \phi(t).$$

Therefore $\{X(t), B(t), \phi(t)\}$ satisfies all conditions in Theorem 4.2. Thus $X = (X(t))$ and $\phi = (\phi(t))$ are characterized as $X = \Gamma_1(X(0), B)$ and $\phi = \Gamma_2(X(0), B)$.

An immediate corollary is the following result due to Lévy.

Corollary. Let $B(t)$ be a one-dimensional Brownian motion such that $B(0) = 0$. Then

(i) the processes $\{|B(t)|\}$ and $\{B(t) - \min_{0 \le s \le t} B(s)\}$ are equivalent in law,

(ii)

$$\lim_{\varepsilon \downarrow 0} \frac{1}{2\varepsilon} \int_0^t I_{[0,\varepsilon)}(B(s) - \min_{0 \le u \le s} B(u))ds = -\min_{0 \le s \le t} B(s).$$

We can give yet another description of the reflecting Brownian motion. Let $x(t)$ be a one-dimensional Brownian motion. Then by (4.1),

$$x(t)^+ - x(0)^+ = \int_0^t I_{(0,\infty)}(x(s))dx(s) + \phi(t, 0).$$

$M(t) = \int_0^t I_{(0,\infty)}(x(s))dx(s)$ is a continuous martingale such that $\langle M \rangle(t) = \int_0^t I_{(0,\infty)}(x(s))ds$. It is easy to see that $\lim_{t \uparrow \infty} \langle M \rangle(t) = \infty$ a.s. Indeed, if we set $\sigma_1 = \min\{t;\ x(t) = 0\}$, $\tau_1 = \min\{t > \sigma_1;\ x(t) = -1\}, \ldots, \sigma_n = \min\{t > \tau_{n-1};\ x(t) = 0\}$, $\tau_n = \min\{t > \sigma_n;\ x(t) = -1\}, \ldots$ and

$$\xi_n = \int_{\sigma_n}^{\tau_n} I_{(0,\infty)}(x(s))ds,$$

then by the strong Markov property of $x(t)$ (Theorem II-6.4), it is easy to see that $\{\xi_n\}$ is independent and identically distributed. By the strong law of large numbers, $\xi_1 + \xi_2 + \cdots + \xi_n \longrightarrow \infty$ a.s. This implies that

$$\lim_{t \uparrow \infty} \int_0^t I_{(0,\infty)}(x(s))ds = \infty \qquad \text{a.s.}$$

Set

$$\tau_t = \inf\{u;\ \int_0^u I_{(0,\infty)}(x(s))ds > t\}.$$

By Theorem II-7.2, $M(\tau_t)$ is a one-dimensional Brownian motion. Also it is easy to see that $X(t) = x(\tau_t)$ is continuous and $X(t) \geq 0$ for all $t \geq 0$, a.s. Therefore

$$\phi(t) := \phi(\tau_t, 0) = X(t) - X(0) - M(\tau_t)$$

is continuous in t and satisfies

$$\int_0^t I_{\{0\}}(X(s))d\phi(s) = \phi(t), \qquad \text{a.s.}$$

Hence $\{X(t) = x(\tau_t),\ B(t) = M(\tau_t),\ \phi(t) = \phi(\tau_t, 0)\}$ is a system satisfying the conditions of Theorem 4.2. Thus we have the following result.

Theorem 4.3. Let $x(t)$ be a one-dimensional Brownian motion and $\tau_t = \inf\{u;\ \int_0^u I_{[0,\infty)}(x(s))ds > t\}$. Set $X(t) = x(\tau_t)$. Then $X(t)$ is a reflecting Brownian motion.

If $x(t)^- = (-x(t)) \vee 0$, then we have similarly

$$x(t)^- - x(0)^- = -\int_0^t I_{(-\infty,0)}(x(s))dx(s) + \phi(t, 0).$$

Let $\eta_t = \inf\{u;\ \int_0^u I_{(-\infty,0)}(x(s))ds > t\}$, $N_t = -\int_0^t I_{(-\infty,0)}(x(s))dx(s)$, $\tilde{B}(t) = N(\eta_t)$ and $Y(t) = -x(\eta_t)$. Then $Y(t)$ is also a reflecting Brownian motion. As we saw above, $X = \Gamma_1(X(0), B)$ and $Y = \Gamma_1(Y(0), \tilde{B})$. Since $\langle M, N\rangle = 0$, B and $\tilde{B}$ are independent by Theorem II-7.3. Therefore, if $X(0)$ and $Y(0)$ are independent (this is the case if $x(0) = x$ a.s. for some $x \in \boldsymbol{R}^1$ or $x(0) \geq 0$ a.s.), then the processes X and Y are independent. This fact indicates roughly that the motion of $x(t)$ on the positive half line $(0, \infty)$ and that on the negative half line $(-\infty, 0)$ are independent. This is seen more clearly by studying the excursions of Brownian motion.

4.3. Excursions of Brownian motion. Let $X = (X(t))$ be a one-dimensional Brownian motion and let $\mathscr{Z} = \{t;\ X(t) = 0\}$. It is well known that, with probability one, $\mathscr{Z}$ is a perfect set of Lebesgue measure 0 and $[0, \infty)\setminus\mathscr{Z} = \bigcup_\alpha e_\alpha$ is a countable union of disjoint open intervals e_α ([73] and [101]). Each interval e_α is called an *excursion interval*; the part of the process $\{X(t),\ t \in e_\alpha\}$ is called an *excursion* of $X(t)$ in $\boldsymbol{R}^1\setminus\{0\}$. In order to study the fine structures of Brownian sample paths, it is sometimes necessary to decompose them into excursions. Here we prefer to proceed in an equivalent but opposite direction: we start with the collection of all excursions and then construct Brownian sample functions.

First of all, we will formulate and construct the collection of all excursions of Brownian motion as a Poisson point process taking values in a function space. This idea is due to Itô [70]. Let $\mathscr{W}^+$ ($\mathscr{W}^-$) be the totality of all continuous functions w: $[0, \infty) \longrightarrow \boldsymbol{R}$ such that $w(0) = 0$ and there exists $\sigma(w) > 0$ such that if $0 < t < \sigma(w)$, then $w(t) > 0$ (resp. $w(t) < 0$), and if $t \geq \sigma(w)$, then $w(t) = 0$. Let $\mathscr{B}(\mathscr{W}^+)$ and $\mathscr{B}(\mathscr{W}^-)$ be the σ-fields on $\mathscr{W}^+$ and $\mathscr{W}^-$ respectively which are generated by Borel cylinder sets. The spaces $\mathscr{W}^+$ and $\mathscr{W}^-$ are called spaces of positive and negative excursions respectively. There exist σ-finite measures n^+ and n^- on $(\mathscr{W}^+, \mathscr{B}(\mathscr{W}^+))$ and on $(\mathscr{W}^-, \mathscr{B}(\mathscr{W}^-))$ respectively such that

$$\begin{aligned} &n^{\pm}(\{w;\, w(t_1) \in A_1,\, w(t_2) \in A_2, \ldots, w(t_n) \in A_n\}) \\ (4.12) \qquad &= \int_{A_1} K^{\pm}(t_1, x_1)dx_1 \int_{A_2} p^0(t_2 - t_1, x_1, x_2)dx_2 \int_{A_3} \cdots \\ &\qquad\qquad \times \int_{A_n} p^0(t_n - t_{n-1}, x_{n-1}, x_n)dx_n \end{aligned}$$

where $0 < t_1 < t_2 < \cdots < t_n$ and $A_i \in \mathscr{B}((0,\infty))$ (resp. $\mathscr{B}((-\infty, 0))$, $i = 1, 2, \ldots, n$,

$$K^+(t, x) = \sqrt{\frac{2}{\pi t^3}}\, x \exp\left(-\frac{x^2}{2t}\right), \qquad t > 0, \quad x \in [0, \infty),$$

$$K^-(t, x) = \sqrt{\frac{2}{\pi t^3}}\,(-x) \exp\left(-\frac{x^2}{2t}\right), \qquad t > 0, \quad x \in (-\infty, 0]$$

and

$$p^0(t, x, y) = \frac{1}{\sqrt{2\pi t}}\left(\exp\left(-\frac{(x-y)^2}{2t}\right) - \exp\left(-\frac{(x+y)^2}{2t}\right)\right),$$
$$t > 0,\ x, y \in [0, \infty) \quad \text{or} \quad x, y \in (-\infty, 0].$$

A way of constructing the measures n^+ and n^- is as follows. In Chapter IV, Example 8.4, we shall see that for every $T > 0$ there exists a probability measure P^T on $\mathscr{W}^+ \cap \{\sigma(w) = T\}$ such that

$$\begin{aligned} &P^T\{w;\, w(t_1) \in dx_1,\, w(t_2) \in dx_2, \ldots, w(t_n) \in dx_n\} \\ &\quad = h(0, 0; t_1, x_1)h(t_1, x_1; t_2, x_2) \cdots h(t_{n-1}, x_{n-1}; t_n, x_n)dx_1dx_2 \cdots dx_n \end{aligned}$$

where

$$h(s, a; t, b) = \begin{cases} \dfrac{K^+(T-t, b)}{K^+(T-s, a)} p^0(t-s, a, b), & 0 < s < t,\ a, b > 0, \\ \sqrt{\dfrac{\pi}{2} T^3}\, K^+(t, b) K^+(T-t, b), & s=0,\ t>0,\ a=0,\ b>0, \end{cases}$$

and $0 < t_1 < t_2 < \cdots < t_n < T$. The σ-finite measure n^+ on $\{\mathscr{W}^+, \mathscr{B}(\mathscr{W}^+)\}$ defined by

$$n^+(B) = \int_0^\infty P^T(B \cap \{\sigma(w) = T\}) \frac{dT}{\sqrt{2\pi T^3}}, \qquad B \in \mathscr{B}(\mathscr{W}^+)$$

satisfies (4.12). n^- can be constructed in a similar way. Let $\mathscr{W} = \mathscr{W}^+ \cup \mathscr{W}^-$ be the sum, $\mathscr{B}(\mathscr{W}) = \mathscr{B}(\mathscr{W}^+) \vee \mathscr{B}(\mathscr{W}^-)$ and n be the σ-finite measure on $\{\mathscr{W}, \mathscr{B}(\mathscr{W})\}$ such that $n|_{\mathscr{W}^\pm} = n^\pm$. By Theorem I-9.1, we can construct a stationary Poisson point process p on $\mathscr{W}$ with characteristic measure n. We call it the *Poisson point process of Brownian excursions*. This Poisson point process is what we intended to be the collection of excursions of Brownian motion. Suppose we are given a Poisson point process of Brownian excursions p on a probability space $(\Omega, \mathscr{F}, P)$. A Brownian motion $X(t)$ is constructed from p by the following steps. Set

$$A(t) = \sum_{s \le t, s \in D_p} \sigma(p(s)) = \int_0^{t+} \int_{\mathscr{W}} \sigma(w) N_p(ds dw)$$

where D_p is the domain of p and $N_p(dsdw)$ is the counting measure of p defined by (9.1) in Chapter I, Section 9. With probability one, $t \longrightarrow A(t)$ is strictly increasing, right continuous and

$$\lim_{t \uparrow \infty} A(t) = \infty.$$

Indeed, it is a time homogeneous Lévy process with increasing paths (cf. Example. II-4.1) with

$$E[\exp(-\lambda A(t))] = \exp(-t\psi(\lambda))$$

where

$$\begin{aligned} \psi(\lambda) &= \int_0^\infty (1 - e^{-\lambda u})\, n(\{w ; \sigma(w) \in du\}) \\ &= \int_0^\infty (1 - e^{-\lambda u}) \frac{2du}{\sqrt{2\pi u^3}} = 2\sqrt{2\lambda}, \end{aligned}$$

i. e., it is a one-sided stable process with exponent 1/2. Let $\phi(t) = A^{-1}(t)$

be the inverse function of $A(t)$. Then, almost surely, $\phi(t)$ is continuous. We now define, for any sample point on which the above mentioned properties of $A(t)$ are satisfied, a function: $[0, \infty] \ni t \longrightarrow X(t) \in R^1$ as follows: For each $t \geq 0$, set $s = \phi(t)$. If $A(s-) < A(s)$, then $s \in D_p$ and we set $X(t) = p(s)(t - A(s-))$. If $A(s-) = A(s)$, we set $X(t) = 0$. *We claim that with probability one, $t \longrightarrow X(t)$ is continuous. Furthermore, we can identify $X(t)$ with a one-dimensional Brownian motion starting at 0 and $\phi(t)$ with the local time at 0:*

$$\phi(t) = \lim_{\varepsilon \downarrow 0} \frac{1}{4\varepsilon} \int_0^t I_{(-\varepsilon, \varepsilon)}(X(s))ds.$$

*Proof**. First, we establish the almost-sure continuity of $t \longrightarrow X(t)$. We have, by (4.21) below,

$$n(\{w;\ \max_{0 \leq t \leq \sigma(w)} |w(t)| \geq \varepsilon\}) = \frac{2}{\varepsilon}, \qquad \varepsilon > 0.$$

Since

$$E[\# \{s \in D_p\ ;\ s \leq T,\ \max_{0 \leq t \leq \sigma[p(s)]} |p(s)[t]| > \varepsilon\}]$$

$$= T \times n(\{w;\ \max_{0 \leq t \leq \sigma(w)} |w(t)| > \varepsilon\}) = \frac{2T}{\varepsilon} < \infty$$

for every $T > 0$ and $\varepsilon > 0$, it holds with probability one that

$$\# \{s \in D_p;\ s \leq T,\ \max_{0 \leq t \leq \sigma[p(s)]} p(s)[t] > \varepsilon\} < \infty$$

for every $T > 0$ and $\varepsilon > 0$. This implies, in particular, that with probability one, for any sequence $s_n \in D_p$ converging to a finite time point t_0 such that $s_n \neq t_0$ for every n,

$$\max_{0 \leq t \leq \sigma[p(s_n)]} p(s_n)[t] \longrightarrow 0 \qquad \text{as } n \longrightarrow \infty.$$

Now the continuity of $t \longrightarrow X(t)$ is easily concluded: It is obvious that $t \longrightarrow X(t)$ is continuous on each interval $(A(s-), A(s))$, $s \in D_p$. If $A(s-) = t = A(s)$, then $X(t) = 0$ and, for intervals $(A(s_n-), A(s_n))$, $s_n \in D_p$, converging to t as $n \longrightarrow \infty$,

$$\max_{t \in (A(s_n-),\ A(s_n))} |X(t)| \longrightarrow 0 \qquad \text{as } n \longrightarrow \infty$$

by the above remark.

* cf.[234] for construction of Brownian motions in more general cases than that treated here.

To identify $X(t)$ with a Brownian motion, we appeal to Theorem II–6.1. We may assume that p is given as an $(\mathscr{F}_t)$-Poisson point process with respect to some reference family $(\mathscr{F}_t)$. Then $A(t)$ is $(\mathscr{F}_t)$-adapted and hence $\phi(t)$ is an $(\mathscr{F}_t)$-stopping time for each t. Let $\mathscr{F}_t = \mathscr{F}_{\phi(t)}$ and $\mathscr{F}_{t-} = \mathscr{F}_{\phi(t)-}$ be the usual σ-fields; in particular $\mathscr{F}_{\phi(t)-}$ is the σ-field generated by sets of the form $A \cap \{s < \phi(t)\}$, $A \in \mathscr{F}_s$, $s \in [0, \infty)$. For each fixed $t > 0$, set $F(s, w, \omega) = w(t - A(s-)) I_{(t \geq A(s-))}$. It is $(\mathscr{F}_s)$-predictable and belongs to $F_p^1 \cap F_p^2$. Indeed

$$\begin{aligned}
&\int_0^u \int_{\mathscr{V}} |w(t - A(s)) I_{(t \geq A(s))}| \hat{N}_p(dsdw) \\
&\quad = \int_0^u ds \int_{\mathscr{V}+} [w(t - A(s)) I_{(t \geq A(s))}] n^+(dw) \\
&\qquad + \int_0^u ds \int_{\mathscr{V}-} [-w(t - A(s)) I_{(t \geq A(s))}] n^-(dw) \\
&\quad = 2 \int_0^u ds I_{(t \geq A(s))} \int_{(0,\infty)} K^+(t - A(s), x) x dx \\
&\quad = 2 \int_0^u I_{(t \geq A(s))} ds
\end{aligned}$$

where $\hat{N}_p(dsdw)$ is the compensator of the point process p. Also,

$$\begin{aligned}
&\int_0^u \int_{\mathscr{V}} |w(t - A(s)) I_{(t \geq A(s))}|^2 \hat{N}_p(dsdw) \\
&\quad = 2 \int_0^u ds\, I_{(t \geq A(s))} \int_{(0,\infty)} K^+(t - A(s), x) x^2 dx \\
&\quad = \frac{8}{\sqrt{2\pi}} \int_0^u (t - A(s))^{1/2} I_{(t \geq A(s))} ds.
\end{aligned}$$

Set $\tilde{N}_p(dsdw) = N_p(dsdw) - \hat{N}_p(dsdw)$. Then, clearly the $X(t)$ defined above is expressed, for each fixed t, as

$$\begin{aligned}
X(t) &= \int_0^{\phi(t)+} \int_{\mathscr{V}} F_t(s, w, \cdot) N_p(dsdw) \\
&= \int_0^{\phi(t)+} \int_{\mathscr{V}} F_t(s, w, \cdot) \tilde{N}_p(dsdw)
\end{aligned}$$

since

$$\int_0^{\phi(t)+} \int_{\mathscr{V}} F_t(s, w, \cdot) \hat{N}_p(dsdw) = 0.$$

Let $\mathscr{H}_t = \mathscr{F}_{t-} \vee \sigma[p(\phi(t))(u - A(\phi(t)-));\ u \leq t] \subset \mathscr{F}_t$.*1 We will show that $X(t)$ is an $(\mathscr{H}_t)$-martingale such that $\langle X \rangle(t) = t$. Set $H(\omega) = H_1(\omega)H_2(\omega)$ where $H_1(\omega)$ is bounded and $\mathscr{F}_{s-}$-measurable and $H_2(\omega) = G(p(\phi(s))^-_{s-A(\phi(s)-)})$. Here $G(w)$ is a bounded $\mathscr{B}(\mathscr{W}^*)$-measurable function on $\mathscr{W}^*$ *2 and for $w \in \mathscr{W}^*$ and $s > 0$, $w_s^- \in \mathscr{W}^*$ is defined by $w_s^-(t) = w(t \wedge s)$. It is sufficient to prove that

$$E(X(t)H(\omega)) = E(X(s)H(\omega)) \tag{4.13}$$

and

$$E([X(t)^2 - t]H(\omega)) = E([X(s)^2 - s]H(\omega)). \tag{4.14}$$

(4.13) is proved as follows.

$$\begin{aligned}
&E(X(t)H(\omega))\\
&= E[\int_0^{\phi(t)+}\int_{\mathscr{V}} F_t(u, w, \omega)\tilde{N}_p(dudw)H(\omega)]\\
&= E[\int_0^{\phi(s)+}\int_{\mathscr{V}} F_t(u, w, \omega)\tilde{N}_p(dudw)H(\omega)]\\
&= E[\sum_{\tau \leq \phi(s),\ \tau \in D_p} F_t(\tau, p(\tau), \omega)H(\omega)]\\
&= E[\sum_{\tau < \phi(s),\ \tau \in D_p} F_t(\tau, p(\tau), \omega)H(\omega)] + E[F_t(\phi(s), p(\phi(s)), \omega)H(\omega)]\\
&:= I_1 + I_2.
\end{aligned}$$

Then $I_1 = 0$ because $F_t(\tau, p(\tau), \omega) = 0$ if $\tau < \phi(s) \leq \phi(t)$. It is known (cf. [15]) that there exists a bounded $(\mathscr{F}_t)$-predictable process $H_u^{(1)}(\omega)$ such that $H_1(\omega) = H_{\phi(s)}^{(1)}(\omega)$. Then

$$\begin{aligned}
I_2 &= E(F_t(\phi(s), p(\phi(s)), \omega)H_1(\omega)H_2(\omega))\\
&= E(\sum_{\tau \leq \phi(s),\ \tau \in D_p} I_{\{s-A(\tau-)<\sigma(p(\tau))\}} F_t(\tau, p(\tau), \omega)H_\tau^{(1)}(\omega)G(p(\tau)^-_{s-A(\tau-)}))\\
&= E(\int_0^{\phi(s)+}\int_{\mathscr{V}} I_{\{s-A(u-)<\sigma(w)\}} F_t(u, w, \omega)H_u^{(1)}(\omega)G(w^-_{s-A(u-)})N_p(dudw))\\
&= E\{\int_0^{\phi(s)} duH_u^{(1)}(\omega)[\int_{\mathscr{V}} I_{\{s-A(u)<\sigma(w)\}}F_t(u, w, \omega)G(w^-_{s-A(u)})n(dw)]\}\\
&= E\{\int_0^{\phi(s)} duH_u^{(1)}(\omega)[\int_{\mathscr{V}} I_{\{s-A(u)<\sigma(w)\}}F_s(u, w, \omega)G(w^-_{s-A(u)})n(dw)]\}.
\end{aligned}$$

---

*1 $p(s) \in \mathscr{W}$, $s \in D_p$, is extended by setting $p(s)(\cdot) \equiv 0$ if $s \notin D_p$.

*2 $\mathscr{W}^* = \mathscr{W} \cup \{\mathbf{0}\}$, where $\mathbf{0}$ is the function $\mathbf{0}(t) \equiv 0$.

Here we used (4.15) below. The following are the fundamental properties of the measure n: if $t > s > 0$ and $g(w)$ is bounded and $\mathscr{B}_s(\mathscr{W})$-measurable* then

$$\int_{\mathscr{V}} w(t)g(w)n(dw) = \int_{\mathscr{V}} w(s)g(w)n(dw) \tag{4.15}$$

and

$$\int_{\mathscr{V}} [w(t)^2 - t\wedge\sigma(w)]g(w)n(dw) = \int_{\mathscr{V}} [w(s)^2 - s\wedge\sigma(w)]g(w)n(dw). \tag{4.16}$$

Now the above argument can be reversed to see that

$$E\{\int_0^{\phi(s)} duH_u^{(1)}(\omega)[\int_{\mathscr{V}} I_{(s-A(u)<\sigma(w))}F_s(u, w, \cdot)G(w^-_{s-A(u)})n(dw)]\}$$
$$= E(X(s)H(\omega)).$$

Equation (4.14) is proved in a similar way by taking $F_t(s, w, \omega) = (w(t - A(s-))^2 - [(t - A(s-))\wedge\sigma(w)])I_{(t\geq A(s-))}$ and by using (4.16). In this case $F_t(\tau, p(\tau), \cdot) \neq 0$ for $\tau < \phi(s)$ but it is easy to see that $F_t(\tau, p(\tau), \cdot) = F_s(\tau, p(\tau), \cdot)$ if $\tau < \phi(s)$. Consequently $X(t)$ is an $(\mathscr{H}_t)$-Brownian motion by Theorem II-6.1.

Next we prove that $\phi(t)$ is the local time at the origin of $X(t)$. It is clear that

$$|X(t)| = \int_0^{\phi(t)+}\int_{\mathscr{V}} |w(t - A(s-))I_{(t>A(s-))}|N_p(dsdw).$$

Noting that $\int_{\mathscr{V}}|w(t)|n(dw) = 2\int_0^\infty xK^+(t,x)dx = 2$ for every $t > 0$, we see that

$$|X(t)| = \int_0^{\phi(t)+}\int_{\mathscr{V}} |w(t - A(s-))I_{(t>A(s-))}|\tilde{N}_p(dsdw) + 2\phi(t).$$

By the same argument as above, we can prove that

$$\int_0^{\phi(t)+}\int_{\mathscr{V}} |w(t - A(s-))I_{(t>A(s-))}|\tilde{N}_p(dsdw)$$

is an $(\mathscr{H}_t)$-martingale, and so by Theorem 4.2 we can conclude that

* $\mathscr{B}_s(\mathscr{W})$ is the σ-field on $\mathscr{W}$ generated by Borel cylinder sets up to time s.

$$2\phi(t) = \lim_{\varepsilon \downarrow 0} \frac{1}{2\varepsilon} \int_0^t I_{[0,\varepsilon)}(|X(s)|)ds.$$

The inverse $A(t)$ of $\phi(t)$ has expression

$$A(t) = \int_0^{t+} \int_{\mathscr{V}} \sigma(w) N_p(dsdw).$$

We know that it is a one-sided stable process with exponent 1/2.

Thus we have established the following formula describing a Brownian sample path $X(t)$ in terms of a Poisson point process of Brownian excursions p:

$$(4.17) \quad \begin{cases} X(t) = \int_0^{\phi(t+)} \int_{\mathscr{V}} w(t - A(s-)) N_p(dsdw) \\ A(t) = \int_0^{t+} \int_{\mathscr{V}} \sigma(w) N_p(dsdw) \\ \text{and } \phi(t) \text{ is the inverse of } t \longmapsto A(t). \end{cases}$$

This expression may be regarded as a formula for the decomposition of a Brownian path into its excursions. Using it, many results on the local time $\phi(t)$ and the zero set $\mathscr{Z}$ of $X(t)$ can be obtained. Before we proceed with some examples, we first introduce the following maps

$$(4.18) \quad T_1 \colon \mathscr{W} \longrightarrow (0, \infty) \text{ defined by } T_1 w = \sigma(w)$$

and

$$(4.19) \quad T_2 \colon \mathscr{W} \longrightarrow (0, \infty) \text{ defined by } T_2 w = \max_{0 \leq t \leq \sigma(w)} |w(t)|.$$

T_1 and T_2 induce stationary Poisson point processes $T_1(p)$ and $T_2(p)$ on $(0, \infty)$ by

$$\boldsymbol{D}_{T_i(p)} = \boldsymbol{D}_p \quad \text{and} \quad T_i(p)(s) = T_i(p(s)), \qquad s \in \boldsymbol{D}_{T_i(p)}, \quad i = 1, 2.$$

The characteristic measure n_1 of $T_1(p)$ is given by

$$(4.20) \quad n_1([x, \infty)) = 2n^+(\{w; \sigma(w) \geq x\}) = 2\int_x^\infty \frac{dt}{\sqrt{2\pi t^3}} = 2\sqrt{\frac{2}{\pi x}}$$

and the characteristic measure n_2 of $T_2(p)$ is given by

$$
\begin{aligned}
n_2([x, \infty)) &= 2n^+(\{w; \max_{0 \le t \le \sigma(w)} w(t) \ge x\}) \\
&= \lim_{\varepsilon \downarrow 0} 2n^+(\{w; \sigma(w) > \varepsilon, \max_{\varepsilon \le t \le \sigma(w)} w(t) \ge x\}) \\
&= \lim_{\varepsilon \downarrow 0} 2 \int_0^\infty K^+(\varepsilon, y) P_y(\sigma_x < \sigma_0) dy \\
&= \lim_{\varepsilon \downarrow 0} 2 \int_0^\infty \sqrt{\frac{2}{\pi \varepsilon^3}}\, y \exp\left(-\frac{y^2}{2\varepsilon}\right) \frac{y \wedge x}{x} dy = \frac{2}{x},
\end{aligned} \tag{4.21}
$$

where P_x is the Wiener measure starting at x and σ_a is the first hitting time to a.*[1]

For a Brownian path $X(s)$ and $\varepsilon > 0$, let

(4.22) $\eta_\varepsilon(t) =$ the number of excursion intervals in $[0, t)$ whose lengths are not less than ε,

and

(4.23) $d_\varepsilon(t) =$ the number of down-crossings of $X(s)$ from ε to 0 and up-crossings of $X(s)$ from $-\varepsilon$ to 0 before time t.

It is immediate from (4.17) that

$$\eta_\varepsilon(t) = N_{T_1(p)}((0, \phi(t)) \times [\varepsilon, \infty))$$

and

$$d_\varepsilon(t) = N_{T_2(p)}((0, \phi(t)) \times [\varepsilon, \infty)).$$

It follows from the strong law of large numbers that

$$P(\lim_{\varepsilon \downarrow 0} \sqrt{\frac{\pi \varepsilon}{2}}\, N_{T_1(p)}((0, a) \times [\varepsilon, \infty)) = 2a \quad \text{for all} \quad a > 0) = 1$$

and

$$P(\lim_{\varepsilon \downarrow 0} \varepsilon N_{T_2(p)}((0, a) \times [\varepsilon, \infty)) = 2a \quad \text{for all} \quad a > 0) = 1.$$

Consequently, we have proved the following results of Lévy *[2]:

*[1] The formula $P_a(\sigma_c < \sigma_b) = (b - a)/(b - c)$, $b < a < c$, is well-known and is easily derived from the fact that $w(t \wedge \sigma_c \wedge \sigma_b) - a$ is a P_a-martingale.

*[2] Lévy only conjectured (4.25); for different proofs, cf. e.g. Itô-McKean [73], Chung-Durret [11] and Williams [179].

$$(4.24)\qquad P\Big(\lim_{\varepsilon\downarrow 0}\sqrt{\frac{\pi\varepsilon}{2}}\eta_\varepsilon(t) = 2\phi(t)\quad\text{for all}\quad t > 0\Big) = 1$$

and

$$(4.25)\qquad P\big(\lim_{\varepsilon\downarrow 0}\varepsilon d_\varepsilon(t) = 2\phi(t)\quad\text{for all}\quad t \geq 0\big) = 1.$$

Let p^+ and p^- be the restriction of p on $\mathscr{W}^+$ and $\mathscr{W}^-$ respectively. Then p^+ and p^- are stationary Poisson point processes on $\mathscr{W}^+$ and $\mathscr{W}^-$ with the characteristic measures n^+ and n^- respectively; moreover, they are mutually independent. We set, respectively,

$$(4.26)\qquad \begin{aligned} A^{\pm}(t) &= \int_0^t\int_{\mathscr{W}^{\pm}}\sigma(w)N_{p^{\pm}}(dsdw) \qquad\text{and} \\ X^{\pm}(t) &= p^{\pm}(s)[t - A^{\pm}(s)] \end{aligned}$$

where $A^{\pm}(s-) \leqq t \leqq A^{\pm}(s)$ under the convention that $p^{\pm}(s) \equiv 0$ if $s \in \boldsymbol{D}_p{}^{\pm}$. The unique s such that $A^{\pm}(s-) \leqq t \leqq A^{\pm}(s)$ are denoted by $\phi^{\pm}(t)$, respectively, so that $\phi^{\pm}(t)$ are the inverse of $t \longrightarrow A^{\pm}(t)$. Also we set

$$\theta^+(t) = \int_0^t I_{[0,\infty)}(X(s))ds \quad\text{and}\quad \theta^-(t) = \int_0^t I_{(-\infty,0]}(X(s))ds$$

where X is the Brownian motion given by (4.17). It is immediately seen that

$$X^+(t) = X(\tau^+(t)) \quad\text{and}\quad X^-(t) = X(\tau^-(t))$$

where $\tau^+(t)$ and $\tau^-(t)$ are the inverse of $t \longrightarrow \theta^+(t)$ and $t \longrightarrow \theta^-(t)$ respectively. Thus X^+ and $-X^-$ are mutually independent reflecting Brownian motions as being functionals of p^+ and p^- respectively and therefore, we recover what was explained at the end of the previous section. If we set

$$(4.27)\qquad \begin{aligned} B^+(t) &= p^+(\phi^+(t)[t - A^+(\phi^+(t))] - \phi^+(t) \qquad\text{and} \\ B^-(t) &= -p^-(\phi^-(t))\,[t - A^-(\phi^-(t))] - \phi^-(t), \end{aligned}$$

then, by the same proof as for (4.17), we can show that B^+ and B^- are Brownian motions and that $X^+(t) = B^+(t) + \phi^+(t)$ and $-X^-(t) = B^-(t) + \phi^-(t)$ are Skorohod equations for X^+ and $-X^-$ respectively. In particular,

$$\phi^+(t) = \lim_{\varepsilon \downarrow 0} \frac{1}{2\varepsilon} \int_0^t I_{[0,\varepsilon)}(X^+(s))ds \quad \text{and}$$

$$\phi^-(t) = \lim_{\varepsilon \downarrow 0} \frac{1}{2\varepsilon} \int_0^t I_{(-\varepsilon,0]}(X^-(s))ds.$$

If $A(t)$ is defined by (4.17), it is immediately seen that

$$A^{\pm}(t) = \theta^{\pm}(A(t)).$$

Let $a < 0$ and set $\sigma_a = \inf\{t; X(t) = a\}$. Then the excursion $p(e) \in \mathscr{W}$ of X such that $A(e-) \leqq \sigma_a \leqq A(e)$ must coincide with $p^-(e') \in \mathscr{W}^-$ where

$$A^-(e'-) \leq \sigma_a^- = \inf\{t;\ X^-(t) = a\} \leq A^-(e').$$

Hence $e = e' = \phi^-(\sigma_a^-)$ and

$$e = \inf\{s \in D_p;\ M_-[p(s)] \leqq a\}$$

where

$$M_-(w) = \inf_{0 \leq t \leq \sigma(w)} w(t), \qquad w \in \mathscr{W}.$$

Then

$$\theta^+(\sigma_a) = \int_0^{A(e)} I_{[0,\infty)}(X(s))ds = A^+(e) = A^+(\phi^-(\sigma_a^-)).$$

e is exponentially distributed with mean $-a$ since

$$\begin{aligned} P[e > u] &= P[N_p((0,u] \times \{w;\ M_-(w) \leq a\}) = 0] \\ &= \exp[-u\, n(\{w;\ M_-(w) \leq a\})] \\ &= \exp[-u/a] \end{aligned}$$

by (4.21). Note that $e = \phi^-(\sigma_a^-)$ and X^+, as being functionals of p^- and p^+ respectively, are independent. Hence we can conclude the following: *the part* $[X^+(t) = X(\tau^+(t));\ \tau^+(t) < \sigma_a]$ *of the reflecting Brownian motion* X^+ *coincides with* $[X^+(t);\ 0 \leqq t < A^+(e) = \inf\{u;\ \phi^+(u) > e\}]$ *where* e *is an exponential holding time independent of the process* X^+ *with mean* $-a$. The process defined in the second way is known as the elastic barrier Brownian motion with parameter $\gamma = -a/(1-a)$, (cf. Example IV-5.5) and the above shows that it can also be constructed from a Brownian motion $X(t)$ by a time change as in the first way.

Let us give one more application of the above consideration. Let $t > 0$ be fixed and set $u = \tau^+(t)$. Then with probability one, $X(u) = X^+(t) > 0$ and $X(u)$ is contained in the excursion $p[\phi(u)] = p^+[\phi^+(t)] \in$

$\mathscr{W}^+$. Hence $\phi(u) = \phi^+(t)$. Now it is obvious that

$$t - A^+(\phi^+(t)-) = u - A(\phi(u)-) = u - A(\phi^+(t)-)$$

and hence

$$\begin{aligned} u &= t - A^+(\phi^+(t)-) + A(\phi^+(t)-) \\ &= t + A^-(\phi^+(t)-) \\ &= t + A^-(\phi^+(t)). \end{aligned}$$

Noting the independence of X^+ and X^- and the fact that $A^-(t)$ is a one-sided stable process with $E[\exp(-\lambda A^-(t))] = \exp[-t\psi(\lambda)]$ where

$$\psi(\lambda) = \int_0^\infty (1 - e^{-\lambda u}) n^-(\{w;\ \sigma(w) \in du\}) = \sqrt{2\lambda},$$

we obtain the following formula due to Williams [235], (cf. McKean [114]):

$$\begin{aligned} &E[\exp(-\lambda\tau^+(t)) \mid X^+(s);\ 0 \leqq s < \infty] \\ &\quad = \exp[-\lambda t - \sqrt{2\lambda}\phi^+(t)]. \end{aligned}$$

This formula can be used to prove the *arcsine law:*

$$(4.28) \qquad P[\theta^+(T) < t] = \frac{2}{\pi}\sin^{-1}\sqrt{\frac{t}{T}}, \qquad 0 \leqq t \leqq T.$$

Indeed

$$E[\exp(-\lambda\tau^+(t))] = E[\exp(-\lambda t - \sqrt{2\lambda}\phi^+(t))]$$

and, noting $\phi^+(t)$ and $\max_{0 \leqq s \leqq t} X(s)$ are equivalent in law (Corollary of Theorem 4.2), this is equal to

$$\begin{aligned} &2e^{-\lambda t}\int_0^\infty e^{-\sqrt{2\lambda}x}(2\pi t)^{-1/2}\exp\left[-\frac{x^2}{2t}\right]dx \\ &= 2e^{-\lambda t}(2\pi t)^{-1/2}\int_0^\infty\left\{\int_0^\infty e^{-\lambda u}(2\pi u^3)^{-1/2}x\exp\left(-\frac{x^2}{2u}\right)du\right\} \\ &\quad \times \exp\left(-\frac{x^2}{2t}\right)dx \\ &= \int_t^\infty \{t(u-t)^3\}^{-1/2}(\pi u)^{-1}e^{-\lambda u}t(u-t)du \end{aligned}$$

from which (4.28) is easily obtained.

In the remainder of this section, we shall present, within our framework of Brownian excursions, several results of Williams on decompositions of Brownian paths, especially on a beautiful description of the Brownian excursion law n or $n^+ = n|_{\mathscr{W}^+}$.

For $w \in \mathscr{W}^+$, we define $\check{w} \in \mathscr{W}^+$ by

$$\check{w}(t) = \begin{cases} w(\sigma(w) - t) & \text{for } 0 \leq t \leq \sigma(w) \\ 0 & \text{for } t \geq \sigma(w). \end{cases}$$

Clearly $\sigma(\check{w}) = \sigma(w)$.

Lemma 4.3. n is invariant under the mapping: $w \longrightarrow \check{w}$.

Proof. P^T on $\mathscr{W}^+ \cap \{w;\ \sigma(w) = T\}$ is invariant under the mapping: $w \longrightarrow \check{w}$ (cf. Example IV-8.4). Since

$$n^+(B) = \int_0^\infty P^T(B \cap \{\sigma(w) = T\})\,(2\pi T^3)^{-\frac{1}{2}}dT, \qquad B \in B(\mathscr{W}^+),$$

the assertion is obvious.

For a continuous path $w \in \boldsymbol{C}([0, \infty) \longrightarrow \boldsymbol{R})$ and $t \geqq 0$, define w_t^+ and w_t^- in $\boldsymbol{C}([0, \infty) \longrightarrow \boldsymbol{R})$ by

$$w_t^+(u) = w(t + u) \quad \text{and} \quad w_t^-(u) = w(t \wedge u).$$

Define $\sigma_a(w)$, $a \in R$, by $\sigma_a(w) = \inf\{t;\ w(t) = a\}$. In the following, a Brownian motion $X(t)$ with $X(0) = a$ is denoted by BM^a and its probability law on $\boldsymbol{C}([0, \infty) \longrightarrow \boldsymbol{R})$ is denoted by P_a. A Bessel diffusion process $Y(t)$ with index 3 (*i.e.*, the radial process of a three-dimensional Brownian motion, cf. Example IV–8.3) such that $Y(0) = a$ ($a \geqq 0$) is denoted by $BES^a(3)$ and its probability law on $\boldsymbol{C}([0, \infty) \longrightarrow \boldsymbol{R}_+)$ is denoted by Q_a where $\boldsymbol{R}_+ = [0, \infty)$. $BES^a(3)$, $a \geqq 0$, is a diffusion process on $[0, \infty)$ with transition probability $q(t, x, y)dy$ where

$$\text{(4.29)} \qquad q(t, x, y) = \begin{cases} \dfrac{1}{x}p^0(t, x, y)y & \text{for } t, x, y > 0 \\ K^+(t, y)y & \text{for } t, y > 0, x = 0 \end{cases}$$

with $p^0(t, x, y)$ and $K^+(t, y)$ given as above.

Lemma 4.4. (i) If $a > 0$,

$$\text{(4.30)} \qquad \begin{aligned} &n^+(\{w;\ w^-_{\sigma_a(w)} \in B_1,\ w^+_{\sigma_a(w)} \in B_2\} \,|\, \sigma_a < \infty) \\ &\quad = Q_0[w^-_{\sigma_a} \in B_1]P_a[w^-_{\sigma_0} \in B_2] \end{aligned}$$

for any Borel subsets B_1, B_2 in $C([0, \infty) \longrightarrow \boldsymbol{R})$.

(ii) If $0 < b < a$,

$$(4.31)\quad \begin{aligned} &P_b[\{w;\ w^-_{\sigma_a(w)} \in B_1,\ w^+_{\sigma_a(w)} \in B_2\} \mid \sigma_a < \sigma_0] \\ &\quad = Q_b[w^-_{\sigma_a} \in B_1]P_a[w \in B_2] \end{aligned}$$

for any Borel subsets B_1, B_2 in $C([0, \infty) \longrightarrow \boldsymbol{R})$.

Proof. (i) We know $n^+(\{w;\ \sigma_a(w) < \infty\}) = 1/a$. Also by (4.12), n^+ has the following property: for any Borel subsets B_1, B_2 in $C([0, \infty) \longrightarrow \boldsymbol{R}_+)$

$$\begin{aligned} &n^+(\{w_t^- \in B_1,\ \sigma(w) > t,\ w_t^+ \in B_2\}) \\ &\quad = E^{n^+}[E^{P_{w(t)}}[w^-_{\sigma_0(w)} \in B_2]\colon\ w_t^- \in B_1,\ \sigma(w) > t] \end{aligned}$$

where E^{n+} (E^{P_a}) denotes the integration by n^+ (resp. P_a). Hence if f_1, $f_2, \ldots, f_m$ and $g_1, g_2, \ldots, g_n$ are bounded Borel functions on $\boldsymbol{R}_+$, $0 < t_1 < t_2 < \ldots < t_m$ and $0 < s_1 < s_2 < \ldots < s_n$,

$$\begin{aligned} &E^{n^+}[f_1(w(t_1))f_2(w(t_2)) \ldots f_m(w(t_m))I_{\{t_m<\sigma_a\}} \\ &\qquad \times g_1(w(\sigma_a + s_1))g_2(w(\sigma_a + s_2)) \ldots g_n(w(\sigma_a + s_n))I_{\{\sigma_a+s_n<\sigma\}}] \\ &\quad = E^{n^+}[f_1(w(t_1))f_2(w(t_2)) \ldots f_m(w(t_m))\, I_{\{\sigma\wedge\sigma_a>t_m\}} \\ &\qquad \times E^{P_{w(t_m)}}[g_1(w(\sigma_a + s_1))g_2(w(\sigma_a + s_2)) \ldots \\ &\qquad\quad g_n(w(\sigma_a + s_n))I_{\{\sigma_a+s_n<\sigma_0\}}]] \\ &\quad = E^{n^+}\{f_1(w(t_1))f_2(w(t_2)) \ldots f_m(w(t_m))I_{\{\sigma\wedge\sigma_a>t_m\}} \\ &\qquad \times E^{P_{w(t_m)}}[E^{P_a}[g_1(w(s_1))g_2(w(s_2)) \ldots g_n(w(s_n))I_{\{s_n<\sigma_0\}}]I_{\{\sigma_a<\sigma_0\}}]\} \\ &\quad = E^{n^+}[f_1(w(t_1))f_2(w(t_2)) \ldots f_m(w(t_m))I_{\{\sigma\wedge\sigma_a>t_m\}}P_{w(t_m)}[\sigma_a < \sigma_0]] \\ &\qquad \times E^{P_a}[g_1(w(s_1))g_2(w(s_2)) \ldots g_n(w(s_n))I_{\{s_n<\sigma_0\}}] \\ &\quad = E^{n^+}[f_1(w(t_1))f_2(w(t_2)) \ldots f_m(w(t_m))I_{\{\sigma_a>t_m\}}w(t_m)] \\ &\qquad \times E^{P_a}[g_1(w(s_1))g_2(w(s_2)) \ldots g_n(w(s_n))]I_{\{s_n<\sigma_0\}} / a \end{aligned}$$

and by using (4.32) below, this is equal to

$$\begin{aligned} &E^{Q_0}[f_1(w(t_1))f_2(w(t_2)) \ldots f_m(w(t_m))I_{\{\sigma_a>t_m\}}] \\ &\qquad \times E^{P_a}[g_1(w(s_1))g_2(w(s_2)) \ldots g_n(w(s_n))I_{\{\sigma_0>s_n\}}] / a. \end{aligned}$$

Then the proof of (i) will be completed. So we prove the following: for any Borel subset B in $C([0, \infty) \longrightarrow \boldsymbol{R}_+)$,

$$(4.32)\quad E^{n^+}[w(t)I_{\{w_t^-\in B\}}] = Q_0\,[w_t^- \in B].$$

For, if $0 < s_1 < s_2 < \ldots < s_n \leqq t$,

$$E^{n+}[g_1(w(s_1))g_2(w(s_2)) \ldots g_n(w(s_n))w(t)]$$
$$= \int_0^\infty \int_0^\infty \ldots \int_0^\infty K^+(s_1, x_1)p^0(s_2 - s_1, x_1, x_2) \ldots$$
$$\times\, p^0(s_n - s_{n-1}, x_{n-1}, x_n)$$
$$\times\, g_1(x_1)g_2(x_2) \ldots g_{n-1}(x_{n-1})g_n(x_n)x_n dx_1 dx_2 \ldots dx_n$$
$$= \int_0^\infty \int_0^\infty \ldots \int_0^\infty q(s_1, 0, x_1)q(s_2-s_1, x_1, x_2) \ldots$$
$$\times\, q(s_n - s_{n-1}, x_{n-1}, x_n)$$
$$\times\, g_1(x_1)g_2(x_2) \ldots g_n(x_n)dx_1 dx_2 \ldots dx_n$$
$$= E^{Q_0}[g_1(w(s_1))g_2(w(s_2)) \ldots g_n(w(s_n))]$$

and this proves (4.32).

The assertions (ii) can be proved in the same way as (i).

Lemma 4.5. Let $a > 0$ and set, for $w \in \boldsymbol{C}([0, \infty) \longrightarrow \boldsymbol{R})$,

(4.33) $$l_a^0(w) = \sup\{t;\ t \leq \sigma_0(w),\ w(t) = a\}$$

(4.34) $$l_a(w) = \sup\{t;\ w(t) = a\}.$$

Then

(4.35) $$P_a[w^-_{l_a^0(w)} \in B] = Q_a[w^-_{l_a(w)} \in B]$$

for any Borel set B in $\boldsymbol{C}([0, \infty) \longrightarrow \boldsymbol{R}_+)$.

Proof. It is sufficient to prove that for bounded Borel functions $f_1, f_2, \ldots, f_n$ on $(0, \infty)$ and $0 < t_1 < t_2 < \ldots < t_n$,

(4.36) $$E^{P_a}[\prod_{j=1}^n f_j(w(t_j))I_{(l_a^0 > t_n)}] = E^{Q_a}[\prod_{j=1}^n f_j(w(t_j))I_{(l_a > t_n)}].$$

We have

the left hand side of (4.36)

$$= E^{P_a}[\prod_{j=1}^n f_j(w(t_j))I_{(\sigma_a(w^+_{t_n}) < \sigma_0(w^+_{t_n}),\ t_n < \sigma_0(w))}]$$
$$= E^{P_a}[\prod_{j=1}^n f_j(w(t_j))g(w(t_n))I_{(t_n < \sigma_0(w))}]$$

$$= \int_0^\infty \int_0^\infty \cdots \int_0^\infty \prod_{j=1}^n \{p^0(t_j - t_{j-1}, x_{j-1}, x_j) f_j(x_j)\} g(x_n) dx_1 dx_2 \ldots dx_n$$

where $t_0 = 0$, $x_0 = a$ and

$$g(x) = P_x[\sigma_a < \sigma_0] = \frac{x \wedge a}{a}, \quad x > 0.$$

Also

the right hand side of (4.36)

$$= E^{Q_a}[\prod_{j=1}^n f_j(w(t_j)) I_{\{\sigma_a(w_{t_n}^+) < \infty\}}]$$

$$= E^{Q_a}[\prod_{j=1}^n f_j(w(t_j)) h(w(t_n))]$$

$$= \int_0^\infty \int_0^\infty \cdots \int_0^\infty \prod_{j=1}^n \{q(t_j - t_{j-1}, x_{j-1}, x_j) f(x_j)\} h(x_n) dx_1 dx_2 \ldots dx_n$$

where $t_0 = 0$, $x_0 = a$ and

$$h(x) = Q_x[\sigma_a < \infty] = \frac{a}{x \vee a}, \quad x > 0.$$

The equality (4.36) is now easily verified by noting $q(t, x, y) = p^0(t, x, y) y/x$.

Corollary 1 (Williams [179]). For given $a > 0$, let $\{X(t)\}$ be BM^a and $\{Y(t)\}$ be $BES^0(3)$. Then $\{X(\sigma_0^X - t), 0 \le t \le \sigma_0^X\}$ is equivalent in law to $\{Y(t), 0 \le t \le l_a^Y\}$ where

$$\sigma_0^X = \inf\{t;\ X(t) = 0\} \quad \text{and}$$
$$l_a^Y = \sup\{t;\ Y(t) = a\}.$$

Proof. From (4.30), we see that $\{w(t);\ 0 \le t \le \sigma(w)\}$ is decomposed, under $n^+(\cdot\ |\sigma_a < \infty)$, into *mutually independent* parts $\{w(t), 0 \le t \le \sigma_a\}$ and $\{w(t + \sigma_a), 0 \le t \le \sigma - \sigma_a\}$, the former being equivalent in law to $\{Y(t), 0 \le t \le \sigma_a^Y\}$ where $\sigma_a^Y = \inf\{t;\ Y(t) = a\}$ and the latter to $\{X(t), 0 \le t \le \sigma_0^X\}$. Hence, the part $\{w(t), 0 \le t \le l_a(w)\}$ ($l_a(w)$ is defined by (4.34)) is decomposed, under $n^+(\cdot|\sigma_a < \infty)$, into mutually independent parts $\{w(t), 0 \le t \le \sigma_a\}$ and $\{w(t + \sigma_a), 0 \le t \le l_a - \sigma_a\}$, and latter being equivalent in law to $\{Z(t);\ 0 \le t \le l_a^Z\}$ where $Z(t)$ is $BES^a(3)$ by virtue of (4.35). Hence we can conclude that the part $\{w(t);\ 0 \le t \le l_a\}$ is, under $n(\cdot\ |\sigma_a < \infty)$, equivalent to $\{Y(t);\ 0 \le t \le l_a^Y\}$. Since $\{\sigma_a(w) < \infty\} = \{\sigma_a(\check{w}) < \infty\}$, it follows from

Lemma 4.3 that $\{\check{w}(t),\ 0 \le t \le l_a(\check{w})\}$ is, under $n(\ \cdot\ |\sigma_a < \infty)$, also equivalent to $\{Y(t),\ 0 \le t \le l_a^Y\}$. But it is obvious that $l_a(\check{w}) = \sigma(w) - \sigma_a(w)$, $\check{w}(t) = w(\sigma(w) - t)$ and we know that $\{w(t + \sigma_a),\ 0 \le t \le \sigma - \sigma_a\}$ is, under $n(\ \cdot\ |\sigma_a < \infty)$, equivalent to $\{X(t),\ 0 \le t \le \sigma_0^X\}$. This clearly implies the assertion.

Let $p^+ = p|_{\mathscr{W}^+}$ be the above point process on $\mathscr{W}^+$ and define $X^+(t)$ and $A^+(t)$ by (4.26). If we define $B^+(t)$ by (4.27) then we know that $B^+(t)$ is BM^0 and hence $X(t) = -B^+(t)$ is also BM^0; that is, the process X defined by

$$
(4.37)\qquad \begin{aligned} X(t) &= \phi^+(t) - [p^+(\phi^+(t))](t - A^+(\phi^+(t)-)) \\ &= \phi^+(t) - X^+(t) \end{aligned}
$$

is BM^0 and

$$
\phi^+(t) = \max_{0 \le s \le t} X(s).
$$

We define a continuous process $Y(t)$ by

$$
(4.38)\qquad \begin{aligned} Y(t) &= \phi^+(t) + [p^+(\phi^+(t))]\,(t - A^+(\phi^+(t)-)) \\ &= 2\max_{0 \le s \le t} X(s) - X(t). \end{aligned}
$$

We fix $a > 0$ and define a point process $\check{p}$ on $\mathscr{W}^+$ by $\boldsymbol{D}_{\check{p}} = \{s \in (0, a);\ a - s \in \boldsymbol{D}_{p^+}\} \cup \{s \in (a, \infty);\ s \in \boldsymbol{D}_{p^+}\}$ and

$$
\check{p}(s) = \begin{cases} \widecheck{p^+(a - s)} & \text{for } s \in \boldsymbol{D}_{\check{p}} \cap (0, a) \\ p^+(s) & \text{for } s \in \boldsymbol{D}_{\check{p}} \cap (a, \infty). \end{cases}
$$

By Lemma 4.3, it is immediately seen that $\check{p}$ has the same law as p^+. If we define $\check{Y}(t)$ from $\check{p}$ in the same way as $Y(t)$ is defined from p^+ by (4.38), then it is immediately seen that

$$
\begin{aligned} A^+(a) &= \check{A}(a) := \sum_{s \le a} \sigma[\check{p}(s)], \\ A^+(a) &= \inf\{t;\ X(t) = a\}, \\ \check{A}(a) &= \sup\{t;\ \check{Y}(t) = a\} \end{aligned}
$$

and

$$
\check{Y}(t) = a - X^+(A^+(a) - t) \qquad \text{for } 0 \le t \le A^+(a).
$$

Therefore, Corollary 1 implies that $\{\check{Y}(t),\ 0 \le t \le l_a^{\check{Y}}\}$ is equivalent to $\{Z(t),\ 0 \le t \le l_a^Z\}$ where Z is $BES^0(3)$. Letting $a \uparrow \infty$, we see that $Y(t)$ defined by (4.38) is actually $BES^0(3)$. Thus we have the following:

Corollary 2 (Pitman [141]). If $X(t)$ is BM^0, $Y(t) = 2 \max_{0 \leq s \leq t} X(s) - X(t)$ is $BES^0(3)$.

Here, we present a useful result for an $(\mathscr{F}_t)$-stationary Poisson point process p on a general state space $(\boldsymbol{X}, \mathscr{B}(\boldsymbol{X}))$ with characteristic measure $n(dx)$. Let $A \in \mathscr{B}([0, \infty)) \times \mathscr{B}(\boldsymbol{X})$ and assume that

$$\int_{\boldsymbol{X}} I_A(s, x) n(dx) = C_A(s) < \infty$$

and

$$\int_0^t C_A(s) ds < \infty \qquad \text{for all} \quad t > 0$$

but

$$\int_0^\infty C_A(s) ds = \infty.$$

Define

$$\theta = \inf\{s \in \boldsymbol{D}_p;\ (s, p(s)) \in A\}. \tag{4.39}$$

Then θ is an $(\mathscr{F}_t)$-stopping time and $P[\theta < \infty] = 1$ because

$$\begin{aligned} P[\theta \geq t] &= P[N_p(\{(s, x) \in A;\ s < t\}) = 0] \\ &= \exp[-\int_0^t C_A(s) ds] \longrightarrow 0 \qquad \text{as} \quad t \longrightarrow \infty. \end{aligned}$$

Lemma 4.6. (i) The point process p^* on $\boldsymbol{X}$ defined by $\boldsymbol{D}_{p^*} = \{t; t + \theta \in \boldsymbol{D}_p\}$ and $p^*(t) = p(t + \theta)$, $t \in \boldsymbol{D}_{p^*}$, is independent of $\mathscr{F}_\theta$.

(ii) $\mathscr{F}_{\theta-}$ and $p(\theta)$ are conditionally independent given θ: To be more precise, if $H(w)$ is bounded $\mathscr{F}_{\theta-}$-measurable and $G(x)$ is bounded $\mathscr{B}(\boldsymbol{X})$-measurable, then we have, for $\lambda > 0$,

$$\begin{aligned} &E[H G[p(\theta)] e^{-\lambda\theta}] \\ &= \int_0^\infty e^{-\lambda s} \Big\{ E[I_{(\theta \geq s)} H_s] \int_{\boldsymbol{X}} G(x) I_A(s, x) n(dx) \Big\} ds \end{aligned}$$

where H_s is a bounded predictable process such that $H = H_\theta$ (cf. [15]). From this formula, we can conclude that

$$P[H \mid \theta = s] = E[I_{(\theta \geq s)} H_s] e^{\int_0^s C_A(u) du} \tag{4.40}$$

$$
(4.41)\qquad \begin{aligned} E[G[p(\theta)]\,|\,\theta = s] &= \int_X G(x)I_A(s, x)n(dx)\{C_A(s)\}^{-1} \\ &= \int_X G(x)I_A(s, x)n(dx)\left[\int_X I_A(s, x)n(dx)\right]^{-1}. \end{aligned}
$$

Proof. (i) This is essentially established in Theorem II–6.5, because $(\mathscr{F}_t)$-stationary Poisson point process p is always independent of $\mathscr{F}_0$.

(ii) We have

$$
\begin{aligned}
&E[H\, G[p(\theta)]e^{-\lambda\theta}] \\
&= E[\sum_{s\in \boldsymbol{D}_p,\, s\leq\theta} H_s G[p(s)]e^{-\lambda s}I_A(s, p(s))] \\
&= E\left[\int_0^{\theta+}\int_X H_s G(x)e^{-\lambda s}I_A(s, x)N_p(dsdx)\right] \\
&= E\left[\int_0^{\theta} e^{-\lambda s}H_s \int_X G(x)I_A(s, x)n(dx)ds\right] \\
&= \int_0^{\infty} e^{-\lambda s}\left\{E[I_{\{\theta\geqq s\}}H_s]\int_X G(x)I_A(s, x)n(dx)\right\}ds
\end{aligned}
$$

and hence the result.

Example 4.1. Let p be the above Poisson point process of Brownian excursions given as an $(\mathscr{F}_t)$-stationary Poisson point process on $\mathscr{W}$ and construct BM^0 X by (4.17). Let $a > 0$ and define $A \in \mathscr{B}((0, \infty)) \times \mathscr{B}(\mathscr{W})$ by

$$
A = (0, \infty) \times \{w;\ M(w) \geqq a\}
$$

where $M(w) = \max_{0\leq s\leq\sigma(w)} w(s)$. If θ is defined by (4.39), then

$$
\theta = \inf\{s \in \boldsymbol{D}_p;\ M[p(s)] \geq a\}
$$

and hence, $A(\theta-) < \sigma_a^X < A(\theta)$ and

$$
A(\theta-) = \sup\{t;\ t < \sigma_a^X,\ X(t) = 0\} : = l_0^{a,X}.
$$

If $G(w) = \Phi(w_{\sigma_a}^-)$, where Φ is a bounded Borel function on $\boldsymbol{C}([0, \infty) \longrightarrow \boldsymbol{R}_+)$, we have

$$
\begin{aligned}
&E[G(p(\theta))\,|\,\theta = s] = E[G(p(\theta))] \\
&= \int_{\mathscr{W}} G(w)I_{\{M(w)\geq a\}}n(dw)\{n(\{M(w) \geq a\})\}^{-1} \\
&= \int_{\mathscr{W}^+} \Phi(w_{\sigma_a}^-)n(dw\,|\,\sigma_a < \infty)
\end{aligned}
$$

$$= \int_{C([0,\,\infty)\to R_+)} \Phi(w^-_{\sigma_a}) Q_0(dw)$$

by (4.30). Thus we obtained the following:

(4.42) *If X is BM^0 and Y is $BES^0(3)$, then $\{X(l_0^{a,X} + t),\ 0 \leq t \leq \sigma_a^X - l_0^{a,X}\}$ is equivalent in law to $\{Y(t),\ 0 \leq t \leq \sigma_a^Y\}$. Furthermore, $\{X(l_0^{a,X} + t),\ 0 \leq t \leq \sigma_a^X - l_0^{a,X}\}$ is independent of $\{X(t),\ 0 \leq t \leq l_0^{a,X}\}$.*

The second assertion follows from the fact that $\{X(t),\ 0 \leq t \leq l_0^{a,X}\}$ is $\mathscr{F}_{\theta-}$-measurable and $E[G[p(\theta)]\,|\,\theta = s]$ is independent of s.

Now we discuss the first decomposition result of Brownian paths due to Williams [179]. Take the $(\mathscr{F}_t)$-point process $p^+ = p|_{\mathscr{W}^+}$ and, for $a > 0$, define BM^a X by

$$X(t) = a + \phi^+(t) - p^+[\phi^+(t)](t - A^+(\phi^+(t)-)).$$

Then

$$a + \phi^+(t) = \max_{0 \leq u \leq t} X(u).$$

Define $A \in \mathscr{B}((0,\ \infty)) \times \mathscr{B}(\mathscr{W}^+)$ by

$$A = \{(s,\ w);\ a + s - M[w] \leq 0\}.$$

If θ is defined by (4.39), it is easy to see that

$$A^+(\theta-) < \sigma_0^X < A^+(\theta), \qquad a + \theta = \max_{0 \leq t \leq \sigma_0^X} X(t)$$

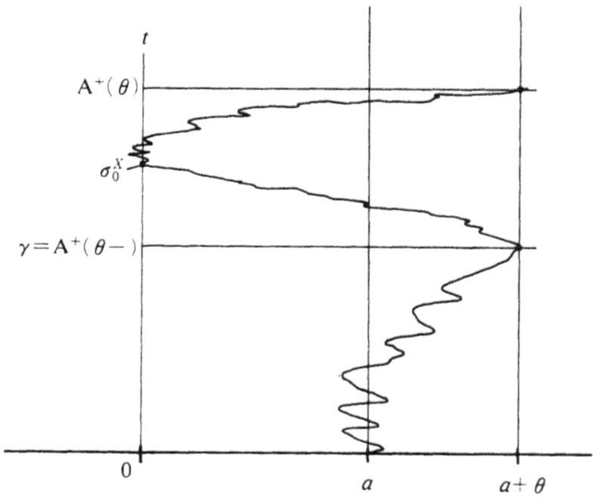

Fig. 1

and $\gamma = A^+(\theta-)$ is the unique time point in $[0, \sigma_0^X]$ which attains this maximum. Then

$$p(\theta)(t) = a + \theta - X(\gamma + t)$$

and by (4.41)

$$\begin{aligned} &E[G[p(\theta)] \mid \theta = s] \\ &= \int_{\mathscr{W}^+} G(w) I_{\{M(w) \geq a+s\}} n^+(dw) \{n^+(\{M(w) \geq a + s\})\}^{-1} \\ &= \int_{\mathscr{W}^+} G(w) n^+(dw \mid \sigma_{a+s} < \infty). \end{aligned}$$

By (4.30) we can conclude that, given θ, $\{X(\gamma + t), 0 \leq t \leq \sigma_0^X - \gamma\}$ is equivalent in law to $\{a + \theta - Y(t), 0 \leq t \leq \sigma_{a+\theta}^Y\}$ where Y is $BES^0(3)$. But, if θ is given, $\{a + \theta - Y(t), 0 \leq t \leq \sigma_{a+\theta}^Y\}$ is equivalent to $\{Y(\sigma_{a+\theta}^Y - t), 0 \leq t \leq \sigma_{a+\theta}^Y\}$ as is easily verified by Corollary 1 above and (4.42).

Next, we consider the part $\{X(t), 0 \leq t \leq \gamma\}$. This part is $\mathscr{F}_{\theta-}$-measurable and, for a bounded Borel function Φ on $C([0, \infty) \longrightarrow \boldsymbol{R}_+)$, we set $H = \Phi(X_\gamma^-)$ where X_γ^- is defined by $X_\gamma^-(t) = X(\gamma \wedge t), t \geq 0$. Then $H = H_\theta$ where H_s is a predictable process defined by $H_s = \Phi(X_{A^+(s-)}^-)$. By (4.40), we have

$$\begin{aligned} &E[\Phi(X_\gamma^-) \mid \theta = s] \\ &= E[I_{\{\theta \geq s\}} \Phi(X_{A^+(s-)}^-)]\{E[I_{\{\theta \geq s\}}]\}^{-1} \\ &= E[\Phi(X_{\sigma_{a+s}^X}^-) \mid \sigma_{a+s}^+ < \sigma_0^X] \end{aligned}$$

since $A^+(s-) = A^+(s) = \sigma_{a+s}^X$ for each fixed $s > 0$. By (4.31) we can conclude that, given θ, $\{X(t), 0 \leq t \leq \gamma\}$ is equivalent to $\{Y(t), 0 \leq t \leq \sigma_{a+\theta}^Y\}$ where Y is $BES^a(s)$. Note that, given θ, $p(\theta)$ and $\mathscr{F}_{\theta-}$ are independent; in particular $p(\theta)$ and $\{X(t), 0 \leq t \leq \gamma\}$ are independent. Finally, we have

$$P[\theta \in ds] = \{\exp[-\int_0^s C_A(u)du]\} C_A(s)ds = \frac{a}{(a+s)^2} ds$$

because

$$C_A(s) = n^+(\{M(w) \geqq a + s\}) = \frac{1}{a+s}.$$

Now we can summarize the above to obtain the following result of Williams: *On a probability space, we set up the following three independent random elements:* $BES^a(3)$ Y^1, $BES^0(3)$ Y^2 *and a random variable* $\theta \in$

(0, ∞) distributed by

$$P(\theta \in ds) = \frac{a}{(a+s)^2} ds.$$

Define

$$Z(t) = \begin{cases} Y^1(t) & , \quad 0 \leq t \leq \sigma_{a+\theta}^{Y1} \\ Y^2(\sigma_{a+\theta}^{Y1} + \sigma_{a+\theta}^{Y2} - t), & \sigma_{a+\theta}^{Y1} \leq t \leq \sigma_{a+\theta}^{Y1} + \sigma_{a+\theta}^{Y2}. \end{cases}$$

Then $\{Z(t), 0 \leq t \leq \sigma_{a+\theta}^{Y1} + \sigma_{a+\theta}^{Y2}\}$ *is equivalent in law to* $\{X(t), 0 \leq t \leq \sigma_0^X\}$ *where X is* BM^a.

This result, combined with (4.30), yields immediately the following description of Brownian excursion law n^+ due to Williams: Let Y^1 and Y^2 be *mutually independent* $BES^0(3)$ *and, for* $a > 0$, *set*

$$X^a(t) = \begin{cases} Y^1(t) & \text{for } 0 \leq t \leq \sigma_a^{Y1} \\ Y^2(\sigma_a^{Y1} + \sigma_a^{Y2} - t) & \text{for } \sigma_a^{Y1} \leq t \leq \sigma_a^{Y1} + \sigma_a^{Y2} \\ 0 & \text{for } t \geq \sigma_a^{Y1} + \sigma_a^{Y2}. \end{cases}$$

Let R_a *be the law on* $\mathscr{W}^+$ *of* X^a. *Then*

$$n^+ = \int_0^\infty (R_a / a^2) da.$$

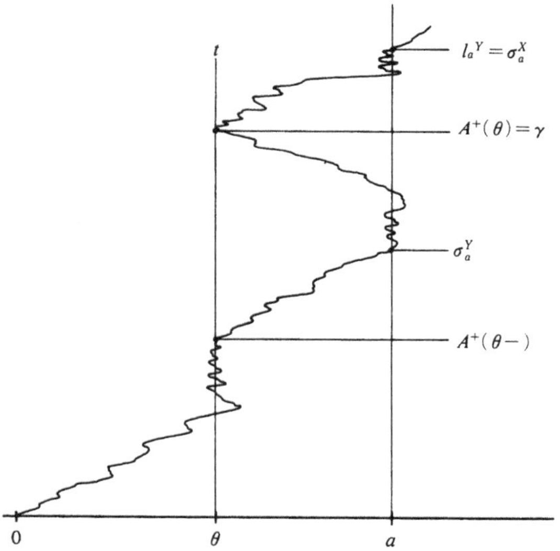

Fig. 2

Next, we discuss the second decomposition result of Brownian paths due to Williams [179]. We start with $BES^0(3)$ Y defined by (4.38) from the $(\mathscr{F}_t)$-point process p^+ on $\mathscr{W}^+$. Let $a > 0$ and define $A \in \mathscr{B}((0, \infty)) \times \mathscr{B}(\mathscr{W}^+)$ by $A = \{(s, w);\ s + M(w) \geq a\}$. If θ is defined by (4.39), it is easy to see that $A^+(\theta-) < \sigma_a^Y < A^+(\theta+)$, $\theta = \min_{\sigma_a^Y \leq t \leq l_a^Y} Y(t)$ and $\gamma = A^+(\theta)$ is the unique time point in (σ_a^Y, l_a^Y) which attains this minimum. Set $X(t) = a - Y(l_a^Y - t)$, $0 \leq t \leq l_a^Y$. Then $l_a^Y = \sigma_a^X$, $l_a^Y - \sigma_a^Y = l_0^{a,X} := \sup\{t;\ t \leq \sigma_a^X,\ X(t) = 0\}$ and, by Corollary 1 above X is a part of a BM^0 up to its hitting to a. The process Y^* defined by $Y^*(t) = Y(t + \gamma) - \theta$ is independent of $\mathscr{F}_\theta$ because Y^* can be constructed from the point process p^* of Lemma 4.6 (i) in the same way as Y from p^+. Then, under the conditional probability $P(\cdot \mid \mathscr{F}_\theta)$, $\{Y(t + \gamma),\ 0 \leq t \leq l_a^Y - \gamma\}$ can be expressed as $\{\theta + Z(t),\ 0 \leq t \leq l_{a-\theta}^Z\}$ where Z is $BES^0(3)$ independent of θ. Since $Y(t + \gamma) = a - X(\sigma_a^X - t - \gamma)$, $0 \leq t \leq \sigma_a^X - \gamma$, we see by Corollary 1 again that $\{X(t), 0 \leq t \leq \sigma_a^X - \gamma\}$, under $P(\cdot \mid \mathscr{F}_\theta)$, can be expressed as $\{W(t), 0 \leq t \leq \sigma_{a-\theta}^W\}$ where W is BM^0 independent of θ.

Also we know by (4.42) that $\{X(l_0^{a,X} + t),\ 0 \leq t \leq \sigma_a^X - l_0^{a,X}\}$ is independent of $\{X(t),\ 0 \leq t \leq l_0^{a,X}\}$ and is equivalent in law to $\{Z(t),\ 0 \leq t \leq \sigma_a^Z\}$ where Z is $BES^0(3)$.

Finally we study the part $\{X(\sigma_a^X - \gamma + t), 0 \leq t \leq l_0^{a,X} - \sigma_a^X + \gamma\}$. Noting that $l_0^{a,X} - \sigma_a^X + \gamma = \gamma - \sigma_a^Y$ and $X(\sigma_a^X - \gamma + t) = Y(\gamma - t)$, $0 \leq t \leq \gamma - \sigma_a^Y$, we study $\{Y(\sigma_a^Y + t),\ 0 \leq t \leq \gamma - \sigma_a^Y\} = \{\theta + p^+(\theta)[\sigma_{a-\theta}(p^+(\theta)) + t],\ \ 0 \leq t \leq \sigma(p^+(\theta)) - \sigma_{a-\theta}(p^+(\theta))\}$. We can compute the joint law of $p^+(\theta)$ and θ as follows:

$$C_A(s) = n^+(\{s + M(w) \geq a\}) = \frac{1}{a - s}, \qquad 0 < s < a,$$

and hence

$$P[\theta \in ds] = \left\{\exp\left(-\int_0^s C_A(u)du\right)\right\} C_A(s)ds = \frac{ds}{a}$$

and

$$\begin{aligned} &E[G(p^+(\theta)) \mid \theta = s] \\ &= \int_{\mathscr{W}^+} G(w) I_{\{M(w) \geq a-s\}} n^+(dw) \{n^+(\{s + M(w) \geq a\})\}^{-1} \\ &= E^{n^+}[G(w) \mid \sigma_{a-s} < \infty]. \end{aligned}$$

Hence, by (4.30),

$$
\begin{aligned}
&E[\Phi([p^+(\theta)]^+_{\sigma_{a-\theta}[p^+(\theta)]}) \mid \theta = s] \\
&\quad = E^{n^+}[\Phi(w^+_{\sigma_{a-s}}) \mid \sigma_{a-s} < \infty] \\
&\quad = E^{P_{a-s}}[\Phi(w^-_{\sigma_0})]
\end{aligned}
$$

for a bounded Borel function Φ on $\boldsymbol{C}([0, \infty) \longrightarrow \boldsymbol{R}_+)$. This implies that $\{Y(\sigma_a^Y + t),\ 0 \leq t \leq \gamma - \sigma_a^Y\}$ is equivalent in law to $\{a - W(t),\ 0 \leq t \leq \sigma_{a-\theta}^W\}$ where W is BM^0 and $\theta \in (0, a)$ is independent of W and uniformly distributed on $(0, a)$. This, combined with Corollary 1 above, implies that $\{X(\sigma_a^X - \gamma + t),\ 0 \leq t \leq l_0^{a,X} - \sigma_a^X + \gamma\}$ is equivalent in law to $\{a - \theta - Z(t),\ 0 \leq t \leq l_{a-\theta}^Z\}$ where Z is $BES^0(3)$ independent of θ. We can now summarize the above to obtain the following result of Williams: *On a probability space, we set up four independent random elements: BM^0 X, $BES^0(3)$ Y^1 and Y^2, a random variable δ uniformly distributed on $(0, a)$. Define*

$$
Z(t) = \begin{cases} X(t) & \text{for } 0 \leq t \leq \sigma_\delta^X \\ \delta - Y^1(t - \sigma_\delta^X) & \text{for } \sigma_\delta^X \leq t \leq \sigma_\delta^X + l_\delta^{Y^1} \\ Y^2(t - \sigma_\delta^X - l_\delta^{Y^1}) & \text{for } \sigma_\delta^X + l_\delta^{Y^1} \leq t \leq \sigma_\delta^X + l_\delta^{Y^1} + \sigma_a^{Y^2}. \end{cases}
$$

Then $\{Z(t),\ 0 \leq t \leq \sigma_\delta^X + l_\delta^{Y^1} + \sigma_a^{Y^2}\}$ is equivalent in law to $\{X(t),$ $0 \leq t \leq \sigma_a^X\}$.

4.4. Some limit theorems for occupation times of Brownian motion. Let $X = (X(t))$ be a one-dimensional Brownian motion such that $X(0) = 0$. Let $f(x)$ be a continuous function with compact support. We are interested in the problem of finding the limit process of $\frac{1}{u(\lambda)}\int_0^{\lambda t} f(X(s))ds$ as $\lambda \longrightarrow \infty$ where $u(\lambda)$ is a some normalizing function. The situation will be quite different accordingly as

$$
\bar{f} = \int_{R^1} f(x)dx \neq 0 \quad \text{or} \quad \bar{f} = 0.
$$

Theorem 4.4.* (i) If $\bar{f} \neq 0$, the family of continuous stochastic processes

$$
t \longmapsto \frac{1}{\sqrt{\lambda}} \int_0^{\lambda t} f(X(s))ds, \qquad \lambda > 0,
$$

* Cf. Papanicolaou-Stroock-Varadhan [137]. The corresponding problem for a two-dimensional Brownian motion was discussed by Kasahara and Kotani [82].

converges in the sense of law on the space of continuous functions to the continuous process $t \longmapsto 2\bar{f}\phi(t, 0)$ as $\lambda \longrightarrow \infty$ where $\phi(t, 0)$ is the local time of $X(t)$ at zero.

(ii) If $\bar{f} = 0$ but f is not identically zero, the family of continuous stochastic processes

$$t \longmapsto \frac{1}{\lambda^{1/4}} \int_0^{\lambda t} f(X(s))ds, \qquad \lambda > 0,$$

converges in the sense of law on the space of continuous functions to the continuous process $t \longmapsto 2\sqrt{\langle f, f\rangle} B(\phi(t, 0))$ as $\lambda \longrightarrow \infty$ where $\phi(t, 0)$ is the local time of $X(t)$ at zero and $B(t)$ is another Brownian motion independent of $X(t)$ with $B(0) = 0$ and $\langle f, f\rangle = -\int_{R^1}\int_{R^1} |x - y| f(x) f(y)dxdy$.

Proof. The proof of (i) is easy. Indeed, by the scaling property of the Brownian motion, $\{X(t)\} \overset{\mathcal{L}}{\approx} \{\lambda^{-1/2}X(\lambda t)\}$ for each $\lambda > 0$. From this it is easy to conclude that $\{\phi(t, a)\} \overset{\mathcal{L}}{\approx} \left\{\frac{1}{\sqrt{\lambda}}\phi(\lambda t, \sqrt{\lambda}a)\right\}$ for each $\lambda > 0$. Then

$$\frac{1}{\sqrt{\lambda}} \int_0^{\lambda t} f(X(s))ds = \frac{2}{\sqrt{\lambda}} \int_{R^1} \phi(\lambda t, a) f(a)da$$

$$\overset{\mathcal{L}}{\approx} 2 \int_{R^1} \phi(t, a/\sqrt{\lambda}) f(a)da.$$

But $2\int_{R^1}\phi(t, a/\sqrt{\lambda})f(a)da$ converges, as $\lambda \longrightarrow \infty$, to $2\phi(t, 0)\int_{R^1} f(a)da$ uniformly in t on each bounded interval a.s. Thus the assertion of (i) is proved.

To prove (ii), set

$$F(x) = \int_{-\infty}^{x} f(y)dy \quad \text{and} \quad G(x) = \int_0^x F(y)dy.$$

Since $f(x)$ is of compact support and null charged, i.e., $\bar{f} = 0$, $F(x)$ is of compact support and $G(x)$ is bounded. By Itô's formula,

$$G(X(t)) = \int_0^t F(X(s))dX(s) + \frac{1}{2}\int_0^t f(X(s))ds.$$

Therefore,

$$\frac{1}{\lambda^{1/4}}\int_0^{\lambda t} f(X(s))ds = \frac{2}{\lambda^{1/4}}G(X(\lambda t)) - \frac{2}{\lambda^{1/4}}\int_0^{\lambda t} F(X(s))dX(s)$$
$$= I_1 + I_2.$$

Clearly $I_1 \longrightarrow 0$ uniformly in t as $\lambda \longrightarrow \infty$ a.s. Set

$$M_\lambda(t) = -\frac{2}{\lambda^{1/4}}\int_0^{\lambda t} F(X(s))dX(s).$$

For each $\lambda > 0$, $M_\lambda(t)$ is a continuous martingale such that

$$\langle M_\lambda\rangle(t) = \frac{4}{\lambda^{1/2}}\int_0^{\lambda t} F(X(s))^2 ds,$$

and by the result (i), this family of processes converges in law to the process $8\overline{F^2}\phi(t,0)$ as $\lambda \longrightarrow \infty$. Define a family of three-dimensional continuous processes $Z_\lambda(t)$, $\lambda > 0$, by

$$Z_\lambda(t) = (M_\lambda(t), \langle M_\lambda\rangle(t), \frac{1}{\lambda^{1/2}}X(\lambda t)).$$

First we show that the family of laws of Z_λ is tight. For this it is clearly sufficient to show that the family of laws of $M_\lambda(t)$ is tight. By Theorem 3.1, we have for $s < t$, that

$$\begin{aligned}
&E([M_\lambda(t) - M_\lambda(s)]^6)\\
&\quad\le \text{const. } E([\langle M_\lambda\rangle(t) - \langle M_\lambda\rangle(s)]^3)\\
&\quad\le \text{const. } \lambda^{-3/2}\int_{\lambda s}^{\lambda t} du\int_{\lambda s}^{u} dv\int_{\lambda s}^{v} dw\int_{R^1} dx\int_{R^1} dy\int_{R^1} dz\\
&\qquad\times \frac{\exp\left(-\frac{x^2}{2w}\right)}{\sqrt{2\pi w}}\frac{\exp\left(-\frac{(y-x)^2}{2(v-w)}\right)}{\sqrt{2\pi(v-w)}}\frac{\exp\left(-\frac{(z-y)^2}{2(u-v)}\right)}{\sqrt{2\pi(u-v)}}F(x)^2F(y)^2F(z)^2\\
&\quad\le \text{const. } \lambda^{-3/2}\int_{\lambda s}^{\lambda t} du\int_{\lambda s}^{u} dv\int_{\lambda s}^{v} dw\frac{1}{\sqrt{2\pi w}\sqrt{2\pi(v-w)}\sqrt{2\pi(u-v)}}\\
&\quad\le \text{const. } (t-s)^{3/2}.
\end{aligned}$$

Hence, we can conclude from Theorem I-4.3 that the laws of $M_\lambda(t)$ constitute a tight family. Next we show that the set of limit points of laws of Z_λ consists of a single point. Let $Z_{\lambda_i} \longrightarrow Z$ in law for some subsequence and $Z = (Z_1(t), Z_2(t), Z_3(t))$. Then $Z_3(t)$ is a Brownian motion and $Z_2(t) = 8\overline{F^2}\phi(t, 0)$ where $\phi(t, 0)$ is the local time of $Z_3(t)$ at zero. By Theorem I-4.2, we may assume that $Z_{\lambda_i}(t) \longrightarrow Z(t)$ uniformly on each compact interval a.s. Then it is easy to prove that (Z_1, Z_3) is a system of continuous martingales such that

$$\langle Z_1, Z_1\rangle_t = Z_2(t) = 8\overline{F^2}\phi(t, 0), \ \langle Z_3, Z_3\rangle_t = t$$

and

$$\langle Z_1, Z_3\rangle_t = \lim_{\lambda\to\infty} \langle M_\lambda, \frac{1}{\lambda^{1/2}} X(\lambda\cdot)\rangle_t = \lim_{\lambda\to\infty}\left(-\frac{2}{\lambda^{3/4}}\int_0^{\lambda t} F(X(s))ds\right) = 0$$

since

$$\frac{1}{\lambda^{1/2}}\int_0^{\lambda t} F(X(s))ds \longrightarrow 2\bar{F}\phi(t, 0) = 0 \quad \text{in law.}$$

By Theorem II-7.3, we can conclude that $Z_1(t) = \sqrt{8\overline{F^2}}B(\phi(t,0))$ where $B(t)$ is a Brownian motion with $B(0) = 0$ which is independent of $Z_3(t)$. This shows that the law of $Z(t)$ is uniquely determined. In particular it has been shown that $M_\lambda(t)$ converges in law to the process $Z_1(t)$. The assertion of (ii) now follows if we notice that $\langle f, f\rangle = 2\overline{F^2}$ but this is immediate since

$$\begin{aligned} -\int_{R^1}\int_{R^1} |x - y| f(x)f(y)dxdy &= -2\iint_{x>y}(x - y)f(x)f(y)dxdy \\ &= -2\iint_{x>y}\left(\int_y^x dz\right)f(x)f(y)dxdy = -2\iiint_{x>z>y} f(x)f(y)dxdydz \\ &= -2\int_{-\infty}^{\infty} dz \int_z^{\infty} f(x)dx \int_{-\infty}^{z} f(y)dy = 2\int_{-\infty}^{\infty}\left(\int_{-\infty}^{z} f(y)dy\right)^2 dz \\ &= 2\int_{-\infty}^{\infty} F(z)^2 dz. \end{aligned}$$

5. Exponential martingales

Let $(\Omega, \mathscr{F}, P)$ and $(\mathscr{F}_t)_{t\geq 0}$ be given as in Section 1 and consider a quasimartingale $X(t)$ such that $X(0) = 0$. We denote by M_X the martingale part of $X(t)$. Then the *exponential quasimartingale* defined by

$$M(t) = \exp\left\{X_t - \frac{1}{2}\langle M_X\rangle_t\right\} \tag{5.1}$$

is the unique solution of the stochastic differential equation

$$(5.2)\qquad \begin{cases} dM = M \cdot dX, \\ M_0 = 1, \end{cases}$$

as is easily seen by Itô's formula and Theorem 2.1.

Define the Hermite polynomials

$$H_n[t, x] = \frac{(-t)^n}{n!} \exp\left(\frac{x^2}{2t}\right) \frac{\partial^n}{\partial x^n} \exp\left(-\frac{x^2}{2t}\right), \quad n \geq 0,$$

that is,

$$(5.3)\qquad \sum_{n=0}^{\infty} \gamma^n H_n[t,x] = \exp\left(\gamma x - \frac{\gamma^2 t}{2}\right) \qquad \text{for every } \gamma \in \boldsymbol{R}.$$

In particular,

$$H_0[t, x] = 1,$$
$$H_1[t, x] = x,$$
$$H_2[t, x] = \frac{x^2}{2} - \frac{t}{2},$$
$$H_3[t, x] = \frac{x^3}{6} - \frac{xt}{2}.$$

For each $\lambda \in \boldsymbol{R}$ set

$$(5.4)\qquad M_\lambda(t) = \exp\left\{\lambda X_t - \frac{\lambda^2}{2}\langle M_X\rangle_t\right\} = \exp\left\{\lambda X_t - \frac{1}{2}\langle M_{\lambda X}\rangle_t\right\}.$$

Thus $M_\lambda(t)$ is the unique solution of

$$(5.5)\qquad \begin{cases} dM_\lambda(t) = \lambda M_\lambda(t) \cdot dX(t) \\ M_\lambda(0) = 1. \end{cases}$$

By (5.3),

$$M_\lambda(t) = \sum_{n=0}^{\infty} \lambda^n H_n[\langle M_X\rangle_t, X_t] := \sum_{n=0}^{\infty} \lambda^n \mathscr{X}_n(t),$$

where we set $\mathscr{X}_n(t) = H_n[\langle M_X\rangle_t, X_t]$. Let N be a positive integer. Letting A_X be the bounded variation part of $X(t)$, we set

$$\eta(t) = \max\{X(t), \langle M_X\rangle(t), \| A_X \|(t)\}, \qquad 0 \leqq t < \infty,$$
$$\sigma_N = \inf\{t;\ \eta(t) \geqq N\}$$

where $\|A_X\|(t)$ is the variation of the mapping: $s \longrightarrow A_X(s)$ on the interval $[0, t]$. We note that $H_n[t, x]$, $n = 0,1,2, \ldots$ can be expressed as

$$H_n[t, x] = \sum_{k=0}^{n} \sum_{j=0}^{n-k} a_n(k, j) t^j x^k.$$

Then we can easily verify by induction on n that

$$\sum_{k=0}^{n} \sum_{j=0}^{n-k} |a_n(k, j)| \leqq 1.$$

This implies that $|H_n[t, x]| \leqq N^n$, $n = 0,1,2, \ldots$, provided $t \leqq N$ and $|x| \leqq N$. Hence

$$|\mathscr{Z}_n(t)| \leqq N^n \qquad \text{for} \quad 0 \leqq t \leqq \sigma_N.$$

Then it is easy to see by (5.5) that for every λ satisfying $0 < \lambda < \dfrac{1}{N}$,

$$\begin{aligned}
M_N(t \wedge \sigma_N) &= 1 + \lambda \int_0^{t \wedge \sigma_N} M_X(s) dX(s) \\
&= 1 + \lambda \int_0^{t \wedge \sigma_N} \Big(\sum_{n=0}^{\infty} \lambda^n \mathscr{Z}_n(s)\Big) dX(s) \\
&= 1 + \lambda \int_0^{t \wedge \sigma_N} \Big(\sum_{n=0}^{\infty} \lambda^n \mathscr{Z}_n(s)\Big) dM_X(s) + \lambda \int_0^{t \wedge \sigma_N} \Big(\sum_{n=0}^{\infty} \lambda^n \mathscr{Z}_n(s)\Big) dA_X(s) \\
&= 1 + \sum_{n=0}^{\infty} \lambda^{n+1} \int_0^{t \wedge \sigma_N} \mathscr{Z}_n(s) dM_X(s) + \sum_{n=0}^{\infty} \lambda^{n+1} \int_0^{t \wedge \sigma_N} \mathscr{Z}_n(s) dA_X(s) \\
&= 1 + \sum_{n=0}^{\infty} \lambda^{n+1} \int_0^{t \wedge \sigma_N} \mathscr{Z}_n(s) dX(s) \\
&= \sum_{n=0}^{\infty} \lambda^n \mathscr{Z}_n(t \wedge \sigma_N).
\end{aligned}$$

Then we have

$$\mathscr{Z}_0(t \wedge \sigma_N) = 1 \text{ and } \mathscr{Z}_n(t \wedge \sigma_N) = \int_0^{t \wedge \sigma_N} \mathscr{Z}_{n-1}(s) dX(s)$$

for $n = 0,1,2, \ldots$. Since

$$\lim_{N \to \infty} \sigma_N = \infty,$$

we deduce from this that

$$\mathscr{Z}_0(t) \equiv 1 \quad \text{and} \quad \mathscr{Z}_n(t) = \int_0^t \mathscr{Z}_{n-1}(s) dX(s)$$

for $n = 1, 2, \ldots$. Hence

$$\mathscr{X}_n(t) = \int_0^t dX(t_1) \int_0^{t_1} dX(t_2) \cdots \int_0^{t_{n-1}} dX(t_n).$$

Thus we have proven the following result ([64] and [113]).

Theorem 5.1.

$$(5.6) \qquad \int_0^t dX(t_1) \int_0^{t_1} dX(t_2) \cdots \int_0^{t_{n-1}} dX(t_n) = H_n[\langle M_X\rangle_t, X_t], \quad n = 1,2, \ldots .$$

Now we will investigate some properties of the exponential quasimartingale $\exp\{X_t - \langle M_X\rangle_t/2\}$.

Theorem 5.2. If $X \in \mathscr{M}$, then the exponential quasimartingale M_t is a continuous local $(\mathscr{F}_t)$-martingale. Furthermore, M_t is a supermartingale, and it is a martingale if and only if

$$E[M_t] = 1 \qquad \text{for every} \quad t \geq 0.$$

Proof. Since M_t is the unique solution of (5.2)

$$M_t - 1 = \int_0^t M_s dX(s) \in \mathscr{M}.$$

Thus, it is easy to see by Fatou's lemma that M_t is a supermartingale.

Theorem 5.3. (Novikov [132]). Let $X \in \mathscr{M}$ and set

$$M_t = \exp\left\{X_t - \frac{1}{2}\langle X\rangle_t\right\}.$$

If

$$(5.7) \qquad E[e^{\langle X\rangle_t/2}] < \infty \qquad \text{for every} \quad t \geq 0,$$

then M_t, $t \geq 0$, is a continuous $(\mathscr{F}_t)$-martingale, i.e.,

$$(5.8) \qquad E[M_t] = 1 \qquad \text{for every } t \geq 0.$$

Proof. By Theorem II-7.2′, on an extension $(\tilde{\Omega}, \tilde{\mathscr{F}}, \tilde{P})$ of Ω with a reference family $(\tilde{\mathscr{F}}_t)$, there exists an $(\tilde{\mathscr{F}}_t)$-Brownian motion $B = (B(t))$ with $B(0) = 0$ such that $X(t) = B(\langle X\rangle(t))$ and $\langle X\rangle(t)$ is an $(\tilde{\mathscr{F}}_t)$-stopping time for each $t \geq 0$. Set

$$\sigma_a = \inf\{t;\, B(t) \leq t - a\}.$$

Then, if $a > 0$, we have for $\lambda > 0$ that

$$E[e^{-\lambda\sigma_a}] = e^{-(\sqrt{1+2\lambda}-1)a}. \tag{5.9}$$

Indeed, setting

$$u(t, x) = e^{-\lambda t}e^{-(\sqrt{1+2\lambda}-1)x},$$

we have by Itô's formula that

$$\begin{aligned} u(t, B_t - t) - u(0, 0) &= \int_0^t \frac{\partial u}{\partial x}(s, B_s - s)dB_s \\ &\quad + \int_0^t \left(\frac{\partial u}{\partial t} - \frac{\partial u}{\partial x} + \frac{1}{2}\frac{\partial^2 u}{\partial x^2}\right)(s, B_s - s)ds \\ &= \int_0^t \frac{\partial u}{\partial x}(s, B_s - s)dB_s. \end{aligned}$$

Thus $t \longmapsto u(t \wedge \sigma_a, B_{t\wedge\sigma_a} - t \wedge \sigma_a)$ is a martingale if $a > 0$ and hence

$$E[u(t \wedge \sigma_a, B_{t\wedge\sigma_a} - t \wedge \sigma_a)] = u(0, 0) = 1.$$

Letting $t \uparrow \infty$, we have by the bounded convergence theorem that

$$E[u(\sigma_a, B_{\sigma_a} - \sigma_a)] = 1.$$

This proves (5.9).

From this we can conclude that

$$E\left[\exp\left(\frac{1}{2}\sigma_a\right)\right] = e^a < \infty.$$

Therefore,

$$E\left[\exp\left(B(\sigma_a) - \frac{1}{2}\sigma_a\right)\right] = e^{-a}E\left[\exp\left(\frac{1}{2}\sigma_a\right)\right] = 1.$$

Combining this result with Theorem 5.2 and setting

$$Y(t) = \exp\left(B(\sigma_a \wedge t) - \frac{1}{2}\sigma_a \wedge t\right),$$

we can show that $Y(t)$, $t \in [0, \infty]$, is a uniformly integrable $(\mathscr{F}_t)$-martingale. Hence for any $(\mathscr{F}_t)$-stopping time σ,

$$E\left[\exp\left(B(\sigma_a \wedge \sigma) - \frac{1}{2}\sigma_a \wedge \sigma\right)\right] = 1.$$

In particular, we have

$$\begin{aligned} 1 &= E\left[\exp\left(B(\sigma_a \wedge \langle X\rangle_t) - \frac{1}{2}\sigma_a \wedge \langle X\rangle_t\right)\right] \\ &= E\left[I_{\{\sigma_a \le \langle X\rangle_t\}} \exp\left(-a + \frac{1}{2}\sigma_a\right)\right] + E\left[I_{\{\sigma_a > \langle X\rangle_t\}} \exp\left(X(t) - \frac{1}{2}\langle X\rangle_t\right)\right]. \end{aligned}$$

Since

$$E\left[I_{\{\sigma_a \le \langle X\rangle_t\}} \exp\left(-a + \frac{1}{2}\sigma_a\right)\right] \le e^{-a} E\left[\exp\left(\frac{1}{2}\langle X\rangle_t\right)\right]$$

we have

$$1 = \lim_{a \to \infty} E\left[I_{\{\sigma_a > \langle X\rangle_t\}} \exp\left(X(t) - \frac{1}{2}\langle X\rangle_t\right)\right] = E\left[\exp\left(X(t) - \frac{1}{2}\langle X\rangle_t\right)\right]$$

and this completes the proof.

Remark 5.1. By a slight modification of the proof, Kazamaki [83] proved the following stronger result: if instead of (5.7) we assume that $E[\exp(X(t)/2)] < \infty$, then the same conclusion (5.8) holds.

The following result is an immediate consequence of Theorem 5.3.

Corollary. Let $X \in \mathscr{M}$. If $\langle X\rangle_t$ is locally bounded, i.e., for every $t > 0$ there exists a positive constant $C(t)$ such that

(5.10) $\qquad \langle X\rangle_t \le C(t) \qquad$ a.s.,

then M_t is a continuous $(\mathscr{F}_t)$-martingale.

6. Conformal martingales

Let $(\Omega, \mathscr{F}, P)$ and $(\mathscr{F}_t)_{t\geq 0}$ be given as usual.

Definition 6.1. By an n-dimensional conformal martingale (local conformal martingale) with respect to $(\mathscr{F}_t)$, we mean a $\boldsymbol{C}^n$-valued continuous $(\mathscr{F}_t)$-adapted process $Z_t = (Z_t^1, Z_t^2, \ldots, Z_t^n)$ such that, writing $Z_t^\alpha = X_t^\alpha + \sqrt{-1}Y_t^\alpha$, $\alpha = 1, 2, \ldots, n$,

(6.1) $\quad X_t^\alpha,\ Y_t^\alpha \in \mathscr{M}_2^c$, (resp. $X_t^\alpha,\ Y_t^\alpha \in \mathscr{M} = \mathscr{M}_2^{c,\mathrm{loc}}$), $\alpha = 1, 2, \ldots, n$

and their stochastic differentials satisfy

$$(6.2)\qquad \begin{aligned} dX_t^\alpha \cdot dX_t^\beta &= dY_t^\alpha \cdot dY_t^\beta \\ dX_t^\alpha \cdot dY_t^\beta &= -dX_t^\beta \cdot dY_t^\alpha \end{aligned} \qquad \alpha, \beta = 1, 2, \ldots, n.$$

(6.2) implies, in particular,

$$(6.3)\qquad dX_t^\alpha \cdot dY_t^\alpha = 0, \quad \alpha = 1, 2, \ldots, n.$$

A one-dimensional conformal martingale (local conformal martingale) is simply called a conformal martingale (resp. local conformal martingale). We complexify the spaces $\mathscr{M}$, $\mathscr{A}$, $\mathscr{Q}$, $d\mathscr{M}$, $d\mathscr{A}$, $d\mathscr{Q}$ in an obvious way and define, for $Z_t = X_t + \sqrt{-1}\,Y_t \in \mathscr{M}$,

$$dZ_t = dX_t + \sqrt{-1}\,dY_t \text{ and } d\bar{Z}_t = dX_t - \sqrt{-1}\,dY_t.$$

It is immediately seen that (6.2) can be equivalently written in the complex form as

$$(6.2)'\qquad dZ_t^\alpha \cdot dZ_t^\beta = 0, \qquad \alpha, \beta = 1, 2, \ldots, n.$$

In other words, $Z_t = (Z_t^1, Z_t^2, \ldots, Z_t^n)$ is an n-dimensional local conformal martingale if and only if all $Z_t^\alpha, Z_t^\alpha Z_t^\beta, \alpha, \beta = 1, 2, \ldots, n$, are $(\mathscr{F}_t)$-local martingales.

Example 6.1. Let $B_t = (B_t^1, B_t^2, \ldots B_t^{2n-1}, B_t^{2n})$ be $2n$-dimensional $(\mathscr{F}_t)$-Brownian motion and set

$$Z_t^\alpha = B_t^{2\alpha-1} + \sqrt{-1}\,B_t^{2\alpha}, \; \alpha = 1, 2, \ldots, n.$$

Then $Z_t = (Z_t^1, Z_t^2, \ldots, Z_t^n)$ is an n-dimensional conformal martingale. Z_t is called an n-dimensional complex Brownian motion.

Example 6.2. Let $(B_t^1, B_t^2, B_t^3, B_t^4)$ be a four-dimensional Brownian motion and let $\{\mathscr{F}_t^{B^1, B^2}\}$ be the proper refence family of (B_t^1, B_t^2) (*i.e.*, the filtration generated by the process (B_t^1, B_t^2)). Let A_t be a continuous $(\mathscr{F}_t^{B1,B2})$-adapted increasing process such that $A_0 = 0$ a.s. Define (Z_t^1, Z_t^2) by

$$Z_t^1 = B_t^1 + \sqrt{-1}B_t^2 \quad \text{and} \quad Z_t^2 = B_{A_t}^3 + \sqrt{-1}B_{A_t}^4.$$

If $\{\mathscr{F}_t\}$ is the proper refence family of (Z_t^1, Z_t^2) (*i.e.*, the filtration generated by the process (Z_t^1, Z_t^2)), then it is not difficult to see that Z_t is a two-dimensional local conformal martingale with respect to $(\mathscr{F}_t)$.

The following two propositions are easily obtained by Doob's optional sampling theorem (Theorem I-6.11) and Knight's theorem (Theorem II-7.3′).

Proposition 6.1. Let $A = (A_t)$ be a continuous $(\mathscr{F}_t)$-adapted process such that $A_0 = 0$ and $t \longrightarrow A_t$ is strictly increasing. Furthermore we assume that

$$\lim_{t\uparrow\infty} A_t = \infty, \quad \text{a. s.}$$

Set $C(t) = \inf\{u;\ A_u > t\}$ and $\tilde{\mathscr{F}}_t = \mathscr{F}_{c(t)}$. Then, for every n-dimensional local conformal martingale Z_t with respect to $(\mathscr{F}_t)$ the time change process $\tilde{Z}_t = Z_{c(t)}$ is an n-dimensional local conformal martingale with respect to $(\tilde{\mathscr{F}}_t)$.

Proposition 6.2. Let $Z_t = (Z_t^1, Z_t^2, \ldots, Z_t^n)$ be an n-dimensional local conformal martingale such that $dZ_t^\alpha \cdot d\bar{Z}_t^\beta = 0$, $\alpha, \beta = 1, 2, \ldots, n$, $\alpha \neq \beta$. Then there exists an n-dimensional complex Brownian motion $\zeta(t) = (\zeta^1(t), \zeta^2(t), \ldots, \zeta^n(t))$ such that

$$Z_t^\alpha = \zeta^\alpha(\langle Z^\alpha\rangle_t), \qquad \alpha = 1, 2, \ldots, n,$$

where $\langle Z^\alpha\rangle_t$ is defined to be the common processes $\langle X^\alpha\rangle_t = \langle Y^\alpha\rangle_t$.

Let $f(x^1, x^2, \ldots, x^n, y^1, y^2, \ldots, y^n)$ be a C^2-function on $\boldsymbol{R}^{2n}$. By setting $z^\alpha = x^\alpha + \sqrt{-1}y^\alpha$, it can be regarded as a C^2-function $f(z^1, z^2, \ldots, z^n)$ on $\boldsymbol{C}^n$. We introduce the differential operator in the usual way:

$$\frac{\partial}{\partial z^\alpha} = \frac{1}{2}\left(\frac{\partial}{\partial x^\alpha} - \sqrt{-1}\frac{\partial}{\partial y^\alpha}\right) \text{ and } \frac{\partial}{\partial \bar{z}^\alpha} = \frac{1}{2}\left(\frac{\partial}{\partial x^\alpha} + \sqrt{-1}\frac{\partial}{\partial y^\alpha}\right).$$

Then the Itô formula for $2n$-dimensional continuous semimartingale $\zeta_t = (X_t^1, X_t^2, \ldots, X_t^n, Y_t^1, Y_t^2 \ldots, Y_t^n)$:

$$df(\zeta_t) = \sum_{\alpha=1}^n \left(\frac{\partial f}{\partial x^i}(\zeta_t)dX_t^\alpha + \frac{\partial f}{\partial y^i}(\zeta_t)dY_t^\alpha\right)$$

$$(6.4)\qquad + \frac{1}{2}\sum_{\alpha,\beta=1}^{n}\left(\frac{\partial^2 f}{\partial x^\alpha \partial x^\beta}(\zeta_t)dX_t^\alpha \cdot dY_t^\beta + 2\frac{\partial^2 f}{\partial x^\alpha \partial y^\beta}(\zeta_t)dX_t^\alpha \cdot dY_t^\beta + \frac{\partial^2 f}{\partial y^\alpha \partial y^\beta}(\zeta_t)dY_t^\alpha \cdot dY_t^\beta\right)$$

can be rewritten, in the complex form, as follows. Setting $Z_t = (Z_t^1, Z_t^2, \ldots, Z_t^n)$, $Z_t^\alpha = X_t^\alpha + \sqrt{-1}\,Y_t^\alpha$,

$$(6.4)'\qquad \begin{aligned} df(Z_t) &= \sum_{\alpha=1}^{n}\left(\frac{\partial f}{\partial z^\alpha}(Z_t)dZ_t^\alpha + \frac{\partial f}{\partial \bar{z}^\alpha}(Z_t)d\bar{Z}_t^\alpha\right) \\ &+ \frac{1}{2}\sum_{\alpha,\beta=1}^{n}\left(\frac{\partial^2 f}{\partial z^\alpha \partial z^\beta}(Z_t)dZ_t^\alpha \cdot dZ_t^\beta + 2\frac{\partial^2 f}{\partial z^\alpha \partial \bar{z}^\beta}(Z_t)dZ_t^\alpha \cdot d\bar{Z}_t^\beta \right. \\ &\left. + \frac{\partial^2 f}{\partial \bar{z}^\alpha \partial \bar{z}^\beta}(Z_t)d\bar{Z}_t^\alpha \cdot d\bar{Z}_t^\beta\right). \end{aligned}$$

If $Z_t = (Z_t^1, Z_t^2, \ldots, Z_t^n)$ is an n-dimensional local conformal martingale, then $dZ_t^\alpha \cdot dZ_t^\beta = 0$ and hence $d\bar{Z}_t^\alpha \cdot d\bar{Z}_t^\beta = \overline{dZ_t^\alpha \cdot dZ_t^\beta} = 0$. Therefore (6.4)′ becomes a simple form:

$$(6.5)\qquad \begin{aligned} df(Z_t) &= \sum_{\alpha=1}^{n}\left(\frac{\partial f}{\partial z^\alpha}(Z_t)dZ_t^\alpha + \frac{\partial f}{\partial \bar{z}^\alpha}(Z_t)d\bar{Z}_t^\alpha\right) \\ &+ \sum_{\alpha,\beta=1}^{n}\frac{\partial^2 f}{\partial z^\alpha \partial \bar{z}^\beta}(Z_t)dZ_t^\alpha \cdot d\bar{Z}_t^\beta. \end{aligned}$$

An important consequence of (6.5) is the following: If $f(z)$ is holomorphic, *i.e.*,

$$\frac{\partial f}{\partial \bar{z}^\alpha} = 0, \qquad \alpha = 1, 2, \ldots, n,$$

we have

$$(6.6)\qquad df(Z_t) = \sum_{\alpha=1}^{n}\frac{\partial f}{\partial z^\alpha}(Z_t)dZ_t^\alpha.$$

In particular, $f(Z_t)$ is a local martingale. Furthermore, it is a local conformal martingale, because $f(z)^2$ being also holomorphic, $f(Z_t)^2$ is also a local martingale. More generally, let $f = (f^1, f^2, \ldots, f^m)$: $\boldsymbol{C}^n \longrightarrow \boldsymbol{C}^m$, be holomorphic, *i.e.*,

$$\frac{\partial f^i}{\partial \bar{z}^\alpha} = 0, \qquad i = 1, 2, \ldots, m \quad \text{and} \quad \alpha = 1, 2, \ldots, n.$$

Then

$$df^i(Z_t) = \sum_{\alpha=1}^{n} \frac{\partial f^i}{\partial z^\alpha}(Z_t) dZ_t^\alpha, \qquad i = 1, 2, \ldots, m$$

and

$$d((f^i f^j)(Z_t)) = \sum_{\alpha=1}^{n} \frac{\partial (f^i f^j)}{\partial z^\alpha}(Z_t) dZ_t^\alpha, \qquad i, j = 1, 2, \ldots, m$$

are local martingales and hence $f(Z_t) = (f^1(Z_t), f^2(Z_t), \ldots, f^n(Z_t))$ is an m-dimensional local conformal martingale. Thus we obtain

Theorem 6.3. Let $Z_t = (Z_t^1, Z_t^2, \ldots, Z_t^n)$ be an n-dimensional local conformal martingale and $f = (f^1, f^2, \ldots, f^n)$: $C^n \longrightarrow C^m$ be holomorphic. Then

$$f(Z_t) = (f^1(Z_t), f^2(Z_t), \ldots, f^m(Z_t))$$

is an m-dimensional local conformal martingale.

Corollary. Let $Z_t = (Z_t^1, Z_t^2, \ldots, Z_t^n)$ be a C^n-valued continuous $(\mathscr{F}_t)$-adapted process. Z_t is an n-dimensional local conformal martingale with respect to $(\mathscr{F}_t)$ if and only if for every holomorphic function f: $C^n \longrightarrow C$, $f(Z_t)$ is a local $(\mathscr{F}_t)$-martingale.

Indeed, "if" part follows by taking $f = z^\alpha$ and $f = z^\alpha z^\beta$, $\alpha, \beta = 1, 2, \ldots, n$. "only if" part follows at once from Theorem 6.3.

Remark 6.1. It $Z_t = (Z_t^1, Z_t^2, \ldots, Z_t^n)$ satisfies that $Z_t \in D$ for all $t \geqq 0$, a.s., where D is a domain in C^n, then the above results remain valid for every f: $D \longrightarrow C^m$ which is holomorphic in D.

CHAPTER IV

Stochastic Differential Equations

1. Definition of solutions

Let $\boldsymbol{R}^d$ be the d-dimensional Euclidean space and let $\boldsymbol{W}^d = \boldsymbol{C}([0, \infty) \longrightarrow \boldsymbol{R}^d)$ be the space of all continuous functions w defined on $[0,\infty)$ with values in $\boldsymbol{R}^d$. For $w_1, w_2 \in \boldsymbol{W}^d$, let

$$\rho(w_1, w_2) = \sum_{k=1}^{\infty} 2^{-k}\big(\max_{0\le t\le k} |w_1(t) - w_2(t)| \wedge 1\big),$$

where $|\cdot|$ denotes the Euclidean metric in $\boldsymbol{R}^d$ (see Chapter I, Section 4). $\boldsymbol{W}^d$ is a complete separable metric space under this metric ρ. Let $\mathscr{B}(\boldsymbol{W}^d)$ be the topological σ-field on $\boldsymbol{W}^d$ and $\mathscr{B}_t(\boldsymbol{W}^d)$ be the sub-σ-field of $\mathscr{B}(\boldsymbol{W}^d)$ generated by $w(s)$, $0 \le s \le t$. In other words, $\mathscr{B}_t(\boldsymbol{W}^d)$ is the inverse σ-field $\rho_t^{-1}[\mathscr{B}(\boldsymbol{W}^d)]$ of $\mathscr{B}(\boldsymbol{W}^d)$ under the mapping $\rho_t\colon \boldsymbol{W}^d \longrightarrow \boldsymbol{W}^d$ defined by

$$(\rho_t w)(s) = w(t \wedge s).$$

Definition 1.1. We shall denote by $\mathscr{A}^{d,r}$ the set of all functions $\alpha(t, w)\colon [0, \infty)\times \boldsymbol{W}^d \longrightarrow \boldsymbol{R}^d\otimes\boldsymbol{R}^r$ such that

(i) it is $\mathscr{B}([0,\infty))\times\mathscr{B}(\boldsymbol{W}^d)/\mathscr{B}(\boldsymbol{R}^d\otimes\boldsymbol{R}^r)$-measurable, and

(ii) for each $t \in [0, \infty)$, $\boldsymbol{W}^d \ni w \longmapsto \alpha(t, w) \in \boldsymbol{R}^d\otimes\boldsymbol{R}^r$ is $\mathscr{B}_t(\boldsymbol{W}^d)/\mathscr{B}(\boldsymbol{R}^d\otimes\boldsymbol{R}^r)$-measurable.

Here we denote by $\boldsymbol{R}^d\otimes\boldsymbol{R}^r$ the totality of real $d\times r$ matrices; $\mathscr{B}(\boldsymbol{R}^d\otimes\boldsymbol{R}^r)$ is the topological σ-field on $\boldsymbol{R}^d\otimes\boldsymbol{R}^r$ obtained by identifying $\boldsymbol{R}^d\otimes\boldsymbol{R}^r$ with dr-dimensional Euclidean space.

We shall denote the (i,j)-entry of the matrix $\alpha(t,w)$ by $\alpha^i_j(t,w)$, $i = 1, 2, \ldots, d$, $j = 1, 2, \ldots, r$.

Suppose we are given $\alpha \in \mathscr{A}^{d,r}$ and $\beta \in \mathscr{A}^{d,1}$. Consider the following *stochastic differential equation* for a d-dimensional continuous process $X=(X(t))_{t\ge0}$:

$$(1.1) \qquad dX_t^i = \sum_{j=1}^{r} \alpha_j^i(t,X)dB^j(t) + \beta^i(t,X)dt \qquad i = 1, 2, \cdots, d$$

or sometimes simply written as

$$(1.1)' \qquad dX_t = \alpha(t,X)dB(t) + \beta(t,X)dt.$$

A precise formulation is as follows.

Definition 1.2. Let $\alpha = (\alpha_j^i(t,w)) \in \mathscr{A}^{d,r}$ and $\beta = (\beta^i(t,w)) \in \mathscr{A}^{d,1}$ be given. By a solution of the equation (1,1), we mean a d-dimensional continuous stochastic process $X=(X(t))_{t\geq 0}$ defined on a probability space $(\Omega,\mathscr{F},P)$ with a reference family $(\mathscr{F}_t)_{t\geq 0}$ such that

(i) there exists an r-dimensional $(\mathscr{F}_t)$-Brownian motion*[1] $B = (B(t))$ with $B(0) = 0$ a.s.;

(ii) $X=(X(t))$ is a d-dimensional continuous process adapted to $(\mathscr{F}_t)_{t\geq 0}$, i.e., X is a mapping: $\omega \in \Omega \longmapsto X(\omega) \in \boldsymbol{W}^d$ such that, for each $t \in [0, \infty)$, it is $\mathscr{F}_t/\mathscr{B}_t(\boldsymbol{W}^d)$-measurable;

(iii) the family of adapted processes $\Phi_j^i(t, \omega)$ and $\psi^i(t, \omega)$ defined by

$$\Phi_j^i(t, \omega) = \alpha_j^i(t,X(\omega))$$

and

$$\psi^i(t, \omega) = \beta^i(t,X(\omega))$$

belong to the spaces $\mathscr{L}_2^{loc}$ *[2] and $\mathscr{L}_1^{loc}$ respectively, where $\mathscr{L}_1^{loc}$ is the set of all measurable $(\mathscr{F}_t)$-adapted processes Ψ such that for every $t \geq 0$, $\int_0^t |\Psi(s,\omega)|\,ds < \infty$ a.s.*[3];

(iv) with probability one, $X(t)=(X^1(t),X^2(t), \ldots ,X^d(t))$ and $B(t) =(B^1(t), B^2(t), \ldots ,B^r(t))$ satisfy

$$(1.2) \qquad X^i(t) - X^i(0) = \sum_{j=1}^{r} \int_0^t \alpha_j^i(s,X)dB^j(s) + \int_0^t \beta^i(s,X)ds,$$
$$i = 1, 2, \ldots, d,$$

where the integral by $dB^j(s)$ is Itô's stochastic integral as defined in Chapter II, Section 1.

In equation (1.2), the first term on the right-hand side is called the *martingale term* and the second term is called the *drift term*.

To emphasize the particular role of the $(\mathscr{F}_t)$-Brownian motion $B =$

*[1] Cf. Chapter I, Section 7.

*[2] Cf. Chapter II, Definition 1.6.

*[3] Ψ, $\Psi' \in \mathscr{L}_1^{loc}$ are identified if $\int_0^t |\Psi(s,\omega)-\Psi'(s,\omega)|\,ds = 0$ for every $t \geq 0$ a.s.

$(B(t))$ in Definition 1.2, we call $X=(X(t))$ a *solution of* (1.1) *with the Brownian motion* $B=(B(t))$, or sometimes we call the pair (X,B) itself a solution of (1.1).

Remark 1.1. The condition (iii) of Definition 1.2 is satisfied if α and β are bounded* or more generally, if

$$\sup\{\|\alpha(t,w)\|+\|\beta(t,w)\|;\ t\in[0,T],\ \|w\|_T\le M\}<\infty$$

for every T and $M>0$, where

$$\|w\|_T=\max_{0\le t\le T}\|w(t)\| \quad\text{and}\quad \|a\|=\sqrt{\sum_{i=1}^{d}\sum_{j=1}^{r}|a_j^i|^2} \qquad \text{for } a\in\boldsymbol{R}^d\otimes\boldsymbol{R}^r.$$

The stochastic differential equations which are most important and which are mainly studied in this book are of the following type.

Definition 1.3. Let $\sigma(t,x)=(\sigma_j^i(t,x))$ be a Borel measurable function $(t,x)\in[0,\infty)\times\boldsymbol{R}^d\longrightarrow\boldsymbol{R}^d\otimes\boldsymbol{R}^r$ and $b(t,x)=(b^i(t,x))$ be a Borel measurable function $(t,x)\in[0,\infty)\times\boldsymbol{R}^d\longrightarrow\boldsymbol{R}^d$. Then $\alpha(t,w)$ and $\beta(t,w)$ defined by $\alpha(t,w)=\sigma(t,w(t))$ and $\beta(t,w)=b(t,w(t))$ clearly satisfy $\alpha\in\mathscr{A}^{d,r}$, $\beta\in\mathscr{A}^{d,1}$. In such a case, the stochastic differential equation (1,1) is said to be of the *Markovian type*. The equation then has the following form:

$$dX(t)=\sigma(t,X(t))dB(t)+b(t,X(t))dt \tag{1.3}$$

or, in terms of its components,

$$dX^i(t)=\sum_{k=1}^{r}\sigma_k^i(t,X(t))dB^k(t)+b^i(t,X(t))dt,\quad i=1,2,\ldots,d. \tag{1.3$'$}$$

Furthermore, if σ and b do not depend on t and are functions of $x\in\boldsymbol{R}^d$ alone, then the equation (1.1) is said to be of the *time-independent* (or *time homogeneous*) *Markovian type*.

Note that an equation of Markovian type reduces to a system of ordinary differential equations (a dynamical system) $\dot{X}_t=b(t,X_t)$ when $\sigma\equiv 0$. Thus a stochastic differential equation generalizes the notion of an ordinary differential equation by adding the effect of random fluctuation.

Now we will present several definitions concerning the uniqueness of solutions. Given $\alpha\in\mathscr{A}^{d,r}$ and $\beta\in\mathscr{A}^{d,1}$, we consider the stochastic differential equation (1.1). We suppose that at least one solution of (1.1) exists.

* i.e., all components of α and β are bounded.

Definition 1.4. We say that the *uniqueness of solutions for* (1.1) *holds* if whenever X and X' are two solutions*[1] whose initial laws*[2] on $\boldsymbol{R}^d$ coincide, then the laws of the processes X and X' on the space $\boldsymbol{W}^d$ coincide.

Remark 1.2. The above definition is equivalent to the following: *the uniqueness of solutions for* (1.1) *holds if whenever X and X' are two solutions of* (1.1) *such that* $X(0) = x$ a.s. *and* $X'(0) = x$ a.s. *for some* $x \in \boldsymbol{R}^d$, *then the laws on the space* $\boldsymbol{W}^d$ *of the processes X and X' coincide.* The equivalence is easily seen if we notice the following fact: if X is a solution of (1.1) on the space $(\Omega,\mathscr{F},P)$ with $(\mathscr{F}_t)_{t\geq 0}$, then, setting $P^\omega = P(\cdot\mid\mathscr{F}_0)$,*[3] we have, for almost all fixed ω, that X is a solution of (1.1) on the space $(\Omega,\mathscr{F},P^\omega)$ with $(\mathscr{F}_t)_{t\geq 0}$ such that $X(0) = X(0,\omega)$ (cf. the corollary of Theorem I-3.2).

The uniqueness defined in Definition 1.4 is sometimes called "*the uniqueness in the sense of probability law*". On the other hand if we consider stochastic differential equations as a tool for defining sample paths of a random process as functionals of Brownian paths, then the following definition might be more natural.

Definition 1.5. (pathwise uniqueness). We say that the *pathwise uniqueness of solutions for* (1.1) *holds* if whenever X and X' are any two solutions defined on the same probability space $(\Omega,\mathscr{F},P)$ with the same reference family $(\mathscr{F}_t)$ and the same r-dimensional $(\mathscr{F}_t)$-Brownian motion such that $X(0) = X'(0)$ a.s., then $X(t) = X'(t)$ for all $t \geq 0$ a.s.

Remark 1.3. We may also consider the following more strict definition of pathwise uniqueness.

We say that the pathwise uniqueness holds (in the strict sense) if whenever X and X' are two solutions such that $X(0) = X'(0)$ a.s. which are defined on the same probability space $(\Omega,\mathscr{F},P)$ with the reference families $(\mathscr{F}_t)$ and $(\mathscr{H}_t)$ respectively, and with the same Brownian motion $B(t)$ which is both an $(\mathscr{F}_t)$- and $(\mathscr{H}_t)$-Brownian motion, then $X(t) = X'(t)$ for all $t \geq 0$ a.s.

Since it is not necessarily true that $B(t)$ is $(\mathscr{F}_t\vee\mathscr{H}_t)$-Brownian motion, the equivalence of this strict definition and Definition 1.5 is not trivial. However it can be proved as an easy consequence of the following Theorem 1.1.

---

*[1] They may be defined on different probability spaces.

*[2] The law of $X(0)$ of a solution X of (1.1) is called the *initial law* or *initial distribution* of the solution.

*[3] $P(\cdot\mid\mathscr{F}_0)$ stands for the regular conditional probability. We can always represent any solution (X,B) on a standard measurable space $(\Omega,\mathscr{F})$ without changing the law of (X, B).

Remark 1.4. Just as in Remark 1.2, we need only consider non-random initial values; i.e., $X(0)=X'(0)=x$ a.s., for some fixed $x\in \boldsymbol{R}^d$.

To understand some of the implications of pathwise uniqueness, it is convenient to introduce the following notion.

In the following, a function $\Phi(x, w): \boldsymbol{R}^d \times \boldsymbol{W}_0^r \longrightarrow \boldsymbol{W}^d$ *1 is called $\hat{\mathscr{E}}(\boldsymbol{R}^d\times \boldsymbol{W}_0^r)$-measurable if, for any Borel probability measure μ on $\boldsymbol{R}^d$, there exists a function $\tilde{\Phi}_\mu(x,w): \boldsymbol{R}^d \times \boldsymbol{W}_0^r \longrightarrow \boldsymbol{W}^d$ which is $\overline{\mathscr{B}(\boldsymbol{R}^d\times \boldsymbol{W}_0^r)}^{\mu\times P^W}/\mathscr{B}(\boldsymbol{W}^d)$-measurable and for almost all x (μ) it holds $\Phi(x, w) = \tilde{\Phi}_\mu(x, w)$, a.a. $w(P^W)$. *2 For such a function $\Phi(x, w)$ and for an $\boldsymbol{R}^d$-valued random variable ξ and an r-dimensional Brownian motion $B = (B(t))$ which are mutually independent, we set $\Phi(\xi, B) := \tilde{\Phi}_\mu(\xi, B)$ where μ is the law of ξ. By this it is a well-defined $\boldsymbol{W}^d$-valued random variable.

Definition 1.6. (strong solution). A solution $X = (X(t))$ of (1.1) with a Brownian motion $B=(B(t))$ is called a *strong solution* if there exists a function $F(x,w)$: $\boldsymbol{R}^d\times \boldsymbol{W}_0^r \longrightarrow \boldsymbol{W}^d$ *1 which is $\hat{\mathscr{E}}(\boldsymbol{R}^d\times \boldsymbol{W}_0^r)$-measurable and, for each $x \in \boldsymbol{R}^d$, $w \longmapsto F(x, w)$ is $\overline{\mathscr{B}_t(\boldsymbol{W}_0^r)}^{P^W}/\mathscr{B}_t(\boldsymbol{W}^d)$-measurable for every $t \geq 0$ and it holds

$$(1.4) \qquad X = F(X(0),B) \quad \text{a.s.}$$

We shall say that *the equation* (1.1) *has a unique strong solution* if there exists a function $F(x, w)$: $\boldsymbol{R}^d\times \boldsymbol{W}_0^r \longrightarrow \boldsymbol{W}^d$ with the same properties as above such that the following is true:

(i) for any r-dimensional $(\mathscr{F}_t)$-Brownian motion $B=(B(t))$ $(B(0) = 0)$ on a probability space with a reference family $(\mathscr{F}_t)$ and any $\boldsymbol{R}^d$-valued random variable ξ which is $\mathscr{F}_0$-measurable, the continuous process $X= F(\xi,B)$ is a solution of (1.1) on this space with $X(0) = \xi$ a.s.;

(ii) for any solution (X, B) of (1.1), $X = F(X(0),B)$ holds a.s.

Thus, a strong solution may be regarded as the function $F(x, w)$ which produces a solution X of (1.1) if we substitute an initial value $X(0)$ and a Brownian motion B.

Theorem 1.1. Given $\alpha \in \mathscr{A}^{d,r}$ and $\beta \in \mathscr{A}^{d,1}$, the equation (1.1) has a unique strong solution if and only if for any Borel probability measure μ on $\boldsymbol{R}^d$, a solution X of (1.1) exists such that the law of initial value $X(0)$ coincides with μ and the pathwise uniqueness of solutions holds.

Proof. If a unique strong solution of (1.1) exists, this means by defi-

*1 $\boldsymbol{W}_0^r = \{w \in \boldsymbol{C}([0, \infty) \longrightarrow \boldsymbol{R}^r);\ w(0) = 0\}$.
*2 P^W is the (r-dimensional) Wiener measure on $\boldsymbol{W}_0^r$ (i.e., the probability law of B).

nition that a function $F(x, w)$: $\boldsymbol{R}^d\times \boldsymbol{W}_0^r \longrightarrow \boldsymbol{W}^d$ exists such that (i) and (ii) above hold. So for a given Borel probability measure μ on $\boldsymbol{R}^d$, let $B=(B(t))$ be an r-dimensional $(\mathscr{F}_t)$-Brownian motion and ξ be an $\mathscr{F}_0$-measurable $\boldsymbol{R}^d$-valued random variable which is distributed as μ defined on some suitable probability space with a reference family $(\mathscr{F}_t)$. Then if we define a continuous process X by $X=F(\xi,B)$, X is a solution of (1.1) such that $X(0)=\xi$ a.s.

Also, if two solutions (X,B) and (X',B') exist on the same probability space such that $B(t)=B'(t)$ and $X(0)=X'(0)$ a.s., then $X=F(X'(0),B) = F(X'(0),B')=X'$. This implies that the pathwise uniqueness of solutions holds.*[1]

Thus what we have to prove is that the existence of a solution for each given initial distribution and the pathwise uniqueness imply the existence of a unique strong solution. So let us assume that for any initial distribution a solution of (1.1) exists and the pathwise uniqueness of solutions holds. Let $x\in \boldsymbol{R}^d$ be fixed and let (X,B) and (X',B') be any solutions of (1.1)*[2] such that $X(0)=x$ and $X'(0)=x$, a.s. Let P_x and P'_x be the probability distributions of (X,B) and (X',B') on the space $\boldsymbol{W}^d\times \boldsymbol{W}_0^r$ respectively. If π: $\boldsymbol{W}^d\times \boldsymbol{W}_0^r \ni (w_1, w_2)\longmapsto w_2 \in \boldsymbol{W}_0^r$ is the projection, then both marginal distributions $\pi(P_x)$ and $\pi(P'_x)$ coincide with P^W, the Wiener measure on $\boldsymbol{W}_0^r$. Let $Q^{w_2}(dw_1)$ and $Q'^{w_2}(dw_1)$ be the regular conditional distributions of w_1 given w_2; that is,

(i) for a fixed $w_2\in \boldsymbol{W}_0^r$, $Q^{w_2}(dw_1)$ is a probability measure on $(\boldsymbol{W}^d, \mathscr{B}(\boldsymbol{W}^d))$,

(ii) for a fixed $A\in\mathscr{B}(\boldsymbol{W}^d)$, $w_2\longmapsto Q^{w_2}(A)$ is $\overline{\mathscr{B}(\boldsymbol{W}_0^r)}^{P^W}$-measurable,

(iii) for every $A_1\in\mathscr{B}(\boldsymbol{W}^d)$ and $A_2\in\mathscr{B}(\boldsymbol{W}_0^r)$,

$$P_x(A_1\times A_2)=\int_{A_2} Q^{w_2}(A_1)P^W(dw_2).$$

Q'^{w_2} is defined similarly from P'_x. We define, on the space $\Omega=\boldsymbol{W}^d\times \boldsymbol{W}^d\times \boldsymbol{W}_0^r$, a Borel probability measure Q by

$$Q(dw_1dw_2dw_3)=Q^{w_3}(dw_1)Q'^{w_3}(dw_2)P^W(dw_3).$$

Let $\mathscr{F}$ be the completion of the topological σ-field $\mathscr{B}(\Omega)$ by Q, and $\mathscr{F}_t = \bigcap_\varepsilon(\mathscr{B}_{t+\varepsilon}\vee\mathscr{N})$, where $\mathscr{B}_t=\mathscr{B}_t(\boldsymbol{W}^d)\times\mathscr{B}_t(\boldsymbol{W}^d)\times\mathscr{B}_t(\boldsymbol{W}_0^r)$ and $\mathscr{N}$ is the set of all Q-null sets. Then clearly (w_1,w_3) and (X,B) have the same distribution and as does (w_2, w_3) and (X',B').

---

*[1] In fact, it implies the pathwise uniqueness in the strict sense of Remark 1.3.

*[2] They may be defined on different probability spaces.

In order to complete our argument, we first need to prove two subsidiary lemmas.

Lemma 1.1. For $A \in \mathscr{B}_t(W^d)$, $w \in W_0^r \longmapsto Q^w(A)$ or $Q'^w(A)$ is $\overline{\mathscr{B}_t(W_0^r)}^{P^W}$-measurable.

Proof. For fixed $t > 0$ and $A \in \mathscr{B}_t(W^d)$, there exists a conditional probability $Q_t^w(A)$ such that $w \in W_0^r \longmapsto Q_t^w(A)$ is $\overline{\mathscr{B}_t(W_0^r)}^{P^W}$-measurable and $P_x(A \times C) = \int_C Q_t^w(A) P^W(dw)$ for every $C \in \mathscr{B}_t(W_0^r)$. If we can show that this equality holds for all $C \in \mathscr{B}(W_0^r)$, then this implies that $Q_t^w(A) = Q^w(A)$ a.a. $w(P^W)$ and the assertion of the lemma holds. We may assume that C is of the form

$$C = \{w \in W_0^r;\, \rho_t w \in A_1,\, \theta_t w \in A_2\}, \quad A_1, A_2 \in \mathscr{B}(W_0^r),$$

where θ_t is defined by $(\theta_t w)(s) = w(t+s) - w(t)$.

Then since $\theta_t w$ and $\mathscr{B}_t(W_0^r)$ are independent with respect to P^W, we have

$$\begin{aligned}
&\int_C Q_t^w(A) P^W(dw) \\
&= \int_{\{\rho_t w \in A_1\}} Q_t^w(A) P^W(dw) P^W(\theta_t w \in A_2) \\
&= P_x(A \times \{\rho_t w \in A_1\}) P^W(\theta_t w \in A_2) \\
&= P_x(A \times \{\rho_t w \in A_1\}) P_x(W^d \times \{\theta_t w \in A_2\}) \\
&= P(X \in A,\, \rho_t(B) \in A_1) P(\theta_t(B) \in A_2) \\
&= P(X \in A,\, \rho_t(B) \in A_1,\, \theta_t(B) \in A_2) \\
&= P(X \in A,\, B \in C) \\
&= P_x(A \times C)
\end{aligned}$$

since $\{X \in A,\, \rho_t(B) \in A_1\} \in \mathscr{F}_t$ and $\theta_t(B)$ and $\mathscr{F}_t$ are independent. This proves the lemma.

Lemma 1.2. $w_3 = (w_3(t))$ is an r-dimensional $(\mathscr{F}_t)$-Brownian motion on $(\Omega, \mathscr{F}, Q)$.

Proof. It is only necessary to prove the independence of $w_3(t) - w_3(s)$ and $\mathscr{F}_s$ for every $t > s$. For this, it is sufficient to prove that

$$E^Q[e^{i\langle \xi, w_3(t) - w_3(s)\rangle} I_{A_1 \times A_2 \times A_3}] \, *$$

* E^Q stands for expectation with respect to the probability Q.

$$= \exp\,[-(|\xi|^2/2)(t-s)]Q(A_1\times A_2\times A_3)$$
$$\text{for } \xi\in \boldsymbol{R}^r,\quad A_1,\ A_2\in\mathscr{B}_s(\boldsymbol{W}^d) \text{ and } A_3\in\mathscr{B}_s(\boldsymbol{W}_0^r).$$

But using Lemma 1.1, we have that

the left hand side

$$= \int_{A_3} e^{i\langle\xi, w_3(t)-w_3(s)\rangle}\, Q^{w_3}(A_1)Q'^{w_3}(A_2)P^W(dw_3)$$
$$= \exp\,[-(|\xi|^2/2)(t-s)]\int_{A_3} Q^{w_3}(A_1)Q'^{w_3}(A_2)P^W(dw_3)$$
$$= \exp\,[-(|\xi|^2/2)(t-s)]Q(A_1\times A_2\times A_3).$$

Now returning to the proof of Theorem 1.1, we conclude from Lemma 1.2 that (w_1,w_3) and (w_2,w_3) are solutions on the *same* space $(\Omega,\mathscr{F},Q)$ with the *same* reference family $(\mathscr{F}_t)$. Hence the pathwise uniqueness implies that $w_1 = w_2$, Q-a.s. This implies that $Q^w\times Q'^w(w_1 = w_2) = 1$ P^W-a.s. Now it is easy to see that there exists a function $w\in \boldsymbol{W}_0^r \longmapsto F_x(w)\in \boldsymbol{W}^d$ such that $Q^w = Q'^w = \delta_{\{F_x(w)\}}$, P^W-a.s. By Lemma 1.1, this function $F_x(w)$ is $\overline{\mathscr{B}_t(\boldsymbol{W}_0^r)}^{P^W}/\mathscr{B}_t(\boldsymbol{W}^d)$-measurable. Clearly $F_x(w)$ is uniquely determined up to P^W-measure 0.

Next, let μ be any given Borel measure on $\boldsymbol{R}^d$ and let (X,B) be any solution of (1.1) such that $X(0)$ is distributed as μ. Then (X,B) is also a solution on $(\Omega,\mathscr{F},P(\cdot\,/\mathscr{F}_0))$ with respect to $(\mathscr{F}_t)$ and hence $P(F_{X(0)}(B) = X|\mathscr{F}_0) = 1$. From this it is easy to conclude that $F_x(w)$ is $\hat{\mathscr{E}}(\boldsymbol{R}^d\times \boldsymbol{W}_0^r)$-measurable and $X = F_{X(0)}(B)$ a.s. Thus the existence of a unique strong solution is now proved.

Corollary. The pathwise uniqueness of solutions implies the uniqueness of solutions (Definition 1.4).

Indeed, in the above proof, we showed that $P_x = P_x'$ which means that the laws of (X,B) and (X',B') coincide. Then, of course, the laws of X and X' coincide. This implies the uniqueness in the sense of Definition 1.4 (cf. Remark 1.2).

Finally, we shall give an example of a stochastic differential equation for which the uniqueness of solutions holds but the pathwise uniqueness does not hold. This example is due to H. Tanaka.

Example 1.1. Consider the following one-dimensional stochastic differential equation of the time-homogeneous Markovian type:

(1.5) $dX(t) = \sigma(X(t))dB(t)$,

where $\sigma(x)=1$ for $x \geq 0$ and $\sigma(x)= -1$ for $x < 0$. For any Borel probability μ on $\boldsymbol{R}^1$ there exists a solution $X(t)$, unique in the law sense, such that the law of $X(0)$ coincides with μ. Indeed, let $B = (B(t))$ be an $(\mathscr{F}_t)$-Brownian motion and let ξ be an $\mathscr{F}_0$-measurable random variable having the distribution μ defined on some suitable probability space with a reference family $(\mathscr{F}_t)$. Set $X(t) = \xi + B(t)$. Then $\tilde{B}(t) = \int_0^t \sigma(X(s))dB(s)$ is an $(\mathscr{F}_t)$-Brownian motion by Theorem II-6.1 and $X(t) = \xi + \int_0^t \sigma(X(s)) d\tilde{B}(s)$, that is, $(X(t), \tilde{B}(t))$ is a solution with the initial value $X(0)=\xi$ distributed as μ. The uniqueness in the sense of law is clear since for any solution $(X(t), B(t))$, $\int_0^t \sigma(X(s))dB(s)$ is a Brownian motion which is independent of $X(0)$.

However the pathwise uniqueness of solutions does not hold for (1.5). For example, if $(X(t), B(t))$ is a solution such that $X(0) = 0$, then $(-X(t), B(t))$ is also a solution. In this case we can prove that $\sigma[B(s); s \leq t] = \sigma[|X(s)|; s \leq t]$; indeed, as we saw in the proof of Theorem III-4.2,

$$|X(t)| = \int_0^t \sigma(X(s))dX(s) + \phi(t) = B(t) + \phi(t),$$

where $\phi(t) = \lim_{\varepsilon \downarrow 0} \frac{1}{2\varepsilon} \int_0^t I_{[0,\varepsilon)}(|X(s)|)ds$. Thus $\sigma[B(s); s \leq t] \subset \sigma[|X(s)|; s \leq t]$. Also it follows from the proof of the same theorem that $|X(t)| = B(t) - \min_{0 \leq s \leq t} B(s)$. This proves the converse inclusion. This relation of σ-fields implies immediately that no strong solution exists for the equation (1.5).

Another example will be given in Example IV-4.1.

2. Existence theorem*[1]

Consider the stochastic differential equation

(1.1) $dX(t) = \alpha(t,X)dB(t) + \beta(t,X)dt$,

where $\alpha \in \mathscr{A}^{d,r}$ and $\beta \in \mathscr{A}^{d,1}$. For $f \in \boldsymbol{C}_b^2(\boldsymbol{R}^d)$,*[2] let

*[1] An existence theorem for stochastic differential equations was first obtained by Skorohod [150].

*[2] $\boldsymbol{C}_b^m(\boldsymbol{R}^d)$ = the set of real m-times continuously differentiable functions which are bounded together with their derivatives up to the m-th order.

(2.1) $$(Af)(t,w) = \frac{1}{2}\sum_{i,j=1}^{d} a^{ij}(t,w)\frac{\partial^2 f}{\partial x^i \partial x^j}(w(t)) + \sum_{i=1}^{d}\beta^i(t,w)\frac{\partial f}{\partial x^i}(w(t)),$$
$$t \in [0, \infty),\ w \in \boldsymbol{W}^d,$$

where

(2.2) $$a^{ij}(t, w) = \sum_{k=1}^{r} \alpha^i_k(t, w)\alpha^j_k(t, w).$$

If $(X(t),B(t))$ is a solution of (1.1) on a probability space $(\Omega,\mathscr{F},P)$ with a reference family $(\mathscr{F}_t)$, then by Itô's formula we have

(2.3) $$\begin{aligned} &f(X(t)) - f(X(0)) - \int_0^t (Af)(s,X)ds \\ &\quad = \sum_{i=1}^{d}\sum_{k=1}^{r}\int_0^t \alpha^i_k(s,X)\frac{\partial f}{\partial x^i}(X(s))dB^k(s) \end{aligned}$$

and hence

(2.4) $$f(X(t)) - f(X(0)) - \int_0^t (Af)(s, X)ds \in \mathscr{M}_2^{c,loc}$$
$$\text{for every}\quad f \in \boldsymbol{C}_b^2(\boldsymbol{R}^d).$$

Conversely, if a d-dimensional continuous adpated process $X=(X(t))$ defined on a probability space $(\Omega,\mathscr{F},P)$ with a reference family $(\mathscr{F}_t)$ satisfies (2.4), then on an extension $(\tilde{\Omega},\tilde{\mathscr{F}},\tilde{P})$ and $(\tilde{\mathscr{F}}_t)$ of $(\Omega, \mathscr{F}, P)$ and $(\mathscr{F}_t)$ we can find an r-dimensional $(\tilde{\mathscr{F}}_t)$-Brownian motion $B = (B(t))$ such that (X, B) is a solution of (1.1). Indeed, let $\boldsymbol{B}_l = \{x \in \boldsymbol{R}^d;\ |x| \leq l\}$ and, for each i, choose $f(x) \in \boldsymbol{C}_b^2(\boldsymbol{R}^d)$ such that $f(x) = x_i$ if $x \in \boldsymbol{B}_l$. Then setting $\sigma_l = \inf\{t;\ X(t) \notin \boldsymbol{B}_l\}$, $l = 1, 2, \ldots$, we see that

$$M_i^{(l)}(t) = X^i(t\wedge\sigma_l) - X^i(0) - \int_0^{t\wedge\sigma_l} \beta^i(s,X)ds \in \mathscr{M}_2^c,$$
$$i = 1, 2, \ldots, d.$$

Thus,

$$M_i(t) = X^i(t) - X^i(0) - \int_0^t \beta^i(s,X)ds \in \mathscr{M}_2^{c,loc},\quad i = 1, 2, \ldots, d.$$

By choosing $f \in \boldsymbol{C}_b^2(\boldsymbol{R}^d)$ such that $f(x) = x^i x^j$, $x \in \boldsymbol{B}_l$, we see similarly that

$$(2.5) \qquad \langle M_i, M_j\rangle(t) = \int_0^t a^{ij}(s,X)ds.$$

By Theorem II-7.1′, we can find an r-dimensional $(\mathscr{F}_t)$-Brownian motion $B = (B(t))$ on an extension $(\tilde{\Omega},\tilde{\mathscr{F}},\tilde{P})$ and $(\tilde{\mathscr{F}}_t)$ such that

$$M_i(t) = \sum_{k=1}^{r} \int_0^t \alpha_k^i(s,X)dB^k(s), \quad i = 1, 2, \ldots, d.$$

Hence (X, B) is a solution of (1.1).

If X satifies (2.4), then its probability law P^X on $(\boldsymbol{W}^d,\mathscr{B}(\boldsymbol{W}^d))$ satisfies

$$(2.6) \qquad f(w(t)) - f(w(0)) - \int_0^t (Af)(s,w)ds \in \mathscr{M}_2^{c,loc} \text{ *1}$$

for every $f \in C_b^2(\boldsymbol{R}^d)$. Clearly, $X(t,w) = w(t)$ is a stochastic process on $(\boldsymbol{W}^d,\mathscr{B}(\boldsymbol{W}^d),P^X)$ with $(\mathscr{B}_{t+}(\boldsymbol{W}^d))$ satisfying (2.4). Thus we have the following result.

Proposition 2.1. The existence of a solution of (1.1) is equivalent to the existence of a d-dimensional continuous process X satisfying (2.4), and this is also equivalent to the existence of a probability P on $(\boldsymbol{W}^d,\mathscr{B}(\boldsymbol{W}^d))$ satisfying (2.6).

Theorem 2.2. Suppose that $\alpha \in \mathscr{A}^{d,r}$ and $\beta \in \mathscr{A}^{d,1}$ are bounded and continuous.*2 Then, for any given probability μ on $(\boldsymbol{R}^d,\mathscr{B}(\boldsymbol{R}^d))$ with compact support, there exists a solution (X, B) of the equation (1.1) such that the law of $X(0)$ coincides with μ, i.e., $P\{X(0) \in A\} = \mu(A)$ for any $A \in \mathscr{B}(\boldsymbol{R}^d)$.

Proof. By Proposition 2.1, it is sufficient to construct a process X with the property (2.4) and $P\{X(0) \in A\} = \mu(A)$ for every $A \in \mathscr{B}(\boldsymbol{R}^d)$. For each $l = 1, 2, \ldots$, let $\phi_l(t)$ be defined by

$$\phi_l(t) = k/2^l \qquad \text{for} \quad k/2^l \le t < (k+1)/2^l \quad (k = 0, 1, 2, \ldots),$$

and set $\alpha_l(t, w) = \alpha(\phi_l(t), w)$ and $\beta_l(t, w) = \beta(\phi_l(t), w)$. Clearly, $\alpha_l \in \mathscr{A}^{d,r}$ and $\beta_l \in \mathscr{A}^{d,1}$. On a probability space $(\Omega,\mathscr{F},P)$ with a reference family $(\mathscr{F}_t)$ we prepare an r-dimensional $(\mathscr{F}_t)$-Brownian motion $B = (B(t))$

---

*1 $\mathscr{M}_2^{c,loc}$ is defined with respect to the reference family $\mathscr{B}_{t+}(\boldsymbol{W}^d)$.

*2 i.e., the function $[0, \infty) \times \boldsymbol{W}^d \ni (t,w) \longmapsto \alpha(t,w) \in \boldsymbol{R}^d \otimes \boldsymbol{R}^r$ or $\beta(t,w) \in \boldsymbol{R}^d$ is bounded and continuous.

and a d-dimensional $\mathscr{F}_0$-measurable random variable ξ such that $P(\xi \in A) = \mu(A)$*[1] for every $A \in \mathscr{B}(\boldsymbol{R}^d)$. Define a d-dimensional continuous process X_l ($l = 1, 2, \ldots$) inductively as follows: $X_l(0) = \xi$, and if $X_l(t)$ is defined for $t \leq k/2^l$, then we define $X_l(t)$, for $k/2^l \leq t \leq (k+1)/2^l$, by

$$X_l(t) = X_l(k/2^l) + \alpha(k/2^l, X_{l,k})(B(t) - B(k/2^l)) \\ + \beta(k/2^l, X_{l,k})(t - k/2^l)$$

where

$$X_{l,k}(t) = \begin{cases} X_l(t), & t \leq k/2^l, \\ X_l(k/2^l), & t \geq k/2^l. \end{cases}$$

Clearly $X_l = (X_l(t))$ is the unique solution of the equation

$$\begin{cases} dX(t) = \alpha_l(t,X)dB(t) + \beta_l(t,X)dt \\ X(0) = \xi. \end{cases} \tag{2.7}$$

Since $\|\alpha_l\|^2 + \|\beta_l\| \leq M$ for some constant $M > 0$, we can apply Theorem III-3.1 to conclude that for every $T > 0$ and $m = 1, 2, \ldots,$

$$\sup_l \sup_{0 \leq t \leq T} E[|X_l(t)|^{2m}] \leq C_m \tag{2.8}$$

and

$$\sup_l E[|X_l(t) - X_l(s)|^{2m}] \leq C_m |t-s|^m, \qquad t, s \in [0, T], \tag{2.9}$$

where C_m ($m = 1, 2, \ldots$) is a constant. Indeed, for (2.9) we have

$$X_l(t) - X_l(s) = \int_s^t \alpha_l(u,X)\,dB(u) + \int_s^t \beta_l(u,X)du$$

and hence*[2]

$$E[|X_l(t) - X_l(s)|^{2m}] \leq C_m^{(1)} E[\|\int_s^t \alpha_l(u,X)dB(u)\|^{2m}] \\ + C_m^{(2)} E[\|\int_s^t \beta_l(u,X)du\|^{2m}]$$

*[1] Note that ξ is bounded since μ has a compact support.

*[2] In the following, $C_m^{(1)}$, $C_m^{(2)}$, ... are positive constants depending on m, T and M.

$$\leq C_m^{(3)} E[(\int_s^t \|\alpha_l(u,X)\|^2 du)^m] + C_m^{(4)} E[(\int_s^t \|\beta_l(u,X)\| du)^m]$$
$$\leq C_m |t-s|^m.$$

Applying Theorem I-4.2 and Theorem I-4.3 for $\alpha = 4$ and $\beta = 2$, we obtain a subsequence $\{l_i\}$, a probability space $(\hat{\Omega}, \hat{\mathscr{F}}, \hat{P})$ and d-dimensional continuous processes $\hat{X}$, $\hat{X}_{l_i}$, $i = 1, 2, \ldots$, such that $X_{l_i} \underset{\approx}{\mathscr{L}} \hat{X}_{l_i}$ and $\hat{X}_{l_i}(t)$ converges to $\hat{X}(t)$ uniformly on each compact interval of $[0, \infty)$ as $i \longrightarrow \infty$ a.s. If $s < t$ and if F is a real bounded continuous and $\mathscr{B}_s(\boldsymbol{W}^d)$-measurable function on $\boldsymbol{W}^d$, then for every $f \in \boldsymbol{C}_b^2(\boldsymbol{R}^d)$,

$$\hat{E}\{(f(\hat{X}(t)) - f(\hat{X}(s)) - \int_s^t (Af)(u,\hat{X})du)F(\hat{X})\}$$
$$(2.10) \qquad = \lim_{i\to\infty} \hat{E}\{(f(\hat{X}_{l_i}(t)) - f(\hat{X}_{l_i}(s)) - \int_s^t (A^{(l_i)}f)(u, \hat{X}_l)du)F(\hat{X}_{l_i})\}$$
$$= 0,$$

where the operator $A^{(l_i)}$ is defined in a similar fashion as A from α_{l_i} and β_{l_i}. Equation (2.10) implies that

$$f(\hat{X}(t)) - f(\hat{X}(0)) - \int_0^t (Af)(u,\hat{X})du$$

is an $(\hat{\mathscr{F}}_t)$-martingale, where

$$\hat{\mathscr{F}}_t = \bigcap_{\varepsilon>0} \sigma[\hat{X}(u); u \leq t+\varepsilon].$$

Thus $\hat{X}$ is a process satisfying (2.4).

Remark 2.1. The condition that μ has compact support is technical and may be removed. Indeed, by what we have shown, for each $x \in \boldsymbol{R}^d$ there exists a solution $X^{(x)}$ such that $\mu = \delta_x$. Let $P_x = P^{X^{(x)}}$. If we can choose P_x in such a way that $x \longmapsto P_x$ is $\mathscr{B}(\boldsymbol{R}^d)/\mathscr{B}(\mathscr{P}(\boldsymbol{W}^d))$-measurable* or $\mathscr{E}(\boldsymbol{R}^d)/\mathscr{B}(\mathscr{P}(\boldsymbol{W}^d))$-measurable then $P(\cdot) = \int_{\boldsymbol{R}^d} \mu(dx)P_x(\cdot)$ is a probability on $(\boldsymbol{W}^d, \mathscr{B}(\boldsymbol{W}^d))$ which satisfies (2.6). Such a selection is always possible because of a general selection theorem ([160], p. 289).

The boundedness assumption on α and β can be weakened, but some kind of restriction on the growth order of α and β is necessary in order to

* $\mathscr{P}(\boldsymbol{W}^d)$ is the set of all probabilities on $(\boldsymbol{W}^d, \mathscr{B}(\boldsymbol{W}^d))$ with the topology of the weak convergence (cf. Chapter I, Section 2).

guarantee the existence of a global solution (i.e., a solution defined for all $t \in [0, \infty)$). We will not, however, discuss this kind of problem in general (see, e.g. [102] and [188]); we shall only discuss it in the case of equations of the Markovian type.

Let $\sigma(x) = (\sigma_k^i(x))$: $\boldsymbol{R}^d \longrightarrow \boldsymbol{R}^d \otimes \boldsymbol{R}^r$ and $b(x) = (b^i(x))$: $\boldsymbol{R}^d \longrightarrow \boldsymbol{R}^d$ be continuous. Consider the following stochastic differential equation

$$(2.11) \quad dX(t) = \sigma(X(t))dB(t) + b(X(t))dt,$$

or, in terms of its components,

$$dX^i(t) = \sum_{k=1}^r \sigma_k^i(X(t))dB^k(t) + b^i(X(t))dt, \quad i = 1, 2, \dots, d.$$

If $x \longmapsto \sigma(x)$ and $x \longmapsto b(x)$ are bounded, then we know by Theorem 2.2 that a solution of (2.11) exists for every given initial distribution with compact support. If we remove this condition of boundedness, then a solution does exist locally but, in general, *blows up* (or *explodes*) in finite time. Therefore it is more convenient to modify the notion of a solution given in Section 1 so as to include *solutions admitting explosions.*

Let $\hat{\boldsymbol{R}}^d = \boldsymbol{R}^d \cup \{\Delta\}$ be the one-point compactification of $\boldsymbol{R}^d$ and

$\hat{\boldsymbol{W}}^d = \{w;\ [0, \infty) \ni t \longmapsto w(t) \in \hat{\boldsymbol{R}}^d$ is continuous and such that if $w(t) = \Delta$, then $w(t') = \Delta$ for all $t' \geq t\}$.

Let $\mathscr{B}(\hat{\boldsymbol{W}}^d)$ be the σ-field generated by Borel cylinder sets. For $w \in \hat{\boldsymbol{W}}^d$, we set

$$(2.12) \qquad e(w) = \inf\{t; w(t) = \Delta\}$$

and call $e(w)$ the *explosion time* of the trajectory w.

Definition 2.1. By a solution $X = (X(t))$ of the equation (2.11) we mean a $(\hat{\boldsymbol{W}}^d, \mathscr{B}(\hat{\boldsymbol{W}}^d))$-valued random variable defined on a probability space $(\Omega, \mathscr{F}, P)$ with a reference family $(\mathscr{F}_t)_{t \geq 0}$ such that

(i) there exists an r-dimensional $(\mathscr{F}_t)$-Brownian motion $B = (B(t))$ with $B(0) = 0$,

(ii) $X = (X(t))$ is adapted to $(\mathscr{F}_t)$, i.e., for each t, $\omega \longmapsto X(t, \omega) \in \hat{\boldsymbol{R}}^d$ is $\mathscr{F}_t$-measurable and

(iii) if $e(\omega) = e(X(\omega))$ is the explosion time of $X(\omega) \in \hat{\boldsymbol{W}}^d$, then for almost all ω,

$$X^i(t) - X^i(0) = \sum_{k=1}^{r} \int_0^t \sigma_k^i(X(s))dB^k(s) + \int_0^t b^i(X(s))ds,$$
$$i = 1, 2, \ldots, d,$$

for all $t \in [0, e(\omega))$.

Remark 2.2. The stochastic integral

$$t \in [0, e(\omega)) \longmapsto \int_0^t \sigma_k^i(X(s))dB^k(s)$$

is well-defined, for if $\sigma_n(\omega) = \inf\{t; |X(t)| \geq n\}$, then $\sigma_k^i(X(s))I_{\{\sigma_n(\omega)\geq s\}}$ is bounded in (s,ω) and hence

$$\int_0^t \sigma_k^i(X(s))I_{\{\sigma_n(\omega)\geq s\}}\,dB^k(s) = \int_0^{t\wedge\sigma_n(\omega)} \sigma_k^i(X(s))dB^k(s)$$

is defined for $t \in [0, \infty)$. Thus $t \in [0, \sigma_n) \longmapsto \int_0^t \sigma_k^i(X(s))dB^k(s)$ is defined for every $n = 1, 2, \ldots$, and hence it is defined on $[0, e(\omega))$ since $e(\omega) = \lim_{n\uparrow\infty} \sigma_n(\omega)$. $e(\omega)$ is called the *explosion time of the solution.*

Now the uniqueness of solutions, the pathwise uniqueness of solutions etc. are defined in the same way as in Section 1: just replace $\boldsymbol{W}^d$ by $\hat{\boldsymbol{W}}^d$.

Theorem 2.3. Given continuous $\sigma(x) = (\sigma_k^i(x))$ and $b(x) = (b^i(x))$, consider the equation (2.11). Then for any probability μ on $(\boldsymbol{R}^d, \mathscr{B}(\boldsymbol{R}^d))$ with compact support, there exists a solution $X = (X(t))$ of (2.11) such that the law of $X(0)$ coincides with μ.

Proof. As in Proposition 2.1, it suffices to show the existence of a $\hat{\boldsymbol{W}}^d$-valued random variable $X = (X(t))$ on a probability space $(\Omega, \mathscr{F}, P)$ with a reference family $(\mathscr{F}_t)$ such that X is $(\mathscr{F}_t)$-adapted, $P(X(0) \in dx) = \mu(dx)$, and for every $f \in \boldsymbol{C}_b^2(\boldsymbol{R}^d)$ and $n = 1, 2, \ldots$,

$$f(X(t\wedge\sigma_n)) - f(X(0)) - \int_0^{t\wedge\sigma_n} (Af)(X(s))ds$$

is an $(\mathscr{F}_t)$-martingale. Here $\sigma_n = \inf\{t; |X(t)| \geq n\}$ and

$$(Af)(x) = \frac{1}{2}\sum_{i,j=1}^{d} a^{ij}(x)\frac{\partial^2 f}{\partial x^i \partial x^j}(x) + \sum_{i=1}^{d} b^i(x)\frac{\partial f}{\partial x^i}(x),$$

$$a^{ij}(x) = \sum_{k=1}^{r} \sigma_k^i(x)\sigma_k^j(x).$$

Let $\rho(x)$ be a continuous function defined on $\boldsymbol{R}^d$ such that $0 < \rho(x) \leq 1$ for every $x \in \boldsymbol{R}^d$ and all $\rho(x)a^{ij}(x)$ and $\rho(x)b^i(x)$ are bounded. Clearly we can choose such a function. Let $(\tilde{A}f)(x) = \rho(x)(Af)(x)$. By Theorem 2.2, there exists a d-dimensional continuous process $\tilde{X} = (\tilde{X}(t))$ (i.e. a $\boldsymbol{W}^d$-valued random variable) on a space $(\Omega,\mathscr{F},P)$ with a reference family $(\mathscr{F}_t)$ such that $P(\tilde{X}(0) \in dx) = \mu(dx)$ and for every $f \in \boldsymbol{C}_b^2(\boldsymbol{R}^d)$,

$$f(\tilde{X}(t)) - f(\tilde{X}(0)) - \int_0^t (\tilde{A}f)(\tilde{X}(s))ds$$

is an $(\mathscr{F}_t)$-martingale. Set

$$A(t) = \int_0^t \rho(\tilde{X}(s))ds \tag{2.13}$$

and

$$e = \int_0^\infty \rho(\tilde{X}(s))ds. \tag{2.14}$$

Since $\rho \leq 1$, $A(t) < \infty$ for every t and $0 < e \leq \infty$. The inverse function $\sigma(t)$ of $t \longmapsto A(t)$ is defined for $t \in [0, e)$ and $\lim_{t \uparrow e} \sigma(t) = \infty$. Since $A(t)$ is $\mathscr{F}_t$-adapted, it is easy to see that $\sigma(t)$ * is an $(\mathscr{F}_t)$-stopping time for each t. Set

$$\mathscr{F}_t = \mathscr{F}_{\sigma(t)} \tag{2.15}$$

and

$$X(t) = \begin{cases} \tilde{X}(\sigma(t)), & t < e, \\ \Delta \quad , & t \geq e. \end{cases} \tag{2.16}$$

By the next lemma, we see that $X = (X(t))$ is an $(\mathscr{F}_t)$-adapted $\hat{\boldsymbol{W}}^d$-valued random variable.

Lemma 2.1. If $e(\omega) < \infty$ then $\lim_{t \uparrow e} X(t) = \Delta$ in $\hat{\boldsymbol{R}}^d$ for a.a. ω.

Proof. It is equivalent to show that, with probability one, if $\int_0^\infty \rho(\tilde{X}(s))ds < \infty$ then $\lim_{t \uparrow \infty} \tilde{X}(t) = \Delta$ in $\hat{\boldsymbol{R}}^d$. Take $0 < a < b$ such that $|\tilde{X}(0)| < b$ a.s. and define

---

* We set $\sigma(t) = \infty$ if $t \geq e$.

$$\begin{aligned} &\tilde{\sigma}_1 = 0 & &\tilde{\tau}_1 = \inf\{t > \tilde{\sigma}_1;\ |\tilde{X}(t)| > b\}, \\ &\tilde{\sigma}_2 = \inf\{t > \tilde{\tau}_1;\ |\tilde{X}(t)| < a\}, & &\tilde{\tau}_2 = \inf\{t > \tilde{\sigma}_2;\ |\tilde{X}(t)| > b\}, \\ &\quad\vdots & &\quad\vdots \quad . \end{aligned}$$

We show that, with probability one, if $\int_0^\infty \rho(\tilde{X}(s))ds < \infty$, then there exists an integer n such that $\tilde{\tau}_n < \infty$ and $\tilde{\sigma}_{n+1} = \infty$. It is sufficient to show that $\int_0^\infty \rho(\tilde{X}(s))ds = \infty$ a.s. on the set $\{\exists n$ such that $\tilde{\sigma}_n < \infty$ and $\tilde{\tau}_n = \infty\} \cup \{\tilde{\sigma}_n < \infty$ for every $n\}$. First, if there exists an integer n such that $\tilde{\sigma}_n < \infty$ and $\tilde{\tau}_n = \infty$, then $|\tilde{X}(t)| \leq b$ for all $t \geq \tilde{\sigma}_n$ and hence $\int_0^\infty \rho(\tilde{X}(s))ds = \infty$ since $\min_{|x|\leq b} \rho(x) > 0$. Next, we show that if $\tilde{\sigma}_n < \infty$ for every n, then $\sum_n (\tilde{\tau}_n - \tilde{\sigma}_n) = \infty$. If we can show this, then

$$\int_0^\infty \rho(\tilde{X}(s))ds \geq \sum_n \int_{\tilde{\sigma}_n}^{\tilde{\tau}_n} \rho(\tilde{X}(s))ds \geq \min_{|x|\leq b} \rho(x) \sum_n (\tilde{\tau}_n - \tilde{\sigma}_n) = \infty.$$

It is clearly equivalent to show that

$$\begin{aligned} \{\prod_n I_{\{\tilde{\sigma}_n<\infty\}}\} \exp[-\sum_n (\tilde{\tau}_n - \tilde{\sigma}_n)] &= \prod_n (I_{\{\tilde{\sigma}_n<\infty\}} \exp[-(\tilde{\tau}_n - \tilde{\sigma}_n)]) \\ &= 0, \quad \text{a.s.,} \end{aligned}$$

i.e.,

$$E(\prod_n I_{\{\tilde{\sigma}_n<\infty\}} \exp[-(\tilde{\tau}_n - \tilde{\sigma}_n)]) = 0. \tag{2.17}$$

We have

$$\begin{aligned} &E(\prod_{n=1}^{m+1} I_{\{\tilde{\sigma}_n<\infty\}} \exp[-(\tilde{\tau}_n - \tilde{\sigma}_n)] | \mathscr{F}_{\tilde{\sigma}_{m+1}}) \\ &\quad = \prod_{n=1}^{m} I_{\{\tilde{\sigma}_n<\infty\}} \exp[-(\tilde{\tau}_n - \tilde{\sigma}_n)] I_{\{\tilde{\sigma}_{m+1}<\infty\}} \\ &\qquad\quad \times E(\exp[-(\tilde{\tau}_{m+1} - \tilde{\sigma}_{m+1})] | \mathscr{F}_{\tilde{\sigma}_{m+1}}). \end{aligned}$$

In the following we assume for simplicity that $d = 1$; a necessary modification of the proof in the general case is left to the reader. Now $\tilde{X}(t)$ is of the form

$$\tilde{X}(t) = \tilde{X}(0) + M(t) + \int_0^t c(s)ds,$$

where $M(t)$ is a continuous $(\mathscr{F}_t)$-martingale such that $\langle M\rangle(t) = \int_0^t d(s)ds$ and $|c(s)|+|d(s)| \leq c$ ($c > 0$ is a constant). We may assume $(\Omega, \mathscr{F})$ is a standard measurable space (e.g we may always take $\Omega = W^d$ and $\mathscr{F} = \mathscr{B}(W^d)$) and let $P(\cdot\ |\mathscr{F}_{\tilde{\sigma}_{m+1}})$ be a regular conditional probability given $\mathscr{F}_{\tilde{\sigma}_{m+1}}$. By Doob's optional sampling theorem, $N_t = M(t+\tilde{\sigma}_{m+1})-M(\tilde{\sigma}_{m+1})$ is a martingale on $\{\tilde{\sigma}_{m+1} < \infty\}$ with respect to the probability $P(\cdot|\mathscr{F}_{\tilde{\sigma}_{m+1}})$ and the reference family $\mathscr{F}_t = \mathscr{F}_{t+\tilde{\sigma}_{m+1}}$ with $\langle N\rangle_t = \int_{\tilde{\sigma}_{m+1}}^{t+\tilde{\sigma}_{m+1}} d(s)ds$. By Theorem II-7.2′, there exists a Brownian motion $b(t)$ ($b(0) = 0$) which is independent of $\mathscr{F}_0 = \mathscr{F}_{\tilde{\sigma}_{m+1}}$ such that $N_t = b(\langle N\rangle_t)$. Then

$$\{|\tilde{X}(t + \tilde{\sigma}_{m+1}) - \tilde{X}(\tilde{\sigma}_{m+1})| > \alpha\} \subseteq \{|M(t + \tilde{\sigma}_{m+1}) - M(\tilde{\sigma}_{m+1})|$$
$$= |b(\langle N\rangle_t)| \geq \alpha/2\} \cup \{\int_{\tilde{\sigma}_{m+1}}^{\tilde{\sigma}_{m+1}+t} |c(s)|ds \geq \alpha/2\}.$$

Consequently,

$$\inf\{t;\ |\tilde{X}(t+\tilde{\sigma}_{m+1}) - \tilde{X}(\tilde{\sigma}_{m+1})| > \alpha\} \geq \frac{\alpha}{2c} \wedge \frac{\sigma_{\alpha/2}}{c}$$

where $\sigma_{\alpha/2} = \inf\{t;\ |b(t)| \geq \alpha/2\}$. Hence

$$I_{\{\tilde{\sigma}_{m+1}<\infty\}} E(\exp[-(\tilde{\tau}_{m+1} - \tilde{\sigma}_{m+1})]|\mathscr{F}_{\tilde{\sigma}_{m+1}})$$
$$\leq I_{\{\tilde{\sigma}_{m+1}<\infty\}} E\left(\exp\left[-\left\{\frac{b-a}{2c} \wedge \frac{\sigma_{(b-a)/2}}{c}\right\}\right]\right)$$
$$\leq E\left(\exp\left[-\left\{\frac{b-a}{2c} \wedge \frac{\sigma_{(b-a)/2}}{c}\right\}\right]\right) := k < 1.$$

Thus

$$E\left(\prod_{n=1}^{m+1} I_{\{\tilde{\sigma}_n<\infty\}} \exp[-(\tilde{\tau}_n - \tilde{\sigma}_n)]\right) \leq kE\left(\prod_{n=1}^{m} I_{\{\tilde{\sigma}_n<\infty\}} \exp[-(\tilde{\tau}_n - \tilde{\sigma}_n)]\right)$$

and (2.17) is now obvious.

Therefore if $\int_0^\infty \rho(\tilde{X}(s))ds < \infty$ then there exists an n such that $\tilde{\tau}_n < \infty$ and $|\tilde{X}(t)| \geq a$ for all $t \geq \tilde{\tau}_n$. Since a was arbitrary, we have $\lim_{t\uparrow\infty} \tilde{X}(t) = \Delta$ in $\hat{\boldsymbol{R}}^d$. The proof of the lemma is now complete.

Now we return to the proof of Theorem 2.3. We have only to show that

$$f(X(t\wedge\sigma_n))-f(X(0))-\int_0^{t\wedge\sigma_n}(Af)(X(s))ds$$

is a martingale. Since

$$f(\tilde{X}(t))-f(\tilde{X}(0))-\int_0^t(\rho Af)(\tilde{X}(s))ds$$

is an $(\mathscr{F}_t)$-martingale, we have by the optional stopping theorem that

$$f(\tilde{X}(t\wedge\tilde{\sigma}_n))-f(\tilde{X}(0))-\int_0^{t\wedge\tilde{\sigma}_n}(\rho Af)(\tilde{X}(s))ds$$

is an $(\mathscr{F}_t)$-martingale for $n=1, 2, \ldots$, where $\tilde{\sigma}_n=\inf\{t;\ |\tilde{X}(t)|\geq n\}$. Again by the optional sampling theorem

$$f(\tilde{X}(\sigma(t)\wedge\tilde{\sigma}_n))-f(\tilde{X}(0))-\int_0^{\sigma(t)\wedge\tilde{\sigma}_n}(\rho Af)(\tilde{X}(s))ds$$

is an $(\mathscr{F}_t)$-martingale. It is easy to see that $\sigma(t)\wedge\tilde{\sigma}_n=\sigma(t\wedge\sigma_n)$ and hence $\tilde{X}(\sigma(t)\wedge\tilde{\sigma}_n)=X(t\wedge\sigma_n)$. Also, we have $t=\int_0^t 1/\rho(\tilde{X}(s))dA(s)$ and hence $\sigma(t)=\int_0^{\sigma(t)}1/\rho(\tilde{X}(s))dA(s)=\int_0^t 1/\rho(X(s))ds$. Consequently,

$$\int_0^{\sigma(t\wedge\sigma_n)}(\rho Af)(\tilde{X}(s))ds=\int_0^{t\wedge\sigma_n}(\rho Af)(X(s))d\sigma(s)=\int_0^{t\wedge\sigma_n}(Af)(X(s))ds.$$

The proof of the theorem is now complete.

Theorem 2.4. If $\sigma(x)=(\sigma_k^i(x))$ and $b(x)=(b^i(x))$ are continuous and satisfy the condition

$$\|\sigma(x)\|^2+\|b(x)\|^2\leq K(1+|x|^2) \tag{2.18}$$

for some positive constant K, then for any solution of (2.11) such that $E(|X(0)|^2)<\infty$, we have $E(|X(t)|^2)<\infty$ for all $t>0$. Thus $e=\infty$ a.s.

Proof. Let $\sigma_n=\inf\{t;|X(t)|\geq n\}$ and $f\in C_b^2(\boldsymbol{R}^d)$ be chosen so that $f(x)=|x|^2$ if $|x|\leq n$. Then since

$$f(X(t\wedge\sigma_n))-f(X(0))-\int_0^{t\wedge\sigma_n}(Af)(X(s))ds$$

is a martingale,

$$E(|X(t\wedge\sigma_n)|^2) = E(|X(0)|^2) + E[\int_0^{t\wedge\sigma_n} (\sum_{i=1}^d a^{ii}(X(s)) + 2\sum_{i=1}^d X^i(s)b^i(X(s)))ds].$$

By (2.18) we have for some constant $c > 0$ that

$$E(|X(t\wedge\sigma_n)|^2) \le E(|X(0)|^2) + c\int_0^t \{1 + E(|X(s\wedge\sigma_n)|^2)\}\, ds.$$

From this we can conclude that

$$E(|X(t\wedge\sigma_n)|^2) \le \{1 + E(|X(0)|^2)\}\, e^{ct} - 1.$$

Letting $n \longrightarrow \infty$, we have

$$E(|X(t)|^2) \le \{1 + E(|X(0)|^2)\}\, e^{ct} - 1,$$

which completes the proof.

Thus the condition (2.18) is a sufficient condition for non-explosion of solutions. A more general criterion for explosion or non-explosion will be given in Chapter VI, Section 4.

3. Uniqueness theorem

In this section we only consider stochastic differential equations of the time homogeneous Markovian type. So suppose we are given $\sigma(x) = (\sigma_k^i(x))$: $\boldsymbol{R}^d \longrightarrow \boldsymbol{R}^d \otimes \boldsymbol{R}^r$ and $b(x) = (b^i(x))$: $\boldsymbol{R}^d \longrightarrow \boldsymbol{R}^d$ which are assumed to be continuous unless otherwise stated. We consider the following stochastic differential equation

(3.1) $$dX(t) = \sigma(X(t))dB(t) + b(X(t))dt$$

or in terms of its components,

$$dX^i(t) = \sum_{k=1}^r \sigma_k^i(X(t))dB^k(t) + b^i(X(t))dt, \quad i = 1, 2, \dots, d.$$

Theorem 3.1. Suppose $\sigma(x)$ and $b(x)$ are locally Lipschitz continuous, i.e., for every $N > 0$ there exists a constant $K_N > 0$ such that

$$(3.2) \qquad \|\sigma(x)-\sigma(y)\|^2+\|b(x)-b(y)\|^2 \leq K_N|x-y|^2 \quad \text{for every} \quad x, y \in \boldsymbol{B}_N.^*$$

Then the pathwise uniqueness of solutions holds for the equation (3.1) and hence it has a unique strong solution.

Proof. Let (X, B) and (X', B') be any two solutions of equation (3.1) defined on some same probability space $(\Omega, \mathscr{F}, P)$ with some same reference family such that $X(0)=X'(0)=x$ and $B(t) \equiv B'(t)$. It is sufficient to show that if $\sigma_N=\inf\{t; |X(t)| \geq N\}$ and $\sigma_N'=\inf\{t; |X'(t)| \geq N\}$ then $\sigma_N=\sigma_N'$ and $X(t)=X'(t)$ for all $t \leq \sigma_N$, $(N=1, 2, \ldots)$. But

$$X(t\wedge\sigma_N\wedge\sigma_N')-X'(t\wedge\sigma_N\wedge\sigma_N') = \int_0^{t\wedge\sigma_N\wedge\sigma_N'}[\sigma(X(s)) - \sigma(X'(s))]dB(s) + \int_0^{t\wedge\sigma_N\wedge\sigma_N'}[b(X(s))-b(X'(s))]ds$$

and hence, if $t \in [0, T]$,

$$\begin{aligned} &E\{|X(t\wedge\sigma_N\wedge\sigma_N') - X'(t\wedge\sigma_N\wedge\sigma_N')|^2\} \\ &\leq 2E\{|\int_0^{t\wedge\sigma_N\wedge\sigma_N'}[\sigma(X(s))-\sigma(X'(s))]dB(s)|^2\} \\ &\quad + 2E\{|\int_0^{t\wedge\sigma_N\wedge\sigma_N'}[b(X(s))-b(X'(s))]ds|^2\} \\ &\leq 2E\{\int_0^{t\wedge\sigma_N\wedge\sigma_N'}\|\sigma(X'(s))-\sigma(X'(s))\|^2 ds\} \\ &\quad + 2TE\{\int_0^{t\wedge\sigma_N\wedge\sigma_N'}|b(X(s))-b(X'(s))|^2 ds\} \\ &\leq 2E\{\int_0^t \|\sigma(X(s\wedge\sigma_N\wedge\sigma_N'))-\sigma(X'(s\wedge\sigma_N\wedge\sigma_N'))\|^2 ds\} \\ &\quad + 2TE\{\int_0^t |b(X(s\wedge\sigma_N\wedge\sigma_N'))-b(X'(s\wedge\sigma_N\wedge\sigma_N'))|^2 ds\} \\ &\leq 2K_N(1+T)\int_0^t E\{|X(s\wedge\sigma_N\wedge\sigma_N')-X'(s\wedge\sigma_N\wedge\sigma_N')|^2\}ds. \end{aligned}$$

It is easy to conclude from this inequality that

$$E\{|X(t\wedge\sigma_N\wedge\sigma_N')-X'(t\wedge\sigma_N\wedge\sigma_N')|^2\}=0 \quad \text{for all} \quad t \in [0, T]$$

* $\boldsymbol{B}_N = \{x; |x| \leq N\}$.

and hence, letting $T \uparrow \infty$, we have

$$X(t\wedge\sigma_N\wedge\sigma'_N) = X'(t\wedge\sigma_N\wedge\sigma'_N) \qquad \text{a.s.} \qquad \text{for all} \quad t \geq 0.$$

Since X and X' are continuous in t a.s., we can conclude that $X(t) = X'(t)$ for all $t \in [0, \sigma_N\wedge\sigma'_N)$ a.s. This clearly implies that $\sigma_N = \sigma'_N$ a.s. and the pathwise uniqueness of solutions of (3.1) is proven. From Theorem 1.1 and Theorem 2.3 we now conclude that the unique strong solution* exists for the equation (3.1).

The existence of a strong solution may be proved more directly by using the method of *successive approximations* as follows. For simplicity we assume that the Lipschitz condition (3.1) holds globally: i.e., there exists a constant $K > 0$ such that

$$(3.3) \qquad \|\sigma(x) - \sigma(y)\|^2 + \|b(x) - b(y)\|^2 \leq K|x-y|^2 \qquad \text{for all } x, y \in \boldsymbol{R}^d.$$

Then, by changing the constant K if necessary, we may assume that

$$(3.4) \qquad \|\sigma(x)\|^2 + \|b(x)\|^2 \leq K(|x|^2 + 1) \qquad \text{for all} \quad x \in \boldsymbol{R}^d.$$

In the following we essentially repeat the same proof as that of Theorem III-2.1. Let $x \in \boldsymbol{R}^d$ be fixed. For a given r-dimensional Brownian motion $B = (B(t))$, we define a sequence $X_n = (X_n(t))$ $(n = 0, 1, 2, \ldots)$ of d-dimensional continuous processes by

$$(3.5) \qquad \begin{cases} X_0(t) = x \\ X_n(t) = x + \int_0^t \sigma(X_{n-1}(s))\,dB(s) + \int_0^t b(X_{n-1}(s))\,ds, \end{cases} \qquad n = 1, 2, \ldots.$$

Set

$$\begin{aligned} X_{n+1}(t) - X_n(t) &= \int_0^t [\sigma(X_n(s)) - \sigma(X_{n-1}(s))]\,dB(s) \\ &\quad + \int_0^t [b(X_n(s)) - b(X_{n-1}(s))]\,ds \\ &= I_1(t) + I_2(t) \qquad \text{say.} \end{aligned}$$

* The space $\boldsymbol{W}^d$ is now replaced by $\hat{\boldsymbol{W}}^d$.

By Doob-Kolmogorov's inequality,

$$E(\sup_{0\le s\le t} |I_1(s)|^2) \le 4E(|I_1(t)|^2)$$
$$= 4E(\int_0^t \|\sigma(X_n(s)) - \sigma(X_{n-1}(s))\|^2 ds)$$
$$\le 4K \int_0^t E(|X_n(s) - X_{n-1}(s)|^2) ds.$$

Also, if $t \in [0, T]$,

$$E(\sup_{0\le s\le t} |I_2(s)|^2) \le E(\{\int_0^t \|b(X_n(s)) - b(X_{n-1}(s))\| ds\}^2)$$
$$\le TE(\int_0^t \|b(X_n(s)) - b(X_{n-1}(s))\|^2 ds)$$
$$\le TK \int_0^t E(|X_n(s) - X_{n-1}(s)|^2) ds.$$

Hence

$$E(\sup_{0\le s\le t} |X_{n+1}(s) - X_n(s)|^2) \le 2K(4+T) \int_0^t E(|X_n(s) - X_{n-1}(s)|^2) ds$$

and therefore,

$$E(\sup_{0\le s\le t} |X_{n+1}(s) - X_n(s)|^2)$$
$$\le \{2K(4+T)\}^n \int_0^t dt_1 \int_0^{t_1} dt_2 \cdots \int_0^{t_{n-1}} dt_n E(|X_1(t_n) - X_0(t_n)|^2)$$

Since

$$E(|X_1(t) - X_0(t)|^2) \le 2E(\|\sigma(x)B(t)\|^2 + \|b(x)\|^2 t^2)$$
$$\le 2(\|\sigma(x)\|^2 t + \|b(x)\|^2 t^2)$$
$$\le 2K(1+T)T(1+|x|^2),$$

we have

$$E(\sup_{0\le t\le T} |X_{n+1}(t) - X_n(t)|^2) \le \text{const. } \{2K(4+T)\}^n T^n/n!.$$

Consequently

$$P\{\sup_{0\le t\le T}|X_{n+1}(t)-X_n(t)|>1/2^n\}\le \text{const. } \{8K(4+T)\}^n T^n/n!$$

By Borel-Cantelli's lemma, we see with probability one that $X_n(t)$ converges uniformly on $[0, T]$, and, since T was arbitrary, $\lim_{n\to\infty} X_n(t) = X(t)$ determines a d-dimensional continuous process which is clearly a solution of (3.1). This solution is of the form $F(x, B)$ for some function $F(x, w)$ on $\boldsymbol{R}^d \otimes \boldsymbol{W}^r$ since each X_n is so. Thus X is a strong solution of (3.1); the uniqueness is clear from Theorem 3.1.

The Lipschitz condition (3.2) for the pathwise uniqueness of solutions of the equation (3.1) can be weakened considerably in the case $d = 1$. For simplicity we state the theorem in the global form assuming that $\sigma(x)$ and $b(x)$ are bounded but it may be localized as Theorem 3.1 in an obvious way.

Theorem 3.2. Let $d = r = 1$*1 and suppose that $\sigma(x)$ and $b(x)$ are bounded. Assume further that the following conditions are satisfied:

(i) there exists a strictly increasing function $\rho(u)$ on $[0, \infty)$ such that $\rho(0) = 0$, $\int_{0+}\rho^{-2}(u)du = \infty$ and $|\sigma(x) - \sigma(y)| \le \rho(|x - y|)$ for all $x, y \in \boldsymbol{R}^1$ *2;

(ii) there exists an increasing and concave function $\kappa(u)$ on $[0, \infty)$ such that $\kappa(0) = 0$, $\int_{0+}\kappa^{-1}(u)du = \infty$ and $|b(x) - b(y)| \le \kappa(|x - y|)$ for all $x, y \in \boldsymbol{R}^1$.

Then the pathwise uniqueness of solutions holds for the equation (3.1) and hence it has the unique strong solution.

Corollary. If σ is Hölder continuous with exponent 1/2 and b is Lipschitz continuous, then the pathwise uniqueness of solutions holds for the equation (3.1) in the case $d = 1$.

Proof of Theorem 3.2. Let $1 > a_1 > a_2 > \cdots > a_n > \cdots > 0$ be defined by

$$\int_{a_1}^1 \rho^{-2}(u)du = 1, \int_{a_2}^{a_1} \rho^{-2}(u)du = 2, \ldots, \int_{a_n}^{a_{n-1}} \rho^{-2}(u)du = n, \ldots.$$

Clearly $a_n \longrightarrow 0$ as $n \longrightarrow \infty$. Let $\psi_n(u), n = 1, 2, \ldots,$ be a continuous function such that its support is contained in (a_n, a_{n-1}),

---

*1 r may be arbitrary. We assume $r = 1$ only for simplicity.

*2 This nice condition for σ was found by T. Yamada ([183]) improving an idea of H. Tanaka in [162].

$$0 \leq \psi_n(u) \leq 2\rho^{-2}(u)/n \quad \text{and} \quad \int_{a_n}^{a_{n-1}} \psi_n(u)du = 1.$$

Such a function obviously exists. Set

$$\varphi_n(x) = \int_0^{|x|} dy \int_0^y \psi_n(u)du, \qquad x \in \boldsymbol{R}^1.$$

It is easy to see that $\varphi_n \in \boldsymbol{C}^2(\boldsymbol{R}^1)$, $|\varphi_n'(x)| \leq 1$ and $\varphi_n(x) \uparrow |x|$ as $n \longrightarrow \infty$. Suppose that we are given two solutions $(X_1(t), B_1(t))$ and $(X_2(t), B_2(t))$ on the same probability space with the same reference family such that $X_1(0) = X_2(0) = x$ and $B_1(t) \equiv B_2(t)$ (: = $B(t)$) a.s. Then we have

$$\begin{aligned} X_1(t) - X_2(t) = & \int_0^t [\sigma(X_1(s)) - \sigma(X_2(s))]dB(s) \\ & + \int_0^t [b(X_1(s)) - b(X_2(s))]ds \end{aligned}$$

and by Itô's formula,

$$\begin{aligned} \varphi_n(X_1(t) - X_2(t)) = & \int_0^t \varphi_n'(X_1(s) - X_2(s))[\sigma(X_1(s)) - \sigma(X_2(s))]dB(s) \\ & + \int_0^t \varphi_n'(X_1(s) - X_2(s))[b(X_1(s)) - b(X_2(s))]ds \\ & + \frac{1}{2}\int_0^t \varphi_n''(X_1(s) - X_2(s))[\sigma(X_1(s)) - \sigma(X_2(s))]^2 ds. \end{aligned}$$

Since the expectation of the first term in the right-hand side is zero, we have

$$\begin{aligned} E[\varphi_n(X_1(t) - X_2(t))] = & E[\int_0^t \varphi_n'(X_1(s) - X_2(s))\{b(X_1(s)) - b(X_2(s))\}ds] \\ & + \frac{1}{2}E[\int_0^t \varphi_n''(X_1(s) - X_2(s))\{\sigma(X_1(s)) - \sigma(X_2(s))\}^2 ds] \\ = & I_1 + I_2. \end{aligned}$$

Now

$$\begin{aligned} |I_1| & \leq \int_0^t E[|b(X_1(s)) - b(X_2(s))|]ds \\ & \leq \int_0^t E[\kappa(|X_1(s) - X_2(s)|)]ds \end{aligned}$$

$$\leq \int_0^t \kappa[E(|X_1(s) - X_2(s)|)]ds$$

by Jensen's inequality, and

$$|I_2| \leq \frac{1}{2} \int_0^t E\left[\frac{2}{n} \rho^{-2}(|X_1(s) - X_2(s)|)\rho^2(|X_1(s) - X_2(s)|)\right] ds$$

$$\leq t/n \longrightarrow 0 \qquad \text{as} \quad n \longrightarrow \infty.$$

Consequently by letting $n \longrightarrow \infty$,

$$E(|X_1(t) - X_2(t)|) \leq \int_0^t \kappa[E(|X_1(s) - X_2(s)|)]ds.$$

Since $\int_{0+} \kappa^{-1}(u)du = +\infty$, the above inequality implies that $E(|X_1(t) - X_2(t)|) = 0$ and hence $X_1(t) = X_2(t)$ a.s. This proves the pathwise uniqueness for (3.1).

The continuity condition for the function σ in the theorem is, in a sense, the best possible. Indeed, suppose for simplicity that $b(x) \equiv 0$ and $\sigma(x)$ is such that $\sigma(x_0) = 0$ and $\int_{x_0-\varepsilon}^{x_0+\varepsilon} \sigma^{-2}(y)dy < \infty$ and $\sigma(x) \geq 1$ for $|x - x_0| > \varepsilon$. Then the equation

$$\text{(3.6)} \qquad \begin{cases} dX(t) = \sigma(X(t))dB(t) \\ X(0) = x_0 \end{cases}$$

has infinitely many solutions. Clearly $X(t) \equiv x_0$ is a solution. Also, for a one-dimensional Brownian motion $b(t)$ with $b(0) = 0$, we set for each $\rho > 0$, $\xi(t) = x_0 + b(t)$, $A_\rho(t) = 2 \int_{-\infty}^{\infty} \phi(t,y)\sigma^{-2}(y)\,dy + \rho\phi(t,x_0) = \int_0^t \sigma^{-2}(\xi(s))ds + \rho\phi(t, x_0)$ and $X_\rho(t) = \xi(A_\rho^{-1}(t))$, where A_ρ^{-1} is the inverse function of $t \longmapsto A_\rho(t)$ and $\phi(t,y)$ is the local time of $\xi(t)$.*[1] Then $X_\rho(t)$ is a solution of (3.6), because $M_t = X_\rho(t) - x_0$ is a continuous martingale with

$$\langle M \rangle_t = A_\rho^{-1}(t) = \int_0^{A_\rho^{-1}(t)} \sigma^2(\xi(s))\,dA_\rho(s) = \int_0^t \sigma^2(X_\rho(s))ds$$

(cf. Proposition 2.1). It is easy to see that the probability law of X_ρ is different for different ρ.

So far we have only presented conditions for pathwise uniqueness.*[2]

*[1] Chapter III, Section 4.

*[2] Except the cases which can be covered by Theorems 3.1 and 3.2, little is known about the pathwise uniqueness and the existence of the strong solutions. Cf. [125] and [189].

There are also several important results for the uniqueness of solutions in the sense of probability laws due mainly to Stroock-Varadhan [157] and Krylov [90]. In particular, Stroock-Varadhan showed that the uniqueness of solutions holds for the equation (1.3) if the matrix $a(t,x) = \sigma(t,x)\sigma(t,x)^*$ (in component form, $a^{ij}(t, x) = \sum_{k=1}^{r} \sigma_k^i(t,x)\sigma_k^j(t,x)$) is continuous, bounded and uniformly positive definite, and if $b(t,x) = (b^i(t,x))$ is bounded and Borel measurable. Here we shall content ourselves with presenting only a particular case of this beautiful and important result.

Theorem 3.3. Consider the equation of the time homogeneous Markovian case (3.1). If $a(x) = \sigma(x)\sigma(x)^*$ is uniformly positive definite, bounded and continuous and $b(x)$ is bounded and Borel measurable, then the uniqueness of solutions holds.

Proof. We assume $b(x) \equiv 0$; the general case is obtained by a transformation of drift which will be discussed in the next section. Set

$$Af(x) = \frac{1}{2}\sum_{i,j=1}^{d} a^{ij}(x)\frac{\partial^2 f}{\partial x^i \partial x^j}(x), \quad f \in \boldsymbol{C}^2(\boldsymbol{R}^d).$$

By Proposition 2.1, it is sufficient to prove that if P_x is a probability on $(\boldsymbol{W}^d, \mathscr{B}(\boldsymbol{W}^d))$ such that

(i) $P_x\{w;\ w(0) = x\} = 1$ and

(ii) $f(w(t)) - f(w(0)) - \int_0^t (Af)(w(s))ds$ is a $(P_x, \mathscr{B}_t(\boldsymbol{W}^d))$-martingale for $f \in \boldsymbol{C}_b^2(\boldsymbol{R}^d)$,

then P_x is uniquely determined. As we shall see in the corollary of Theorem 5.1, it suffices to prove that $E_x[\int_0^\infty e^{-\lambda t} f(w(t))dt]$ is uniquely determined for every $f \in \boldsymbol{C}_b(\boldsymbol{R}^d)$; that is, for any two probabilities P_x and P_x' on $(\boldsymbol{W}^d, \mathscr{B}(\boldsymbol{W}^d))$ satisfying (i) and (ii),

$$E_x\Big[\int_0^\infty e^{-\lambda t} f(w(t))dt\Big] = E_x'\Big[\int_0^\infty e^{-\lambda t} f(w(t))dt\Big] \tag{3.7}$$

for every $f \in \boldsymbol{C}_b(\boldsymbol{R}^d)$. We shall obtain this result by a perturbation argument on Wiener measure. First we shall show that there exists a positive constant ε such that if $a(x) = (a^{ij}(x))$ satisfies

$$|a^{ij}(x) - \delta_{ij}| \le \varepsilon \qquad \text{for all} \quad x \in \boldsymbol{R}^d \text{ and } i,j = 1, 2, \ldots, d, \tag{3.8}$$

then for any x and for any two probabilities P_x and P_x' satisfying (i) and (ii), (3.7) holds.

We now list some analytical properties of operators related to Brownian motion which we shall need in the subsequent discussion. Set

$$g_t(x) = (2\pi t)^{-d/2} \exp\left(-\frac{|x|^2}{2t}\right), \qquad t > 0, \quad x \in \boldsymbol{R}^d,$$

$$v_\lambda(x) = \int_0^\infty e^{-\lambda t} g_t(x)dt, \qquad \lambda > 0, \quad x \in \boldsymbol{R}^d$$

and

$$(V_\lambda f)(x) = \int_{\boldsymbol{R}^d} v_\lambda(x-y)f(y)dy.$$

Then the following facts hold. Let $\lambda > 0$ be fixed.

(1) V_λ is a bounded operator on $\mathscr{L}_p(\boldsymbol{R}^d)$ into itself such that $\|V_\lambda\|_p \leq \|v_\lambda\|_1 = 1/\lambda$. This is a consequence of the well-known Hausdorff-Young's inequality.

(2) If $p > \frac{d}{2} \vee 1$, then there exists a constant A_p depending on p and d only such that

$$|V_\lambda f(x)| \leq A_p\|f\|_p \qquad \text{for every } f \in \mathscr{L}_p(\boldsymbol{R}^d) \text{ and } x \in \boldsymbol{R}^d.$$

Indeed, by Hölder's inequality,

$$|V_\lambda f(x)| \leq \|v_\lambda\|_q\|f\|_p \qquad \text{where} \quad 1/p + 1/q = 1,$$

and

$$\|v_\lambda\|_q \leq \int_0^\infty e^{-\lambda t} \|g_t\|_q dt \leq \text{const.} \int_0^\infty e^{-\lambda t} t^{-\frac{d}{2}\left(1-\frac{1}{q}\right)} dt < \infty$$

if $\frac{d}{2}\left(1 - \frac{1}{q}\right) = \frac{d}{2p} < 1$.

(3) For each i and j, $\frac{\partial^2 V_\lambda f}{\partial x^i \partial x^j}$ for $f \in C_K^\infty(\boldsymbol{R}^d)$ * can be extended to a bounded operator on $\mathscr{L}_p(\boldsymbol{R}^d)$ into itself for every $p > 1$; that is, there exists a constant C_p depending on p and d only such that

$$\left\|\frac{\partial^2 V_\lambda f}{\partial x^i \partial x^j}\right\|_p \leq C_p\|f\|_p.$$

* $C_K^\infty(\boldsymbol{R}^d)$ is the space of C^∞-functions on $\boldsymbol{R}^d$ with compact supports.

The proof in the case of $p = 2$ is easily furnished by the Fourier transform, but in the general case we have to appeal to the deep $\mathscr{L}_p$-theory for singular integrals (cf. [152]).

Suppose $a(x) = (a^{ij}(x))$ satisfies (3.8) and let P_x be a probability on $(\boldsymbol{W}^d, \mathscr{B}(\boldsymbol{W}^d))$ satisfying the above conditions (i) and (ii). Then

$$E_x[f(w(t))] = f(x) + \int_0^t E_x[(Af)(w(s))]ds \qquad \text{for } f \in \boldsymbol{C}_b^2(\boldsymbol{R}^d)$$

and hence for fixed $\lambda > 0$ and $x \in \boldsymbol{R}^d$,

$$\lambda E_x[\int_0^\infty e^{-\lambda t} f(w(t))dt] = f(x) + E_x[\int_0^\infty e^{-\lambda t}(Af)(w(t))dt].$$

Consequently, denoting $E_x[\int_0^\infty e^{-\lambda t} g(w(t))dt]$ by $\mu_\lambda(g)$, we have that

$$\mu_\lambda\Big(\lambda f - \frac{1}{2}\Delta f\Big) = f(x) + \frac{1}{2}\mu_\lambda\Big(\sum_{i,j=1}^d c^{ij}(\cdot)\frac{\partial^2 f}{\partial x^i \partial x^j}(\cdot)\Big),$$

where $c^{ij}(y) = a^{ij}(y) - \delta_{ij}$. Letting $f = V_\lambda h$ for $h \in \boldsymbol{C}_K^\infty(\boldsymbol{R}^d)$,

$$\mu_\lambda(h) = V_\lambda h(x) + \frac{1}{2}\mu_\lambda\Big(\sum_{i,j=1}^d c^{ij}(\cdot)\frac{\partial^2 V_\lambda h}{\partial x^i \partial x^j}(\cdot)\Big).$$

Using the above properties (2) and (3) for $p > d/2 \vee 1$,

$$|\mu_\lambda(h)| \leq A_p\|h\|_p + \frac{\varepsilon}{2}d^2 C_p \sup_{\|f\|_p \leq 1} |\mu_\lambda(f)|\ \|h\|_p.$$

Therefore, if $\sup_{\|f\|_p\leq 1} |\mu_\lambda(f)| = \|\mu_\lambda\|_q < \infty$, then we can conclude that $\|\mu_\lambda\|_q \leq A_p/\Big(1 - \frac{\varepsilon}{2}d^2 C_p\Big)$ for all $\varepsilon > 0$ such that $1 - \frac{\varepsilon}{2}d^2 C_p > 0$. This can be verified as follows. Set

$$Y_m(t,w) = w\Big(\frac{k}{2^m} \wedge m\Big), \quad t \in [k/2^m, (k+1)/2^m), \quad k = 0, 1, \ldots$$

and

$$X_m(t,w) = x + \int_0^t \sigma(Y_m(s))dB(s),$$

where $B = (B(t,w))$ is a d-dimensional $(\mathscr{B}_t(\boldsymbol{W}^d))$-Brownian motion such that

$$w(t) = x + \int_0^t \sigma(w(s))dB(s)$$

(cf. Theorem II-7.1). Then it is easy to see that the probability law $P^{(m)}$ of the process $(X_m(t))$ converges weakly to P_x and, in particular,

$$\mu_\lambda^{(m)}(f) := E_x\Big[\int_0^\infty e^{-\lambda t} f(X_m(t))dt\Big] \longrightarrow \mu_\lambda(f), \quad f \in C_b(\boldsymbol{R}^d).$$

Since $\{w(t),\, t \in [k/2^m, (k+1)/2^m)\}$ is, with respect to the regular conditional probability $P^m(\cdot \mid \mathscr{B}_{k/2m}(\boldsymbol{W}^d))$, a linear transformation of the d-dimensional Brownian motion by the constant matrix $\sigma(w(k/2^m))$, we see from (2) that

$$\|\mu_\lambda^{(m)}\|_q = \sup_{\|f\|_p \le 1} \Big| E_x\Big[\int_0^\infty e^{-\lambda t} f(X_m(t))dt\Big]\Big| < \infty.$$

Then by the same argument as above we have

$$\|\mu_\lambda^{(m)}\|_q \le A_p \Big/ \Big(1 - \frac{\varepsilon}{2} d^2 C_p\Big).$$

Finally, by letting $m \longrightarrow \infty$

$$\|\mu_\lambda\|_q \le \varliminf_{m\to\infty} \|\mu_\lambda^{(m)}\|_q \le A_p \Big/ \Big(1 - \frac{\varepsilon}{2} d^2 C_p\Big).$$

Let P'_x be another probability on $(\boldsymbol{W}^d, \mathscr{B}(\boldsymbol{W}^d))$ satisfying the above conditions (i) and (ii). Then for each fixed x and $f \in C_K^2(\boldsymbol{R}^d)$,

$$\mu_\lambda(f) = V_\lambda f(x) + \mu_\lambda(K_\lambda f) \quad \text{and} \quad \mu'_\lambda(f) = V_\lambda f(x) + \mu'_\lambda(K_\lambda f)$$

where

$$K_\lambda f(y) = \frac{1}{2}\sum_{i,j=1}^d c^{ij}(y)\frac{\partial^2 V_\lambda f}{\partial x^i \partial x^j}(y)$$

and μ'_λ is defined in the same way as μ_λ from P'_x. Therefore

$$(\mu_\lambda - \mu'_\lambda)(f) = (\mu_\lambda - \mu'_\lambda)(K_\lambda f).$$

But we have $\|K_\lambda f\|_p \le \frac{d^2}{2}\varepsilon C_p \|f\|_p$. Consequently

$$\sup_{\|f\|_p \le 1} |(\mu_\lambda - \mu'_\lambda)(f)| \le \frac{d^2}{2}\varepsilon C_p \sup_{\|f\|_p \le 1} |(\mu_\lambda - \mu'_\lambda)(f)|$$

and hence, if we choose $\varepsilon > 0$ such that $\frac{d^2}{2}\varepsilon C_p < 1$, then $\mu_\lambda(f) = \mu'_\lambda(f)$.

Thus we have proved the uniqueness of the probabilities $\{P_x\}_{x \in R^d}$ provided $a^{ij}(x)$ satisfies the condition (3.8) for $\varepsilon > 0$ such that $\frac{d^2}{2}\varepsilon C_p < 1$. Clearly (δ_{ij}) may be replaced by any positive definite constant matrix $C = (C^{ij})$, and $\varepsilon > 0$ can be chosen independent of C if $A \le \underline{\lambda}(C) \le \bar{\lambda}(C) \le B$ for some positive constants A, B, where $\underline{\lambda}(C)$ and $\bar{\lambda}(C)$ are the least and the largest eigenvalues of C respectively. Now we remove the restriction (3.8) by the following standard localization argument. Set

$$\tau(w) = \inf\{t;\ \max_{1 \le i,j \le d} |a^{ij}(w(t)) - a^{ij}(w(0))| \ge \varepsilon\}.$$

Then the result we obtained implies that for any $\{P'_x\}$ satisfying (i) and (ii),

$$(3.9) \qquad P_x\{w_\tau^- \in A\} = P'_x\{w_\tau^- \in A\} \quad \text{for every } x \text{ and } A \in \mathscr{B}(\boldsymbol{W}^d)$$

where $w_\tau^- \in \boldsymbol{W}^d$ is defined by $w_\tau^-(u) = w(\tau \wedge u)$. We denote $P_x\{w_\tau^- \in A\}$, $x \in \boldsymbol{R}^d$, $A \in \mathscr{B}(\boldsymbol{W}^d)$, by $p\{x, A\}$. By Doob's optional sampling theorem, it is easy to see that if θ is a $(\mathscr{B}_t(\boldsymbol{W}^d))$-stopping time such that $P_x(\theta < \infty) = 1$, then for P_x-almost all w, $P'_w(A) = P_x(w_\theta^+ \in A \,|\, \mathscr{B}_\theta(\boldsymbol{W}^d))$, $A \in \mathscr{B}(\boldsymbol{W}^d)$ satisfies the above condition (ii) and

$$P'_w(w';\ w'(0) = w(\theta(w))) = 1.^*$$

Here $w_\theta^+ \in \boldsymbol{W}^d$ is defined by $(w_\theta^+)(u) = w(\theta + u)$. Using this fact, it is easy to conclude the following: letting

$$\begin{aligned} \tau_0(w) &= 0 \\ \tau_1(w) &= \tau(w) \\ \tau_2(w) &= \tau_1(w) + \tau(w_{\tau_1}^+) \\ &\vdots \end{aligned}$$

* Cf. the corollary of Theorem I-3.2.

$$\tau_{n+1}(w) = \tau_n(w) + \tau(w^+_{\tau_n})$$

and $\Phi_n w = (w^+_{\tau_n})^-_{\tau(w^+_{\tau_n})}$, for $n = 0, 1, \ldots,$

$$P_x\{w;\ \Phi_0 w \in A_0,\ \Phi_1 w \in A_1, \ldots, \Phi_n w \in A_n\}$$

$$= \int_{A_0} p\{x, dw_0\} \int_{A_1} p\{w_0(\tau(w_0)), dw_1\} \int \cdots$$

$$\times \int_{A_n} p\{w_{n-1}(\tau(w_{n-1})), dw_n\}$$

$$= P'_x\{w;\ \Phi_0 w \in A_0,\ \Phi_1 w \in A_1, \ldots, \Phi_n w \in A_n\}.$$

Since $\tau_n(w) \longrightarrow \infty$ for every w, this implies $P_x = P'_x$.

4. Solution by transformation of drift and by time change

Certain stochastic differential equations can be solved (i.e., we can show the existence and the uniqueness of solutions) by some probabilistic methods. These methods sometimes apply even for equations which are not covered by the theorems obtained so far.

4.1. *The transformation of drift.* Let $(\Omega,\mathscr{F},P)$ be a probability space with a reference family $(\mathscr{F}_t)$. In the following, we assume that $(\Omega,\mathscr{F},P)$ and $(\mathscr{F}_t)_{t>0}$ possess the below property:

(4.1) *Suppose, for every $t \geq 0$, that μ_t is an absolutely continuous probability measure on $(\Omega, \mathscr{F}_t)$ with respect to P such that μ_t restricted on $\mathscr{F}_s$ coincides with μ_s for any $t > s \geq 0$. Then there exists a probability measure μ on $(\Omega,\mathscr{F})$ such that μ restricted on $\mathscr{F}_t$ coincides with μ_t for every $t \geq 0$.*

That is, we assume that any consistent family of absolutely continuous probability measures with respect to P can be extended to a probability measure on $(\Omega,\mathscr{F})$. For example, if $(\Omega,\mathscr{F})$ is a standard measurable space, then the above condition (4.1) is satisfied.

For $X \in \mathscr{M}_2^{c,loc}$ we set

$$M(t) = \exp[X(t) - \langle X\rangle(t)/2]. \tag{4.2}$$

We assume that M is a martingale. This is true, for example, by Theorem III-5.3 if $E(\exp(\frac{1}{2}\langle X\rangle_t)) < \infty$ for every $t \geq 0$. Then if we define $\hat{P}_t$ for each $t \geq 0$ by

(4.3) $\hat{P}_t(A) = E[M(t): A], \qquad A \in \mathscr{F}_t,$

$\hat{P}_t$ is a measure on $(\Omega, \mathscr{F}_t)$ and $\hat{P}_t|_{\mathscr{F}_s} = \hat{P}_s$ for $t > s \geq 0$. In fact, if $A \in \mathscr{F}_s$,

$$\hat{P}_t(A) = E[M(t): A] = E[E(M(t)|\mathscr{F}_s): A] = E[M(s): A].$$

By assumption (4.1), there exists a probability $\hat{P}$ on $(\Omega, \mathscr{F})$ such that $\hat{P}|_{\mathscr{F}_t} = \hat{P}_t$.

Definition 4.1. $\hat{P}$ is called the probability measure *which has density* M *with respect to* P. We denote $\hat{P}$ as $\hat{P} = M \cdot P$.

In this way we obtain a new system $(\Omega, \mathscr{F}, \hat{P})$ and $(\mathscr{F}_t)_{t \geq 0}$. The spaces of martingales with respect to this system are denoted by $\hat{\mathscr{M}}_2$, $\hat{\mathscr{M}}_2^c$, etc. The following theorem was established by Girsanov [40]* in the case when $X(t)$ is a Brownian motion.

Theorem 4.1. (i) Let $Y \in \mathscr{M}_2^{c, loc}$. If we define $\tilde{Y}$ by

(4.4) $\tilde{Y}(t) = Y(t) - \langle Y, X \rangle(t),$

then $\tilde{Y} \in \hat{\mathscr{M}}_2^{c, loc}$.

(ii) Let $Y_1, Y_2 \in \mathscr{M}_2^{c, loc}$ and define $\tilde{Y}_1$ and $\tilde{Y}_2$ by (4.4). Then

(4.5) $\langle Y_1, Y_2 \rangle = \langle \tilde{Y}_1, \tilde{Y}_2 \rangle.$

Remark 4.1. Since $P|_{\mathscr{F}_t}$ and $\hat{P}|_{\mathscr{F}_t}$ are mutually absolutely continuous for each fixed $t \geq 0$, there is no ambiguity in the statement of (ii): (4,5) holds P-a.s. and $\hat{P}$-a.s.

Proof. Assume first that $\tilde{Y}(t)$ is bounded in the sense that, for each $t \geq 0$, $\tilde{Y}(t) \in \mathscr{L}^\infty(\Omega, \mathscr{F})$. By Itô's formula,

$$\begin{aligned} &d(M(t)\tilde{Y}(t)) \\ &\quad = d\{M(t)(Y(t) - \langle Y, X \rangle(t))\} \\ &\quad = \tilde{Y}(t)dM(t) + M(t)dY(t) - M(t)d\langle Y, X \rangle(t) + dM(t) \cdot dY(t) \\ &\quad = \tilde{Y}(t)dM(t) + M(t)dY(t) \end{aligned}$$

since $dM(t) = M(t)dX(t)$ and hence $dM(t) \cdot dY(t) = M(t)d\langle X, Y \rangle(t)$. This implies that $M(t)\tilde{Y}(t)$ is a martingale and thus

* Cf. Cameron-Martin [9], Maruyama [110] and Motoo [123].

$$\hat{E}[\tilde{Y}(t)\,|\,\mathscr{F}_s] = E[M(t)\tilde{Y}(t)\,|\,\mathscr{F}_s]M(s)^{-1} = \tilde{Y}(s).$$

In general, we choose a sequence $\{\sigma_n\}$ of $(\mathscr{F}_t)$-stopping times such that, for each n, $t \longmapsto \tilde{Y}(t\wedge\sigma_n)$ is bounded in the above sense. Since $\tilde{Y}(t\wedge\sigma_n) = Y(t\wedge\sigma_n) - \langle X, Y\rangle(t\wedge\sigma_n) = \widetilde{Y^{\sigma_n}}(t)$, where $Y^{\sigma_n} \in \mathscr{M}_2^{c,\,loc}$ is defined by $Y^{\sigma_n}(t) = Y(t\wedge\sigma_n)$, $\widetilde{Y^{\sigma_n}} \in \hat{\mathscr{M}}_2^{c,\,loc}$ by the above proof and hence $\tilde{Y} \in \hat{\mathscr{M}}_2^{c,\,loc}$.

The proof of (ii) can be given in a similar way.

Corollary. Let $Y \in \mathscr{M}_2^{c,\,loc}$, $\Phi \in \mathscr{L}_2^{loc}(Y)$ and $Z(t) = \int_0^t \Phi(s)dY(s)$. Then $\Phi \in \mathscr{L}_2^{loc}(\tilde{Y})$ and

$$\tilde{Z}(t) = Z(t) - \langle Z, X\rangle(t) = \int_0^t \Phi(s)d\tilde{Y}(s). \tag{4.6}$$

Proof. By the definition of stochastic integrals, it suffices to prove (4.6) for $\Phi \in \mathscr{L}_0$; but in this case the assertion is clear.

Theorem 4.1 implies that if we transform the probability measure P into $\hat{P} = M\cdot P$, every continuous local martingale Y with respect to P is transformed under the probability $\hat{P}$ to $Y =$ *a continuous local martingale*$+\langle Y, X\rangle$. That is, the transformation of the probability measures $P \longrightarrow \hat{P} = M\cdot P$ induces a drift $\langle Y, X\rangle$ for every local martingale Y. For this reason, the transformation of measures $P \longrightarrow \hat{P}$ is called *transformation of drift*. It is also often called the *Girsanov transformation.*

The method of transformation of drift can be applied to solve a class of stochastic differential equations. Suppose $\alpha \in \mathscr{A}^{d,r}$ and $\beta \in \mathscr{A}^{d,1}$ are given and let us consider the stochastic differential equation

$$dX(t) = \alpha(t,X)dB(t) + \beta(t,X)dt. \tag{1.1}$$

Suppose that a solution (X, B) is given on a probability space $(\Omega,\mathscr{F},P)$ with a reference family $(\mathscr{F}_t)$. Without loss of generality, we may assume that this system satisfies condition (4.1). Choose $\gamma \in \mathscr{A}^{r,1}$ such that it is bounded or, more generally, satisfies

$$E\left[\exp\left(\frac{1}{2}\int_0^t \|\gamma(s,X)\|^2 ds\right)\right] < \infty \qquad \text{for every} \quad t > 0.$$

Then

$$M(t) = \exp\left\{\int_0^t \gamma(s,X)dB(s) - \frac{1}{2}\int_0^t \|\gamma(s,X)\|^2 ds\right\} \tag{4.7}$$

is an $(\mathscr{F}_t)$-martingale. Let $\hat{P} = M\cdot P$. By Theorem 4.1,

$$(4.8)\qquad \tilde{B}(t) = B(t) - \int_0^t \gamma(s,X)ds$$

is an r-dimensional $(\mathscr{F}_t)$-Brownian motion on the probability space $(\Omega, \mathscr{F}, \hat{P})$ with the reference family $(\mathscr{F}_t)_{t\geq 0}$. Indeed, since $X(t)$ in (4.2) is now

$$\int_0^t \gamma(s,X)dB(s)\ (= \sum_{i=1}^r \int_0^t \gamma^i(s,X)dB^i(s)) := Y(t),$$

we have

$$\tilde{B}^i(t) = B^i(t) - \langle B^i, Y\rangle(t) = B^i(t) - \int_0^t \gamma^i(s,X)ds \in \hat{\mathscr{M}}_2^{c,loc}$$

and $\langle \tilde{B}^i, \tilde{B}^j\rangle(t) = \langle B^i, B^j\rangle(t) = \delta_{ij}t$, implying that $\tilde{B}$ is an r-dimensional $(\mathscr{F}_t)$-Brownian motion. By the corollary of Theorem 4.1, we have

$$(4.9)\qquad dX(t) = \alpha(t,X)d\tilde{B}(t) + [\beta(t,X) + \alpha(t,X)\gamma(t,X)]dt.$$

This implies that $(X(t), \tilde{B}(t))$ is a solution of the stochastic differential equation (4.9) on the probability space $(\Omega,\mathscr{F},\hat{P})$ with the reference family $(\mathscr{F}_t)_{t\geq 0}$. *Thus we get a solution of* (4.9) *by applying the transformation of drift to a solution of* (1.1). Furthermore, if the uniqueness of solutions (cf. Definition 1.4) holds for the equation (1.1), then it also holds for the equation (4.9). To show this, we may assume without loss of generality that $\gamma \in \mathscr{A}^{r,1}$ is of the form $\gamma(t,w) = \alpha^*(t,w)\eta(t,w)$ with some $\eta \in \mathscr{A}^{d,1}$. ($\alpha^*(t,w) \in \mathscr{A}^{r,d}$ is the transposed of $\alpha(t,w)$; as usual, we regard it as a linear mapping $\boldsymbol{R}^d \longrightarrow \boldsymbol{R}^r$.) Indeed, let $\boldsymbol{R}^r = \alpha^*(t,w)(\boldsymbol{R}^d) \oplus \alpha^*(t,w)(\boldsymbol{R}^d)^\perp$ be the orthogonal decomposition and $\gamma(t,w) = \gamma_1(t,w) + \gamma_2(t,w)$ under this decomposition. Then $\gamma_1 \in \mathscr{A}^{r,1}$. γ_1 is clearly bounded if γ is bounded and $\alpha(t,w)\gamma(t,w) = \alpha(t,w)\gamma_1(t,w)$ since $\alpha(t,w)\gamma_2(t,w) = 0$ as we see by $(\alpha(t,w)\gamma_2(t,w),y) = (\gamma_2(t,w),\ \alpha^*(t,w)y) = 0$ for any $y \in \boldsymbol{R}^d$. Finally choose $\eta(t,w)$ from the affine space $\{y;\ \alpha^*(t,w)y = \gamma_1(t,w)\}$ in $\boldsymbol{R}^d$ such that $\eta \in \mathscr{A}^{d,1}$; for example, let $\eta(t,w)$ be the unique element in the affine space which attains the minimal distance from the origin. By replacing γ by γ_1 we can conclude the above assertion.

Then $M(t)$ given by (4.7) is a well determined functional of X:

$$M(t) = \exp\left\{\int_0^t \gamma(s,X)dB(s) - \frac{1}{2}\int_0^t \|\gamma(s,X)\|^2 ds\right\}$$

$$= \exp\left\{\int_0^t \eta(s,X)dM_X(s) - \frac{1}{2}\int_0^t \|\alpha^*(s,X)\eta(s,X)\|^2 ds\right\}$$

where

$$M_X(t) = X(t) - X(0) - \int_0^t \beta(s,X)ds = \int_0^t \alpha(s,X)dB(s).$$

Thus the uniqueness of solutions for the equation (1.1) implies that the law of the joint process $(X(t),M(t))$ is uniquely determined by the initial law of X.

Now we shall show that the uniqueness of solutions for the equation (1.1) implies that for the equation (4.9). In fact, starting with any solution $(\hat{X}(t), \hat{B}(t))$ of the equation (4.9) on a probability space $(\Omega,\mathscr{F},\hat{P})$ with a reference family $(\mathscr{F}_t)_{t\geq 0}$ satisfying the condition (4.1), we define

$$\hat{M}(t) = \exp\left[-\int_0^t \gamma(s, \hat{X})d\hat{B}(s) - \frac{1}{2}\int_0^t \|\gamma(s, \hat{X})\|^2 ds\right],$$

$$B(t) = \hat{B}(t) + \int_0^t \gamma(s,\hat{X})ds,$$

and

$$P = \hat{M}\cdot\hat{P}.$$

Then $(\hat{X}(t), B(t))$ is a solution of equation (1.1) with respect to the probability P and it is easy to see that if we apply the above transformation of drift to the solution $(\hat{X}(t), B(t))$, then we come back to $(\hat{X}(t), \hat{B}(t))$. Thus *any* solution of (4.9) is obtained by the transformation of drift from *some* solution of (1.1), and hence combined with the above remark, the uniqueness of the solutions of (1.1) implies that of (4.9). Summarizing, we have the following result.

Theorem 4.2. If the equation (1.1) has a unique solution, then the equation (4.9) also has a unique solution; moreover a solution of (4.9) is obtained from a solution of (1.1) by the transformation of drift.

Corollary. Suppose $\beta \in \mathscr{A}^{d,1}$ is bounded. Then the stochastic differential equation

$$(4.10) \qquad dX(t) = dB(t) + \beta(t,X)dt$$

(i.e., the stochastic differential equation (1.1) with $\alpha = I$, the identity matrix) has a unique solution and is constructed as follows. We choose on some probability space $(\Omega,\mathscr{F},P)$ with a reference family $(\mathscr{F}_t)_{t\geq 0}$ satisfying the condition (4.1) a d-dimensional $(\mathscr{F}_t)$-Brownian motion $B(t)$ with $B(0)=0$ and a d-dimensional, $\mathscr{F}_0$-measurable random variable $X(0)$ with given distribution μ on $\boldsymbol{R}^d$. We set

$$X(t) = X(0) + B(t),$$
$$M(t) = \exp\left[\int_0^t \beta(s,X)dB(s) - \frac{1}{2}\int_0^t \|\beta(s,X)\|^2 ds\right],$$
$$\hat{P} = M\cdot P$$

and

$$\tilde{B}(t) = B(t) - \int_0^t \beta(s,X)ds.$$

Then $(X(t), \tilde{B}(t))$ is a solution of (4.10) on the probability space $(\Omega,\mathscr{F},\hat{P})$ with the reference family $(\mathscr{F}_t)_{t\geq 0}$.

In this way, the stochastic differential equation (4.10) is always solved uniquely by the method of the transformation of drift, (cf. Maruyama [110]). But the solution thus obtained is not necessarily a strong solution in general. In fact, Cirel'son [12] gave an example of a stochastic differential equation of the form (4.10) for which the solution is not strong.

Example 4.1. (Cirel'son [12]). Let $d=1$ and $\beta(t,w)\in\mathscr{A}^{1,1}$ be defined as follows: let $\theta(x) = x \pmod 1 = x - [x] \in [0,1)$, $x\in\boldsymbol{R}^1$, and let, $\{t_k;\ k=0,\ -1,\ -2,\ \dots\}$ be a sequence such that $0 < t_{k-1} < t_k$ and $\lim_{k\to-\infty} t_k = 0$. For $w\in\boldsymbol{W}^1$, set

$$\beta(t,w) = \begin{cases} \theta\left(\dfrac{w(t_k)-w(t_{k-1})}{t_k - t_{k-1}}\right), & \text{if } t\in[t_k, t_{k+1}), k=-1,-2,\dots, \\ 0, & \text{if } t=0 \text{ or } t\geq t_0. \end{cases} \tag{4.11}$$

It is clear that $\beta(t,w)\in\mathscr{A}^{1,1}$ and is bounded. However, the one-dimensional stochastic differential equation (4.10) does not have a strong solution.

Proof.* Suppose on the contrary, this stochastic differential equation has a strong solution $(X(t), B(t))$. We may assume that $X(0)=x$ a.s. for

* We learned the following simple proof from Shiryaev by private communication.

some constant $x \in \boldsymbol{R}^1$. Then, by the definition of strong solution, we have $\mathscr{F}_t^X = \sigma[X(s); s \leq t] \subset \mathscr{F}_t^B = \sigma[B(s); s \leq t]$. Now

$$X(t_{k+1}) - X(t_k) = B(t_{k+1}) - B(t_k) + \int_{t_k}^{t_{k+1}} \beta(t,X)dt$$
$$= B(t_{k+1}) - B(t_k) + \theta\Big(\frac{X(t_k) - X(t_{k-1})}{t_k - t_{k-1}}\Big)(t_{k+1} - t_k)$$
$$, \text{ for } k = -1, -2, \dots,$$

and hence if we set

$$\eta_k = \frac{X(t_k) - X(t_{k-1})}{t_k - t_{k-1}}, \; \xi_k = \frac{B(t_k) - B(t_{k-1})}{t_k - t_{k-1}}, \quad k = -1, -2, \dots,$$

then

$$\eta_{k+1} = \xi_{k+1} + \theta(\eta_k). \tag{4.12}$$

Consequently

$$e^{2\pi i \eta_{k+1}} = e^{2\pi i \xi_{k+1}} e^{2\pi i \eta_k}, \qquad (i = \sqrt{-1}),$$

and therefore, for every $l = 0, 1, 2, \cdots$,

$$e^{2\pi i \eta_{k+1}} = e^{2\pi i \xi_{k+1}} e^{2\pi i \xi_k} \cdots e^{2\pi i \xi_{k-l+1}} e^{2\pi i \eta_{k-l}}. \tag{4.13}$$

Since $B(t)$ is an $(\mathscr{F}_t)$-Brownian motion and $(X(t), B(t))$ is $(\mathscr{F}_t)$-adapted, we have that ξ_{k+1} is independent of $\sigma[\xi_k, \xi_{k-1}, \xi_{k-2}, \dots, \eta_k, \eta_{k-1}, \eta_{k-2}, \dots]$ and so from (4.13)

$$E[e^{2\pi i \eta_{k+1}}] = \prod_{n=0}^{l} E[e^{2\pi i \xi_{k-n+1}}] E[e^{2\pi i \eta_{k-l}}].$$

Since $|E[e^{2\pi i \eta_{k-l}}]| \leq 1$ and $E[e^{2\pi i \xi_n}] = \exp[-2\pi^2/(t_n - t_{n-1})]$, we have

$$|E[e^{2\pi i \eta_{k+1}}]| \leq \exp[-2\pi^2 \sum_{n=0}^{l} (t_{k-n+1} - t_{k-n})^{-1}].$$

Letting $l \longrightarrow \infty$, we deduce that

$$E[e^{2\pi i \eta_{k+1}}] = 0, \quad k = -1, -2, \dots. \tag{4.14}$$

Set $\mathscr{B}_l^{k+1} = \sigma[B(t) - B(s);\ t_{k-l} \leq s < t \leq t_{k+1}]$. Then $\sigma[X(u), B(u), u \leq t_{k-l}]$ $(\subset \mathscr{F}_{t_{k-l}})$ is independent of $\mathscr{B}_l^{k+1}$ and hence, by (4.13) and (4.14),

$$E[e^{2\pi i\eta_{k+1}} \mid \mathscr{B}_l^{k+1}] = e^{2\pi i\xi_{k+1}}\, e^{2\pi i\xi_k} \cdots e^{2\pi i\xi_{k-l+1}} E[e^{2\pi i\eta_{k-l}}] = 0.$$

Since $\bigvee_{l=0}^{\infty} \mathscr{B}_l^{k+1} = \mathscr{F}_{t_{k+1}}^{B} \supset \mathscr{F}_{t_{k+1}}^{X}$, it follows by letting $l \uparrow \infty$ that

$$e^{2\pi i\eta_{k+1}} = E[e^{2\pi i\eta_{k+1}} \mid \mathscr{F}_{t_{k+1}}^{B}] = \lim_{l\to\infty} E[e^{2\pi i\eta_{k+1}} \mid \mathscr{B}_l^{k+1}] = 0,$$

(cf. Theorem I-6.6). But this is clearly a contradiction.

4.2. *Time change*. Another probabilistic method which is sometimes useful in solving stochastic differential equations is method of random time change. The general theory of time changes in the martingale theory is well known; we have already discussed it somewhat in Chapter II and in Chapter III, Section 1. In order to avoid undesirable complications, we restrict the class of time changes and formulate it as follows.

Let $\boldsymbol{I}$ be the class of functions

$$\phi: t \in [0, \infty) \longmapsto \phi_t \in [0, \infty)$$

satisfying

(i) $\phi_0 = 0$,
(ii) ϕ is continuous and strictly increasing and
(iii) $\phi_t \uparrow \infty$ as $t \uparrow \infty$.

Clearly, if ϕ^{-1} is the inverse function of $\phi \in \boldsymbol{I}$, then $\phi^{-1} \in \boldsymbol{I}$. $\boldsymbol{I}$ is a subset of $\boldsymbol{W}^1$ and the Borel fields induced by $\mathscr{B}(\boldsymbol{W}^1)$, $\mathscr{B}_t(\boldsymbol{W}^1)$ are denoted by $\mathscr{B}(\boldsymbol{I})$ and $\mathscr{B}_t(\boldsymbol{I})$ respectively. Each $\phi \in \boldsymbol{I}$ defines a transformation T^ϕ of $\boldsymbol{W}^d$ into itself by

$$(4.15) \qquad T^\phi: w \in \boldsymbol{W}^d \longrightarrow (T^\phi w) \in \boldsymbol{W}^d$$

where $(T^\phi w)(t) = w(\phi_t^{-1})$, $t \in [0, \infty)$. T^ϕ is called the *time change* defined by $\phi \in \boldsymbol{I}$.

Let a probability space $(\Omega, \mathscr{F}, P)$ with a reference family $(\mathscr{F}_t)_{t\geq 0}$ be given. Consider a mapping $\phi: \Omega \ni \omega \longmapsto \phi(\omega) \in \boldsymbol{I}$ which is $\mathscr{F}_t/\mathscr{B}_t(\boldsymbol{I})$-measurable for each t. Such a $\phi = (\phi_t(\omega))$ is called a *process of the time change*. Clearly $\phi = (\phi_t(\omega))$ is an $(\mathscr{F}_t)$-adapted increasing process, and hence if $\phi_t^{-1}(\omega)$ is the inverse function of $t \longmapsto \phi_t(\omega)$ then ϕ_t^{-1} is an

$(\mathscr{F}_t)$-stopping time for each fixed $t \in [0, \infty)$. If $X = (X(t))$ is a continuous $(\mathscr{F}_t)$-adapted process, then $T^\phi X = ((T^\phi X)(t))$ defined by $(T^\phi X)(t) = X(\phi_t^{-1})$ is a continuous $(\mathscr{F}_{\phi_t^{-1}})$-adapted process. The transformation $X \longmapsto T^\phi X$ defined above is called the *time change of X by the process of time change* ϕ.

Given a process of time change ϕ, we define a new reference family $(\tilde{\mathscr{F}}_t)$ by $\tilde{\mathscr{F}}_t = \mathscr{F}_{\phi_t^{-1}}$, $t \in [0, \infty)$. The class $\mathscr{M}_2^{c,loc}$ with respect to $\tilde{\mathscr{F}}_t$ is denoted by $\tilde{\mathscr{M}}_2^{c,loc}$. It is an important consequence of Doob's optional sampling theorem that if $X \in \mathscr{M}_2^{c,loc}$, then $T^\phi X \in \tilde{\mathscr{M}}_2^{c,loc}$, and if $X, Y \in \mathscr{M}_2^{c,loc}$, then

$$(4.16) \qquad \langle T^\phi X, T^\phi Y \rangle = T^\phi \langle X, Y \rangle.$$

Now we apply the method of time change to solve stochastic differential equations. In the following we consider the case of $d = 1$, $r = 1$ and $\beta = 0$. Thus we consider for given $\alpha(t,w) \in \mathscr{A}^{1,1}$ the equation

$$(4.17) \qquad dX(t) = \alpha(t,X)dB(t).$$

We assume for simplicity that there exist positive constants C_1 and C_2 such that

$$(4.18) \qquad C_1 \le \alpha(t,w) \le C_2.$$

As we saw in Theorem 4.2, if we can solve the equation (4.17), then we can solve the general equation having drift term $\beta(t,w)dt$ by the method of transformation of drift.

Theorem 4.3. (i) Let $b = (b(t))$ be a one-dimensional $(\mathscr{F}_t)$-Brownian motion with $b(0) = 0$ given on a probability space $(\Omega, \mathscr{F}, P)$ with a reference family $(\mathscr{F}_t)_{t \ge 0}$ and let $X(0)$ be an $\mathscr{F}_0$-measurable random variable. Define a continuous process $\xi = (\xi(t))$ by $\xi(t) = X(0) + b(t)$. Let $\phi = (\phi_t)$ be a process of time change such that

$$(4.19) \qquad \phi_t = \int_0^t \alpha(\phi_s, T^\phi \xi)^{-2} ds$$

holds a.s. Then if we set $X = T^\phi \xi$ (i.e., $X(t) = \xi(\phi_t^{-1}) = X(0) + b(\phi_t^{-1})$) and $\tilde{\mathscr{F}}_t = \mathscr{F}_{\phi_t^{-1}}$, there exists an $(\tilde{\mathscr{F}}_t)$-Brownian motion $B = (B(t))$ such that $(X(t), B(t))$ is a solution of (4.17) on the probability space $(\Omega, \mathscr{F}, P)$ with the reference family $(\tilde{\mathscr{F}}_t)_{t \ge 0}$.

(ii) Conversely, if $(X(t), B(t))$ is a solution of (4.17) on a probability space $(\Omega,\mathscr{F},P)$ with a reference family $(\mathscr{F}_t)_{t\geq 0}$, then there exist a reference family $(\hat{\mathscr{F}}_t)_{t\geq 0}$, an $(\hat{\mathscr{F}}_t)$-Brownian motion $b=(b(t))$ with $b(0)=0$, and a process of time change $\phi=(\phi_t)$ with respect to the family $(\hat{\mathscr{F}}_t)$ such that, if we set $\xi(t)=X(0)+b(t)$, then (4.19) holds a.s. and $X=T^{\phi}\xi$. That is, any solution of (4.17) is given as in (i).

Corollary. Suppose we are given a one-dimensional $(\mathscr{F}_t)$-Brownian motion $b=(b(t))$ and an $\mathscr{F}_0$-measurable random variable $X(0)$. Define $\xi=(\xi(t))$ by $\xi(t)=X(0)+b(t)$. If there exists a process of time change ϕ such that (4.19) holds and if such a ϕ is unique, (i.e., if ψ is another process of time change satisfying (4.19), then $\phi(t)\equiv\psi(t)$ a.s.), then the solution of (4.17) with the initial value $X(0)$ exists and is unique. Moreover, the solution is given by $X=T^{\phi}\xi$.

Proof. (i) If $b=(b(t))$, $X(0)$ and $\phi=(\phi_t)$ are given as in (i) of the theorem, then $M=T^{\phi}b\in\tilde{\mathscr{M}}_2^{c,loc}$ and $\langle M\rangle(t)=\phi_t^{-1}$. If $X=T^{\phi}\xi$, then by (4.19), $t=\int_0^t\alpha(\phi_s,X)^2d\phi_s$ and hence

$$\langle M\rangle(t)=\phi_t^{-1}=\int_0^{\phi_t^{-1}}\alpha(\phi_s,X)^2d\phi_s=\int_0^t\alpha(s,X)^2ds.$$

We set $B(t)=\int_0^t(\alpha(s,X))^{-1}dM(s)$. Then $B\in\tilde{\mathscr{M}}_2^{c,loc}$ and $\langle B\rangle(t)=\int_0^t\alpha(s,X)^{-2}d\langle M\rangle(s)=t$. This implies that B is an $(\hat{\mathscr{F}}_t)$-Brownian motion. Since $M(t)=X(t)-X(0)=\int_0^t\alpha(s,X)dB(s)$, (X,B) is a solution of (4.17).

(ii) Let $(X(t), B(t))$ be a solution of (4.17) on a probability space $(\Omega,\mathscr{F},P)$ with a reference family $(\mathscr{F}_t)_{t\geq 0}$. Then $M(t)=X(t)-X(0)\in\mathscr{M}_2^{c,loc}$ and $\langle M\rangle(t)=\int_0^t\alpha(s,X)^2ds$. Set $\psi(t)=\langle M\rangle(t)$, $\phi_t=\psi_t^{-1}$ and $\hat{\mathscr{F}}_t=\mathscr{F}_{\psi_t^{-1}}$. Clearly $\phi=(\phi_t)$ is a process of time change with respect to $(\hat{\mathscr{F}}_t)$ and the process $b=(b(t))=(M(\phi_t))$ is an $(\hat{\mathscr{F}}_t)$-Brownian motion (Theorem II-7.2). If we now set $\xi(t)=X(0)+b(t)$, then $T^{\phi}\xi=X$. Furthermore, since $t=\int_0^t\alpha(s,X)^{-2}d\psi_s$, it follows that

$$\begin{aligned}\phi_t&=\int_0^{\phi_t}\alpha(s,X)^{-2}d\psi_s=\int_0^t\alpha(\phi_u,X)^{-2}du\\&=\int_0^t\alpha(\phi_u,T^{\phi}\xi)^{-2}du\end{aligned}$$

and consequently $\phi=(\phi_t)$ satisfies (4.19). The proof of the theorem is now complete.

Example 4.2. Let $a(x)$ be a bounded Borel measurable function on $\boldsymbol{R}^1$ such that $a(x) \geq C$ for some positive constant C. Set $\alpha(t,w) = a(w(t))$. Thus we are considering a stochastic differential equation of time-independent Markovian type. If $X(0)$ and $b = b(t)$ are given, then equation (4.19) may be written as

$$(4.20) \qquad \phi_t = \int_0^t a[T^\phi \xi(\phi_s)]^{-2} ds = \int_0^t a(\xi(s))^{-2} ds,$$

where $\xi(t) = X(0) + b(t)$. Consequently ϕ_t is uniquely determined from $\xi(t)$ and hence the stochastic differential equation

$$dX(t) = a(X(t))dB(t)$$

is solved as $X(t) = \xi(\phi_t^{-1})$.

Example 4.3. Let $a(t,x)$ be a bounded Borel measurable function on $[0,\infty)\times\boldsymbol{R}^1$ such that $a(t,x) \geq C$ for some positive constant C. Set $\alpha(t,w) = a(t,w(t))$. In this case, equation (4.19) is given as

$$(4.21) \qquad \phi_t = \int_0^t a[\phi_s, T^\phi \xi(\phi_s)]^{-2} ds = \int_0^t a[\phi_s, \xi(s)]^{-2} ds.$$

This is equivalent to the following differential equation for ϕ_t along each fixed sample path of $\xi(t)$:

$$(4.21)' \qquad \begin{cases} \dot{\phi}_t = 1/a[\phi_t, \xi(t)]^2,^* \\ \phi_0 = 0. \end{cases}$$

One simple sufficient condition that (4.21)′ has the unique solution is that $a(t, x)$ be Lipschitz continuous in t; in this case the stochastic differential equation

$$dX(t) = a(t,X(t))dB(t)$$

is solved uniquely as $X(t) = \xi(\phi_t^{-1})$ (cf. Yershov [187]). On the other hand, Stroock-Varadhan [159] proved that the above stochastic differential equation always has a unique solution, and this fact, in turn, can be used

---

* $\dot{\phi}_t = \dfrac{d}{dt}\phi_t$.

to show that (4.21)′ has a unique solution ϕ_t along each fixed sample path of $\xi(t)$. (cf. Watanabe [167]).

Example 4.4. (Nisio [130]). Let $f(x)$ be a locally bounded Borel measurable function on $\boldsymbol{R}^1$, $a(x)$ be a bounded Borel measurable function on $\boldsymbol{R}^1$ such that $a(x) \geq C$ for some positive constant C, and $y \in \boldsymbol{R}^1$. Set $\alpha(t,w) = a[y+\int_0^t f(w(s))ds]$, $w \in \boldsymbol{W}^1$. The corresponding stochastic differential equation is

$$(4.22) \qquad dX(t) = a[y+\int_0^t f(X(s))ds]dB(s)$$

and the equation (4.19) is given as

$$(4.23) \qquad \phi_t = \int_0^t a(y+\int_0^{\phi_s} f[(T^\phi\xi)(u)]du)^{-2}ds.$$

Now

$$\begin{aligned}\int_0^{\phi_s} f[(T^\phi\xi)(u)]du &= \int_0^{\phi_s} f[\xi(\phi_u^{-1})]du \\ &= \int_0^s f[\xi(u)]d\phi_u = \int_0^s f(\xi(u))\dot{\phi}_u du.\end{aligned}$$

Thus (4.23) is equivalent to

$$(4.24) \qquad \begin{cases} \dot{\phi}_t = 1/a(y+\int_0^t f(\xi(u))\dot{\phi}_u du)^2, \\ \phi_0 = 0. \end{cases}$$

and this can be solved uniquely for each given $\xi(t)$ as follows. Set

$$Z(t) = \int_0^t f(\xi(u))\dot{\phi}_u du.$$

Then

$$\dot{Z}(t) = f(\xi(t))\dot{\phi}_t = f(\xi(t))/a(y+Z(t))^2$$

and hence

$$\int_0^t a(y+Z(s))^2\dot{Z}(s)ds = \int_0^t f(\xi(s))ds.$$

Consequently, if we define $A(x)$ by

$$A(x) = \begin{cases} \int_0^x a(y+z)^2 dz, & \text{for} \quad x > 0, \\ -\int_x^0 a(y+z)^2 dz, & \text{for} \quad x \leq 0, \end{cases}$$

then $A(Z(t)) = \int_0^t f(\xi(s))ds$ and therefore $Z(t) = A^{-1}[\int_0^t f(\xi(s))ds]$ where $A^{-1}(x)$ is the inverse function of $x \longmapsto A(x)$. Thus ϕ_t is solved uniquely as

$$\phi_t = \int_0^t a(y + Z(s))^{-2}ds = \int_0^t a(y+A^{-1}[\int_0^s f(\xi(u))du])^{-2}ds$$

and so the equation (4.22) is solved uniquely as $X(t) = \xi(\phi_t^{-1})$.

Note that in the special case of $f(x) = x$, equation (4.22) is equivalent to the following equation of motion with random acceleration:

$$(4.25) \qquad \begin{cases} dY(t) = X(t)dt \\ dX(t) = a(Y(t))dB(t) \\ Y(0) = y. \end{cases}$$

5. Diffusion processes

Diffusion processes constitute a class of stochastic processes which are characterized by two properties: the Markovian property and the continuity of trajectories. Since it is beyond the scope of this book to discuss diffusion processes in full generality,*[1] we restrict our attention to the class of diffusion processes which can be described by stochastic differential equations. This class of diffusions is known to be sufficiently wide both in theory and applications*[2]; furthermore, the stochastic calculus provides us with a very powerful tool for studying such diffusions.

First, we give a formal definition of diffusion processes. Let S be a topological space. We sometimes find it convenient to attach an extra point

---

*[1] Only in the one dimensional case is a satisfactory theory known, cf. Itô-McKean [73] and Dynkin [22].

*[2] There are, however, many recent results on multi-dimensional diffusion processes which can not be covered by the method of stochastic differential equations; see, e.g. Fukushima [30], Ikeda-Watanabe [53], Motoo [124] and Orey [135].

Δ to S, either as an isolated point or as point at infinity if S is locally compact. Thus we set $S' = S \cup \{\Delta\}$. Δ is called the *terminal point*. Either S or S' is called the *state space*. Let $\bar{W}(S)$ be the set of all functions w: $[0, \infty) \ni t \longmapsto w(t) \in S'$ such that there exists $0 \leq \zeta(w) \leq \infty$ with the following properties:

(i) $w(t) \in S$ for all $t \in [0, \zeta(w))$ and the mapping $t \in [0, \zeta(w)) \longmapsto w(t)$ is continuous;

(ii) $w(t) = \Delta$ for all $t \geq \zeta(w)$.

$\zeta(w)$ is called the *life time* of the trajectory w. For convenience, we set $w(\infty) = \Delta$ for every $w \in \bar{W}(S)$. A Borel cylinder set in $\bar{W}(S)$ is defined for some integer n, a sequence $0 \leq t_1 < t_2 < \cdots < t_n$ and a Borel subset A in $S'^n = S' \times S' \times \cdots \times S'$ as

$$\pi^{-1}_{t_1, t_2, \ldots, t_n}(A)$$

where $\pi_{t_1, t_2, \ldots, t_n}: \bar{W}(S) \longrightarrow S'^n$ is given by

$$\pi_{t_1, t_2, \ldots, t_n}(w) = (w(t_1), w(t_2), \ldots, w(t_n)).$$

Here, of course, a Borel subset in a topological space is any set in the smallest σ-field containing all open sets. Let $\mathscr{B}(\bar{W}(S))$ be the σ-field in $\bar{W}(S)$ generated by all Borel cylinder sets and let $\mathscr{B}_t(\bar{W}(S))$ be that generated by all cylinder sets up to time t, i.e., sets expressed in the form $\pi^{-1}_{t_1, t_2, \ldots, t_n}(A)$ where $t_n \leq t$. A family of probabilities $\{P_x, x \in S'\}$ on $(\bar{W}(S), \mathscr{B}(\bar{W}(S)))$ is called *Markovian*[*1] if

(i) $P_x\{w; w(0) = x\} = 1$ for every $x \in S'$,

(ii) $x \in S \longmapsto P_x(A)$ is Borel measurable (or more generally, universally measurable)[*2] for each $A \in \mathscr{B}(\bar{W}(S))$, and

(iii) for every $t > s \geq 0$, $A \in \mathscr{B}_s(\bar{W}(S))$ and a Borel subset Γ in S',

$$(5.1) \qquad P_x(A \cap \{w; w(t) \in \Gamma\}) = \int_A P_{w'(s)}[w; w(t-s) \in \Gamma] P_x(dw')$$

for every $x \in S'$.

A Markovian system $\{P_x, x \in S'\}$ is called *conservative* if

$$(5.2) \qquad P_x\{w; \zeta(w) = \infty\} = 1 \qquad \text{for every} \quad x \in S.$$

In this case we need not consider the point Δ since almost all sample paths

---

*1 We only consider the time homogeneous case.

*2 Cf. Chapter I, §1.

lie in S. For a Markovian system $\{P_x, x \in S'\}$, $t \in [0, \infty)$, $x \in S'$ and a Borel subset Γ of S', we set

$$P(t,x,\Gamma) = P_x\{w;\ w(t) \in \Gamma\}. \tag{5.3}$$

The family $\{P(t,x,\Gamma)\}$ is called the *transition probability* of a Markovian system. By the successive application of (5.1), it is easy to see that

$$\begin{aligned} &P_x[w(t_1)\in A_1, w(t_2)\in A_2, \dots, w(t_n)\in A_n] \\ &= \int_{A_1} P(t_1, x, dx_1) \int_{A_2} P(t_2 - t_1, x_1, dx_2) \int_{A_3} \cdots \\ &\qquad \times \int_{A_n} P(t_n - t_{n-1}, x_{n-1}, dx_n) \end{aligned} \tag{5.4}$$

$$\text{for } 0 < t_1 < t_2 < \dots < t_n,\ A_i \in \mathscr{B}(S'),$$

and thus we see that *two Markovian systems on the same state space with the same transition probability coincide.*

Let a *Markovian* system $\{P_x\}$ be given. For each $t \geq 0$, we set $\mathscr{F}_t(\overline{W}(S)) = \bigcap_{\varepsilon>0} \bigcap_{x\in S'} \overline{\mathscr{B}_{t+\varepsilon}(\overline{W}(S))}^{P_x}$*[1] and $\mathscr{F}_\infty(\overline{W}(S)) = \bigvee_{t>0} \mathscr{F}_t(\overline{W}(S))$. A mapping $w \in \overline{W}(S) \longmapsto \sigma(w) \in [0, \infty)$ is called a *stopping time* if it is an $(\mathscr{F}_t(\overline{W}(S)))$-stopping time, i.e., if for every $t \geq 0$, $\{w;\ \sigma(w) \leq t\} \in \mathscr{F}_t(\overline{W}(S))$. For a stopping time σ, we set $\mathscr{F}_\sigma(\overline{W}(S)) = \{A \in \mathscr{F}_\infty(\overline{W}(S));\ A \cap \{w;\ \sigma(w) \leq t\} \in \mathscr{F}_t(\overline{W}(S))$ for every $t \geq 0\}$. The Markovian system $\{P_x\}$ is called a *strongly Markovian system* if for every $t \geq 0$, stopping time σ, $A \in \mathscr{F}_\sigma(\overline{W}(S))$ and a Borel subset Γ of S', we have that

$$\begin{aligned} &P_x(A \cap \{w;\ w(t + \sigma(w)) \in \Gamma\}) \\ &= \int_A P_{w'(\sigma(w'))}[w;\ w(t) \in \Gamma] P_x(dw') \end{aligned} \tag{5.5}$$

for every $x \in S'$.*[2]

Definition 5.1. A family of probabilities $\{P_x\}_{x\in S'}$ on $(\overline{W}(S), \mathscr{B}(\overline{W}(S)))$ is called a *system of diffusion measures*, or simply a *diffusion* if it is a strongly Markovian system.

Definition 5.2. A stochastic process $X = \{X(t)\}$ on S' defined on a

*[1] It follows from this definition that $\mathscr{F}_t(\overline{W}(S))$ is right continuous, i.e., $\mathscr{F}_{t+0}(\overline{W}(S)) = \mathscr{F}_t(\overline{W}(S))$ for every $t \geq 0$.

*[2] We need only assume that (5.5) is valid for bounded stopping time; otherwise, replace σ by $\sigma \wedge n$, A by $A \cap \{\sigma \leq n\}$ and then let $n \uparrow \infty$.

probability space $(\Omega,\mathscr{F},P)$ is called a *diffusion process on* S if there exists a system of diffusion measures $\{P_x\}_{x\in S'}$ such that, for almost all ω, $[t \longmapsto X(t)] \in \overline{\boldsymbol{W}}(S)$ and the probability law (i.e. the image measure) on $\overline{\boldsymbol{W}}(S)$ of $[t \longmapsto X(t)]$ coincides with $P_\mu(\cdot)=\int_{S'} P_x(\cdot)\mu(dx)$ where μ is the Borel measure on S' defined by $\mu(dx) = P\{\omega; X(0, \omega) \in dx\}$ and is called the *initial distribution* of X.

If $X = (X(t))$ is a diffusion process and if we set $\zeta(\omega) = \inf\{t; X(t) = \Delta\}$, then it is clear that, with probability one, $[0, \zeta) \ni t \longmapsto X(t) \in S$ is continuous and $X(t) = \Delta$ for all $t \geq \zeta$. ζ is called the *life time* of the diffusion process X. X is called *conservative* if $\zeta(\omega) = \infty$ a.s.

Now, let $\boldsymbol{C}(S')$ be the Banach space of all real or complex valued bounded continuous functions defined on S' and $(A,\mathscr{D}(A))$ be a linear operator on $\boldsymbol{C}(S')$ into itself with the domain of definitions $\mathscr{D}(A)$. Let $\{P_x, x\in S'\}$ be a system of probability measures on $(\overline{\boldsymbol{W}}(S),\mathscr{B}(\overline{\boldsymbol{W}}(S)))$ such that $x \longmapsto P_x(A)$ is Borel measurable (or universally measurable) for $A\in \overline{\boldsymbol{W}}(S)$. The following definition is based on an idea of Stroock and Varadhan [160].

Definition 5.3. $\{P_x\}$ is called a *diffusion measure generated* (or *determined*) *by the operator* A (or simply as *an* A*-diffusion*) if it is a strongly Markovian system satisfying

(i)

$$P_x\{w;\ w(0) = x\} = 1 \qquad \text{for every} \quad x,$$

(ii)

$$f(w(t)) - f(w(0)) - \int_0^t (Af)(w(s))ds$$

is a $(P_x, \mathscr{B}_t(\overline{\boldsymbol{W}}(S)))$-martingale for every $f \in \mathscr{D}(A)$ and every x.

Theorem 5.1. Suppose that $\{P_x, x \in S'\}$ is a system of probability measures on $(\overline{\boldsymbol{W}}(S),\mathscr{B}(\overline{\boldsymbol{W}}(S)))$ satisfying conditions (i) and (ii) of Definition 5.3. Suppose further that $\{P_x\}$ is unique; i.e.,

(iii) if $\{P_x'\}$ is any other system of probability measures on $(\overline{\boldsymbol{W}}(S), \mathscr{B}(\overline{\boldsymbol{W}}(S)))$ satisfying the conditions (i) and (ii) of Definition 5.3, then $P_x' = P_x$ for every x.

Then $\{P_x\}$ is a system of diffusion measures generated by the operator A.

Proof. It is only necessary to show that $\{P_x\}$ is a strongly Markovian system, i.e., it satisfies (5.5). Since

$$X_f(t) = f(w(t)) - f(w(0)) - \int_0^t (Af)(w(s))ds$$

is a $(P_x,\mathscr{B}_t(\overline{W}(S)))$-martingale for every $x\in S'$ and $t\longmapsto X_f(t)$ is right continuous, it is clear that it is also a $(P_x,\mathscr{F}_t(\overline{W}(S)))$-martingale. If σ is a bounded stopping time, then by Doob's optional sampling theorem $t\longmapsto X_f(t+\sigma)$ is a $(P_x,\mathscr{F}_{t+\sigma}(\overline{W}(S)))$-martingale. In particular, for every $t>s$, $A\in\mathscr{F}_{s+\sigma}(\overline{W}(S))$ and $C\in\mathscr{F}_\sigma(\overline{W}(S))$, we have

$$E_x(X_f(t+\sigma)-X_f(s+\sigma)\colon A\cap C)=0.$$

This implies that

$$E_x(X_f(t+\sigma)-X_f(s+\sigma)\colon A\,|\,\mathscr{F}_\sigma(\overline{W}(S)))=0,\quad \text{a.a. } w(P_x).$$

Therefore, if $\tilde{P}^w(A)=P_x(\theta_\sigma^{-1}(A)\,|\,\mathscr{F}_\sigma(\overline{W}(S)))$, $A\in\mathscr{B}(\overline{W}(S))$ is the regular conditional probability given $\mathscr{F}_\sigma(\overline{W}(S))$ where $\theta_\sigma\colon\overline{W}(S)\longrightarrow\overline{W}(S)$ is defined by $(\theta_\sigma w)(t)=w(\sigma(w)+t)$, then $\tilde{P}^w(w';\ w'(0)=w(\sigma(w)))=1$ for a.a. $w(P_x)$ and $X_f(t)$ is a $(\tilde{P}^w,\mathscr{B}_t(\overline{W}(S)))$-martingale. By assumption (iii), we have $\tilde{P}^w=P_{w(\sigma(w))}$ which clearly implies (5.5).

Corollary. Let $\{P_x,\ x\in S'\}$ be a system of probability measures on $(\overline{W}(S),\mathscr{B}(\overline{W}(S)))$ which satisfies the conditions (i) and (ii) of Definition 5.3. The uniqueness condition (iii) of Theorem 5.1 then follows from the weaker condition (iv):

(iv) if $\{P'_x\}$ is any other system of probability measures on $(\overline{W}(S),\mathscr{B}(\overline{W}(S)))$ satisfying (i) and (ii), then

$$(5.6)\qquad \int_{\overline{W}(S)} f(w(t))P_x(dw)=\int_{\overline{W}(S)} f(w(t))P'_x(dw)$$

for every $t\geq 0$, $x\in S'$ and $f\in\mathscr{T}$, where $\mathscr{T}$ is some total family in $\boldsymbol{C}(S')$.* The equality (5.6) can also be replaced by

$$(5.7)\qquad \int_{\overline{W}(S)}\int_0^\infty e^{-\lambda t}f(w(t))dtP_x(dw)=\int_{\overline{W}(S)}\int_0^\infty e^{-\lambda t}f(w(t))dtP'_x(dw)$$

for every $\lambda>0$, $x\in S'$ and $f\in\mathscr{T}$, where $\mathscr{T}$ is some total family in $\boldsymbol{C}(S')$.

Proof. Just as in the above proof, we have

---

* $\mathscr{T}\subset\boldsymbol{C}(S')$ is called *total* if for any two Borel probability measures μ and ν on S', $\int_{S'}f(x)\mu(dx)=\int_{S'}f(x)\nu(dx)$ for any $f\in\mathscr{T}$ implies that $\mu=\nu$.

$$\int_{\hat{W}(S)} f(w'(t))\tilde{P}^w(dw') = \int_{\hat{W}(S)} f(w'(t))P_{w(\sigma(w))}(dw')$$

for every bounded stopping time σ and hence $\{P_x\}$ is a strongly Markovian system. Since $\mathscr{T}$ is total, our assumption (5.6) implies that the transition probability $P(t, x, \Gamma) = P_x(w(t) \in \Gamma)$ is uniquely determined. Consequently, by (5.4), P_x is uniquely determined as a measure on $\mathscr{B}(\hat{W}(S))$, i.e. (iii) is satisfied. The equivalence of (5.6) and (5.7) is obvious.

The following theorem is an easy consequence of Theorem 5.1.

Theorem 5.2. Let $(A, \mathscr{D}(A))$ be a linear operator on $C(S')$ and let $(\Omega, \mathscr{F}, P)$ and $(\mathscr{F}_t)_{t \geq 0}$ be given as usual. Suppose that for each $x \in S'$ there exists an S'-valued, $(\mathscr{F}_t)$-adapted stochastic process $X_x = (X(t))$ such that

(i) with probability one, $X(0) = x$, $[0, \zeta(\omega) = \zeta(X_x)) \ni t \longmapsto X(t) \in S$ is continuous and $X(t) = \Delta$ for $t \geq \zeta$ and

(ii) for every $f \in \mathscr{D}(A)$,

$$X_f(t) = f(X(t)) - f(X(0)) - \int_0^t (Af)(X(s))ds$$

is an $(\mathscr{F}_t)$-martingale.

Let P_x be the probability law of the process X_x on $(\hat{W}(S), \mathscr{B}(\hat{W}(S)))$ and suppose that $x \longmapsto P_x$ is universally measurable and, furthermore, that P_x is uniquely determined for every $x \in S'$. Then $\{P_x\}_{x \in S'}$ is the system of diffusion measures determined by the operator A and the stochastic process X_x is a diffusion process satisfying $X(0) = x$.

We call X_x the *A-diffusion process starting at* $x \in S'$.

Example 5.1. Let $S = \mathbf{R}^d$ and $S' = \mathbf{R}^d \cup \{\Delta\}$, where Δ is attached to $\mathbf{R}^d$ as an isolated point. Let $\mathscr{D}(A) = \{f \in C(S');\ f|_{\mathbf{R}^d} \in C_b^2(\mathbf{R}^d)\}$ and define A on $\mathscr{D}(A)$ by

$$(5.8) \qquad Af(x) = \begin{cases} \frac{1}{2}\Delta f(x), & x \in \mathbf{R}^d, \\ 0, & x = \Delta. \end{cases}$$

Then the operator A generates a unique diffusion $\{P_x\}$; This is just *d-dimensional Brownian motion* i.e., P_x is the Wiener measure on $\mathbf{W}^d$ starting at $x \in \mathbf{R}^d$.

To prove this, we first show that P_x, $x \in \mathbf{R}^d$, is conservative. Indeed, $f(x)$ defined by $f(x)|_{\mathbf{R}^d} \equiv 0$ and $f(\Delta)=1$ belongs to $\mathscr{D}(A)$ and $Af(x) \equiv 0$. Thus $I_\Delta(w(t))$ is a martingale with respect to P_x for every x, and hence if

$x \in \boldsymbol{R}^d$, $I_\Delta(w(t)) = 0$, a.a. $w(P_x)$, that is, $P_x(\zeta(w) = \infty) = 1$. Consequently for every $f \in C_b^2(\boldsymbol{R}^d)$ and $x \in \boldsymbol{R}^d$, $f(w(t)) - f(w(0)) - \int_0^t \frac{1}{2}(\Delta f)(w(s))ds$ is a P_x-martingale and we can apply the same proof as that of Theorem II-6.1 to show that P_x is the Wiener measure starting at x. Thus the A-diffusion is just d-dimensional Brownian motion.

Example 5.2. Let S' and $\mathscr{D}(A)$ be as in Example 5.1. Define A on $\mathscr{D}(A)$ by

$$Af(x) = \begin{cases} \frac{1}{2}\Delta f(x) + \sum_{i=1}^{d} c_i \frac{\partial f}{\partial x^i}(x), & x \in \boldsymbol{R}^d, \\ 0, & x = \Delta, \end{cases} \tag{5.9}$$

where $c = (c_i) \in \boldsymbol{R}^d$ is a constant. Then the operator A generates a unique diffusion $\{P_x\}$; P_x is the probability law of the process $X(t) = x + B(t) + ct$ where $B(t)$ is a d-dimensional Brownian motion with $B(0) = 0$. This diffusion is called *the d-dimensional Brownian motion with drift c.*

This can be proved in the same way as in Example 5.1.

Example 5.3. Let $S' = \boldsymbol{R}^d \cup \{\Delta\}$ be as in the previous examples. Let $\mathscr{D}(A) = \{f \in C(S'); f|_{\boldsymbol{R}^d} \in C_b^2(\boldsymbol{R}^d)$ and $f(\Delta) = 0\}$ and define A on $\mathscr{D}(A)$ by

$$Af(x) = \begin{cases} \frac{1}{2}(\Delta f)(x) - cf(x), & x \in \boldsymbol{R}^d, \\ 0, & x = \Delta, \end{cases} \tag{5.10}$$

where $c > 0$ is a constant. Then, the operator A generates a unique diffusion $\{P_x\}$ on $\boldsymbol{R}^d$; P_x is the probability law of the process X_x defined as follows. Let $(B(t))$ $(B(0) = 0)$ be a d-dimensional Brownian motion and e be an independent exponentially distributed random variable with mean $1/c$. We define

$$X_x(t) = \begin{cases} x + B(t), & \text{if} \quad t < e, \\ \Delta, & \text{if} \quad t \geq e. \end{cases}$$

This diffusion is called *the d-dimensional Brownian motion with the random absorption rate c.* To prove this assertion, we first note that the system $\{P_x\}$ clearly satisfies the conditions (i) and (ii) of Definition 5.3. Secondly the function $f_\xi(x)$, $\xi \in \boldsymbol{R}^d$, defined by

$$f_\xi(x) = \begin{cases} e^{i\langle \xi, x\rangle}, & x \in \boldsymbol{R}^d, \\ 0, & x = \Delta, \end{cases}$$

is in $\mathscr{D}(A)$, and hence if $\{P'_x\}$ satisfies (i) and (ii) of Definition 5.3, then

$$f_\xi(w(t)) - f_\xi(w(0)) - \int_0^t (Af_\xi)(w(s))ds$$

is a P'_x-martingale. Thus, if $x \in \boldsymbol{R}^d$,

$$E'_x[e^{i\langle \xi, w(t)\rangle}: \zeta > t] = e^{i\langle \xi, x\rangle} - \left(\frac{|\xi|^2}{2} + c\right)\int_0^t E'_x[e^{i\langle \xi, w(s)\rangle}: \zeta > s]ds^*$$

and so $E'_x[f_\xi(w(t))] = E'_x[e^{i\langle \xi, w(t)\rangle}: \zeta > t] = e^{i\langle \xi, x\rangle - (|\xi|^2/2 + c)t}$. Since $\{f_\xi;\ \xi \in \boldsymbol{R}^d\}$ is total, we conclude that $\{P'_x\}$ coincides with $\{P_x\}$ by the corollary of Theorem 5.1.

Example 5.4. Let $S = \boldsymbol{R}^d_+ := \{x = (x^1, x^2, \ldots, x^d) \in \boldsymbol{R}^d;\ x^d \geq 0\}$, $S' = \boldsymbol{R}^d_+ \cup \{\Delta\}$ where Δ is attached to $\boldsymbol{R}^d_+$ as an isolated point. Let $\mathscr{D}(A) = \{f \in \boldsymbol{C}(S'); f|_{\boldsymbol{R}^d_+} \in \boldsymbol{C}^2_b(\boldsymbol{R}^d_+)$ and $\left.\frac{\partial f}{\partial x^d}\right|_{\partial S} = 0\}$ where $\partial S = \{x \in \boldsymbol{R}^d_+; x^d = 0\}$ and define A on $\mathscr{D}(A)$ by

$$(5.11) \qquad Af(x) = \begin{cases} \frac{1}{2}\Delta f(x), & x \in \boldsymbol{R}^d_+, \\ 0, & x = \Delta. \end{cases}$$

Then the operator A generates a unique diffusion $\{P_x\}$ on $\boldsymbol{R}^d_+$; P_x is the probability law of the process $X_x(t)$ defined as follows. Let $B(t) = (B^1(t), B^2(t), \ldots, B^d(t))$ be a d-dimensional Brownian motion with $B(0) = 0$ and

$$X_x(t) = (x^1 + B^1(t), x^2 + B^2(t), \ldots, x^{d-1} + B^{d-1}(t), |x^d + B^d(t)|).$$

This diffusion is called the ***reflecting barrier Brownian motion on*** $\boldsymbol{R}^d_+$. To prove this, we first note that by Chapter III, Section 4.2 $X^k_x(t) = x^k + \tilde{B}^k(t) + \delta_{kd}\phi(t)$ for $k = 1, 2, \ldots, d$ where $\tilde{B}(t)$ is a d-dimensional

* E'_x stands for expectation with respect to P'_x.

Wiener martingale* and $\phi(t)$ is the local time of $X_x^d(t)$ at 0: $\phi(t) = \lim_{\varepsilon \downarrow 0} \frac{1}{2\varepsilon} \int_0^t I_{(0,\varepsilon)}(X_x^d(s))ds$. Hence, by Itô's formula,

$$\begin{aligned}
f(X_x(t)) - f(x) - \int_0^t \frac{1}{2}\Delta f(X_x(s))ds \\
&= \sum_{k=1}^{d} \int_0^t \frac{\partial f}{\partial x^k}(X_x(s))d\tilde{B}^k(s) + \int_0^t \frac{\partial f}{\partial x^d}(X_x(s))I_{\partial S}(X_x(s))d\phi(s) \\
&= \sum_{k=1}^{d} \int_0^t \frac{\partial f}{\partial x^k}(X_x(s))d\tilde{B}^k(s) \qquad \text{if } f \in \mathscr{D}(A).
\end{aligned}$$

Thus $\{P_x\}$ satisfies the conditions (i) and (ii) of Definition 5.3. To prove the uniqueness of any such system $\{P_x'\}$ we set, for $\xi = (\xi^1, \xi^2, \ldots, \xi^d) \in \boldsymbol{R}^d$,

$$f_\xi(x) = \begin{cases} \prod_{k=1}^{d-1} e^{i\xi^k x^k} \cos \xi^d x^d, & x \in \boldsymbol{R}_+^d, \\ 0, & x = \Delta. \end{cases}$$

Then $f_\xi \in \mathscr{D}(A)$ and hence

$$f_\xi(w(t)) - f_\xi(x) - \int_0^t (Af_\xi)(w(s))ds$$

is a P_x'-martingale. Noting that $Af_\xi = -\frac{|\xi|^2}{2} f_\xi$, we have

$$E_x'[f_\xi(w(t))] = f_\xi(x) - \frac{|\xi|^2}{2} \int_0^t E_x'[f_\xi(w(s))]ds.$$

This equation implies that $E_x'[f_\xi(w(t))] = \exp\left(-\frac{|\xi|^2}{2}t\right) f_\xi(x)$, and so since $\{f_\xi, \xi \in \boldsymbol{R}^d\}$ is total, we can apply the corollary of Theorem 5.1 to conclude that $P_x' = P_x$.

Example 5.5 ([73]). For simplicity, we consider the one-dimensional case only. Let $S=[0, \infty)$ and $S'=[0, \infty) \cup \{\Delta\}$, where Δ is attached to S as an isolated point. For a given parameter γ $(0 \leq \gamma \leq 1)$, let $\mathscr{D}(A) = \{f \in \boldsymbol{C}(S'); f|_{[0,\infty)} \in \boldsymbol{C}_b^2([0,\infty))$ and $(1-\gamma)f(0) = \gamma f'(0)\}$ and define A on $\mathscr{D}(A)$ by

* Indeed, $\tilde{B}^i(t) = B^i(t)$ for $i \neq d$, and $\tilde{B}^d(t) = \int_0^t \operatorname{sgn}(x^d + B^d(s))dB^d(s)$.

$$(5.12) \qquad Af(x) = \begin{cases} \dfrac{1}{2}\dfrac{d^2}{dx^2}f(x), & x \in [0,\infty), \\ 0, & x = \Delta. \end{cases}$$

The operator A generates a unique diffusion $\{P_x\}$ on $[0, \infty)$ which is called the *elastic barrier Brownian motion with parameter* γ. P_x is the probability law of the following process $X_x(t)$. Let $\xi_x(t)$ be the reflecting Brownian motion starting at x and $\phi(t)$ be its local time at 0: $\phi(t) = \lim_{\varepsilon \downarrow 0} \frac{1}{2\varepsilon} \int_0^t I_{[0,\varepsilon)}$ $(\xi_x(s))ds$. Let e be a random variable which is exponentially distributed with mean $\gamma/(1-\gamma)$ and is independent of ξ_x. Set

$$X_x(t) = \begin{cases} \xi_x(t), & \text{if } \ t < \zeta := \inf\{t;\, \phi(t) > e\}, \\ \Delta, & \text{if } \ t \geq \zeta. \end{cases}$$

The proof follows by a similar argument as in Examples 5.3 and 5.4: we take for $\xi \in \boldsymbol{R}$

$$f_\xi(x) = \begin{cases} \gamma\xi \cos \xi x + (1-\gamma) \sin \xi x, & x \in [0, \infty), \\ 0, & x = \Delta. \end{cases}$$

Example 5.6. Let S be a bounded smooth domain in $\boldsymbol{R}^d$ and $S' = S \cup \{\Delta\}$, where Δ is attached to S as a *point at infinity*.*[1] Let $\mathscr{D}(A) = C_0^2(S)$ $(:= \{f;$ twice continuously differentiable in S and tends to 0 at $\Delta\})$ and define A on $\mathscr{D}(A)$ by

$$(5.13) \qquad Af(x) = \begin{cases} \dfrac{1}{2}\Delta f(x), & x \in S, \\ 0, & x = \Delta. \end{cases}$$

Then the operator A generates a unique diffusion $\{P_x\}$ on S' which is called the *absorbing barrier Brownian motion on S* or the *minimal Brownian motion on S*. P_x is the probability law of a Brownian motion starting at $x \in S$ stopped at the first instant when it hits the boundary of S (which is identified with Δ). Again it is clear that $\{P_x\}$ satisfies the conditions (i) and (ii) of Definition 5.3. Suppose $\{P'_x\}$ also satisfies these conditions. Let $G_\alpha(x,y)$, $\alpha > 0$, be the Green function of S for the operator $L_\alpha = \alpha - \Delta/2$ with Dirichlet boundary conditions: i.e., if $f \in C_K^\infty(S)$ *[2] then $G_\alpha f(x) = \int_S G_\alpha(x, y) f(y) dy$ is the unique solution of

*[1] i.e., S' is the one point compactification of S.
*[2] $C_K^\infty(S)$ is the set of all C^∞-functions f with the support $S(f) \subset S$.

$$\begin{cases} \alpha u - \frac{1}{2}\Delta u = f, \\ \quad u|_{\partial S} = 0. \end{cases}$$

Let $f \in C_K^\infty(S)$ and set $u = G_\alpha f$. Then $u \in \mathscr{D}(A)$ and $Au = \alpha u - f$. Hence

$$u(w(t)) - u(w(0)) - \frac{1}{2}\int_0^t \Delta u(w(s))ds$$

is a P_x'-martingale and, therefore for every $t \geq 0$,

$$E_x'[u(w(t))] - u(x) = \alpha \int_0^t E_x'[u(w(s))]ds - \int_0^t E_x'[f(w(s))]ds.$$

Consequently

$$\begin{aligned} &\int_0^\infty e^{-\alpha t} E_x'[u(w(t))]dt - \frac{u(x)}{\alpha} \\ &\quad = -\int_0^\infty [\int_0^t E_x'[u(w(s))]ds] d(e^{-\alpha t}) \\ &\qquad + \frac{1}{\alpha}\int_0^\infty [\int_0^t E_x'[f(w(s))]ds] d(e^{-\alpha t}) \\ &\quad = \int_0^\infty e^{-\alpha t} E_x'[u(w(t))]dt - \frac{1}{\alpha}\int_0^\infty e^{-\alpha t} E_x'[f(w(t))]dt \end{aligned}$$

and hence

$$u(x) = \int_0^\infty e^{-\alpha s} E_x'[f(w(s))]ds = G_\alpha f(x).$$

Now apply the corollary of Theorem 5.1.

6. Diffusion processes generated by differential operators and stochastic differential equations

Suppose that we are given a second order differential operator A on $\boldsymbol{R}^d$:

$$(6.1) \qquad Af(x) = \frac{1}{2}\sum_{i,j=1}^d a^{ij}(x) \frac{\partial^2 f}{\partial x^i \partial x^j}(x) + \sum_{i=1}^d b^i(x) \frac{\partial f}{\partial x^i}(x),$$

where $a^{ij}(x)$ and $b^i(x)$ are real continuous functions on $\boldsymbol{R}^d$ and $(a^{ij}(x))$ is symmetric and non-negative definite; i.e., $a^{ij}(x) = a^{ji}(x)$ and $\sum_{i,j=1}^d a^{ij}(x)$

$\xi^i\xi^j \geq 0$ for all $\xi = (\xi^i) \in \boldsymbol{R}^d$ and all $x \in \boldsymbol{R}^d$. We take, as the domain of the definition of A, the space $\boldsymbol{C}_K^2(\boldsymbol{R}^d)$ consisting of all twice continuously differentiable functions having compact support. The notion of the diffusion process generated by the operator A (A-diffusion) was defined in the previous section. To be precise, we formulate it again as follows. Let $\hat{\boldsymbol{R}}^d = \boldsymbol{R}^d \cup \{\Delta\}$ be the one-point compactification of $\boldsymbol{R}^d$. Every function f on $\boldsymbol{R}^d$ is regarded as a function on $\hat{\boldsymbol{R}}^d$ by extending it as $f(\Delta) = 0$. Let $\hat{\boldsymbol{W}}^d$ be defined as in Section 2 and $e(w)$ be defined by (2.12).

Definition 6.1. By a *diffusion measure* generated by the operator A (or simply an A-diffusion), we mean a system $\{P_x, x \in \boldsymbol{R}^d\}$ * of probabilities on $(\hat{\boldsymbol{W}}^d, \mathscr{B}(\hat{\boldsymbol{W}}^d))$ which is strongly Markovian and satisfies

(i)

$$P_x\{w;\ w(0) = x\} = 1 \qquad \text{for every } x \in \boldsymbol{R}^d,$$

(ii)

$$f(w(t)) - f(w(0)) - \int_0^t (Af)(w(s))ds$$

is a $(P_x, \mathscr{B}_t(\hat{\boldsymbol{W}}^d))$-martingale for every $f \in \boldsymbol{C}_K^2(\boldsymbol{R}^d)$ and every $x \in \boldsymbol{R}^d$.

Remark 6.1. By Theorem 5.1, we know that any system $\{P_x, x \in \boldsymbol{R}^d\}$ of probabilities on $(\hat{\boldsymbol{W}}^d, \mathscr{B}(\hat{\boldsymbol{W}}^d))$ satisfying (i) and (ii) and that $x \longmapsto P_x$ is universally measurable is strongly Markovian and hence is an A-diffusion if it satisfies further the following uniqueness condition:

(iii) if $\{P'_x\}$ is another system of probabilities on $(\hat{\boldsymbol{W}}^d, \mathscr{B}(\hat{\boldsymbol{W}}^d))$ satisfying the above conditions (i) and (ii), then $P'_x = P_x$ for all x.

Also, by the corollary of Theorem 5.1, (iii) may be replaced by the following weaker condition:

(iii)′ if $\{P'_x\}$ is another system of probabilities on $(\hat{\boldsymbol{W}}^d, \mathscr{B}(\hat{\boldsymbol{W}}^d))$ satisfying the above conditions (i) and (ii), then

$$\int_{\hat{\boldsymbol{W}}^d} f(w(t))P_x(dw) = \int_{\hat{\boldsymbol{W}}^d} f(w(t))P'_x(dw)$$

for every $t \geq 0$, $x \in \boldsymbol{R}^d$ and f in a total family of functions on $\boldsymbol{R}^d$.

Definition 6.2. A stochastic process $X = (X(t))$ on $\hat{\boldsymbol{R}}^d$ is called *a diffusion process generated by the operator A* or simply *A-diffusion process*, if almost all samples $[t \longmapsto X(t)]$ belong to $\hat{\boldsymbol{W}}^d$ and the probability law of X coincides with $P_\mu(\cdot) = \int_{\boldsymbol{R}^d} P_x(\cdot)\mu(dx)$ where $\{P_x\}$ is a diffusion measure

* To be precise, P_Δ should be included in the system but, since P_Δ is the trivial measure δ_{w_Δ} where $w_\Delta(t) \equiv \Delta$, we usually omit it.

generated by A and μ is the probability law of $X(0)$.

Our problem now is to consider the existence and uniqueness of A-diffusions.

Let $\sigma = (\sigma^i_k(x)) \in \boldsymbol{R}^d \otimes \boldsymbol{R}^r$ be such that

$$(6.2) \qquad x \longmapsto \sigma(x) \ \textit{is continuous and}\ a^{ij}(x) = \sum_{k=1}^{r} \sigma^i_k(x)\sigma^j_k(x) \quad \text{for } i\ j = 1, 2, \dots, d.$$

Clearly such a σ exists for some r. We choose one such σ and fix it. We now consider the following stochastic differential equation

$$(6.3) \qquad dX^i(t) = \sum_{k=1}^{r} \sigma^i_k(X(t))dB^k(t) + b^i(X(t))dt, \qquad i = 1, 2, \dots, d.$$

According to Theorem 2.3, we know that for every $x \in \boldsymbol{R}^d$ there exists a solution $X(t)$ of (6.3) such that $X(0) = x$. By Itô's formula (Theorem II-5.1),

$$f(X(t)) - f(X(0)) = \sum_{k=1}^{r}\sum_{i=1}^{d}\int_0^t \frac{\partial f}{\partial x^i}(X(s))\sigma^i_k(X(s))dB^k(s) + \int_0^t (Af)(X(s))ds$$

for every $f \in C^2_K(\boldsymbol{R}^d)$. From this, it is clear that the law P_x on $\hat{\boldsymbol{W}}^d$ of the process X satisfies the conditions (i) and (ii) of Definition 6.1. We shall now show that the uniqueness of solutions for the stochastic differential equation (6.3) is equivalent to the uniqueness condition (iii) in Remark 6.1. Indeed, it is obvious that (iii) implies the uniqueness of solutions of (6.3). On the other hand, if $\{P_x\}$ is the system of probabilities on $\hat{\boldsymbol{W}}^d$ satisfying the conditions (i) and (ii) of Definition 6.1, then we can conclude by the same proof as in Section 2 that there exists some extension $(\Omega, \mathscr{F}, P)$ with a reference family $(\mathscr{F}_t)$ of the probability space $(\hat{\boldsymbol{W}}^d, \mathscr{B}(\hat{\boldsymbol{W}}^d), P_x)$ with the reference family $(\mathscr{B}_t(\hat{\boldsymbol{W}}^d))$,* and an $(\mathscr{F}_t)$-Brownian motion $B = ((B(t))$ such that setting $X(t) = w(t)$ and $e = e(w)$, we have for $t \in [0, e)$,

$$X^i(t) = x^i + \sum_{k=1}^{r}\int_0^t \sigma^i_k(X(s))dB^k(s) + \int_0^t b^i(X(s))ds, \ i = 1, 2, \dots, d.$$

This implies that $(X(t), B(t))$ is a solution of (6.3) such that $X(0) = x$. Since

---

* $\mathscr{B}_t(\hat{\boldsymbol{W}}^d)$ is defined in the same way as $\mathscr{B}_t(\boldsymbol{W}^d)$.

the probability law of the process $X(t)$ is clearly P_x, the uniqueness of solutions of (6.3) implies the uniqueness condition (iii). Thus we have the following result.

Theorem 6.1. Let the differential operator (6.1) be given as above and choose any $\sigma = (\sigma_k^i(x))$ such that (6.2) holds. Then the A-diffusion $\{P_x, x \in \boldsymbol{R}^d\}$ exists uniquely if and only if the uniqueness of solutions holds for the stochastic differential equation (6.3). In this case, P_x is the probability law on $(\hat{\boldsymbol{W}}^d, \mathscr{B}(\hat{\boldsymbol{W}}^d))$ of a solution $X = (X(t))$ of (6.3) such that $X(0) = x$.*1

Stroock-Varadhan's result (Theorem 3.3) implies that if $(a^{ij}(x))$ is bounded, continuous and uniformly positive definite and $(b^i(x))$ is bounded, then the A-diffusion exists uniquely; moreover, it is a conservative diffusion. In the general case when $(a^{ij}(x))$ may degenerate, we have by Theorem 3.1 that if we can choose the above $\sigma(x) = (\sigma_k^i(x))$ such that $\sigma(x)$ and $b(x) = (b^i(x))$ are locally Lipschitz continuous, then the A-diffusion exists uniquely. Furthermore, this A-diffusion is conservative (i.e., $P_x(e = \infty) = 1$, for every $x \in \boldsymbol{R}^d$) if $\sigma(x)$ and $b(x)$ satisfy the growth condition $\|\sigma(x)\| + \|b(x)\| \leq K(1+|x|)$ for some positive constant K. In the one-dimensional case, the Lipschitz condition on $\sigma(x)$ may be weakened as in Theorem 3.2. In particular, if we can choose $\sigma(x)$ such that it is locally Hölder continuous of exponent 1/2 and if $b(x)$ is locally Lipschitz continuous, then the A-diffusion exists uniquely. An important question now is to determine when we can choose a sufficiently smooth σ such that (6.2) holds for a given matrix a. For this we have the following result.

Proposition 6.2. (i) Let $\mathfrak{S}^r$ be the set of all $r \times r$ symmetric, non-negative definite matrices. If $a(x)\colon \boldsymbol{R}^d \longrightarrow \mathfrak{S}^r$ is in the class $\boldsymbol{C}_b^2(\boldsymbol{R}^d)$,*2 then the square root $\sigma(x)$ (i.e., $\sigma(x)\colon \boldsymbol{R}^d \longrightarrow \mathfrak{S}^r$ satisfying the property $\sigma(x)\sigma(x)^* = a(x)$) is uniformly Lipschitz continuous on $\boldsymbol{R}^d$.

(ii) If $a(x)\colon \boldsymbol{R}^d \longrightarrow \mathfrak{S}^r$ is twice continuously differentiable, then the square root $\sigma(x)$ is locally Lipschitz continuous.

Proof. Clearly it is only necessary to prove (i). We shall prove this in the case $d = 1$; the general case follows from the fact that a function is uniformly Lipschitz continuous on $\boldsymbol{R}^d$ if it is uniformly Lipschitz continuous in each variable $x^i \in \boldsymbol{R}^1$ for fixed $\hat{x} = (x^1, x^2, \ldots, x^{i-1}, x^{i+1}, \ldots, x^d)$ with the Lipschitz constant independent of $\hat{x}$. Fix $x_0 \in \boldsymbol{R}^1$. We

---

*1 The universal measurability of $x \longmapsto P_x$ is seen as in the proof of Theorem 1.1.

*2 We say that $a(x)$ is in $\boldsymbol{C}_b^2(\boldsymbol{R}^d)$ if every component of $a(x)$ is in $\boldsymbol{C}_b^2(\boldsymbol{R}^d)$.

can choose an orthogonal matrix P such that $Pa(x_0)P^*$ is a diagonal matrix. Set $\tilde{a}(x) = Pa(x)P^*$. For a positive constant $\varepsilon > 0$, let

$$a_\varepsilon(x) = a(x) + \varepsilon I \quad \text{and} \quad \tilde{a}_\varepsilon(x) = Pa_\varepsilon(x)P^* = \tilde{a}(x) + \varepsilon I.$$

The square roots of $a_\varepsilon(x) = (a_\varepsilon^{ij}(x))$ and $\tilde{a}_\varepsilon(x) = (\tilde{a}_\varepsilon^{ij}(x))$ are denoted by $\sigma_\varepsilon(x) = (\sigma_\varepsilon^{ij}(x))$ and $\tilde{\sigma}_\varepsilon(x) = (\tilde{\sigma}_\varepsilon^{ij}(x))$ respectively. Then $\sigma_\varepsilon(x)$ and $\tilde{\sigma}_\varepsilon(x) = P\sigma_\varepsilon(x)P^*$ are clearly in $C_b^2(R^1)$. By differentiating both sides of $\tilde{a}_\varepsilon^{ij}(x) = \sum_{k=1}^{r} \tilde{\sigma}_\varepsilon^{ik}(x)\tilde{\sigma}_\varepsilon^{jk}(x)$ at $x = x_0$, we have

$$\dot{\tilde{a}}_\varepsilon^{ij}(x_0) = (\tilde{\sigma}_\varepsilon^{ii}(x_0) + \tilde{\sigma}_\varepsilon^{jj}(x_0))\dot{\tilde{\sigma}}_\varepsilon^{ij}(x_0).^* \tag{6.4}$$

Let $K = \sup_{1 \le i, j \le r, x \in R^1} |\ddot{a}^{ij}(x)| = \sup_{1 \le i, j \le r, x \in R^1} |\ddot{\tilde{a}}_\varepsilon^{ij}(x)|$ and set $f(x) = \langle \lambda, \tilde{a}_\varepsilon(x)\lambda \rangle$ for $\lambda \in R$. Then since $f(x) \ge 0$ for all $x \in R^1$, we have

$$\begin{aligned} 0 \le f(x+h) &= f(x) + \dot{f}(x)h + \frac{1}{2}\ddot{f}(x+\theta h)h^2 \\ &\le f(x) + \dot{f}(x)h + \frac{1}{2}\Big(\sum_{i=1}^{r} |\lambda_i|\Big)^2 K h^2 \end{aligned}$$

and hence

$$\dot{f}(x)^2 \le 2f(x)\Big(\sum_{i=1}^{r} |\lambda_i|\Big)^2 K.$$

Letting λ be $\delta_i = (\delta_{ik})_{k=1}^{r}$, $\delta_j = (\delta_{jk})_{k=1}^{r}$ and $\delta_i + \delta_j$ respectively, we obtain that

$$\dot{\tilde{a}}_\varepsilon^{ii}(x)^2 \le 2K\tilde{a}_\varepsilon^{ii}(x), \quad \dot{\tilde{a}}_\varepsilon^{jj}(x)^2 \le 2K\tilde{a}_\varepsilon^{jj}(x)$$

and

$$(\dot{\tilde{a}}_\varepsilon^{ii}(x) + 2\dot{\tilde{a}}_\varepsilon^{ij}(x) + \dot{\tilde{a}}_\varepsilon^{jj}(x))^2 \le 8K(\tilde{a}_\varepsilon^{ii}(x) + 2\tilde{a}_\varepsilon^{ij}(x) + \tilde{a}_\varepsilon^{jj}(x)).$$

Consequently there exists a constant $c(K)$ independent of $\varepsilon > 0$ such that

$$|\dot{\tilde{a}}_\varepsilon^{ij}(x)| \le c(K)(\tilde{a}_\varepsilon^{ii}(x)^{1/2} + \tilde{a}_\varepsilon^{jj}(x)^{1/2})$$

for all $x \in R^1$. If we set $x = x_0$, then

---

* For every $f \in C_b^2(R^1)$, we set $\dot{f}(x) = \frac{d}{dx}f(x)$ and $\ddot{f}(x) = \frac{d^2}{dx^2}f(x)$.

$$|\dot{\tilde{a}}_\varepsilon^{ij}(x_0)| \leq c(K)(\tilde{\sigma}_\varepsilon^{ii}(x_0) + \tilde{\sigma}_\varepsilon^{jj}(x_0))$$

and combining this with (6.4), we have $|\dot{\tilde{\sigma}}_\varepsilon^{ij}(x_0)| \leq c(K)$. Since $\sigma_\varepsilon^{ij}(x_0) = (P^*\tilde{\sigma}_\varepsilon(x_0)P)_{ij}$, it is easy to see that $|\dot{\sigma}_\varepsilon^{ij}(x_0)| \leq rc(K)$. But x_0 is arbitrary and hence,

$$|\dot{\sigma}_\varepsilon^{ij}(x)| \leq rc(K) \qquad \text{for all} \quad x \in \boldsymbol{R}^1.$$

Thus $|\sigma_\varepsilon^{ij}(x) - \sigma_\varepsilon^{ij}(y)| \leq rc(K)|x-y|$. Letting $\varepsilon \downarrow 0$, we conclude that

$$|\sigma^{ij}(x) - \sigma^{ij}(y)| \leq rc(K)|x-y|.$$

Corollary. If the coefficients of the differential operator (6.1) satisfy (i) $a^{ij}(x)$ is twice continuously differentiable, $i, j = 1, 2, \ldots, d$, and (ii) $b^i(x)$ is continuously differentiable, $i = 1, 2, \ldots, d$, then the A-diffusion exists uniquely.

7. Stochastic differential equations with boundary conditions

In the previous section we discussed a class of diffusion processes described by second order differential operators. If we consider the case of a domain with boundary, a diffusion is usually described by a *second order differential operator plus a boundary condition.* A general class of boundary conditions was found by Wentzell [175]. Here we will discuss the construction of such diffusions by means of stochastic differential equations.*1 For simplicity we only consider diffusion processes on the upper half space $\boldsymbol{R}_+^d$, $d \geq 2$.

So let $D = \boldsymbol{R}_+^d = \{x = (x^1, x^2, \ldots, x^d);\ x^d \geq 0\}$, $\partial D = \{x \in D;\ x^d = 0\}$ be the boundary of D and $\mathring{D} = \{x \in D;\ x^d > 0\}$ be the interior of D. Suppose we are given a second order differential operator on D acting on $\boldsymbol{C}_K^2(D)$:*2

$$(7.1) \qquad Af(x) = \frac{1}{2}\sum_{i,j=1}^d a^{ij}(x)\frac{\partial^2 f}{\partial x^i \partial x^j}(x) + \sum_{i=1}^d b^i(x)\frac{\partial f}{\partial x^i}(x)$$

where $a^{ij}(x)$ and $b^i(x)$ are bounded continuous functions on D and $(a^{ij}(x))$ is symmetric and non-negative definite. Suppose also we are given a *boundary operator of the Wentzell type*; i.e., a mapping from $\boldsymbol{C}_K^2(D)$ into the space of continuous functions on ∂D given as follows:

---

*1 [49], [166] and [169].

*2 $\boldsymbol{C}_K^2(D)$ = the set of all twice continuously differentiable functions with compact support.

(7.2)
$$Lf(x) = \frac{1}{2}\sum_{i,j=1}^{d-1} \alpha^{ij}(x)\frac{\partial^2 f}{\partial x^i \partial x^j}(x) + \sum_{i=1}^{d-1}\beta^i(x)\frac{\partial f}{\partial x^i}(x)$$
$$+ \delta(x)\frac{\partial f}{\partial x^d}(x) - \rho(x)Af(x), \qquad x \in \partial D,$$

where $\alpha^{ij}(x)$, $\beta^i(x)$, $\delta(x)$ and $\rho(x)$ are bounded continuous functions on ∂D such that $(\alpha^{ij}(x))_{i,j=1}^{d-1}$ is symmetric and non-negative definite, $\delta(x) \geq 0$ and $\rho(x) \geq 0$.

Definition 7.1. By a *diffusion measure generated by the pair* (A,L) of operators given above, or simply (A,L)*-diffusion*, we mean a system $\{P_x, x \in D\}$ of probabilities on $(W(D), \mathscr{B}(W(D)))$ *1 which is strongly Markovian (cf. Section 5) and satisfies the following two conditions:
(i) $P_x\{w; w(0) = x\} = 1$ for every $x \in D$;
(ii) there exists a function $\phi(t,w)$ defined on $[0, \infty) \times W(D)$ such that

a) for a.a. $w(P_x)$, $\phi(0,w) = 0$, $t \longmapsto \phi(t,w)$ is continuous and non-decreasing, and

$$\int_0^t I_{\partial D}(w(s))d\phi(s,w) = \phi(t,w), \quad \text{for all} \quad t \geq 0,$$

b) for each $t \geq 0$, $w \longmapsto \phi(t,w)$ is $\mathscr{B}_t(W(D))$-measurable,*2

(7.3)
$$f(w(t)) - f(w(0)) - \int_0^t (Af)(w(s))ds - \int_0^t (Lf)(w(s))d\phi(s,w)$$

is a $(P_x, \mathscr{B}_t(W(D)))$*-martingale for every* $f \in C_K^2(D)$ and

(7.4)
$$\int_0^t I_{\partial D}(w(s))ds = \int_0^t \rho(w(s))d\phi(s,w) \quad \text{a.s. } (P_x).$$

Remark 7.1. Suppose (A,L) and (A',L') are two pairs of operators as above such that $Af(x) = A'f(x)$ and $Lf(x) = c(x)L'f(x)$ for all $f \in C_K^2(D)$, where $c(x)$ is a positive continuous function on ∂D. Then an (A',L')-diffusion is also an (A,L)-diffusion since

$$\int_0^t (L'f)(w(s))d\phi'(s,w) = \int_0^t (Lf)(w(s))d\tilde{\phi}(s,w)$$

*1 $W(D) = C([0, \infty) \to D)$ = the space of all continuous functions w: $[0, \infty) \ni t \longmapsto w(t) \in D$ with the topology of uniform convergence on bounded intervals.
*2 $\mathscr{B}_t(W(D))$ is the σ-field on $W(D)$ generated by cylinder sets up to time t.

where $\tilde{\phi}(t, w) = \int_0^t c(w(s))d\phi'(s, w)$ and $\tilde{\phi}$ satisfies the conditions a) and b) of Definition 7.1. Consequently, there is one degree of freedom in defining the operator L.

Remark 7.2. If we define another boundary operator L' by

$$
\begin{aligned}
L'f(x) &= Lf(x) + \rho(x)Af(x) \\
&= \frac{1}{2}\sum_{i,j=1}^{d-1} \alpha^{ij}(x)\frac{\partial^2 f}{\partial x^i \partial x^j}(x) + \sum_{i=1}^{d-1} \beta^i(x)\frac{\partial f}{\partial x^i}(x) + \delta(x)\frac{\partial f}{\partial x^d}(x)
\end{aligned}
\tag{7.5}
$$

then the expression (7.3) is given as follows:

$$
\begin{aligned}
& f(w(t)) - f(w(0)) - \int_0^t (Af)(w(s))ds - \int_0^t (Lf)(w(s))d\phi(s,w) \\
&= f(w(t)) - f(w(0)) - \int_0^t I_{\mathring{D}}(w(s))(Af)(w(s))ds \\
&\qquad - \int_0^t I_{\partial D}(w(s))(Af)(w(s))ds - \int_0^t (Lf)(w(s))d\phi(s,w) \\
&= f(w(t)) - f(w(0)) - \int_0^t I_{\mathring{D}}(w(s))(Af)(w(s))ds \\
&\qquad - \int_0^t \rho(w(s))(Af)(w(s))d\phi(s,w) - \int_0^t (Lf)(w(s))d\phi(s,w) \\
&\hspace{20em} \text{(by (7.4))}, \\
&= f(w(t)) - f(w(0)) - \int_0^t I_{\mathring{D}}(w(s))(Af)(w(s))ds \\
&\hspace{10em} - \int_0^t (L'f)(w(s))d\phi(s,w).
\end{aligned}
$$

Thus (7.3) is equivalent (under (7.4)) to the statement that

$$
\begin{aligned}
& f(w(t)) - f(w(0)) - \int_0^t I_{\mathring{D}}(w(s))(Af)(w(s))ds \\
& \hspace{8em} - \int_0^t (L'f)(w(s))d\phi(s,w)
\end{aligned}
\tag{7.3$'$}
$$

is a $(P_x, \mathscr{B}_t(\boldsymbol{W}(D)))$*-martingale for every* $f \in \boldsymbol{C}_K^2(D)$.

Definition 7.2. A continuous stochastic process $X=(X(t))$ on D is called a *diffusion process generated by the pair of operators* (A,L), or *simply* (A,L)*-diffusion process*, if the probability law of X on $(\boldsymbol{W}(\boldsymbol{D}), \mathscr{B}(\boldsymbol{W}(\boldsymbol{D})))$

coincides with $P_\mu(\cdot) = \int_D P_x(\cdot)\mu(dx)$, where $\{P_x\}$ is a diffusion measure generated by (A,L) and μ is the probability law of $X(0)$.

Remark 7.3. By Theorem 5.1, we know that any system $\{P_x, x \in D\}$ of probabilities on $(\boldsymbol{W}(D), \mathscr{B}(\boldsymbol{W}(D)))$ satisfying (i) and (ii) of Definition 7.1 and that $x \longmapsto P_x$ is universally measurable is strongly Markovian and hence is an (A,L)-diffusion if it satisfies further the following uniqueness condition:

(iii) if $\{P'_x\}$ is another system of probabilities on $(\boldsymbol{W}(D), \mathscr{B}(\boldsymbol{W}(D)))$ satisfying the above conditions (i) and (ii), then $P'_x = P_x$ for every $x \in D$.

Now we will discuss the existence and uniqueness of (A, L)-diffusions by the method of stochastic differential equations. First we will formulate a stochastic differential equation which describes an (A, L)-diffusion process. For this, we choose $\sigma(x) = (\sigma^i_k(x)): D \longrightarrow \boldsymbol{R}^d \otimes \boldsymbol{R}^r$ and $\tau(x) = (\tau^i_l(x)): \partial D \longrightarrow \boldsymbol{R}^{d-1} \otimes \boldsymbol{R}^s$ which are continuous and

$$a^{ij}(x) = \sum_{k=1}^{r} \sigma^i_k(x)\sigma^j_k(x), \quad i,j = 1, 2, \cdots, d, \tag{7.6}$$

and

$$\alpha^{ij}(x) = \sum_{l=1}^{s} \tau^i_l(x)\tau^j_l(x), \quad i,j = 1, 2, \cdots, d-1. \tag{7.7}$$

Consider the following stochastic differential equation

$$\begin{cases} dX^i(t) = \sum_{k=1}^{r} \sigma^i_k(X(t))I_{\mathring{D}}(X(t))dB^k(t) + b^i(X(t))I_{\mathring{D}}(X(t))dt \\ \qquad + \sum_{l=1}^{s} \tau^i_l(X(t))I_{\partial D}(X(t))dM^l(t) + \beta^i(X(t))I_{\partial D}(X(t))d\phi(t), \\ \qquad\qquad i = 1, 2, \cdots, d-1, \\ dX^d(t) = \sum_{k=1}^{r} \sigma^d_k(X(t))I_{\mathring{D}}(X(t))dB^k(t) + b^d(X(t))I_{\mathring{D}}(X(t)dt \\ \qquad + \delta(X(t))d\phi(t), \\ I_{\partial D}(X(t))dt = \rho(X(t))d\phi(t). \end{cases} \tag{7.8}$$

An intuitive meaning of this equation is as follows. $\phi(t)$ is an increasing process which increases only when $X(t)$ is on the boundary ∂D and is called the *local time of* $X(t)$ *on* ∂D. $d\phi(t)$ acts only when $X(t) \in \partial D$ and causes the reflection at ∂D. $\{B^k(t), M^l(t)\}$ is a mutually orthogonal*[1] system of martingales such that $d\langle B^k\rangle(t) = dt$, $k = 1, 2, \ldots, r$, and $d\langle M^l\rangle(t) = d\phi(t)$, $l = 1, 2, \ldots, s$, i.e., B is an r-dimensional Brownian motion in the ordinary time, and M is an s-dimensional Brownian motion if the time is measured by the local time $\phi(t)$. The function $\rho(x)$ represents the rate of sojourn of $X(t)$ on the boundary: note that $\rho(x) \equiv 0$ if and only if $\int_0^t I_{\partial D}(X(s))ds = 0$ for every $t \geq 0$ a.s. In this case, we say that the boundary ∂D is *non-sticky*; otherwise it is called *sticky*.

A precise formulation of (7.8) is as follows.

Definition 7.3. By a solution of equation (7.8)*[2] we mean a system of stochastic processes $\mathfrak{X} = [X(t) = (X^1(t), X^2(x), \ldots, X^d(t)), B(t) = (B^1(t), B^2(t), \ldots, B^r(t)), M(t) = (M^1(t), M^2(t), \ldots, M^s(t)), \phi(t)]$ defined on a probability space $(\Omega, \mathscr{F}, P)$ with a reference family $(\mathscr{F}_t)_{t\geq 0}$ such that

(i) $X(t)$ is a D-valued continuous $(\mathscr{F}_t)$-adapted process,

(ii) $\phi(t)$ is a continuous $(\mathscr{F}_t)$-adapted increasing process such that $\phi(0) = 0$ and

$$\int_0^t I_{\partial D}(X(s))d\phi(s) = \phi(t), \quad t \geq 0, \quad \text{a.s.},$$

(iii) $\{B(t), M(t)\}$ is a system of elements in $\mathscr{M}_2^{c,loc}$ such that $\langle B^k, B^j\rangle(t) = \delta_{kj}t$, $\langle B^k, M^l\rangle = 0$ and $\langle M^l, M^m\rangle(t) = \delta_{lm}\phi(t)$, and

(iv) with probability one,

$$\begin{aligned} X^i(t) = X^i(0) &+ \sum_{k=1}^r \int_0^t \sigma_k^i(X(s))I_{\mathring{D}}(X(s))dB^k(s) + \int_0^t b^i(X(s))I_{\mathring{D}}(X(s))ds \\ &+ \sum_{l=1}^s \int_0^t \tau_l^i(X(s))I_{\partial D}(X(s))dM^l(s) + \int_0^t \beta^i(X(s))I_{\partial D}(X(s))d\phi(s) \end{aligned}$$

(7.8′) $$i = 1, 2, \ldots, d-1,$$

$$\begin{aligned} X^d(t) = X^d(0) &+ \sum_{k=1}^r \int_0^t \sigma_k^d(X(s))I_{\mathring{D}}(X(s))dB^k(s) \\ &+ \int_0^t b^d(X(s))I_{\mathring{D}}(X(s))ds + \int_0^t \delta(X(s))d\phi(s), \end{aligned}$$

$$\int_0^t I_{\partial D}(X(s))ds = \int_0^t \rho(X(s))d\phi(s).$$

Definition 7.4. We say that the *uniqueness of solutions* for (7.8) holds

*[1] With respect to the random inner product $\langle\ ,\ \rangle$ of Definition II-2.1.

*[2] Also we call it a solution corresponding to the coefficients $[\sigma, b, \tau, \beta, \delta, \rho]$.

if whenever $\mathfrak{X}$ and $\mathfrak{X}'$ are any two solutions of (7.8) whose initial laws coincide, then the probability laws of $X=(X(t))$ and $X'=(X'(t))$ on $(\boldsymbol{W}(D), \mathscr{B}(\boldsymbol{W}(D)))$ coincide.*

The following theorem can be proved in almost the same way as Theorem 6.1.

Theorem 7.1. Let the differential operator A and the boundary operator L be given as above and choose continuous σ and τ satisfying (7.6) and (7.7). Then the (A,L)-diffusion $\{P_x, x \in D\}$ exists uniquely if and only if, for every probability μ on $(D, \mathscr{B}(D))$ there exists a solution of (7.8) such that the probability law of $X(0)$ coincides with μ and the uniqueness of solutions holds for (7.8). P_x is the probability law on $(W(D), \mathscr{B}(W(D)))$ of a solution $X(t)$ of (7.8) such that $X(0) = x$.

Theorem 7.2. We assume for the stochastic differential equation (7.8) that σ, b, τ, β, δ, ρ satisfy the following: σ and b are bounded and Lipschitz continuous on D, τ, β and δ are bounded and Lipschitz continuous on ∂D and ρ is bounded and continuous on ∂D. Furthermore, we assume that σ satisfies

$$a^{dd}(x) = \sum_{k=1}^{r} \sigma_k^d(x)\sigma_k^d(x) \geq c, \quad x \in \partial D, \tag{7.9}$$

and

$$\delta(x) \geqq c, \qquad x \in \partial D \tag{7.10}$$

for some positive constant c. Then for any probability μ on $(D, \mathscr{B}(D))$ there exists a solution $X(t)$ such that the probability law of $X(0)$ coincides with μ. Furthermore, the uniqueness of solutions holds for the equation (7.8).

Corollary. For a given pair of operators (A,L) satisfying (7.9) and (7.10), suppose that we can choose σ and τ for some r and s such that (7.6) and (7.7) hold and $\sigma, b, \tau, \beta, \delta, \rho$ satisfy the assumption of Theorem 7.2. Then (A,L)-diffusion exists uniquely.

Proof of Theorem 7.2. Let $c(x)$ be a continuous function on ∂D such that $c_1 \leqq c(x) \leqq c_2$, $x \in \partial D$, for some positive constants c_1 and c_2. Then it is easy to see that $\mathfrak{X} = [X(t), B(t), M(t), \phi(t)]$ is a solution corresponding to $[\sigma, b, \tau, \beta, \delta, \rho]$ if and only if $\tilde{\mathfrak{X}} = [\tilde{X}(t), \tilde{B}(t), \tilde{M}(t), \tilde{\phi}(t)]$ with $\tilde{X}(t) \equiv X(t)$, $\tilde{B}(t) \equiv B(t)$,

$$\tilde{M}(t) = \int_0^t \{\sqrt{c(X(s))}\}^{-1} d\, M(s), \qquad \tilde{\phi}(t) = \int_0^t \{c(X(s))\}^{-1} ds$$

* We sometimes call the process $X = (X(t))$ itself a solution of (7.8).

is a solution corresponding to $[\sigma, b, \sqrt{c\tau}, c\beta, c\delta, c\rho]$. Under the assumption (7.10), therefore, we may always assume that $\delta(x)$ is normalized to be $\delta(x) \equiv 1$.

We will prove the theorem in the following three steps.

(1°) The case of non-sticky boundary; i.e., $\rho(x) \equiv 0$ and $\sigma_1^d(x) \equiv 1$, $\sigma_k^d(x) \equiv 0$, $k = 2, 3, \cdots, r$, and $b^d(x) \equiv 0$.

(2°) The case of non-sticky boundary i.e.; $\rho(x) \equiv 0$.

(3°) The general case.

(1°) *The case of* $\rho(x) \equiv 0$, $\sigma_1^d(x) \equiv 1$, $\sigma_k^d(x) \equiv 0$, $k = 2, 3, \ldots, r$, *and* $b^d(x) \equiv 0$.

First we show the existence of solutions. Let μ be a given Borel probability on D. On a probability space we construct the following three objects such that they are mutually independent.

(i) $x(0) = (x^1(0), x^2(0), \ldots, x^d(0))$, a D-valued random variable with the distribution μ,

(ii) $B(t) = (B^1(t), B^2(t), \ldots, B^r(t))$, an r-dimensional Brownian motion with $B(0) = 0$ and

(iii) $\hat{B}(t) = (\hat{B}^1(t), \hat{B}^2(t), \ldots, \hat{B}^s(t))$, an s-dimensional Brownian motion with $\hat{B}(0) = 0$.

Define $\phi(t)$ and $X^d(t)$ by

$$\phi(t) = \begin{cases} 0, & t \leq \sigma_0 := \min\{t;\, B^1(t) + x^d(0) = 0\}, \\ -\min\limits_{\sigma_0 \leq s \leq t}(B^1(s) + x^d(0)), & t > \sigma_0 \end{cases} \tag{7.11}$$

and

$$X^d(t) = x^d(0) + B^1(t) + \phi(t). \tag{7.12}$$

As we saw in Chapter III, Section 4.2, $X^d(t)$ is a reflecting Brownian motion on $[0,\infty)$ and $\phi(t)$ is the local time of $X^d(t)$ at 0: $\phi(t) = \lim\limits_{\varepsilon \downarrow 0} \frac{1}{2\varepsilon} \int_0^t I_{[0,\varepsilon)}(X^d(s))ds$. Next define $M(t) = (M^1(t), M^2(t), \ldots, M^s(t))$ by $M(t) = \hat{B}(\phi(t))$. Set $\mathscr{F}_t = \bigcap\limits_n \mathscr{F}'_{t+1/n}$, where $\mathscr{F}_t'$ is the σ-field generated by $x(0)$ and $\{B(u), M(u)\}_{u \leq t}$. It is then clear that $\{B(t), M(t)\}$ is a system of elements in $\mathscr{M}_2^{c,loc}$ satisfying the conditions in (iii) of Definition 7.3. Consider the following stochastic differential equation for $\tilde{X}(t) = (X^1(t), X^2(t), \ldots, X^{d-1}(t))$:

$$dX^i(t) = \sum_{k=1}^r \sigma_k^i(\tilde{X}(t), X^d(t))dB^k(t) + b^i(\tilde{X}(t), X^d(t))dt$$

$$(7.13) \qquad + \sum_{l=1}^{s} \tau_l^i(\tilde{X}(t), 0)dM^l(t) + \beta^i(\tilde{X}(t),0)d\phi(t),$$

$$X^i(0) = x^i(0), \qquad i = 1, 2, \ldots, d-1.$$

By Theorem III-2.1, the solution $\tilde{X}(t)$ exists uniquely. $X(t) = (\tilde{X}(t), X^d(t))$ is a continuous D-valued process satisfying $\int_0^t I_{\partial D}(X(s))ds = \int_0^t I_{\{0\}}(X^d(s))ds = 0$ for every $t \geq 0$ a.s. and $\int_0^t I_{\partial D}(X(s))d\phi(s) = \int_0^t I_{\{0\}}(X^d(s))d\phi(s) = \phi(t)$ for every $t \geq 0$ a.s. In particular,

$$I_{\mathring{D}}(X(t))dB^k(t) = dB^k(t), \qquad k = 1, 2, \ldots, r,$$

and

$$I_{\mathring{D}}(X(t))dt = dt.$$

Consequently $\mathfrak{X} = [X(t), B(t), M(t), \phi(t)]$ is a solution of (7.8).

Next we show the uniqueness of solutions. The equation (7.8) implies that

$$dX^d(t) = dB^1(t) + d\phi(t)$$

and by Theorem III-4.2, $X^d(t)$ and $\phi(t)$ are uniquely determined from $X^d(0)$ and $B^1(t)$ as (7.12) and (7.11). By Theorem II-7.3, $\{B(t), \hat{B}(t) = M(\phi^{-1}(t))\}$ is an $(r+s)$-dimensional Brownian motion which is independent of $X(0)$. Therefore, the probability law of $[X(0), (B(t)), (M(t))]$ is uniquely determined from the law μ of $X(0)$. Since the solution $\tilde{X}(t)$ of (7.13) is unique and is constructed as in Theorem III-2.1, it is clear that the law of $\mathfrak{X} = [X(t) = (\tilde{X}(t), X^d(t)), B(t), M(t), \phi(t)]$ is uniquely determined from the law μ.

(2°) *The general non-sticky case*: $\rho(x) \equiv 0$.

First we discuss some transformations of solutions.

(a) *Transformation of Brownian motion.*

Let $\mathfrak{X} = [X(t), B(t), M(t), \phi(t)]$ be a solution on a space $(\Omega, \mathscr{F}, P)$ with $(\mathscr{F}_t)$ corresponding to the coefficients $[\sigma, b, \tau, \beta, 0]$. Let $p(x) = (p_j^k(x))$: $D \longrightarrow O(r)$ be a continuous function defined on D with values in the r-dimensional orthogonal group $O(r)$. Set

$$\tilde{B}^k(t) = \sum_{j=1}^{r} \int_0^t p_j^k(X(u))dB^j(u), \qquad k = 1, 2, \ldots, r.$$

Then $\tilde{B}(t) = (\tilde{B}^k(t))$ is an r-dimensional $(\mathscr{F}_t)$-Brownian motion (Example II-6.1) and $\tilde{\mathfrak{X}} = [X(t), \tilde{B}(t), M(t), \phi(t)]$ is a solution on $(\Omega, \mathscr{F}, P)$ with $(\mathscr{F}_t)$ corresponding to the coefficients $[\tilde{\sigma}, b, \tau, \beta, 0]$, where $\tilde{\sigma} = \sigma p^{-1}$. The transformation $\mathfrak{X} \longrightarrow \tilde{\mathfrak{X}}$ is called a *transformation of Brownian motion deter-*

mined by p and is denoted by $\mathfrak{X} \xrightarrow[p]{(a)} \tilde{\mathfrak{X}}$. Clearly $\mathfrak{X}$ is also obtained from $\tilde{\mathfrak{X}}$ by transformation of the same type determined by p^{-1}: $\tilde{\mathfrak{X}} \xrightarrow[p^{-1}]{(a)} \mathfrak{X}$.

(b) *Time change.*

Let $\mathfrak{X} = [X(t), B(t), M(t), \phi(t)]$ be a solution on a space $(\Omega,\mathcal{F},P)$ with $(\mathcal{F}_t)$ corresponding to the coefficients $[\sigma, b, \tau, \beta, 0]$. Let $c(x)$ be a continuous function on D such that $c_1 \leq c(x) \leq c_2$ for some positive constants c_1, c_2. Set $A(t) = \int_0^t c(X(u))du$ and denote by $A^{-1}(u)$ the inverse of $t \longmapsto A(t)$. Let $\tilde{X}(t) = X(A^{-1}(t))$, $\tilde{B}(t) = (\tilde{B}^k(t))$ where $\tilde{B}^k(t) = \int_0^t \sqrt{c(\tilde{X}(u))}\, dB^k(A^{-1}(u))$, $\tilde{M}(t) = M(A^{-1}(t))$ and $\tilde{\phi}(t) = \phi(A^{-1}(t))$. Also set $\tilde{\mathcal{F}}_t = \mathcal{F}_{A^{-1}(t)}$. Then we see at once (cf. Chapter III, Section 1 or Section 4.2) that $\tilde{\mathfrak{X}} = [\tilde{X}(t),\tilde{B}(t),\tilde{M}(t),\tilde{\phi}(t)]$ is a solution on $(\Omega,\mathcal{F},P)$ with $(\tilde{\mathcal{F}}_t)$ corresponding to the coefficients $[c^{-1/2}\sigma,c^{-1}b,\tau,\beta,0]$. The transformation $\mathfrak{X} \longrightarrow \tilde{\mathfrak{X}}$ is called a *transformation of time change determined by c* and is denoted by $\mathfrak{X} \xrightarrow[c]{(b)} \tilde{\mathfrak{X}}$. Clearly $\mathfrak{X}$ is also obtained from $\tilde{\mathfrak{X}}$ by transformation of the same type determined by c^{-1}: $\tilde{\mathfrak{X}} \xrightarrow[c^{-1}]{(b)} \mathfrak{X}$.

(c) *Transformation of drift.*

Let $\mathfrak{X} = [X(t), B(t), M(t), \phi(t)]$ be a solution on $(\Omega,\mathcal{F},P)$ with $(\mathcal{F}_t)$* corresponding to the coefficients $[\sigma, b, \tau, \beta, 0]$. Let $d(x) = (d^1(x), d^2(x), \ldots, d^r(x))$ be a bounded $\boldsymbol{R}^r$-valued continuous function defined on D and set

$$\mu(t) = \exp[\sum_{k=1}^{r} \int_0^t d^k(X(s))dB^k(s) - \frac{1}{2}\sum_{k=1}^{r} \int_0^t d^k(X(s))^2 ds].$$

Then $\mu(t)$ is a positive $(\mathcal{F}_t)$-martingale and $\tilde{P} = \mu \cdot P$ is defined by Definition 4.1. Set $\tilde{\mathfrak{X}}(t) = [X(t), \tilde{B}(t), M(t), \phi(t)]$, where $\tilde{B}^k(t) = B^k(t) - \int_0^t d^k(X(s))ds$, $k = 1, 2, \ldots, r$. It is easy to see from Theorem 4.1 that $\tilde{\mathfrak{X}}(t)$ is a solution on $(\Omega,\mathcal{F},\tilde{P})$ with $(\mathcal{F}_t)$ corresponding to the coefficients $[\sigma, \tilde{b} = b+\sigma d, \tau, \beta, 0]$. The transformation $\mathfrak{X} \longrightarrow \tilde{\mathfrak{X}}$ is called a *transformation of drift determined by d* and is denoted by $\mathfrak{X} \xrightarrow[d]{(c)} \tilde{\mathfrak{X}}$. Clearly $\mathfrak{X}$ is also obtained from $\tilde{\mathfrak{X}}$ by transformation of the same type determined by $-d$; $\tilde{\mathfrak{X}} \xrightarrow[-d]{(c)} \mathfrak{X}$.

With these preparations completed, we will now show the existence and uniqueness of solutions in the case $\rho \equiv 0$. Let σ, b, τ and β satisfy the assumptions in Theorem 7.2. Then there exists $p(x)$: $D \longrightarrow O(r)$ such that each component of $p(x)$ is Lipschitz continuous and

* We may assume without loss of generality that $(\Omega, \mathcal{F})$ is a standard probability space.

$$\sigma(x)p(x)^{-1} = \begin{pmatrix} * & * & *, \ldots, * \\ * & * & *, \ldots, * \\ |\sigma^d(x)|, & 0, & 0, \ldots, 0 \end{pmatrix},$$

where $\sigma^i(x) = (\sigma^i_k(x))_{k=1}^r$ is the i-th row of $\sigma(x)$ and

$$|\sigma^i(x)| = \sqrt{\sum_{k=1}^r \{\sigma^i_k(x)\}^2}, \quad i = 1, 2, \ldots, d.$$

Indeed, $p_1(x) = \sigma^d(x)/|\sigma^d(x)| : D \longrightarrow \boldsymbol{S}^{r-1} = \{x \in R^r; |x| = 1\}$ is Lipschitz continuous. We choose $p_k(x)$: $D \longrightarrow \boldsymbol{S}^{r-1}$, $k = 2, 3, \ldots, r$ such that $p_k(x)$ is Lipschitz continuous and the system $[p_1(x), p_2(x), \ldots, p_r(x)]$ is orthonormal in $\boldsymbol{R}^r$ for every $x \in D$. Such a selection of $p_k(x)$ is always possible. Then $p(x)$: $D \longrightarrow O(r)$ whose k-th row is $p_k(x)$, $k = 1, 2, \ldots, r$, is what we want.

Next set $c(x) = |\sigma^d(x)|^2$ and define $d(x) = (d^1(x), d^2(x), \ldots, d^r(x))$ by

$$d^1(x) = -b^d(x)/c(x) \quad \text{and} \quad d^i(x) \equiv 0, \quad i = 2, 3, \ldots, r.$$

Let $\mathfrak{X}$ be a solution corresponding to the coefficients $[\sigma, b, \tau, \beta, 0]$. If we operate on $\mathfrak{X}$ by the successive transformations

$$\mathfrak{X} \xrightarrow[p]{\text{(a)}} \mathfrak{X}_1 \xrightarrow[c]{\text{(b)}} \mathfrak{X}_2 \xrightarrow[d]{\text{(c)}} \mathfrak{X}_3,$$

then $\mathfrak{X}_3$ is a solution corresponding to the coefficients $[\tilde{\sigma}, \tilde{b}, \tilde{\tau}, \tilde{\beta}, 0]$, where $\tilde{\sigma} = (\sigma p^{-1})/\sqrt{c}$, $\tilde{b} = c^{-1}b + \tilde{\sigma}d$, $\tilde{\tau} = \tau$ and $\tilde{\beta} = \beta$.

Clearly $\tilde{\sigma}^d_1(x) \equiv 1$, $\tilde{\sigma}^d_k(x) \equiv 0$, $k = 2, 3, \ldots, r$, and $\tilde{b}^d(x) \equiv 0$. By the result of case (1°), the law of $\mathfrak{X}_3$ is uniquely determined from the law μ of $X(0)$. Since

$$\mathfrak{X}_3 \xrightarrow[-d]{\text{(c)}} \mathfrak{X}_2 \xrightarrow[c^{-1}]{\text{(b)}} \mathfrak{X}_1 \xrightarrow[p^{-1}]{\text{(a)}} \mathfrak{X},$$

the law of $\mathfrak{X}$ is uniquely determined from μ. This completes the proof of the uniqueness. The existence of solutions is also clear. We know the existence of a solution $\mathfrak{X}_3$ corresponding to $[\tilde{\sigma}, \tilde{b}, \tilde{\tau}, \tilde{\beta}, 0]$. Consequently $\mathfrak{X}$ is obtained by the above transformation.

(3°) *The general case.*

Let $[\sigma, b, \tau, \beta, \rho]$ satisfy the conditions in Theorem 7.2. Construct a solution $\mathfrak{X} = [X(t), B(t), M(t), \phi(t)]$ on a space $(\Omega, \mathscr{F}, P)$ with $(\mathscr{F}_t)$ corresponding to $[\sigma, b, \tau, \beta, 0]$. Taking an extension of Ω if necessary, we may assume that there exists an r-dimensional Brownian motion $B^* = (B^*(t))$ on Ω which is independent of $\mathfrak{X}$. Let $A(t) = t + \int_0^t \rho(X(s))d\phi(s)$ and $A^{-1}(t)$ be the inverse of $t \longmapsto A(t)$. Set $\tilde{X}(t) = X(A^{-1}(t))$, $\tilde{M}(t) = M(A^{-1}(t))$, $\tilde{\phi}(t) = \phi(A^{-1}(t))$ and $\tilde{\mathscr{F}}_t = \mathscr{F}_{A^{-1}(t)} \vee \sigma\{B^*(s);\ s \leq t\}$. Also set

$$\tilde{B}(t) = B(A^{-1}(t)) + \int_0^t I_{\partial D}(\tilde{X}(s))dB^*(s).$$

Then $\tilde{\mathfrak{X}} = [\tilde{X}(t),\tilde{B}(t),\tilde{M}(t),\tilde{\phi}(t)]$ is a solution corresponding to $[\sigma,b,\tau,\beta,\rho]$. This can be proved easily if we note the following relations

$$A^{-1}(t) = \int_0^t I_{\mathring{D}}(\tilde{X}(s))ds \quad \text{and} \quad \int_0^t I_{\partial D}(\tilde{X}(s))ds = \int_0^t \rho(\tilde{X}(s))d\tilde{\phi}(s).$$

These are the consequences of

$$\int_0^t I_{\mathring{D}}(X(s))dA_s = \int_0^t I_{\mathring{D}}(X(s))ds = t$$

and

$$\int_0^t I_{\partial D}(X(s))dA_s = \int_0^t \rho(X(s))d\phi_s.$$

Next we show the uniqueness of solutions. Let $\tilde{\mathfrak{X}} = [\tilde{X}(t), \tilde{B}(t), \tilde{M}(t), \tilde{\phi}(t)]$ be any solution corresponding to $[\sigma, b, \tau, \beta, \rho]$. Set $\tilde{A}(t) = \int_0^t I_{\mathring{D}}(\tilde{X}(s))ds$. Then $t \longmapsto \tilde{A}(t)$ is strictly increasing a.s. Indeed if this is not true, there exist $0 < r_1 < r_2$ such that if we set $\Omega_{r_1,r_2} = \{\omega; \tilde{A}(r_1) = \tilde{A}(r_2)\}$, then $P(\Omega_{r_1,r_2}) > 0$. But

$$\begin{aligned}\Omega_{r_1,r_2} &\underset{a.s.}{\subset} \{r_2 - r_1 = \int_{r_1}^{r_2} I_{\partial D}(\tilde{X}(s))ds = \int_{r_1}^{r_2} \rho(\tilde{X}(s))d\tilde{\phi}(s)\} \\ &\underset{a.s.}{\subset} \{\int_{r_1}^{r_2} d\tilde{\phi}(s) > 0\}\end{aligned}$$

and

$$\begin{aligned}\Omega_{r_1,r_2} &\underset{a.s.}{\subset} \{I_{\mathring{D}}(\tilde{X}(s)) = 0 \quad \text{for all} \quad s \in [r_1, r_2]\} \\ &\underset{a.s.}{\subset} \{\sum_{k=1}^{r} \int_{r_1}^{r_2} \sigma_k^d(\tilde{X}(s))I_{\mathring{D}}(\tilde{X}(s))d\tilde{B}^k(s) = 0 \\ &\qquad \text{and} \int_{r_1}^{r_2} b^d(\tilde{X}(s))I_{\mathring{D}}(\tilde{X}(s))ds = 0\}.\end{aligned}$$

Therefore,

$$\Omega_{r_1,r_2} \underset{a.s.}{\subset} \{\tilde{X}^d(r_2) = \tilde{X}^d(r_1) + \tilde{\phi}(r_2) - \tilde{\phi}(r_1) > \tilde{X}^d(r_1)\}$$
$$\underset{a.s.}{\subset} \{\tilde{X}(r_2) \in \mathring{D}\}.$$

But this is clearly a contradiction.

Thus the inverse $\tilde{A}^{-1}(t)$ of $t \longmapsto \tilde{A}(t)$ is continuous. Set $\mathfrak{X} = [X(t) = \tilde{X}(\tilde{A}^{-1}(t)),\ B(t) = \int_0^{\tilde{A}^{-1}(t)} I_{\mathring{D}}(\tilde{X}(s))d\tilde{B}(s),\ M(t) = \tilde{M}(\tilde{A}^{-1}(t)),\ \phi(t) = \tilde{\phi}(\tilde{A}^{-1}(t))]$. Then it is easy to see that $\mathfrak{X}$ is a solution corresponding to $[\sigma, b, \tau, \beta, 0]$. Also,

$$t = \int_0^t I_{\mathring{D}}(\tilde{X}(s))ds + \int_0^t I_{\partial D}(\tilde{X}(s))ds = \tilde{A}(t) + \int_0^t \rho(\tilde{X}(s))d\tilde{\phi}(s)$$

and hence

$$\tilde{A}^{-1}(t) = t + \int_0^t \rho(X(s))d\phi(s).$$

This implies that $\tilde{\mathfrak{X}}$ is obtained from $\mathfrak{X}$ as above. Since the law of $\mathfrak{X}$ is unique, this implies that the law of $\tilde{\mathfrak{X}}$ is also unique.

Thus, we have constructed a general class of (A,L)-diffusion processes by means of stochastic differential equations. But we assumed that $\delta(x) > 0$ everywhere on ∂D and normalized it so that $\delta(x) \equiv 1$. From a probabilistic point of view, this assumption is too restrictive and should be weakened to the condition that $\delta(x) + \rho(x) > 0$ everywhere on ∂D. Roughly speaking, $\delta(x) > 0$ implies that there is reflection at x and $\rho(x) > 0$ implies that there is sojourn at x. Therefore it is intuitively clear that $\delta(x) = \rho(x) = 0$ is impossible but $\delta(x) = 0$ and $\rho(x) > 0$ may be allowed. We can give another method of constructing (A,L)-diffusion processes which covers the general case of $\delta(x) + \rho(x) > 0$. This method, which is similar to the one given in Chapter III, Section 4.3, consists in piecing together excursions from the boundary to the boundary. As we shall see, the probabilistic structure of the diffusion is clearly revealed by this type of construction ([170], [174] and [229]).

For simplicity, we consider the case of $A = \Delta/2$ leaving the general case to [174] and [229]. So let $D = \boldsymbol{R}^d_+$, $A = \Delta/2$ (i.e., $a^{ij}(x) = \delta_{ij}$, $b^i(x) = 0$) and let L be given by (7.2). We assume that

(7.14) $$\inf_{x \in \partial D} [\rho(x) + \delta(x)] > 0.$$

Assume furthermore that there exists $\tau(x) = (\tau^i_l(x)) \colon \partial D \longrightarrow \boldsymbol{R}^{d-1} \otimes \boldsymbol{R}^s$ such that

(7.15) $$\alpha^{ij}(x) = \sum_{l=1}^{s} \tau_l^i(x)\tau_l^j(x), \quad x \in \partial D, \quad i, j = 1, 2, \ldots, d-1,$$

and assume that all the functions $\tau_l^i(x)$, $\beta^i(x)$, $\mu(x)$ are Lipschitz continuous on ∂D.

Let $\mathscr{W}_0(D)$ be the totality of all continuous functions $w: [0, \infty) \longrightarrow D$ with $w(0) = 0$ such that there exists $\sigma(w) > 0$ having the property that if $0 < t < \sigma(w)$, then $w(t) \in \mathring{D}$ and if $t \geq \sigma(w)$, then $w(t) = w(\sigma(w)) \in \partial D$. Let $\mathscr{B}(\mathscr{W}_0(D))$ be the σ-field generated by Borel cylinder sets and let n be the σ-finite measure on $(\mathscr{W}_0(D), \mathscr{B}(\mathscr{W}_0(D)))$ defined as follows. In Chapter III, Section 4.3, we defined the path space $\mathscr{W}^+$ and the σ-finite measure n^+ on $(\mathscr{W}^+, \mathscr{B}(\mathscr{W}^+))$. Let P_0 be the Wiener measure on W_0^{d-1} starting at 0, and define n as the image measure of $P_0 \times n^+$ under the map

$$W_0^{d-1} \times \mathscr{W}^+ \ni (\omega, w) \longmapsto (\omega_{\sigma(w)}^-, w) \in \mathscr{W}_0(D)$$

where $\omega_{\sigma(w)}^-$ is defined by $\omega_{\sigma(w)}^-(t) = \omega(t \wedge \sigma(w))$. If we set

$$K^+(t,x) = \prod_{i=1}^{d-1} \frac{1}{\sqrt{2\pi t}} \exp\left(-\frac{(x^i)^2}{2t}\right) \sqrt{\frac{2}{\pi t^3}}\, x^d \exp\left(-\frac{(x^d)^2}{2t}\right)$$
$$\text{for } t > 0, \ x = (x^1, x^2, \ldots, x^d) \in D,$$

and

$$p^\circ(t,x,y) = \prod_{i=1}^{d-1} \frac{1}{\sqrt{2\pi t}} \exp\left(-\frac{(x^i - y^i)^2}{2t}\right) \frac{1}{\sqrt{2\pi t}} \left(\exp\left(-\frac{(x^d - y^d)^2}{2t}\right) - \exp\left(-\frac{(x^d + y^d)^2}{2t}\right)\right) \quad \text{for} \quad t > 0, \quad x, y \in D,$$

then n is the unique measure on $(\mathscr{W}_0(D), \mathscr{B}(\mathscr{W}_0(D)))$ such that

$$n(\{w; w(t_1) \in A_1, w(t_2) \in A_2, \ldots, w(t_n) \in A_n\})$$
$$= \int_{A_1} K^+(t_1, x_1)dx_1 \int_{A_2} p^\circ(t_2 - t_1, x_1, x_2)dx_2 \int_{A_3} \cdots \times \int_{A_n} p^\circ(t_n - t_{n-1}, x_{n-1}, x_n)dx_n$$

for $0 < t_1 < t_2 < \cdots < t_n$ and $A_i \in \mathscr{B}(\dot{D})$. Let $\mathscr{W}(D)$ be the totality of continuous paths w: $[0, \infty) \longrightarrow D$ such that $w(0) \in \partial D$ and $w(t) = w(t \wedge \sigma(w))$, where $\sigma(w) = \inf\{t > 0;\ w(t) = \partial D\}$. Let $\mathbf{0}$ denote the constant path $\mathbf{0}(t) = 0 \in \partial D$. Let us define a mapping $T_c: \mathscr{W}_0(D) \longrightarrow \mathscr{W}_0(D) \cup \{\mathbf{0}\}$ for every $c \geq 0$ by

$$(7.16)\qquad (T_c w)(t) = \begin{cases} cw(t/c^2), & c > 0, \\ 0, & c = 0 \end{cases}$$

and let us also define a map

$$\Phi\colon \partial D \times \mathscr{W}_0(D) \ni (x, w) \longmapsto \Phi(x, w) \in \mathscr{W}(D)$$

by

$$(7.17)\qquad \Phi(x,w)(t) = x + (T_{\delta(x)} w)(t), \quad t \geq 0.$$

Clearly $\Phi(x,w)(0) = x$ and $\sigma[\Phi(x,w)] = \delta(x)^2\sigma(w)$. Let us define $\phi\colon \partial D \times \mathscr{W}_0(D) \ni (x, w) \longmapsto \phi(x,w) \in \partial D$ by

$$(7.18)\qquad \begin{aligned} \phi(x,w) &= \Phi(x,w)(\sigma[\Phi(x,w)]) - x \\ &= \delta(x)w(\sigma(w)). \end{aligned}$$

It is easy to see that for every $x, y \in \partial D$

$$(7.19)\qquad \begin{aligned} &\int_{\mathscr{W}_0(D) \cap \{\sigma(w) \leq 1\}} |\phi(x,w) - \phi(y,w)|^2\, n(dw) \\ &= |\delta(x) - \delta(y)|^2 \int_{\mathscr{W}_0(D) \cap \{\sigma(w) \leq 1\}} |w(\sigma(w))|^2 n(dw) \\ &= (d-1)\sqrt{\frac{2}{\pi}}\, |\delta(x) - \delta(y)|^2\, {}^{*} \\ &\leq K|x - y|^2. \end{aligned}$$

Let us take the following on an appropriate probability space $(\Omega, \mathscr{F}, P)$ with a reference family $(\mathscr{F}_t)$;

(i) $\mathscr{G}_t \subset \mathscr{F}_0$; an increasing family of sub σ-fields of $\mathscr{F}_0$ and a d-dimensional $(\mathscr{G}_t)$-Brownian motion $B(t) = (B^i(t))$ with $B(0) = 0$,

(ii) an s-dimensional $(\mathscr{F}_t)$-Brownian motion $B^*(t) = (B^l(t)^*)$

and

(iii) an $(\mathscr{F}_t)$-stationary Poisson point process p on $(\mathscr{W}_0(D), \mathscr{B}(\mathscr{W}_0(D)))$ with the characteristic measure n.

We shall now construct a path function of an (A,L)-diffusion process. Let $x \in D$ be given as the initial point. Firstly, we set

---

* Note that $n(\{w;\ \sigma(w) \in dt,\ w(\sigma(w)) \in dx\})$
$= (2\pi t^3)^{-1/2} dt (2\pi t)^{-(d-1)/2} \exp\left(-\dfrac{|x|^2}{2t}\right) dx, \quad t > 0,\ x \in \partial D.$

$$(7.20) \qquad X^x(t) = x + B(t) \qquad \text{for} \quad t \leq \sigma_0,$$

where $\sigma_0 = \inf\{t \geq 0;\ x + B(t) \in \partial D\}$. Set $\xi_0 = X^x(\sigma_0)$. Then ξ_0 is an $(\mathscr{F}_0)$-measurable ∂D-valued random variable. Secondly, we solve the following stochastic differential equation of jump type for the process $\xi(t) = (\xi^i(t))_{i=1}^d$ on ∂D:

$$(7.21) \qquad \begin{cases} \xi^d(t) \equiv 0, \\ \xi^i(t) = \xi_0^i + \sum_{l=1}^{s} \int_0^t \tau_l^i(\xi(s))dB^l(s)^* + \int_0^t \beta^i(\xi(s))ds \\ \qquad + \int_0^{t+}\int_{\mathscr{W}_0(D)} \phi^i(\xi(s-), w)I_{\{\sigma(w)\leq 1\}}\tilde{N}_p(dsdw) \\ \qquad + \int_0^{t+}\int_{\mathscr{W}_0(D)} \phi^i(\xi(s-), w)I_{\{\sigma(w)>1\}}N_p(dsdw), \\ \qquad\qquad i = 1, 2, \ldots, d-1. \end{cases}$$

(7.21) is a stochastic differential equation of the jump type which will be discussed in Section 9. Noting the Lipschitz continuity of τ and β, (7.19) and that $n(\{w;\ \sigma(w) > 1\}) = \int_1^\infty (2\pi t^3)^{-1/2}dt < \infty$, we can apply Theorem 9.1 to conclude that $\xi(t)$ is determined uniquely as an $(\mathscr{F}_t)$-adapted right-continuous process on ∂D with left-hand limits.

Thirdly, set

$$(7.22) \qquad \begin{aligned} A(t) &= \sigma_0 + \int_0^{t+}\int_{\mathscr{W}_0(D)} \sigma[\Phi(\xi(s-),w)]N_p(dsdw) + \int_0^t \rho(\xi(s))ds \\ &= \sigma_0 + \sum_{s\leq t,\, s\in D_p} \delta(\xi(s-))^2\sigma[p(s)] + \int_0^t \rho(\xi(s))ds. \end{aligned}$$

It is easy to show using (7.14) that $A(t)$ is an $(\mathscr{F}_t)$-adapted right-continuous process such that $t \longmapsto A(t)$ is strictly increasing and $\lim_{t\uparrow\infty} A(t) = \infty$ a.s. For every $t \geq 0$, there exists a unique $s \geq 0$ such that $A(s-) \leq t \leq A(s)$. If $s = 0$, i.e., $0 \leq t \leq \sigma_0$, $X^x(t)$ was already defined by (7.20). If $s > 0$ and $A(s-) < A(s)$, then this implies that $s \in \boldsymbol{D}_p$ and we set

$$(7.23) \qquad X^x(t) = \Phi(\xi(s-), p(s))(t - A(s-)).$$

If $s > 0$ and $A(s-) = A(s)$, then $\xi(s) = \xi(s-)$ and we set

$$(7.24) \qquad X^x(t) = \xi(s).$$

In this way we have defined a stochastic process $X^x(t)$; it is obvious by the way of construction that $t \longmapsto X^*(t)$ is continuous a.s. The remaining problem is to show that $X^*(t)$ is an (A,L)-diffusion process and it is unique. We can show that this $X^*(t)$ satisfies (7.8) by using the results in [234] or, we may argue as follows: We can show the existence of solution $X(t)$ to (7.8) by a similar tightness argument as in Section 2. We can show, by decomposing any such solution $X(t)$ into excursions $\{X(t),\ t \in e_\alpha\}$ where e_α is one of intervals $(A(s-),\ A(s))$ with $A(t) = \phi^{-1}(t)$, that $X(t)$ is obtained as explained above from a Poisson point process of Brownian excursions p and auxiliary Brownian motions $B(t)$ and $B^*(t)$. From this we can conclude the uniqueness of solutions to (7.8) and, at the same time, that the above constructed process $X^*(t)$ actually satisfies (7.8). For the details, we refer to [229].

8. Examples

Example 8.1. (*Linear* or *Gaussian diffusions*). Let $\sigma = (\sigma^i_k)$ be a constant $d\times r$-matrix and $\beta = (\beta^i_k)$ be a constant $d\times d$-matrix. Set $b^i(x) = \sum_{k=1}^{d} \beta^i_k x^k$, $x = (x^1, x^2, \ldots, x^d) \in \mathbf{R}^d$. Consider the following stochastic differential equation

$$(8.1) \qquad dX^i_t = \sum_{k=1}^{r} \sigma^i_k dB^k_t + b^i(X_t)dt, \quad i = 1, 2, \ldots, d,$$

or in matrix notation,

$$(8.1) \qquad dX_t = \sigma dB_t + \beta X_t dt.$$

We know by the general theory (Theorem 3.1) that the solution exists uniquely; it is given explicitly as follows. Let

$$e^{t\beta} = \sum_{k=0}^{\infty} \frac{t^k}{k!} \beta^k.$$

Then the solution $X(t)$ of (8.1) is given as

$$(8.2) \qquad X(t) = e^{\beta t}(X(0) + \int_0^t e^{-\beta s}\, \sigma dB(s)),$$

or in component form,

$$X^i(t) = \sum_{j=1}^{d} (e^{\beta t})^i_j \{X^j(0) + \sum_{l=1}^{d}\sum_{k=1}^{r} \int_0^t (e^{-\beta s})^j_l \sigma^l_k dB^k(s)\}.$$

The proof is easily seen from the relation

$$d(e^{-\beta t}X(t)) = e^{-\beta t}(dX(t) - \beta X(t)dt) = e^{-\beta t}\sigma dB(t).$$

In particular, if the initial value $X(0)$ is Gaussian distributed, then $X(t)$ is a Gaussian process.* For example, if $d = 1$ and

$$dX(t) = dB(t) - \gamma X(t)dt \qquad (\gamma > 0), \tag{8.3}$$

$X(t)$ is solved as

$$X(t) = e^{-\gamma t}(X(0) + \int_0^t e^{\gamma s}dB(s)). \tag{8.4}$$

(see also Example 2.1 of Chapter III, Section 2). Suppose that $X(0)$ is Gaussian distributed with mean 0 and variance σ^2. Then the covariance of $X(t)$ is given as

$$\begin{aligned} E(X(t)X(s)) &= e^{-\gamma(t+s)}\sigma^2 + \int_0^s e^{-\gamma(t-u)}e^{-\gamma(s-u)}du \\ &= \left(\sigma^2 - \frac{1}{2\gamma}\right)e^{-\gamma(t+s)} + \frac{1}{2\gamma}e^{-\gamma(t-s)}, \qquad \text{if} \quad t > s. \end{aligned}$$

In particular, if $\sigma^2 = 1/2\gamma$, $X(t)$ is a stationary Gaussian process [18].

The equation (8.3) is known as *Langevin's equation* and the solution $X(t)$ in (8.4) is known as *Ornstein-Uhlenbeck's Brownian motion.*

A slight general equation

$$dX(t) = (aX(t) + b)dB(t) + (cX(t) + d)dt \tag{8.5}$$

can be solved in a similar way where a, b, c and d are real constants. First, we note that (8.5) is equivalent to

$$\begin{aligned} dX(t) &= aX(t)\circ dB(t) - \frac{1}{2}a(aX(t) + b)dt + bdB(t) \\ &\qquad + (cX(t) + d)dt \\ &= aX(t)\circ dB(t) + \left(c - \frac{1}{2}a^2\right) X(t)dt + bdB(t) \end{aligned} \tag{8.6}$$

* By the definition of solutions, $X(0)$ and $B(t)$ are always independent.

$$+\left(d-\frac{1}{2}ab\right)dt.$$

If $M(t)=\exp\left\{-aB(t)-\left(c-\frac{1}{2}a^2\right)t\right\}$, then

$$dM(t)=-aM(t)\circ dB(t)-\left(c-\frac{1}{2}a^2\right)M(t)dt$$

and consequently

$$M(t)^{-1}\circ dM(t)=-\left[adB(t)+\left(c-\frac{1}{2}a^2\right)dt\right].$$

Now (8.6) is equivalent to

$$dX(t)=-X(t)M(t)^{-1}\circ dM(t)+bdB(t)+\left(d-\frac{1}{2}ab\right)dt$$

i.e.,

$$d(M(t)X(t))=bM(t)\circ dB(t)+\left(d-\frac{1}{2}ab\right)M(t)dt.$$

Therefore $X(t)$ is solved uniquely as

$$X(t)=M(t)^{-1}[X(0)+b\int_0^t M(s)\circ dB(s)+\left(d-\frac{1}{2}ab\right)\int_0^t M(s)ds] \tag{8.7}$$

where $M(t)=\exp\left\{-aB(t)-\left(c-\frac{1}{2}a^2\right)t\right\}$.

Similarly if we consider a multi-dimensional stochastic differential equation

$$dX^i(t)=\sum_{p=1}^{r}\left(\sum_{j=1}^{d}L^i_{pj}X^j(t)+c^i_p\right)\circ dB^p(t)+\left(\sum_{j=1}^{d}L^i_{0j}X^j(t)+c^i_0\right)dt$$
$$X^i(0)=x^i, \qquad i=1,2,\ldots,d,$$

where L^i_{pj}, $i,j=1,2,\ldots,d$, $p=0,1,\ldots,r$, are constants such that $L^i_{pj}=0$ for $i>j$, then the solution is given by

$$X^d(t) = M^d(t)^{-1}(x^d + \int_0^t M^d(s)\circ d\eta^d(s))$$

with $\eta^d(t) = \sum_{p=1}^{d} c_p^d B^p(t) + c_0^d t$ and

$$X^i(t) = M^i(t)^{-1}(x^i + \int_0^t M^i(s)\circ d\eta^i(s))$$

with

$$\eta^i(t) = \sum_{j=i+1}^{d} \int_0^t X^j(s)\circ d\xi_j^i(s) + \sum_{p=1}^{r} c_p^i B^p(t) + c_0^i t,$$

$$i = d-1, d-2, \ldots, 1.$$

Here

$$\xi_j^i(t) = \sum_{p=1}^{r} L_{pj}^i B^p(t) + L_{0j}^i t$$

and

$$M^i(t) = \exp(-\xi_i^i(t)), \qquad i, j = 1, 2, \ldots, d.$$

In this case, the Lie algebra $\mathfrak{L}(L_0, L_1, \ldots, L_r)$ generated by vector fields

$$L_p = \sum_{i=1}^{d}\sum_{j=1}^{d} (L_{pj}^i x^j + c_p^i)\frac{\partial}{\partial x^i}, \qquad p = 0, 1, \ldots, d,$$

is solvable. A general result for the representation of solutions in such a case was obtained by H. Kunita [96].

Example 8.2. Let a, c, d be real constants such that $a > 0$. Consider the following one-dimensional stochastic differential equation:

$$\text{(8.8)} \qquad dX(t) = (2aX(t)\vee 0)^{1/2}dB(t) + (cX(t)+d)dt.$$

Since the coefficients $\sigma(x) = (2ax\vee 0)^{1/2}$ and $b(x) = cx + d$ satisfy the condition of Theorem 3.2 and also the growth condition (2.18), a global strong solution $X(t)$ exists uniquely for every given initial value $X(0)$. If $d \geq 0$ and $X(0) \geq 0$ a.s., then $X(t) \geq 0$ for all $t \geq 0$ a.s. Indeed, in the case $d = 0$, it is obvious that $X(t) \equiv 0$ a.s. if $X(0) \equiv 0$ a.s. by the uniqueness of solutions. Setting $\sigma = \inf\{t; X(t) = 0\}$ we see that $\tilde{X}(t) = X(t+\sigma)$ is a solution of (8.8) with $\tilde{X}(0) = 0$ on the space ($\tilde{\Omega} = \{\omega; \sigma(\omega) < \infty\}$, $\tilde{\mathscr{F}} = \mathscr{F}|_{\tilde{\Omega}}$, $\tilde{P} = P(\cdot\,|\tilde{\Omega})$) and hence $\tilde{X}(t) \equiv 0$ a.s. on $\tilde{\Omega}$. This implies that $X(t) \equiv X(t\wedge\sigma)$ a.s. and consequently, $X(t) \geq 0$ a.s. if $X(0) \geq 0$ a.s. In the case $d > 0$, set $\sigma_{-\varepsilon} = \inf\{t; X(t) = -\varepsilon\}$ where $\varepsilon > 0$ is such that $-c\varepsilon + d > 0$. Assume that $P(\sigma_{-\varepsilon} < \infty) > 0$. Then, with probability one, if we take any $r < \sigma_{-\varepsilon}$ such that $X(t) < 0$ if $t \in (r, \sigma_{-\varepsilon})$, we have

$$dX(t) = (cX(t) + d)dt$$

on the interval $(r, \sigma_{-\varepsilon})$ and hence $t \longmapsto X(t)$ is increasing on this interval. This is clearly impossible.

Thus the solution of (8.8) defines a conservative diffusion process $\{P_x\}$ on $[0,\infty)$ in the case $d \geq 0$. It is the L-diffusion process where L is the operator

$$Lf(x) = ax\frac{d^2}{dx^2}f(x) + (cx + d)\frac{d}{dx}f(x) \tag{8.9}$$

acting on $C_K^2([0, \infty))$.

We shall now prove the following formula:

$$E_x(e^{-\lambda w(t)}) = \left[\frac{a\lambda}{c}(e^{ct} - 1) + 1\right]^{-d/a} \exp\left\{-\frac{\lambda e^{ct}x}{\frac{a\lambda}{c}(e^{ct} - 1) + 1}\right\}. \tag{8.10}$$

(If $c = 0$, we understand that $\frac{1}{c}(e^{ct} - 1) = t$.) Indeed by Itô's formula, with respect to P_x we have that

$$u(t,w(t)) - u(0,x) = \text{a martingale} + \int_0^t \left[\frac{\partial u}{\partial t} + Lu\right](s,w(s))ds$$

for every $u(t,x) \in C_b^{1,2}([0, \infty)\times[0, \infty))$. * Noting that the function $v(t, x)$ in the right-hand side of (8.10) satisfies that $\dfrac{\partial v}{\partial t} = Lv$ and $v(0+, x) = e^{-\lambda x}$, we set $u(t,x) = v(t_0 - t,x)$ for fixed t_0 and apply Itô's formula for $u(t, x)$. Then $v(t_0 - t,w(t)) - v(t_0,x)$ is a P_x-martingale and hence $E_x[v(t_0-t,w(t))] = v(t_0,x)$. Letting $t = t_0$, we have $E_x(e^{-\lambda w(t_0)}) = v(t_0, x)$.

* $C_b^{1,2}([0, \infty)\times[0, \infty)) \ni u(t, x)$ implies that all the derivatives of u up to the first order in t and up to the second order in x are continuous and bounded.

Let $x > 0$ and set $\sigma_0 = \inf\{t; w(t) = 0\}$. Then

$$(8.11) \qquad \begin{aligned} P_x(\sigma_0 < \infty) &> 0 \quad \text{if} \quad 0 \leq d < a \\ &= 1 \quad \text{if} \quad 0 \leq d < a \quad \text{and} \quad c \leq 0 \end{aligned}$$

and

$$(8.12) \qquad P_x(\sigma_0 = \infty) = 1 \qquad \text{if} \quad d \geq a.$$

For the proof of (8.11) and (8.12), set

$$s(x) = \int_1^x \exp\left\{-\int_1^y \frac{cz+d}{az}dz\right\}dy = e^{c/a}\int_1^x \exp\left[-\frac{c}{a}y\right]y^{-d/a}dy$$

and

$$\kappa(x) = \int_1^x \exp\left[-\int_1^y \frac{cz+d}{az}dz\right]\left\{\int_1^y \exp\left[\int_1^z \frac{c\eta+d}{a\eta}d\eta\right]\frac{dz}{az}\right\}dy.$$

It is easy to see that $s(0+) = -\infty$ if and only if $d \geq a$ and $s(\infty) = \infty$ if and only if $c < 0$ or $c = 0$ and $d \leq a$. Note also that $\kappa(0+) < \infty$ if $d < a$. Now the assertions follows from Theorem VI-3.1 and Theorem VI-3.2. We also remark that the boundary $x = 0$ is regular* and reflecting if $0 < d < a$, and exit and absorbing if $d = 0$. If $d \geq a$ the boundary is entrance, and it is easy to conclude that

$$(8.13) \qquad P_0(w(t) > 0 \quad \text{for all} \quad t > 0) = 1.$$

Indeed, letting $\lambda \uparrow \infty$ in (8.10), we have $P_0(w(t) > 0) = 1$ for every $t > 0$. Combining this with (8.12), we can conclude that

$$P_0(w(t+s) > 0 \quad \text{for all} \quad s \geq 0) = 1 \quad \text{for every } t > 0.$$

Since t is arbitrary, this implies (8.13).

Example 8.3. (*Bessel diffusions*). For $\alpha > 0$, let L_α be the differential operator on $[0, \infty)$ defined by

$$(8.14) \qquad L_\alpha f(x) = \frac{1}{2}\left[\frac{d^2}{dx^2}f(x) + \frac{\alpha-1}{x}\frac{d}{dx}f(x)\right]$$

* Itô-McKean [73].

with the domain

$$\mathscr{D}(L_\alpha) = \{f \in C_b^2([0,\infty));\quad \text{for some constants } 0 < a_1 < a_2 \text{ and}$$
(8.15)
$$f(x) = cx^2 \quad \text{if } x \in [0, a_1]$$
$$\text{and} \quad f(x) = 0 \quad \text{if} \quad x \in [a_2,\infty)\}.$$

For $f \in \mathscr{D}(L\alpha)$, we set $L\alpha f(0) = c(\alpha-1)$ so that $L\alpha f \in C_b([0,\infty))$. There exists a unique conservative diffusion process generated by the operator L_α which is called the *Bessel diffusion process with index* α. Bessel diffusion processes are essentially a particular case of the diffusions discussed in the previous example. Let

(8.16)
$$\tilde{L}_\alpha f(x) = 2x\frac{d^2}{dx^2}f(x) + \alpha\frac{d}{dx}f(x) \quad \text{and} \quad c$$
$$\mathscr{D}(\tilde{L}_\alpha) = \{\tilde{f}(x) = f(\sqrt{x}\,);\, f \in \mathscr{D}(L\alpha)\}\ .$$

Then an $\tilde{L}_\alpha$-diffusion $\{P_x\}_{x\in[0,\infty)}$ exists uniquely. Indeed, the diffusion of Example 8.2 in the case $a = 2$, $c = 0$ and $d = \alpha$ is clearly an $\tilde{L}_\alpha$-diffusion. Conversely, by the same proof as in Theorem 6.1, we can prove that if $\{\tilde{P}_x^{(\alpha)}\}_{x\in[0,\infty)}$ is $\tilde{L}_\alpha$-diffusion then $\{X(t, w) = w(t)\}$ is a solution of equation (8.8) for $a = 2$, $c = 0$ and $d = \alpha$ with $X(0) = x$. By the uniqueness of solutions of (8.8), we can conclude that the $\tilde{L}_\alpha$-diffusion is unique. It is easy to see that

$$(\tilde{L}_\alpha\tilde{f})(x^2) = (L_\alpha f)(x), \qquad f \in \mathscr{D}(L_\alpha) \tag{8.17}$$

where $\tilde{f}(x) = f(\sqrt{x})$. Now we can conclude that the L_α-diffusion $\{P_x^{(\alpha)}\}_{x\in[0,\infty)}$ is unique and $P_x^{(\alpha)}$ is the image measure on $W([0,\infty))$ of the measure $\tilde{P}_{x^2}^{(\alpha)}$ under the mapping

$$W([0,\infty)) \ni w \longmapsto \sqrt{w} \in W([0,\infty)),$$

where the path $\sqrt{w}$ is defined by $\sqrt{w}(t) = \sqrt{w(t)}$. That is, the Bessel diffusion $X_\alpha^x(t)$ of index α starting at x is obtained as $X_\alpha^x(t) = \sqrt{Y_\alpha^{x^2}(t)}$, where $Y_\alpha^{x^2}$ is the unique solution of

$$\begin{cases} dY(t) = 2(Y(t)\vee 0)^{1/2}dB(t) + \alpha dt \\ Y(0) = x^2. \end{cases} \tag{8.18}$$

By (8.10), we see that

(8.19) $$E_x^{(\alpha)}(e^{-\lambda w(t)^2}) = (2\lambda t + 1)^{-\alpha/2} \exp\left(-\frac{\lambda x^2}{2\lambda t + 1}\right).$$

Inverting the Laplace transform, we see that

$$E_x^{(\alpha)}(f(w(t))) = \int_0^\infty p^{(\alpha)}(t, x, y) f(y) dy,$$

where

(8.20) $$p^{(\alpha)}(t, x, y) = \frac{\exp[-(x^2+y^2)/2t]}{t(xy)^{\alpha/2-1}} y^{\alpha-1} I_{\frac{\alpha}{2}-1}\left(\frac{xy}{t}\right)$$

and $I_\nu(x) = \left(\frac{x}{2}\right)^\nu \sum_{n=0}^\infty \frac{\left(\frac{x}{2}\right)^{2n}}{n!\Gamma(\nu+n+1)}$ is the modified Bessel function.

We can prove an interesting property of family of Bessel diffusions by using equation (8.18). Let B_1 and B_2 be two independent Brownian motions and α_1 and α_2 be positive constants. Consider the equations

$$\begin{cases} dY_1(t) = 2(Y_1(t) \vee 0)^{1/2} dB_1(t) + \alpha_1 dt \\ Y_1(0) = y_1 \in [0, \infty) \end{cases}$$

and

$$\begin{cases} dY_2(t) = 2(Y_2(t) \vee 0)^{1/2} dB_2(t) + \alpha_2 dt \\ Y_2(0) = y_2 \in [0, \infty). \end{cases}$$

Set $Y_3(t) = Y_1(t) + Y_2(t)$ and

$$B_3(t) = \int_0^t \sqrt{\frac{Y_1(s)}{Y_1(s) + Y_2(s)}} dB_1(s) + \int_0^t \sqrt{\frac{Y_2(s)}{Y_1(s) + Y_2(s)}} dB_2(s).^*$$

Then $B_3(t)$ is a Brownian motion by Theorem II-6.1 and

$$\begin{cases} dY_3(t) = 2(Y_3(t) \vee 0)^{1/2} dB_3(t) + (\alpha_1 + \alpha_2) dt \\ Y_3(0) = y_1 + y_2. \end{cases}$$

Thus the law of Y_3 is $\tilde{P}_{y_1+y_2}^{(\alpha_1+\alpha_2)}$. Consequently, if $X^\alpha(t)$ and $X^\beta(t)$ are mutually

* We know as a consequence of (8.10) that $P(Y_i(t) > 0) = 1$ for every $t > 0$.

independent Bessel diffusions of index α and β respectively, $\sqrt{|X^\alpha(t)|^2 + |X^\beta(t)|^2}$ is a Bessel diffusion of index $\alpha + \beta$. In particular, if $\alpha = d$, $d = 1, 2, \ldots$, $X^\alpha(t)$ can be identified with the radial process of d-dimensional Brownian motion. See [73], [147], [168] for further information on Bessel diffusions.

Example 8.4. (*Brownian excursions*).*[1] Let $T > 0$ be fixed. Consider the following stochastic differential equation

$$(8.21) \qquad \begin{cases} dX(t) = 2(X(t) \vee 0)^{1/2} dB(t) + \left(3 - \dfrac{2X(t)}{T-t}\right) dt \\ X(0) = 0. \end{cases}$$

This is an equation similar to (8.18). Hence it can be shown that there exists a unique solution $X(t)$ for $t \in [0, T)$ and that

$$(8.22) \qquad P(X(t) > 0 \quad \text{for all } t \in (0, T)) = 1.^{*2}$$

Furthermore, $X(t)$ defines a time-dependent Markov process. To be precise, for $0 \leq s < T$ and $x \in [0, \infty)$, let $\boldsymbol{W}_{s,x}$ be the totality of all continuous paths $w: [s, T) \ni t \longmapsto w(t) \in [0, \infty)$ such that $w(s) = x$ and $w(t) > 0$ for all $t \in (s, T)$, $\mathscr{B}(\boldsymbol{W}_{s,x})$ be the σ-field on $\boldsymbol{W}_{s,x}$ generated by Borel cylinder sets and $\mathscr{B}_t(\boldsymbol{W}_{s,x})$, $s < t < T$, be the sub σ-field generated by Borel cylinder sets depending only on the interval $[s, t]$. Let $\hat{\boldsymbol{P}}_{0,0}$ be the probability law on $(\boldsymbol{W}_{0,0}, \mathscr{B}(\boldsymbol{W}_{0,0}))$ of the solution $X(t)$ of (8.21) and more generally $\hat{\boldsymbol{P}}_{s,x}$ be the probability law on $(\boldsymbol{W}_{s,x}, \mathscr{B}(\boldsymbol{W}_{s,x}))$ of the unique solution $\{X(t)\}_{t \in [s,T)}$ of

$$(8.23) \qquad \begin{cases} dX(t) = 2(X(t) \vee 0)^{1/2}\, dB(t) + \left(3 - \dfrac{2X(t)}{T-t}\right) dt \\ X(s) = x. \end{cases}$$

The Markovian property of $\hat{\boldsymbol{P}}_{0,0}$ is now formulated as follows;
for $0 \leq s < t$ and $f \in \boldsymbol{B}([0, \infty))$,*[3]

$$(8.24) \qquad \hat{E}_{0,0}[f(w(t)) \,|\, \mathscr{B}_s(\boldsymbol{W}_{0,0})] = \hat{E}_{s,w(s)}[f(w'(t))] \quad \text{a.a. } w \ (\hat{\boldsymbol{P}}_{0,0}).$$

More generally, for $0 \leq u < s < t$, $x \in [0, \infty)$ and $f \in \boldsymbol{B}([0, \infty))$,

---

*[1] [73].

*[2] To prove (8.22) rigorously, apply the comparison theorem (Theorem VI-1.1) to equation (8.8) with $a = 2$, $c = 0$, $2 < d < 3$ and $X(0) = 0$ and equation (8.21).

*[3] $\boldsymbol{B}([0, \infty))$ is the totality of all bounded Borel measurable functions on $[0, \infty)$.

(8.25) $$\hat{E}_{u,x}[f(w(t))\,|\,\mathscr{B}_s(\boldsymbol{W}_{u,x})] = \hat{E}_{s,w(s)}[f(w'(t))] \quad \text{a.a. } w\ (\hat{P}_{u,x}).$$

The proof of the above can be given as in Section 5 using the uniqueness of solutions of (8.23) for every s and x.

Let $P_{s,x}$ be the image measure on $(\boldsymbol{W}_{s,x}, \mathscr{B}(\boldsymbol{W}_{s,x}))$ of the measure $\hat{P}_{s,x^2}$ under the mapping $\boldsymbol{W}_{s,x^2} \ni w \longrightarrow \sqrt{w} \in \boldsymbol{W}_{s,x}$ where the path $\sqrt{w}$ is defined, of course, by $\sqrt{w}(t) = \sqrt{w(t)}$. Then it is obvious that the Markovian property (8.25) also holds for $\{P_{s,x}\}$.

Set

(8.26) $$p^0(t, x, y) = \frac{1}{\sqrt{2\pi t}}\left(\exp\left(-\frac{(x-y)^2}{2t}\right) - \exp\left(-\frac{(x+y)^2}{2t}\right)\right)$$
$$t > 0, \quad x, y \in [0, \infty),$$

(8.27) $$K(t, x) = \sqrt{\frac{2}{\pi t^3}}\, x \exp\left(-\frac{x^2}{2t}\right), \qquad t > 0, \quad x \in [0, \infty),$$

and

(8.28) $$p(s, x; t, y) = \begin{cases} \dfrac{K(T-t, y)}{K(T-s, x)}\, p^0(t-s, x, y), \\ \qquad \text{if } 0 \le s < t < T, x, y \in (0, \infty), \\ \sqrt{\dfrac{\pi(T-s)^3}{2}}\, K(T-t, y) K(t-s, y) \\ \qquad \text{if } 0 \le s < t < T;\ x = 0 \text{ and } y > 0. \end{cases}$$

We shall show that for every $x \ge 0$ and $s \ge 0$,

(8.29) $$\begin{aligned} &P_{s,x}\{w;\, w(t_1) \in dx_1,\, w(t_2) \in dx_2, \ldots, w(t_n) \in dx_n\} \\ &= p(s, x; t_1, x_1) p(t_1, x_1; t_2, x_2) \ldots p(t_{n-1}, x_{n-1}; t_n, x_n) \\ &\qquad \times dx_1 dx_2 \cdots dx_n, \end{aligned}$$

for every $s < t_1 < t_2 < \cdots < t_n < T$. It is sufficient to prove that

(8.30) $$E_{s,x}[f(w(t))] = \int_{[0,\infty)} p(s, x; t, y) f(y) dy,$$
$$0 \le s < t,\ x \in [0, \infty),\ f \in \boldsymbol{B}([0, \infty)),$$

since (8.29) is obtained from (8.30) by successive applications of the Markovian property (8.25) for $\{P_{s,x}\}$. Set, for $0 \le s < t < T$,

$$u(s, x; t) = \int_0^\infty p(s, x; t, y) f(t, y) dy,$$

where $f(t,y)$ is a bounded smooth function. Then we can verify by direct calculation that $u(s,x; t)$ satisfies

$$\text{(8.31)} \quad \begin{cases} -\dfrac{\partial u}{\partial s}(s, x; t) = \left\{\dfrac{1}{2}\dfrac{\partial^2}{\partial x^2} + \left(\dfrac{1}{x} - \dfrac{x}{T-s}\right)\dfrac{\partial}{\partial x}\right\} u(s,x; t), \\ \qquad\qquad\qquad\qquad\qquad s \in (0, t), \quad x \in (0, \infty), \\ \lim\limits_{s \uparrow t, x \to y} u(s, x; t) = f(t, y). \end{cases}$$

If we set

$$\tilde{u}(s, x; t) = \int_0^\infty p(s, \sqrt{x}; t, y) f(t, y^2) dy,$$

$\tilde{u}(s, x; t)$ satisfies

$$\text{(8.32)} \quad \begin{cases} -\dfrac{\partial \tilde{u}}{\partial s}(s, x; t) = \left\{2x\dfrac{\partial^2}{\partial x^2} + \left(3 - \dfrac{2x}{T-s}\right)\dfrac{\partial}{\partial x}\right\} \tilde{u}(s, x; t) \\ \lim\limits_{s \uparrow t, x \to y} \tilde{u}(s, x; t) = f(t, y). \end{cases}$$

By (8.32) and Itô's formula, we see for each $\tau < T$ that $[s, \tau) \ni t \longmapsto \tilde{u}(t, X(t); \tau)$ is a martingale if $X(t)$ is the solution of (8.23). Hence

$$E(\tilde{u}(t, X(t); \tau)) = \tilde{u}(s, x; \tau)$$

for every $t \in [s, \tau)$. Letting $t \uparrow \tau$

$$E[f(\tau, X(\tau))] = \hat{E}_{s,x}[f(\tau, w(\tau))] = \tilde{u}(s, x; \tau).$$

Consequently

$$E_{s,x}[f(\tau, w(\tau))] = \hat{E}_{s,x^2}[f(\tau, \sqrt{w(\tau)})] = \int_0^\infty p(s, x; \tau, y) f(\tau, y) dy$$

and (8.30) is proved.

It is immediately seen from (8.29) that

$$P_{0,0}\{w; w(t_1) \in dx_1, w(t_2) \in dx_2, \ldots, w(t_n) \in dx_n\}$$

$$= P_{0,0}\{w;\, w(T-t_1) \in dx_1,\, w(T-t_2) \in dx_2,\, \ldots,\, w(T-t_n) \in dx_n\}$$

for every $0 < t_1 < t_2 < \cdots < t_n < T$. This shows that $P_{0,0}$ is invariant under the time inversion $w \longmapsto \hat{w}$, where $\hat{w}$ is defined by $\hat{w}(t) = w(T-t)$. From this, we can conclude that

$$P_{0,0}\{w;\, \lim_{t \uparrow T} w(t) = 0\} = 1. \tag{8.33}$$

Hence $P_{0,0}$ may be regarded as a probability on the space $\boldsymbol{W}_{0,0} = \{w;\, [0, T] \ni t \longmapsto w(t)$ is coninuous, $w(0) = w(T) = 0$ and $w(t) > 0$ for $t \in (0, T)\}$.

Example 8.5. (*Pinned Brownian motion*).

Let $X(t)$ be a one-dimensional Brownian motion such that $X(0) = 0$. For fixed $t_0 > 0$ and $x, y \in \boldsymbol{R}^1$, define the process $X_x^{t_0,y} = (X_x^{t_0,y}(t))_{0 \le t \le t_0}$ by

$$\begin{aligned} X_x^{t_0,y}(t) &= x + X(t) + \frac{t}{t_0}(-X(t_0) + (y-x)) \\ &= x + \frac{t}{t_0}(y-x) + X_0^{t_0,0}(t). \, * \end{aligned} \tag{8.34}$$

It is easy to verify that the probability law of $X_x^{t_0,y}$ coincides with $P_x(\cdot \,|\, w(t_0) = y)$, where P_x is the Wiener measure starting at x. The process $X_x^{t_0,y}$ is called a ***pinned Brownian motion***. Consider the following stochastic differential equation

$$\begin{cases} dX(t) = dB(t) - \dfrac{y - X(t)}{t - t_0}\, dt \\ X(0) = x. \end{cases} \tag{8.35}$$

Clearly the solution $X(t)$ exists uniquely for $t \in [0, t_0)$. By (8.35) we have

$$(t - t_0)d\left(\frac{X(t) - y}{t - t_0}\right) = dB(t),$$

and hence $X(t)$ is solved as

---

* The process $X_0^{t_0,0}$ is sometimes called the *Brownian bridge*.

$$(8.36) \qquad X(t) = x + \frac{t}{t_0}(y - x) + (t - t_0)\int_0^t \frac{dB(s)}{s - t_0}, \quad t < t_0.$$

It is now easy to identify the process $X(t)$ with $X_x^{t_0,y}(t)$. Both $X_0^{t_0,0}(t)$ and $(t - t_0)\int_0^t \frac{dB(s)}{s - t_0}$ are centered Gaussian processes with the covariance

$$\Gamma(s, t) = t \wedge s - \frac{ts}{t_0}.$$

Thus the equation (8.35) is the stochastic differential equation determining the pinned Brownian motion $X_x^{t_0,y}$.

9. Stochastic differential equations with respect to Poisson point processes

So far we have only considered stochastic differential equations with respect to Brownian motions. For such equations, the solutions are always continuous processes. We can also consider more general stochastic differential equations* which include Poisson point processes as well as Brownian motions; in this case, however, the solutions are usually discontinuous processes. For simplicity, we consider such general equations in the case of the time-homogeneous Markovian type.

Let $\{\boldsymbol{U}, \mathscr{B}_U\}$ be a measurable space and $n(du)$ be a σ-finite measure on it. Let U_0 be a set in $\mathscr{B}_U$ such that $n(U \setminus U_0) < \infty$. Let $\sigma(x) = (\sigma_k^i(x))$ be a Borel measurable function $\boldsymbol{R}^d \longrightarrow \boldsymbol{R}^d \otimes \boldsymbol{R}^r$, $b(x) = (b^i(x))$ be a Borel measurable function $\boldsymbol{R}^d \longrightarrow \boldsymbol{R}^d$, and $f(x, u) = (f^i(x, u))$ be a $\mathscr{B}(\boldsymbol{R}^d) \times \mathscr{B}_U$-measurable function $\boldsymbol{R}^d \times \boldsymbol{U} \longrightarrow \boldsymbol{R}^d$ such that for some positive constant K,

$$(9.1) \qquad \|\sigma(x)\|^2 + \|b(x)\|^2 + \int_{U_0} \|f(x, u)\|^2 n(du) \leq K(1 + |x|^2), \; x \in \boldsymbol{R}^d.$$

Consider the following stochastic differential equation

$$\begin{aligned} X^i(t) = X^i(0) &+ \sum_{k=1}^r \int_0^t \sigma_k^i(X(s))dB^k(s) + \int_0^t b^i(X(s))ds \\ (9.2) \qquad &+ \int_0^{t+}\int_U f^i(X(s-), u) I_{U_0}(u) \tilde{N}_p(dsdu) \\ &+ \int_0^{t+}\int_U f^i(X(s-), u) I_{U \setminus U_0}(u) N_p(dsdu), \quad i = 1, 2, \ldots, d, \end{aligned}$$

* They are also called *stochastic differential equations of the jump type.*

where $B = (B^k(t))$ is an r-dimensional Brownian motion, p is a stationary Poisson point process on $\boldsymbol{U}$ with characteristic measure n and N_p and $\tilde{N}_p$ are defined in Chapter II, Section 3. A precise formulation is as follows. By a solution of the equation (9.2), we mean a right continuous process $X = (X(t))$ with left hand limits on $\boldsymbol{R}^d$ defined on a probability space $(\Omega, \mathscr{F}, P)$ with a reference family $(\mathscr{F}_t)$ such that X is $(\mathscr{F}_t)$-adapted and there exist an r-dimensional $(\mathscr{F}_t)$-Brownian motion $B = (B^k(t))$ and an $(\mathscr{F}_t)$-stationary Poisson point process p on $\boldsymbol{U}$ with characteristic measure n such that the equation (9.2) holds a.s.

Theorem 9.1. If $\sigma(x)$, $b(x)$ and $f(x, u)$ satisfy in addition to (9.1) the Lipschitz condition

$$\|\sigma(x) - \sigma(y)\|^2 + \|b(x) - b(y)\|^2 + \int_{U_0} \|f(x, u) - f(y, u)\|^2 n(du) \leq K|x - y|^2, \qquad x, y \in \boldsymbol{R}^d, \tag{9.3}$$

then for any given r-dimensional $(\mathscr{F}_t)$-Brownian motion $B = (B^k(t))$, any $(\mathscr{F}_t)$-stationary Poisson point process p with characteristic measure n and any $\boldsymbol{R}^d$-valued $\mathscr{F}_0$-measurable random variable ξ defined on a probability space with a reference family $(\mathscr{F}_t)$, there exists a unique d-dimensional $(\mathscr{F}_t)$-adapted right-continuous process $X(t)$ with left-hand limits which satisfies equation (9.2) and such that $X(0) = \xi$ a.s.

Proof. Suppose $B = (B^k(t))$, p and ξ are given as above. Let $\boldsymbol{D} = \{s \in \boldsymbol{D}_p;\, p(s) \in \boldsymbol{U} \setminus U_0\}$. Since $n(\boldsymbol{U} \setminus U_0) < \infty$, $\boldsymbol{D}$ is a discrete set in $(0, \infty)$ a.s. Let $\sigma_1 < \sigma_2 < \cdots < \sigma_n < \cdots$ be the enumeration of all elements in $\boldsymbol{D}$. It is easy to see that σ_n is an $(\mathscr{F}_t)$-stopping time for each n and $\lim_{n \uparrow \infty} \sigma_n = \infty$ a.s.* First we shall show the existence and uniqueness of solutions in the time interval $[0, \sigma_1]$. For this, consider the following equation

$$\begin{aligned} Y^i(t) = \xi^i &+ \sum_{k=1}^{r} \int_0^t \sigma_k^i(Y(s))dB^k(s) + \int_0^t b^i(Y(s))ds \\ &+ \int_0^{t+} \int_U f^i(Y(s-), u) I_{U_0}(u) \tilde{N}_p(dsdu), \quad i = 1, 2, \ldots, d. \end{aligned} \tag{9.4}$$

Noting the following general formula

$$E[\{\int_0^{t+} \int_U g(Y(s-), u) I_{U_0}(u) \tilde{N}_p(dsdu)\}^2]$$

* We disregard the trivial case of $n(\boldsymbol{U} \setminus U_0) = 0$.

$$= \int_0^t ds \int_{U_0} E[g^2(Y(s),u)]n(du)$$

and the assumption (9.3), we can show by the same argument as in the proof of Theorem 3.1 that the solution $Y(t)$ of (9.4) exists uniquely and is constructed as follows: if $\xi = y$, a constant point in $\boldsymbol{R}^d$, the solution is constructed by the successive approximation as in the proof of Theorem 3.1. The solution is a measurable function of y, B and p in the obvious sense. The solution for a general initial value ξ is obtained by replacing the variable y of this function with ξ. Set

$$X_1(t) = \begin{cases} Y(t), & 0 \leq t < \sigma_1, \\ Y(\sigma_1-) + f(Y(\sigma_1-), p(\sigma_1)), & t = \sigma_1. \end{cases}$$

The process $\{X_1(t)\}_{t\in[0,\sigma_1]}$ is clearly the unique solution of (9.2) in the time interval $[0, \sigma_1]$. Next, set $\tilde{\xi} = X_1(\sigma_1)$, $\tilde{B} = (\tilde{B}^k(t))$ where $\tilde{B}^k(t) = B^k(t+\sigma_1) - B^k(\sigma_1)$, and $\tilde{p} = (\tilde{p}(t))$ where $\boldsymbol{D}_{\tilde{p}} = \{s;\ s+\sigma_1 \in \boldsymbol{D}_p\}$ and $\tilde{p}(s) = p(s+\sigma_1)$. We can determine the process $\tilde{X}_2(t)$ on $[0, \tilde{\sigma}_1]$ with respect to $\tilde{\xi}$, $\tilde{B}$ and $\tilde{p}$ in the same way as $X_1(t)$. Clearly $\tilde{\sigma}_1$, defined with respect to $\tilde{p}$, coincides with $\sigma_2 - \sigma_1$. Define $\{X(t)\}_{t\in[0,\sigma_2]}$ by

$$X(t) = \begin{cases} X_1(t), & t \in [0, \sigma_1], \\ \tilde{X}_2(t - \sigma_1), & t \in [\sigma_1, \sigma_2]. \end{cases}$$

It is easy to see that $\{X(t)\}_{t\in[0,\sigma_2]}$ is the unique solution of (9.2) in the time interval $[0, \sigma_2]$. Continuing this process successively, $X(t)$ is determined uniquely in the time interval $[0, \sigma_n]$ for every n and hence $X(t)$ is determined globally.

We have actually proved, under the assumption (9.3), the unique existence of the strong solution of (9.2). The uniqueness in law is obvious from this stronger result.

CHAPTER V

Diffusion Processes on Manifolds

1. Stochastic differential equations on manifolds

Let M be a d-dimensional C^∞-manifold i.e., M is a Hausdorff topological space with an open covering $\{U_\alpha\}_{\alpha\in\Lambda}$ of M, each U_α provided with a homeomorphism ϕ_α with an open subset $\phi_\alpha(U_\alpha)$ of $\boldsymbol{R}^d$ such that, if $U_\alpha\cap U_\beta\neq\phi$ the function $\phi_\beta\circ\phi_\alpha^{-1}$ from $\phi_\alpha(U_\alpha\cap U_\beta)$ into $\phi_\beta(U_\alpha\cap U_\beta)$ is a C^∞-function. U_α is called a *coordinate neighborhood* and for $x\in U_\alpha$, $\phi_\alpha(x)=(x^1,x^2,\ldots,x^d)\in\boldsymbol{R}^d$ is called a *local coordinate of* x. In this book we always assume that M is connected and σ-compact. It is well-known then that M is paracompact and has a countable open base.*

A function $f(x)$ defined on an open subset D of M is called C^∞ (or smooth) if it is C^∞ as a function of the local coordinate, i.e., $f\circ\phi_\alpha^{-1}$ is C^∞ on $\phi_\alpha(U_\alpha\cap D)$ for every α. Let $F(M)$ be the totality of all real valued C^∞-functions on M and $F_0(M)$ be the subclass of $F(M)$ consisting of all functions in $F(M)$ with compact support. $F(M)$ and $F_0(M)$ are algebras over the field of real numbers $\boldsymbol{R}$ with the usual rules of $f+g$, fg and λf ($f,g\in F(M)$ or $F_0(M)$, $\lambda\in\boldsymbol{R}$).

Let $x\in M$. By a *tangent vector at* x we mean a linear mapping V of $F(M)$ into $\boldsymbol{R}$ such that

$$V(fg)=V(f)g(x)+f(x)V(g).$$

The set of all tangent vectors at x forms a linear space $T_x(M)$, called the *tangent space at* x, with the rules

$$(V+V')(f)=V(f)+V'(f) \qquad \text{and} \qquad (\lambda V)(f)=\lambda V(f).$$

* Cf. [112].

Let $(x^1,x^2,\dots,x^d)$ be a local coordinate in a coordinate neighborhood U of x. Every $f\in F(M)$ is expressed on U as a C^∞-function $f(x^1,x^2,\dots,x^d)$. Then $f\longmapsto \left(\frac{\partial f}{\partial x^i}\right)(x)$ is a tangent vector at x for every $i=1,2,\dots,d$. This is denoted by $\left(\frac{\partial}{\partial x^i}\right)_x$. It is easy to see that $\left\{\left(\frac{\partial}{\partial x^i}\right)_x\right\}_{i=1,2,\dots,d}$ forms a base for $T_x(M)$.

By a *vector field* we mean a mapping $V: x\in M\longmapsto V(x)\in T_x(M)$. V is called a C^∞*-vector field* if for every $f\in F(M)$, $(Vf)(x):=V(x)f$ is a C^∞-function. Thus V is a C^∞-vector field if and only if V is a linear mapping of $F(M)$ into $F(M)$ (or $F_0(M)$ into $F_0(M)$) such that $V(fg)=V(f)g+fV(g)$. In this book we only consider C^∞-vector fields unless otherwise stated. The totality of C^∞-vector fields is denoted by $\mathfrak{X}(M)$.

Let $A_0,A_1,\dots,A_r\in\mathfrak{X}(M)$. We consider the following stochastic differential equation given in an intuitive form

$$dX(t) = A_\alpha(X(t))\circ dB^\alpha(t) + A_0(X(t))dt.^{*1} \tag{1.1}$$

A precise formulation is as follows. Let $\hat{M}=M$ or $M\cup\{\Delta\}$ (= the one-point compactification of M) accordingly as M is compact or non-compact. Let $\hat{W}(M)$ be the path space defined by

$$\hat{W}(M) = \{w;\ w \text{ is a continuous mapping } [0,\infty)\longrightarrow\hat{M} \text{ such that } w(0)\in M \text{ and if } w(t)=\Delta \text{ then } w(t')=\Delta \text{ for all } t'\ge t\}$$

and let $\mathscr{B}(\hat{W}(M))$ be the σ-field generated by the Borel cylinder sets. The *explosion time* $e(w)$ is defined by

$$e(w) = \inf\{t;\ w(t)=\Delta\}.$$

Definition 1.1. A solution $X=(X(t))$ of (1.1) is any $(\mathscr{F}_t)$-adapted $\hat{W}(M)$-valued random variable (i.e., a continuous process on $\hat{M}$ with Δ as a trap) defined on a probability space with a reference family $(\mathscr{F}_t)$ and an r-dimensional $(\mathscr{F}_t)$-Brownian motion $B=(B(t))$ with $B(0)=0$ such that the following is satisfied: for every $f\in F_0(M)$,*2

$$f(X(t)) - f(X(0)) = \int_0^t (A_\alpha f)(X(s))\circ dB^\alpha(s) + \int_0^t (A_0 f)(X(s))ds, \tag{1.2}$$

*1 According to the usual convention, the summation sign is abbreviated for repeated indices appearing once at the top and once at the bottom.

*2 We define $f(\Delta)=0$ for every $f\in F_0(M)$.

where the first term on the right-hand side is understood in the sense of the Fisk-Stratonovich integral defined in Chapter III, Section 1.

The results of Chapter IV applied to each coordinate neighborhood enable us to obtain a unique strong solution of (1.1). Namely we have the following result.

Theorem 1.1. There exists a function $F: M\times W_o^r \longrightarrow \hat{W}(M)$ which is $\bigcap_{\mu} \overline{\mathscr{B}(M)\times\mathscr{B}_t(W_o^r)}^{\mu\times P^W}/\mathscr{B}_t(\hat{W}(M))$-measurable[*1] for every $t\geq 0$ such that

(i) for every solution $X=(X(t))$ with respect to the Brownian motion $B=(B(t))$, it holds that

$$X=F(X(0),B) \qquad \text{a.s.,}$$

and

(ii) for every r-dimensional $(\mathscr{F}_t)$-Brownian motion $B=(B(t))$ with $B(0)=0$ defined on a probability space with a reference family $(\mathscr{F}_t)$ and an M-valued $(\mathscr{F}_0)$-measurable random variable ξ, $X=F(\xi,B)$ is a solution of (1.1) with $X(0)=\xi$ a.s.

Proof. Take a coordinate neighborhood [*2] U and express $A_\alpha=\sigma_\alpha^i(x)\dfrac{\partial}{\partial x^i}$, $\alpha=0,1,\ldots,r$, under the local coordinates $(x^1,x^2,\ldots,x^d)$ in U. Extend the functions $\sigma_\alpha^i(x)$ to bounded smooth functions on $\boldsymbol{R}^d$ and then consider the following stochastic differential equation

$$\text{(1.3)} \qquad \begin{cases} dX_t^i=\sigma_\alpha^i(X_t)\circ dB^\alpha(t)+\sigma_0^i(X_t)dt \\ X_0^i=x^i, \qquad\qquad i=1,2,\ldots,d. \end{cases}$$

Note that (1.3) is equivalent to

$$\text{(1.3)}' \qquad \begin{cases} dX_t^i=\sigma_\alpha^i(X_t)dB^\alpha(t)+\bar{\sigma}_0^i(X_t)dt \\ X_0^i=x^i, \qquad\qquad i=1,2,\ldots,d, \end{cases}$$

where

[*1] Here μ runs over all probabilities on $(M,\mathscr{B}(M))$. W_0^r, P^W, $\mathscr{B}_t(W_0^r)$ have the same meaning as in Chapter IV: W_0^r is the space of continuous paths in $\boldsymbol{R}^r$ starting at 0, P^W is the Wiener measure on W_0^r and $\mathscr{B}_t(W_0^r)$ is the σ-field generated by the Borel cylinder sets up to time t. $\mathscr{B}_t(\hat{W}(M))$ is defined similarly.

[*2] Here we choose a relatively compact coordinate neighbourhood. Such a remark will sometimes be necessary in the future but usually we do not mention it.

$$(1.4)\qquad \tilde{\sigma}_0^i(x) = \sigma_0^i(x) + \frac{1}{2}\sum_{\alpha=1}^{r}\Big(\frac{\partial}{\partial x^k}\sigma_\alpha^i(x)\Big)\sigma_\alpha^k(x).$$

It follows from the results of Chapter IV that the unique strong solution of (1.3) exists; i.e., there exists a mapping F: $\boldsymbol{R}^d \times \boldsymbol{W}_0^r \longrightarrow \hat{\boldsymbol{W}}^d$ with the properties as in Theorem 1.1 such that any solution X of (1.3) is given as $X=F(x,B)$ where $x=(x^1,x^2,\ldots,x^d)$. $F(x,w) \equiv (X(t,x,w))$ itself is the solution of (1.3) with respect to the canonical realization $w=(w(t))$ of Brownian motion on $\{\boldsymbol{W}_0^r, P^W\}$ with the reference family $\{\mathscr{F}_t^0\}$ defined by $\mathscr{F}_t^0 = \mathscr{B}_t^{P^W}(\boldsymbol{W}_0^r)$, $t \geq 0$. Take $x = (x^1, x^2, \ldots, x^d) \in U$ and set $\tau_U(w) = \inf\{t; X(t,x,w) \notin U\}$. Define $X_U = (X_U(t,x,w))$ by

$$X_U(t,x,w) = X(t\wedge\tau_U(w),x,w).$$

For each $x \in M$ and coordinate neighborhood U containing x we construct the local solution X_U as above. It is easy to see that if U and $\tilde{U}$ are two coordinate neighborhoods and $x \in U \cap \tilde{U}$, then $X_U(t,x,w) = X_{\tilde{U}}(t,x,w)$ for all $t \leq \tau_U(w)\wedge\tau_{\tilde{U}}(w)$. Indeed, if $A_\alpha = \tilde{\sigma}_\alpha^i(\tilde{x})\dfrac{\partial}{\partial\tilde{x}^i}$ under the local coordinate $\tilde{x} = (\tilde{x}^1,\tilde{x}^2,\ldots,\tilde{x}^d)$ in $\tilde{U}$, then we have

$$(1.5)\qquad \tilde{\sigma}_\alpha^i(\tilde{x}(x)) = \sigma_\alpha^k(x)\frac{\partial\tilde{x}^i}{\partial x^k}$$

and the equation for $X_{\tilde{U}}(t,x,w)$ is of the form

$$(1.6)\qquad d\tilde{X}_t^i = \tilde{\sigma}_\alpha^i(\tilde{X}_t)\circ dw^\alpha(t) + \tilde{\sigma}_0^i(\tilde{X}_t)dt.$$

On the other hand, it follows from the chain rule (Theorem III-1.3) that the process X_U under the local coordinate $\tilde{x}$ in $\tilde{U}$, i.e., $\tilde{x}(X_U(t,x,w)) = (\tilde{x}_t^i)$, satisfies

$$\begin{aligned} d\tilde{x}_t^i &= \frac{\partial\tilde{x}^i}{\partial x^k}(X(t))\circ dX^k(t) \\ &= \frac{\partial\tilde{x}^i}{\partial x^k}(X(t))\sigma_\alpha^k(X(t))\circ dw^\alpha(t) + \frac{\partial\tilde{x}^i}{\partial x^k}(X(t))\sigma_0^k(X(t))dt \\ &= \tilde{\sigma}_\alpha^i(\tilde{x}_t)\circ dw^\alpha(t) + \tilde{\sigma}_0^i(\tilde{x}_t)dt. \end{aligned}$$

Thus $\tilde{x}_t = \tilde{x}(X_U(t,x,w))$ satisfies the same equation (1.6) as $\tilde{X}_t = X_{\tilde{U}}(t,x,w)$, and so by the uniqueness of solutions, we conclude that $X_U(t,x,w) = X_{\tilde{U}}(t,x,w)$ for all $t \leq \tau_U(w)\wedge\tau_{\tilde{U}}(w)$.

Now we will patch together the local solutions into a global solution.

We first choose two systems of coordinate neighborhoods $\{U_\alpha\}$ and $\{V_\alpha\}$ which form locally finite coverings of M such that $\bar{U}_\alpha \subset V_\alpha$ and and $\bar{V}_\alpha$ is also contained in another coordinate neighborhood W_α. Let $x \in M$ and $U_1, U_2, \cdots, U_l$ be the totality of coordinate neighborhoods in the system $\{U_\alpha\}$ containing x. Then the process $\hat{X}(t,x,w) = X_{V_i}(t,x,w)$ is well-defined for $t \in [0, \hat{t}_x(w)]$ where $\hat{t}_x(w) = \max_{1\le i\le l} \{\tau_{V_i}(w)\}$. Here we set $\hat{X}(\infty,x,w) = \Delta$, for covenience. Define $\tau_1(w) = \hat{t}_x(w)$ and $X(t) = \hat{X}(t)$ for $t \in [0, \tau_1]$. Inductively, if $\tau_n(w)$ and $X(t) = (X(t,x,w))$ are defined for $t \in [0, \tau_n(w)]$, then on the set $\{w;\ \tau_n(w) < \infty\}$, we define

$$x_n = X(\tau_n), \quad w_n = \theta_{\tau_n} w,^* \quad \tau_{n+1} = \tau_n + \hat{t}_{x_n}(w_n)$$

and $X(t) = \hat{X}(t - \tau_n, x_n, w_n)$ for $t \in [\tau_n, \tau_{n+1}]$. In this way, $X(t)$ is defined for $t \in [0, \tau_\infty)$ where $\tau_\infty = \lim_{n\to\infty} \tau_n$. We now show that

$$\text{(1.7)} \qquad \lim_{t\uparrow\tau_\infty} X(t) = \Delta \qquad \text{on the set } \{w;\ \tau_\infty(w) < \infty\}.$$

We can choose an increasing sequence $\{M_n\}$ of finite sums of $\{V_\alpha\}$ such that

$$\bigcup_{n=1}^{\infty} M_n = M \quad \text{and} \quad \bar{M}_n \subset M_{n+1} \qquad \text{for every} \quad n \ge 1.$$

Furthermore we may assume that if $\bar{V}_k \cap \partial M_n \neq \phi$, then $\bar{V}_k \subset M_{n+1}$. For each n we define

$$\begin{aligned}
&\sigma_1 = 0 && \tilde{\sigma}_1 = \inf\{t > \sigma_1;\ X(t) \in M_{n+1}^c\}\\
&\sigma_2 = \inf\{t > \tilde{\sigma}_1;\ X(t) \in M_n\} && \tilde{\sigma}_2 = \inf\{t > \sigma_2;\ X(t) \in M_{n+1}^c\}\\
&\sigma_3 = \inf\{t > \tilde{\sigma}_2;\ X(t) \in M_n\} && \tilde{\sigma}_3 = \inf\{t > \sigma_3;\ X(t) \in M_{n+1}^c\}\\
&\quad\vdots && \quad\vdots
\end{aligned}$$

It is sufficient for (1.7) to show that for every n, on the set $\{w; \tau_\infty(w) < \infty\}$, there exists an integer k such that

$$\text{(1.8)} \qquad \tilde{\sigma}_k(w) < \infty \quad \text{and} \quad \sigma_{k+1}(w) = \infty.$$

To show this, it suffices to prove that $\tau_\infty(w) = \infty$ on the set

* $\theta_t\colon W_0^r \longrightarrow W_0^r$ is defined as in Chapter IV: $(\theta_t w)(s) = w(t+s) - w(t)$.

$$\{w;\ \exists\ k \text{ such that } \sigma_k(w) < \infty \text{ and } \tilde{\sigma}_k(w) = \infty\}$$
$$\cup\ \{w;\ \sigma_k(w) < \infty \text{ for every } k\}.$$

First if $\sigma_k(w) < \infty$ for every k, then we can show

$$\sum_{k=1}^{\infty} \{\tilde{\sigma}_k(w) - \sigma_k(w)\} = \infty \tag{1.9}$$

by the same argument as in the proof of Lemma IV-2.1. This clearly implies that $\tau_\infty(w) = \infty$. Next, consider the case when there exists k such that $\sigma_k(w) < \infty$ and $\tilde{\sigma}_k(w) = \infty$. Then we have

$$X(t) \in M_{n+1} \qquad \text{for all} \quad t \geq \sigma_k. \tag{1.10}$$

We can now conclude by the same argument as in the proof of (1.9) that $\tau_\infty(w) = \infty$.

We set $X(t) = \Delta$ for $t \geq \tau_\infty$ on the set $\{w; \tau_\infty < \infty\}$. Thus we have defined $X(t) = (X(t,x,w))$ as a mapping

$$M \times W_0^r \ni (x, w) \longmapsto X = (X(t,x,w)) \in \hat{W}(M).$$

It is easy to see that it is a solution of (1.1). Indeed, it is obvious that for every $f \in F_0(M)$,

$$\begin{aligned} f(X(t \wedge \tau_1)) - f(x) &= \int_0^{t\wedge\tau_1} \frac{\partial f}{\partial x^i}(X(s))\sigma_\alpha^i(X(s))\circ dw^\alpha(s) \\ &\quad + \int_0^{t\wedge\tau_1} \frac{\partial f}{\partial x^i}(X(s))\sigma_0^i(X(s))ds \\ &= \int_0^{t\wedge\tau_1} (A_\alpha f)(X(s))\circ dw^\alpha(s) + \int_0^{t\wedge\tau_1} (A_0 f)(X(s))ds. \end{aligned}$$

Similarly, on the set $\{w; \tau_n(w) < \infty\}$,

$$\begin{aligned} & f(X(t \wedge \tau_{n+1})) - f(X(t \wedge \tau_n)) \\ &= \int_0^{(t - t\wedge\tau_n)\wedge \hat{t}_{x_n}(w_n)} (A_\alpha f)(\hat{X}(s,x_n,w_n))\circ dw_n^\alpha(s) \\ &\quad + \int_0^{(t - t\wedge\tau_n)\wedge \hat{t}_{x_n}(w_n)} (A_0 f)(\hat{X}(s,x_n,w_n))ds \\ &= \int_{t\wedge\tau_n}^{t\wedge\tau_{n+1}} (A_\alpha f)(X(s))\circ dw^\alpha(s) + \int_{t\wedge\tau_n}^{t\wedge\tau_{n+1}} (A_0 f)(X(s))ds. \end{aligned}$$

Summing up, we have

$$\begin{aligned} f(X(t)) - f(x) &= f(X(t\wedge\tau_\infty)) - f(x) \\ &= \int_0^{t\wedge\tau_\infty} (A_\alpha f)(X(s))\circ dw^\alpha(s) + \int_0^{t\wedge\tau_\infty} (A_0 f)(X(s))ds \\ &= \int_0^t (A_\alpha f)(X(s))\circ dw^\alpha(s) + \int_0^t (A_0 f)(X(s))ds. \end{aligned}$$

The uniqueness of solutions is also easily proved.

Remark 1.1. We can also construct the solution of (1.1) more directly by appealing to Whitney's imbedding theorem ([176]). M is imbedded into $\boldsymbol{R}^{2d+1}$ as a closed submanifold of $\boldsymbol{R}^{2d+1}$ and the vector fields $A_\alpha(x)$ are restrictions on M of smooth vector fields $\tilde{A}_\alpha(x)$ on $\boldsymbol{R}^{2d+1}$. The stochastic differential equation corresponding to $\tilde{A}_\alpha(x)$ is defined globally in the Euclidean coordinate system and the solution is constructed as in Chapter IV. If the initial value is on M, then it is easy to see that the solution remains on M. Thus this solution actually defines the solution of (1.1). A construction of the Brownian motion on a sphere given in Chapter III, Section 2 is a typical example of the method of imbedding.

Theorem 1.2. Let P_x be the probability law on $\hat{\boldsymbol{W}}(M)$ of the solution $X=(X(t))$ of (1.1) with the initial value $X(0)=x$. Then $\{P_x\}_{x\in M}$ is a diffusion generated by the second order differential operator

$$(7.11)\qquad Af = \frac{1}{2}\sum_{\alpha=1}^{r} A_\alpha(A_\alpha f) + A_0 f, \qquad f\in F_0(M).$$

Proof. Using the uniqueness of solutions we can show that $\{P_x\}$ has the strong Markov property. Actually, we can prove the following stronger result: for any $(\mathscr{F}_t^0)$-stopping time $\sigma(w)$, we have $X(t+\sigma(w),x,w) = X(t,X(\sigma(w),x,w),\theta_\sigma w)$ for all $t\geq 0$ and almost all w such that $\sigma(w)<\infty$.
Since for $f\in F_0(M)$,

$$\begin{aligned} df(X(t)) &= (A_\alpha f)(X(t))\circ dw^\alpha(t) + (A_0 f)(X(t))dt \\ &= (A_\alpha f)(X(t))dw^\alpha(t) + (A_0 f)(X(t))dt \\ &\quad + \frac{1}{2} d(A_\alpha f)(X(t))\cdot dw^\alpha(t) \end{aligned}$$

and

$$d(A_\beta f)(X(t)) = A_\alpha(A_\beta f)(X(t))\circ dw^\alpha(t) + (A_0 A_\beta f)(X(t))dt,$$

we have

$$d(A_\alpha f)(X(t))\cdot dw^\alpha(t) = \sum_{\alpha=1}^{r} A_\alpha(A_\alpha f)(X(t))dt.$$

Consequently,

$$df(X(t)) = (A_\alpha f)(X(t))dw^\alpha(t) + (Af)(X(t))dt,$$

where (Af) is defined by(7.11). This proves that $X=(X(t))$ is an A-diffusion.

Remark 1.2. By a similar argument as in Chapter IV, we can deduce the uniqueness of the A-diffusion $\{P_x\}_{x\in M}$ on $\hat{W}(M)$ from the uniqueness of solutions of (1.1).

2. Flow of diffeomorphisms

Given vector fields $A_\alpha \in \mathfrak{X}(M)$, $\alpha = 0, 1, \ldots, r$, we constructed in Section 1 a mapping $X = (X(t,x,w))$: $M \times W_0^r \ni (x,w) \longmapsto X(\cdot,x,w) \in \hat{W}(M)$. This may also be regarded as a mapping: $[0,\infty)\times M \times W_0^r \ni (t, x, w) \longmapsto X(t,x,w) \in \hat{M}$. The main purpose of this section is to show that the mapping $x \in M \longmapsto X(t,x,w) \in \hat{M}$ is a local diffeomorphism of M for each fixed $t \geq 0$ and for almost all w such that $X(t,x,w) \in M$.

First we discuss the case of $M = \boldsymbol{R}^d$. Let $\sigma(x) = (\sigma_k^i(x)) \in \boldsymbol{R}^d \otimes \boldsymbol{R}^r$ and $b(x) = (b^i(x)) \in \boldsymbol{R}^d$ be given such that they are smooth functions (i.e., C^∞-functions) on $\boldsymbol{R}^d$, $\|\sigma(x)\| + |b(x)| \leq K(1+|x|)$ for some positive constant K and all the derivatives of σ_k^i and b^i are bounded. Let $X = (X(t, x, w))$ be the unique solution of

$$\text{(2.1)} \qquad \begin{cases} dX_t^i = \sigma_\alpha^i(X_t)dw^\alpha(t) + b^i(X_t)dt \\ X_0 = x, \qquad\qquad i = 1, 2, \ldots, d, \end{cases}$$

defined on the space (W_0^r, P^W) with the reference family $(\mathscr{F}_t^0)$. As we saw in Chapter IV, the solution $X=(X(t,x,w))$ exists uniquely and $E\{|X(t)|^2\} < \infty$ for all $t \geq 0$.* This result will be strengthened below as $E\{|X(t)|^p\} < \infty$ for all $p \geq 1$. In particular, $e = \infty$ a.s.

First of all we shall prove a lemma on an approximation of solutions

* E stands for the expectation on the space (W_0^r, P^W).

by polygonal paths (cf. [111]). Let

$$(2.2)\qquad \phi_n(s) = k/2^n \qquad \text{if} \quad s \in [k/2^n, (k+1)/2^n), \quad k = 0, 1, \ldots .$$

Lemma 2.1. Let $A(x) = (A^i_\alpha(x)) \in \boldsymbol{R}^m \otimes \boldsymbol{R}^r$ and $\beta(x) = (\beta^i(x)) \in \boldsymbol{R}^m$ be given and satisfy the following conditions;

(i) there exists a positive constant K such that

$$\|A(x)\| + |\beta(x)| \leq K(1+|x|) \qquad \text{for every } x \in \boldsymbol{R}^m,$$

(ii) for every $N > 0$, there exists a positive constant K_N such that

$$\|A(x) - A(y)\| + |\beta(x) - \beta(y)| \leq K_N |x-y|$$

for every $x,y \in \boldsymbol{R}^m$ such that $|x| \leq N$ and $|y| \leq N$. Let $\alpha(t)$ and $\alpha_n(t)$, $n = 1, 2, \ldots$, be continuous $\boldsymbol{R}^m$-valued $(\mathscr{F}_t^0)$-adapted *1 processes such that for some $p \geq 2$

$$(2.3)\qquad \sup_n E\{\sup_{0\leq t\leq T} |\alpha_n(t)|^{p+1}\} < \infty \quad \text{and} \quad E\{\sup_{0\leq t\leq T} |\alpha_n(t) - \alpha(t)|^p\} \longrightarrow 0$$

as $n \longrightarrow \infty$.*2 Let $Y(t)$ and $Y_n(t)$, $n = 1, 2, \ldots$, be continuous $\boldsymbol{R}^m$-valued $(\mathscr{F}_t^0)$-adapted processes such that

$$(2.4)\qquad Y^i(t) = \alpha^i(t) + \int_0^t A^i_\alpha(Y(s))dw^\alpha(s) + \int_0^t \beta^i(Y(s))ds \quad {}^{*3}$$

and

$$(2.4)'\qquad Y^i_n(t) = \alpha^i_n(t) + \int_0^t A^i_\alpha(Y_n(\phi_n(s)))dw^\alpha(s) + \int_0^t \beta^i(Y_n(\phi_n(s)))ds$$

$$\text{for } i = 1, 2, \ldots, m, \quad n = 1, 2, \ldots .$$

Then, for every $T > 0$,

$$(2.5)\qquad E\{\sup_{0\leq t\leq T} |Y_n(t) - Y(t)|^p\} \longrightarrow 0 \qquad \text{as} \quad n \longrightarrow \infty.$$

---

*1 As in Section 1, we consider the Wiener space $(W_0^r, \mathscr{B}(W_0^r), P^W)$ and $\mathscr{F}_t^0 = \mathscr{B}_t^{P^W}(W_0^r)$, $t \geq 0$.

*2 E stands for the expectation on the Wiener space. $T > 0$ is any fixed constant.

*3 $w(t) = (w^\alpha(t))$ is the canonical realization of r-dimensional Brownian motion on the space (W_0^r, P^W).

Proof. Let $T>0$ be arbitrary but fixed. First we remark that (2.3) implies

$$E\{\sup_{0\le t\le T}|\alpha(t)|^{p+1}\}<\infty \tag{2.6}$$

and

$$E\{\sup_{0\le t\le T}|\alpha_n(t)-\alpha_n(\phi_n(t))|^p\}\longrightarrow 0 \quad \text{as} \quad n\longrightarrow\infty. \tag{2.7}$$

(2.6) follows easily from Fatou's lemma and (2.7) follows from *

$$\begin{aligned}E\{\sup_{0\le t\le T}&|\alpha_n(t)-\alpha_n(\phi_n(t))|^p\}\\ &\le K_1[E\{\sup_{0\le t\le T}|\alpha(t)-\alpha(\phi_n(t))|^p\}+E\{\sup_{0\le t\le T}|\alpha_n(t)-\alpha(t)|^p\}],\end{aligned}$$

the right-hand side tending to zero as $n\longrightarrow\infty$ by the dominated convergence theorem and (2.3).

In the following we assume for simplicity of notation that $m=1$ and $r=1$. We shall show that

$$\sup_n E\{\sup_{0\le t\le T}|Y_n(t)|^{p+1}\}<\infty. \tag{2.8}$$

From (2.4)′ we have

$$\begin{aligned}E\{\sup_{0\le s\le t}|Y_n(s)|^{p+1}\}\le K_2[&E\{\sup_{0\le s\le t}|\alpha_n(s)|^{p+1}\}\\ &+E\{\sup_{0\le s\le t}|\int_0^s A(Y_n(\phi_n(u)))dw(u)|^{p+1}\}\\ &+E\{\sup_{0\le s\le t}|\int_0^s \beta(Y_n(\phi_n(u)))du|^{p+1}\}]\end{aligned}$$

for $t\in[0,T]$. By Theorem III-3.1 and Hölder's inequality,

$$\begin{aligned}E\{\sup_{0\le s\le t}&|\int_0^s A(Y_n(\phi_n(u)))dw(u)|^{p+1}\}\\ &\le K_3E\{|\int_0^t A(Y_n(\phi_n(s)))^2ds|^{(p+1)/2}\}\end{aligned}$$

* In the following K_1, K_2, ... are positive constants independent of n (which may depend on T).

$$\leq K_4 \int_0^t E\{|A(Y_n(\phi_n(s)))|^{p+1}\} ds$$

$$\leq K_5 \int_0^t [1 + E\{|Y_n(\phi_n(s))|^{p+1}\}]ds$$

and

$$E\{\sup_{0\leq s\leq t} |\int_0^s \beta(Y_n(\phi_n(u)))du|^{p+1}\}$$

$$\leq K_6 \int_0^t E\{|\beta(Y_n(\phi_n(s)))|^{p+1}\} ds$$

$$\leq K_7 \int_0^t [1 + E\{|Y_n(\phi_n(s))|^{p+1}\}]ds.$$

Consequently

$$E\{\sup_{0\leq s\leq t} |Y_n(s)|^{p+1}\} \leq K_8(1 + \int_0^t E\{|Y_n(\phi_n(s))|^{p+1}\} ds). \tag{2.9}$$

Then obviously

$$E\{|Y_n(\phi_n(t))|^{p+1}\} \leq K_8(1 + \int_0^t E\{|Y_n(\phi_n(s))|^{p+1}\} ds)$$

and we can deduce from this (using here a similar truncation argument as in the proof of Theorem III-3.1 or Theorem IV-2.4) that

$$E\{|Y_n(\phi_n(t))|^{p+1}\} \leq K_8 \exp\{K_8 t\}$$

Substituting this inequality into (2.9), we obtain (2.8). Similarly we can prove

$$E\{\sup_{0\leq t\leq T} |Y(t)|^{p+1}\} < \infty. \tag{2.10}$$

Next we set

$$\sigma_n^N = \inf\{t;\ |Y_n(t)| \geq N\}$$

and

$$\sigma^N = \inf\{t;\ |Y(t)| \geq N\}$$

for every $N > 0$. Then, for $t \in [0, T]$,

$$E\{\sup_{0\le s\le t\wedge\sigma_n^N\wedge\sigma^N} |Y_n(s) - Y(s)|^p\}$$
$$\le K_9[E\{\sup_{0\le s\le t} |\alpha_n(s) - \alpha(s)|^p\}$$
$$+ E\{\sup_{0\le s\le t\wedge\sigma_n^N\wedge\sigma^N} |\int_0^s \{A(Y_n(\phi_n(u))) - A(Y(u))\} dw(u)|^p\}$$
$$+ E\{\sup_{0\le s\le t\wedge\sigma_n^N\wedge\sigma^N} |\int_0^s \{\beta(Y_n(\phi_n(u))) - \beta(Y(u))\} du|^p\}]$$

and by estimating similarly as above, this is dominated by *

$$K_{10}[E\{\sup_{0\le s\le T} |\alpha_n(s) - \alpha(s)|^p\}$$
$$+ E\{\int_0^{t\wedge\sigma_n^N\wedge\sigma^N} |A(Y_n(\phi_n(s))) - A(Y(s))|^p ds\}$$
$$+ E\{\int_0^{t\wedge\sigma_n^N\wedge\sigma^N} |\beta(Y_n(\phi_n(s))) - \beta(Y(s))|^p ds\}]$$
$$\le o(1) + L_N E\{\int_0^{t\wedge\sigma_n^N\wedge\sigma^N} |Y_n(\phi_n(s)) - Y(s)|^p ds\}$$
$$\le o(1) + L_N \int_0^t E\{|Y_n(\phi_n(s\wedge\sigma_n^N\wedge\sigma^N)) - Y(s\wedge\sigma_n^N\wedge\sigma^N)|^p\} ds.$$

By writing $s' = s\wedge\sigma_n^N\wedge\sigma^N$ for simplicity we have

$$E\{|Y_n(\phi_n(s')) - Y(s')|^p\}$$
$$\le K_{11}(E\{|Y_n(\phi_n(s')) - Y_n(s')|^p\} + E\{|Y_n(s') - Y(s')|^p\})$$

and

$$E\{|Y_n(\phi_n(s')) - Y_n(s')|^p\}$$
$$\le K_{12}[E\{|\alpha_n(\phi_n(s')) - \alpha_n(s')|^p\}$$
$$+ E\{|A(Y_n(\phi_n(s')))(w(s') - w(\phi_n(s')))|^p\}$$
$$+ E\{|\beta(Y_n(\phi_n(s')))(s' - \phi_n(s'))|^p\}]$$
$$= o(1)$$

by (2.7). Consequently

* L_N and L'_N are positive constants which may depend on N. $o(1)$ denotes quantities which tend to zero as $n \longrightarrow \infty$ uniformly in $t \in [0, T]$.

$$E\{\sup_{0\leq s\leq t\wedge\sigma_n^N\wedge\sigma^N}|Y_n(s)-Y(s)|^p\}$$

$$\leq o(1)+L'_N\int_0^t E\{|Y_n(s\wedge\sigma_n^N\wedge\sigma^N)-Y(s\wedge\sigma_n^N\wedge\sigma^N)|^p\}\,ds.$$

We can conclude from this that

$$E\{\sup_{0\leq s\leq T\wedge\sigma_n^N\wedge\sigma^N}|Y_n(s)-Y(s)|^p\}\longrightarrow 0 \quad\text{as}\quad n\longrightarrow\infty \tag{2.11}$$

for every $N>0$. Finally

$$\begin{aligned}
E\{\sup_{0\leq s\leq T}|Y_n(s)-Y(s)|^p\}
&\leq E\{\sup_{0\leq s\leq T\wedge\sigma_n^N\wedge\sigma^N}|Y_n(s)-Y(s)|^p\}\\
&\quad+E\{\sup_{0\leq s\leq T}(|Y_n(s)|+|Y(s)|)^p:\ \sigma_n^N\leq T\}\\
&\quad+E\{\sup_{0\leq s\leq T}(|Y_n(s)|+|Y(s)|)^p:\ \sigma^N\leq T\}\\
&\leq E\{\sup_{0\leq s\leq T\wedge\sigma_n^N\wedge\sigma^N}|Y_n(s)-Y(s)|^p\}\\
&\quad+2E\{\sup_{0\leq s\leq T}(|Y_n(s)|+|Y(s)|)^p:\sup_{0\leq s\leq T}(|Y_n(s)|+|Y(s)|)\geq N\}\\
&\leq E\{\sup_{0\leq s\leq T\wedge\sigma_n^N\wedge\sigma^N}|Y_n(s)-Y(s)|^p\}\\
&\qquad+\frac{2}{N}E\{\sup_{0\leq s\leq T}(|Y_n(s)|+|Y(s)|)^{p+1}\}.
\end{aligned}$$

Now we can easily deduce from (2.8), (2.10) and (2.11) that

$$E\{\sup_{0\leq s\leq T}|Y_n(s)-Y(s)|^p\}\longrightarrow 0 \qquad\text{as}\quad n\longrightarrow\infty.$$

q.e.d.

For given $\sigma(x)$ and $b(x)$ as above, let $X_n(t)=(X_n^i(t,x,w))$ be the solution, on the Wiener space $(W_0^r,\ P^W)$ with respect to the canonical realization $w(t)=(w^\alpha(t))$ of Brownian motion, of the equation

$$X_n(t)=x+\int_0^t\sigma_\alpha(X_n(\phi_n(s)))\,dw^\alpha(s)+\int_0^t b(X_n(\phi_n(s)))ds, \tag{2.12}$$

in component form,

$$X_n^i(t)=x^i+\int_0^t\sigma_\alpha^i(X_n(\phi_n(s)))dw^\alpha(s)+\int_0^t b^i(X_n(\phi_n(s)))ds, \tag{2.12$'$}$$

$$i=1,2,\ldots,d.$$

$X_n(t)$ is uniquely determined. Indeed, $X_n(0)=x$ and if $X_n(t)$ is determined for $t\in[0,(k-1)/2^n]$, then

$$X_n(t) = X_n((k-1)/2^n) + \sigma_\alpha(X_n((k-1)/2^n))(w^\alpha(t)-w^\alpha((k-1)/2^n)) \\ +b(X_n((k-1)/2^n))(t-(k-1)/2^n)$$

for $t\in[(k-1)/2^n, k/2^n]$. It is also clear from this that $X_n(t)$ is expressed as

$$X_n(t) = F(x,w(1/2^n),w(2/2^n), \dots ,w([2^n t]/2^n),w(t))$$

for some C^∞-function F: $\boldsymbol{R}^d\times(\boldsymbol{R}^r)^{[2^n t]+1} \longmapsto \boldsymbol{R}^d$. In particular, $x\longmapsto X_n(t,x,w)$ is C^∞ for every $t\geq 0$ and $w\in \boldsymbol{W}_0^r$. Let $D^\alpha = \dfrac{\partial^{|\alpha|}}{\partial x_1^{\alpha_1}\cdots\partial x_d^{\alpha_d}}$ for $\alpha=(\alpha_1,\alpha_2,\dots,\alpha_d)$, $|\alpha|=\alpha_1+\alpha_2+\cdots+\alpha_d$ and set

$$Y^i_{\alpha,(n)}(t,x,w) = D^\alpha X^i_n(t,x,w).$$

Proposition 2.1. For each $x\in\boldsymbol{R}^d$, $1\leq i\leq d$, and each α, there exists a unique process $Y^i_\alpha(t)=(Y^i_\alpha(t,x,w))$ such that

$$E\{\sup_{0\leq t\leq T}|Y^i_{\alpha,(n)}(t) - Y^i_\alpha(t)|^p\} \longrightarrow 0 \quad \text{as} \quad n\longrightarrow\infty \tag{2.13}$$

for all $T>0$ and $p\geq 1$. Furthermore, the convergence (2.13) is uniform in x on each compact set in $\boldsymbol{R}^d$.

Proof. First consider the case of $|\alpha|=1$. If we set $Y^i_{j,(n)}(t,x,w) = \dfrac{\partial}{\partial x^j}X^i_n(t,x,w)$, $Y_{(n)}(t)=(Y^i_{j,(n)}(t,x,w))$ is determined by the equation (in matrix notation)

$$\begin{aligned} Y_{(n)}(t) &= I+\int_0^t \sigma'_\alpha(X_n(\phi_n(s)))Y_{(n)}(\phi_n(s))dw^\alpha(s) \\ &\quad+\int_0^t b'(X_n(\phi_n(s)))Y_{(n)}(\phi_n(s))ds, \end{aligned} \tag{2.14}$$

where $\sigma'_\alpha(x)^i_j = \dfrac{\partial}{\partial x^j}\sigma^i_\alpha(x)$, $b'(x)^i_j = \dfrac{\partial}{\partial x^j}b^i(x)$ and $I=(\delta^i_j)$. Applying Lemma 2.1 to the system of equations (2.12) and (2.14) combined together, we have that

$$E\{\sup_{t\in[0,T]}|X_n(t)-X(t)|^p\} + E\{\sup_{t\in[0,T]}\|Y_{(n)}(t)-Y(t)\|^p\}\longrightarrow 0 \quad \text{as} \quad n\longrightarrow\infty \tag{2.15}$$

for all $T > 0$ and $p \geq 1$, where $X(t)$ is the solution of (2.1) and $Y(t)$ $= (Y_j^i(t, x, w))$ is the solution of

$$(2.16) \qquad Y(t) = I + \int_0^t \sigma'_\alpha(X(s)) Y(s) dw^\alpha(s) + \int_0^t b'(X(s)) Y(s) ds.$$

As is immediately seen in the proof of Lemma 2.1, the convergence (2.15) is uniform in x on every compact set.

Next, set

$$Y_{j,k,(n)}^i(t) = \frac{\partial^2}{\partial x^j \partial x^k} X_n^i(t,x,w).$$

Then $Y_{j,k,(n)}^i(t) = (Y_{j,k,(n)}^i(t,x,w))$ is determined by the equation

$$
\begin{aligned}
Y_{j_1,j_2,(n)}^i(t) = &\int_0^t \sigma'_\alpha(X_n(\phi_n(s)))_k^i Y_{j_1,j_2,(n)}^k(\phi_n(s)) dw^\alpha(s) \\
&+ \int_0^t b'(X_n(\phi_n(s)))_k^i Y_{j_1,j_2,(n)}^k(\phi_n(s)) ds \\
&+ \int_0^t \sigma''_\alpha(X_n(\phi_n(s)))_{k,l}^i Y_{j_1,(n)}^k(\phi_n(s)) Y_{j_2,(n)}^l(\phi_n(s)) dw^\alpha(s) \\
&+ \int_0^t b''(X_n(\phi_n(s)))_{k,l}^i Y_{j_1,(n)}^k(\phi_n(s)) Y_{j_2,(n)}^l(\phi_n(s)) ds,
\end{aligned}
$$
(2.17)

where

$$\sigma''_\alpha(x)_{k,l}^i = \frac{\partial^2}{\partial x^k \partial x^l} \sigma_\alpha^i(x) \quad , \quad b''(x)_{k,l}^i = \frac{\partial^2}{\partial x^k \partial x^l} b^i(x) .$$

If we denote by $\alpha_{j_1,j_2,(n)}^i(t)$ the sum of last two terms on the right hand side of (2.17), then noting Theorem III-3.1 and (2.15) it is easy to conclude that

$$(2.18) \qquad E\{ \sup_{t \in [0,T]} |\alpha_{j_1,j_2,(n)}^i(t) - \alpha_{j_1,j_2}^i(t)|^p \} \longrightarrow 0 \qquad \text{as} \quad n \longrightarrow \infty$$

for all $T > 0$ and $p \geq 1$. Here

$$(2.19) \qquad \alpha_{j_1,j_2}^i(t) = \int_0^t \sigma''_\alpha(X(s))_{k,l}^i Y_{j_1}^k(s) Y_{j_2}^l(s) dw^\alpha(s)$$

$$+\int_0^t b''(X(s))^i_{k,l} Y^k_{j_1}(s) Y^l_{j_2}(s) ds$$

and this convergence is uniform in x on each compact set. Now we can apply Lemma 2.1 * to conclude that

$$E\{\sup_{t\in[0,T]} |Y^i_{j_1,j_2,(n)}(t) - Y^i_{j_1,j_2}(t)|^p\} \longrightarrow 0 \quad \text{as} \quad n \longrightarrow \infty \tag{2.20}$$

for every $T > 0$ and $p \geq 1$, where $Y^i_{j_1,j_2}(t)$ is determined by the equation

$$\begin{aligned} Y^i_{j_1,j_2}(t) &= \int_0^t \sigma'_\alpha(X(s))^i_k Y^k_{j_1,j_2}(s) dw^\alpha(s) \\ &\quad + \int_0^t b'(X(s))^i_k Y^k_{j_1,j_2}(s) ds + \alpha^i_{j_1,j_2}(t). \end{aligned} \tag{2.21}$$

Furthermore, the convergence is uniform in x on each compact set.

By continuing this process step by step for higher order derivatives we complete the proof of the proposition.

Now we shall introduce the following seminorms for smooth functions ϕ on $\boldsymbol{R}^d$. For a bounded domain $\Omega \subset \boldsymbol{R}^d$, $p \geq 1$ and $m = 1, 2, \ldots$, we set

$$||\phi||^\Omega_{p,m} = \sum_{|\alpha|\leq m} \{\int_\Omega |D^\alpha \phi(x)|^p dx\}^{1/p}, \quad ||\phi||^\Omega_{\infty,m} = \sum_{|\alpha|\leq m} \sup_{x\in\Omega} |D^\alpha \phi(x)|.$$

For each Ω and m, we can find $\Omega' \supset \Omega$, $m' > m$, $p \geq 1$ and a constant $K > 0$ such that the following inequality holds for all smooth functions ϕ on $\boldsymbol{R}^d$:

$$||\phi||^\Omega_{\infty,m} \leq K ||\phi||^{\Omega'}_{p,m'}. \tag{2.22}$$

This is an obvious consequence of the well-known inequalities of Sobolev [151].

Now (2.13) clearly implies that

$$E\{\sup_{0\leq t\leq T} |Y^i_{\alpha,(n)}(t,x,w) - Y^i_{\alpha,(m)}(t,x,w)|^p\} \longrightarrow 0$$

as $n, m \longrightarrow \infty$ for all $T > 0$ and $p \geq 1$ uniformly on each compact set in x. Consequently,

* We apply this lemma to the system of equations (2.12) and (2.17) combined together.

$$E\{\sup_{0\le t\le T}\int_\Omega |Y^t_{\alpha,(n)}(t,x,w) - Y^t_{\alpha,(m)}(t,x,w)|^p\, dx\}$$

$$\le \int_\Omega dx\, E\{\sup_{0\le t\le T} |Y^t_{\alpha,(n)}(t,x,w) - Y^t_{\alpha,(m)}(t,x,w)|^p\}$$

$$\longrightarrow 0 \qquad\qquad \text{as} \quad n, m \longrightarrow \infty$$

for every bounded domain Ω. This implies that

$$E\{\sup_{0\le t\le T}\|X_n(t,\cdot,w) - X_m(t,\cdot,w)\|^{\Omega}_{p,l}\} \longrightarrow 0$$

as $n, m \longrightarrow \infty$ for all $T > 0, p \ge 1, l = 1, 2, \ldots$ and bounded domain Ω. By a standard argument we can extract a subsequence of X_n, denoted by X_n again, such that for almost all w (P^W),

$$\sup_{0\le t\le T}\|X_n(t,\cdot,w) - X_m(t,\cdot,w)\|^{\Omega}_{p,l} \longrightarrow 0$$

as $n, m \longrightarrow \infty$ for all bounded domain Ω, $p \ge 1$, $l = 1, 2, \ldots$ and $T > 0$. The inequality (2.22) then implies that, for almost all w,

$$\text{(2.23)} \qquad \sup_{0\le t\le T}\|X_n(t,\cdot,w) - X_m(t,\cdot,w)\|^{\Omega}_{\infty,l} \longrightarrow 0$$

as $n, m \longrightarrow \infty$ for all bounded domain Ω, $l = 1, 2, \ldots$ and $T > 0$. Consequently, for almost all w,

$$\lim_{n\to\infty} X_n(t,x,w) = \hat{X}(t,x,w)$$

exists uniformly in (t, x) on each compact set in $[0, \infty)\times \boldsymbol{R}^d$; furthermore $(t, x) \longmapsto \hat{X}(t,x,w) \in \boldsymbol{R}^d$ is continuous and for each $t \in [0, \infty)$, $x \longmapsto \hat{X}(t,x,w)$ is a $\boldsymbol{C}^\infty$-mapping from $\boldsymbol{R}^d$ into $\boldsymbol{R}^d$. We also have by (2.15) that

$$P^W[w;\, \hat{X}(t,x,w) = X(t,x,w) \quad \text{for all} \quad t \ge 0] = 1,$$

for all x, i.e., $(\hat{X}(t,x,w))$ is a modification of the family of solutions $(X(t,x,w))$ of the stochastic differential equation (2.1).

Thus we have obtained the following result.

Proposition 2.2. Let $X=(X(t,x,w))$ be the solution of (2.1). Then a modification of $X(t,x,w)$, denoted by $X(t,x,w)$ again, can be chosen so that the mapping $x \in \boldsymbol{R}^d \longmapsto X(t, x, w) \in \boldsymbol{R}^d$ is $\boldsymbol{C}^\infty$ for each fixed t, a.s.

Next we shall show that the mapping $x \longmapsto X(t,x,w)$ is a diffeomorphism of $\boldsymbol{R}^d$. In doing this, it is more convenient to rewrite the equation using Fisk-Stratonovich differentials. So instead of the equation (2.1), we consider the following equation

$$(2.24)\qquad \begin{cases} dX_t^i = \sigma_\alpha^i(X_t)\circ dw^\alpha(t) + b^i(X_t)dt \\ X_0 = x, \qquad\qquad\qquad i = 1, 2, \dots, d. \end{cases}$$

Note that (2.24) may be transformed into an equation of the form (2.1) with $b^i(x)$ replaced by $\bar{b}^i(x)$ where

$$(2.25)\qquad \bar{b}^i(x) = b^i(x) + \frac{1}{2}\sum_{\alpha=1}^{r}\left(\frac{\partial}{\partial x^k}\sigma_\alpha^i(x)\right)\sigma_\alpha^k(x).$$

Clearly $\sigma_\alpha^i(x)$ and $\bar{b}^i(x)$ satisfy the same assumptions as $\sigma_\alpha^i(x)$ and $b^i(x)$ and therefore, it is just a matter of convenience as to whether we write the equation in the form (2.1) or in the form (2.24). We shall now show that the C^∞-mapping $x \longmapsto X(t,x,w)$ defined by the solutions of (2.24) is a diffeomorphism. By (2.16), the Jacobian matrix $(Y_j^i(t)) = \left(\frac{\partial}{\partial x^j}X^i(t, x, w)\right)$ satisfies (in matrix notation)

$$(2.26)\qquad Y(t) = I + \int_0^t \sigma_\alpha'(X(s))Y(s)dw^\alpha(s) + \int_0^t \bar{b}'(X(s))Y(s)ds$$

where $\bar{b}$ is given by (2.25). It is easy to see that (2.26) is equivalent to

$$(2.26)'\qquad Y(t) = I + \int_0^t \sigma_\alpha'(X(s))Y(s)\circ dw^\alpha(s) + \int_0^t b'(X(s))Y(s)ds.$$

Now let $Z(t)=(Z_j^i(t))$ be the solution of

$$(2.27)\qquad Z(t) = I - \int_0^t Z(s)\sigma_\alpha'(X(s))\circ dw^\alpha(s) - \int_0^t Z(s)b'(X(s))ds.^*$$

Then

$$d(Z(t)Y(t)) = Z(t)\circ dY(t) + dZ(t)\circ Y(t) = 0$$

and therefore $Z(t)Y(t) \equiv I$. This proves that $Y(t)$ is invertible and $Y(t)^{-1} = Z(t)$. Consequently, the C^∞-mapping $x \longmapsto X(t,x,w)$ is a local diffeomorphism into R^d. By using the next Lemma it is easy to see that it is a bijection.

* To be precise we consider the system of equations (2.24) and (2.27) combined together. It is a stochastic differential equation on $R^d \times R^{d^2}$ whose coefficients satisfy the same kind of regularity and growth conditions.

Indeed, $x = X(t, \hat{X}(t,x,\hat{w}),w)$ and $x = \hat{X}(t,X(t,x,w),\hat{w})$ for every x a.s. Thus the mapping is a diffeomorphism of $\boldsymbol{R}^d$.

Lemma 2.2.* Let $\hat{X}(t,x,w)$ be constructed as above from the equation

$$
(2.28) \qquad \begin{cases} dX_t^i = \sigma_\alpha^i(X_t)\circ dw^\alpha(t) - b^i(X_t)dt \\ X_0 = x. \end{cases}
$$

Then, for every fixed $T > 0$, we have

$$
(2.29) \qquad X(T - t, x, w) = \hat{X}(t, X(T, x, w), \hat{w})
$$

for every $0 \leq t \leq T$ and x, a.s. (P^W). Here $\hat{w}$ is another Wiener process defined by

$$
(2.30) \qquad \hat{w}(t) = w(T-t) - w(T), \qquad 0 \leq t \leq T.
$$

Proof. Let

$$
B_n^\alpha(t, w) = \frac{(\bar{t}_n - t)}{(\bar{t}_n - \underline{t}_n)} w^\alpha(\underline{t}_n) + \frac{(t - \underline{t}_n)}{(\bar{t}_n - \underline{t}_n)} w^\alpha(\bar{t}_n), \qquad \alpha = 1, 2, \ldots, r,
$$

where $\bar{t}_n = \dfrac{(k+1)}{2^n} T$ and $\underline{t}_n = \dfrac{k}{2^n} T$ if $\dfrac{kT}{2^n} \leq t < \dfrac{(k+1)T}{2^n}$, $k = 1, 2, \ldots, 2^n - 1$, $n = 1, 2, \ldots$. It is clear that

$$
B_n^\alpha(T-t, w) - w^\alpha(T) = B_n^\alpha(t, \hat{w}), \qquad 0 \leq t \leq T.
$$

Consider the following two systems of ordinary differential equation for every $n = 1, 2, \ldots$

$$
\begin{cases} \dfrac{dX_n^i}{dt}(t) = \displaystyle\sum_{\alpha=1}^r \sigma_\alpha^i(X_n(t)) \frac{dB_n^\alpha}{dt}(t,w) + b^i(X_n(t)) \\ X_n(0) = x \end{cases} \qquad i = 1, 2, \ldots, d,
$$

and

* Malliavin [106].

$$\begin{cases} \dfrac{dX_n^i}{dt}(t) = \sum_{\alpha=1}^{r} \sigma_\alpha^i(X_n(t)) \dfrac{dB_n^\alpha}{dt}(t,w) - b^i(X_n(t)) \\ X_n(0) = x. \end{cases} \qquad i = 1, 2, \ldots, d,$$

Solutions will be denoted by $X_n(t,x,w)$ and $\hat{X}_n(t,x,w)$ respectively. It is immediately seen by the uniqueness of solutions that

$$X_n(T-t,x,w) = \hat{X}_n(t,X_n(T,x,w),\hat{w}), \qquad n = 1, 2, \ldots .$$

We can apply the corollary of Theorem VI-7.3 to obtain (2.29) by taking the limits.

Summarizing we state the following result.

Theorem 2.3.*1 Let $X(t,x,w)$ be the solution of (2.24) (or (2.1)) on the Wiener space $(\boldsymbol{W}_0^r, P^W)$. Then a modification of $X(t,x,w)$ can be chosen so that the mapping $x \longmapsto X(t,x,w)$ is a diffeomorphism of $\boldsymbol{R}^d$ a.s., for each $t \in [0, \infty)$. The Jacobian matrix $(Y_j^i(t,x,w)) = \left(\dfrac{\partial}{\partial x^j} X^i(t,x,w)\right)$ is determined by the equation (2.26)′ (resp. (2.16)).

Thus we have a one-parameter family of diffeomorphisms $X_t(w)$: $x \longmapsto X(t,x,w)$ for $t \in [0,\infty)$. Clearly $X_0(w)$=the identity and $X_s(\theta_t w) \circ X_t(w) = X_{t+s}(w)$ for almost all w.

Furthermore it actually holds a stronger statement: a.s., $x \longmapsto X(t,x,w)$ is a diffeomorphism of $\boldsymbol{R}^d$ for all $t \geq 0$. For details, see Kunita [209].

Now we return to the case of a compact manifold M. The solution $(X(t,x,w))$ of (1.1) may be considered as a family of mappings X_t: $x \longmapsto X(t,x,w)$ from M into M.

Theorem 2.4. Assume that M is a compact manifold. $X(t,x,w)$ has a modification,*2 which is denoted by $X(t,x,w)$ again, such that the mapping $X_t(w)$: $x \longmapsto X(t,x,w)$ is $\boldsymbol{C}^\infty$ in the sense that $x \longmapsto f(X(t,x,w))$ is $\boldsymbol{C}^\infty$ for every $f \in F(M)$ and all fixed $t \in [0, \infty)$, a.s. Furthermore, for each $x \in M$ and $t \in [0, \infty)$, the differential $X(t,x,w)_*$ of the mapping $x \longmapsto X(t,x,w)$;

$$X(t,x,w)_*: T_x(M) \longmapsto T_{X(t,x,w)}(M)$$

*1 Cf. Funaki [31], Malliavin [106] and Elworthy [25]

*2 By a modification of $X(t,x,w)$ we mean a process $\hat{X}(t,x,w)$ such that $P^W\{\hat{X}(t,x,w) = X(t,x,w)$ for all $t \geq 0\} = 1$ for all x.

is an isomorphism a.s. on the set $\{w;\ X(t,x,w) \in M\}$.

Proof. Let $x_0 \in M$ and $t \in [0, \infty)$ be fixed. Then for almost all w such that $X(t,x_0,w) \in M$, there exist an integer $n > 0$ and a sequence of coordinate neighborhoods $U_1, U_2, \cdots, U_n$ such that

$$\{X(s,x_0,w);\ s \in [(k-1)t/n, kt/n]\} \subset U_k,\ k = 1, 2, \cdots, n.$$

By Theorem 2.3, we can easily conclude that if U is a coordinate neighborhood and $\{X(s,y_0,w);\ s \in [0, t_0]\} \subset U$, then $y \longmapsto X(t_0,y,w)$ is a diffeomorphism in a neighborhood of y_0. Consequently, since

$$X(t,x_0,w) = [X_{t/n}(\theta_{(n-1)t/n}w) \circ \cdots \circ X_{t/n}(\theta_{t/n}w) \circ X_{t/n}](x_0)$$

the assertions of the theorem follow at once.

In case of non-compact manifolds, the solution $(X(t,x,w))$ of (1.1) may be considered as a family of mappings X_t: $x \to X(t,x,w)$ from M into $\hat{M} = M \cup \{\Delta\}$ and local results can be deduced from the compact case. However to obtain global results about the family of mappings X_t we must assume some additional conditions. Elworthy [197], Chapter VIII can be consulted for detailed information about these.

Now let us introduce the *bundle of linear frames* $GL(M)$ *on* M.* By a *frame* $e = [e_1, e_2, \ldots, e_d]$ at x we mean a linearly independent system of vectors $e_i \in T_x(M)$, $i = 1, 2, \ldots, d$, i.e., a basis of $T_x(M)$. $GL(M)$ is defined as the collection of all frames at all points $x \in M$:

$$GL(M) = \{r = (x, e);\ x \in M \text{ and } e \text{ is a frame at } x\}.$$

$GL(M)$ can be given a structure of a C^∞-manifold as follows. Let $\{U_\alpha, \phi_\alpha\}$ be a coordinate system of M. Set

$$\tilde{U}_\alpha = \{r = (x, e) \in GL(M);\ x \in U_\alpha \text{ and } e \text{ is a frame at } x\}$$

and define the mapping $\tilde{\phi}_\alpha$ from $\tilde{U}_\alpha$ onto $\phi_\alpha(U_\alpha) \times GL(d, R) \subset \boldsymbol{R}^d \times \boldsymbol{R}^{d^2}$ by

$$\tilde{\phi}_\alpha(r) = (\phi_\alpha(x) = (x^1, x^2, \ldots, x^d),\ e^i_j,\ i, j = 1, 2, \ldots, d)$$

where

$$e_j = e^i_j \left(\frac{\partial}{\partial x^i}\right)_x.$$

* [7] and [131].

Clearly $(\tilde{U}_\alpha, \tilde{\phi}_\alpha)$ defines a coordinate system of $GL(M)$ and so $GL(M)$ has the structure of C^∞-manifold of dimension $d + d^2$. An element a of the group $GL(d, \boldsymbol{R})$ acts on $GL(M)$ from the right by

(2.31) $$r \cdot a = (x, ea), \qquad r = (x, e)$$

where $ea = [(ea)_1, (ea)_2, \cdots, (ea)_d]$ is a frame at x defined by

$$(ea)_j = a^i_j e_i, \qquad j = 1, 2, \cdots, d.$$

Thus $GL(M)$ is a principal fibre bundle with the structural group $GL(d, \boldsymbol{R})$. The projection $\pi: GL(M) \longrightarrow M$ is defined, as usual, by $\pi(x, e) = x$.

Every vector field $L \in \mathfrak{X}(M)$ induces a vector field $\tilde{L}$ on $GL(M)$ as follows. Let $f \in F(GL(M))$. Then $\tilde{L}f$ is given by

(2.32) $$(\tilde{L}f)(r) = \frac{d}{dt} f((\exp tL)x, (\exp tL)_* e)\,|_{t=0}$$

where $r = (x, e)$ and $(\exp tL)_* e = [(\exp tL)_* e_1, (\exp tL)_* e_2, \ldots, (\exp tL)_* e_d]$. Here, of course, $\exp tL$ is the local diffeomorphism $x \longmapsto x(t,x)$ defined by the differential equation

$$\begin{cases} \dfrac{dx^i}{dt}(t,x) = a^i(x(t,x)), & \left(L = a^i(x)\dfrac{\partial}{\partial x^i}\right), \\ x(0,x) = x & \end{cases}$$

and $(\exp tL)_*$ is its differential which is an isomorphism $T_x(M) \longrightarrow T_{(\exp tL)x}(M)$ for each $x \in M$. Let $A_0, A_1, \ldots, A_r \in \mathfrak{X}(M)$ and $X_t = (X(t,x,w))$ be the flow of diffeomorphisms on M constructed above. Then $\tilde{A}_0, \tilde{A}_1, \ldots, \tilde{A}_r \in \mathfrak{X}(GL(M))$ defines a flow of diffeomorphisms $r_t = (r(t, r, w))$ on $GL(M)$, and it is easy to see from the definition that

$$r(t,r,w) = (X(t,x,w), e(t,r,w))$$

where $r = (x, e)$ and $e(t,r,w) = X(t,x,w)_* e$.* The expressions under a local coordinate are as follows: $X(t,x,w)$ is determined by the equation (2.24) where $A_\alpha(x) = \sigma^i_\alpha(x)\dfrac{\partial}{\partial x^i}$, $\alpha = 1, 2, \ldots, d$, $A_0(x) = b^i(x)\dfrac{\partial}{\partial x^i}$ and

(2.33) $$e^i_j(t,x,w) = Y^i_k(t,x,w) e^k_j$$

* $X(t, x, w)_*$ is the differential of $x \longmapsto X(t,x,w)$ and, of course, $X(t, x, w)_* e = [X(t,x,w)_* e_1, X(t,x,w)_* e_2, \ldots, X(t,x,w)_* e_d]$.

where $Y_k^i(t,x,w)$ is determined by the equation (2.26′).

Let $L \in \mathfrak{X}(M)$. We shall define a function $f_L^i(r) \in F(GL(M))$ for each, $i = 1, 2, \ldots, d$ by

$$f_L^i(r) = (e^{-1})_k^i a^k(x) \tag{2.34}$$

in a local coordinate $(x^i, e = (e_j^i))$ of $GL(M)$, where $L = a^i(x)\dfrac{\partial}{\partial x^i}$ and e^{-1} is the inverse matrix of e. It is easy to see that (2.34) is independent of a choice of local coordinates and thus defines a global function on $GL(M)$. It is also easy to prove that for $L_1, L_2 \in \mathfrak{X}(M)$,

$$(\tilde{L}_1 f_{L_2}^i)(r) = f_{[L_1, L_2]}^i(r), \qquad i = 1, 2, \ldots, d. \tag{2.35}$$

Here $\tilde{L}_1$ is defined by (2.32) for L_1 and $[L_1, L_2] = L_1L_2 - L_2L_1$ is the usual Poisson bracket. Therefore, we have that

$$\begin{aligned} & f_L^i(r(t,r,w)) - f_L^i(r) \\ & = \int_0^t f_{[A_\alpha, L]}^i(r(s,r,w)) \circ dw^\alpha(s) + \int_0^t f_{[A_0, L]}^i(r(s, r, w))ds \end{aligned} \tag{2.36}$$

for every $L \in \mathfrak{X}(M)$ and $i = 1, 2, \ldots, d$.

3. Heat equation on a manifold

Let M be a C^∞-manifold and $A_0, A_1, \ldots, A_r \in \mathfrak{X}(M)$. Let $X_t = (X(t,x,w))$ be the flow of diffeomorphisms on M constructed in the preceding section. Then $x \longmapsto f(X(t,x,w))$ is C^∞ for any $f \in F_0(M)$. *Throughout this section, we shall assume that the vector fields A_k, $k = 0, 1, \ldots, r$, have the property that*

$$E[\sup_{t\in[0,T]} \sup_{x\in U} |D^\alpha\{f(X(t,x,w))\}|] < \infty$$

for all $f \in F_0(M)$, every coordinate neighbourhood U such that $\bar{U}$ is compact, every $T > 0$ and every multi-index α.

This condition is satisfied if $M = \boldsymbol{R}^d$ and if the coefficients $(b^i(x))$ and $(\sigma_\alpha^i(x))$ in $A_0(x) = b^i(x)\dfrac{\partial}{\partial x^i}$ and $A_\alpha(x) = \sigma_\alpha^i(x)\dfrac{\partial}{\partial x^i}$, $\alpha = 1, 2, \ldots, r$, satisfy the condition of the preceding section. It is also satisfied if M is imbedded into $\boldsymbol{R}^m$ (cf. Remark 1.1) such that A_k, $k = 0, 1, \ldots, r$ are restrictions of vector fields $\tilde{A}_k$ on $\boldsymbol{R}^m$ which themselves satisfy the condition. In particular, if M is compact the above condition is always satisfied.

Define the second order differential operator A acting on $F(M)$ by

$$(3.1) \qquad Af(x) = \frac{1}{2}\sum_{\alpha=1}^{r} A_\alpha(A_\alpha f)(x) + (A_0 f)(x).$$

We shall show that the function $u(t,x)$ defined by

$$(3.2) \qquad u(t,x) = E[f(X(t,x,w))], \qquad f \in F_0(M)$$

is a smooth solution of the following heat equation

$$(3.3) \qquad \begin{cases} \dfrac{\partial v}{\partial t}(t,x) = (Av)(t,x) \\ \lim\limits_{t\downarrow 0, y\to x} v(t,y) = f(x). \end{cases}$$

First, we shall prove that the function $u(t,x)$ defined by (3.2) belongs to $C^\infty([0, \infty)\times M)$.* It is clear that $u(t, x)$ is a C^∞-function of $x \in M$ since $x \longmapsto f(X(t,x,w))$ is C^∞ and the differentiation under the expectation sign is justified by the above condition. By (1.2), we have

$$f(X(t,x,w)) - f(x) = \text{a martingale} + \int_0^t (Af)(X(s,x,w))ds$$

for each $x \in M$, and hence

$$(3.4) \qquad u(t,x) = f(x) + \int_0^t E[(Af)(X(s,x,w))]ds.$$

Since $A^n f \in F_0(M)$, $n = 1, 2, \ldots$, we have

$$\begin{aligned} u(t,x) &= f(x) + t(Af)(x) + \int_0^t dt_1 \int_0^{t_1} E[(A^2 f)(X(t_2,x,w))]dt_2 \\ &= f(x) + t(Af)(x) + \frac{t^2}{2}(A^2 f)(x) \\ &\quad + \int_0^t dt_1 \int_0^{t_1} dt_2 \int_0^{t_2} E[(A^3 f)(X(t_3,x,w))]dt_3 \\ &= f(x) + t(Af)(x) + \frac{t^2}{2}(A^2 f)(x) \\ &\quad + \cdots + \int_0^t dt_1 \int_0^{t_1} dt_2 \cdots \int_0^{t_{n-1}} E[(A^n f)(X(t_n,x,w))]dt_n. \end{aligned}$$

* $C^\infty([0, \infty)\times M)$ is the class of C^∞-functions on $[0, \infty)\times M$.

Now it is clear that $u(t,x) \in C^\infty([0, \infty) \times M)$.

Next we shall show that

$$(3.5) \qquad (Au_t)(x) = E[(Af)(X(t,x,w))],$$

where for each fixed $t \geq 0$, we set $u_t(x) = u(t,x)$. Applying Itô's formula to the smooth function $u_t(x)$, we have (writing $X_s = X(s,x,w)$)

$$u_t(X_s) - u_t(x) = \int_0^s (A_\alpha u_t)(X_u) dw^\alpha(u) + \int_0^s (Au_t)(X_u) du.$$

Let U be a relatively compact neighborhood of x and let $\sigma = \inf\{t; X_t \notin U\}$. Then we have

$$E[u_t(X_{s\wedge\sigma})] - u_t(x) = E[\int_0^{s\wedge\sigma} (Au_t)(X_u) du],$$

and hence

$$\left| (Au_t)(x) - \frac{E[u_t(X_{s\wedge\sigma})] - u_t(x)}{E[s\wedge\sigma]} \right|$$

$$= \left| \frac{E[\int_0^{s\wedge\sigma} \{(Au_t)(X_u) - (Au_t)(X_0)\} du]}{E[s\wedge\sigma]} \right|$$

$$= \left| \frac{E[\int_0^{s\wedge\sigma} Y_u du] + E[\int_0^{s\wedge\sigma} du \int_0^u (A^2 u_t)(X_\xi) d\xi]}{E[s\wedge\sigma]} \right|$$

where $Y_u = \int_0^u (A_\alpha A u_t)(X_\xi) dw^\alpha(\xi)$. Clearly,

$$\left| \frac{E[\int_0^{s\wedge\sigma} du \int_0^u (A^2 u_t)(X_\xi) d\xi]}{E[s\wedge\sigma]} \right| \leq s \max_{x \in \bar{U}} |(A^2 u_t)(x)|.$$

Also

$$\left| \frac{E[\int_0^{s\wedge\sigma} Y_u du]}{E[s\wedge\sigma]} \right|$$

$$\leq \frac{E[(s\wedge\sigma)\sup_{0\leq u\leq s\wedge\sigma}|Y_u|]}{E[s\wedge\sigma]} \leq \frac{(E[(s\wedge\sigma)^2])^{1/2}(E[\sup_{0\leq u\leq s\wedge\sigma}|Y_u|^2])^{1/2}}{E[s\wedge\sigma]}$$

$$\leq K\frac{(E[(s\wedge\sigma)^2])^{1/2}(E[\langle Y\rangle_{s\wedge\sigma}])^{1/2}}{E[s\wedge\sigma]}, \quad \text{(by Theorem III-3.1)},$$

$$\leq K\frac{(E[(s\wedge\sigma)^2])^{1/2}(E[s\wedge\sigma])^{1/2}\,(\max_{x\in \bar{U}}\sum_{\alpha=1}^{r}((A_\alpha Au_t)(x))^2)^{1/2}}{E[s\wedge\sigma]}$$

$$\leq K\,s^{1/2}(\max_{x\in \bar{U}}\sum_{\alpha=1}^{r}((A_\alpha Au_t)(x))^2)^{1/2},$$

where K is a positive constant. Consequently,

$$\lim_{s\downarrow 0}\frac{E[u_t(X_{s\wedge\sigma})]-u_t(x)}{E[s\wedge\sigma]} = (Au_t)(x). \tag{3.6}$$

On the other hand, by the strong Markov property of Brownian motion,

$$\begin{aligned}
&E[u_t(X_{s\wedge\sigma})]-u_t(x)\\
&\quad = E[f(X(t,X_{s\wedge\sigma},\theta_{s\wedge\sigma}w))]-E[f(X(t,x,w))]\\
&\quad = E[f(X(t+s\wedge\sigma,x,w))]-E[f(X(t,x,w))]\\
&\quad = E[\int_t^{t+s\wedge\sigma}(Af)(X_u)du]\\
&\quad = E[\int_0^{s\wedge\sigma}(Af)(X_{u+t})du].
\end{aligned}$$

Therefore,

$$\lim_{s\downarrow 0}\frac{E[u_t(X_{s\wedge\sigma})]-u_t(x)}{E[s\wedge\sigma]} = \lim_{s\downarrow 0}\frac{E[\int_0^{s\wedge\sigma}(Af)(X_{u+t})du]}{E[s\wedge\sigma]}. \tag{3.7}$$

It is immediately seen that $E[s\wedge\sigma] = s+o(s)$ since $s-E(s\wedge\sigma)\leq sP(\sigma\leq s) = o(s)$. So

$$E[\int_0^{s\wedge\sigma}(Af)(X_{u+t})du] = E[\int_0^{s}(Af)(X_{u+t})du] + o(s)$$

since Af is bounded. Therefore we can conclude that the limit in (3.7) is equal to

$$\lim_{s \downarrow 0} \frac{1}{s} E[\int_0^s (Af)(X_{u+t})du] = E[(Af)(X_t)].$$

This completes the proof of (3.5).

From (3.4) and (3.5), it follows that

$$u(t,x) = f(x) + \int_0^t (Au)(s,x)ds$$

and therefore we have

$$\frac{\partial u}{\partial t}(t,x) = Au(t,x),$$

i.e., $u(t,x)$ satisfies the heat equation (3.3).

Conversely, let $v(t,x) \in C^{1,2}([0, \infty) \times M)$ * be a bounded solution of (3.3). We shall further assume that $v(t,x)$ satisfies the following condition:

$$\lim_{n \uparrow \infty} E[v(t-\sigma_n, X(\sigma_n, x, w)) : \sigma_n \leq t] = 0 \tag{3.8}$$

for every $t > 0$ and $x \in M$, where $\sigma_n = \inf\{t; X(t, x, w) \notin D_n\}$ and D_n is an increasing sequence of relatively compact open sets in M such that $\bigcup_n D_n = M$.

Clearly (3.8) is satisfied for any bounded v if, for instance, $(X(t,x,w))$ is conservative, i.e., $P(e[X(\cdot,x,w)] = \infty) = 1$ for every $x \in M$. This is true since

$$e[X(\cdot,x,w)] = \lim_{n \uparrow \infty} \sigma_n.$$

Now by Itô's formula, we have for each $t_0 > 0$ and $0 \leq t \leq t_0$,

$$\begin{aligned} &E[v(t_0 - t \wedge \sigma_n, X(t \wedge \sigma_n, x, w))] - v(t_0, x) \\ &= E[\int_0^{t \wedge \sigma_n} \left\{(Av)(t_0 - s, X(s, x, w)) - \frac{\partial v}{\partial t}(t_0 - s, X(s,x,w))\right\} ds] \\ &= 0. \end{aligned}$$

Letting $n \uparrow \infty$ and noting (3.8), we obtain

* $C^{1,2}([0, \infty) \times M)$ is the totality of all functions $f(t, x)$ on $[0, \infty) \times M$ which are continuously differentiable in t and twice continuously differentiable in x.

$$E\{v(t_0 - t, X(t,x,w)) : e[X(\cdot,x,w)] > t\} = v(t_0,x).$$

Now letting $t \uparrow t_0$, we see by bounded convergence theorem that the left hand side tends to $E[f(X(t_0,x,w))] = u(t_0,x)$. Therefore $v(t,x)$ must coincide with $u(t,x)$.

The above results are summarized in the following theorem.

Theorem 3.1. For $f \in F_0(M)$, define $u(t,x)$ by (3.2). Then $u(t,x) \in \boldsymbol{C}^\infty([0, \infty)\times M)$ and satisfies the heat equation (3.3). Conversely, if $v(t,x) \in \boldsymbol{C}^{1,2}([0,\infty)\times M)$ is bounded and satisfies the heat equation (3.3) and the condition (3.8), then $v(t,x)$ must coincide with $u(t,x)$.

Remark 3.1. Generally a bounded solution of (3.3) is not unique and a kind of conditions like (3.8) is necessary in order to assure the uniqueness. For example, if $M=(0,\infty)$ and the operator $Au = u''/2$, i.e., $A_0 = 0$ and $A_1 = \frac{d}{dx}$, then, for $f \in F_0(M)$, both

$$v_1(t,x) = \int_0^\infty \frac{1}{\sqrt{2\pi t}}\left(\exp\left(-\frac{(x-y)^2}{2t}\right) - \exp\left(-\frac{(x+y)^2}{2t}\right)\right) f(y)dy$$

and

$$v_2(t,x) = \int_0^\infty \frac{1}{\sqrt{2\pi t}}\left(\exp\left(-\frac{(x-y)^2}{2t}\right) + \exp\left(-\frac{(x+y)^2}{2t}\right)\right) f(y)dy$$

are solutions of the heat equation. $v_1(t,x)$ is the solution which satisfies the condition (3.8).

Now let $c(x) \in F(M)$ such that it is bounded from above: $\sup_{x\in M} c(x) < \infty$ and

$$x \longmapsto E[\exp\{\int_0^t c(X(s,x,w))ds\} f(X(t,x,w))]$$

is $\boldsymbol{C}^\infty$ for any $f \in F_0(M)$ and $t \geq 0$. This condition is always satisfied if $c \in F_0(M)$.

Theorem 3.2. The function (Feynman-Kac formula) $u(t,x)$ defined by

$$(3.9) \qquad u(t,x) = E[\exp\{\int_0^t c(X(s,x,w))ds\} f(X(t,x,w))], \quad f \in F_0(M)$$

is a solution in $C^\infty([0, \infty)\times M)$ of the heat equation

$$(3.10) \qquad \begin{cases} \dfrac{\partial v}{\partial t}(t,x) = (Av)(t,x) + c(x)v(t,x) \\ \lim\limits_{t\downarrow 0, y\to x} v(t,y) = f(x) \end{cases}$$

which is bounded on $[0, T]\times M$ for each $T > 0$.

Conversely if $v(t,x)$ is a solution in $C^{1,2}([0, \infty)\times M)$ of the equation (3.10) such that it is bounded on $[0, T]\times M$ for each $T > 0$ and

$$(3.11) \qquad \lim_{n\uparrow\infty} E[\exp\{\int_0^{\sigma_n} c(X(s,x,w))ds\}\, v(t-\sigma_n,\ X(\sigma_n,x,w)) : \sigma_n \le t] = 0$$

then $v(t,x)$ coincides with $u(t,x)$ given by (3.9).

The proof is given in the same way as in Theorem 3.1. This time, however, we use Itô's formula as follows:

$$d[\exp\{\int_0^t c(X_s)ds\} f(X_t)] = \exp\{\int_0^t c(X_s)ds\}\,(A_\alpha f)(X_t)dw^\alpha(t)$$
$$+ \exp\{\int_0^t c(X_s)ds\}\,(Af(X_t)+c(X_t)f(X_t))dt$$

and

$$d[\exp\{\int_0^t c(X_s)ds\}\, v(t_0-t,\ X_t)]$$
$$= \exp\{\int_0^t c(X_s)ds\}\,(A_\alpha v)(t_0-t,\ X_t)dw^\alpha(t)$$
$$+ \exp\{\int_0^t c(X_s)ds\}\,\Big(-\frac{\partial v}{\partial t}(t_0-t,\ X_t) + (Av)(t_0-t,\ X_t)$$
$$+ c(X_t)v(t_0-t,\ X_t)\Big)dt.$$

4. Non-degenerate diffusions on a manifold and their horizontal lifts

Consider a two dimensional Brownian motion on a plane and suppose that the trajectory of a Brownian particle is traced in ink. We roll a sphere on the plane along the Brownian curve without slipping. The resulting path which is thus transferred defines a random curve on the sphere, and, indeed, it defines a Brownian motion on the sphere. This idea of constructing spherical Brownian motion was first proposed by Bochner. Such a method can also be carried over to a general Riemannian manifold. By making use of the connection in the sense of Levi-Civita, we can "*roll*" the manifold along a smooth curve in Euclidean space and, although a Brownian curve in Euclidean space is not smooth, stochastic calculus enables us to "*roll*" the manifold along such a curve. A Brownian motion on a Riemannian manifold can be obtained in this way. By generalizing the connection of Levi-Civita to a class of affine connections, we can obtain more general diffusions on a manifold; indeed, the most general non-degenerate smooth diffusions can be obtained in this manner. Such a process will be carried out below by constructing a flow of diffeomorphisms on the bundle of orthonormal frames over the manifold. This method is due to Eells and Elworthy [23].*[1] As we shall see in the next section, it is closely related to stochastic parallel displacement which was first introduced by Itô [68].

Before going into details, let us quickly recall several fundamental notions in differential geometry. Let M be a C^∞-manifold. A *tensor of type* (p,q) *at* x is an element in the tensor product $T_x(M)^p_q$. Here $T_x(M)^p_q = \underbrace{T_x(M)\otimes T_x(M)\otimes\cdots\otimes T_x(M)}_{p}\otimes\underbrace{T_x(M)^*\otimes T_x(M)^*\otimes\cdots\otimes T_x(M)^*}_{q}$ is the linear space formed of all multi-linear mappings u:
$\underbrace{T_x(M)^*\times T_x(M)^*\times\cdots\times T_x(M)^*}_{p}\times\underbrace{T_x(M)\times T_x(M)\times\cdots\times T_x(M)}_{q}\longrightarrow \boldsymbol{R}$
with the usual rules of addition and scalar multiplication.*[2] Choosing a local coordinate $(x^1,x^2,\ldots,x^d)$ introduces a basis $\left(\frac{\partial}{\partial x^1}\right)_x, \left(\frac{\partial}{\partial x^2}\right)_x, \ldots, \left(\frac{\partial}{\partial x^d}\right)_x$ in $T_x(M)$; its dual basis is denoted by $(dx^1)_x, (dx^2)_x, \ldots, (dx^d)_x$. We denote by $\left(\frac{\partial}{\partial x^{i_1}}\right)_x\otimes\cdots\otimes\left(\frac{\partial}{\partial x^{i_p}}\right)_x\otimes(dx^{j_1})_x\otimes\cdots\otimes(dx^{j_q})_x$ the element $u\in T_x(M)^p_q$ such that

$$u\left((dx^{k_1})_x,\ldots,(dx^{k_p})_x,\left(\frac{\partial}{\partial x^{l_1}}\right)_x,\ldots,\left(\frac{\partial}{\partial x^{l_q}}\right)_x\right)$$
$$=\delta^{k_1}_{i_1}\cdots\delta^{k_p}_{i_p}\,\delta^{j_1}_{l_1}\cdots\delta^{j_q}_{l_q}$$

for every $k_1,k_2,\ldots,k_p,l_1,l_2,\ldots,l_q$. Clearly the system

$$(4.1)\qquad \left\{\left(\frac{\partial}{\partial x^{i_1}}\right)_x\otimes\left(\frac{\partial}{\partial x^{i_2}}\right)_x\otimes\cdots\otimes\left(\frac{\partial}{\partial x^{i_p}}\right)_x\otimes(dx^{j_1})_x\otimes(dx^{j_2})_x\right.$$

*[1] Cf. Also their works given in the references of [23] and Malliavin [104].
*[2] $T_x(M)^*$ is the dual space of $T_x(M)$.

$$\otimes \cdots \otimes (dx^{j_q})_x\Big\} \quad i_1, i_2, \ldots, i_p, j_1, j_2, \ldots, j_q = 1, 2, \ldots, d$$

forms a basis of $T_x(M)^p_q$. A C^∞*-tensor field of type* (p, q) *1 is a mapping

$$u\colon M \ni x \longmapsto u(x) \in T_x(M)^p_q$$

whose components $u^{i_1 i_2 \cdots i_p}_{j_1 j_2 \cdots j_q}(x)$ with respect to the basis (4.1) are C^∞ in every coordinate neighborhood. The $u^{i_1 i_2 \cdots i_p}_{j_1 j_2 \cdots j_q}$ obey the usual rule under a change of coordinates:

$$\tilde{u}^{i_1 i_2 \cdots i_p}_{j_1 j_2 \cdots j_q}(x) = \frac{\partial \tilde{x}^{i_1}}{\partial x^{k_1}} \frac{\partial \tilde{x}^{i_2}}{\partial x^{k_2}} \cdots \frac{\partial \tilde{x}^{i_p}}{\partial x^{k_p}} \frac{\partial x^{l_1}}{\partial \tilde{x}^{j_1}} \frac{\partial x^{l_2}}{\partial \tilde{x}^{j_2}} \cdots \frac{\partial x^{l_q}}{\partial \tilde{x}^{j_q}} u^{k_1 k_2 \cdots k_p}_{l_1 l_2 \cdots l_q}(x). \tag{4.2}$$

Conversely, if a system of C^∞-functions $\{u^{i_1 i_2 \cdots i_p}_{j_1 j_2 \cdots j_q}(x)\}$ is defined in every coordinate neighborhood and satisfies (4.2), then there exists a unique (p, q)-tensor field whose components coincide with it.

A (1,0)-tensor field is just a vector field. A (0,1)-tensor field is called a *differential 1-form.* Generally, a *(differential) p-form* is a $(0,p)$-tensor which is alternate, i.e., its components satisfy

$$u_{\sigma(i_1)\sigma(i_2)\cdots\sigma(i_p)}(x) = \operatorname{sgn}(\sigma) u_{i_1 i_2 \cdots i_p}(x)$$

for every permutation σ.*2 If we set $dx^{i_1} \wedge dx^{i_2} \wedge \cdots \wedge dx^{i_p} = \frac{1}{p!} \sum_\sigma \operatorname{sgn}(\sigma)\, dx^{\sigma(i_1)} \otimes dx^{\sigma(i_2)} \otimes \cdots \otimes dx^{\sigma(i_p)}$, then a p-form $u(x)$ is expressed as

$$\begin{aligned} u(x) &= u_{i_1 i_2 \cdots i_p}(x)\, dx^{i_1} \wedge dx^{i_2} \wedge \cdots \wedge dx^{i_p} \\ &= p! \sum_{i_1 < i_2 < \cdots < i_p} u_{i_1 i_2 \cdots i_p}(x) dx^{i_1} \wedge dx^{i_2} \wedge \cdots \wedge dx^{i_p}. \end{aligned}$$

The *exterier product* $\alpha \wedge \beta$ of a p-form α and q-form β is a $(p+q)$-form defined by

$$(\alpha \wedge \beta)(x) = \alpha_{k_1 k_2 \cdots k_p}(x) \beta_{k_{p+1} k_{p+2} \cdots k_{p+q}}(x) dx^{k_1} \wedge dx^{k_2} \wedge \ldots \wedge dx^{k_{p+q}}. \tag{4.3}$$

The *exterior derivative* $d\alpha$ of p-form α is a $(p+1)$-form defined by

*1 Also we call it simply (p, q)-tensor field.

*2 Such a property is clearly independent of the choice of coordinates. The notion of a symmetric tensor can be defined similarly.

$$(4.4)\qquad (d\alpha)(x) = \frac{\partial}{\partial x^i} \alpha_{i_1 i_2 \ldots i_p}(x)\, dx^i \wedge dx^{i_1} \wedge \cdots \wedge dx^{i_p}.$$

By an *affine connection* ∇ we mean a rule which associates to every $X \in \mathfrak{X}(M)$ a linear mapping ∇_X: $\mathfrak{X}(M) \longrightarrow \mathfrak{X}(M)$ having the following properties:

$$(4.5)\qquad \begin{aligned} &\text{(i) } \nabla_X Y \text{ is bilinear in } X \text{ and } Y;\\ &\text{(ii) } \nabla_{fX+gY} = f\nabla_X + g\nabla_Y;\\ &\text{(iii) } \nabla_X(fY) = f\nabla_X Y + (Xf)Y.^* \end{aligned}$$

The operator ∇_X is called *covariant differentiation with respect to X*. A system of functions $\{\Gamma^i_{jk}(x)\}$ is defined in a coordinate neighborhood by

$$\nabla_{\partial_i}\partial_j = \Gamma^k_{ij}(x) \frac{\partial}{\partial x^k} \qquad \left(\partial_i = \frac{\partial}{\partial x^i}\right).$$

The $\Gamma^k_{ij}(x)$ are called the *components of the connection* ∇. In a local coordinate system, $\nabla_X Y$ may be expressed as

$$(4.6)\qquad \nabla_X Y = \left[X^i(x)\, Y^j(x) \Gamma^k_{ij}(x) + X^i(x) \frac{\partial}{\partial x^i} Y^k(x)\right] \frac{\partial}{\partial x^k}$$

where $X = X^i(x)\dfrac{\partial}{\partial x^i}$ and $Y = Y^i(x)\dfrac{\partial}{\partial x^i}$. The components of the connection obey the following transformation rule under a change of coordinates

$$(4.7)\qquad \tilde{\Gamma}^k_{ij} = \frac{\partial x^p}{\partial \tilde{x}^i}\frac{\partial x^q}{\partial \tilde{x}^j}\frac{\partial \tilde{x}^k}{\partial x^r}\Gamma^r_{pq} + \frac{\partial^2 x^r}{\partial \tilde{x}^i \partial \tilde{x}^j}\frac{\partial \tilde{x}^k}{\partial x^r}.$$

Conversely, any system of smooth functions $\Gamma^k_{ij}(x)$ defined in each coordinate neighborhood and satisfying the rule (4.7) determines an affine connection by (4.6).

Given a tensor field $u(x) = (u^{i_1 i_2 \cdots i_p}_{j_1 j_2 \ldots j_q}(x))$ of type (p, q), a tensor field $(\nabla u)(x) = (u^{i_1 i_2 \cdots i_p}_{j_1 j_2 \ldots j_q;k}(x))$ of type $(p,q+1)$ is defined by

$$\begin{aligned} &u^{i_1 i_2 \cdots i_p}_{j_1 j_2 \ldots j_q;k}(x)(:= \nabla_k u^{i_1 i_2 \cdots i_p}_{j_1 j_2 \ldots j_q}(x))\\ (4.8)\qquad &= \frac{\partial}{\partial x^k} u^{i_1 i_2 \cdots i_p}_{j_1 j_2 \ldots j_q}(x) + \sum_{\alpha=1}^{p} \Gamma^{i_\alpha}_{kl}(x) u^{i_1 i_2 \ldots l \ldots i_p}_{j_1 j_2 \ldots j_q}(x)\\ &\quad - \sum_{\beta=1}^{q} \Gamma^m_{k, j_\beta}(x) u^{i_1 i_2 \cdots i_p}_{j_1 j_2 \ldots m \ldots j_q}(x) \end{aligned}$$

* For $f \in F(M)$ and $X \in \mathfrak{X}(M)$, $fX \in \mathfrak{X}(M)$ is defined by $(fX)g = fX(g)$ for every $g \in F(M)$.

where the indices l and m in u take the place of i_α and j_β respectively. By the transformation rule (4.7), it is easy to see that $\nabla u(x)$ is a tensor field. $\nabla u(x)$ is called the *covariant derivative of* $u(x)$. For $X = X^i \dfrac{\partial}{\partial x^i} \in \mathfrak{X}(M)$, the (p,q)-tensor $\nabla_X u$ defined by

$$(\nabla_X u)^{i_1 i_2 \cdots i_p}_{j_1 j_2 \cdots j_q} = X^k u^{i_1 i_2 \cdots i_p}_{j_1 j_2 \cdots j_q ; k}$$

is called the *covariant derivaitive of* $u(x)$ *in the direction* X. Note that if $u = Y \in \mathfrak{X}(M)$, the above coincides with the original $\nabla_X Y$.

Let c: $I \ni t \longmapsto c(t) \in M$ * be a (piecewise) smooth curve in M and $u(t) = (u^{i_1 i_2 \cdots i_p}_{j_1 j_2 \cdots j_q}(t))$ be a tensor field along c; that is, $u(t) \in T_{c(t)}(M)^p_q$ for $t \in I$ and $t \longmapsto u(t)$ is (piecewise) smooth. $u(t)$ is said to be *parallel along* c (with respect to the connection ∇) if

$$\begin{aligned}
&\frac{d}{dt} u^{i_1 i_2 \cdots i_p}_{j_1 j_2 \cdots j_q}(t) + \sum_{\alpha=1}^{p} \Gamma^{i_\alpha}_{kl}(c(t)) u^{i_1 i_2 \cdots l \cdots i_p}_{j_1 j_2 \cdots j_q}(t) \frac{dc^k(t)}{dt} \\
&\qquad - \sum_{\beta=1}^{q} \Gamma^{m}_{k j_\beta}(c(t)) u^{i_1 i_2 \cdots i_p}_{j_1 j_2 \cdots m \cdots j_q}(t) \frac{dc^k(t)}{dt} = 0, \qquad t \in I.
\end{aligned} \tag{4.9}$$

In particular, a tensor field $u(x)$ is parallel along c if and only if $(\nabla_{\dot c(t)} u)\,(c(t))\left(:= u^{i_1 i_2 \cdots i_p}_{j_1 j_2 \cdots j_q ; k}(c(t)) \dfrac{dc^k(t)}{dt}\right) = 0$, $t \in I$. For $t_0, t_1 \in I$, $t_0 \leq t_1$, $u(t_1)$ is uniquely determined from $u(t_0)$ as a solution of (4.9) and we say that $u(t_1)$ is obtained from $u(t_0)$ by *parallel displacement along the curve* $c(t)$.

Consider a manifold M with an affine connection $\nabla = \{\Gamma^i_{jk}(x)\}$. Let $GL(M)$ be the bundle of linear frames. For each $r \in GL(M)$,

$$H_r = \left\{ X = a^i \left(\frac{\partial}{\partial x^i}\right)_x - \Gamma^i_{kl}(x) e^l_j a^k \frac{\partial}{\partial e^i_j}\,; \quad (a^i) \in \mathbf{R}^d \right\} \tag{4.10}$$

is a linear subspace of $T_r(GL(M))$ which is clearly independent of the choice of local coordinates (x^i, e^i_j). A tangent vector X in H_r is called *horizontal.* An affine connection may also be defined as a rule which assigns a linear subspace H_r of $T_r(GL(M))$ at each $r \in GL(M)$ (cf. Nomizu [131]). H_r is called the *horizontal subspace*. Let $\xi \in T_x(M)$. $\tilde\xi \in T_r(GL(M))$ is called a *horizontal lift* of ξ if $\tilde\xi$ is horizontal, $\pi(r) = x$ and $(d\pi)_r \tilde\xi = \xi$. $\tilde\xi$ is unique if r such that $\pi(r) = x$ is given. Given $X \in \mathfrak{X}(M)$, there exists a unique $\tilde X \in \mathfrak{X}(GL(M))$ such that $\tilde X_r$ is the horizontal lift of $X_{\pi(r)}$ for every $r \in GL(M)$. $\tilde X$ is called the *horizontal lift* of X. In a local coordinate,

* I denotes an interval of $\mathbf{R}^1$.

(4.11) $$\tilde{X} = X^i(x)\frac{\partial}{\partial x^i} - \Gamma^q_{ij}(x)X^i(x)e^j_p\frac{\partial}{\partial e^q_p}$$

if $X = X^i(x)\frac{\partial}{\partial x^i}$. Given a smooth curve $c: I \ni t \longmapsto c(t) \in M$, a smooth curve $\tilde{c}: I \ni t \longmapsto \tilde{c}(t) \in GL(M)$ is called a *horizontal lift* of c if (i) $\frac{d\tilde{c}}{dt}(t)$ is horizontal and (ii) $\pi(\tilde{c}(t)) = c(t)$ for $t \in I$. Clearly, if $r = (x, e = [e_1, e_2, \ldots, e_d])$ is given where x is the starting point of c, then a horizontal lift $\tilde{c}$ starting at r exists and is unique. Indeed, $\tilde{c}(t) = (c(t), e(t) = [e_1(t), e_2(t), \ldots, e_d(t)])$, where $e_i(t) \in T_{c(t)}(M)$ is obtained from e_i by parallel displacement along the curve c. For each $j = 1, 2, \ldots, d$, there exists a unique vector field $\tilde{L}_j \in \mathfrak{X}(GL(M))$ such that $(\tilde{L}_j)_r$ is the horizontal lift of $e_j \in T_x(M)$ for every $r = (x, e = [e_1, e_2, \ldots, e_d])$. In a local coordinate (x^i, e^i_j), $\tilde{L}_j$ may be expressed as

(4.12) $$\tilde{L}_j = e^i_j\frac{\partial}{\partial x^i} - \Gamma^q_{kl}e^k_j e^l_p\frac{\partial}{\partial e^q_p}.$$

$\{\tilde{L}_1, \tilde{L}_2, \ldots, \tilde{L}_d\}$ is called the *system of canonical horizontal vector fields* or *basic vector fields* ([7] and [131]).

Let $u(x) = (u^{i_1 i_2 \ldots i_p}_{j_1 j_2 \ldots j_q}(x))$ be a (p,q)-tensor field. Define a system of smooth functions $F_u(r) = \{F^{i_1 i_2 \ldots i_p}_{u j_1 j_2 \ldots j_q}(r)\}$ on $GL(M)$ by

(4.13) $$u(x) = F^{i_1 i_2 \ldots i_p}_{u j_1 j_2 \ldots j_q}(r) e_{i_1} \otimes e_{i_2} \otimes \cdots \otimes e_{i_p} \otimes e^{j_1}_* \otimes e^{j_2}_* \otimes \cdots \otimes e^{j_q}_*$$

for $r = (x, e = [e_1, e_2, \ldots, e_d])$ and $e_* = [e^1_*, e^2_*, \ldots, e^d_*]$ is the dual base of e. In a local coordinate (x^i, e^i_j), it may be expressed as

(4.14) $$F^{i_1 i_2 \ldots i_p}_{u j_1 j_2 \ldots j_q}(r) = u(x)^{k_1 k_2 \ldots k_p}_{l_1 l_2 \ldots l_q} e^{l_1}_{j_1} e^{l_2}_{j_2} \cdots e^{l_q}_{j_q} f^{i_1}_{k_1} f^{i_2}_{k_2} \cdots f^{i_p}_{k_p}$$

where (f^i_j) is the inverse matrix of (e^i_j). $F_u(r) = \{F^{i_1 i_2 \ldots i_p}_{u j_1 j_2 \ldots j_q}(r)\}$ is called the *scalarization of the tensor field* $u(x)$ or the *system of components of the tensor field* $u(x)$ *read in the frame* e. $F_u(r)$ is $GL(d, \boldsymbol{R})$-*equivariant* in the sense that

(4.15) $$F^{i_1 i_2 \cdots i_p}_{u j_1 j_2 \cdots j_q}(r) = F^{k_1 k_2 \cdots k_p}_{u l_1 l_2 \cdots l_q}(r \cdot a) a^{i_1}_{k_1} a^{i_2}_{k_2} \cdots a^{i_p}_{k_p} b^{l_1}_{j_1} b^{l_2}_{j_2} \cdots b^{l_q}_{j_q}$$

for every $a = (a^i_j) \in GL(d, \boldsymbol{R})$ where $r \cdot a$ is the action of a on r defined by (2.31); (b^i_j) is the inverse of (a^i_j). Conversely, every $GL(d, \boldsymbol{R})$-equivariant system $F(r) = \{F^{i_1 i_2 \cdots i_p}_{j_1 j_2 \cdots j_q}(r)\}$ of smooth functions on $GL(M)$ is given as $F = F_u$ for some uniquely determined tensor field u.

Proposition 4.1.

$$\tilde{L}_m(F_{u j_1 j_2 \ldots j_q}^{i_1 i_2 \ldots i_p})(r) = (F_{\nabla u})^{i_1 i_2 \ldots i_p}_{j_1 j_2 \ldots j_q; m}(r) \tag{4.16}$$

for every $i_1, i_2, \ldots, i_p, j_1, j_2, \ldots, j_q$ and $m = 1, 2, \ldots, d$, where $\{\tilde{L}_m\}$ is the system of canonical horizontal vector fields.

The proof is left to the reader.

For an affine connection $\nabla = \{\Gamma_{ij}^k\}$, $T_{ij}^k = \Gamma_{ij}^k - \Gamma_{ji}^k$ are the components of a tensor T of type (1,2). T is called the *torsion tensor*. An intrinsic definition of the torsion tensor T is given by

$$T(X,Y) = \nabla_X Y - \nabla_Y X - [X, Y], \qquad X, Y \in \mathfrak{X}(M).$$

An affine connection $\nabla = \{\Gamma_{ij}^k\}$ is called *torsion free* or *symmetric* if the torsion tensor is zero, i.e., $\Gamma_{ij}^k = \Gamma_{ji}^k$.

A C^∞-manifold M is called a *Riemannian manifold* if a tensor field $g = (g_{ij})$ of type (0, 2) is given on M such that

(i) g is symmetric, i.e., $g_{ij}(x) = g_{ji}(x)$;

(ii) g is positive definite, i.e., $g_{ij}(x)\xi^i\xi^j > 0$ for all x and $\xi \neq 0 \in \mathbf{R}^d$.* g is called *fundamental tensor field* or *Riemannian metric* (tensor field). It defines an inner product on each tangent space $T_x(M)$ by

$$\langle \xi, \eta \rangle = g_{ij}(x)\xi^i\eta^j, \quad \xi = \xi^i \left(\frac{\partial}{\partial x^i}\right)_x \quad \text{and} \quad \eta = \eta^i \left(\frac{\partial}{\partial x^i}\right)_x. \tag{4.17}$$

An affine connection $\nabla = \{\Gamma_{ij}^k\}$ is called *compatible with the Riemannian metric* g if the inner product is preserved during a parallel displacement of tangent vectors. That is, for every smooth curve $c(t)$ and tangent vectors $\xi^i(t)\dfrac{\partial}{\partial x^i}$ and $\eta^i(t)\dfrac{\partial}{\partial x^i}$ at $c(t)$

$$\frac{d\xi^i}{dt} + \Gamma_{jk}^i(c(t))\frac{dc^j(t)}{dt}\xi^k(t) = 0 \quad \text{and} \quad \frac{d\eta^i}{dt} + \Gamma_{jk}^i(c(t))\frac{dc^j(t)}{dt}\eta^k(t) = 0$$

imply that

$$\frac{d}{dt}(g_{ij}(c(t))\xi^i(t)\eta^j(t)) = 0.$$

From this it is easy to conclude that ∇ is compatible with g if and only if

* Properties (i) and (ii) are clearly independent of the choice of coordinates.

$$\text{(4.18)} \qquad \frac{\partial}{\partial x^k} g_{ij} = g_{lj} \Gamma^l_{ki} + g_{il} \Gamma^l_{kj} \quad \text{for all} \quad i, j, k = 1, 2, \ldots, d.$$

An affine connection compatible with g is not unique (see Proposition 4.3 below), but if we assume further that it is symmetric then it is unique. Indeed, (4.18) together with $\Gamma^k_{ij} = \Gamma^k_{ji}$ implies that

$$\text{(4.19)} \qquad \Gamma^k_{ij} = \{{}_i{}^k{}_j\} := \frac{1}{2}\left(\frac{\partial}{\partial x^i} g_{mj} + \frac{\partial}{\partial x^j} g_{im} - \frac{\partial}{\partial x^m} g_{ij}\right) g^{km}.$$

This connection is called the *Riemannian connection* or the *connection of Levi-Civita*. The $\{{}_i{}^k{}_j\}$ are called the *Christoffel symbols*.

Let $O(M)$ be a submanifold of $GL(M)$ defined by

$$\text{(4.20)} \qquad O(M) = \{r = (x,e) \in GL(M);\ e \text{ is an orthonormal base of } T_x(M)\}.$$

In a local coordinate (x^i, e^i_j) of $GL(M)$, $r \in O(M)$ if and only if

$$\text{(4.21)} \qquad g_{kl} e^k_i e^l_j = \delta_{ij},$$

or equivalently,

$$\text{(4.22)} \qquad \sum_{m=1}^{d} e^i_m e^j_m = g^{ij}$$

where (g^{ij}) is the inverse matrix of (g_{ij}). The equivalence of (4.21) and (4.22) is easily verified as follows. Set $e = (e^i_j)$ and $G = (g_{ij})$. Then

$$\begin{aligned} \text{(4.21)} &\Longleftrightarrow e^*Ge = I \Longleftrightarrow G = (e^*)^{-1}e^{-1} \\ &\Longleftrightarrow G^{-1} = ee^* \\ &\Longleftrightarrow \text{(4.22)}. \end{aligned}$$

Now the orthogonal group $O(d)$ acts on $O(M)$, and $O(M)$ is a principal fibre bundle over M with the structural group $O(d)$. $O(M)$ is called the *bundle of orthonormal frames on M*. Let ∇ be an affine connection compatible with g and $c: [a, b] \longrightarrow M$ be a smooth curve in M. If $r = (c(0), e) \in O(M)$, then the horizontal lift $\tilde{c}(t) = (c(t), e(t))$ of $c(t)$ lies in $O(M)$ since $e(t)$ is an orthonormal frame at $c(t)$. Similarly, a horizontal vector field $\tilde{X}$ of $X \in \mathfrak{X}(M)$, if restricted to $O(M)$, is a vector field on $O(M)$, and the canonical horizontal vector fields $\tilde{L}_m$, $m = 1, 2, \ldots, d$, are vector fields on $O(M)$.

Let M be a Riemannian manifold and $\nabla = \{\Gamma^k_{ij}\}$ be an affine connection compatible with the Riemannian metric g. The connection ∇ enables us to "*roll*" M along a curve $\gamma(t)$ in $\boldsymbol{R}^d$ to obtain a curve $c(t)$ in M as the trace of $\gamma(t)$. Intutitively the infinitesimal motion of $c(t)$ is that of $\gamma(t)$ in the tangent space which can be identified with $\boldsymbol{R}^d$ by choosing an orthonormal frame and the infinitesimal motion of the frame is given by the connection i.e., parallel displacement along the curve $c(t)$. To be precise, let γ: $[0, \infty) \ni t \longmapsto \gamma(t) \in \boldsymbol{R}^d$ be a smooth curve in $\boldsymbol{R}^d$. Let $r = (x, e) \in O(M)$ and define a curve $\tilde{c}(t) = (c(t), e(t))$ in $O(M)$ by

$$(4.23) \qquad \begin{cases} \dfrac{dc}{dt}(t) = \dfrac{d\gamma^\alpha}{dt}(t) e_\alpha(t) \\ \nabla_{\dot{c}(t)} e(t) = 0 \\ c(0) = x \\ e(0) = e. \end{cases}$$

In local coordinates,

$$(4.24) \qquad \begin{cases} \dfrac{dc^i}{dt}(t) = e^i_\alpha(t) \dfrac{d\gamma^\alpha}{dt}(t), & i = 1, 2, \ldots, d \\ \dfrac{de^i_\alpha}{dt}(t) = -\Gamma^i_{ml}(c(t)) e^l_\alpha(t) \dfrac{dc^m}{dt}(t) & \\ c^i(0) = x^i & \\ e^i_\alpha(0) = e^i_\alpha & i, \alpha = 1, 2, \ldots, d. \end{cases}$$

The equation (4.23) may be written as

$$(4.25) \qquad \begin{cases} \dfrac{d\tilde{c}}{dt}(t) = \tilde{L}_\alpha(\tilde{c}(t)) \dfrac{d\gamma^\alpha}{dt}(t) \\ \tilde{c}(0) = r, \end{cases}$$

where $\{\tilde{L}_1, \tilde{L}_2, \ldots, \tilde{L}_d\}$ is the system of canonical horizontal vector fields. The curve $c(t) = \pi(\tilde{c}(t))$ in M depends on the choice of the initial frame e at x; we shall denote it as $c(t) = c(t, r, \gamma)$, $r = (x, e)$. It follows at once that

$$(4.26) \qquad c(t, r \cdot a, \gamma) = c(t, r, a\gamma), \qquad t \in [0, \infty),\ a \in O(d),$$

where $r \cdot a$ is defined by (2.31) and the curve $a\gamma$ in $\boldsymbol{R}^d$ is defined by

(4.27) $(a\gamma)(t) = a\gamma(t), \quad$ *i.e.* $\quad (a\gamma)^i\,(t) = a^i_j\,\gamma^j(t).$

Now let $w(t) = (w^\alpha(t))$ be the canonical realization of a d-dimensional Wiener process. Stochastic calculus enables us to define a random curve $X(t)$ in M in the same way. Let $r(t) = (r(t, r, w))$ be the solution of the stochastic differential equation

$$(4.28)\qquad \begin{cases} dr(t) = \tilde{L}_\alpha(r(t))\circ dw^\alpha(t) \\ r(0) = r. \end{cases}$$

$r(t,r,w)$ is the flow of diffeomorphisms on $O(M)$ corresponding to the canonical horizontal vector fields $\tilde{L}_1, \tilde{L}_2, \ldots, \tilde{L}_d$ and the drift vector field * $\tilde{L}_0 \equiv 0$. In local coordinates, (4.28) is equivalent to

$$(4.29)\qquad \begin{cases} dX^i(t) = e^i_\alpha(t)\circ dw^\alpha(t) & i = 1, 2, \ldots, d \\ de^i_\alpha(t) = -\,\Gamma^i_{mk}(X(t))e^k_\alpha(t)\circ dX^m(t), & i, \alpha = 1, 2, \ldots, d, \end{cases}$$

where $r(t) = (X^i(t), e^i_\alpha(t))$. That the solution $r(t) = (X^i(t), e^i_\alpha(t))$ lies on $O(M)$ if $r(0) \in O(M)$ is clear since $\tilde{L}_\alpha$ is a vector field on $O(M)$. Of course, one can also verify directly that $d(g_{ij}(X(t))e^i_\alpha(t)e^j_\beta(t)) = 0$ by using (4.18). Now a stochastic curve $X(t) = (X^i(t))$ on M is defined by $X(t) = \pi[r(t)]$. By (4.26) we have (writing $X(t) = (X(t,r,w))$)

(4.30) $\quad X(t, r\cdot a, w) = X(t,r,aw) \quad$ for $t \geq 0$, $a \in O(d)$ and $w \in W^d_0$.

But $aw = (aw(t))$ is another d-dimensional Wiener process and hence the probability law of $X(\cdot, r\cdot a, w)$ is independent of $a \in O(d)$. In other words, the probability law of $X(\cdot, r, w)$ depends only on $x = \pi(r)$. We denote it by P_x. It is now easy to deduce the strong Markov property of the system $\{P_x\}$ from that of $r(\cdot, r, w)$.

Remark 4.1. Of course $r(t, r, w)$ can be defined as a flow of diffeomorphisms on $GL(M)$ for any affine connection but its projection to M is not strong Markov in general because it is not usually true that the law of $\pi[r(\cdot, r, w)]$ depends only on $x = \pi(r)$. This is the reason why we restrict ourselves to an affine connection compatible with g and to a flow of diffeomorphisms on $O(M)$.

Thus we have a diffusion $\{P_x\}$ on M. We shall now show that it is A-diffusion process where the differential operator A is given by

* In the stochastic differential equation (1.1), the vector field A_0 is called the *drift vector field.*

(4.31) $$A = \frac{1}{2}\Delta_M + b.$$

Here Δ_M is the Laplace-Beltrami operator on M given by

(4.32) $$\Delta_M f = g^{ij}\nabla_j^R \nabla_i^R f\ {}^* = g^{ij}\frac{\partial^2 f}{\partial x^i \partial x^j} - g^{ij}\{{}_i{}^k{}_j\}\frac{\partial f}{\partial x^k},$$

and $b = b^i(x)\dfrac{\partial}{\partial x^i}$ is the vector field given by

(4.33) $$b^i = \frac{1}{2} g^{mk}(\{{}_m{}^i{}_k\} - \Gamma^i_{mk}).$$

Indeed, considering $f(r) \equiv f(x)$ for $r = (x, e)$, we have

$$\begin{aligned} f(X(t)) - f(X(0)) &= f(r(t)) - f(r(0)) \\ &= \int_0^t (\tilde{L}_\alpha f)(r(s)) \circ dw^\alpha(s) \\ &= \int_0^t \tilde{L}_\alpha f(r(s)) dw^\alpha(s) + \frac{1}{2}\int_0^t \sum_{\alpha=1}^d \tilde{L}_\alpha(\tilde{L}_\alpha f)(r(s)) ds. \end{aligned}$$

Therefore it is sufficient to show that

$$\frac{1}{2}\sum_{\alpha=1}^d \tilde{L}_\alpha(\tilde{L}_\alpha f) = Af.$$

Note that A given by (4.31) may also be written as

$$A = \frac{1}{2} g^{ij}\nabla_i \nabla_j = \frac{1}{2}\left\{g^{ij}\frac{\partial f^2}{\partial x^i \partial x^j} - g^{ij}\Gamma^k_{ij}\frac{\partial f}{\partial x^k}\right\}.$$

By Proposition 4.1,

$$\tilde{L}_\alpha(\tilde{L}_\alpha f) = \tilde{L}_\alpha(F_{\nabla f})_\alpha = (F_{\nabla\nabla f})_{\alpha\alpha} = (\nabla_i \nabla_j f) e^i_\alpha e^j_\alpha.$$

Hence

$$\sum_{\alpha=1}^d \tilde{L}_\alpha(\tilde{L}_\alpha f) = \sum_{\alpha=1}^d (\nabla_i \nabla_j f) e^i_\alpha e^j_\alpha = g^{ij}\nabla_i \nabla_j f$$

---

* $\nabla^R = (\{{}_i{}^k{}_j\})$ is the Riemannian connection.

by (4.22). The above results are summarized below.

Theorem 4.2. Let M be a Riemannian manifold with an affine connection ∇ which is compatible with the Riemannian metric g, and let $\tilde{L}_1, \tilde{L}_2, \ldots, \tilde{L}_d$ be the system of canonical horizontal vector fields. Consider the stochastic differential equation (4.28) on $O(M)$. The solution defines a flow of diffeomorphisms $r(t) = (r(t,r,w))$ on $O(M)$ and its projection $X(t) = \pi[r(t)]$ defines a diffusion process on M corresponding to the differential operator A given by (4.31).

Definition 4.1. The process $r(t)$ in Theorem 4.2 is called the *horizontal lift* of the A-diffusion $X(t)$.

Definition 4.2. In the case $A = \Delta_M/2$, the A-diffusion $X(t)$ is called the *Brownian motion on M.*

Thus the horizontal lift $r(t)$ of a Brownian motion $X(t)$ on M is constructed by means of the Riemannian connection. Now we shall prove the following

Proposition 4.3. (i) For every vector field $b = b^i(x)\dfrac{\partial}{\partial x^i}$ on a Riemannian manifold M, there exists an affine connection $\nabla = \{\Gamma^k_{ij}\}$ on M compatible with the Riemannian metric g such that (4.33) holds.

(ii) Two affine connections $\nabla = \{\Gamma^k_{ij}\}$ and $\nabla' = \{\Gamma'^k_{ij}\}$ compatible with the Riemannian metric g satisfy

$$g^{jk}\Gamma^i_{jk} = g^{jk}\Gamma'^i_{jk} \qquad \text{for all} \quad i = 1, 2, \ldots, d \tag{4.34}$$

if and only if

$$T^n_{in} = T'^n_{in}, \quad i = 1, 2, \ldots, d, \tag{4.35}$$

where $\{T^k_{ij}\}$ and $\{T'^k_{ij}\}$ are the torsion tensors of ∇ and ∇' respectively.

Proof. (i) Define

$$\Gamma^i_{jk} = \{{}_j{}^i{}_k\} + \frac{2}{d-1}(\delta^i_j b_k - g_{jk}b^i)$$

where $b_i = g_{ij}b^j$. Then, since $\delta^i_j b_k - g_{jk}b^i$ are the components of a (1, 2)-tensor, $\nabla = \{\Gamma^i_{jk}\}$ satisfies (4.7) and hence defines an affine connection ([185]). It satisfies (4.18) since

$$\begin{aligned}
&\frac{\partial}{\partial x^l} g_{pq} - g_{mq}\Gamma^m_{lp} - g_{pm}\Gamma^m_{lq} \\
&= \frac{\partial}{\partial x^l} g_{pq} - g_{mq}\{{}_l{}^m{}_p\} - g_{pm}\{{}_l{}^m{}_q\} \\
&\quad - \frac{2}{d-1}(g_{mq}\delta^m_l b_p - g_{mq}g_{lp}b^m + g_{pm}\delta^m_l b_q - g_{pm}g_{lq}b^m) \\
&= -\frac{2}{d-1}(g_{lq}b_p - b_q g_{lp} + g_{pl}b_q - g_{lq}b_p) \\
&= 0.
\end{aligned}$$

Also,

$$\begin{aligned}
\frac{1}{2}g^{lk}(\{{}_l{}^i{}_k\} - \Gamma^i_{lk}) &= -\frac{1}{d-1}g^{lk}(\delta^i_l b_k - g_{lk}b^i) \\
&= -\frac{1}{d-1}(g^{ik}b_k - b^i d) = b^i.
\end{aligned}$$

(ii). First we note the following identity for any affine connection $\{\Gamma^i_{jk}\}$ compatible with g: if $\{T^i_{jk}\}$ is the torsion tensor and $S^i_{jk} = g^{im}g_{jn}T^n_{mk}$, then

$$\Gamma^i_{jk} = \{{}_j{}^i{}_k\} + \frac{1}{2}T^i_{jk} + \frac{1}{2}(S^i_{jk} + S^i_{kj}). \tag{4.36}$$

Indeed, by (4.18),

$$\begin{aligned}
&\frac{\partial}{\partial x^k} g_{sj} - g_{mj}\Gamma^m_{ks} - g_{sm}\Gamma^m_{kj} = 0, \\
&\frac{\partial}{\partial x^j} g_{sk} - g_{mk}\Gamma^m_{js} - g_{sm}\Gamma^m_{jk} = 0 \qquad \text{and} \\
&-\frac{\partial}{\partial x^s} g_{jk} + g_{mk}\Gamma^m_{sj} + g_{jm}\Gamma^m_{sk} = 0.
\end{aligned}$$

Hence by (4.19),

$$\begin{aligned}
\{{}_j{}^i{}_k\} &= \frac{1}{2}g^{is}\left(\frac{\partial}{\partial x^k} g_{sj} + \frac{\partial}{\partial x^j} g_{sk} - \frac{\partial}{\partial x^s} g_{jk}\right) \\
&= \frac{1}{2}g^{is}g_{mj}(\Gamma^m_{ks} - \Gamma^m_{sk}) + \frac{1}{2}g^{is}g_{mk}(\Gamma^m_{js} - \Gamma^m_{sj}) + \frac{1}{2}(\Gamma^i_{kj} + \Gamma^i_{jk})
\end{aligned}$$

$$= -\frac{1}{2}(S^i_{jk} + S^i_{kj}) + \frac{1}{2}(\Gamma^i_{kj} + \Gamma^i_{jk})$$

$$= -\frac{1}{2}(S^i_{jk} + S^i_{kj}) + \Gamma^i_{jk} - \frac{1}{2}T^i_{jk}.$$

This proves (4.36). Since $g^{jk}T^i_{jk} = 0$, (4.34) holds if and only if

$$g^{jk}S^i_{jk} = g^{jk}S'^i_{jk} \qquad \text{for all} \quad i = 1, 2, \ldots, d,$$

that is

$$g^{jk}g^{im}g_{jn}(T^n_{mk} - T'^n_{mk}) = g^{im}(T^n_{mn} - T'^n_{mn}) = 0, \quad i = 1, 2, \ldots, d.$$

Finally this is equivalent to

$$g_{ki}g^{im}(T^n_{mn} - T'^n_{mn}) = (T^n_{kn} - T'^n_{kn}) = 0, \quad k = 1, 2, \ldots, d. \qquad \text{q.e.d.}$$

From this we easily see that the correspondence between $\nabla = \{\Gamma^i_{jk}\}$ and b defined by (4.33) is a bijection if $d = 2$, while it is a many to one surjection if $d > 2$.

Let M be a differentiable manifold and A be a second order differential operator on M which is expressed in local coordinates as

$$(4.37) \qquad Af(x) = \frac{1}{2}a^{ij}(x)\frac{\partial^2 f}{\partial x^i \partial x^j}(x) + b^i(x)\frac{\partial f}{\partial x^i}(x), \quad f \in F(M)$$

where $(a^{ij}(x))$ is symmetric and non-negative definite.*[1] If $(a^{ij}(x))$ is strictly positive definite, i.e.,

$$(4.38) \qquad a^{ij}(x)\xi_i\xi_j > 0 \qquad \text{for all} \quad x \quad \text{and} \quad \xi = (\xi_i) \in \boldsymbol{R}^d \setminus \{0\},$$ *[2]

we say that the operator A is *non-degenerate*. The corresponding A-diffusion is also called *non-degenerate*. Now any non-degenerate diffusion on M can be constructed by Theorem 4.2. Indeed, let A be a non-degenerate differential operator. In a change of local coordinates, $(a^{ij}(x))$ and $(b^i(x))$ transform accordingly as

$$(4.39) \qquad \tilde{a}^{ij}(x) = a^{kl}(x)\frac{\partial \tilde{x}^i}{\partial x^k}\frac{\partial \tilde{x}^j}{\partial x^l}$$

*[1], *[2] These properties of $(a^{ij}(x))$ are independent of the choice of local coordinates.

and

$$(4.40)\qquad \bar{b}^i(x) = b^k(x)\frac{\partial \tilde{x}^i}{\partial x^k} + \frac{1}{2}a^{kl}(x)\frac{\partial^2 \tilde{x}^i}{\partial x^k \partial x^l}.$$

(4.39) implies that $(a^{ij}(x))$ is a tensor field of type (2,0) and hence its inverse matrix $(g_{ij}(x))$ defines a tensor field of type (0,2) which is also symmetric and positive definite. Thus, it defines a Riemannian metric g on M and so M is a Riemannian manifold. Then we have

$$(4.41)\qquad A = \frac{1}{2}\Delta_M + \bar{b}$$

where $\bar{b} = \bar{b}^i(x)\dfrac{\partial}{\partial x^i}$ is given by

$$(4.42)\qquad \bar{b}^i(x) = b^i(x) + \frac{1}{2}g^{ij}\left\{{}_{i}{}^{i}{}_{j}\right\}.$$

Obviously $\bar{b}$ is a vector field on M. Now choose an affine connection $\nabla = \{\Gamma^i_{jk}\}$ on M compatible with g such that

$$\bar{b}^i(x) = \frac{1}{2}g^{mk}\left(\left\{{}_{m}{}^{i}{}_{k}\right\} - \Gamma^i_{mk}\right).*$$

This is always possible by Proposition 4.3. The A-diffusion is now constructed as in Theorem 4.2.

Remark 4.2. In this construction of an A-diffusion we have made use of an affine connection. An A-diffusion can also be constructed using only the Riemannian connection ∇^R. In this case, we first construct a flow of diffeomorphisms $r(t) = (r(t, r, w))$ on $O(M)$ as the solution of the stochastic differential equation

$$(4.43)\qquad dr(t) = \tilde{L}_\alpha(r(t))\circ dw^\alpha(t) + \tilde{L}_0(r(t))dt,$$

where $(\tilde{L}_1, \tilde{L}_2, \ldots, \tilde{L}_d)$ is the system of canonical horizontal vector fields corresponding to ∇^R, and the drift vector field $\tilde{L}_0$ is the horizontal lift (with respect to ∇^R) of the vector field $\bar{b}$. The A-diffusion $X(t)$ is then obtained by the projection: $X(t) = \pi[r(t)]$.

Finally, we shall study some problems related to the invariant measure

* $g^{mk} = a^{mk}$ by definition.

of a non-degenerate A-diffusion. For simplicity, we shall assume that M is compact and orientable. As we saw above, we may assume without loss of generality that M is a Riemannian manifold and A is of the form

$$A = \frac{1}{2}\Delta_M + b, \tag{4.44}$$

where Δ_M is the Laplace-Beltrami operator and $b \in \mathfrak{X}(M)$. Let $\{P_x\}$ be the system of diffusion measures determined by A (i.e., the A-diffusion). P_x is a probability measure on $\hat{W}(M)$.*[1] The *transition semigroup T_t of the A-diffusion* is defined by

$$(T_t f)(x) = \int_{W(M)} f(w(t)) P_x(dw), \qquad f \in C(M). \tag{4.45}$$

Let Ω be a domain (i.e., a connected open set) in M and define $\rho^\Omega w \in \hat{W}(\Omega)$, $w \in \hat{W}(M)$ by

$$(\rho^\Omega w)(t) = \begin{cases} w(t) & \text{if } t < \tau_\Omega(w) \\ \Delta & \text{if } t \geq \tau_\Omega(w) \end{cases}$$

where $\tau_\Omega(w) = \inf\{t;\ w(t) \notin \Omega\}$. The image measure of P_x $(x \in \Omega)$ under the mapping ρ^Ω is denoted by P_x^Ω. It is a probability measure on $\hat{W}(\Omega)$. As is easily seen, $\{P_x^\Omega\}_{x\in\Omega}$ defines a diffusion on Ω and it is called the *minimal A-diffusion on Ω*. Its transition semigroup is defined by

$$\begin{aligned} (T_t^\Omega f)(x) &= \int_{W(\Omega)} f(w(t)) P_x^\Omega(dw) \\ &= \int_{W(M)} f(w(t)) I_{\{\tau_\Omega(w) > t\}} P_x(dw), \quad f \in C_b(\Omega).^{*2} \end{aligned} \tag{4.46}$$

Definition 4.3. A Borel measure $\mu(dx)$ on M is called an *invariant measure* of the A-diffusion $\{P_x\}$ if

$$\int_M T_t f(x)\mu(dx) = \int_M f(x)\mu(dx) \qquad \text{for all } f \in C(M). \tag{4.47}$$

Definition 4.4. (i). An A-diffusion $\{P_x\}$ is said to be *symmetrizable* if there exists a Borel measure $\nu(dx)$ on M such that

*[1] Since M is compact, $\hat{W}(M) = W(M)$: the set of all continuous paths in M.
*[2] We set $f(\Delta) = 0$ for $f \in C_b(\Omega)$.

(4.48) $$\int_M T_t f(x)g(x)\nu(dx) = \int_M f(x)(T_t g)(x)\nu(dx)$$
for all $f, g \in C(M)$.

(ii). An A-diffusion $\{P_x\}$ is said to be *locally symmetrizable* if $\{P_x^{\Omega}\}$ is symmetrizable for every simply connected domain $\Omega \subset M$, i.e., if there exists a Borel measure $\nu^{\Omega}(dx)$ on Ω such that

(4.49) $$\int_{\Omega} T_t^{\Omega} f(x)g(x)\nu^{\Omega}(dx) = \int_{\Omega} f(x)T_t^{\Omega} g(x)\nu^{\Omega}(dx)$$
for all $f, g \in C_b(\Omega)$.

It is clear that if $\{P_x\}$ is symmetrizable, the measure ν in (4.48) is an invariant measure.

A differential 1-form ω_b is defined from the vector field b by

(4.50) $$\omega_b = b_i(x)dx^i,$$

where $b = b^i \dfrac{\partial}{\partial x^i}$ and $b_i = g_{ij}b^j$ in local coordinates. By a well-known theorem of de Rham-Kodaira ([145]), ω_b has the following orthogonal decomposition:*

(4.51) $$\omega_b = dF + \delta\beta + \alpha$$

where $F \in F(M)$, β is a 2-form and α is a harmonic 1-form. Here we briefly recall some of necessary notions. An inner product is defined on the totality $A_p(M)$ of all p-forms on M by

(4.52) $$(\alpha, \beta)_p = \int_M \langle \alpha, \beta \rangle dx,$$

where $\alpha = \sum_{i_1<i_2<\cdots<i_p} \alpha_{i_1,i_2,\ldots,i_p} dx^{i_1} \wedge dx^{i_2} \wedge \cdots \wedge dx^{i_p}$, $\beta = \sum_{i_1<i_2<\cdots<i_p} \beta_{i_1,i_2,\ldots,i_p} dx^{i_1} \wedge dx^{i_2} \wedge \cdots \wedge dx^{i_p}$, $\beta^{i_1,i_2,\ldots,i_p} = g^{i_1j_1}g^{i_2j_2} \cdots g^{i_pj_p}\beta_{j_1,j_2,\ldots,j_p}$, $\langle \alpha, \beta \rangle = \sum_{i_1<i_2<\cdots<i_p} \alpha_{i_1,i_2,\ldots,i_p}(x)\beta^{i_1,i_2,\ldots,i_p}(x)$ and dx is a volume element defined by

$$dx = \sqrt{\det(g_{ij}(x))}\, dx^1dx^2 \cdots dx^d.$$

The operator $\delta\colon A_p(M) \longrightarrow A_{p-1}(M)$ is defined by

* With respect to the inner product defined by (4.52) below.

(4.53) $\quad (d\alpha, \beta)_p = (\alpha, \delta\beta)_{p-1}, \quad \alpha \in A_{p-1}(M), \beta \in A_p(M).$

De Rham-Kodaira's Laplacian $\square$: $A_p(M) \longrightarrow A_p(M)$ is defined by

(4.54) $\quad \square = -(d\delta + \delta d).$

$\alpha \in A_p(M)$ is called *harmonic* if $\square\alpha = 0$. The totality of all harmonic p-forms is denoted by $H_p(M)$. It is known that $\square\alpha = 0$ if and only if $d\alpha = 0$ and $\delta\alpha = 0$. For $f \in F(M)$, grad $f \in \mathfrak{X}(M)$ is defined by

(4.55) $$\operatorname{grad} f = g^{ij} \frac{\partial f}{\partial x^j} \frac{\partial}{\partial x^i},$$

and for $X \in \mathfrak{X}(M)$, div$X \in F(M)$ is defined by

(4.56) $$\begin{aligned} \operatorname{div} X &= -\delta\omega_X \\ &= \frac{1}{\sqrt{\det G}} \frac{\partial}{\partial x^i} (X^i \sqrt{\det G}).^* \end{aligned}$$

Then the Laplace-Beltrami operator (4.32) is also given as

(4.57) $$\Delta_M f = \operatorname{div}(\operatorname{grad} f) = -\delta df \quad \text{for} \quad f \in F(M).$$

It is easy to see that μ is an invariant measure of the A-diffusion $\{P_x\}$ if and only if

(4.58) $$\int_M Af(x)\mu(dx) = 0 \qquad \text{for all} \quad f \in F(M).$$

Indeed, we know by Theorem 3.1 that $u(t,x) = T_t f(x)$, $f \in F(M)$, is the unique solution of

$$\begin{cases} \dfrac{\partial u}{\partial t} = Au(t,x) \\ \lim\limits_{t \downarrow 0, y \to x} u(t,y) = f(x). \end{cases}$$

Also, we saw in the proof of Theorem 3.1 that $Au(t,x) = T_t(Af)(x)$. Hence if (4.47) holds, we have by differentiating with respect to t that

* $G = (g_{ij})$ and $X = X^i \dfrac{\partial}{\partial x^i}$.

$$\int_M T_t(Af)(x)\mu(dx) = 0 \qquad \text{for all} \quad f \in F(M).$$

We obtain (4.58) by letting $t \downarrow 0$. Conversely, if (4.58) is satisfied, then

$$0 = \int_M (Au)(t,x)\mu(dx) = \frac{d}{dt}\int_M T_t f(x)\mu(dx).$$

Consequently (4.47) holds for $f \in F(M)$ and hence for $f \in \boldsymbol{C}(M)$.

Proposition 4.4.

$$(Af, h)_0 = (f, A^*h)_0,^{*1} \qquad f, h \in F(M),$$

where

$$(4.59) \qquad A^*h = -\frac{1}{2}\delta dh + \delta(h\omega_b) = \frac{1}{2}\Delta_M h - \operatorname{div}(hb).^{*2}$$

The proof is immediate from the definition.

Proposition 4.5. An invariant measure $\mu(dx)$ of the A-diffusion exists and is unique up to a multiplicative constant; moreover, $\mu(dx)$ is given as $v(x)dx$ where $v \in F(M)$ is a solution of

$$(4.60) \qquad A^*v = 0.$$

Proof. The equation (4.58) is equivalent to $A^*\mu = 0$ in the sense of Schwartz distributions on M. Since A^* is elliptic, any solution must be of the form $\mu = vdx$, $v \in F(M)$, by Weyl's lemma ([1]). Furthermore, $\lambda = 0$ is the largest eigenvalue of the eigenvalue problem $(A - \lambda)\phi = 0$, it is simple and $\{c\phi_0;\ c \in \boldsymbol{R}\}$ is the eigenspace where $\phi_0(x) \equiv 1$. Therefore, $\lambda = 0$ is also the largest eigenvalue of the eigenvalue problem $(A^* - \lambda)\phi = 0$, it is simple and its associated eigenspace is given as $\{c\phi_0^*;\ c \in \boldsymbol{R}\}$ where we can choose ϕ_0^* such that $\phi_0^*(x) > 0$ for all $x \in M$.*3 Hence all invariant measures of the A-diffusion are given as $\mu = c\phi_0^* dx$ for some constant $c > 0$.

---

*1 $(f, g)_0 = \int_M f(x)g(x)dx$, $dx = \sqrt{\det G}\, dx^1 dx^2 \cdots dx^d$.

*2 For $f \in F(M)$ and $\alpha \in \Lambda_p(M)$, $\alpha = \alpha_{i_1,i_2,\ldots,i_p} dx^{i_1} \wedge dx^{i_2} \wedge \cdots \wedge dx^{i_p}$, $f\alpha$ is defined by $f\alpha = (f\alpha_{i_1,i_2,\ldots,i_p})dx^{i_1} \wedge dx^{i_2} \wedge \cdots \wedge dx^{i_p}$.

*3 This fact is well known. It is also a consequence of the theory of positive operators [89].

We choose $U(x) \in F(M)$ so that $A^*(e^{-U}) = 0$, i.e., $e^{-U}dx$ is an invariant measure. Since $A^*(e^{-U}) = -\frac{1}{2}\delta d(e^{-U}) + \delta(e^{-U}\omega_b) = 0$, $e^{-U}\omega_b - \frac{1}{2}d(e^{-U}) = \delta\beta_1 + \alpha_1$ for some $\beta_1 \in \Lambda_2(M)$ and $\alpha_1 \in H_1(M)$, i.e.,

$$(4.61) \qquad e^{-U}\omega_b = \frac{1}{2}d(e^{-U}) + \delta\beta_1 + \alpha_1,$$

where $\beta_1 \in \Lambda_2(M)$ and $\alpha_1 \in H_1(M)$. Conversely, if U satisfies (4.61) with some β_1 and α_1 then

$$A^*(e^{-U}) = \delta\left(e^{-U}\omega_b - \frac{1}{2}d(e^{-U})\right) = \delta(\delta\beta_1 + \alpha_1) = 0$$

and hence $e^{-U(x)}dx$ is an invariant measure. (4.61) gives the de Rham-Kodaira decomposition of the 1-form $e^{-U}\omega_b$.

Theorem 4.6. (i). The A-diffusion is symmetrizable if and only if

$$(4.62) \qquad \delta\beta = \alpha = 0 \qquad \text{in} \quad (4.51);^*$$

and this is equivalent to

$$(4.63) \qquad \delta\beta_1 = \alpha_1 = 0 \qquad \text{in} \quad (4.61).$$

The condition (4.62) or (4.63) is equivalent to the condition that b be given as

$$(4.64) \qquad b = \text{grad } F, \qquad F \in F(M),$$

and in this case the invariant measures are of the form constant $\times\, e^{2F(x)}\, dx$.

(ii). The A-diffusion is locally symmetrizable if and only if

$$(4.65) \qquad \delta\beta = 0 \quad \text{in} \quad (4.51);$$

this condition is equivalent to

$$(4.66) \qquad d\omega_b = 0.$$

---

* [88] and [127].

(iii). The A-diffusion has a measure cdx ($c > 0$: constant) as its invariant measure if and only if $dF = 0$ in (4.51) and this condition is equivalent to

$$\delta\omega_b = -\text{div } b = 0. \tag{4.67}$$

Corollary. The A-diffusion is symmetric with respect to the Riemannian volume dx (i.e., it is symmetrizable and the measure ν in (4.48) is dx) if and only if it is the Brownian motion on M.

Proof. Let $U_0(x) \in F(M)$ be determined by

$$\int_M e^{-U_0(x)}dx = 1 \quad \text{and} \quad A^*(e^{-U_0}) = 0.$$

Then the measure $e^{-U_0(x)}dx$ is the unique invariant probability measure of the A-diffusion. If we introduce another inner product on $F(M)$ by

$$\langle u, v\rangle = \int_M u(x)v(x)\, e^{-U_0(x)}dx, \tag{4.68}$$

then it is easy to see that

$$\langle Au, v\rangle = \langle u, \hat{A}v\rangle$$

where

$$\hat{A}v = \frac{1}{2}\Delta_M v - (b + \text{grad } U_0)v. \tag{4.69}$$

Suppose the A-diffusion is symmetrizable. As we remarked before, the measure ν in (4.48) is an invariant measure, and hence if ν is normalized so that it is a probability measure, we must have $\nu(dx) = e^{-U_0(x)}dx$. Thus $\langle T_t u, v\rangle = \langle u, T_t v\rangle$ for $u, v \in F(M)$ by (4.48), and so by differentiating with respect t we obtain

$$\langle T_t Au, v\rangle = \langle u, T_t Av\rangle.$$

Letting $t \downarrow 0$, we have $\langle Au, v\rangle = \langle u, Av\rangle$, and consequently $\hat{A}v = Av$ for all $v \in F(M)$, that is,

$$b = -(b + \text{grad } U_0). \tag{4.70}$$

Hence

$$b = -\frac{1}{2}\text{grad } U_0.$$

Conversely, if $b = \text{grad } F$, $F \in F(M)$, then $\omega_b = dF$ and so $e^{2F}\omega_b = \frac{1}{2}d(e^{2F})$. Therefore $U = -2F$ satisfies the equation (4.61) with $\delta\beta_1 = \alpha_1 = 0$. Hence $e^{-U}dx$ is an invariant measure and $U = U_0 + c$ for some constant c. Consequently we have $b = -\frac{1}{2}\text{grad } U_0$ and this implies that $\hat{A}v = Av$, $v \in F(M)$. But generally, the A-diffusion $\{P_x\}$ and $\hat{A}$-diffusion $\{\hat{P}_x\}$ are connected to each other, through their transition semigroups T_t and $\hat{T}_t$ respectively, by the relation

$$\langle T_t u, v\rangle = \langle u, \hat{T}_t v\rangle, \qquad u, v \in F(M).$$

Indeed,

$$\begin{aligned}\frac{d}{ds}\langle T_{t-s}u, \hat{T}_s v\rangle &= \langle -AT_{t-s}u, \hat{T}_s v\rangle + \langle T_{t-s}u, \hat{A}\hat{T}_s v\rangle \\ &= -\langle T_{t-s}u, \hat{A}\hat{T}_s v\rangle + \langle T_{t-s}u, \hat{A}\hat{T}_s v\rangle \\ &= 0,\end{aligned}$$

and hence

$$\begin{aligned}\langle T_t u, v\rangle - \langle u, \hat{T}_t v\rangle &= -\int_0^t \frac{d}{ds}\langle T_{t-s}u, \hat{T}_s v\rangle ds \\ &= 0.\end{aligned}$$

Therefore $\hat{A} = A$ implies that $\langle T_t u, v\rangle = \langle u, T_t v\rangle$ and consequently the A-diffusion is symmetrizable. Now the proof of (i) is completed. The proof of (ii) can be given similarly. Finally, the measure dx is an invariant measure if and only if the function U in (4.61) is a constant, that is, $\omega_b = \delta\beta + \alpha$, for some $\beta \in \Lambda_2(M)$ and $\alpha \in H_1(M)$.

Remark 4.3. It might be interesting to give an probabilistic interpretation for conditions like "$\alpha = 0$ in (4.51)", "$\alpha_1 = 0$ in (4.61)" or "$\beta_1 = 0$ in (4.61)". In this connection, Manabe [109] obtained the following result. He first defined "the stochastic Kronecker index" $I(X[0, t], c)$ between a $d-1$ chain c in M and an orbit $X[0, t] = \{X(s); s \in [0, t]\}$ of A-diffusion

$X(t)$ as a stochastic process. He then showed that $\alpha_1 = 0$ in (4.61) if and only if for every $d-1$ cycle c

$$\lim_{t\uparrow\infty} I(X[0, t], c)/t = 0 \qquad \text{a.s.}$$

5. Stochastic parallel displacement and heat equation for tensor fields

Let M be a Riemannian manifold and Δ be the Laplace-Beltrami operator. Then, as we saw above, the solution of the heat equation

$$(5.1) \qquad \begin{cases} \dfrac{\partial u}{\partial t} = \dfrac{1}{2}\Delta u \\ u|_{t=0} = f \end{cases}$$

can be solved uniquely as

$$(5.2) \qquad u(t,x) = E[f(X(t,x))]$$

where $X(t,x)$ is a Brownian motion on M starting at x. In order to generalize this fact to the case of heat equation for tensor fields, Itô ([68], [72]) introduced the notion of *stochastic parallel displacement*. His idea is as follows. Consider the equation (5.1) where u and f are now tensor fields on M and $\Delta u = g^{ij}\nabla_i\nabla_j u$, ∇ being the covariant differentiation with respect to the Riemannian connection. Then the solution u is given by

$$(5.3) \qquad u(t,x) = E[f(X(t,x))_*],$$

where $f(X(t,x))_*$ is a tensor at x obtained from the tensor $f(X(t, x))$ at $X(t, x)$ by parallel displacement along the *time reversed* Brownian curve. The difficulty in this procedure is in obtaining $f(X(t,x))_*$ as the parallel translate of $f(X(t,x))$ along the time reversed Brownian curve. Itô defined it as a limit of the parallel translate along a piece wise geodesic curve from $X(t,x)$ to x which approximates the time reversed Brownian curve. But as we saw in the previous section, we can use stochastic calculus to perform a parallel displacement along the Brownian curve from x to $X(t,x)$ in the usual sense of time. Moreover, we can actually realize Itô's idea using only such parallel displacements: instead of translating a tensor at $X(t,x)$ to a tensor at x, we translate an orthonormal frame e at x to an orthonormal frame $e(t)$ at $X(t,x)$ along the Brownian curve and *read the tensor at* $X(t, x)$ *using this frame* $e(t)$. This approach is due to Malliavin [104]. It is now clear that the process $r(t) = (X(t,x), e(t))$ is the horizontal lift on $O(M)$ of the Brownian curve $X(t,x)$ constructed in the previous section. The heat equation (5.1) for tensor fields is solved in this manner. By modifying the expectation with a Feynman-Kac type weight we can also solve the heat equation for differential forms

$$(5.4) \qquad \begin{cases} \dfrac{\partial \alpha}{\partial t} = \dfrac{1}{2} \square \alpha \\ \alpha \,|_{t=0} = f \end{cases}$$

where $\square$ is the Laplacian of de Rham-Kodaira (4.54).

Let M be a compact Riemannian manifold and $O(M)$ be the bundle of orthonormal frames over M. Let $\{\tilde{L}_1, \tilde{L}_2, \ldots, \tilde{L}_d\}$ be the system of canonical horizontal vector fields on $O(M)$ with respect to the Riemannian connection $\{{}_j{}^i{}_k\}$. As in the previous section, the flow of diffeomorphisms $r(t) = (r(t,r,w))$ on $O(M)$ is defined by the solution of the stochastic differential equation

$$(5.5) \qquad \begin{cases} dr(t) = \tilde{L}_\alpha(r(t)) \circ dw^\alpha(t) \\ r(0) = r. \end{cases}$$

$r(t)$ defines a diffusion process on $O(M)$ which corresponds to the differential operator $\frac{1}{2} \Delta_{O(M)}$, where

$$(5.6) \qquad \Delta_{O(M)} = \sum_\alpha \tilde{L}_\alpha(\tilde{L}_\alpha).$$

$r = (r(t))$ is called the *horizontal Brownian motion on* $O(M)$ and $\Delta_{O(M)}$ is called the *horizontal Laplacian of Bochner*. As we saw in the previous section, the projection $X(t) = \pi[r(t)]$ is the Brownian motion on M. We write $r(t, r, w) = (X(t, r, w), e(t, r, w))$. We know that

$$(5.7) \qquad r(t, r\cdot a, w) = r(t, r, aw)\cdot a$$

i.e.

$$(5.8) \qquad X(t, r\cdot a, w) = X(t, r, aw) \quad \text{and} \quad e(t, r\cdot a, w) = e(t, r, aw)a$$

for every $a \in O(d)$.*

* See (2.31) for the definition of the action $r\cdot a$ of a on r.

Let $T(x) = (T^{i_1 i_2 \cdots i_p}_{j_1 j_2 \cdots j_q}(x))$ be a (p, q)-tensor field and $F_T(r) = (F_T{}^{i_1 i_2 \cdots i_p}_{j_1 j_2 \cdots j_q}(r))$ be its scalarization. It is a system of smooth functions on $O(M)$. Let Δ be the Laplace-Beltrami operator acting on tensor fields defined by

$$(5.9) \qquad (\Delta T)^{i_1 i_2 \cdots i_p}_{j_1 j_2 \cdots j_q} = g^{ij}(\nabla(\nabla T))^{i_1 i_2 \cdots i_p}_{j_1 j_2 \cdots j_q ij} = g^{ij} T^{i_1 i_2 \cdots i_p}_{j_1 j_2 \cdots j_q ; i ; j},$$

where ∇T is the covariant derivative of T with respect to the Riemannian connection. By Proposition 4.1, we have

$$(5.10) \qquad (\Delta_{O(M)} F_T{}^{i_1 i_2 \cdots i_p}_{j_1 j_2 \cdots j_q})(r) = F_{\Delta T}{}^{i_1 i_2 \cdots i_p}_{j_1 j_2 \cdots j_q}(r)$$
$$\text{for every} \quad i_1, i_2, \ldots, i_p, j_1, j_2, \ldots, j_q.$$

Indeed

$$\begin{aligned}
&\Delta_{O(M)}(F_T{}^{i_1 i_2 \cdots i_p}_{j_1 j_2 \cdots j_q}) \\
&= \sum_\alpha \tilde{L}_\alpha \tilde{L}_\alpha (F_T{}^{i_1 i_2 \cdots i_p}_{j_1 j_2 \cdots j_q}) \\
&= \sum_\alpha F_{\nabla\nabla T}{}^{i_1 i_2 \cdots i_p}_{j_1 j_2 \cdots j_q \alpha\alpha} \\
&= \sum_\alpha e^i_\alpha e^j_\alpha e^{l_1}_{j_1} e^{l_2}_{j_2} \cdots e^{l_q}_{j_q} f^{i_1}_{k_1} f^{i_2}_{k_2} \cdots f^{i_p}_{k_p} (\nabla\nabla T)^{k_1 k_2 \cdots k_p}_{l_1 l_2 \cdots l_q ij} \\
&= e^{l_1}_{j_1} e^{l_2}_{j_2} \cdots e^{l_q}_{j_q} f^{i_1}_{k_1} f^{i_2}_{k_2} \cdots f^{i_p}_{k_p} (\Delta T)^{k_1 k_2 \cdots k_p}_{l_1 l_2 \cdots l_q} \\
&= (F_{\Delta T})^{i_1 i_2 \cdots i_p}_{j_1 j_2 \cdots j_q}
\end{aligned}$$

since $\sum_\alpha e^i_\alpha e^j_\alpha = g^{ij}$.

For a given (p, q)-tensor field $f = (f(x))$, we define a system of functions $U^{i_1 i_2 \cdots i_p}_{j_1 j_2 \cdots j_q}(t, r)$ on $[0, \infty) \times O(M)$ by

$$(5.11) \qquad U^{i_1 i_2 \cdots i_p}_{j_1 j_2 \cdots j_q}(t, r) = E[(F_f)^{i_1 i_2 \cdots i_p}_{j_1 j_2 \cdots j_q}(r(t, r, w))]$$

for every $i_1, i_2, \ldots, i_p, j_1, j_2, \ldots, j_q = 1, 2, \ldots, d$. By Theorem 3.1 $U^{i_1 i_2 \cdots i_p}_{j_1 j_2 \cdots j_q}$ is the unique solution of the heat equation

$$(5.12) \qquad \begin{cases} \dfrac{\partial V}{\partial t} = \dfrac{1}{2} \Delta_{O(M)} V \\ V|_{t=0} = (F_f)^{i_1 i_2 \cdots i_p}_{j_1 j_2 \cdots j_q}. \end{cases}$$

Since $F_f = \{(F_f)^{i_1 i_2 \cdots i_p}_{j_1 j_2 \ldots j_q}\}$ is $O(d)$-equivariant, we can easily prove by (5.7) and (5.8) that $U(t) = \{U^{i_1 i_2 \cdots i_p}_{j_1 j_2 \ldots j_q}(t, \cdot)\}$ is $O(d)$-equivariant for every $t \geq 0$. Therefore there exists a unique (p,q)-tensor field $u(t, \cdot) = \{u^{i_1 i_2 \cdots i_p}_{j_1 j_2 \ldots j_q}(t, \cdot)\}$ such that

$$U^{i_1 i_2 \cdots i_p}_{j_1 j_2 \ldots j_q}(t,r) = (F_{u(t,\cdot)})^{i_1 i_2 \cdots i_p}_{j_1 j_2 \ldots j_q}(r).$$

By (5.10), it is clear that $u(t, \cdot)$ is the unique solution of (5.1). In this way the heat equation (5.1) for tensor fields can be solved uniquely.

Next we consider the equation (5.4) for differential forms. Let

$$\alpha(x) = \sum_{i_1 < i_2 < \cdots < i_p} \alpha_{i_1 i_2 \cdots i_p}(x) dx^{i_1} \wedge dx^{i_2} \wedge \cdots \wedge dx^{i_p}$$

be a p-form. We agree to define $\alpha_{i_1 i_2 \cdots i_p}(x)$ for all system of indices by the alternation property, so that $\alpha_{i_1 i_2 \cdots i_p}(x) = 0$ unless the indices i_1, $i_2, \cdots, i_p$ are all distinct and $\alpha_{i_1 i_2 \cdots i_p}(x) = \mathrm{sgn}(\sigma)\alpha_{\sigma(i_1)\sigma(i_2)\cdots\sigma(i_p)}(x)$ for any permutation σ. Then.

$$\alpha(x) = \frac{1}{p!} \alpha_{i_1 i_2 \cdots i_p}(x) dx^{i_1} \wedge dx^{i_2} \wedge \cdots \wedge dx^{i_p}.$$

Following [145] *or* [128], *we identify* $\alpha(x)$ (*in a coordinate neighborhood*) *with the system of functions* $(\alpha_{i_1 i_2 \cdots i_p}(x))$ *and write* $\alpha(x) = (\alpha_{i_1 i_2 \cdots i_p}(x))$. *Note that these components* $(\alpha_{i_1 i_2 \cdots i_p}(x))$ *of p-form* α *are* $p!$ *times of those when* α *is regarded as an alternate* $(0, p)$ *tensor.*

Since the Riemannian connection is torsion free, we have

$$(5.13) \qquad (d\alpha)_{i_1 i_2 \ldots i_{p+1}} = \sum_{\nu=1}^{p+1} (-1)^{\nu-1} \nabla_{i_\nu} \alpha_{i_1 \ldots \hat{i}_\nu \ldots i_{p+1}}$$

where the circumflex denotes omission. Combining this with (4.53) it is easy to see that

$$(5.14) \qquad (\delta\alpha)_{i_1 i_2 \ldots i_{p-1}} = - g^{ik} \nabla_k \alpha_{i\, i_1 \ldots i_{p-1}} := - \nabla^i \alpha_{i\, i_1 \ldots i_{p-1}} \cdot$$

By (5.13) and (5.14),

$$(d\delta\alpha)_{i_1 i_2 \ldots i_p} = \sum_{\nu=1}^{p} (-1)^{\nu} \nabla_{i_\nu} \nabla^i \alpha_{i\, i_1 \ldots \hat{i}_\nu \ldots i_p}$$

and

$$(\delta d\alpha)_{i_1 i_2 \cdots i_p} = -\nabla^i \nabla_i \alpha_{i_1 i_2 \cdots i_p} - \sum_{\nu=1}^{p} (-1)^\nu \nabla^i \nabla_{i_\nu} \alpha_{i\, i_1 \cdots \hat{i}_\nu \cdots i_p} \,.$$

Then it is clear that

$$(5.15) \qquad (\Box \alpha)_{i_1 i_2 \ldots i_p} = \nabla^i \nabla_i \alpha_{i_1 i_2 \ldots i_p} - \sum_{\nu=1}^{p} (-1)^\nu (\nabla_{i_\nu} \nabla^i - \nabla^i \nabla_{i_\nu}) \alpha_{i i_1 \ldots \hat{i}_\nu \ldots i_p} \,.$$

Notice Ricci's identity

$$(\nabla_l \nabla^k - \nabla^k \nabla_l) \alpha_{i_1 i_2 \ldots i_p} = \sum_{\nu=1}^{p} R^{h \cdot k \cdot}_{\cdot i_\nu \cdot l} \, \alpha_{i_1 \ldots i_{\nu-1} h i_{\nu+1} \ldots i_p}$$

where

$$R^i{}_{jkl} = \frac{\partial}{\partial x^k} \left\{ {}_l{}^i{}_j \right\} - \frac{\partial}{\partial x^l} \left\{ {}_k{}^i{}_j \right\} + \left(\left\{ {}_l{}^a{}_j \right\} \left\{ {}_k{}^i{}_a \right\} - \left\{ {}_k{}^a{}_j \right\} \left\{ {}_l{}^i{}_a \right\} \right)$$

is the component of the curvature tensor and $R^{i \cdot k \cdot}_{\cdot j \cdot l} = g^{ka} R^i{}_{jal}$ [145]. Applying this identity to (5.15) we see that (5.15) can be rewritten as

$$(5.16) \qquad \begin{aligned} (\Box \alpha)_{i_1 i_2 \ldots i_p} &= (\Delta \alpha)_{i_1 i_2 \ldots i_p} - \sum_{\nu=1}^{p} (-1)^\nu R^{h \cdot k \cdot}_{\cdot k \cdot i_\nu} \, \alpha_{h\, i_1 \ldots \hat{i}_\nu \ldots i_p} \\ &\quad - 2 \sum_{\mu < \nu} (-1)^{\mu+\nu} R^{h \cdot k \cdot}_{\cdot i_\nu \cdot i_\mu} \, \alpha_{k h i_1 \ldots \hat{i}_\mu \ldots \hat{i}_\nu \ldots i_p} \,. \end{aligned}$$

The relation (5.16) is called *Weitzenböck's formula* ([128], [145]). For the 1-form $\alpha = \alpha_i(x) dx^i$,

$$(5.17) \qquad (\Box \alpha)_i = (\Delta \alpha)_i + R_i^j \alpha_j$$

where $R_i^j = R^{j \cdot k \cdot}_{\cdot k \cdot i}$. In exactly the same way as above, we see that the equation (5.4) is equivalent to the following equation for alternate and $O(d)$-equivariant systems $V(t, \cdot) = \{V_{i_1 i_2 \ldots i_p}(t, r)\}$ of functions on $O(M)$

$$(5.18) \qquad \begin{aligned} &\frac{\partial}{\partial t} V_{i_1 i_2 \ldots i_p}(t, r) \\ &\quad = \frac{1}{2} \Delta_{O(M)} V_{i_1 i_2 \ldots i_p}(t, r) - \frac{1}{2} \sum_{\nu=1}^{p} (-1)^\nu J^h_{i_\nu}(r) V_{h i_1 \ldots \hat{i}_\nu \ldots i_p}(t, r) \\ &\qquad - \sum_{\mu < \nu} (-1)^{\mu+\nu} J^{hk}_{i_\nu i_\mu}(r) V_{k h i_1 \cdots \hat{i}_\mu \cdots \hat{i}_\nu \cdots i_p}(t, r) \end{aligned}$$

$$: = \frac{1}{2}(\square_{O(M)} V)_{i_1 i_2 \cdots i_p}(t, r),$$

$$V_{i_1 i_2 \cdots i_p}(0, r) = (F_f)_{i_1 i_2 \cdots i_p}(r),$$

where $\{J^j{}_i(r)\}$ is the scalarization of $R^j{}_i = R^{j\cdot k}_{\cdot k\cdot i}$ and $\{J^i{}_j{}^k{}_l(r)\}$ is the scalarization of $\{R^{i\cdot k}_{\cdot j\cdot l}\}$. Just note that $\square_{O(M)} F_\alpha = F_{\square\alpha}$ for $\alpha \in \Lambda_p(M)$ by (5.16).

By using the tensor $\{R^i{}_{jkl}\}$ we can form various mixed tensors by raising and lowering indices via

$$R_{ijkl} = g_{ia} R^a{}_{jkl}, \quad R_{i\cdot k\cdot l}^{\cdot j\cdot\cdot} = g^{ja} R_{iakl} \quad \text{etc.}$$

Then a straightforward calculation in local coordinates shows that

$$R_{ijkl} = R_{klij} = - R_{ijlk}.$$

The Ricci tensor R_{ij} is defined by

$$R_{ij} = R_{ik\cdot j}^{\cdot\cdot k\cdot} (= R^k{}_{ijk})$$

Then we have $R_{ij} = R_{ji}$. We note that $J^j{}_i(r) = J_{ji}(r)$ and $J^i{}_j{}^k{}_l(r) = J_{ijkl}(r)$ for all i, j, k, l where $(J_{ji}(r))$ and $(J_{ijkl}(r))$ are scalarizations of $(R_{ji}(x))$ and $(R_{ijkl}(x))$, respectively: It is a general fact that raising and lowering indices of components are irrelevant for scalarizations of tensors.

The equation (5.18) is apparently complicated and so following [194] we now rewrite it into more compact and manageable form by introducing several algebraic notions, especially exterior and interior multiplications (creation and annihilation operators) over exterior (Grassman) algebra.

Let $\boldsymbol{R}^d$ be the d-dimensional Euclidean space and $\delta^1, \delta^2, \cdots, \delta^d$ be the canonical basis, i.e.,

$$\delta^i = (0, \cdots, 0, \overset{i-th}{1}, 0, \cdots, 0).$$

Let $\Lambda\boldsymbol{R}^d = \sum_{p=0}^{d} \oplus \overset{p}{\Lambda}\boldsymbol{R}^d$ be the exterior (= Grassman) algebra over $\boldsymbol{R}^d$: $\overset{p}{\Lambda}\boldsymbol{R}^d$ is the Euclidean space of dimension $\binom{d}{p}$ with basis

$$\delta^{i_1} \wedge \delta^{i_2} \wedge \cdots \wedge \delta^{i_p}, \qquad 1 \leq i_1 < i_2 < \cdots < i_p \leq d$$

and hence, $\Lambda\boldsymbol{R}^d$ is the Euclidean space of dimension 2^d. An element

ω of $\overset{p}{\Lambda}\boldsymbol{R}^d$ is uniquely expressed as

$$\omega = \sum_{i_1<i_2<\cdots<i_p} \omega_{i_1 i_2\cdots i_p}\, \delta^{i_1}\wedge\delta^{i_2}\wedge \cdots \wedge\delta^{i_p}$$

and we agree to define $\omega_{i_1 i_2\cdots i_p}$ for all system of indices by the alternation property so that $\omega_{i_1 i_2\cdots i_p} = 0$ unless $i_1, i_2, \cdots, i_p$ are all distinct and $\omega_{i_1 i_2\cdots i_p} = \operatorname{sgn}(\sigma)\omega_{\sigma(i_1)\sigma(i_2)\cdots\sigma(i_p)}$ for any permutation σ. Then we have

$$\omega = \frac{1}{p!}\omega_{i_1 i_2\cdots i_p}\delta^{i_1}\wedge\delta^{i_2}\wedge \cdots \wedge\delta^{i_p}.$$

We denote $\omega = (\omega_{i_1 i_2\cdots i_p})$ and call $\omega_{i_1 i_2\cdots i_p}$ components of ω.

If V_1, V_2 are vector spaces, the space of all linear mappings from V_1 and V_2 is denoted by $\operatorname{Hom}(V_1, V_2)$. For a vector space V, $\operatorname{Hom}(V, V)$ is denoted simply by $\operatorname{End}(V)$. Thus $\operatorname{End}(V)$ is the algebra of all linear mappings on V. For each $i = 1, 2, \cdots, d$, let $a_i^* \in \operatorname{End}(\Lambda\boldsymbol{R}^d)$ be defined by

$$a_i^*(\omega) = \delta^i \wedge \omega, \qquad \omega \in \Lambda\boldsymbol{R}^d. \tag{5.19}$$

Thus a_i^* is a linear mapping on $\Lambda\boldsymbol{R}^d$ sending $\overset{p}{\Lambda}\boldsymbol{R}^d$ into $\overset{p+1}{\Lambda}\boldsymbol{R}^d$ for each p (creation operator). The dual of a_i^* is denoted by a_i. Hence $a_i \in \operatorname{End}(\Lambda\boldsymbol{R}^d)$ which sends $\overset{p}{\Lambda}\boldsymbol{R}^d$ into $\overset{p+1}{\Lambda}\boldsymbol{R}^d$ (annihilation operator). It is easy to see that

$$\begin{aligned} &\{a_i, a_j\} = \{a_i^*, a_j^*\} = 0 \qquad \text{and} \\ &\{a_i^*, a_j\} = \delta_{ij} I \end{aligned} \tag{5.20}$$

where

$$\{a, b\} = ab + ba \qquad \text{for } a, b \in \operatorname{End}(\Lambda\boldsymbol{R}^d).$$

In terms of components, these linear mappings satisfy the following: If $\omega = (\omega_{i_1 i_2\cdots i_p})$, i.e.,

$$\omega = \frac{1}{p!}\,\omega_{i_1 i_2\cdots i_p}\delta^{i_1}\wedge\delta^{i_2}\cdots \wedge\delta^{i_p},$$

then, for $i = 1,2, \cdots, d,$

$$(5.21)\qquad [a_i^*(\omega)]_{i_1 i_2 \cdots i_{p+1}} = -\sum_{\nu=1}^{p+1} (-1)^\nu \, \omega_{i_1 i_2 \cdots \hat{i}_\nu \cdots i_{p-1}} \delta_{i, i_\nu}$$

$$(5.22)\qquad [a_i(\omega)]_{i_1 i_2 \cdots i_{p-1}} = \omega_{i i_1 \cdots i_{p-1}}.$$

For $\alpha = (\alpha_{ij}) \in \boldsymbol{R}^d \otimes \boldsymbol{R}^d$, define $D_1[\alpha] \in \mathrm{End}(\Lambda \boldsymbol{R}^d)$ by

$$(5.23)\qquad D_1[\alpha] = \sum_{i,j=1}^{d} \alpha_{ij} a_i^* a_j.$$

$D_1[\alpha]$ sends $\overset{p}{\Lambda} \boldsymbol{R}^d$ into $\overset{p}{\Lambda} \boldsymbol{R}^d$ for every p and we see by (5.21) and (5.22) that

$$(5.24)\qquad (D_1[\alpha](\omega))_{i_1 i_2 \cdots i_p} = -\sum_{\nu=1}^{p} \sum_{j=1}^{d} (-1)^\nu \, \omega_{j i_1 \cdots \hat{i}_\nu \cdots i_p} \alpha_{i_\nu j}$$

if $\omega = (\omega_{i_1 i_2 \cdots i_p})$. For $\beta = (\beta_{ijkl}) \in \boldsymbol{R}^d \otimes \boldsymbol{R}^d \otimes \boldsymbol{R}^d \otimes \boldsymbol{R}^d$, define $D_2[\beta] \in \mathrm{End}\,(\Lambda \boldsymbol{R}^d)$ by

$$(5.25)\qquad D_2[\beta] = \sum_{i,j,k,l=1}^{d} \beta_{ijkl} a_i^* a_k^* a_j a_l.$$

$D_2[\beta]$ also sends $\overset{p}{\Lambda} \boldsymbol{R}^d$ into $\overset{p}{\Lambda} \boldsymbol{R}^d$ for every p. Furthermore, if β satisfies

$$(5.26)\qquad \beta_{ijkl} = \beta_{klij}$$

we see easily that

$$(5.27)\qquad \begin{aligned} &(D_2[\beta](\omega))_{i_1 i_2 \cdots i_p} \\ &\quad = 2 \sum_{1 \le \nu < \mu \le p} \sum_{l,k=1}^{d} (-1)^{\nu+\mu} \, \omega_{l k i_1 \cdots \hat{i}_\nu \cdots \hat{i}_\mu \cdots i_p} \beta_{i_\nu l i_\mu k} \end{aligned}$$

if $\omega = (\omega_{i_1 i_2 \cdots i_p})$. Hence, if β satisfies furthermore

$$(5.28)\qquad \beta_{ijkl} = -\beta_{jikl} = -\beta_{ijlk}\ ,$$

then we have

(5.29)
$$(D_2[\beta](\omega))_{i_1 i_2 \cdots i_p} = 2 \sum_{1\le\nu<\mu\le p} \sum_{l,k=1}^{d} (-1)^{\nu+\mu}\, \omega_{lki_1\cdots\hat{i}_\nu\cdots\hat{i}_\mu\cdots i_p}\beta_{ki_\mu l i_\nu}.$$

Noting that $J_{ij}(r) = J_{ji}(r)$ and that $J_{ijkl}(r)$ satisfies both (5.26) and (5.28), we can now rewrite (5.18) in the form

(5.18)′
$$\begin{aligned}\frac{\partial}{\partial t} V &= \frac{1}{2}\Box_{O(M)} V \\ &= \frac{1}{2}\Delta_{O(M)} V + \frac{1}{2} D_1[J_{ij}(r)]\, V - \frac{1}{2} D_2[J_{ijkl}(r)]\, V \\ V(0,r) &= F_f(r).\end{aligned}$$

If we further define, for $\beta = (\beta_{ijkl}) \in \boldsymbol{R}^d \otimes \boldsymbol{R}^d \otimes \boldsymbol{R}^d \otimes \boldsymbol{R}^d$, $\tilde{D}_2[\beta] \in \mathrm{End}(\Lambda\boldsymbol{R}^d)$ by

(5.30)
$$\tilde{D}_2[\beta] = \sum_{i,j,k,l=1}^{d} \beta_{ijkl} a_i^* a_j a_k^* a_l,$$

then it is immediately seen from (5.20) that

$$\tilde{D}_2[\beta] = -\, D_2[\beta] + D_1[\alpha]$$

where

$$\alpha = (\alpha_{ij}) \in \boldsymbol{R}^d \otimes \boldsymbol{R}^d, \qquad \alpha_{ij} = \sum_{k=1}^{d} \beta_{ikkj}.$$

Hence (5.18) can finally be rewritten in the following simple form ([194]):

(5.18)″
$$\begin{aligned}\frac{\partial V}{\partial t} &= \frac{1}{2}\Box_{O(M)} V \\ &= \frac{1}{2}\Delta_{O(M)} V + \frac{1}{2}\tilde{D}_2[J_{ijkl}(r)]\, V \\ V(0,r) &= F_f(r).\end{aligned}$$

Thus, *to find a solution* $\alpha(t,x)$ *of* (5.4) *is equivalent to find a* $\Lambda\boldsymbol{R}^d$*-valued,* $O(d)$*-equivariant function* $U(t,r) = (U_{i_1 i_2 \cdots i_p}(t,r))$ *defined on* $(0, \infty) \times O(M)$ *which satisfies* (5.18)″.

Now we are going to show that a $\Lambda\boldsymbol{R}^d$-valued smooth function $U(t,r)$ on $[0, \infty) \times O(M)$ which satisfies (5.18)″ exists and is unique; furthermore, this $U(t,r)$ is $O(d)$-equivariant. In order to construct this

$U(t,r)$, we modify the expectation in (5.11) with Feynman-Kac type weight. So let $r(t) = r(t,r,w)$ be the solution of (5.5) defined on the d-dimensional Wiener space $(\boldsymbol{W}_0^d, P^W)$ and, for each $w \in \boldsymbol{W}_0^d$ and $r \in O(M)$, consider the following ordinary differential equation for $\mathrm{End}(\Lambda \boldsymbol{R}^d)$-valued function $M(t)$:

$$\begin{aligned}(5.31)\qquad &\frac{dM}{dt}(t) = \frac{1}{2}M(t)\,\tilde{D}_2[J_{ijkl}(r(t,r,w))]\\ &M(0) = I\ (:= \text{the identity in } \mathrm{End}(\Lambda \boldsymbol{R}^d)).\end{aligned}$$

Clearly the unique solution $M(t) \in \mathrm{End}(\Lambda \boldsymbol{R}^d)$ of (5.31) exists. We denote $M(t) = M(t,r,w)$ to clarify the dependence of $M(t)$ on $r = r(0)$ and w.

For a given p-form

$$f(x) = \frac{1}{p!} f_{i_1 i_2 \cdots i_p}(x) dx^{i_1} \wedge dx^{i_2} \wedge \cdots \wedge dx^{i_p},$$

let $F_f(r) = \{(F_f)_{i_1 i_2 \cdots i_p}(r)\}$ be its scalarization and define a $\overset{p}{\Lambda} \boldsymbol{R}^d$-valued function $U(t,r)$ on $[0, \infty) \times O(M)$ by

$$(5.32)\qquad U(t,r) = E[M(t,r,w)\, F_f(r(t,r,w))].$$

We can conclude as in Section 3 that $U(t,r)$ is smooth on $[0, \infty) \times O(M)$. Note that the following Itô formula holds for any smooth $\overset{p}{\Lambda} \boldsymbol{R}^d$-valued smooth function $V(t,r)$ on $[0, \infty) \times O(M)$:

$$\begin{aligned}&M(t)V(t,r(t)) - V(0,r)\\ &= \int_0^t M(s)\,(\tilde{L}_\alpha V)(s,r(s))dw^\alpha(s)\\ (5.33)\qquad &+ \int_0^t M(s)\,\{(\frac{\partial V}{\partial s}(s,r(s)) + \frac{1}{2}\Delta_{O(M)}\, V(s, r(s))\\ &\qquad + \frac{1}{2}\tilde{D}_2[J_{ijkl}(r(s))]\, V(s,r(s))\}ds\\ &= a\ martingale + \int_0^t M(s)(\frac{\partial V}{\partial s} + \frac{1}{2}\Box_{O(M)} V)(s,r(s))ds.\end{aligned}$$

Then, by the same argument as in Theorem 3.1, we can show that $U(t,r)$ defined by (5.32) is the unique solution of (5.18)″.

Finally, we show that $U(t,r)$ is $O(d)$-equivariant. For this, we introduce some notations. Let $g = (g^i_j) \in O(d)$ and define $\lambda(g) \in \mathrm{End}(\Lambda \boldsymbol{R}^d)$ by

$$(5.34) \qquad \lambda(g)\,[\delta^{i_1} \wedge \delta^{i_2} \wedge \cdots \wedge \delta^{i_p}] = (\delta^{i_1}g) \wedge (\delta^{i_2}g) \wedge \cdots \wedge (\delta^{i_p}g)$$

where $\delta^i g = g^i_\beta \delta^\beta$, $i = 1,2, \cdots, d$. In other words, $\lambda(g) \in \mathrm{End}(\Lambda \boldsymbol{R}^d)$ which sends $\overset{p}{\Lambda}\boldsymbol{R}^d$ into $\overset{p}{\Lambda}\boldsymbol{R}^d$ for every p and satisfies

$$(5.34)' \qquad [\lambda(g)\,(\omega)]_{i_1 i_2 \cdots i_p} = g^{\beta_1}_{i_1} g^{\beta_2}_{i_2} \cdots g^{\beta_p}_{i_p}\, \omega_{\beta_1 \beta_2 \cdots \beta_p}$$

if $\omega = (\omega_{i_1 i_2 \cdots i_p}) \in \overset{p}{\Lambda}\boldsymbol{R}^d$. Note that $\lambda(gh) = \lambda(h)\lambda(g)$ for $g,h \in O(d)$ and hence, λ defines an action of $O(d)$ on $\Lambda \boldsymbol{R}^d$ from the *right*. The fact that $F_f(r)$ is $O(d)$-equivariant can now be stated as

$$(5.35) \qquad F_f(r \cdot g) = \lambda(g)F(r) \quad \text{for every } g \in O(d) \text{ and } r \in O(M),$$

where the action $r \to r \cdot g$ of $g \in O(d)$ is defined by (2.31). Also, for $g = (g^i_j) \in O(d)$, define an $\boldsymbol{R}^d \otimes \boldsymbol{R}^d$-valued function $[\tau(g)J]_{ij}(r)$ and an $\boldsymbol{R}^d \otimes \boldsymbol{R}^d \otimes \boldsymbol{R}^d \otimes \boldsymbol{R}^d$-valued function $[\tau(g)J]_{ijkl}(r)$, both defined on $O(M)$, by

$$(5.36) \qquad [\tau(g)J]_{ij}(r) = g^\alpha_i g^\beta_j J_{\alpha\beta}(r)$$

and

$$(5.37) \qquad [\tau(g)J]_{ijkl}(r) = g^\alpha_i g^\beta_j g^\gamma_k g^\delta_l J_{\alpha\beta\gamma\delta}(r).$$

The fact that $\{J_{ij}(r)\}$ and $\{J_{ijkl}(r)\}$ are $O(d)$-equivariant can now be stated as

$$(5.38) \qquad J_{ij}(r \cdot g) = [\tau(g)J]_{ij}(r)$$

and

$$(5.39) \qquad J_{ijkl}(r \cdot g) = [\tau(g)J]_{ijkl}(r)$$

for every i,j,k,l, $g \in O(d)$ and $r \in O(M)$. It follows from (5.24) and (5.29) that

$$D_1[[\tau(g)J]_{ij}(r)] = \lambda(g)D_1[J_{ij}(r)]\,\lambda(g)^{-1}$$

and

$$D_2[[\tau(g)J]_{ijkl}(r)] = \lambda(g)D_2[J_{ijkl}(r)]\,\lambda(g)^{-1}.$$

Hence, combining these with (5.38) and (5.39), we have

$$(5.40)\qquad D_1[J_{ij}(r\cdot g)] = \lambda(g)\,D_1[J_{ij}(r)]\,\lambda(g)^{-1}$$

and

$$(5.41)\qquad D_2[J_{ijkl}(r\cdot g)] = \lambda(g)D_2[J_{ijkl}(r)]\,\lambda(g)^{-1}.$$

We can conclude by (5.40), (5.41) and (5.7) that

$$(5.42)\qquad M(t,r\cdot g,w) = \lambda(g)M(t,r,gw)\,\lambda(g)^{-1},\qquad g \in O(d).$$

From this and (5.35), we see immediately that $U(t,r)$ satisfies

$$\begin{aligned} U(t,rg) &= \lambda(g)E[M(t,r,gw)F_f(r(t,r,gw))] \\ &= \lambda(g)E[M(t,r,w)F_f(r(t,r,w))] \\ &= \lambda(g)U(t,r). \end{aligned}$$

This means that $U(t,r)$ is $O(d)$-equivariant.

Malliavin [104] used the above to obtain an interesting generalization of a vanishing theorem of Bochner ([185]) for harmonic 1-forms.

6. The case with boundary conditions

We shall now discuss a similar probabilistic construction of the solution of the heat equation (5.4) for differential forms in the case of a manifold with boundary.*

Let M be a Riemannian manifold of dimension d with smooth boundary. The interior and boundary of M are denoted by $\mathring{M}$ and ∂M respectively. Near the boundary, we can choose a coordinate neighborhood U and a local coordinate $x = (x^1, x^2, \ldots, x^d)$ in U such that $x^d \geq 0$ for all $x \in U$ and $x \in U \cap \partial M$ if and only if $x^d = 0$. The tangent vector $n(x) = n^i(x)\dfrac{\partial}{\partial x^i}$ at $x \in \partial M$ given by

$$(6.1)\qquad n^i(x) = g^{id}(x)/\sqrt{g^{dd}(x)},\qquad i = 1, 2, \ldots, d$$

is called the *inward pointing unit normal vector at* x. For a smooth function f defined in U,

* The material in this section is adapted from [55] and [172].

$$\frac{\partial f}{\partial n}(x) = n^i(x)\frac{\partial f}{\partial x^i}(x), \qquad x \in \partial M, \tag{6.2}$$

is called the *normal derivative* of f *at* x. Let

$$\delta^{i_1 i_2 \cdots i_p}_{j_1 j_2 \cdots j_p} = \begin{vmatrix} \delta^{i_1}_{j_1}\delta^{i_1}_{j_2} \cdots \delta^{i_1}_{j_p} \\ \cdots\cdots\cdots \\ \delta^{i_p}_{j_1}\delta^{i_p}_{j_2} \cdots \delta^{i_p}_{j_p} \end{vmatrix}$$

and $\varepsilon_{i_1 i_2 \cdots i_d}$ be the skew symmetric $(0, d)$ tensor field defined by

$$\varepsilon_{i_1 i_2 \cdots i_d}(x) = \sqrt{\det(g_{ij}(x))}\, \delta^{12 \cdots d}_{i_1 i_2 \cdots i_d}. \tag{6.3}$$

For a p-form α, its adjoint $*\alpha$ is a $(d-p)$-form defined as follows. If α is expressed as

$$\alpha(x) = \sum_{i_1 < i_2 < \cdots < i_p} \alpha_{i_1 i_2 \cdots i_p}(x)\, dx^{i_1} \wedge dx^{i_2} \wedge \cdots \wedge dx^{i_p}, \tag{6.4}$$

then

$$*\alpha(x) = \sum_{j_1 < j_2 < \cdots < j_{d-p}} \alpha^*{}_{j_1 j_2 \cdots j_{d-p}}(x) dx^{j_1} \wedge dx^{j_2} \wedge \cdots \wedge dx^{j_{d-p}}, \tag{6.5}$$

where

$$\alpha^*_{j_1 j_2 \cdots j_{d-p}}(x) = \sum_{i_1 < i_2 < \cdots < i_p} \varepsilon_{i_1 i_2 \cdots i_p j_1 j_2 \cdots j_{d-p}}(x) \alpha^{i_1 i_2 \cdots i_p}(x) \tag{6.6}$$

and

$$\alpha^{i_1 i_2 \cdots i_p}(x) = g^{i_1 j_1}(x) g^{i_1 j_2}(x) \cdots g^{i_p j_p}(x) \alpha_{j_1 j_2 \cdots j_p}(x).$$

Let θ be a p-form. We denote by $\theta_{\tan}$ the restriction of θ to ∂M and call it the tangent component of θ. We define the normal component of θ by $\theta_{\text{norm}} = \theta - \theta_{\tan}$.

Let $\eta(x)$ be a differential 1-form such that

$$\eta|_{\partial M}(x) = \eta_i(x) dx^i, \qquad x \in \partial M \tag{6.7}$$

where $\eta_i(x) = g_{ij}(x) n^j(x)$ and $n^i(x)$ is defined by (6.1). Then it is easy to see that the restriction $(-1)^{p(d-p)+p-1}\, [*(*\theta \wedge \eta)] \wedge \eta|_{\partial M}$ of the p-form $(-1)^{p(d-p)+p-1}\, [*(*\theta \wedge \eta)] \wedge \eta$ is uniquely determined from θ and coincides with θ_{norm}.

Definition 6.1. A differential form θ is said to satisfy the *absolute boundary conditions* if $\theta_{\mathrm{norm}} = 0$ and $(d\theta)_{\mathrm{norm}} = 0$ ([144]).

We want to solve the heat equation (5.4) with the absolute boundary conditions, namely,

$$\begin{cases} \dfrac{\partial \alpha}{\partial t} = \dfrac{1}{2} \square \alpha \\ \alpha|_{t=0} = f \\ \alpha_{\mathrm{norm}} = 0, \qquad (d\alpha)_{\mathrm{norm}} = 0 \end{cases} \tag{6.8}$$

where $f \in A_p(M)$ is given. In order to avoid undesirable complications, we shall restrict ourselves to the case of 1-forms. Near the boundary, we can choose a coordinate neighborhood U and a local coordinate $x = (x^1, x^2, \ldots, x^d)$ in U such that the following holds:*

(i) $x^d \geq 0$ for all $x \in U$;
(ii) $x \in U \cap \partial M$ if and only if $x^d = 0$;
(iii) the metric tensor $g(x) = (g_{ij}(x))$ satisfies $g_{id}(x) = 0$ for $i = 1, 2, \ldots, d-1$.

In this local coordinate, it is easy to see that $\theta \in A_1(M)$ satisfies the absolute boundary conditions if and only if

$$\theta_d(x) = 0 \quad \text{and} \quad \frac{\partial}{\partial x^d} \theta_i(x) = 0, \qquad i = 1, 2, \ldots, d-1, \quad x \in U \cap \partial M. \tag{6.9}$$

Indeed, we have

$$\theta_{\mathrm{norm}} = \theta_d(x) dx^d \quad \text{and} \quad (d\theta)_{\mathrm{norm}} = \sum_{i=1}^{d-1} \left(\frac{\partial \theta_i}{\partial x^d} - \frac{\partial \theta_d}{\partial x^i} \right) dx^d \wedge dx^i,$$

and so (6.9) follows at once.

Let $O(M)$ be the bundle of orthonormal frames over M. If $F_\theta(r) = ([F_\theta]_i(r))$ is the scalarization of θ, i.e.,

$$[F_\theta]_i(r) = \theta_j(x) e_i^j, \qquad r = (x^i, e_j^i),$$

then (6.9) holds if and only if

$$f_d^j [F_\theta]_j(r) = 0 \quad \text{and} \quad f_i^j \frac{\partial}{\partial x^d} [F_\theta]_j(r) = 0, \quad i = 1, 2, \ldots, d-1, \tag{6.10}$$

---

* Cf. [13], [144].

where (f_j^i) is the inverse of (e_j^i). Thus, the initial value problem (6.8) for differential 1-forms is equivalent to the following initial value problem for $\boldsymbol{R}^d$-valued functions $(U_i(t,r))$ on $O(M)$ which are $O(d)$-equivariant:

$$
(6.11)\quad \begin{cases} \dfrac{\partial U_i}{\partial t}(t,r) = \dfrac{1}{2}\{\Delta_{O(M)} U_i(t,r) + J_i^j(r) U_j(t,r)\} \\ U_i(0,r) = (F_f)_i(r) \\ f_i^j \dfrac{\partial}{\partial x^d} U_j(t,r)|_{\partial O(M)} = 0, \qquad i = 1, 2, \ldots, d-1, \\ f_d^j U_j(t,r)|_{\partial O(M)} = 0 \end{cases}
$$

where $\{J_i^j(r)\}$ is the scalarization of the tensor $\{R_i^j(x)\}$ and $\partial O(M) = \{r = (x, e) \in O(M);\ x \in \partial M\}$. We note that by using same notation as in Section 5, (6.11) can also be rewritten in the following form:

$$
\begin{aligned}
\frac{\partial}{\partial t} U &= \frac{1}{2}\Box_{O(M)} U \\
&= \frac{1}{2}[\Delta_{O(M)} + D_1[J_j^i]]U \\
U(0,r) &= F_f(r) \\
f_i^j \frac{\partial}{\partial x^d} & U_j(t,r)|_{\partial O(M)} = 0, \quad i = 1, 2, \cdots, d-1 \\
f_d^j & U_j(t,r)|_{\partial O(M)} = 0.
\end{aligned}
$$

We will now solve this initial value problem using the horizontal Brownian motion $(r(t))$ on $O(M)$. $(r(t))$ can be obtained by solving stochastic differential equations with boundary conditions (Chapter IV, Section 7). Since the construction of solutions can be localized, we do not hesitate to put the following assumptions.

Assumption (A). M is the upper half space of $\boldsymbol{R}^d$:

$$M = \{x;\ x = (x^1, x^2, \ldots, x^d) \in \boldsymbol{R}^d,\ x^d \geq 0\}$$

and

$$\partial M = \{x \in M;\ x^d = 0\}.$$

Assumption (B). The Riemannian metric tensor $(g_{ij}(x))$ under the coordinate $x = (x^1, x^2, \ldots, x^d)$ consists of $\boldsymbol{C}^\infty$-functions which are bounded together with all of their partial derivatives. Furthermore, $(g_{ij}(x))$ is uniformly positive definite and satisfies $g_{id}(x) \equiv 0$ for $i = 1, 2, \ldots, d-1$.

We consider the following stochastic differential equation for the process $(X(t), e(t))$ on $\boldsymbol{R}^d_+ \times \boldsymbol{R}^{d^2}$:

$$
(6.12) \quad \begin{cases} dX^i_t = e^i_k(t) \circ dB^k(t) + \delta_{id} d\phi(t) \\ de^i_\alpha(t) = -\left\{ {}_l{}^i{}_k \right\} (X(t)) e^k_\alpha(t) \circ dX^l(t) \\ \quad = -\left\{ {}_l{}^i{}_k \right\} (X(t)) e^k_\alpha(t) e^l_m(t) \circ dB^m(t) - \left\{ {}_d{}^i{}_k \right\} (X(t)) e^k_\alpha(t) d\phi(t) \\ \qquad\qquad i, \alpha = 1, 2, \ldots, d. \end{cases}
$$

Here $\left\{ {}_l{}^i{}_k \right\}(x)$ are the Christoffel symbols and $B(t) = (B^i(t))$ is a d-dimensional Brownian motion. $\phi(t)$ is a continuous non-decreasing process which increases only when $X(t) \in \partial M$. (6.12) is a particular case of the stochastic differential equations discussed in Chapter IV, Section 7, and so by Theorem IV-7.2, we know that for any Borel probability measure μ on $\boldsymbol{R}^d_+ \times \boldsymbol{R}^{d^2}$, the solution $(X(t), e(t))$ of (6.12) with the initial law μ exists uniquely. In the same way as in Section 4, we see that if

$$g_{kl}(X(0)) e^k_i(0) e^l_j(0) = \delta_{ij}$$

then for all $t \geq 0$

$$g_{kl}(X(t)) e^k_i(t) e^l_j(t) = \delta_{ij}$$

holds almost surely; i.e., if $(X(0), e(0)) \in O(M)$, then $(X(t), e(t)) \in O(M)$ for all $t \geq 0$ a.s. Thus we have a diffusion process $r(t) = (X(t), e(t))$ on $O(M)$. It is called the *horizontal Brownian motion on the bundle of orthonormal frames $O(M)$ with reflecting boundary*. Let $\tilde{L}_1, \tilde{L}_2, \ldots, \tilde{L}_d$ be the canonical horizontal vector fields:

$$
(6.13) \quad (\tilde{L}_m F)(r) = e^i_m \frac{\partial F}{\partial x^i}(r) - \left\{ {}_k{}^i{}_l \right\} (x) e^l_m e^k_j \frac{\partial F}{\partial e^i_j}(r), \quad (r = (x, e = (e^i_j))),
$$
$$m = 1, 2, \ldots, d,$$

and define the horizontal Laplacian of Bochner by

$$
(6.14) \quad \Delta_{O(M)} = \sum_{m=1}^{d} \tilde{L}_m(\tilde{L}_m).
$$

Set

(6.15) $\alpha^{dd}(r) = g^{dd}(x), \quad \alpha^{d,i}_m(r) = -e^k_m \left\{{}^{\,i}_{d\,k}\right\}(x) g^{dd}(x).$

Theorem 6.1. Let $r(t) = (X(t), e(t) = (e^i_m(t)))$ be the horizontal Brownian motion with reflecting boundary constructed above from the solution of (6.12).

(i) For any smooth function $F(t, r)$ on $[0, \infty) \times O(M)$,

$$dF(t,r(t)) = (\tilde{L}_m F)(t,r(t))dB^m(t) + \left\{\frac{1}{2}(\Delta_{O(M)}F)(t,r(t)) + \frac{\partial F}{\partial t}(t,r(t))\right\}dt + (\tilde{X}_d F)(t,r(t))d\phi(t),$$

where $\tilde{X}_d$ is the horizontal lift of the vector field $X_d = \frac{\partial}{\partial x^d}$ which is given explicitly as

$$(6.16) \qquad (\tilde{X}_d F)(t,r) = \frac{\partial F}{\partial x^d}(t,r) + \frac{\alpha^{d,i}_m(r)}{\alpha^{dd}(r)} \frac{\partial F}{\partial e^i_m}(t,r).$$

(ii)

$$(6.17) \qquad \begin{cases} dX^d(t)\cdot dX^d(t) = \alpha^{dd}(r(t))dt \\ dX^d(t)\cdot de^i_m(t) = \alpha^{d,i}_m(r(t))dt, \qquad m, i = 1, 2, \ldots, d. \end{cases}$$

Proof. (i) is immediately obtained from Itô's formula. (ii) is easily proved once we notice

$$\sum_\alpha e^i_\alpha(t) e^j_\alpha(t) = g^{ij}(X(t)).$$

Theorem 6.1 implies that the process $(r(t))$ on $O(M)$ is determined by the differential operator $\frac{1}{2}\Delta_{O(M)}$ with the boundary condition $\tilde{X}_d F = 0$ on $\partial O(M)$. (6.17) implies that the process $(r(t))$ is a *normally reflecting diffusion process* in the sense of Definition 6.2 given below.

In order to obtain a solution of (6.11), we consider the canonical realization $\{r(t, w), w \in \boldsymbol{W}(O(M)), P_r\}$ of the horizontal Brownian motion on $O(M)$ with reflecting boundary: $\boldsymbol{W}(O(M)) = \boldsymbol{C}([0, \infty) \longrightarrow O(M))$, P_r is the probability law on $\boldsymbol{W}(O(M))$ of the solution $r(t)$ of (6.12) with $r(0) = r$, and $r(t, w) = w(t)$ for $w \in \boldsymbol{W}(O(M))$. For a Borel probability measure μ on $O(M)$, P_μ is defined by $P_\mu(B) = \int_{O(M)} P_r(B) \times \mu(dr)$, $B \in \mathscr{B}(\boldsymbol{W}(O(M)))$.* Let $\mathscr{F} = \bigcap_\mu \overline{\mathscr{B}(\boldsymbol{W}(O(M)))}^{P_\mu}$ and $\mathscr{F}_t =$

* $\mathscr{B}(\boldsymbol{W}(O(M)))$ and $\mathscr{B}_t(\boldsymbol{W}(O(M)))$ are defined as usual.

$\{A \in \bar{\mathscr{F}};$ for any μ there exists B_μ such that $B_\mu \in \mathscr{B}_t(W(O(M)))$ and $P_\mu(A \triangle B_\mu) = 0\}$. We now fix μ and give the following discussions on the probability space $(W(O(M)), \bar{\mathscr{F}}, P_\mu)$. Writing $r(t, w) = (X(t) = X(t, w), e(t) = e(t, w))$, we set

$$\phi(t) = \lim_{\varepsilon \downarrow 0} \frac{1}{2\varepsilon} \int_0^t I_{[0,\varepsilon)}(X^d(s)) g^{dd}(X(s)) ds$$

and

$$B^i(t) = \int_0^t [e(s)^{-1}]^i_k \circ [dX^k(s) - \delta^k_d d\phi(s)].$$

Then $\{B^i(t)\}$ is a d-dimensional $(\mathscr{F}_t)$-Brownian motion and $\{r(t) = r(t,w), \phi(t)\}$ satisfies the equation (6.12).

Lemma 6.1. $\{P_r\}$ is invariant under the action T_a of a, $a \in O(d)$, from the right; that is, if $w \cdot a \in W(O(M))$ is defined for $w \in W(O(M))$ by $(w \cdot a)(t) = w(t) \cdot a$ and if $T_a(P_r)$ is the image measure of P_r under the mapping $w \longmapsto w \cdot a$, then

(6.18) $\qquad T_a(P_r) = P_{r \cdot a}\,.$

Proof. Let $r(t)$ be a solution of (6.12) with $B(t)$ and $\phi(t)$ such that $r(0) = r$. Then for $a \in O(d)$, $\tilde{r}(t) = r(t) \cdot a$ is a solution of (6.12) with $\tilde{B}(t) = a^{-1}B(t)$ and $\tilde{\phi}(t) = \phi(t)$ such that $\tilde{r}(0) = r \cdot a$. $\tilde{B}(t)$ is another d-dimensional Brownian motion and hence (6.18) holds by the uniqueness of solutions.

By this lemma we see that $X(t)$ defines a diffusion process on M and it is easily seen as in Section 4 that $X(t)$ is determined by $\Delta_M/2$ with boundary condition $\frac{\partial f}{\partial n} = 0$ on ∂M. This diffusion is called the *Brownian motion on M with reflecting boundary* or simply, *reflecting Brownian motion* on M.

Let $R^d \otimes R^d$ be the algebra of all $d \times d$ real matrices $a = (a^i_j)$ endowed with the norm $\|a\|^2 = \sum_{i,j=1}^{d} |a^i_j|^2$. *It is convenient for our purpose to define the multiplication in* $R^d \otimes R^d$ *by the following rule:** *for* $a = (a^i_j)$ *and* $b = (b^i_j)$, $ab = (c^i_j)$ *where*

(6.19) $\qquad c^i_j = a^k_j b^i_k.$

* This convention is used only in this section.

Let $P = (p^i_j)$ where

$$p^i_j = \begin{cases} 1 & \text{if } i = d \text{ and } j = d \\ 0 & \text{otherwise} \end{cases}$$

and $Q = I - P$.

In the following we fix a Borel probability measure μ on $O(M)$ and restrict our attention to the probability space $(\boldsymbol{W}(O(M)), \mathcal{F}, P_\mu)$. Then $r(t, w) = (X(t, w), e(t, w) = (e^i_j(t, w)))$ (also denoted simply by $r(t) = (X(t), e(t) = (e^i_j(t))))$ is the solution of (6.12) with $B(t)$ and $\phi(t)$ defined as above. Following H.Airault [2], we consider the following stochastic differential equation for an $\boldsymbol{R}^d \otimes \boldsymbol{R}^d$-valued process $K(t) = (K^i_j(t,w))$ as described below:

(6.20) (i) *for any* $t \geq 0$ *such that* $X(t) \in \mathring{M}$,

$$(6.20)_a \qquad dK^1(t) := dK(t)P = K(t)\{e(t)^{-1}de(t) + \frac{1}{2} R(X(t))dt\} P$$

where $R(x) = (R^i_j(x))$ *is the tensor defines by* $R^j_i = R^{j \cdot k \cdot}_{\cdot k \cdot i}$, (cf. Section 5)

(ii) *for any* $t \geq 0$,

$$(6.20)_b \qquad \begin{aligned} dK^2(t) &:= dK(t)Q \\ &= K(t)\{e(t)^{-1}de(t) + \frac{1}{2}R(X(t))dt\} I_{\mathring{M}}(X(t))Q; \end{aligned}$$

(iii) *with probability one,* $t \longmapsto K^1(t) = K(t)P$ *is right-continuous with left-hand limits: furthermore*

$$K^1(t) = 0 \qquad \textit{if} \quad X(t) \in \partial M,$$

and the initial valued is given by

$$(6.21) \qquad K^1(0) = I_{\mathring{M}}(X(0))e(0)P, \qquad K^2(0) = e(0)Q.$$

Remark 6.1. (a) A precise formulation of $(6.20)_a$ is as follows: if a continuous process $Y(t)$ is defined by a semimartingale integral

$$Y(t) = \int_0^t K(s)\left\{e(s)^{-1}de(s) + \frac{1}{2}R(X(s))ds\right\}P,$$

then, with probability one,

$$K^1(t) - K^1(s) = Y(t) - Y(s)$$

for all $s \leq t$ such that $X(u) \in \mathring{M}$ for every $u \in [s, t]$.

(b) (6.20), (ii) automatically implies that $t \longmapsto K^2(t)$ is continuous with probability one.

The above stochastic differential equation may also be expressed in the equivalent form of a stochastic integral equation as the next lemma shows.

Lemma 6.2. An $\boldsymbol{R}^d \otimes \boldsymbol{R}^d$-valued process $K(t)$ adapted to $(\mathscr{F}_t)$ is a solution of the above stochastic differential equation (6.20) with the initial condition (6.21) if and only if

$$\text{(6.22)} \quad \begin{cases} K^1(t) := K(t)P \\ \quad = I_{\{t<\sigma\}}(e(0) + \int_0^t K(u)[e(u)^{-1}de(u) + \frac{1}{2} R(X(u))du])P \\ \quad + I_{\{t \geq \sigma\}} \int_{\tau(t)}^t K(u)[e(u)^{-1} de(u) + \frac{1}{2} R(X(u))du]P \\ K^2(t) := K(t)Q = e(0)Q + \int_0^t K(u)[e(u)^{-1}de(u) \\ \quad + \frac{1}{2} R(X(u))du] I_{\mathring{M}}(X(u))Q, \end{cases}$$

where

$$\text{(6.23)} \qquad \sigma = \begin{cases} \inf\{s;\ X(s) \in \partial M\} \\ \infty \quad \text{if} \quad \{\ \} = \phi, \end{cases}$$

is the *first hitting time* of $X(t)$ to ∂M and

$$\text{(6.24)} \qquad \tau(t) = \begin{cases} \sup\{s;\ s \leq t,\ X(s) \in \partial M\} \\ 0 \quad \text{if} \quad \{\ \} = \phi \end{cases}$$

is the *last exit time* before t from ∂M.

Remark 6.2. $\int_{\tau(t)}^t \cdot$ is understood, of course, as $Y(t) - Y(\tau(t))$ where $Y(t)$ is the continuous process given by $Y(t) = \int_0^t \cdot$.

The proof is easy and omitted.

Let Ξ be the totality of all $\boldsymbol{R}^d \otimes \boldsymbol{R}^d$-valued processes $\xi(t) = (\xi(t,w)^i_j)$ defined on $(\boldsymbol{W}(O(M)), \mathscr{F}, P)$ adapted to $(\mathscr{F}_t)$ such that $t \longmapsto \xi(t)$ is right continuous with left-hand limits a.s. and satisfies

$$\sup_{t \in [0,T]} E_\mu[\|\xi(t)\|^2] < \infty \qquad \text{for all} \quad T > 0. \tag{6.25}$$

Define a mapping Φ: $\Xi \longrightarrow \Xi$ by

$$\begin{aligned} \Phi^1(\xi)(t) :&= \Phi(\xi)(t)P \\ &= I_{\{\sigma > t\}}(e(0) + \int_0^t \xi(u)[e(u)^{-1} de(u) + \frac{1}{2} R(X(u))du])P \\ &\quad + I_{\{\sigma \le t\}} \int_{\tau(t)}^t \xi(u)[e(u)^{-1} de(u) + \frac{1}{2} R(X(u))du]P \end{aligned} \tag{6.26$_a$}$$

$$\begin{aligned} \Phi^2(\xi)(t) :&= \Phi(\xi)(t)Q \\ &= eQ + \int_0^t \xi(u)[e(u)^{-1} de(u) + \frac{1}{2} R(X(u))du] I_{\mathring{M}}(X(u))Q. \end{aligned} \tag{6.26$_b$}$$

Let $A(t)$ be the right-continuous inverse of $t \longmapsto \phi(t)$ and set

$$\boldsymbol{D} = \{s \ge 0;\ A(s-) < A(s)\}. \tag{6.27}$$

If $t > 0$ is fixed, then $\tau(t) = A(\phi(t)-)$ a.s. By Theorem 6.6 given below, we see that if $g(t)$ is an $(\mathscr{F}_t)$-well measurable process such that $t \longmapsto E_\mu[g(t)^2]$ is locally bounded then

$$\begin{aligned} &E_\mu[\{\int_{\tau(t)}^t g(s)dB^k(s)\}^2] \\ &\le E_\mu[\sum_{u \in \boldsymbol{D}} \{\int_{A(u-)\wedge t}^{A(u)\wedge t} g(s)dB^k(s)\}^2] = E_\mu[\int_0^t g(s)^2 ds]. \end{aligned} \tag{6.28}$$

It is easy to show from this that for every $T > 0$ there is a constant $K = K(T) > 0$ such that

$$E_\mu[\|\Phi(\xi)(t)\|^2] \le K(1 + \int_0^t E_\mu[\|\xi(u)\|^2]du) \qquad \text{for all } t \in [0, T]. \tag{6.29}$$

This proves that $\Phi(\xi) \in \Xi$ if $\xi \in \Xi$. Again using (6.28), we have that, for $\xi, \eta \in \Xi$,

(6.30) $$E_\mu[\|\Phi(\xi)(t) - \Phi(\eta)(t)\|^2] \leq K\int_0^t E_\mu[\|\xi(s) - \eta(s)\|^2]ds, \quad t \in [0, T].$$

Theorem 6.2. The stochastic differential equation (6.20) with the initial condition (6.21) has one and only one solution $K(t) \in \Xi$.

Proof. Let $\xi_n \in \Xi$, $n = 0, 1, \ldots$ be defined by $\xi_0 \equiv 0$ and $\xi_n = \Phi(\xi_{n-1})$, $n = 1, 2, \ldots$. Using (6.30) we can show that there exists $\xi \in \Xi$ such that

$$E_\mu[\sup_{0 \leq t \leq T} \|\xi_n(t) - \xi(t)\|^2] \longrightarrow 0.$$

Then clearly ξ is a solution of (6.22). The uniqueness also follows from (6.30). The arguments are standard and similar to those given in Chapter III or Chapter IV and so we omit the details.

Let $K(t) = (K^i_j(t, w))$ be the solution of (6.20) with the initial condition (6.21) and define $M(t) = (M^i_j(t, w))$ by

(6.31) $$M(t,w) = K(t,w)e(t,w)^{-1}, \qquad t \geq 0.$$

Theorem 6.3. $M = \{M(t,w)\}$ is an $\boldsymbol{R}^d \otimes \boldsymbol{R}^d$-valued MOF of the horizontal Brownian motion on $O(M)$ with reflecting boundary;* i.e.,
(i) $M(t,w)$ is $(\mathscr{F}_t)$-adapted,
(ii) for every $t,s \geq 0$, $M(t+s,w) = M(s,w)M(t,\theta_s w)$ a.s., where the shift operator θ_s: $\boldsymbol{W}(O(M)) \longrightarrow \boldsymbol{W}(O(M))$ is defined by $(\theta_s w)(t) = w(t+s)$.

Proof. (i) is obvious. To prove (ii), we fix s and set $\tilde{K}(t) = K(t+s,w)$, $\tilde{X}(t) = X(t+s,w)$ and $\tilde{e}(t) = e(t+s,w)$. Then it is clear that $\tilde{K}(t)$ satisfies the above stochastic differential equation (6.20) with respect to $(\tilde{X}(t), \tilde{e}(t))$. On the other hand, by applying the shift operator θ_s to $K(t)$, we see that $\bar{K}(t) = K(t, \theta_s w)$ satisfies the same equation with respect to $(X(t, \theta_s w), e(t, \theta_s w)) = (\tilde{X}(t), \tilde{e}(t))$. If we set

$$\tilde{K}'(t) = K(s,w)e(s,w)^{-1}\bar{K}(t),$$

then

$$\tilde{K}'(0) = K(s,w)e(s,w)^{-1}(I_{\partial M}(X(s,w))e(s,w)P + e(s,w)Q)$$

---
* [139].

$$= K(s,w)(I_{\dot{M}}(X(s,w))P + Q)$$
$$= K(s,w)$$

by (6.20) (iii). Hence $\tilde{K}(t)$ and $\tilde{K}'(t)$ satisfy the same equation and the same initial condition. Consequently $\tilde{K}(t) \equiv \tilde{K}'(t)$ by the uniqueness of solutions. That is,

$$K(t+s,w) = K(s,w)e(s,w)^{-1}K(t,\theta_s w).$$

Multiplying by $e(t+s,w)^{-1} = e(t, \theta_s w)^{-1}$ from the right yields (ii).

The following lemmas delineate some properties of MOF $M = \{M(t,w)\}$.

Lemma 6.3. If $X(0) \in \partial M$, then

$$Pe(0)^{-1}M(t) = 0 \qquad \text{for all} \quad t \geq 0.$$

Proof. It is enough to prove that $Pe(0)^{-1}K(t) = 0$ for all $t \geq 0$. If $X(0) \in \partial M$, then

$$Pe(0)^{-1}K(0) = Pe(0)^{-1}(I_{\dot{M}}(X(0))e(0)P + e(0)Q) = PQ = 0.$$

Since $\tilde{K}(t) = Pe(0)^{-1}K(t)$ satisfies (6.20), $\tilde{K}(t) = 0$ by the uniqueness of solutions.

Lemma 6.4 $M(t, w \cdot a) = aM(t, w)a^{-1}, \qquad t \geq 0,\ a \in O(d).$

Proof. Since $X(t, w \cdot a) = X(t,w)$ and $e(t, w \cdot a) = ae(t,w)$,* we see at once that $K(t, w \cdot a) = aK(t,w)$ by the uniqueness of solutions of (6.20). Thus $M(t, w \cdot a) = aK(t, w)[ae(t,w)]^{-1} = aK(t,w)e(t,w)^{-1}a^{-1} = aM(t,w)a^{-1}$, which completes the proof.

Now we can solve the equation (6.11). First of all, however, we shall adopt the following convention in addition to the multiplication rule (6.19): *for a d-dimensional* $b = (b_i)$ *and* $a = (a_i^j) \in \boldsymbol{R}^d \otimes \boldsymbol{R}^d$, $ab = c$ *is the d-dimensional vector* $c = (c_i)$ *defined by*

$$(6.32) \qquad c_i = a_i^j b_j .$$

* This was denoted by $e(t, w)a$ in (2.31). Here we are adopting the multiplication rule (6.19) and hence it should be written as $ae(t, w)$.

Under this convention, (6.11) is rewritten as

$$(6.11)\qquad \begin{cases} \dfrac{\partial U}{\partial t} = \dfrac{1}{2}\{\Delta_{O(M)}U + JU\} \\ U|_{t=0} = F_f \\ \left(Pe^{-1}U + Qe^{-1}\dfrac{\partial U}{\partial x^d}\right) = 0 \qquad \text{if} \quad r = (x, e) \in \partial O(M). \end{cases}$$

More generally, we consider the following initial value problem for the heat equation of $\boldsymbol{R}^d$-valued functions $U(t, r) = (U_i(t, r))$ on $O(M)$:

$$(6.33)\qquad \begin{cases} \dfrac{\partial U}{\partial t} = \dfrac{1}{2}\{\Delta_{O(M)}U + JU\} \\ U|_{t=0} = F \\ Pe^{-1}U + Qe^{-1}\left[\dfrac{\partial U}{\partial x^d} - \dfrac{\partial U}{\partial e^i_m}\left\{{}_d{}^i{}_k\right\}(x)e^k_m + e\Gamma_d(x)e^{-1}U\right] = 0 \\ \qquad\qquad\qquad\qquad\qquad \text{if} \quad r = (x, e) \in \partial O(M), \end{cases}$$

where $\Gamma_d(x) \in \boldsymbol{R}^d \otimes \boldsymbol{R}^d$ is defined by

$$\Gamma_d(x) = (\left\{{}_d{}^i{}_j\right\}(x)).$$

If $U(t,r)$ is $O(d)$-equivariant, this implies that $U(t,r) = e\tilde{U}(x)$, $r = (x,e)$, where $\tilde{U}(x)$ is a smooth $\boldsymbol{R}^d$-valued function on M, and hence

$$\frac{\partial U}{\partial e^i_m}\left\{{}_d{}^i{}_k\right\}(x)e^k_m = e\Gamma_d(x)e^{-1}U.$$

Thus the initial value problem (6.33) is reduced to the initial value problem (6.11) in the case of $O(d)$-equivariant functions. We now construct a semigroup corresponding to the initial value problem (6.33) by using the *MOF M*.

Let $\boldsymbol{C}_0(O(M) \longrightarrow \boldsymbol{R}^d)$ be the set of all bounded continuous functions $F(r)$ on $O(M)$ taking values in $\boldsymbol{R}^d$ such that

$$(6.34)\qquad Pe^{-1}F(r) = 0 \qquad \text{if} \quad r = (x, e) \in \partial O(M).$$

For $F \in \boldsymbol{C}_0(O(M) \longrightarrow \boldsymbol{R}^d)$ and $t \geq 0$, set

$$(6.35)\qquad (H_tF)(r) = E_r[M(t,w)F(r(t,w))].$$

Theorem 6.4. (i) $\{H_t\}$ defines a one-parameter semigroup of operators on $C_0(O(M) \longrightarrow \boldsymbol{R}^d)$.

(ii) If F is $O(d)$-equivariant, then so is H_tF for all $t \geq 0$.

Proof. If $r \in \partial O(M)$, then by Lemma 6.3 $Pe^{-1}(H_tF)(r) = 0$. The continuity in r of the functions $H_tF(r)$, $t \geq 0$ follows from the continuity of $r \longmapsto P_r \in \mathscr{P}(W(O(M)))$,* which in turn is a consequence of the uniqueness of solutions of the stochastic differential equations. The semigroup property of $\{H_t\}$ is obvious since M is an MOF. Finally, (ii) follows from Lemmas 6.1 and 6.4.

Theorem 6.5. Let $F(t, r) = (F_i(t, r))$ be a smooth function on $[0, \infty) \times O(M)$ taking values in $\boldsymbol{R}^d$ such that for each $t \geq 0$, $r \longmapsto F(t, r)$ is a function in $C_0(O(M) \longrightarrow \boldsymbol{R}^d)$. Then, with probability one,

$$
\begin{aligned}
&M(t)F(t, r(t)) - M(0)F(0, r(0)) \\
&= \int_0^t M(u)(\tilde{L}_\alpha F)(u, r(u))dB^\alpha(u) \\
&\quad + \int_0^t M(u)\left(\left[\frac{\partial F}{\partial t}(u, r(u)) + \frac{1}{2}\{\Delta_{O(M)}F(u, r(u))\right.\right. \\
&\qquad\qquad \left.\left. + J(r(u))F(u, r(u))\}\right]\right) du \qquad (6.36) \\
&\quad + \int_0^t M(u)e(u)\,Qe(u)^{-1}\left[\frac{\partial F}{\partial x^d}(u, r(u)) - \frac{\partial F}{\partial e^l_m}(u, r(u))\right. \\
&\qquad \left. \times \left\{{}_{d}{}^{l}{}_{k}\right\}(X(u))e^k_m(u) + e(u)\Gamma_d(X(u))e(u)^{-1}F(u, r(u))\right] d\phi(u).
\end{aligned}
$$

Proof. As we remarked above, the diffusion $(r(t))$ is a normally reflecting diffusion on $O(M)$ in the sense of Definition 6.2 given below. As we shall see, a characteristic feature of such a diffusion is that if $f(t, r)$ is a smooth function on $[0, \infty) \times O(M)$, and if $g(t)$ is an $(\mathscr{F}_t)$-adapted process such that $s \longmapsto g(s)$ is right continuous with left-hand limits and $s \longmapsto E_\mu(g(s)^2)$ is locally bounded, then the following identity holds:

$$
\sum_{\substack{s \leq \phi(t) \\ s \in D}}^{*} \int_{A(s-)\wedge t}^{A(s)\wedge t} g(s)df(s, r(s)) = \int_0^t g(s)df(s, r(s)), \tag{6.37}
$$

* $\mathscr{P}(W(O(M)))$ is the totality of all probabilities on $W(O(M))$ with the topology of weak convergence.

where the integral is understood in the sense of stochastic integral by the semimartingale $s \longmapsto f(s, r(s))$, $\int_{A(s-)\wedge t}^{A(s)\wedge t}\cdot$ is defined similarly as in Remark 6.2 and the sum $\sum\limits_{\substack{s\le\phi(t)\\ s\in D}}^{*}$ is understood as the limit in probability of the finite sum $\sum\limits_{\substack{s\le\phi(t)\\ A(s)-A(s-)>\varepsilon}}\cdot$ as $\varepsilon\downarrow 0$.

First we prove the following lemma.

Lemma 6.5. For any t such that $X_t \in \mathring{M}$,

$$dM(t) = \frac{1}{2} M(t)J(r(t))dt, \tag{6.38}$$

i.e., if $u \in \boldsymbol{D}$, then for every $s \le t$ such that $[s, t] \subset (A(u-), A(u))$ we have

$$M(t) - M(s) = \frac{1}{2}\int_s^t M(u)J(r(u))du. \tag{6.39}$$

Proof. First we note that $J = eRe^{-1}$ by the convention (6.19). Then by (6.20) and Itô's formula, if $X(t) \in \mathring{M}$,

$$\begin{aligned} dM(t) &= K(t)e(t)^{-1}\{de(t)e(t)^{-1} + e(t)d(e(t)^{-1}) + de(t)\cdot d(e(t)^{-1})\} \\ &\quad + \frac{1}{2} K(t)e(t)^{-1}J(r(t))dt \\ &= K(t)e(t)^{-1}d(e(t)e(t)^{-1}) + \frac{1}{2} K(t)e(t)^{-1}J(r(t))dt \\ &= \frac{1}{2} M(t)J(r(t))dt. \end{aligned}$$

Now we return to the proof of (6.36). By Lemma 6.5 and Itô's formula,

$$\begin{aligned} &\sum_{\substack{s\le\phi(t)\\ s\in D}}^{*} \{M(t\wedge A(s))F(t\wedge A(s), r(t\wedge A(s))) \\ &\qquad\qquad - M(A(s-))F(A(s-), r(A(s-)))\} \\ &= \sum_{\substack{s\le\phi(t)\\ s\in D}}^{*} \int_{A(s-)}^{A(s)\wedge t} M(u)dF(u,r(u)) \\ &\quad + \frac{1}{2}\sum_{\substack{s\le\phi(t)\\ s\in D}}^{*} \int_{A(s-)}^{A(s)\wedge t} M(u)J(r(u))F(u,r(u))du. \end{aligned} \tag{6.40}$$

Using (6.37) and Theorem 6.1 (i), (6.40) is equal to

$$\int_0^t M(u)dF(u,r(u)) + \frac{1}{2}\int_0^t M(u)J(r(u))F(u,r(u))du$$

$$= \int_0^t M(u)(\tilde{L}_m F)(u,r(u))dB^m(u)$$

$$+ \int_0^t M(u)\left[\frac{\partial F}{\partial t}(u,r(u)) + \frac{1}{2}\{\Delta_{O(M)}F(u,r(u)) + J(r(u))F(u,r(u))\}\right]du \tag{6.41}$$

$$+ \int_0^t M(u)\left[\frac{\partial F}{\partial x^d}(u,r(u)) - \frac{\partial F}{\partial e^i_m}(u,r(u))\{{}_d{}^i{}_k\}(X(u))e^k_m(u)\right]d\phi(u).$$

On the otherhand, for every $s < t$,

$$M(t)F(t,r(t)) - M(s)F(s,r(s))$$
$$= K^1(t)Pe(t)^{-1}F(t,r(t)) - K^1(s)Pe(s)^{-1}F(s,r(s))$$
$$+ K^2(t)Qe(t)^{-1}F(t,r(t)) - K^2(s)Qe(s)^{-1}F(s,r(s)).$$

Noting that $Pe(t)^{-1}F(t,r) = 0$ if $r \in \partial O(M)$ and that $r(A(u)) \in \partial O(M)$ if $u \in \boldsymbol{D}$, and $r(A(u-)) \in \partial O(M)$ if $u \in \boldsymbol{D}$ and $u > 0$, we see that the first line of (6.40) is equal to

$$[K^1(t)Pe(t)^{-1}F(t,r(t)) - K^1(0)Pe(0)^{-1}F(0,r(0))]$$
$$+ \sum_{\substack{s \le \phi(t)\\ s \in \boldsymbol{D}}}^{*} \{K^2(t\wedge A(s))Qe(t\wedge A(s))^{-1}F(t\wedge A(s), r(t\wedge A(s))) \tag{6.42}$$
$$- K^2(A(s-))Qe(A(s-))^{-1}F(A(s-), r(A(s-)))\}.$$

By (6.20) and the fact that $\int_0^t I_{\partial M}(X(u))du \equiv 0$,

$$dK^2(u) = K(u)\{e(u)^{-1}de(u) + \frac{1}{2}R(X(u))du\}Q$$
$$- K(u)e(u)^{-1}de(u)I_{\partial M}(X(u))Q.$$

Hence it is clear that $d\{K^2(u)Qe(u)^{-1}F(u,r(u))\}$ has the form

$$g_1(u)de(u) + g_2(u)du + g_3(u)d(e(u)^{-1}) + g_4(u)dF(u,r(u))$$
$$-I_{\partial M}(X(u))K(u)e(u)^{-1}de(u)Qe(u)^{-1}F(u,r(u))$$

where $u \longmapsto g_i(u)$, $i = 1, 2, 3, 4$, are all right continuous $(\mathscr{F}_t)$-adapted processes with left-hand limits. Using again the general formula (6.37), the second term of (6.42) is equal to

$$\sum_{\substack{s\leq\phi(t)\\ s\in D}}^{*} \int_{A(s-)}^{A(s)\wedge t} d\{K^2(u)Qe(u)^{-1}F(u, r(u))\}$$
$$= \int_0^t g_1(u)de(u) + \int_0^t g_2(u)du + \int_0^t g_3(u)d(e(u)^{-1})$$
$$+ \int_0^t g_4(u)dF(u, r(u))$$
$$= K^2(t)Qe(t)^{-1}F(t, r(t)) - K^2(0)Qe(0)^{-1}F(0, r(0))$$
$$+ \int_0^t I_{\partial M}(X(u))K(u)e(u)^{-1}de(u)Qe(u)^{-1}F(u, r(u)).$$

Because of (6.12) and since $I_{\partial M}(X(u))Pe(u)^{-1}F(u, r(u)) = 0$,

$$\int_0^t I_{\partial M}(X(u))K(u)e(u)^{-1}de(u)Qe(u)^{-1}F(u, r(u))$$
$$= -\int_0^t K(u)e(u)^{-1}e(u)\Gamma_d(X(u))e(u)^{-1}F(u, r(u))d\phi(u)$$
$$= -\int_0^t K(u)\Gamma_d(X(u))e(u)^{-1}F(u, r(u))d\phi(u).$$

Thus we have

$$M(t)F(t, r(t)) - M(0)F(0, r(0))$$
$$- \int_0^t M(u)e(u)\Gamma_d(X(u))e(u)^{-1}F(u, r(u))d\phi(u)$$
$$= \int_0^t M(u)(\tilde{L}_m F)(u, r(u))dB^m(u)$$

(6.43)
$$+ \int_0^t M(u)\Big[\frac{\partial F}{\partial t}(u, r(u)) + \frac{1}{2}\{\Delta_{O(M)}F(u, r(u)) + J(r(u))F(u, r(u))\}\Big]du$$
$$+ \int_0^t M(u)\Big[\frac{\partial F}{\partial x^d}(u, r(u)) - \frac{\partial F}{\partial e^i_m}(u, r(u))\left\{{}_{d}{}^{i}{}_{k}\right\}(X(u))e^k_m(u)\Big]d\phi(u).$$

Finally we remark that if $g(u)$ is $(\mathscr{F}_t)$-well measurable process, then

$$\int_0^t M(u)g(u)d\phi(u)$$

$$(6.44)\qquad \begin{aligned} &= \int_0^t K^1(u)e(u)^{-1}g(u)d\phi(u) + \int_0^t K^2(u)e(u)^{-1}g(u)d\phi(u) \\ &= \int_0^t K^2(u)e(u)^{-1}g(u)d\phi(u) \\ &= \int_0^t M(u)e(u)Qe(u)^{-1}g(u)d\phi(u) \end{aligned}$$

since $I_{\partial M}(X(u))K^1(u) = 0$. Now (6.36) follows from (6.43). q.e.d.

Theorem 6.5 may be regarded as a martingale version of the statement that $u(t, r) = H_t F(r)$ solves (6.33).

In the remainder of this section, we shall elaborate on the above mentioned notion of *normally reflecting diffusions* and especially on the formula (6.37). Let D be the upper half space of $\boldsymbol{R}^d$ and $\sigma(x) = (\sigma_k^i(x))$, $b(x) = (b^i(x))$, $\tau(x)=(\tau_l^i(x))$, $\beta(x)=(\beta^i(x))$ be given as in Chapter IV, Section 7. Consider the non-sticky stochastic differential equation (7.8) in Chapter IV, Section 7 corresponding to $[\sigma,b,\tau,\beta,1,0]$. Let $a^{ij}(x)$ and $\tau^{ij}(x)$ be defined by (7.6) and (7.7) in Chapter IV, Section 7. Let $\mathfrak{X} = (X(t), B(t), M(t), \phi(t))$ be a solution. We know that $X(t)$ is a diffusion process on D determined by the differential operator

$$(6.45)\qquad Af(x) = \frac{1}{2}\sum_{i,j=1}^{d} a^{ij}(x)\frac{\partial^2 f}{\partial x^i \partial x^j}(x) + \sum_{i=1}^{d} b^i(x)\frac{\partial f}{\partial x^i}(x)$$

with the boundary condition

$$(6.46)\qquad Lf(x) = \frac{1}{2}\sum_{i,j=1}^{d-1} \alpha^{ij}(x)\frac{\partial^2 f}{\partial x^i \partial x^j}(x) + \sum_{i=1}^{d-1} \beta^i(x)\frac{\partial f}{\partial x^i}(x) + \frac{\partial f}{\partial x^d}(x) = 0 \quad \text{on } \partial D.$$

Set

$$(6.47)\qquad \mathscr{Z} = \{s \geq 0;\ X(s) \in \partial D\}.$$

Since $\int_0^t I_{\partial D}(X(s))ds = 0$ a.s., $\mathscr{Z}$ has Lebesgue measure 0 a.s. and $(0, \infty) \setminus \mathscr{Z} = \bigcup_\alpha e_\alpha$, where $e_\alpha = (l_\alpha, r_\alpha)$ are mutually disjoint open intervals. Each e_α is called an *excursion interval* and the part $(X(t), t \in e_\alpha)$ is called an *excursion of* $X(t)$. Let $A(t)$ be the right-continuous inverse function of $t \longmapsto \phi(t)$. Set $\boldsymbol{D} = \{s \in [0, \infty);\ A(s) - A(s-) > 0\}$.* Then it is easy to see that the totality of excursion intervals coincides with the set of intervals $\{(A(u-), A(u)),\ u \in \boldsymbol{D}\}$. Let $\mathscr{F}_t^0$ be the completion of $\sigma[X(u),$

* $A(0-) = 0$.

$B(u), u \leq t]$ and let $g(s)$ be an $\mathscr{F}_t^0$-well measurable process such that $s \longmapsto E[g(s)^2]$ is bounded on each finite interval. Then $Y^k(t) = \int_0^t g(s)dB^k(s)$ is a continuous $\mathscr{F}_t^0$-martingale. For each excursion interval $e_\alpha = (l_\alpha, r_\alpha)$,

$$\int_{e_\alpha \cap [0,t]} g(s)dB^k(s) = Y^k(r_\alpha \wedge t) - Y^k(l_\alpha \wedge t)$$

by definition. It is also denoted by

$$\int_{A(u-)\wedge t}^{A(u)\wedge t} g(s)dB^k(s) \qquad \text{if} \quad e_\alpha = (A(u-), A(u)).$$

Theorem 6.6. (i).

$$E(\sum_{u\in D}[\int_{A(u-)\wedge t}^{A(u)\wedge t} g(s)dB^k(s)]^2) = E[\int_0^t g(s)^2 ds], \tag{6.48}$$

$$k = 1, 2, \ldots, r, \quad t \geq 0.$$

(ii). Assume further that $s \longmapsto g(s)$ is right-continuous with left-hand limits and that $a^{ij}(x)$ is C^3 on D. Then

$$\begin{aligned} \sum_{u\in D}{}^{*} \int_{A(u-)\wedge t}^{A(u)\wedge t} g(s)dB^k(s) &= \int_0^t g(s)dB^k(s) \\ &+ \int_0^t g(s)\frac{\sigma_k^d(X(s))}{a^{dd}(X(s))}d\phi(s), \qquad k = 1, 2, \ldots, r, \quad t \geq 0. \end{aligned} \tag{6.49}$$

Here

$$\sum_{u\in D}{}^{*} \int_{A(u-)\wedge t}^{A(u)\wedge t} \cdot$$

is defined as the limit in probability of finite sum

$$\sum_{\substack{u\in D \\ A(u)-A(u-)>\varepsilon}} \int_{A(u-)\wedge t}^{A(u)\wedge t} \cdot$$

if and only if the limit exists.

Proof. We shall first prove the special case of the reflecting Brownian motion and reduce the general case to this special case.

(a) *The case of the reflecting Brownian motion*: i.e., the case $\sigma_k^i(x) = \delta_k^i$ with $r = d$, $b(x) \equiv 0$, $\tau(x) \equiv 0$, $\beta(x) \equiv 0$ and $\delta(x) \equiv 1$.

In this case, the system $\mathfrak{X}(t) = (X(t), B(t), \phi(t))$ is determined by the equation

$$(6.50) \qquad \begin{cases} X^i(t) = X^i(0) + B^i(t), & i = 1, 2, \dots, d-1 \\ X^d(t) = X^d(0) + B^d(t) + \phi(t). \end{cases}$$

Let the path spaces $\boldsymbol{W}(D)$, $\mathscr{W}(D)$, $\mathscr{W}_0(D)$, the corresponding σ-fields $\mathscr{B}(\boldsymbol{W}(D))$, $\mathscr{B}(\mathscr{W}(D))$, $\mathscr{B}(\mathscr{W}_0(D))$ and the σ-finite measure n on $(\mathscr{W}_0(D), \mathscr{B}(\mathscr{W}_0(D)))$ be defined as in Chapter IV, Section 7. If we set $\boldsymbol{D}_p = \{u \in (0, \infty); A(u) - A(u-) > 0\} = \boldsymbol{D} \setminus \{0\}$ and

$$p(u) = \begin{cases} X(t + A(u-)) - X(A(u-)), & 0 \leq t \leq A(u) - A(u-) := \sigma[p(u)] \\ X(A(u)) - X(A(u-)), & t \geq A(u) - A(u-) \end{cases}$$
$$\text{for} \quad u \in \boldsymbol{D}_p,$$

then we know that p: $\boldsymbol{D}_p \ni u \longmapsto p(u) \in \mathscr{W}_0(D)$ is an $(\mathscr{F}_t)$-stationary Poisson point process on $(\mathscr{W}_0(D), \mathscr{B}(\mathscr{W}_0(D)))$ with characteristic measure n, where $\mathscr{F}_t = \mathscr{F}_{A(t)}$. Let n^ξ, $\xi \in D$, be the image measure on $\mathscr{W}(D)$ of n under the mapping $w \longmapsto \xi + w$.* We shall now introduce the following notations

(6.51) ρ_t: $\boldsymbol{W}(D) \longrightarrow \boldsymbol{W}(D)$ defined by $(\rho_t w)(s) = w(t \wedge s)$ (*stopped path*)

(6.52) θ_t: $\boldsymbol{W}(D) \longrightarrow \boldsymbol{W}(D)$ defined by $(\theta_t w)(s) = w(t+s)$; (*shifted path*)

(6.53) $\rho_{\partial D}$: $\boldsymbol{W}(D) \longrightarrow \boldsymbol{W}(D)$ defined by $(\rho_{\partial D} w)(s) = w(s \wedge \sigma(w))$

where

(6.54) $\sigma(w) = \inf\{t > 0;\, w(t) \in \partial D\}$

(*stopped path on reaching the boundary*).

Let, by X, be denoted the path $t \longmapsto X(t)$. Clearly this is a $\boldsymbol{W}(D)$-valued random variable.

* $(\xi + w)(t) = \xi + w(t)$.

Excursion formula I. Let $Z(s)$ be an $(\mathscr{F}_t)$-predictable non-negative process and $f(s,w,w')$ be a non-negative Borel function on $(0, \infty)\times W(D)\times \mathscr{W}(D)$. Then

$$E\{\sum_{s\leq t, s\in D_p} Z(s)f(A(s-), \rho_{A(s-)}X, \rho_{\partial D}[\theta_{A(s-)}X])\} \tag{6.55}$$
$$= E\{\int_0^t Z(s)[\int_{\mathscr{W}(D)} f(A(s), \rho_{A(s)}X, w)n^{X(A(s))}(dw)]ds\}.$$

This follows immediately from the fact that the sum under the expectation on the left is just

$$\int_0^{t+}\int_{\mathscr{W}_0(D)} Z(s)f(A(s-), \rho_{A(s-)}X, X(A(s-)) + w)N_p(dsdw)$$

(cf. Chapter II, Section 3).

By a random time change $t \longmapsto \phi(t)$ in (6.55) we have

Excursion formula II. Let $Z(s)$ be an $(\mathscr{F}_t)$-well measurable non-negative process and $f(s, w, w')$ be as above. Then

$$E\{\sum_{s\leq \phi(t), s\in D_p} Z(A(s-))f(A(s-),\rho_{A(s-)}X,\rho_{\partial D}[\theta_{A(s-)}X])\} \tag{6.56}$$
$$= E\{\int_0^t Z(s)[\int_{\mathscr{W}(D)} f(s, \rho_s X,w)n^{X(s)}(dw)]d\phi(s)\}.$$

Let $\tau(t)$ be defined by (6.24). By setting

$$f(s, w, w') = g(t - s, s, w, w')I_{\{\sigma(w')>t-s\}}$$

in (6.56),*[1] we have

Last exit formula. Let $Z(s)$ be an $(\mathscr{F}_t)$-well measurable non-negative process and $g(s,s',w,w')$ be a non-negative Borel function on $(0, \infty)\times(0, \infty)\times W(D)\times \mathscr{W}(D)$. Then

$$E\{Z(\tau(t))g(t - \tau(t), \tau(t), \rho_{\tau(t)}X, \rho_{\partial D}[\theta_{\tau(t)}X])I_{\{\tau(t)>0\}}\} \tag{6.57}$$
$$= E\{\int_0^t Z(s)[\int_{\mathscr{W}(D)} g(t-s, s, \rho_s X, w)I_{\{\sigma(w)>t-s\}}n^{X(s)}(dw)]d\phi(s)\}.$$

For each $i = 1, 2, \ldots, r$ and $t_0 > 0$, $w^i(t+t_0) - w^i(t_0)$ is a continuous $\mathscr{B}_{t+t_0}(\mathscr{W}(D))$-martingale with respect to $n^{\xi}(\cdot \mid \sigma(w) > t_0)$.*[2]

*[1] This idea is due to Maisonneuve [103].

*[2] $\xi \in \partial D$ is fixed. $\sigma(w)$ is defined by (6.54).

Hence, for any $\mathscr{B}_t(\mathscr{W}(D))$-well measurable process $\Phi(s, w)$ such that

$$\int_{\mathscr{W}(D)} [\int_0^{\sigma(w)\wedge t} |\Phi(s, w)|^2 ds] n^{\xi}(dw) < \infty \qquad \text{for each} \quad t > 0,$$

we can define the stochastic integral

$$\int_{t_0\wedge\sigma(w)}^{t\wedge\sigma(w)} \Phi(s,w)dw^i(s).$$

It is easy to see that

$$\lim_{t_0 \downarrow 0} \int_{t_0\wedge\sigma(w)}^{t\wedge\sigma(w)} \Phi(s,w)dw^i(s) = \int_0^{t\wedge\sigma(w)} \Phi(s,w)dw^i(s)$$

exists in $\mathscr{L}_2(\mathscr{W}(D), n^{\xi})$,

$$\int_{\mathscr{W}(D)} [(\int_0^{t\wedge\sigma(w)} \Phi(s,w)dw^i(s))^2] n^{\xi}(dw) = \int_{\mathscr{W}(D)} [\int_0^{t\wedge\sigma(w)} \Phi(s,w)^2 ds] n^{\xi}(dw) \tag{6.58}$$

and for every $\mathscr{B}_{t_0}(\mathscr{W}(D))$-measurable $H(w)$ in $\mathscr{L}_2(\mathscr{W}(D), n^{\xi})$,

$$\begin{aligned}\int_{\mathscr{W}(D)} [\int_0^t \Phi(s, w)dw^i(s)] H(w) n^{\xi}(dw)\\ = \int_{\mathscr{W}(D)} [\int_0^{t_0} \Phi(s,w)dw^i(s)] H(w) n^{\xi}(dw), \qquad t > t_0.\end{aligned} \tag{6.59}$$

First we shall prove (6.48). Without loss of generality we may assume that $X(u)$ is given in canonical form, i.e., $X(u,w) = w(u)$, $w \in \boldsymbol{W}(D)$. Now let $g(s) = g(s,w)$ be $(\mathscr{B}_t(\boldsymbol{W}(D)))$-well measurable process such that $s \longmapsto E(g(s)^2)$ is bounded on each finite interval. For given $s > 0$, $w \in \boldsymbol{W}(D)$ and $w' \in \mathscr{W}(D)$, set

$$\Phi_g^s(u, w, w') = \begin{cases} g(s+u, [w, w']_s) & \text{if} \quad w(s) = w'(0), \\ 0 & \text{otherwise} \end{cases} \tag{6.60}$$

where $[w, w']_s$ is defined by

$$[w, w']_s(u) = \begin{cases} w(u), & 0 \le u \le s, \\ w'(u-s), & u > s. \end{cases} \tag{6.61}$$

Let $t > 0$ be given and fixed. Let

$$(6.62) \qquad f_1^t(s,w,w') = \begin{cases} [\int_0^{(t-s)\wedge\sigma(w')} \Phi_g^s(u,w,w')dw'^i(u)]^2, & t \geq s > 0 \\ 0, & t < s \end{cases}$$

in the sense explained above. By (6.58),

$$\int_{\mathscr{W}(D)} f_1^t(s, w, w')n^{w(s)}(dw') = \begin{cases} \int_{\mathscr{W}(D)} [\int_{s\wedge t}^{(\sigma(w')+s)\wedge t} g(u, [w, w']_s)^2 du] n^{w(s)}(dw') \\ \qquad\qquad \text{if} \quad w(s) = w'(0), \\ 0 \qquad\qquad \text{otherwise.} \end{cases}$$

It is clear from the definition that

$$(6.63) \qquad f_1^t(A(s-), p_{A(s-)}X, p_{\partial D}[\theta_{A(s-)}X])$$

$$= (\int_{A(s-)\wedge t}^{A(s)\wedge t} g(u)dX^i(u))^2 = (\int_{A(s-)\wedge t}^{A(s)\wedge t} g(u)dB^i(u))^2.$$

By the excursion formula and (6.62),

$$E\{\sum_{s\in D} (\int_{A(s-)\wedge t}^{A(s)\wedge t} g(u)dB^i(u))^2\}$$

$$= E\{(\int_0^{\sigma\wedge t} g(u)dB^i(u))^2\} + E\{\sum_{s\in D, s\leq\phi(t)} (\int_{A(s-)\wedge t}^{A(s)\wedge t} g(u)dB^i(u))^2\}$$

$$= E\{\int_0^{\sigma\wedge t} g(u)^2 du\} + E\{\int_0^t d\phi(s) \int_{\mathscr{W}(D)} [\int_{s\wedge t}^{(\sigma(w')+s)\wedge t} g(u,[w,w']_s)^2 du] n^{w(s)}(dw')\}$$

$$= E\{\int_0^{\sigma\wedge t} g(u)^2 du\} + E\{\sum_{s\in D_p, s\leq\varphi(t)} \int_{A(s-)}^{A(s)\wedge t} g(u)^2 du\}$$

$$= E[\int_0^t g(u)^2 du].$$

Now we shall prove (ii). In the case (a), (6.49) is given as

$$(6.64) \qquad \sum_{s\in D}^* \int_{A(s-)\wedge t}^{A(s)\wedge t} g(u)dB^i(u) = \int_0^t g(u)dB^i(u), \quad i = 1, 2, \ldots, d-1$$

and

(6.65) $$\sum_{s\in D}^{*} \int_{A(s-)\wedge t}^{A(s)\wedge t} g(u)dB^d(u) = \int_0^t g(u)dB^d(u) + \int_0^t g(u)d\phi(u).$$

First we prove (6.65). In the following, we shall call $g(s)$ a *step process* if there exists a sequence of $(\mathscr{B}_t(W(D)))$-stopping times $\sigma_0 \equiv 0 < \sigma_1 < \sigma_2 < \cdots < \sigma_n < \cdots \longrightarrow \infty$ and $\{\mathscr{B}_{\sigma_i}(W(D))\}$-measurable random variable g_i such that $g(s) = g_i$ if $s \in [\sigma_i, \sigma_{i+1})$ for $i = 0, 1, \ldots$.

Lemma 6.6. Let $g(s)$ be a step process. Then (6.65) holds.

Proof. If, for example, $g(s) \equiv 1$, then (6.65) is trivially true: we have

$$\int_{A(s-)\wedge t}^{A(s)\wedge t} dB^d(u) = \int_{A(s-)\wedge t}^{A(s)\wedge t} dX^d(t) = X^d(A(s)\wedge t) - X^d(A(s-)\wedge t)$$

$$= \begin{cases} 0, & s \in \boldsymbol{D}_p,\ \ A(s) \le t \ \text{ or } \ A(s-) > t \\ X^d(t), & s \in \boldsymbol{D}_p,\ \ A(s-) \le t < A(s) \\ X^d(\sigma \wedge t) - X^d(0), & s = 0 \end{cases}$$

and hence the left-hand side of (6.65) is equal to $X^d(t) - X^d(0) = B^d(t) + \phi(t)$. A similar argument applies if $g(s)$ is a step process.

Lemma 6.7. Let $g(s)$ be a $\mathscr{B}_t(W(D))$-adapted process such that $s \longrightarrow g(s)$ is right-continuous with left-hand limits. Then for every $\varepsilon > 0$, there exists a step process $g_\varepsilon(s)$ such that

(6.66) $$|g_\varepsilon(s) - g(s)| \le \varepsilon \qquad \text{for every } s.$$

Proof. Let $\{\sigma_n\}$ be defined by $\sigma_0 = 0$ and

$$\sigma_n = \inf\{t > \sigma_{n-1};\ |g(t) - g(\sigma_{n-1})| \ge \varepsilon\} \wedge n.$$

Then $\sigma_n \uparrow \infty$, and $g_\varepsilon(s) = \sum_{n=0}^{\infty} g(\sigma_n) I_{[\sigma_n, \sigma_{n+1})}(s)$ has the desired properties.

Lemma 6.8. For $\xi \in \partial D$ and $t > 0$, let

(6.67) $$\mu^{\xi,t}(B) = n^\xi(B \mid \sigma > t) = \frac{n^\xi(B: \sigma > t)}{n^\xi(\sigma > t)} \qquad \text{for} \quad B \in \mathscr{B}(\mathscr{W}(D)).$$

Then $\mu^{\xi,t}$ is a Markovian measure on $\mathscr{W}(D)$ concentrated on $\{w \in \mathscr{W}(D);\ w(0) = \xi,\ \sigma(w) > t\}$ such that

$$\begin{aligned}(6.68)\qquad &\mu^{\xi,t}\{w; w(t_1)\in E_1,\ w(t_2)\in E_2,\ \ldots,\ w(t_n)\in E_n\}\\ &= \int_{E_1} dx_1 \int_{E_2} dx_2 \cdots \int_{E_n} dx_n K^{\xi}(t_1, x_1) \prod_{i=1}^{n-1} p(t_i, x_i; t_{i+1}, x_{i+1})\\ &\text{for}\quad 0 < t_1 < t_2 < \cdots < t_n < t \quad\text{and}\quad E_i \in \mathscr{B}(\mathring{D})\end{aligned}$$

In the above,

$$K^{\xi}(s, x) = \frac{K^+(s, x-\xi)h(t-s, x)}{K(t)}, \qquad s > 0,\quad x \in \mathring{D}$$

$$p(s, x; u, y) = \frac{h(t-u, y)}{h(t-s, x)} p^0(u-s, x, y),\ 0 < s < u < t,\ x, y \in \mathring{D}$$

$$h(s, x) = \int_D p^0(s, x, y)dy = \frac{2}{\sqrt{2\pi s}} \int_0^{x^d} \exp\left\{-\frac{\eta^2}{2s}\right\} d\eta$$

and

$$K(t) = \int_D K^+(t, x)dx = \sqrt{\frac{2}{\pi t}},$$

where $K^+(t,x)$ and $p^0(t, x, y)$ are given as in Chapter IV, Section 7.

This lemma is easily proved from the properties of the measure n.

Corollary. The process

$$b^d(s) = w^d(s) - \int_0^s A(t, u, w(u))du \qquad (0 \leq s < t)$$

is a one-dimensional Brownian motion with respect to the probability measure $\mu^{\xi,t}$ on $\mathscr{W}(D)$, where

$$(6.69)\qquad A(t, u, x) = \frac{\left(\frac{\partial}{\partial x^d} h(t-u, x)\right)}{h(t-u, x)} = \frac{\exp\left\{\frac{-(x^d)^2}{2(t-u)}\right\}}{\int_0^{x^d} \exp\left\{-\frac{\eta^2}{2(t-u)}\right\} d\eta}.$$

Lemma 6.9. Let $g(s)$ be a bounded $\{\mathscr{B}_t(\boldsymbol{W}(D))\}$-well measurable process. Then for fixed $s, w \in \boldsymbol{W}(D)$ and for any $\varepsilon \geq \varepsilon' > 0$,

$$\begin{aligned}(6.70)\qquad &\int_{\mathscr{V}(D)} \left|\int_0^{\varepsilon'} \Phi_g^s(u, w, w')dw'^d(u)\right| I_{\{\sigma(w')>\varepsilon\}} n^{w(s)}(dw')\\ &\leq \left(\sqrt{\frac{2}{\pi}} + 1\right) \|g\|_\infty\end{aligned}$$

where $\|g\|_\infty = \sup_{u,w} |g(u,w)|$ and Φ_g^s is defined by (6.60).

Proof. By the corollary to Lemma 6.8,

$$\begin{aligned}
&\int_{\mathscr{W}(D)} \Big| \int_0^{\varepsilon'} \Phi_g^s(u, w, w')dw'^d(u) \Big| I_{\{\sigma(w')>\varepsilon\}} n^{w(s)}(dw') \\
&\le \int_{\mathscr{W}(D)} \{ \Big| \int_0^{\varepsilon'} \Phi_g^s(u, w, w')db^d(u) \Big| \} \mu^{w(s),\varepsilon}(dw') n^{w(s)}(\sigma(w') > \varepsilon) \\
&+ \int_{\mathscr{W}(D)} \{ \int_0^{\varepsilon'} |\Phi_g^s(u,w,w')A(\varepsilon,u,w'(u))| du \} \mu^{w(s),\varepsilon}(dw') n^{w(s)}(\sigma(w') > \varepsilon) \\
&\le \|g\|_\infty \varepsilon'^{1/2} \sqrt{\frac{2}{\pi\varepsilon}} + \|g\|_\infty \\
&\le \left(\sqrt{\frac{2}{\pi}} + 1\right) \|g\|_\infty .
\end{aligned}$$

Here we used the following facts:

$$n^\xi(\sigma(w) > \varepsilon) = \int_D K^+(\varepsilon, x)dx = \sqrt{\frac{2}{\pi\varepsilon}}$$

and

$$\begin{aligned}
&\int_{\mathscr{W}(D)} [\int_0^{\varepsilon'} A(\varepsilon, u, w(u))du] \mu^{\xi,\varepsilon}(dw) n^\xi(\sigma(w) > \varepsilon) \\
&\le \int_{\mathscr{W}(D)} [\int_0^{\varepsilon} A(\varepsilon, u, w(u))du] \mu^{\xi,\varepsilon}(dw) n^\xi(\sigma(w) > \varepsilon) \\
&= \int_{\mathscr{W}(D)} w^d(\varepsilon) n^\xi(dw) \\
&= \int_D x^d K^+(\varepsilon, x - \xi)dx = 1.
\end{aligned}$$

Now let $g(s)$ be a $\{\mathscr{B}_t(W(D))\}$-well measurable process such that $s \longmapsto E(g(s)^2)$ is bounded on each bounded interval. We introduce the following notation:

$$S_\varepsilon(t) = \sum_{s \le \phi(t),\, A(s)-A(s-)>\varepsilon} \int_{A(s-)}^{A(s)\wedge t} g(s)dB^d(s), \tag{6.71}$$

$$Y_\varepsilon(t) = \int_0^t d\phi(s) [\int_{\mathscr{W}(D)} \{ \int_0^{\sigma(w')\wedge(t-s)} \Phi_g^s(u, w, w')dw'^d(u) I_{\{\sigma(w')>\varepsilon\}} \} \times n^{w(s)}(dw')], \tag{6.72}$$

(6.73) $\quad M_\varepsilon(t) = S_\varepsilon(t) - Y_\varepsilon(t).$

Lemma 6.10. $M_\varepsilon(t) \longrightarrow \int_0^t g(s)dB^d(s)$ in $\mathscr{L}_2(P)$ as $\varepsilon \downarrow 0$.

Proof. Set

(6.74) $$f_2^t(s, w, w') = \int_0^{(t-s)\wedge\sigma(w')} \Phi_g^s(u, w, w(s)+w')dw'^d(u)$$

for $0 \leq s \leq t$, $w \in \boldsymbol{W}(D)$ and $w' \in \mathscr{W}_0(D)$.

Then clearly

$$M_\varepsilon(t) = \int_0^{\sigma\wedge t} g(u)dB^d(u)I_{\{\sigma>\varepsilon\}}$$
$$+ \int_0^{\phi(t)}\int_{\mathscr{W}_0(D)} f_2^t(A(s-), \rho_{A(s-)}X, w')I_{\{\sigma(w')>\varepsilon\}}\tilde{N}_p(dsdw')$$

in the sense of stochastic integrals (Chapter II, Section 3). Consequently for each fixed t, $M_\varepsilon(t) \longrightarrow M(t)$ in $\mathscr{L}_2(P)$ as $\varepsilon \longrightarrow 0$, where

$$M(t) = \int_0^{\sigma\wedge t} g(u)dB^d(u) + \int_0^{\phi(t)}\int_{\mathscr{W}_0(D)} f_2^t(A(s-), \rho_{A(s-)}X, w')\tilde{N}_p(dsdw')$$

and

$$E(M(t)^2) = E[\int_0^{\sigma\wedge t} g(u)^2du] + E[\int_0^t d\phi(s)\int_{\mathscr{W}_0(D)} \{f_2^t(s, \rho_s X, w')\}^2 n(dw')]$$
$$= E[\int_0^t g(u)^2du].$$

Next we assume that $g(s)$ is bounded and prove that $M(t)$ is an $\{\mathscr{B}_t(\boldsymbol{W}(D))\}$-martingale. It is sufficient to show that for any bounded Borel measurable functions $F_1(w)$, $F_2(w)$ on $\boldsymbol{W}(D)$ and $0 < t_1 < t_2$,

(6.75) $\quad E(M(t_2)H) = E(M(t_1)H)$

where

$$H(w) = F_1(\rho_{A(\phi(t_1)-)}w)F_2(\rho_{t_1-A(\phi(t_1)-)}[\theta_{A(\phi(t_1)-)}w]).$$

We prove the following estimate

(6.76) $$E(M_\varepsilon(t_2)H) = E(M_\varepsilon(t_1)H) + o(1) \qquad (\varepsilon \downarrow 0)$$

from which (6.75) follows. First, it is clear that

$$E\Big(\int_0^{\sigma\wedge t_2} g(u)dB^d(u)I_{\{\sigma>\varepsilon\}}H\Big) = E\Big(\int_0^{\sigma\wedge t_1} g(u)dB^d(u)I_{\{\sigma>\varepsilon\}}H\Big) + o(1).$$

By the martingale property of the stochastic integral with respect to $\tilde{N}_p$, we have

$$E\Big[\int_0^{\phi(t_2)}\int_{\mathscr{W}_0(D)} f_2^{t_2}(A(s-), \rho_{A(s-)}X, w')I_{\{\sigma(w')>\varepsilon\}}\tilde{N}_p(dsdw')H\Big]$$
$$= E\Big[\int_0^{\phi(t_1)}\int_{\mathscr{W}_0(D)} f_2^{t_2}(A(s-), \rho_{A(s-)}X, w')I_{\{\sigma(w')>\varepsilon\}}\tilde{N}_p(dsdw')H\Big].$$

Now

$$\int_0^{\phi(t_1)}\int_{\mathscr{W}_0(D)} f_2^{t_2}(A(s-), \rho_{A(s-)}X, w')I_{\{\sigma(w')>\varepsilon\}}\tilde{N}_p(dsdw')$$
$$= \sum_{s\in D_{\mathscr{p}},\, s\leq\phi(t_1),\, \sigma(\theta_{A(s-)}X)>\varepsilon} f_2^{t_2}(A(s-), \rho_{A(s-)}X, \hat{\rho}_{\partial D}[\theta_{A(s-)}X])$$
$$- \int_0^{t_1} d\phi(s)\int_{\mathscr{W}_0(D)} f_2^{t_2}(s, \rho_s X, w')I_{\{\sigma(w')>\varepsilon\}}n(dw')$$
$$:= I_1^\varepsilon - I_2^\varepsilon,$$

where $\hat{\rho}_{\partial D}[\theta_{A(s-)}X] = \rho_{\partial D}[\theta_{A(s-)}X] - X(A(s-))$. If $\varepsilon < t_2 - t_1$ and $s \leq t_1$,

$$\int_{\mathscr{W}_0(D)} f_2^{t_2}(s, \rho_s X, w')I_{\{\sigma(w')>\varepsilon\}}n(dw')$$
$$= \int_{\mathscr{W}(D)}\Big[\int_0^{(t_2-s)\wedge\sigma(w')} \Phi_g^s(u, \rho_s X, w')dw'^d(u)\Big]I_{\{\sigma(w')>\varepsilon\}}n^{X(s)}(dw')$$
$$= \int_{\mathscr{W}(D)}\Big[\int_0^{\varepsilon} \Phi_g^s(u, \rho_s X, w')dw'^d(u)\Big]I_{\{\sigma(w')>\varepsilon\}}n^{X(s)}(dw')$$

and hence

$$I_2^\varepsilon = \int_0^{t_1} d\phi(s)\int_{\mathscr{W}(D)} f_2^{t_1}(s, \rho_s X, w')I_{\{\sigma(w')>\varepsilon\}}n^{X(s)}(dw')$$
$$+ \int_{t_1-\varepsilon}^{t_1} d\phi(s)\Big\{\int_{\mathscr{W}(D)}\Big[\int_0^{\varepsilon} \Phi_g^s(u, \rho_s X, w')dw'^d(u)$$
$$- \int_0^{(t_1-s)} \Phi_g^s(u, \rho_s X, w')dw'^d(u)\Big]I_{\{\sigma(w')>\varepsilon\}}n^{X(s)}(dw')\Big\}.$$

If we denote the second term by $\delta(\varepsilon)$, then by Lemma 6.9, $E(\delta(\varepsilon)) = o(1)$. Next,

$$\begin{aligned} I_1^\varepsilon &= \sum_{\substack{s \in D_p,\, s<\phi(t_1) \\ \sigma(\theta_{A(s-)}X)>\varepsilon}} f_2^{t_2}(A(s-), \rho_{A(s-)}X, \hat{\rho}_{\partial D}[\theta_{A(s-)}X]) \\ &\quad + f_2^{t_2}(\tau(t_1), \rho_{\tau(t_1)}X, \hat{\rho}_{\partial D}[\theta_{\tau(t_1)}X]) \\ &:= I_{11}^\varepsilon + I_{12}^\varepsilon \end{aligned}$$

where $\tau(t_1) = A(\phi(t_1)-)$. By the last exit formula,

$$\begin{aligned} E(I_{12}^\varepsilon H) &= E[\int_0^{t_1} F_1(\rho_s X)d\phi(s)\{\int_{\mathscr{W}(D)}[\int_0^{\sigma(w')\wedge(t_2-s)} \Phi_g^s(u, \rho_s X, w')dw'^d(u)] \\ &\quad \times I_{\{\sigma(w')>\varepsilon\vee(t_1-s)\}} F_2(\rho_{t_1-s}w')n^{X(s)}(dw')\}] \\ &= E[\int_0^{t_1} F_1(\rho_s X)d\phi(s)\{\int_{\mathscr{W}(D)}[\int_0^{\varepsilon\vee(t_1-s)} \Phi_g^s(u, \rho_s X, w')dw'^d(u)] \\ &\quad \times I_{\{\sigma(w')>\varepsilon\vee(t_1-s)\}} F_2(\rho_{t_1-s}w')n^{X(s)}(dw')\}] \\ &= E[\int_0^{t_1} F_1(\rho_s X)d\phi(s)\{\int_{\mathscr{W}(D)}[\int_0^{(t_1-s)} \Phi_g^s(u, \rho_s X, w')dw'^d(u)] \\ &\quad \times I_{\{\sigma(w')>\varepsilon\vee(t_1-s)\}} F_2(\rho_{t_1-s}w')n^{X(s)}(dw')\}] \\ &\quad + E[\int_{t_1-\varepsilon}^{t_1} F_1(\rho_s X)d\phi(s)\{\int_{\mathscr{W}(D)}[\int_0^{\varepsilon} \Phi_g^s(u, \rho_s X, w')dw'^d(u) \\ &\quad - \int_0^{t_1-s} \Phi_g^s(u, \rho_s X, w')dw'^d(u)] I_{\{\sigma(w')>\varepsilon\}} F_2(\rho_{t_1-s}w')n^{X(s)}(dw')\}] \\ &:= a_1(\varepsilon) + a_2(\varepsilon). \end{aligned}$$

Then clearly

$$E(I_{11}^\varepsilon H) + a_1(\varepsilon) = E(\{\sum_{\substack{s \in D_p,\, s\le\phi(t_1) \\ \sigma(\theta_{A(s-)}X)>\varepsilon}} f_2^{t_1}(A(s-), \rho_{A(s-)}X, \hat{\rho}_{\partial D}[\theta_{A(s-)}X])\} H).$$

By Lemma 6.9, $a_2(\varepsilon) = o(1)$. Now the proof of (6.76) is complete.

Let $g(s)$ be a bounded step process. By Lemma 6.6,

$$S_\varepsilon(t) \longrightarrow \int_0^t g(s)dB^d(s) + \int_0^t g(s)d\phi(s) \quad \text{a.s.} \quad \text{as} \quad \varepsilon \longrightarrow 0$$

and hence

$$Y_\varepsilon(t) = S_\varepsilon(t) - M_\varepsilon(t) \longrightarrow \int_0^t g(s)dB^d(s) + \int_0^t g(s)d\phi(s) - M(t)$$

in probability as $\varepsilon \longrightarrow 0$. By Lemma 6.9, we can easily conclude that this limit is of the form $\int_0^t h(s)d\phi(s)$ for some bounded adapted process $h(s)$. Therefore,

$$\int_0^t h(s)d\phi(s) = \int_0^t g(s)dB^d(s) + \int_0^t g(s)d\phi(s) - M(t).$$

We conclude from this that

$$\int_0^t h(s)d\phi(s) = \int_0^t g(s)d\phi(s)$$

and

$$\int_0^t g(s)dB^d(s) = M(t).$$

Let $g(s)$ be general. Then we take a sequence $\{g_k(s)\}$ of bounded step processes such that

$$E[\int_0^t |g_k(s) - g(s)|^2 ds] \longrightarrow 0 \quad (k \longrightarrow \infty).$$

It is easy to see that $M_k(t)$ corresponding to $g_k(s)$ converges to $M(t)$ and hence

$$M(t) = \int_0^t g(s)dB^d(s).$$

Lemma 6.11. Let $g(s)$ be a $\{\mathscr{B}_t(W(D))\}$-adapted process such that $s \longmapsto g(s)$ is right-continuous with left limits and $s \longmapsto E[g(s)^2]$ is locally bounded. Let $Y_\varepsilon(t)$ be defined by (6.72). Then

$$Y_\varepsilon(t) \longrightarrow \int_0^t g(u)d\phi(u) \quad \text{in probability} \quad \text{as} \quad \varepsilon \longrightarrow 0. \tag{6.77}$$

Proof. First assume that $g(s)$ is a step process. Then by Lemma 6.6

$$S_\varepsilon(t) \longrightarrow \int_0^t g(s)dB^d(u) + \int_0^t g(u)d\phi(u) \qquad \text{a.s.},$$

and by Lemma 6.10,

$$M_\varepsilon(t) \longrightarrow \int_0^t g(u)dB^d(u) \qquad \text{in} \quad \mathscr{L}_2(P).$$

Thus (6.77) holds. Now let $g(s)$ be general. By Lemma 6.7, we can choose a sequence $\{g_k(s)\}$ of step processes such that

$$g_k(s) - \frac{1}{k} \leq g(s) \leq g_k(s) + \frac{1}{k}.$$

Since $Y_\varepsilon(t)$ may also be expressed as

$$Y_\varepsilon(t) = \int_0^t d\phi(s) \int_{\mathscr{W}(D)} \Big[\int_0^{\sigma(w')\wedge(t-s)\wedge\varepsilon} \Phi_g^s(u,w,w')A(\varepsilon,u,w'(u))du \times I_{\{\sigma(w')>\varepsilon\}}\Big] n^{X(s)}(dw'),$$

we have

$$Y_\varepsilon^k(t) - \frac{1}{k}\phi(t) \leq Y_\varepsilon(t) \leq Y_\varepsilon^k(t) + \frac{1}{k}\phi(t)$$

where Y_ε^k corresponds to g_k. The desired conclusion then follows by first letting $\varepsilon \longrightarrow 0$ and then $k \longrightarrow \infty$.

Now we are ready to conclude the proof of (6.65). By Lemma 6.10 and Lemma 6.11

$$S_\varepsilon(t) = M_\varepsilon(t) + Y_\varepsilon(t)$$

converges in probability to

$$\int_0^t g(s)dB^d(s) + \int_0^t g(s)d\phi(s)$$

and this proves (6.65).

The proof of (6.64) is immediate from the following reasoning. Suppose that a family of disjoint open non-random intervals $\{e_\alpha\}$ in $[0, \infty)$ is given such that $[0, \infty)\setminus \bigcup_\alpha e_\alpha$ has zero measure. Then it is obvious that

$$\sum_\alpha{}^* \int_{[0,t]\cap e_\alpha} g(u)dB^i(u) = \int_0^t g(u)dB^i(u)$$

Indeed, if E is the union of all e_α such that $|e_\alpha| > \varepsilon$,

$$\sum_{\substack{\alpha \\ |e_\alpha|>\varepsilon}} \int_{[0,t]\cap e_\alpha} g(u)dB^i(u) = \int_0^t I_E(u)g(u)dB^i(u)$$

since the e_α are non-random intervals; moreover, it is clear that

$$\int_0^t I_E(u)g(u)dB^i(u) \longrightarrow \int_0^t g(u)dB^i(u) \qquad \text{in} \quad \mathscr{L}_2(P) \quad \text{as} \quad \varepsilon \longrightarrow 0.$$

Since $\{A(u)\}$ is defined only through $\{B^d(t)\}$, it is independent of $\{B^i(t), i=1,2,\ldots,d-1\}$. By Fubini's theorem, the intervals $(A(s-), A(s))$ can be treated as non random intervals and hence (6.64) follows.

(b) *General case.*

The proof of (i) is similar to the proof in case (a). As for (ii), we first remark that we may assume $d=r$. Indeed, if $r<d$, we set

$$\sigma_k^d(x) \equiv 0 \qquad \text{for} \quad r<k\leq d$$

and then adjoin $d-r$ independent Wiener processes $B^{r+1}(t), B^{r+2}(t), \ldots, B^d(t)$. If $r>d$, we consider the r-dimensional process $(Y^1(t), Y^2(t), \ldots, Y^{r-d}(t), X^1(t), \ldots, X^d(t))$ by setting e.g., $Y^1(t)=B^1(t)$, $Y^2(t)=B^2(t), \ldots, Y^{r-d}(t)=B^{r-d}(t)$.

First, we consider the case $\sigma_k^d(s) \equiv \delta_k^d$ and $b^d(x) \equiv 0$. Then $[B^1(t), B^2(t), \ldots, B^{d-1}(t), X^d(t) = X^d(0)+B^d(t)+\phi(t)]$ is a refelcting Brownian motion and the proof in case (i) applies. Secondly, we consider the case $\sigma_k^d(x) \equiv \delta_k^d$. Then by a change of drift (Chapter IV, Section 7), it is reduced to the first case. Thirdly, we consider the general case. It is reduced to the second case by the following change of coordinates and transformation of Brownian motion. Since the $a^{ij}(x)$ are C^3 by assumption, we can find a C^2-function $f(x)$ on D such that $f(x)\geq 0$, $f(x)=0$ if and only if $x\in\partial D$ and

$$a^{ij}(x)\frac{\partial f}{\partial x^i}\frac{\partial f}{\partial x^j} \equiv 1 \qquad \text{on} \quad \partial D.$$

For $\mathfrak{X}=(X(t),B(t),M(t),\phi(t))$, set $\hat{\mathfrak{X}}=(\hat{X}(t),B(t),M(t)\,\phi(t))$ where $\hat{X}^i(t) = X^i(t)$, $i=1,2,\ldots,d-1$ and $\hat{X}^d(t)=f(X(t))$. $\hat{X}$ corresponds to $[\hat{\sigma},\hat{b},\hat{\tau},\hat{\beta},1,0]$ with $\hat{a}^{dd}(x)\equiv 1$. By a transformation of Brownian motion from $B(t)$ to $\hat{B}(t)$ (Chapter IV, Section 7), we may assume that $\hat{\sigma}_k^d(x) = \delta_k^d$.

The proof of Theorem 6.6 is now complete.

Let $\mathfrak{X}=(X(t),B(t),M(t),\phi(t))$ be given as above and $f(t,x)$ be a smooth function on $[0, \infty)\times D$. Then $f(t, X(t))$ is a continuous $(\mathscr{F}_t^0)$-semimartingale.

Theorem 6.7. Let $g(t)$ be an $(\mathscr{F}_t^0)$-adapted process such that $t \longmapsto g(t)$ is right-continuous with left-hand limits and $t \longmapsto E[g(t)^2]$ is locally bounded. Then we have

$$
\begin{aligned}
\sum_{s\in D}^{*} \int_{A(s-)\wedge t}^{A(s)\wedge t} & g(u)df(u,X(u)) \\
&= \int_0^t g(u)df(u,X(u)) - \sum_{i=1}^{d-1}\sum_{l=1}^{s}\int_0^t g(u)\frac{\partial f}{\partial x^i}(u,X(u))\tau_l^i(X(u))dM^l(u) \\
&\quad - \int_0^t g(u)\Big\{\sum_{i=1}^{d-1}\Big(\beta^i((X(u)) - \frac{a^{di}(X(u))}{a^{dd}(X(u))}\Big)\frac{\partial f}{\partial x^i}(u,X(u)) \\
&\quad + \frac{1}{2}\sum_{i,j=1}^{d-1}\alpha^{ij}(X(u))\frac{\partial^2 f}{\partial x^i \partial x^j}(u,X(u))\Big\}d\phi(u).
\end{aligned}
\tag{6.78}
$$

Proof. Let A and L be defined by (6.45) and (6.46). By Itô's formula,

$$
\begin{aligned}
g(u)df(u,X(u)) &= g(u)\frac{\partial f}{\partial t}(u,X(u))du \\
&\quad + \sum_{i=1}^{d}\sum_{k=1}^{r} g(u)\frac{\partial f}{\partial x^i}(u,X(u))\sigma_k^i(X(u))dB^k(u) \\
&\quad + \sum_{i=1}^{d-1}\sum_{l=1}^{s} g(u)\frac{\partial f}{\partial x^i}(u,X(u))\tau_l^i(X(u))dM^l(u) \\
&\quad + g(u)(A_x f)(u,X(u))du + g(u)(L_x f)(u,X(u))d\phi(u).
\end{aligned}
$$

By Theorem 6.6, the left-hand side of (6.78) is equal to

$$
\begin{aligned}
&\int_0^t g(u)\frac{\partial f}{\partial t}(u,X(u))du + \sum_{i=1}^{d}\sum_{k=1}^{r}\int_0^t g(u)\frac{\partial f}{\partial x^i}(u,X(u))\sigma_k^i(X(u))dB^k(u) \\
&\quad + \sum_{i=1}^{d}\sum_{k=1}^{r}\int_0^t g(u)\frac{\partial f}{\partial x^i}(u,X(u))\frac{\sigma_k^i(X(u))\sigma_k^d(X(u))}{a^{dd}(X(u))}d\phi(u) \\
&\quad + \int_0^t g(u)(A_x f)(u,X(u))du \\
&= \int_0^t g(u)df(u,X(u)) - \sum_{i=1}^{d-1}\sum_{l=1}^{s}\int_0^t g(u)\frac{\partial f}{\partial x^i}(u,X(u))\tau_l^i(X(u))dM^l(u) \\
&\quad - \int_0^t g(u)[L_x f(u,X(u)) - \sum_{i=1}^{d}\frac{a^{di}(X(u))}{a^{dd}(X(u))}\frac{\partial f}{\partial x^i}(u,X(u))]d\phi(u).
\end{aligned}
$$

Since

$$L_x f(u,x) - \sum_{i=1}^{d} \frac{a^{di}(x)}{a^{dd}(x)} \frac{\partial f}{\partial x^i}(u,x)$$
$$= \sum_{i=1}^{d-1} \left\{ \beta^i(x) - \frac{a^{di}(x)}{a^{dd}(x)} \right\} \frac{\partial f}{\partial x^i}(u,x) + \frac{1}{2} \sum_{i,j=1}^{d-1} \alpha^{ij}(x) \frac{\partial^2 f}{\partial x^i \partial x^j}(u,x),$$

we have obtained the conclusion.

Corollary. The identity

$$\sum_{s \in D}^{*} \int_{A(s-)\wedge t}^{A(s)\wedge t} g(u) df(u,X(u)) = \int_0^t g(u) df(u,X(u)) \tag{6.79}$$

holds for every smooth $f(u, x)$ on $[0, \infty) \times D$ if and only if

$$\alpha^{ij}(x) = 0 \quad \text{and} \quad \beta^i(x) = \frac{a^{di}(x)}{a^{dd}(x)} \qquad \text{identically on } \partial D, \tag{6.80}$$
$$i, j = 1, 2, \ldots, d-1.$$

Definition 6.2. We say that X is a *normally reflecting diffusion process* if (6.80) is satisfied.

7. Kähler diffusions

Let ϕ be a mapping from an open set D of C^n into C^m. Then we can write $\phi(z) = (\phi^1(z), \phi^2(z), \cdots, \phi^m(z))$ where ϕ^i is a complex-valued function defined on D. As in the Section III-6 if each ϕ^i is holomorphic on D, then ϕ is called a holomorphic mapping from D into C^m. A Hausdorff topological space M is called a *d-dimensional complex manifold* if M has an open covering $\{U_\alpha\}_{\alpha \in A}$ such that for each U_α there is a homeomorphism ϕ_α from U_α onto an open set D_α of C^d satisfying the following property: if $U_\alpha \cap U_\beta \neq \phi$ the mapping $\phi_\beta \circ \phi_\alpha^{-1}$ from $\phi_\alpha(U_\alpha \cap U_\beta)$ into $\phi_\beta(U_\alpha \cap U_\beta)$ is a holomorphic mapping. $\{(U_\alpha, \phi_\alpha)\}_{\alpha \in A}$ is called a *system of holomorphic coordinate neighborhoods of* M. Let U be an open set of M provided with a homeomorphism ϕ from U onto an open set D of C^d. (U, ϕ) is called a *holomorphic coordinate neighborhood* of M if the following property holds: If $U \cap U_\alpha \neq \phi$, $\alpha \in A$, then the mapping $\phi_\alpha \circ \phi^{-1}$ from $\phi(U \cap U_\alpha)$ to $\phi_\alpha(U \cap U_\alpha)$ and the mapping $\phi \circ \phi_\alpha^{-1}$ from $\phi_\alpha(U \cap U_\alpha)$ to $\phi(U \cap U_\alpha)$ are both holomorphic. For a holomorphic coordinate neighborhood (U, ϕ), we set

$$\phi(z) = (z^1(z), z^2(z), \cdots, z^d(z)), \quad z \in U$$

and we call $(z^1, z^2, \ldots, z^d)$ the *system of complex local coordinates* on (U, ϕ). Since we can identify $\boldsymbol{C}^d$ with $\boldsymbol{R}^{2d}$, a d-dimensional complex manifold M can be considered as a $2d$-dimensional real manifold. Hence a holomorphic coordinate neighborhood (U, ϕ) of M is a C^∞ coordinate neighborhood. For a system of complex local coordinates $(z^1, z^2, \cdots, z^d)$ on (U, ϕ), we denote by x^k and y^k the real and imaginary parts of z^k respectively. Then $(x^1, y^1, x^2, y^2, \cdots, x^d, y^d)$ is a system of local coordinates of M. Hence

$$\left\{\left(\frac{\partial}{\partial x^1}\right)_z, \left(\frac{\partial}{\partial y^1}\right)_z, \cdots, \left(\frac{\partial}{\partial x^d}\right)_z, \left(\frac{\partial}{\partial y^d}\right)_z\right\},$$

is a basis of $T_z(M)$ and

$$\{(dx^1)_z, (dy^1)_z, \cdots, (dx^d)_z, (dy^d)_z\}$$

is a basis of $T_z^*(M)$. As in the Section III-6, we set

$$\left(\frac{\partial}{\partial z^k}\right)_z = \frac{1}{2}\left\{\left(\frac{\partial}{\partial x^k}\right)_z - \sqrt{-1}\left(\frac{\partial}{\partial y^k}\right)_z\right\}$$

$$\left(\frac{\partial}{\partial \bar{z}^k}\right)_z = \frac{1}{2}\left\{\left(\frac{\partial}{\partial x^k}\right)_z + \sqrt{-1}\left(\frac{\partial}{\partial y^k}\right)_z\right\}$$

and also set

$$(dz^k)_z = (dx^k)_z + \sqrt{-1}(dy^k)_z$$

$$(d\bar{z}^k)_z = (dx^k)_z - \sqrt{-1}(dy^k)_z.$$

Then we see immediately that

$$\left\{\left(\frac{\partial}{\partial z^1}\right)_z, \left(\frac{\partial}{\partial \bar{z}^1}\right)_z, \cdots, \left(\frac{\partial}{\partial z^d}\right)_z, \left(\frac{\partial}{\partial \bar{z}^d}\right)_z\right\}$$

is a basis of the complexification $T_z^C(M)$ of $T_z(M)$ and

$$\{(dz^1)_z, (d\bar{z}^1)_z, \cdots, (dz^d)_z, (d\bar{z}^d)_z\}$$

is a basis of the complexification $T_z^{*C}(M)$ of $T_z^*(M)$.

Define a linear transformation J_z of $T_z(M)$ by

$$J_z\left(\frac{\partial}{\partial x^k}\right)_z = \left(\frac{\partial}{\partial y^k}\right)_z, \; J_z\left(\frac{\partial}{\partial y^k}\right)_z = -\left(\frac{\partial}{\partial x^k}\right)_z, \quad k = 1, 2, \cdots, d.$$

It is easy to see that J_z can be defined independently of the choice of the system of complex local coordinates $(z^1, z^2, \cdots, z^d)$. The mapping J: $z \to J_z$ is called the almost complex structure attached to M. A Riemannian metric g on a complex manifold M is called a *Hermitian metric* on M if g satisfies

$$(7.1) \qquad g_z(J_z X, J_z Y) = g_z(X, Y) \qquad \text{for } X, Y \in T_z(M)$$

at each point z of M. We extend the value of g at $z \in M$ to a symmetric bilinear form on $T_z^C(M)$ by defining

$$\begin{aligned} &g_z(X + \sqrt{-1}\,Y, X' + \sqrt{-1}\,Y') \\ &= (g_z(X, X') - g_z(Y, Y')) + \sqrt{-1}(g_z(X, Y') + g_z(Y, X')) \end{aligned}$$

for $X + \sqrt{-1}\,Y$, $X' + \sqrt{-1}\,Y' \in T_z^C(M)$. We set

$$g_{\alpha\beta}(z) = g_z\left(\left(\frac{\partial}{\partial z^\alpha}\right)_z, \left(\frac{\partial}{\partial z^\beta}\right)_z\right), \qquad g_{\alpha\bar\beta}(z) = g_z\left(\left(\frac{\partial}{\partial z^\alpha}\right)_z, \left(\frac{\partial}{\partial \bar z^\beta}\right)_z\right)$$

$$g_{\bar\alpha\beta}(z) = g_z\left(\left(\frac{\partial}{\partial \bar z^\alpha}\right)_z, \left(\frac{\partial}{\partial z^\beta}\right)_z\right), \qquad g_{\bar\alpha\bar\beta}(z) = g_z\left(\left(\frac{\partial}{\partial \bar z^\alpha}\right)_z, \left(\frac{\partial}{\partial \bar z^\beta}\right)_z\right)$$

$$, \alpha, \beta = 1, 2, \cdots, d,$$

and call $g_{\alpha\beta}$, $g_{\alpha\bar\beta}$, $g_{\bar\alpha,\beta}$, $g_{\bar\alpha,\bar\beta}$ the components of g with respect to $(z^1, z^2, \cdots, z^d)$. It is easy to see that

$$g_{\alpha\beta} = g_{\beta\alpha}, \quad g_{\bar\alpha\bar\beta} = g_{\bar\beta\bar\alpha}, \quad g_{\alpha\bar\beta} = g_{\bar\beta\alpha}, \quad \bar g_{\alpha\beta} = g_{\bar\alpha\bar\beta}, \quad \bar g_{\alpha\bar\beta} = g_{\bar\alpha\beta}.$$

In terms of components of g the condition (7.1) is expressed as

$$(7.2) \qquad g_{\alpha\beta} = g_{\bar\alpha\bar\beta} = 0, \quad \alpha, \beta = 1, 2, \cdots, d.$$

For a Hermitian metric g, we set

$$(7.3) \qquad \omega_z(X, Y) = g_z(J_z X, Y) \qquad \text{for} \quad X, Y \in T_z(M).$$

The mapping ω: $z \to \omega_z$ defines a real differential 2-form on M. ω is called the differential form on M of degree 2 attached to the Hermitian metric g. ω is expressed as

$$\omega = \sqrt{-1} \sum_{\alpha, \beta = 1}^{d} g_{\alpha\bar\beta}\, dz^\alpha \wedge d\bar z^\beta.$$

If ω is a closed form, i.e.

(7.4) $$d\omega = 0,$$

then the Hermitian metric g is called a *Kähler metric*. A complex manifold endowed with a Kähler metric is called a *Kähler manifold.* We note that in terms of components of g the condition (7.4) is expressed as

(7.5) $$\frac{\partial g_{\alpha\bar\beta}}{\partial z^\gamma} = \frac{\partial g_{\gamma\bar\beta}}{\partial z^\alpha} \qquad \alpha, \beta, \gamma = 1, 2, \cdots, d.$$

We now note that if (M, g) is a Kähler manifold, the Laplace-Beltrami operator Δ on M may be written as

(7.6) $$\Delta = \sum_{\alpha,\beta=1}^{d} g^{\alpha\bar\beta}(z)\frac{\partial^2}{\partial z^\alpha \partial \bar z^\beta}\left(= \sum_{\alpha,\beta=1}^{d} g^{\bar\beta\alpha}(z)\frac{\partial^2}{\partial z^\alpha \partial \bar z^\beta}\right)$$

where $(g^{\alpha\bar\beta}(z))$ and $(g^{\bar\beta\alpha}(z))$ are given by

(7.7) $$\sum_\gamma g_{\alpha,\bar\gamma}(z) g^{\bar\gamma\beta}(z) = \delta^\beta_\alpha, \quad \sum_\gamma g^{\bar\alpha\gamma}(z) g_{\gamma\bar\beta}(z) = \delta^{\bar\alpha}_{\bar\beta} \qquad \alpha, \beta = 1, 2, \cdots, d,$$

respectively. Indeed a straightforward calculation in local coordinates shows that if g is a Hermitian metric, then the principal part of Δ is equal to the right hand of (7.6). Combining (7.5) with V-(4.32) we can show that the coefficients of the lower order terms of Δ identically vanish. These imply (7.6).

Definition 7.1. The Brownian motion on a Kähler manifold M, i.e. the diffusion process on M generated by the differential operator given by (7.6) is called the *Kähler diffusion* on M.

We note that the Riemannian connection ∇ defined by IV-(4.19) is extended by complex linearity to act on complex vector fields, i.e., for complex vector fields

$$Z_k = X_k + \sqrt{-1}Y_k \qquad k = 1, 2$$

we set

$$\nabla_{Z_1} Z_2 = \nabla_{X_1} X_2 - \nabla_{Y_1} Y_2 + \sqrt{-1}(\nabla_{X_1} Y_2 + \nabla_{Y_1} X_2).$$

Define *the components of the connection* ∇ *with respect to* $\dfrac{\partial}{\partial z^\alpha}$ and $\dfrac{\partial}{\partial \bar z^\beta}$ by

$$\nabla_{\frac{\partial}{\partial z^\alpha}} \frac{\partial}{\partial z^\beta} = \sum_\gamma \left(\Gamma^\gamma_{\alpha\beta} \frac{\partial}{\partial z^\gamma} + \Gamma^{\bar\gamma}_{\alpha\beta} \frac{\partial}{\partial \bar z^\gamma} \right)$$

$$\nabla_{\frac{\partial}{\partial z^\alpha}} \frac{\partial}{\partial \bar z^\beta} = \sum_\gamma \left(\Gamma^\gamma_{\alpha\bar\beta} \frac{\partial}{\partial z^\gamma} + \Gamma^{\bar\gamma}_{\alpha\bar\beta} \frac{\partial}{\partial \bar z^\gamma} \right)$$

$$\nabla_{\frac{\partial}{\partial \bar z^\alpha}} \frac{\partial}{\partial z^\beta} = \sum_\gamma \left(\Gamma^\gamma_{\bar\alpha\beta} \frac{\partial}{\partial z^\gamma} + \Gamma^{\bar\gamma}_{\bar\alpha\beta} \frac{\partial}{\partial \bar z^\gamma} \right)$$

$$\nabla_{\frac{\partial}{\partial \bar z^\alpha}} \frac{\partial}{\partial \bar z^\beta} = \sum_\gamma \left(\Gamma^\gamma_{\bar\alpha\bar\beta} \frac{\partial}{\partial z^\gamma} + \Gamma^{\bar\gamma}_{\bar\alpha\bar\beta} \frac{\partial}{\partial \bar z^\gamma} \right)$$

If (M, g) is a Kähler manifold, we obtain

$$\begin{aligned} &\Gamma^\alpha_{\beta\bar\gamma} = \Gamma^\alpha_{\bar\beta\bar\gamma} = \Gamma^{\bar\alpha}_{\beta\gamma} = \Gamma^{\bar\alpha}_{\bar\beta\gamma} = 0 \\ (7.8) \qquad &\Gamma^\alpha_{\bar\beta\gamma} = \Gamma^{\bar\alpha}_{\beta\bar\gamma} = 0 \\ &\Gamma^\alpha_{\beta\gamma} = \Gamma^\alpha_{\gamma\beta}, \quad \Gamma^\alpha_{\beta\gamma} = \Gamma^{\bar\alpha}_{\bar\beta\bar\gamma}. \end{aligned}$$

Furthermore, for a Kähler manifold the coefficients of ∇ are determined by

$$\sum_\gamma g_{\gamma\bar\delta}\, \Gamma^\gamma_{\alpha\beta} = \frac{\partial g_{\delta\alpha}}{\partial z^\beta}, \quad \sum_\gamma g_{\bar\gamma\bar\delta}\, \Gamma^{\bar\gamma\beta}_{\bar\alpha} = \frac{\partial g_{\delta\bar\alpha}}{\partial \bar z^\beta}.$$

In the rest of this section, unless otherwise stated we *assume* that (M, g) is a Kähler manifold. We introduce the bundle of unitary frames over M: By a *unitary frame* $e = [e_1, e_2, \cdots, e_d]$ at z we mean a system of complex tangent vectors $e_\alpha \in T^C_z(M)$ such that

$$g_z(e_\alpha, \bar e_\beta) = \delta_{\alpha\beta} \qquad \alpha, \beta = 1, 2, \cdots, d$$

and each e_α is of holomorphic type, i.e., it has no component of $\frac{\partial}{\partial \bar z^\beta}$, $\beta = 1, 2, \cdots, d$. $U(M)$ is defined as the collection of all unitary frames at all points $z \in M$:

$$U(M) = \{ r = (z, e);\ z \in M \text{ and } e \text{ is a unitary frame at } z \}.$$

Then $U(M)$ is a principal fibre bundle with the structure group $U(d)$. This bundle is called the *bundle of unitary frames* over M. For details, see [207]. Since M is Kählerian, $U(M)$ is invariant under the parallel displacement with respect to the connection ∇.

As in the Section 4, the Kähler diffusion on M is constructed by the solution of the stochastic differential equation corresponding to the canonical horizontal vector fields. In this case, it is described more conveniently by using the complex structure. Let (W_0^{2d}, P^W) be the standard $2d$-dimensional Wiener space and define a system of complex-valued martingales $\zeta(t) = (\zeta^\alpha(t))_{\alpha=1}^d$ by

$$\zeta^\alpha(t) = \frac{1}{\sqrt{2}}(w^{2\alpha-1}(t) + \sqrt{-1}w^{2\alpha}(t)), \quad \alpha = 1, 2,\cdots, d.$$

As stated in the Example III-6.1, $\zeta(t)$ is called a d-dimensional complex Brownian motion and it is a typical example of d-dimensional conformal martingale. We consider the following stochastic differential equation

$$dr(t) = L_\alpha(r(t)) \circ d\zeta^\alpha(t) \tag{7.9}$$

where $\{L_1, L_2,\cdots, L_d\}$ is the system of the canonical horizontal vector field on $U(M)$, i.e. $L_\alpha(r) \in T_r^C(U(M))$ is defined to be the horizontal lift of $e_\alpha \in T_z^C(M)$, $\alpha = 1, 2,\cdots, d$ where $r = (z, e = [e_1, e_2,\cdots, e_d]) \in U(M)$. The meaning of (7.9) is as follows: we say that *$r(t)$ is a solution of (7.9) if $r(t)$ is a continuous process on $U(M)$ such that, for every $F \in F(U(M))$, $F(r(t))$* is a semimartingale and satisfies

$$F(r(t)) - F(r(0)) = \int_0^t L_\alpha F(r(s)) \circ d\zeta^\alpha(s) + \int_0^t (\bar{L}_\alpha F)(r(s)) \circ d\bar{\zeta}^\alpha(s).$$

In a complex local coordinate, it is given as follows:

$$(7.9)' \quad \begin{aligned} dZ^i(t) &= e^i_\alpha(t) \circ d\zeta^\alpha(t) \qquad i = 1, 2,\cdots, d \\ de^i_\alpha(t) &= -\Gamma^i_{\gamma\beta}(Z(t))e^\beta_\alpha(t) \circ dZ^\gamma(t), \qquad i, \alpha = 1, 2,\cdots, d. \end{aligned}$$

Since $d\zeta^\alpha \cdot d\zeta^\beta = 0$ and $dZ^\alpha \cdot dZ^\beta = 0$, (7.9)′ is equivalent to

$$(7.9)'' \quad \begin{aligned} dZ^i(t) &= e^i_\alpha(t)d\zeta^\alpha(t) \qquad i = 1, 2,\cdots, d \\ de^i_\alpha(t) &= -\Gamma^i_{\gamma\beta}(Z(t))e^\beta_\alpha(t)dZ^\gamma(t), \qquad i, \alpha = 1, 2,\cdots, d. \end{aligned}$$

If $r(0) \in U(M)$, then the solution $r(t) = (Z^i(t), e^i_\alpha(t))$ lies on $U(M)$ and the $Z(t) = (Z^1(t), Z^2(t), \cdots, Z^d(t))$ is a d-dimensional local conformal margingale. Combining these results with III-(6.5), we obtain the following: *The solution $r(t)$ to (7.9) with $r(0) \in U(M)$ lies on $U(M)$ and its projection $Z(t) = \pi(r(t))$ defines the Kähler diffusion* on M where $\pi\colon U(M) \to M$ is the natural projection given by $\pi(r) = z$ for $r = (z, e) \in U(M)$.

Example 7.1. Let $D = \{z \in C;\ |z| < 1\}$ be the unit disc in the complex plane $\boldsymbol{C}$ endowed with the Riemannian metric

$$ds^2 = |dz|^2/(1 - |z|^2)^2 \qquad z \in D. \tag{7.10}$$

Then D is a Kähler manifold and the metric g given by (7.10) is called the *Poincaré metric*. Since (D, g) is a realization of the Lobachevskii plane, the Kähler diffusion on (D, g) is called the *Lobachevskii Brownian motion in the unit disc*. This diffusion is obtained from the one-dimensional complex Brownian motion by a transformation of time change determined by the function $c(z) = (1 - |z|^2)^{-2}$. Let $\zeta(t)$ be the one-dimensional complex Brownian motion defined above. For $z \in D$, we set

$$\zeta^z(t) = z + \zeta(t)$$

and let

$$\sigma = \inf\{t;\ |\zeta^z(t)| = 1\}.$$

We also set

$$A_t = \int_0^t c(\zeta^z(s))ds, \qquad t < \sigma.$$

Then it holds that $\lim_{t\uparrow\sigma} A_t = \infty$ a.s. and hence its inverse $C_t = \inf\{u;\ A_u > t\}$ can be defined for $t \in [0, \infty)$. To see this, we note that $r(t) = |\zeta^z(t)|$ is $BES^{|z|}(2)$: *the Bessel diffusion process with index* 2 (Example IV-8.3) and $r(C_t)$ is a one-dimensional diffusion on $[0, 1)$ starting at $|z|$ with generator

$$\frac{(1 - r^2)^2}{2}\left(\frac{d^2}{dr^2} + \frac{1}{r}\frac{d}{dr}\right).$$

By Theorem VI-3.2 we can see that if $|z| \neq 0$, $r(C_t)$ can not hit 0 nor 1 in a finite time almost surely. Noting that $\lim_{t\uparrow\sigma} A_t := e < \infty$ implies

$\lim_{t\uparrow e} r(C_t) = 1$, we can conclude that $\lim_{t\uparrow \sigma} A_t = \infty$ a.s.

Then $Z_t^z(t) = \zeta^z(C_t)$ is a local conformal martingale by Proposition III-6.1 and it possesses the properties that $Z^z(t) \in D$, $t \geqq 0$ a.s. and $\lim_{t\uparrow\infty} Z_t^z(t) := Z_\infty^z$ exists and $Z_\infty^z \in \partial D = \{z;\ |z| = 1,\ z \in \boldsymbol{C}\}$ a.s. It is clear that $\{Z^z(t),\ t \geqq 0\}$ defines the Lobachevskii Brownian motion starting at z in the unit disc.

Remark 7.1. Let $p(t, z_1, z_2)$, $t > 0$, $z_1, z_2 \in D$ be the transition density function with respect to the Riemannian volume

$$m(dz) = \frac{dxdy}{(1 - |z|^2)^2}, \qquad z = x + \sqrt{-1}y \in D$$

on D of the Lobachevskii Brownian motion. Then it holds that

$$p(t, z_1, z_2) = p(t, \rho(z_1, z_2)), \qquad t > 0,\ z_1, z_2 \in D$$

where $\rho(z_1, z_2)$, $z_1, z_2 \in D$ is the Poincaré distance between z_1 and z_2, i.e.

$$\rho(z_1, z_2) = \frac{1}{2}\log\frac{1+r}{1-r}, \quad r = \left|\frac{z_1 z_2}{1 - \bar{z}_1 z_2}\right|$$

and

$$p(t, \rho) = \frac{e^{-\frac{t}{8}}}{(2\pi t)^{3/2}}\int_\rho^\infty \frac{re^{-\frac{r^2}{2t}}dr}{\sqrt{\mathrm{ch} r - \mathrm{ch}\rho}}.$$

Finally we give one of simplest examples in which conformal martingales can be applied to problems in complex analysis.

Example 7.2. Let D be a polydisc in $\boldsymbol{C}^2$ given by

$$D = \{z = (z^1, z^2) \in \boldsymbol{C}^2;\ |z^i| < 1, i = 1, 2\}$$

and let $z_0 = (z_0^1, z_0^2) \in D$. Let $Z(t) = (Z^1(t),\ Z^2(t))$ where $Z^1(t)$ and $Z^2(t)$ are mutually independent Lobachevskii Brownian motions in the unit disc in $\boldsymbol{C}$ as given in the Example 7.1 such that $Z^i(0) = z_0^i$, $i = 1, 2$. Then $Z(t)$ is a two-dimensional conformal martingale such that $Z(0) = z_0$ and with probability one, $Z(t) \in D$ for all $t \geqq 0$, $\lim_{t\uparrow\infty} Z(t) := Z_\infty$ exists and

$$Z_\infty \in \widehat{\partial D} = \{z = (z^1, z^2);\ |z^i| = 1,\quad i = 1, 2\}.$$

Note that the usual topological boundary ∂D of D is given by

$$\partial D = \{z = (z^1, z^2);\ |z^1| \leqq 1 \text{ and } |z^2| = 1 \text{ or } |z^1| = 1 \text{ and } |z^2| \leqq 1\}$$

and $\widehat{\partial D}$ is much smaller than ∂D. $\widehat{\partial D}$ is called the *distingushed boundary* of D. Now if ϕ: $D \to \boldsymbol{C}$ is holomorphic, $\phi(Z(t))$ is a local conformal martingale. As a simple application of this fact, we show the following: *Let* ϕ: $\bar{D} = D \cup \partial D \to \boldsymbol{C}$ *be continuous and* ϕ, *restricted on D, be holomorphic. Then* $\max_{z\in\bar{D}}|\phi(z)|$ *is always attained on* $\widehat{\partial D}$:

(7.11) $$\max_{z\in\bar{D}}|\phi(z)| = \max_{z\in\widehat{\partial D}}|\phi(z)|.$$

Indeed $\phi(Z(t))$ is a bounded martingale and $\lim_{t\uparrow\infty} \phi(Z(t)) = \phi(Z_\infty)$. Hence

$$\phi(z_0) = E[\phi(Z(0))] = E[\phi(Z_\infty)]$$

and therefore,

$$|\phi(z_0)| \leqq E[|\phi(Z_\infty)|].$$

We know that $Z_\infty \in \widehat{\partial D}$ a.s. and hence

$$|\phi(z_0)| \leqq \max_{z\in\widehat{\partial D}}|\phi(z)|.$$

Since z_0 can be choosen as any point in D, this clearly implies the assertion (7.11)

For these topics, we refer the reader to Debiard-Gaveau [195] and Kaneko-Taniguchi [205]. For further applications of conformal diffusions to complex analysis, see Gaveau [200], Malliavin [108], Durrett [196], Fukushima-Okada [198] etc.

8. Malliavin's stochastic calculus of variation for Wiener functionals

As we have seen above, strong solutions of stochastic differential equations are functions of Brownian motions. Such functions are often called *Wiener functionals* or *Brownian functionals* and they have been studied by many people with a variety of motivations (cf. e.g. [10], [64],

[79], and [178]). Recently P. Malliavin ([106], [107]) gave a new approach to the analysis of Wiener functionals, especially to the analysis of strong solutions of stochastic differential equations. We follow the assmptions and notations of Section 2 and let $X = (X(t, x, w))$ be the solution of (2.1) realized on the r-dimensional Wiener space $(\boldsymbol{W}_0^r, P^W)$. Recall that $\boldsymbol{W}_0^r$ is the totality of continuous functions w: $[0, \infty) \to \boldsymbol{R}^r$ such that $w(0) = 0$ with the topology of uniformly convergence on every bounded interval, i.e. the topology induced by a countable system of norms

$$||w||_n = \max_{0 \leq t \leq n} |w(t)|, \quad n = 1, 2, \cdots, \quad w \in \boldsymbol{W}_0^r.$$

If $t\ (> 0)$ and x are fixed, the mapping: $w \to X(t, x, w)$ is a d-dimensional Wiener functional, i.e. a P^W-measurable function of w, but it is not in a class of functionals to which the classical calculus of variations or Fréchet differential calculus on a countably normed space $\boldsymbol{W}_0^r$ can be applied. Generally, it is not even continuous in w. It is an important discovery of Malliavin that this functional, however, can be differentiated in w as many times as we want if the differentiation is understood properly. Moreover he showed that these derivatives can actually be used to produce fruitful results. Examples of such successful applications initiated by Malliavin, then followed by Kusuoka-Stroock, Bismut, Watanabe, Léandre and so on, are among others, in the problems of regularity, estimates and asymptotics of heat kernels. As we saw in Sections 3 and 5, the initial value problems of heat equations can be solved by probabilistic method of representing solutions as expectations of certain Wiener functionals. With a help of the Malliavin calculus, we can proceed one step further and represent the heat kernel, i.e. the fundamental solution of a heat equation, as a *generalized expectation* (in the same sense as in the Schwartz distribution theory) of a certain *generalized* Wiener functional. This method, as we shall see in the subsequent sections, is quite useful in the above mentioned problems of heat kernels.

We start with the r-dimensional Wiener space $(\boldsymbol{W}_0^r, P^W)$. We write simply $\boldsymbol{W}_0^r = \boldsymbol{W}$ and $P^W = P$ when there is no confusion. As usual, P-measurable functions defined on the Wiener space $(\boldsymbol{W}, P)$ are called *Wiener functionals* and two Wiener functionals with the same range space are identified whenever they coincide P-almost everywhere. Needless to say that the most important function spaces of Wiener functionals are L_p-spaces. In the following, we denote by E a real separable Hilbert space. As usual, we denote by $\boldsymbol{L}_p(P; E)$ $(1 \leq p < \infty)$ or simply by $\boldsymbol{L}_p(E)$ the real L_p-space formed of all E-valued Wiener functionals F such that $|F(w)|_E$ is p-th integrable and endowed with the norm

$$\|F\|_p = (\int_W |F(w)|_E^p P(dw))^{1/p}$$

where $|e|_E = \langle e, e \rangle_E^{1/2}$ is the norm of $e \in E$ and $\langle\ ,\ \rangle_E$ is the inner product of E. In the case $E = \boldsymbol{R}$, $\boldsymbol{L}_p(E)$ is denoted simply by $\boldsymbol{L}_p$. In the Malliavin calculus, we introduce, besides L_p-spaces of Winer functionals, a system of Sobolev spaces of Wiener functionals so that we can develop a differential calculus of Wiener functionals. Malliavin defined the notion of derivatives and Sobolev spaces in terms of Ornstein-Uhlenbeck processes over Wiener spaces. These notions have been studied further by Shigekawa [148]; Meyer [218], Kusuoka-Stroock [211] and Sugita [225], [226], among others. Sugita [226], in particular, showed the equivalence of apparently different approaches of these authors. Here we develop the theory along the line of [204] and [232].

Let H be the Hilbert space formed of all $h \in \boldsymbol{W}$ such that each component of $h(t) = (h^1(t), h^2(t), \cdots, h^r(t))$ is absolutely continuous in t and has square-integrable derivatives. Endow H with the Hilbertian norm

$$|h|_H^2 = \int_0^\infty |\dot{h}(t)|^2 dt, \qquad h \in H$$

where $\dot{h}(t) = (\dot{h}^1(t), \dot{h}^2(t), \cdots, \dot{h}^r(t))$ and

$$\dot{h}^i(t) = \frac{d}{dt} h^i(t), \qquad i = 1, 2, \cdots, r.$$

This H is often called the *Cameron-Martin subspace* of $\boldsymbol{W}$. For $h \in H$, the stochastic integral (often called Wiener integral)

$$(8.1) \qquad [h](w) = \sum_{i=1}^r \int_0^\infty \dot{h}^i(t) dw^i(t)$$

is well defined and $[h] \in \boldsymbol{L}_2$ as a function of w. Indeed, L_2-norm of $[h]$ coincides with $|h|_H$ i.e. the linear mapping: $H \ni h \rightarrow [h] \in \boldsymbol{L}_2$ is an isometry.

Definition 8.1. (i) A function F: $\boldsymbol{W} \rightarrow \boldsymbol{R}$ is called a *polynomial functional* if there exist an $n \in \boldsymbol{N}, h_1, h_2, \cdots, h_n \in H$ and a real polynomial $p(x_1, x_2, \cdots, x_n)$ of n-variables such that

$$(8.2) \qquad F(w) = p([h_1](w), [h_2](w), \cdots, [h_n](w)).$$

The totality of all polynomial functionals is denoted by $\boldsymbol{P}$.

(ii) A functional F: $\boldsymbol{W} \to \boldsymbol{R}$ is called a *smooth functional* if there exist an $n \in \boldsymbol{N}$, $h_1, h_2, \cdots, h_n \in H$ and a tempered $\boldsymbol{C}^\infty$-function $f(x_1, x_2, \cdots, x_n)$ on $\boldsymbol{R}^n$ such that

$$(8.3) \qquad F(w) = f([h_1](w), [h_2](w), \cdots, [h_n](w)).$$

Here, f is called a tempered $\boldsymbol{C}^\infty$-function if it is $\boldsymbol{C}^\infty$ and all derivatives of f are of polynomial growth order: i.e. for each multi-index $\alpha = (\alpha_1, \alpha_2, \cdots, \alpha_n)$, $\alpha_i \in \boldsymbol{Z}^+$, positive constants K_α and N_α exist such that

$$|D_\alpha f(x)| \leqq K_\alpha(1 + |x|^2)^{N_\alpha} \qquad \text{for all} \quad x \in \boldsymbol{R}^n,$$

where D_α is the differential operator defined by

$$D_\alpha = \left(\frac{\partial}{\partial x^1}\right)^{\alpha_1}\left(\frac{\partial}{\partial x^2}\right)^{\alpha_2} \cdots \left(\frac{\partial}{\partial x^d}\right)^{\alpha_d}$$

(previously denoted by D^α in Section 2). The totality of smooth functionals is denoted by $\boldsymbol{S}$

(iii) An E-valued functional F: $\boldsymbol{W} \to E$ is called an *E-valued polynomial functional* (*smooth functional*) if there exist an $m \in \boldsymbol{N}$, $e_1, e_2, \cdots, e_m \in E$ and $F_1, F_2, \cdots, F_m \in \boldsymbol{P}$ (resp. $\boldsymbol{S}$) such that

$$F(w) = F_1(w)e_1 + F_2(w)e_2 + \cdots + F_m(w)e_m.$$

The totality of E-valued polynomial functionals and that of E-valued smooth functionals are denoted by $\boldsymbol{P}(E)$ and $\boldsymbol{S}(E)$ respectively.

Remark 8.1. In (8.2) and (8.3), $h_1, h_2, \cdots, h_n \in H$ can be chosen to satisfy the orthonormality condition: $(h_i, h_j) = \delta_{ij}$, $i, j = 1, 2, \cdots, n$, if n and p or f are suitably modified. In this case, the degree of polynomial p is uniquely determined from F and it is called the *degree of the polynomial functional* F. Note also that the joint law of $([h_1](w), [h_2](w), \cdots, [h_n](w))$ is the n-dimensional standard Gaussian distribution N_n, i.e.

$$N_n(dx) = (2\pi)^{-\frac{n}{2}} \exp\{-\frac{1}{2}\sum_{i=1}^n x_i^2\} dx^1 dx^2 \cdots dx^n.$$

The totality of $F \in \boldsymbol{P}$ of degree at most n is denoted by $\boldsymbol{P}_n$.

In the following, we often state definitions, properties, etc. in the case $E = \boldsymbol{R}$ for simplicity: All the statements are valid in the case of general E with obvious modifications.

Remark 8.2. $\boldsymbol{P} \subset \boldsymbol{S} \subset \boldsymbol{L}_p$ for every $1 \leqq p < \infty$ and $\boldsymbol{P}$ is dense in $\boldsymbol{L}_p$. Gnerally, $\boldsymbol{P}(E) \subset \boldsymbol{S}(E) \subset \boldsymbol{L}_p(E)$ and $\boldsymbol{P}(E)$ is dense in $\boldsymbol{L}_p(E)$.

Proof. We fix an *ONB* $\{h_i\}$ in H. Note that, for every $c > 0$,

$$(8.4) \qquad E[\exp\{c\sum_{k=1}^{n} |\lambda^k[h_k](w)|\}] < \infty$$
$$\text{for every } n = 1, 2, \cdots \quad \text{and } \lambda = (\lambda^1, \lambda^2, \cdots, \lambda^n) \in \boldsymbol{R}^n$$

and $\mathscr{B}(\boldsymbol{W}) = \sigma[[h_i](w),\ i = 1, 2, \cdots]$, where E denotes the expectation with respect to the Wiener measure P. It immediately follows that

$$\boldsymbol{P} \subset \boldsymbol{S} \subset \bigcap_{1 \leqq p < \infty} \boldsymbol{L}_p \quad \text{and} \quad \mathscr{B}(W) = \bigvee_{n} \mathscr{B}_n(W)$$

where $\mathscr{B}_n(W) = \sigma[[h_i](w),\ i = 1, 2, \cdots, n]$. We prove that $\boldsymbol{P}$ is dense in $\boldsymbol{L}_p$, $1 \leqq p < \infty$. Suppose the contrary. Then there exists a non-zero $X \in \boldsymbol{L}_q$, $(1/p + 1/q = 1)$ such that

$$(8.5) \qquad E[XF] = 0 \qquad \text{for every} \quad F \in \boldsymbol{P}.$$

Noting (8.4) and that $1 < q \leqq \infty$, we have

$$E[|X| \exp\{\sum_{k=1}^{n} |\lambda^k[h_k](w)|\}] < \infty$$

and, combining this with (8.5), we can conclude that

$$E[X \exp\{\sqrt{-1}\sum_{k=1}^{n}\lambda^k[h_k](w)\}]$$
$$(8.6) \qquad = \sum_{m=0}^{\infty} \frac{(\sqrt{-1})^m}{m!} E[X(\sum_{K=1}^{n}\lambda^k[h_k](w))^m]$$
$$= 0$$
$$\text{for every } n = 1, 2, \cdots \text{ and } \lambda = (\lambda^1, \lambda^2, \cdots, \lambda^n) \in \boldsymbol{R}^n.$$

If $X_n = E[X | \mathscr{B}_n(W)]$, then there exists $f(x^1, x^2, \cdots, x^n) \in \mathscr{L}_1(\boldsymbol{R}^n, N_n)$ such that $X_n(w) = f([h_1](w), [h_2](w), \cdots, [h_n](w))$ and (8.6) implies that

$$E[X_n \exp\{\sqrt{-1}\sum_{k=1}^{n} \lambda^k[h_k](w)|\}]$$
$$= E[X \exp\{\sqrt{-1}\sum_{k=1}^{n}\lambda^k[h_k](w)\}]$$

$$= 0,$$

i.e.

$$\int_{R^n} f(x) e^{\sqrt{-1}(\lambda, x)} N_n(dx) = 0$$

$$\text{for every } \lambda = (\lambda^1, \lambda^2, \cdots, \lambda^n) \in \boldsymbol{R}^n.$$

By the uniqueness of Fourier transform, we can conclude that $f(x)N_n(dx) = 0$ i.e. $f(x) = 0$ a.e. $x(N_n)$, implying that $X_n = 0$ a.s. By Theorem I–6.6, $X = \lim_{n \to \infty} X_n$ a.s. and hence $X = 0$, leading us therefore to a contradiction.

$\boldsymbol{L}_2$ is a Hilbert space and is decomposed into the direct sum of mutually orthogonal subspaces known as *Wiener's homogeneous chaos:*

$$\boldsymbol{L}_2 = \boldsymbol{C}_0 \oplus \boldsymbol{C}_1 \oplus \cdots \oplus \boldsymbol{C}_n \oplus \cdots$$

where $\boldsymbol{C}_0 = \{\text{constants}\}$ and

$$\boldsymbol{C}_n = \bar{\boldsymbol{P}}_n \cap [\boldsymbol{C}_0 \oplus \boldsymbol{C}_1 \oplus \cdots \oplus \boldsymbol{C}_{n-1}]^{\perp}.$$

Here $-$ and $\perp$ denote the closure and the orthogonal complement in $\boldsymbol{L}_2$, respectively. The projection operator $\boldsymbol{L}_2 \to \boldsymbol{C}_n$ is denoted by J_n.

Let $H_n(x)$, $n = 0, 1, \cdots$, be Hermite polynomials defined by

$$H_n(x) = \frac{(-1)^n}{n!} \exp\left(\frac{x^2}{2}\right) \frac{d^n}{dx^n} \exp\left(-\frac{x^2}{2}\right), \quad n = 0, 1, \cdots \tag{8.7}$$

(previously denoted by $H_n[1, x]$ in Section III-5). Let

$$\Lambda = \{\boldsymbol{a} = (a_1, a_2, \cdots);\ a_i \in \boldsymbol{Z}^+,\ a_i = 0 \text{ except for a finite number of } i\text{'s}\}$$

and, for $\boldsymbol{a} \in \Lambda$, set

$$\boldsymbol{a}! = \prod_{i=1}^{\infty} a_i! \quad \text{and} \quad |\boldsymbol{a}| = \sum_{i=1}^{\infty} a_i.$$

Let $\{h_i\}_{i=1}^{\infty}$ be an orthonormal base (ONB) of H and define $H_{\boldsymbol{a}}(w) \in \boldsymbol{P}$, $\boldsymbol{a} \in \Lambda$, by

(8.8) $$H_{\boldsymbol{a}}(w) = \prod_{i=1}^{\infty} H_{a_i}([h_i](w)).$$

Since $H_0(x) = 1$ and $a_i = 0$ except for finite number of i's, this product is actually finite and defines a polynomial functional. We note that

$$H_{\boldsymbol{a}} \in \boldsymbol{P}_n \quad \text{if} \quad |\boldsymbol{a}| \leqq n.$$

Proposition 8.1. (i) $\{\sqrt{\boldsymbol{a}!}H_{\boldsymbol{a}}(w);\ \boldsymbol{a} \in \Lambda\}$ is an *ONB* in $\boldsymbol{L}_2$.

(ii) $\{\sqrt{\boldsymbol{a}!}H_{\boldsymbol{a}}(w);\ \boldsymbol{a} \in \Lambda,\ |\boldsymbol{a}| = n\}$ is an *ONB* in $\boldsymbol{C}_n$.

Proof. First recall that $\{\sqrt{n!}H_n(x)\}\}_{n=0}^{\infty}$ is an *ONB* in $\mathscr{L}_2(\boldsymbol{R}, N_1)$, where N_1 is the one-dimensional standard Gaussian measure, i.e.

$$N_1(dx) = \frac{1}{\sqrt{2\pi}}\exp\left\{-\frac{|x|^2}{2}\right\}dx.$$

Since $\{[h_i](w)\}$ are independent identically distributed random variables (*i.i.d.*) with the one-dimensional standard Gaussian distribution, it is immediatly seen that

$$\int_W H_{\boldsymbol{a}}(w)H_{\boldsymbol{b}}(w)P(dw) = \frac{1}{\boldsymbol{a}!}\delta_{\boldsymbol{a},\boldsymbol{b}}, \qquad \boldsymbol{a}, \boldsymbol{b} \in \Lambda.$$

Since $\boldsymbol{P}$ is dense in $\boldsymbol{L}_2$, the assertion of the proposition is almost obvious.

Corollary. If $F \in \boldsymbol{P}$, then $J_nF \in \boldsymbol{P}$ and $F = \sum_n J_nF$ is a finite sum.

Indeed, if F is represented as (8.2) with orthonormal $\{h_i\}$ extend $\{h_i\}$ to an *ONB* of H and apply the proposition. Since $p(x_1, x_2, \cdots, x_n)$ can be expressed as a linear combination of $\prod_{i=1}^{n} H_{a_i}(x_i)$, the assertion is obvious.

Hence, for a given real sequence $\phi = (\phi_n)_{n=0}^{\infty}$, we can define an operator T_ϕ on $\boldsymbol{P}$ by

(8.9) $$T_\phi F = \sum_{n=0}^{\infty} \phi_n J_n F, \qquad F \in \boldsymbol{P}.$$

T_ϕ can be extended to a self adjoint operator on $\boldsymbol{L}_2$ (denoted again by

T_ϕ) by setting

$$\textit{the domain } \mathscr{D}(T_\phi) = \{F \in \boldsymbol{L}_2;\ \sum_{n=0}^{\infty} \phi_n^2 \|J_n F\|_2^2 < \infty\}$$

and

$$T_\phi F = \sum_{n=0}^{\infty} \phi_n J_n F, \qquad F \in \mathscr{D}(T_\phi).$$

If $\phi_n = -n$, T_ϕ is denoted by L and called the *Ornstein-Uhlenbeck operator* or *number operator*. Also $(I - L)^s$, for $s \in \boldsymbol{R}$, is defined to be T_ϕ with $\phi_n = (1+n)^s$, $n = 0, 1, \cdots$. If $\phi_n = \exp\{-nt\}$, $t \geqq 0$, $n = 0, 1, \cdots$, we denote T_ϕ by T_t. Namely, we define

$$T_t F = \sum_{n=0}^{\infty} e^{-nt} J_n F, \qquad F \in \boldsymbol{P},\ t \geqq 0. \tag{8.10}$$

It is clear that $\{T_t\}$ defines a one-parameter semigroup of operators on $\boldsymbol{P}$. If (8.10) is extended to $F \in \boldsymbol{L}_2$, it define a contraction operators on $\boldsymbol{L}_2$, i.e.,

$$\|T_t F\|_2 \leqq \|F\|_2, \qquad F \in \boldsymbol{L}_2.$$

Hence $\{T_t\}$ defines a one parameter semigroup of symmetric and contraction operators on $\boldsymbol{L}_2$ and is called the *Ornstein-Uhlenbeck semigroup.*

Proposition 8.2. The Ornstein-Uhlenbeck semigroup T_t on $\boldsymbol{P}$ is also given by

$$\begin{aligned} T_t F(w) &= \int_W F(e^{-t}w + \sqrt{1-e^{-2t}}v)P(dv) \\ &= \int_W T_t(w, dv)F(v) \end{aligned} \quad , F \in \boldsymbol{P}, \tag{8.11}$$

where $T_t(w, dv)$, $w \in \boldsymbol{W}$, is the image measure on $\boldsymbol{W}$ of P under the mapping: $\boldsymbol{W} \ni v \to e^{-t}w + \sqrt{1-e^{-2t}}v \in \boldsymbol{W}$.

Proof. We denote the operator defined by the right-hand side of (8.11) by $\hat{T}_t$. Let $h \in H$ and

$$F(w) = \exp\{\sqrt{-1}[h](w) + |h|_H^2/2\}.$$

Then noting that $[h](w)$ is Gaussian distributed with mean 0 and variance $|h|_H^2$, we have

$$\hat{T}_t F(w) = \int_W \exp\{\sqrt{-1}e^{-t}[h](w) + \sqrt{-1}\sqrt{1-e^{-2t}}[h](v) + \frac{1}{2}|h|_H^2\}P(dv)$$

$$(8.12) \qquad = \exp\{\sqrt{-1}e^{-t}[h](w) + \frac{1}{2}|h|_H^2\} \times \int_W \exp\{\sqrt{-1}\sqrt{1-e^{-2t}}[h](v)\}P(dv)$$

$$= \exp\{\sqrt{-1}e^{-t}[h](w) + \frac{1}{2}e^{-2t}|h|_H^2\}.$$

Let $\lambda = (\lambda^1, \lambda^2, \cdots, \lambda^n)$, $n \in \boldsymbol{N}$, and $h = \lambda^1 h_1 + \lambda^2 h_2 + \cdots + \lambda^n h_n$ where $\{h^i\} \in H$ is orthonormal. Setting

$$F(w) = \prod_{i=1}^n \exp\{\sqrt{-1}\lambda^i[h_i](w) - \frac{1}{2}(\sqrt{-1}\lambda^i)^2\}$$

we have, by II-(5.3) and (8.7)

$$F(w) = \sum_{m_1, m_2, \ldots, m_n = 0}^{\infty} \prod_{j=1}^n (\sqrt{-1}\lambda^j)^{m_j} \prod_{j=1}^n H_{m_j}([h_j](w)).$$

Applying $\hat{T}_t$ to both sides, we obtain

$$\exp\{\sqrt{-1}e^{-t}[h](w) + \frac{1}{2}e^{-2t}|h|_H^2\}$$
$$= \hat{T}_t h(w)$$
$$= \sum_{m_1, m_2, \ldots, m_n = 0}^{\infty} \prod_{j=1}^n (\sqrt{-1}\lambda^j)^{m_j} \hat{T}_t(\prod_{i=1}^n H_{m_i}([h_i](\cdot))(w).$$

Since the left-hand side is

$$\prod_{j=1}^n \exp\{\sqrt{-1}\lambda^j e^{-t}[h_j](w) - \frac{1}{2}(\lambda^j e^{-t}\sqrt{-1})^2\}$$
$$= \prod_{j=1}^n \{\sum_{k=0}^{\infty}(\sqrt{-1}\lambda^j e^{-t})^k H_k([h_j](w))\}$$

$$= \sum_{m_1, m_2, \cdots, m_n=0}^{\infty} e^{-(m_1+m_2+\cdots+m_n)t} \prod_{j=1}^{n} (\sqrt{-1}\lambda^j)^{m_j} \prod_{j=1}^{n} H_{m_j}([h_j](w)),$$

we can conclude that

$$\hat{T}_t(\prod_{j=1}^{n} H_{m_j}([h_j](\cdot))(w) = e^{-(m_1+m_2+\cdots+m_n)t} \prod_{j=1}^{n} H_{m_j}([h_j](w)),$$

that is

$$(\hat{T}_t H_{\boldsymbol{a}})(w) = e^{-|\boldsymbol{a}|t} H_{\boldsymbol{a}}(w) \qquad \text{for any} \quad \boldsymbol{a} \in \Lambda.$$

This, combined with Proposition 8.1, implies

$$\hat{T}_t = T_t$$

which completes the proof.

Thus, the Ornstein-Uhlenbeck semigroup is a symmetric Markovian semigroup, i.e., a semigroup given by a probability kernel $T_t(w, dv)$ as (8.11) which is symmetric in the sense of (4.48) with respect to its invariant measure $\nu = P$. It can be extended to F in $\boldsymbol{S}$ by (8.10) and it is easy to see that $T_t F \in \boldsymbol{S}$, $t \geqq 0$. Furthermore, it can be extended to any Borel function $F(w)$ on $\boldsymbol{W}$ with growth order

$$|F(w)| \leqq C_1 \exp\{C_2 \|w\|_n\}$$

for some positive constants C_1, C_2 and n. As we have noticed, $\{T_t\}$ is an L_2-contraction symmetric semigroup on $\boldsymbol{L}_2$. Furthermore it is a strongly continuous contraction semigroup on $\boldsymbol{L}_p$, $1 \leqq p < \infty$, i.e., for $t > 0$

$$\text{(8.13)} \qquad \begin{matrix} \text{(i)} & \|T_t F\|_p \leqq \|F\|_p \\ \text{(ii)} & \lim_{t \downarrow 0} \|T_t F - F\|_p = 0 \end{matrix} \qquad \text{for any } F \in \boldsymbol{L}_p.$$

In fact, it is easy to see that for any $F \in \boldsymbol{L}_p$ satisfying $|F|^p \in \boldsymbol{L}_2$,

$$\begin{aligned} &\int_{\boldsymbol{W}} |T_t F|^p(w) P(dw) \\ &= \int_{\boldsymbol{W}} \left| \int_{\boldsymbol{W}} T_t(w, dv) F(v) \right|^p P(dw) \\ &\leqq \int_{\boldsymbol{W}} \left\{ \int_{\boldsymbol{W}} T_t(w, dv) |F(v)|^p \right\} P(dw) \end{aligned}$$

$$= \int_W T_t(|F|^p)(w)P(dw)$$

$$= \int_W |F|^p(w)(T_t 1)(w)P(dw) \qquad \text{(by the symmetry of } T_t \text{ in } L_2)$$

$$= \int_W |F|^p(w)P(dw).$$

In the general case, for any $F \in L_p$, we choose $F_n \in L_p$ such that $|F_n|^p \in L_2$ and $F_n \to F$ in L_p.

(ii) of (8.13) is easily concluded first for $F \in P$ and then by a standard limiting argument.

We have defined the Ornstein-Uhlenbeck operator L on P by

$$(8.14) \qquad LF = \sum_{n\geqq 0}(-n)J_nF, \qquad F \in P.$$

Clearly

$$(8.15) \qquad \frac{d}{dt}T_tF = T_t(LF) = L(T_tF), \quad F \in P.$$

We now determine the expression of L. Let $F \in P$ be given by (8.2) with $\{h_i\}$ orthonormal. Then, writing, $\xi(w) = (\xi^1(w), \xi^2(w), \cdots, \xi^n(w)) = ([h_1](w), [h_2](w), \cdots, [h_n](w)) \in R^n$ and $\partial_i p = \partial p/\partial x^i$, $i = 1,2,\cdots, n$ for simplicity,

$$\frac{d}{dt}(T_tF)(w) = \frac{d}{dt}\int_{R^n} p(e^{-t}\xi(w) + \sqrt{1-e^{-2t}}\eta)\left(\frac{1}{\sqrt{2\pi}}\right)^n e^{-\frac{|\eta|^2}{2}}d\eta$$

$$= -\int_{R^n}\sum_{i=1}^n e^{-t}\xi^i(w)(\partial_i p)(e^{-t}\xi(w) + \sqrt{1-e^{-2t}}\eta)\left(\frac{1}{\sqrt{2\pi}}\right)^n e^{-\frac{|\eta|^2}{2}}d\eta$$

$$+ \int_{R^n}\sum_{i=1}^n \frac{\eta^i e^{-2t}}{\sqrt{1-e^{-2t}}}(\partial_i p)(e^{-t}\xi(w) + \sqrt{1-e^{-2t}}\eta)\left(\frac{1}{\sqrt{2\pi}}\right)^n e^{-\frac{|\eta|^2}{2}}d\eta$$

$$= -\int_{R^n}\sum_{i=1}^n e^{-t}\xi^i(w)(\partial_i p)(e^{-t}\xi(w) + \sqrt{1-e^{-2t}}\eta)\left(\frac{1}{\sqrt{2\pi}}\right)^n e^{-\frac{|\eta|^2}{2}}d\eta$$

$$+ \int_{R^n}\sum_{i=1}^n e^{-t}((\partial_i^2 p)(e^{-t}\xi(w) + \sqrt{1-e^{-2t}}\eta))\left(\frac{1}{\sqrt{2\pi}}\right)^n e^{-\frac{|\eta|^2}{2}}d\eta$$

by an integration by parts. Hence

$$\text{(8.16)}\qquad \begin{aligned} LF(w) &= \frac{d}{dt}(T_tF)|_{t=0}(w) \\ &= \sum_{i=1}^{n}((\partial_i^2 p)(\xi(w)) - \xi^i(w)(\partial_i p)(\xi(w)). \end{aligned}$$

Similarly, if $F \in \boldsymbol{S}$ is given by (8.3), $LF(w) = \frac{d}{dt}(T_tF)|_{t=0}(w)$ is also given by (8.16) with p replaced by f. For $F \in \boldsymbol{P}$ (or $\boldsymbol{S}$) and $h \in H$, we define the derivative of F in the direction of h by

$$D_hF(w) = \lim_{\varepsilon\downarrow 0}\frac{F(w+\varepsilon h) - F(w)}{\varepsilon}.$$

It is clear that there exists unique $DF \in \boldsymbol{P}(H)$ (resp. $DF \in \boldsymbol{S}(H)$) satisfying

$$D_hF(w) = \langle DF(w), h\rangle_H \qquad h \in H.$$

Indeed, if $F \in \boldsymbol{P}$ is given by (8.2)

$$\text{(8.17)}\qquad DF(w) = \sum_{i=1}^{n}(\partial_i p)(\xi(w))h_i,$$

and similarly in the case of $F \in \boldsymbol{S}$. More generally, if $F \in \boldsymbol{P}(E)$ then $DF \in \boldsymbol{P}(H \otimes E)$, $H \otimes E$ being the tensor product of H and E. Remember that the tensor product $E_1 \otimes E_2$ of two separable Hilbert spaces E_1 and E_2 is a Hilbert space formed of all linear operators A: $E_1 \to E_2$ of Hilbert-Schmidt type endowed with the Hilbert-Schmidt norm

$$\|A\|_{HS} = \{\sum_{i,j=1}^{\infty} \langle Ae_i, e_j'\rangle_{E_2}^2\}^{1/2}$$

for some (= any) ONB's $\{e_i\}$ in E_1 and $\{e_j'\}$ in E_2. Indeed, if

$$F(w) = \sum_{k=1}^{m} p_k([h_1](w), [h_2](w), \cdots, [h_n](w))e_k, \qquad e_k \in E$$

then

$$DF(w) = \sum_{k=1}^{m}\sum_{i=1}^{n}(\partial_i p_k)([h_1](w), [h_2](w), \cdots, [h_n](w))h_i \otimes e_k$$

where $h_i \otimes e_k \in H \otimes E$ is defined by

$$[h_i \otimes e_k](h) = \langle h_i, h\rangle_H\, e_k \qquad h \in H.$$

If $F \in \boldsymbol{P}$, then $D^2F \in \boldsymbol{P}(H \otimes H)$ and D^2F, as a linear operator on H, is of trace class for almost all $w \in \boldsymbol{W}$. Indeed, if $F \in \boldsymbol{P}$ is given by (8.2) with $\{h_i\}$ orthonormal, then extending $\{h_i\}$ to an ONB in H,

$$\text{trace } D^2F(w) = \sum_{i=1}^{\infty} \langle D^2F(w)(h_i), h_i \rangle_H = \sum_{i=1}^{n} (\partial_i^2 p)(\xi(w)). \tag{8.18}$$

From (8.16), (8.17) and (8.18), we have

$$LF(w) = \text{trace } D^2F(w) - [DF](w). \tag{8.19}$$

Here, for $G \in \boldsymbol{P}(H)$ generally, $[G] \in \boldsymbol{P}$ is defined as follows: If G is represented by $G = \sum_{i=1}^{n} G_i(w) h_i$ with $G_i \in \boldsymbol{P}$ and $h_i \in H$,

$$[G](w) = \sum_{i=1}^{m} G_i(w)[h_i](w). \tag{8.20}$$

This definition of $[G]$ is well-defined independently of a way of representing G. If we introduce a linear operator D^*: $\boldsymbol{P}(H) \to \boldsymbol{P}$ by

$$D^*F(w) = -\text{ trace } DF(w) + [F](w), \qquad F \in \boldsymbol{P}(H), \tag{8.21}$$

then we have

$$L = -D^*D. \tag{8.22}$$

All the above definitions and formulas extend to the case of $\boldsymbol{S}$ and $\boldsymbol{S}(H)$. In the same way as we obtained (8.16), we can prove

$$\int_W \langle DF(w), G(w) \rangle_H P(dw) = \int_W F(w) D^*G(w) P(dw) \quad \text{for all } F \in \boldsymbol{S} \text{ and } G \in \boldsymbol{S}(H). \tag{8.23}$$

Also it is easy to prove the following chain rules: If $F_i \in \boldsymbol{S}$, $i = 1, 2, \cdots, n$ and $g(x_1, x_2, \cdots, x_n)$ is a tempered C^∞-function on $\boldsymbol{R}^n$, then

$$D(g(F_1, F_2, \cdots, F_n)) = \sum_{i=1}^{n} \frac{\partial g}{\partial x^i}(F_1, F_2, \cdots, F_n) DF_i \tag{8.24}$$

$$
\begin{aligned}
&L(g(F_1, F_2,\cdots, F_n)) \\
(8.25)\qquad &= \sum_{i,j=1}^{n} \frac{\partial^2 g}{\partial x^i \partial x^j}(F_1, F_2,\cdots, F_n) \langle DF_i, DF_j\rangle_H \\
&+ \sum_{i=1}^{n} \frac{\partial g}{\partial x_i}(F_1, F_2,\cdots, F_n) LF_i.
\end{aligned}
$$

In particular, if $F, G \in \boldsymbol{S}$,

$$(8.26)\qquad D(FG) = FDG + GDF$$

and

$$(8.27)\qquad \langle DF, DG\rangle_H = \frac{1}{2}\{L(FG) - (LF)G - FLG\}.$$

If, $F,G,J \in \boldsymbol{S}$, then

$$
\begin{aligned}
&\langle D\langle DF, DG\rangle_H, DJ\rangle_H \\
&= \langle D^2F, DG \otimes DJ\rangle_{H\otimes H} + \langle D^2G, DF \otimes DJ\rangle_{H\otimes H}.
\end{aligned}
$$

If $F \in \boldsymbol{S}$ and $G \in \boldsymbol{S}(H)$, then $FG \in \boldsymbol{S}(H)$ and

$$(8.29)\qquad D^*(FG) = -\langle DF, G\rangle_H + FD^*G.$$

Hence, if $F, G \in \boldsymbol{S}$

$$(8.30)\qquad D^*(FDG) = -\langle DF, DG\rangle_H - FLG.$$

As before, let E be a separable real Hilbert space.

Definition 8.2. For $1 < p < \infty$ and $s \in \boldsymbol{R}$, we define a norm $\|\cdot\|_{p,s}$ on $\boldsymbol{P}(E)$ by

$$(8.31)\qquad \|F\|_{p,s} = \|(I - L)^{s/2}F\|_p.$$

Note that $(I - L)^{s/2}F \in \boldsymbol{P}(E)$ is defined by the Wiener chaos decomposition as

$$(I - L)^{s/2}F = \sum_{n=0}^{\infty}(1 + n)^{s/2}J_nF.$$

Proposition 8.3. These norms possess the following basic properties:

(i) (Monotonicity) if $s \leqq s'$ and $1 < p' \leqq p < \infty$, then,

$$||F||_{p,s} \leqq ||F||_{p',s'}, \qquad F \in \boldsymbol{P}(E).$$

(ii) (Compatibility) if $s,s' \in \boldsymbol{R}$ and $p,p' \in (1, \infty)$ and if $F_n \in \boldsymbol{P}(E)$ satisfy $||F_n||_{p,s} \to 0$ as $n \to \infty$ and $||F_n - F_m||_{p',s'} \to 0$ as $n,m \to \infty$, then $||F_n||_{p',s'} \to 0$ as $n \to \infty$.

(iii) (Duality) if $s \in \boldsymbol{R}$, $p \in (1, \infty)$ and $G \in \boldsymbol{P}(E)$, then

$$||G||_{q,-s} = \sup\{ \,|\int_W <F(w), G(w)>_E P(dw)|\,; F \in \boldsymbol{P}(E), ||F||_{p,s} \leqq 1\}$$

where $1/p + 1/q = 1$.

Proof. (i) Since T_t is a contraction semigroup on $\boldsymbol{L}_p(E)$, $1 < p < \infty$, the operator

$$(I - L)^{-s} = \frac{1}{\Gamma(s)} \int_0^\infty e^{-t} t^{s-1} T_t dt$$

is also a contraction on $\boldsymbol{L}_p(E)$ if $s > 0$:

$$||(I - L)^{-s} F||_p \leqq ||F||_p, \qquad F \in \boldsymbol{P}(E).$$

From this, (i) is easily concluded.

(ii) Let $G_n = (I - L)^{s'/2} F_n \in \boldsymbol{P}(E)$. Then $||F_n - F_m||_{p',s'} \to 0$ as n, $m \to \infty$ means that $||G_n - G_m||_{p'} \to 0$ as n, $m \to \infty$ and hence, there exists $G \in \boldsymbol{L}_p(E)$, such that $||G_n - G||_{p'} \to 0$ as $n \to \infty$. But $||F_n||_{p,s} \to 0$ as $n \to \infty$ implies that

$$||(I - L)^{(s-s')/2} G_n||_p \to 0 \qquad \text{as } n \to \infty.$$

Let $J \in \boldsymbol{P}(E)$. Then

$$(I - L)^{\frac{s'-s}{2}} J \in \boldsymbol{P}(E) \subset \boldsymbol{L}_q(E)$$

for every $1 < q < \infty$. Hence

$$\begin{aligned}\int_W <G(w), J(w)>_E P(dw) &= \lim_{n\to\infty} \int_W <G_n(w), J(w)>_E P(dw) \\ &= \lim_{n\to\infty} \int_W <(I - L)^{\frac{s-s'}{2}} G_n(w), (I - L)^{\frac{s'-s}{2}} J>_E P(dw) \\ &= 0\end{aligned}$$

and this implies that $G = 0$. Then

$$\|F_n\|_{p', s'} = \|G_n\|_{p'} \to 0 \qquad \text{as } n \to \infty.$$

(iii) This can be easily obtained by the well-known duality between $L_p(E)$ and $L_q(E)$ if $1/p + 1/q = 1$.

Definition 8.3. Let $1 < p < \infty$ and $s \in \boldsymbol{R}$. Define $\boldsymbol{D}_{p,s}(E)$ to be the completion of $\boldsymbol{P}(E)$ by the norm $\| \quad \|_{p,s}$.

It is easy to see that $\boldsymbol{D}_{p,s}(E)$ coincides with the completion of $\boldsymbol{S}(E)$ by the norm $\| \quad \|_{p,s}$. Now it is clear that

$$\boldsymbol{D}_{p,0}(E) = \boldsymbol{L}_p(E) \tag{8.32}$$

and by (i) and (ii) of Proposition 8.3, we have

$$\boldsymbol{D}_{p', s'}(E) \hookrightarrow \boldsymbol{D}_{p,s}(E) \qquad \text{for } s \leqq s' \text{ and } 1 < p' \leqq p < \infty \tag{8.33}$$

where $\hookrightarrow$ denotes the continuous inclusion. By (iii) of Proposition 8.3, we have

$$\boldsymbol{D}'_{p, s}(E) = \boldsymbol{D}_{q,-s}(E), \qquad s \in \boldsymbol{R}, \qquad p \in (1, \infty) \tag{8.34}$$

under the obvious identification where $1/p + 1/q = 1$. Set

$$\boldsymbol{D}_\infty(E) = \bigcap_{1<p<\infty,\ s>0} \boldsymbol{D}_{p,s}(E) \tag{8.35}$$

and

$$\boldsymbol{D}_{-\infty}(E) = \bigcup_{1<p<\infty,\ s>0} \boldsymbol{D}_{p,-s}(E). \tag{8.36}$$

Then $\boldsymbol{D}_\infty(E)$ is a complete countably normed space (Fréchet space) and $\boldsymbol{D}_{-\infty}(E)$ is its dual. The spaces $\boldsymbol{D}_{p,s}(E)$, $\boldsymbol{D}_\infty(E)$, $\boldsymbol{D}_{-\infty}(E)$,$\cdots$ are denoted by $\boldsymbol{D}_{p,s}$, $\boldsymbol{D}_\infty$, $\boldsymbol{D}_{-\infty}$,$\cdots$ if $E = \boldsymbol{R}$. Since

$$\boldsymbol{D}_{p,s}(E) \subset \boldsymbol{D}_{p,0}(E) = \boldsymbol{L}_p(E) \qquad \text{if } s > 0,$$

elements in $\bigcup_{1<p<\infty,\ s>0} \boldsymbol{D}_{p,s}(E)$ are Wiener functionals in the usual sense, but elements in $\boldsymbol{D}_{p,s}(E)$ for $s < 0$ are usually not so. We have a clear analogy with the Schwartz distribution theory and it may be natural to call such elements as *generalized Wiener functionals.* The coupling

$${}_{\boldsymbol{D}_{-\infty}(E)}(\Phi, F)_{\boldsymbol{D}_\infty(E)}, \qquad \Phi \in \boldsymbol{D}_{-\infty}(E),\ F \in \boldsymbol{D}_\infty(E).$$

is also denoted by $E(<\Phi, F>_E)$ and it is called a *generalized expectation.* In particular, $\mathbf{1}$ (= the functional identically equal to 1) $\in \boldsymbol{D}_\infty$ and, for $\Phi \in \boldsymbol{D}_{-\infty}$, the coupling ${}_{\boldsymbol{D}_{-\infty}}(\Phi, \mathbf{1})_{\boldsymbol{D}_\infty}$ is called the generalized expectation of Φ and is denoted by the usual notation $E(\Phi)$ because it is compatible when

$$\Phi \in \bigcup_{1<p<\infty} \boldsymbol{L}_p.$$

It is clear from the definition of our Sobolev spaces $\boldsymbol{D}_{p,s}(E)$ that the Ornstein-Uhlenbeck operator L: $\boldsymbol{P}(E) \to \boldsymbol{P}(E)$ can be extended uniquely to L: $\boldsymbol{D}_{-\infty}(E) \to \boldsymbol{D}_{-\infty}(E)$ and it is continuous as a operator $\boldsymbol{D}_{p,s+2}(E) \to \boldsymbol{D}_{p,s}(E)$ for every $p \in (1, \infty)$ and $s \in \boldsymbol{R}$. The following important result is due to Meyer [218]. Unfortunatly, we can not have enough space here to give its complete proof. For the proof, see [225] and [232].

Theorem 8.4. (Meyer [218]). For $p \in (1, \infty)$ and $k \in \boldsymbol{Z}^+$, there exist positive constants $c_{p,k}$ and $C_{p,k}$ such that

$$c_{p,k}\|D^k F\|_p \leqq \|F\|_{p,k} \leqq C_{p,k}\sum_{l=0}^{k}\|D^l F\|_p \tag{8.37}$$

for every $F \in \boldsymbol{P}(E)$.

By using this we obtain the following theorem.

Theorem 8.5. (Meyer [218], Sugita [225], P. Kree-M. Kree [208]). The operator D: $\boldsymbol{P}(E) \to \boldsymbol{P}(H \otimes E)$ (or $\boldsymbol{S}(E) \to \boldsymbol{S}(H \otimes E)$) can be extended uniquely to an operator D: $\boldsymbol{D}_{-\infty}(E) \to \boldsymbol{D}_{-\infty}(H \otimes E)$ so that its restriction D: $\boldsymbol{D}_{p,s+1}(E) \to \boldsymbol{D}_{p,s}(H \otimes E)$ is continuous for every $p \in (1, \infty)$ and $s \in \boldsymbol{R}$.

By the duality, we immediately have the following

Corollary. The operator D^*: $\boldsymbol{P}(H \otimes E) \to \boldsymbol{P}(E)$ (or $\boldsymbol{S}(H \otimes E) \to \boldsymbol{S}(E)$) can be extended uniquely to an operator D^*: $\boldsymbol{D}_{-\infty}(H \otimes E) \to \boldsymbol{D}_{-\infty}(E)$ so that its restriction D^*: $\boldsymbol{D}_{p,s+1}(H \otimes E) \to \boldsymbol{D}_{p,s}(E)$ is continuous for every $p \in (1, \infty)$ and $s \in \boldsymbol{R}$.

It is now clear that $L = -D^*D$ in this extended sense. These results imply, in particular, that operators D, D^* and L, first defined on the space of polynomial functionals or smooth functionals, are closable and therefore can be extended to Sobolev space $\boldsymbol{D}_{p,s}(E)$. It may be interesting to note that the following operator on $\boldsymbol{P}$;

$$\boldsymbol{P} \ni F \rightarrow \text{trace } D^2F \in \boldsymbol{P}$$

is not closable. Indeed, let $r = 1$ and $F_k \in \boldsymbol{P}$, $k = 1, 2, \cdots$, be defined by

$$F_k(w) = \sum_{n=1}^{2^k} \left| w\left(\frac{n}{2^k}\right) - w\left(\frac{n-1}{2^k}\right) \right|^2 - 1.$$

It is easy to show that $F_k \rightarrow 0$ in $\boldsymbol{L}_p$ for every $1 < p < \infty$ but trace $D^2F_k = 2$ for every $k = 1, 2, \cdots$.

To prove Theorem 8.5, we prepare the following lemmas and propositions.

Lemma 8.1. (Commutation relations involving D). For a real sequence $\phi = (\phi_n)_{n=0}^{\infty}$,

$$DT_\phi = T_{\phi^+}D \qquad \text{on} \qquad \boldsymbol{P}(E)$$

where

$$\phi^+(n) = \phi(n+1), \qquad n = 0, 1, \cdots.$$

Proof. For simplicity, we only show the lemma in case of $E = \boldsymbol{R}$. By Proposition 8.1, it suffices to prove

$$DT_\phi H_{\boldsymbol{a}}(w) = T_{\phi^+}DH_{\boldsymbol{a}}(w) \qquad \text{for every} \qquad \boldsymbol{a} \in \Lambda,$$

where

$$H_{\boldsymbol{a}}(w) = \prod_{j=1}^{\infty} H_{a_i}([h_i](w))$$

and $\{h_i\}_{i=1}^{\infty}$ is an *ONB* of H. Since

$$T_\phi H_{\boldsymbol{a}}(w) = \phi(|a|)H_{\boldsymbol{a}}(w) \qquad \boldsymbol{a} \in \Lambda,$$

we have

$$DT_\phi H_{\boldsymbol{a}}(w) = \phi(|\boldsymbol{a}|)DH_{\boldsymbol{a}}(w). \tag{8.38}$$

Furthermore, noting that $H_n'(x) = H_{n-1}(x)$, we easily see

$$DH_{\boldsymbol{a}}(w) = \sum_{a_j>0} H_{a_j-1}([h_j](w)) \prod_{i\neq j} H_{a_i}([h_j](w))h_j$$

and hence

$$DH_{\boldsymbol{a}}(w) \in \boldsymbol{C}_{|\boldsymbol{a}|-1}.$$

This implies

$$T_{\phi+}DH_{\boldsymbol{a}}(w) = \sum_{n=1}^{\infty} \phi(n+1)J_n DH_{\boldsymbol{a}}(w) = \phi(|\boldsymbol{a}|)DH_{\boldsymbol{a}}(w).$$

Combining this with (8.38) yields the conclusion.

Proposition 8.6. (Hypercontractivity of the Ornstein-Uhlenbeck semigroup, Nelson [220]). Let $p \in (1, \infty)$, $t > 0$ and $q(t) = (p-1)e^{2t} + 1$. Then

$$\|T_t G\|_{q(t)} \leqq \|G\|_p \qquad \text{for every } G \in \boldsymbol{L}_p.$$

Proof. The following proof is due to Neveu [221]. Let $\{B_t^{(i)}, t \geqq 0\}$, $i = 1, 2$, be two n-dimensional Brownian motions with $B_0^{(i)} = 0$ defined on a probability space $\{\Omega, \mathscr{F}, Q\}$. We assume that $\{B_t^{(i)}, t \geqq 0\}$, $i = 1,2$ are independent. For given $\lambda \in (0, 1)$, we set

$$q = q(\lambda) = (p-1)\lambda^{-2} + 1$$

and define $q' = q'(\lambda)$ by

$$\frac{1}{q} + \frac{1}{q'} = 1.$$

It is easy to see that $\{(p-1)(q'-1)\}^{1/2} = \lambda$. We define

$$\xi_t^{(1)} = B_t^{(1)} \quad \text{and} \quad \xi_t^{(2)} = \lambda B_t^{(1)} + \sqrt{1-\lambda^2}B_t^{(2)}, \qquad 0 \leqq t \leqq 1$$

and $\mathscr{F}_t^{(i)}$, $i = 1, 2$, the proper reference families of $\{\xi_t^{(i)}\}$, i.e. $\mathscr{F}_t^{(i)} = \sigma[\xi_s^{(i)}, 0 \leqq s \leqq t]$, $i = 1, 2$. Let f and g be C^∞-functions defined on $\boldsymbol{R}^n$ such that

$$a \leqq f(x), g(x) \leqq b \qquad \text{for some } 0 < a < b.$$

Define martingale $\{M_t, 0 \leqq t \leqq 1\}$ and $\{N_t, 0 \leqq t \leqq 1\}$ by

$$M_t = E[f(\xi_1^{(1)})^{q'} | \mathscr{F}_t^{(1)}],\ N_t = E[g(\xi_1^{(2)})^p | \mathscr{F}_t^{(2)}], \quad 0 \leqq t \leqq 1.$$

Then these can be written in the following forms (Theorem II-6.6):

$$M_t = M_0 + \int_0^t \phi_s d\xi_s^{(1)}, \qquad N_t = N_0 + \int_0^t \psi_s d\xi_s^{(2)}.$$

By Itô's formula, we have

$$\begin{aligned}
&E^Q[M_1^{1/q'} N_1^{1/p}] - E^Q[M_0^{1/q'} N_0^{1/p}] \\
&= -\frac{1}{2} E^Q\Big[\int_0^1 M_t^{\frac{1}{q'}-2} N_t^{\frac{1}{p}-2}\Big[\frac{1}{p}\Big(1-\frac{1}{p}\Big)M_t^2\psi_t^2 - 2\frac{\lambda}{pq'}M_t N_t \phi_t \psi_t \\
&\qquad\qquad + \frac{1}{q'}\Big(1-\frac{1}{q'}\Big)N_t^2\phi_t^2\Big]dt\Big] \\
&= -\frac{1}{2} E^Q\Big[\int_0^1 M_t^{\frac{1}{q'}-2} N_t^{\frac{1}{p}-2}\Big\{\Big(\frac{\sqrt{p-1}}{p}M_t\psi_t - \frac{\sqrt{q'-1}}{q'}N_t\phi_t\Big)^2\Big\}dt\Big]
\end{aligned}$$

$$(\text{ by } \sqrt{(p-1)(q'-1)} = \lambda\ )$$

where E^Q denotes the expectation with respect to Q. This implies that

$$\begin{aligned}
E^Q[f(\xi_1^{(1)})g(\xi_1^{(2)})] &= E^Q[M_1^{1/q'} N_1^{1/p}] \\
&\leqq E^Q[M_0^{1/q'} N_0^{1/p}] = (E^Q[f(\xi_1^{(1)})^{q'}])^{1/q'}(E^Q[g(\xi_1^{(2)})^p])^{1/p}.
\end{aligned}$$

Hence we obtain

$$\begin{aligned}
&\int_{R^n}\int_{R^n} g(\lambda\xi + \sqrt{1-\lambda^2}\eta)f(\xi)\Big(\frac{1}{\sqrt{2\pi}}\Big)^n e^{-\frac{|\xi|^2}{2}}\Big(\frac{1}{\sqrt{2\pi}}\Big)^n e^{-\frac{|\eta|^2}{2}} d\xi d\eta \\
&\leqq \Big\{\int_{R^n} f(\xi)^{q'}\Big(\frac{1}{\sqrt{2\pi}}\Big)^n e^{-\frac{|\xi|^2}{2}} d\xi\Big\}^{1/q'}\Big\{\int_{R^n} g(\xi)^p\Big(\frac{1}{\sqrt{2\pi}}\Big)^n e^{-\frac{|\xi|^2}{2}} d\xi\Big\}^{1/p}
\end{aligned}$$

Taking $\lambda = e^{-t}$, $t > 0$, we have

$$\begin{aligned}
&\int_{R^n}\Big\{\int_{R^n} g(\lambda\xi + \sqrt{1-e^{-2t}}\eta)\Big(\frac{1}{\sqrt{2\pi}}\Big)^n e^{-\frac{|\eta|^2}{2}} d\eta\Big\} f(\xi)\Big(\frac{1}{\sqrt{2\pi}}\Big)^n e^{-\frac{|\xi|^2}{2}} d\xi) \\
&\leqq \Big\{\int_{R^n} g(\xi)^p\Big(\frac{1}{\sqrt{2\pi}}\Big)^n e^{-\frac{|\xi|^2}{2}} d\xi\Big\}^{1/p}\Big\{\int_{R^n} f(\xi)^{\hat{q}(t)'}\Big(\frac{1}{\sqrt{2\pi}}\Big)^n e^{-\frac{|\xi|^2}{2}} d\xi\Big\}^{1/\hat{q}(t)'}
\end{aligned}$$

where $\hat{q}(t)' = q(e^{-t})'$. By combining this with Remark 8.2 and Proposition 8.2, we easily obtain

$$\Big|\int_W T_t G(w)F(w)P(dw)\Big| \leqq \|G\|_p \|F\|_{\hat{q}(t)},$$

$$\text{for every} \quad G \in \boldsymbol{L}_p \quad \text{and} \quad F \in \boldsymbol{L}_{\hat{q}(t)'},$$

which completes the proof.

Proposition 8.7. J_n: $L_2 \to C_n$ is a bounded operator on L_p, $p \in (1, \infty)$.

Proof. Let $p > 2$ and t be the positive constant such that

$$e^{2t} = p - 1.$$

Then by Proposition 8.6 we have

$$\|T_t F\|_p \leqq \|F\|_2 \qquad F \in L_p \ .$$

In particular

$$\|T_t J_n F\|_p \leqq \|J_n F\|_2 \leqq \|F\|_2 \leqq \|F\|_p \ .$$

Since

$$\|T_t J_n F\|_p = e^{-nt}\|J_n F\|_{p'}$$

this implies that

$$\|J_n F\|_p \leqq e^{nt}\|F\|_p \qquad \text{for } F \in L_p \ . \tag{8.39}$$

For $1 < p < 2$, consider the adjoint J_n^* of J_n. Then (8.39), applied to p' such that $1/p + 1/p' = 1$, yields that

$$\|J_n^* F\|_p \leqq e^{-nt}\|F\|_p \qquad \text{for } F \in P \ .$$

But, for $F \in P$, $J_n^* F = J_n F$. By the denseness of P, the result follows.

Lemma 8.2. (L_p-multiplier theorem).* Let $\phi = \{\phi(n)\}_{n=0}^{\infty}$ be a real positive sequence such that

$$\phi(n) = \sum_{k=0}^{\infty} a_k \left(\frac{1}{n^\alpha}\right)^k, \qquad \text{for} \qquad n \geqq n_0$$

for some $n_0 \in N$, $\quad 1 \geq \alpha > 0$ and

$$\sum_{k=0}^{\infty} |a_k| \left(\frac{1}{n_0^\alpha}\right)^k < \infty.$$

Then for every $1 < p < \infty$ there exists a positive constant C_p such that

*) First obtained by Meyer [218]. Proof given here is due to Shigekawa (Private communication).

$$\|T_\phi F\|_p \leqq C_p\|F\|_p \qquad \text{for every} \qquad F \in \boldsymbol{P}.$$

Proof. First we consider the case $\alpha = 1$. We set

$$T_\phi = \sum_{n=0}^{n_0-1} \phi(n)J_n + \sum_{n=n_0}^{\infty} \phi(n)J_n \qquad := T_\phi^{(1)} + T_\phi^{(2)} .$$

By Proposition 8.7, $T_\phi^{(1)}$ is L_p-bounded, i.e., there exists a positive constant C_p,

$$\|T_\phi^{(1)}F\|_p \leqq C_p\|F\|_p \qquad \text{for} \qquad F \in \boldsymbol{L}_p. \tag{8.40}$$

We now show that there exists $C > 0$ such that

$$\|T_t(I - J_0 - J_1 - \cdots - J_{n_0-1})F\|_p \leqq Ce^{-n_0 t}\|F\|_p, \qquad \text{for all } t > 0 \text{ and } F \in \boldsymbol{L}_p. \tag{8.41}$$

First we consider the case $p = 2$. For $F \in \boldsymbol{L}_2$, we have

$$\begin{aligned} &\|T_t(I - J_0 - J_1 - \cdots - J_{n_0-1})F\|_2^2 \\ &= \|\sum_{k=n_0}^{\infty} e^{-kt}J_kF\|_2^2 = \sum_{k=n_0}^{\infty} \|e^{-kt}J_kF\|_2^2 \\ &\leqq e^{-2n_0 t}\sum_{k=n_0}^{\infty} \|J_kF\|_2^2 \leqq e^{-2n_0 t}\|F\|_2^2. \end{aligned} \tag{8.42}$$

Next we consider the case $p > 2$. Take t_0 such that $p = e^{2t_0} + 1$. By Proposition 8.6, we obtain

$$\begin{aligned} &\|T_{t_0+t}(I - J_0 - J_1 - \cdots - J_{n_0-1})F\|_p^2 \\ &\quad\leqq \|T_t(I - J_0 - J_1 - \cdots - J_{n_0-1})F\|_2^2 \\ &\quad\leqq \|\sum_{n=n_0}^{\infty} e^{-nt}J_nF\|_2^2 \\ &\quad\leqq e^{-2n_0 t}\sum_{n=n_0}^{\infty} \|J_nF\|_2^2 \\ &\quad\leqq e^{-2n_0 t}\|F\|_2^2 \\ &\quad\leqq e^{-2n_0 t}\|F\|_p^2 = e^{2n_0 t_0}e^{-2n_0(t+t_0)}\|F\|_p^2. \end{aligned} \tag{8.43}$$

Furthermore for $t \leqq t_0$ we obtain, by (8.13),

$$\|T_t\,(I - J_0 - J_1 - \cdots - J_{n_0-1})F\|_p$$

$$(8.44) \qquad \begin{aligned} &\leqq \|(I - J_0 - J_1 - \cdots - J_{n_0-1})F\|_p \\ &\leqq \|I - J_0 - J_1 - \cdots - J_{n_0-1}\|_p \|F\|_p \end{aligned}$$

where $\|I - J_0 - J_1 - \cdots - J_{n_0-1}\|_p$ is the operator norm of $(I - J_0 - J_1 - \cdots - J_{n_0-1})$ in L_p.

Combining (8.42), (8.43) and (8.44) we obtain (8.41) with $p \geqq 2$. For $1 < p < 2$, the (8.41) follows by duality.

We now set

$$R_{n_0} = \int_0^\infty T_t(I - J_0 - J_1 - \cdots - J_{n_0-1})dt.$$

Then, from (8.41) we obtain

$$\|R_{n_0}F\|_p \leqq C\frac{1}{n_0}\|F\|_p.$$

Since

$$R_{n_0}^2 F = \int_0^\infty \int_0^\infty T_{t+s}(I - J_0 - J_1 - \cdots - J_{n_0-1})F \, dtds$$

we obtain

$$\|R_{n_0}^2 F\|_p \leqq C\frac{1}{n_0^2}\|F\|_p$$

and repeating this we have

$$(8.45) \qquad \|R_{n_0}^k F\|_p \leqq C\frac{1}{n_0^k}\|F\|_p, \qquad k = 1, 2, \cdots .$$

Note that

$$R_{n_0}^k F = \frac{1}{n^k} J_n F, \qquad k = 1, 2, \cdots, \text{ if } F \in C_n,\ n \geqq n_0.$$

Hence we obtain

$$T_\phi^{(2)} F = \sum_{n=n_0}^\infty \sum_{k=0}^\infty a_k R_{n_0}^k J_n F = \sum_{k=0}^\infty a_k R_{n_0}^k F.$$

By (8.45),

$$\|T_\phi^{(2)}F\|_p \leqq C\Big(\sum_{k=0}^{\infty}|a_k|\Big(\frac{1}{n_0}\Big)^k\Big)\|F\|_p.$$

Combining this with (8.40), we can conclude that there exists $C_p > 0$, such that

$$\|T_\phi F\| \leqq C_p\|F\|_p \qquad F \in \boldsymbol{P}.$$

For the general case i.e., $0 < \alpha < 1$, define

$$Q_t^{(\alpha)} = \sum_{n=0}^{\infty} e^{-n^\alpha t} J_n F = \int_0^\infty (T_s F)\mu_t^{(\alpha)}(ds)$$

where $\mu_t^{(\alpha)}$ is the probability measure defined by

$$\int_0^\infty e^{-\lambda s}\mu_t^{(\alpha)}(ds) = \exp(-\lambda^\alpha t) \qquad \text{for every } \lambda > 0.$$

As in the case $\alpha = 1$, write

$$T_\phi = T_\phi^{(1)} + T_\phi^{(2)}.$$

By the same reason, $T_\phi^{(1)}$ is L_p-bounded. From (8.41), we have

$$\|Q_t^{(\alpha)}(I - J_0 - J_1 - \cdots - J_{n_0-1})F\|_p$$
$$\leqq C\int_0^\infty \|F\|_p e^{-n_0 s}\mu_t^{(\alpha)}(ds) = C\exp(-n_0^\alpha t)\|F\|_p.$$

Define

$$R_{n_0} = \int_0^\infty Q_t^{(\alpha)}(I - J_0 - J_1 - \cdots - J_{n_0-1})dt$$

and proceed as in the case $\alpha = 1$. Then we obtain that $T_\phi^{(2)}$ is L_p-bounded. This concludes the proof.

Proof of Theorem 8.5. In case of $E = \boldsymbol{R}$, we show the theorem. By Lemma 8.1, we obtain

$$|(I - L)^{s/2}DF|_H = |DR(I - L)^{s/2}F|_H \qquad F \in \boldsymbol{P}$$

where $\boldsymbol{R}$ is the operator given by

$$R = \sum_{n=1}^{\infty}\Big(\frac{n}{n+1}\Big)^{s/2} J_n.$$

Note that $R = T_\phi$ with

$$\phi(n) = \begin{cases} 0, & n = 0 \\ \left(\dfrac{n}{n+1}\right)^{s/2} = \left(\dfrac{1}{1+\frac{1}{n}}\right)^{s/2}, & n \geqq 1 \end{cases}$$

and

$$h(x) = \left(\frac{1}{1+x}\right)^{\frac{s}{2}}$$

is analytic near $x = 0$. By Theorem 8.4 and Lemma 8.2, there exist positive constants C_p and C_p' such that

$$\begin{aligned} \|(I-L)^{s/2}DF\|_p &= \|DR(I-L)^{s/2}F\|_p \\ &\leqq C_p\|(I-L)^{\frac{1}{2}}R(I-L)^{\frac{s}{2}}F\|_p \\ &= C_p\|R(I-L)^{(s+1)/2}F\|_p \\ &\leqq C_p'\|(I-L)^{(s+1)/2}F\|_p \\ &= C_p'\|F\|_{p,s+1}. \end{aligned}$$

Therefore

$$\|DF\|_{p,s} \leqq C_p'\|F\|_{p,s+1} \qquad F \in \boldsymbol{P}$$

from which the result follows by a limiting argument.

Proposition 8.8. Let E_1 and E_2 be real separable Hilbert spaces. For any $p, q \in (1, \infty)$ and $k = 0, 1, 2, \cdots$ such that

$$\frac{1}{p} + \frac{1}{q} = \frac{1}{r} < 1$$

there exists a constant $C_{p,q,k} > 0$ such that

$$\|F \otimes G\|_{r,k} \leqq C_{p,q,k}\|F\|_{p,k}\|G\|_{q,k} \tag{8.46}$$

for every $F \in \boldsymbol{P}(E_1)$ and $G \in \boldsymbol{P}(E_2)$.

Proof. Indeed,

$$D(F \otimes G) = DF \otimes G + F \otimes DG$$

and hence

$$|D(F \otimes G)|_{H\otimes E_1 \otimes E_2} \leqq |DF|_{H\otimes E_1}|G|_{E_2} + |F|_{E_1}|DG|_{H\otimes E_2}$$

from which (8.46) in the case of $k = 1$ is easily obtained by noting the equivalence of norms $\|F\|_p + \|DF\|_p$ and $\|F\|_{p,1}$ (which is a consequence of Meyer's theorem). We can obtain (8.46) by the same argument successively.

From (8.46) and the duality (Proposition 8.3, (iii)), it is easy to obtain the following.

Corollary. For every $p,q \in (1, \infty)$ such that

$$\frac{1}{p} + \frac{1}{q} = \frac{1}{r} < 1$$

and $k = 0, 1, \cdots$, there exists a constant $C'_{p,q,k} > 0$ such that

$$\|F \otimes G\|_{r,-k} \leqq C'_{p,q,k}\|F\|_{p,k}\|G\|_{q,-k} \tag{8.47}$$

for every $F \in \boldsymbol{P}(E_1)$ and $G \in \boldsymbol{P}(E_2)$.

By (8.46) and (8.47), we see that $F \otimes G \in \boldsymbol{D}_{r,k}(E_1 \otimes E_2)$ is well-defined for every $F \in \boldsymbol{D}_{p,k}(E_1)$ and $G \in \boldsymbol{D}_{q,k}(E_2)$ and, $F \otimes G \in \boldsymbol{D}_{r,-k}(E_1 \otimes E_2)$ is well-defined for every $F \in \boldsymbol{D}_{p,k}(E_1)$ and $G \in \boldsymbol{D}_{q,-k}(E_2)$ provided $\frac{1}{p} + \frac{1}{q} = \frac{1}{r} < 1$. In particular, $\boldsymbol{D}_\infty$ is an algebra, i.e., if $F,G \in \boldsymbol{D}_\infty$ then $FG \in \boldsymbol{D}_\infty$. Furthermore, $\boldsymbol{D}_\infty(E)$ and $\boldsymbol{D}_{-\infty}(E)$ are $\boldsymbol{D}_\infty$-modules, i.e., if $F \in \boldsymbol{D}_\infty$ and $G \in \boldsymbol{D}_\infty(E)$ ($\boldsymbol{D}_{-\infty}(E)$), then $FG \in \boldsymbol{D}_\infty(E)$ (resp. $\boldsymbol{D}_{-\infty}(E)$). Note that if $F \in \boldsymbol{D}_\infty$ and $G \in \boldsymbol{D}_{-\infty}$, the generalized expectation $E(FG)$ of $FG \in \boldsymbol{D}_{-\infty}$ coincides with ${}_{\boldsymbol{D}_{-\infty}}\langle G, F\rangle_{\boldsymbol{D}_\infty}$.

We note that chain rules (8.24)–(8.30) can be easily extended to Wiener functionals belonging to Sobolev spaces in an obvious way: In particular these formulas hold if $\boldsymbol{P}$ and $\boldsymbol{P}(E)$ are replaced by $\boldsymbol{D}_\infty$ and $\boldsymbol{D}_\infty(E)$ respectively. Also the integration by parts formula (8.23) can be extended to the case $F \in \boldsymbol{D}_{p,1}$ and $G \in \boldsymbol{D}_{q,1}(H)$ such that

$$\frac{1}{p} + \frac{1}{q} = 1.$$

Of course, (8.23) holds in the case $F \in \boldsymbol{D}_{p,s+1}$ and $G \in \boldsymbol{D}_{q,-s}(H)$,

$$\frac{1}{p}+\frac{1}{q}=1, \qquad s \in \boldsymbol{R},$$

if the both sides of (8.23) are understood as the natural coupling between $\boldsymbol{D}_{p,s}(H)$ and $\boldsymbol{D}_{q,-s}(H)$ $(=(\boldsymbol{D}_{p,s}(H))')$ and $\boldsymbol{D}_{p,s+1}$ and $\boldsymbol{D}_{q,-s-1}(=(\boldsymbol{D}_{p,s+1})')$, respectively.

9. Pull-back of Schwartz distributions under Wiener mappings and the regularity of induced measures (probability laws)

We now apply the results of the previous section to the study of regularity and asymptotics of probability density of finite dimensional Wiener functionals. For this, we regard a d-dimensional Wiener functional as a d-dimensional Wiener mapping F: $\boldsymbol{W} \to \boldsymbol{R}^d$ and discuss the pull-back of Schwartz tempered distributions on $\boldsymbol{R}^d$ under this mapping. This pull-back can be defined as an element in $\boldsymbol{D}_{-\infty}$, i.e., as a generalized Wiener functionals, if the Wiener mapping satisfies certain conditions on regularity and non-degeneracy usually referred to as Malliavin's conditions. If the Dirac delta function δ_x at $x \in \boldsymbol{R}^d$ is pulled back by a Wiener mapping F: $\boldsymbol{W} \to \boldsymbol{R}^d$, then the generalized expectation $E[\delta_x(F)]$ of this pull-back $\delta_x(F)$ coincides with the density $p_F(x)$ of the probability law of functional F with respect to the Lebesgue measure on $\boldsymbol{R}^d$. We can discuss the continuous or differentiable dependence on parameters of the pull-back thereby deduce the regularity of the probability density.

So let $(\boldsymbol{W}, P)$ be the r-dimensional Wiener space and let $F(w) = (F^1(w), F^2(w), \cdots, F^d(w))$ be a mapping $\boldsymbol{W} \to \boldsymbol{R}^d$. We assume that it is smooth in the following sense of Malliavin:

(A.1) $\quad F \in \boldsymbol{D}_\infty(\boldsymbol{R}^d)$, i.e. $F^i \in \boldsymbol{D}_\infty, \qquad i=1,2,\cdots,d.$

Then we can define

(9.1) $\quad \sigma^{ij}(w) = \langle DF^i(w), DF^j(w)\rangle_H, \qquad i,j=1,2,\cdots,d$

and we know that $\sigma^{ij} \in \boldsymbol{D}_\infty$. The matrix $\sigma(w) = (\sigma^{ij}(w))$ is non-negative definite and hence det $\sigma(w) \geqq 0$ for almost all $w(P)$. $\sigma(w)$ is called the *Malliavin convariance of F.*

Definition 9.1. We say that F is non-degenerate in the sense of Malliavin if it satisfies

$$\text{(A.2)} \qquad (\det \sigma(w))^{-1} \in \bigcap_{1<p<\infty} \boldsymbol{L}_p.$$

Here it is understood that $1/0 = \infty$.

Assume that F: $\boldsymbol{W} \to \boldsymbol{R}^d$ satisfies Malliavin's conditions (A.1) and (A.2). For every $\varepsilon > 0$, define $\sigma_\varepsilon = (\sigma_\varepsilon^{ij})$ by

$$\sigma_\varepsilon^{ij} = \sigma^{ij} + \varepsilon\delta^{ij}$$

where δ^{ij} is Kronecker's delta. Then, setting $(\gamma_{ij}^\varepsilon) = (\sigma_\varepsilon^{ij})^{-1}$, we have $\gamma_{ij}^\varepsilon \in \boldsymbol{D}_\infty$, $i, j = 1, 2, \cdots, d$ because there exist tempered C^∞-functions $f_{ij}(x)$, $i, j = 1, 2, \cdots, d$ on $\boldsymbol{R}^{d^2}$ such that

$$\gamma_{ij}^\varepsilon = f_{ij}(\sigma_\varepsilon), \qquad i, j = 1, 2, \cdots, d.$$

It is easy to prove by the chain rule that

$$\text{(9.2)} \qquad D\gamma_{ij}^\varepsilon = -\sum_{k,l=1}^{d} \gamma_{ik}^\varepsilon \gamma_{jl}^\varepsilon D\sigma_\varepsilon^{kl}.$$

Furthermore it is easy to deduce from (A.2) that

$$D^k\gamma_{ij}^\varepsilon \to D^k\gamma_{ij} \qquad \text{in} \quad \boldsymbol{L}_p(H^{\otimes k})$$

for every $k = 0, 1, \cdots$, $1 < p < \infty$ and $i, j = 1, 2, \cdots, d$ where $H^{\otimes k} = \underbrace{H \otimes H \otimes \cdots \otimes H}_{k}$ and $(\gamma_{ij}) = (\sigma^{ij})^{-1}$. This shows that

$$\gamma_{ij} \in \boldsymbol{D}_\infty, \qquad i, j = 1, 2, \cdots, d.$$

Also by (9.2), we have

$$\text{(9.3)} \qquad D\gamma_{ij} = -\sum_{k,l=1}^{d} \gamma_{ik}\gamma_{jl} D\sigma^{kl}.$$

Let ϕ: $\boldsymbol{R}^d \to \boldsymbol{R}$ be a tempered C^∞-function. Then, by (8.24),

$$D(\phi \circ F) = \sum_{i=1}^{d} (\partial_i \phi \circ F) DF^i.$$

Hence

$$<D(\phi \circ F), DF^j>_H = \sum_{i=1}^{d} (\partial_i \phi \circ F) <DF^i, DF^j>_H$$

$$= \sum_{i=1}^{d} \partial_i \phi \circ F \cdot \sigma^{ij}$$

and therefore,

$$\partial_i \phi \circ F = \sum_{j=1}^{d} \gamma_{ij} \langle D(\phi \circ F), DF^j \rangle_H. \tag{9.4}$$

Let $g \in \boldsymbol{D}_\infty$. Then, we obtain by (8.23) that

$$\begin{aligned}
&\int_W \partial_i \phi \circ F(w) \cdot g(w) P(dw) \\
&= \sum_{j=1}^{d} \int_W \langle D(\phi \circ F)(w), \gamma_{ij}(w) g(w) DF^j(w) \rangle_H P(dw) \\
&= \sum_{j=1}^{d} \int_W \phi \circ F(w) \cdot D^*[\gamma_{ij}(w) g(w) DF^j(w)] P(dw) \\
&= \int_W \phi \circ F(w) \cdot \Phi_i(w; g) P(dw)
\end{aligned}$$

where

$$\begin{aligned}
\Phi_i(w; g) &= \sum_{j=1}^{d} D^*[\gamma_{ij}(w) g(w) DF^j(w)] \\
&= - \sum_{j=1}^{d} \{ g(w) \langle D\gamma_{ij}(w), DF^j(w) \rangle_H \\
&\quad + \gamma_{ij}(w) \langle Dg(w), DF^j(w) \rangle_H + \gamma_{ij}(w) g(w) LF^j(w) \}
\end{aligned} \tag{9.5}$$

by (8.26) and (8.30). Hence

$$\begin{aligned}
&\int_W (\partial_{i_1} \partial_{i_2} \cdots \partial_{i_k} \phi) \circ F(w) \cdot g(w) P(dw) \\
&= \int_W \phi \circ F(w) \cdot \Phi_{i_1, i_2, \ldots, i_k}(w; g) P(dw)
\end{aligned} \tag{9.6}$$

where $\Phi_{i_1, i_2, \cdots, i_k}(w; g) \in \boldsymbol{D}_\infty$ is determined successively by

$$\Phi_{i_1, i_2, \ldots, i_\nu}(w; g) = \Phi_{i_\nu}(w; \Phi_{i_1, i_2, \ldots, i_{\nu-1}}(\cdot\,; g)).$$

We can conclude from (9.5) that $\Phi_{i_1, i_2, \ldots, i_k}(w; g)$ has a form

$$\begin{aligned}(9.7)\quad &\Phi_{i_1, i_2, \ldots, i_k}(w; g)\\ &= \Psi_0(w)g(w) + \langle \Psi_1(w), Dg(w)\rangle_H + \cdots\\ &\quad + \langle \Psi_k(w), D^k g(w)\rangle_{H^{\otimes k}}\end{aligned}$$

where

$$\Psi_\nu \in \boldsymbol{D}_\infty(H^{\otimes \nu}), \qquad \nu = 0, 1, 2, \cdots$$

which are obtained as polynomials in γ_{ij}, F^i and their derivatives. Similarly we have for $\Delta = \sum_{i=1}^{d} (\partial/\partial x^i)^2$,

$$\begin{aligned}(9.8)\quad &\int_W \left(\left(1 + |x|^2 - \frac{\Delta}{2}\right)^k \phi\right) \circ F(w) \cdot g(w) P(dw)\\ &= \int_W \phi \circ F(w) \cdot \eta_{2k}(w; g) P(dw)\end{aligned}$$

where $\eta_{2k}(w; g) \in \boldsymbol{D}_\infty$ has a similar expression as (9.7). Indeed, $\eta_{2k}(w; g)$ is obtained in the same way as $\Phi_{i_1, i_2, \cdots, i_k}(w; g)$ with some more polynomials of F multiplied in each step of the integration by parts. In particular, for every $q > 1$,

$$(9.9)\qquad \sup\{\|\eta_{2k}(\,\cdot\,; g)\|_1;\ g \in \boldsymbol{D}_{q, 2k},\ \|g\|_{q, 2k} \leqq 1\} < \infty$$

where $\| \ \|_1$ denotes the $\boldsymbol{L}_1$-norm.

We introduce a system of Sobolev spaces on $\boldsymbol{R}^d$ as follows: Let $\mathscr{S}(\boldsymbol{R}^d)$ be the real Schwartz space of rapidly decreasing C^∞-functions on $\boldsymbol{R}^d$. For $\phi \in \mathscr{S}(\boldsymbol{R}^d)$ and $k = 0, \pm 1, \pm 2, \cdots$, introduce the norm

$$(9.10)\qquad \|\phi\|_{2k} = \|(1 + |x|^2 - \Delta/2)^k \phi\|_\infty$$

where $\| \ \|_\infty$ is the sup-norm on $\boldsymbol{R}^d$. Let $\mathscr{S}_{2k}$, $k = 0, \pm 1, \pm 2, \cdots$ be the completion of $\mathscr{S}(\boldsymbol{R}^{2d})$ by the norm $\| \ \|_{2k}$. Then we have

$$\begin{aligned}&\cdots \subset \mathscr{S}_2 \subset \mathscr{S}_0 =: \{f(x);\ \text{continuous on } \boldsymbol{R}^d \text{ and } \lim_{|x|\to\infty} |f(x)|\\ &\qquad = 0\} \subset \mathscr{S}_{-2} \subset \cdots,\\ &\bigcap_{k>0} \mathscr{S}_{2k} = \mathscr{S}(\boldsymbol{R}^d) \quad \text{and} \quad \bigcup_{k>0} \mathscr{S}_{-2k} = \mathscr{S}'(\boldsymbol{R}^d).\end{aligned}$$

Theorem 9.1. Let F: $\boldsymbol{W} \to \boldsymbol{R}^d$ be a d-dimensional Wiener mapping satisfying Malliavin's conditions of (A.1) and (A.2). Then, for every $p > 1$ and $k = 0, 1, 2, \cdots$, there exists a positive constant $C =$

$C(p, k; F)$ such that the following estimate holds: For every $\phi \in \mathscr{S}(\boldsymbol{R}^d)$,

$$(9.11) \qquad \|\phi \circ F\|_{p,-2k} \leqq C\|\phi\|_{-2k}.$$

(Note that $\phi \circ F \in \boldsymbol{D}_\infty$).

Proof. Define q by $1/p + 1/q = 1$. Then by the duality

$$\|\phi \circ F\|_{p,-2k} = \sup\{\int_W \phi \circ F(w) \cdot g(w)P(dw);\ g \in \boldsymbol{D}_{q,2k}, \\ \|g\|_{q,2k} \leqq 1\}$$

and, by (9.8) and (9.9),

$$\begin{aligned} &|\int_W \phi \circ F(w) \cdot g(w)P(dw)| \\ &= |\int_W \{(1 + |x|^2 - \Delta/2)^k(1 + |x|^2 - \Delta/2)^{-k}\phi\} \circ F(w) \cdot \\ &\quad \times g(w)P(dw) \\ &= |\int_W \{(1 + |x|^2 - \Delta/2)^{-k}\phi\} \circ F(w) \cdot \eta_{2k}(w; g)P(dw)| \\ &\leqq \|(1 + |x|^2 - \Delta/2)^{-k}\phi\|_\infty \|\eta_{2k}(\cdot\ ; g)\|_1 \\ &\leqq \|\phi\|_{-2k}\|\eta_{2k}(\cdot\ ; g)\|_1. \end{aligned}$$

Now (9.11) is established by setting $C = C(p, k; F)$ to be the supremum in (9.9)

Corollary. The mapping $\mathscr{S}(\boldsymbol{R}^d) \ni \phi \to \phi \circ F \in \boldsymbol{D}_\infty$ can be extended uniquely to a continuous linear mapping

$$\mathscr{S}_{-2k} \ni T \to T \circ F \in \boldsymbol{D}_{p,-2k}$$

for every $1 < p < \infty$ and $k = 0, 1, 2, \cdots$. In particular, for every $T \in \mathscr{S}'(\boldsymbol{R}^d)$, $T \circ F \in \boldsymbol{D}_{-\infty}$ is well-defined and

$$(9.12) \qquad T \circ F \in \bigcup_{k=1}^{\infty} \bigcap_{1<p<\infty} \boldsymbol{D}_{p,-2k}.$$

This $T \circ F$, denoted also by $T(F)$, is called the *pull-back of Schwartz distribution $T \in S'(\boldsymbol{R}^d)$ on $\boldsymbol{R}^d$ by the Wiener mapping F: $\boldsymbol{W} \to \boldsymbol{R}^d$*, or the *composite of distribution T on $\boldsymbol{R}^d$ and d-dimensional Wiener functional F.*

The fact (9.12) is important because $T\circ F$ can be coupled with test functionals in a class much larger than $\boldsymbol{D}_\infty$: By introducing the notations

$$(9.13)\qquad \tilde{\boldsymbol{D}}_\infty = \bigcap_{k=1}^{\infty} \bigcup_{1<p<\infty} \boldsymbol{D}_{p,k} \qquad \text{and}$$

$$(9.14)\qquad \tilde{\boldsymbol{D}}_{-\infty} = \bigcup_{k=1}^{\infty} \bigcap_{1<p<\infty} \boldsymbol{D}_{p,-k},$$

if $G \in \tilde{\boldsymbol{D}}_\infty$, then $G\cdot T\circ F \in \boldsymbol{D}_{-\infty}$ is well-defined and hence the generalized expectation

$$E[G\cdot T\circ F] = {}_{D_{-\infty}}\langle 1, G\cdot T\circ F\rangle_{D_\infty} = {}_{\tilde{D}_{-\infty}}\langle G, T\circ F\rangle_{\tilde{D}_\infty}$$

is well-defined.

Example 9.1. Let $r = 1$ and so $\boldsymbol{W} = \boldsymbol{W}_0^1$. For $w \in \boldsymbol{W}$, set

$$G(w) = \exp\{\sigma \int_0^1 w(s)^2 ds\}, \qquad \sigma \in \boldsymbol{R}.$$

Then $G(w) \in \tilde{\boldsymbol{D}}_\infty$ if and only if $\sigma < \pi^2/8$ and if $f(x)$ is a $\boldsymbol{C}^\infty$-function on $\boldsymbol{R}$ with compact support, $G(w)f(w(1)) \in \tilde{\boldsymbol{D}}^\infty$ if $\sigma < \pi^2/2$. The proof of these facts can be easily provided by the formulas (6.9) and (6.10) of Section VI–6. Note that $G(w) \in \boldsymbol{D}_\infty$ if and only if $\sigma \leqq 0$.

The above result can be applied to show the existence of smooth density of the probability law $F_*(P)$ of F, i.e. the image measure of P under F. For this we first note the following fact.

Lemma 9.1. Let $\delta_y \in \mathscr{S}'(\boldsymbol{R}^d)$ be the Dirac δ-function at $y \in \boldsymbol{R}^d$ and m be a positive integer. Then

$$\delta_y \in \mathscr{S}_{-2m} \quad \text{and} \quad D_\alpha \delta_y \in \mathscr{S}_{-2m-2k} \qquad \text{if } m > \frac{d}{2} \text{ and } |\alpha| \leqq 2k,$$

where $\alpha = (\alpha_1, \alpha_2, \cdots, \alpha_d)$, $\alpha_i \in \boldsymbol{Z}^+$, is a multi-index, $|\alpha| = \alpha_1 + \alpha_2 + \cdots + \alpha_d$ and D_α is the differential operator:

$$D_\alpha = \left(\frac{\partial}{\partial x^1}\right)^{\alpha_1}\left(\frac{\partial}{\partial x^2}\right)^{\alpha_2}\cdots\left(\frac{\partial}{\partial x^d}\right)^{\alpha_d}.$$

Furthermore if $m > d/2$, then the mapping:

$$\boldsymbol{R}^d \ni y \to \delta_y \in \mathscr{S}_{-2m-2k}$$

is $\boldsymbol{C}^{2k}$, i.e. $2k$-times continuously differentiable.

Proof. First we note that the operator $A =: (1 + |x|^2 - \Delta/2)^{-1}$ is defined by a kernel $A(x, y)$:

$$Af(x) = \int_{\boldsymbol{R}^d} A(x, y) f(y) dy$$

and $A(x, y)$ is given by

$$A(x, y) = \int_0^\infty e^{-t} p_1(t, x, y) dt$$

where $p_1(t, x, y)$ is defined in (6.12) of Section VI-6 (in the case $d=2$, but similarly in general). From this, we see that $A(x, y) \geqq 0$, $A(x, y)$ is $\boldsymbol{C}^\infty$ in $(x, y) \in (\boldsymbol{R}^d \times \boldsymbol{R}^d) \backslash \{(x, y);\ x = y\}$, and

$$A(x, y) \sim \text{const.} \times |x - y|^{-(d-2)} \qquad \text{as } |x - y| \downarrow 0.$$

Also it is immediately seen that

$$A(x, y) \leqq \tilde{A}(x, y), \qquad x, y \in \boldsymbol{R}^d$$

where $\tilde{A}(x, y)$ is the kernel of the operator $\tilde{A} =: (1 - \Delta/2)^{-1}$. Hence

$$\begin{aligned}
&((1 + |x|^2 - \Delta/2)^{-m} \delta_y)(x) \\
&= \int_{\boldsymbol{R}^d} \int_{\boldsymbol{R}^d} \cdots \int_{\boldsymbol{R}^d} A(x, y_1) A(y_1, y_2) \cdots A(y_{m-1}, y) dy_1 dy_2 \cdots dy_{m-1} \\
&\leqq \int_{\boldsymbol{R}^d} \int_{\boldsymbol{R}^d} \cdots \int_{\boldsymbol{R}^d} \tilde{A}(x, y_1) \tilde{A}(y_1, y_2) \cdots \tilde{A}(y_{m-1}, y) dy_1 dy_2 \cdots dy_{m-1} \\
&= \left(\frac{1}{2\pi}\right)^d \int_{\boldsymbol{R}^d} \frac{e^{\sqrt{-1}\langle \xi,\, x-y \rangle}}{\left(1 + \frac{|\xi|^2}{2}\right)^m} d\xi
\end{aligned}$$

and the right-hand side is in $\mathscr{S}_0$ as a function of x if $m > \frac{d}{2}$. From this we can easily deduce that

$$(1 + |x|^2 - \Delta/2)^{-m}\delta_y \in \mathscr{S}_0 \qquad \text{for each } y$$

and the mapping $\boldsymbol{R}^d \ni y \to (1 + |x|^2 - \Delta/2)^{-m}\delta_y \in \mathscr{S}_0$ is continuous provided $m > d/2$. This proves the first assertion and the second assertion can be proved similarly. The last assertion is obvious if we notice that

$$D_\alpha \delta_y = (-1)^{|\alpha|}(D_y)_\alpha \delta_y$$

where $(D_y)_\alpha$ is the differentiation with respect to the parameter y.

Theorem 9.2. Let F: $\boldsymbol{W} \to \boldsymbol{R}^d$ satisfy the conditions (A.1) and (A.2) and let $F_*(P)$ be the probability law of F. Then $F_*(P)$ has the smooth density $p_F(y)$ with respect to the Lebesgue measure dy on $\boldsymbol{R}^d$ and $p_F(y)$ can be given by

$$p_F(y) = E[\delta_y(F)] \tag{9.15}$$

as generalized expectation of $\delta_y(F)$. More generally, for every multi-index α,

$$(D_\alpha p_F)(y) = E[((D_y)_\alpha \delta_y)(F)] = E[(-1)^{|\alpha|}(D_\alpha \delta_y)(F)]. \tag{9.16}$$

Proof. Let k be a non-negative integer. By the lemma above and the continuity of the mapping: $\mathscr{S}_{-2m} \ni T \to T \circ F \in \boldsymbol{D}_{p,-2m}$, we can deduce at once that the mapping:

$$\boldsymbol{R}^d \ni y \to \delta_y(F) \in \boldsymbol{D}_{p,-2m_0-2k}$$

is $\boldsymbol{C}^{2k}$ where $m_0 = [d/2] + 1$ and hence, for every $g \in \boldsymbol{D}_{q,2m_0+2k}$, $1/p + 1/q = 1$, the mapping: $\boldsymbol{R}^d \ni y \to E[g\delta_y(F)]$ is $\boldsymbol{C}^{2k}$. In particular, the mapping: $\boldsymbol{R}^d \ni y \to E[g\delta_y(F)] \in \boldsymbol{R}$ is $\boldsymbol{C}^\infty$ if $g \in \tilde{\boldsymbol{D}}_\infty$. Thus the mapping: $\boldsymbol{R}^d \ni y \to E[\delta_y(F)] \in \boldsymbol{R}$ is $\boldsymbol{C}^\infty$. (9.15) can be verified by noting that for every $\phi \in \mathscr{S}(\boldsymbol{R}^d)$

$$\begin{aligned}\int_{\boldsymbol{R}^d} \phi(y) E[\delta_y(F)]dy &= E[\int_{\boldsymbol{R}^d} \phi(y)\delta_y(F)dy] \\ &= E[(\int_{\boldsymbol{R}^d} \phi(y)\delta_y(\,\cdot\,)dy) \circ F] \\ &= E[\phi \circ F].\end{aligned}$$

Now (9.16) is obvious.

In the above proof, it was shown that if $g \in \bigcup_{p>1} \boldsymbol{D}_{p,\,2m_0+2k}$ where m_0

$= [d/2] + 1$ and k is a non-negative integer, then the mapping: $\boldsymbol{R}^d \ni y \to E[g\delta_y(F)]$ is $\boldsymbol{C}^{2k}$. It is easy to see that

$$E[g\delta_y(F)] = p_F(y)E[g \mid F = y] \tag{9.17}$$

for almost all y such that $p_F(y) > 0$. Thus we have obtained a regularity result for conditional expectation:

Corollary. If $g \in \bigcup_{p>1} \boldsymbol{D}_{p,2m_0+2k}$ where $m_0 = [d/2] + 1$ and k is a non-negative integer, the conditional expectation $E[g \mid F = y]$ has a $\boldsymbol{C}^{2k}$-version on the set $\{y; p_F(y) > 0\}$. In particular, it has a $\boldsymbol{C}^\infty$-version on the same set provided $g \in \tilde{\boldsymbol{D}}_\infty$.

Example 9.2. Let $r = d$ and F: $\boldsymbol{W}(= \boldsymbol{W}_0^d) \to \boldsymbol{R}^d$ be defined, for fixed $t > 0$ and $x \in \boldsymbol{R}^d$, by $F(w) = x + w(t)$. Then $F \in \boldsymbol{D}_\infty(\boldsymbol{R}^d)$ (actually $\in \boldsymbol{P}(\boldsymbol{R}^d)$) and $\sigma^{ij}(w) = \delta^{ij}t$, $i, j = 1, 2, \cdots, d$. Hence $\gamma_{ij} = \delta_{ij}/t$ where δ_{ij} is Kronecker's delta and hence (A.1) and (A.2) are clearly satisfied. Let $k \in \boldsymbol{C}^{2m}(\boldsymbol{R}^d)$ which is tempered in the sense $D^\alpha k$, $|\alpha| \leqq 2m$, are all polynomial growth order and

$$\varlimsup_{|x|\to\infty} k(x)/|x|^2 = \alpha < \infty.$$

If α is sufficiently small (actually it is sufficient to assume $\alpha < 1/2t$), we can easily see that

$$g(w) = \exp\{\int_0^t k(x + w(s))ds\} \in \bigcup_{p>1} \boldsymbol{D}_{p,2m}.$$

Hence

$$p(t, x, y) := p_F(y) = E[(\exp\{\int_0^t k(x + w(s))ds\})\delta_y(x + w(t))]$$

is $\boldsymbol{C}^{2k}$ in y if $k = m - m_0 \geqq 0$. By the Feynman-Kac formula (Theorem 3.2), it is easy to identify $p(t, x, y)$ with the fundamental solution of heat equation

$$\frac{\partial u}{\partial t} = \frac{1}{2}\Delta u + ku.$$

The generalized Wiener functional $\delta_y(x + w(t))$ is known as Donsker's δ-function (cf. [210]).

Now we consider a family $\{F_\alpha(w)\}_{\alpha\in I}$ of d-dimensional Wiener functionals depending on a parameter α, the parameter set being assumed to be an m-dimensional interval $I \subset \boldsymbol{R}^m$, for simplicity. We assume that $F_\alpha \in \boldsymbol{D}_\infty(\boldsymbol{R}^d)$ for all $\alpha \in I$ and also that, denoting by $\sigma(\alpha) = (\sigma^{ij}(\alpha))$, $\sigma^{ij}(\alpha) = \langle DF^i_\alpha, DF^j_\alpha\rangle$ the Malliavin covariance of F_α,

$$\sup_{\alpha\in I}\|\det \sigma(\alpha)^{-1}\|_p < \infty \qquad \text{for } 1 < p < \infty.$$

Then it is clear that the constant $C_\alpha = C(p, k;\ F_\alpha)$ in Theorem 9.1 can be chosen such that

$$\sup_{\alpha\in I} C_\alpha < \infty.$$

Hence, for $T \in \mathscr{S}_{-2k}$, if we choose $\phi_n \in \mathscr{S}(\boldsymbol{R}^d)$ such that $\phi_n \to T$ in $\mathscr{S}_{-2k}$ as $n \to \infty$, then for every $1 < p < \infty$, $\phi_n \circ F_\alpha \to T\circ F_\alpha$ in $\boldsymbol{D}_{p,-2k}$ as $n \to \infty$ *uniformly in* $\alpha \in I$. From this we can easily conclude the following: If the mapping: $I \ni \alpha \to F_\alpha \in \boldsymbol{D}_\infty$ is continuous ($2m$-times differentiable) then, for $T \in \mathscr{S}_{-2k}$, the mapping: $I \ni \alpha \to T\circ F_\alpha \in \boldsymbol{D}_{p,-2k}$ is continuous for every $1 < p < \infty$ (resp. the mapping: $I \ni \alpha \to T\circ F_\alpha \in \boldsymbol{D}_{p,-2k-2m}$ is $2m$-times differentiable for every $1 < p < \infty$). We can deduce from these facts the continuity or differentiability in α of the generalized expectation $E[g_\alpha \cdot T\circ F_\alpha]$ if the mapping: $I \ni \alpha \to g_\alpha \in \boldsymbol{D}_{q,2k}$ is continuous or differentiable where $1/p + 1/q = 1$. Note that, in the differentiable case, we have

$$\frac{\partial}{\partial\alpha_\nu}(T\circ F_\alpha) = \sum_{i=1}^d \frac{\partial F^i_\alpha}{\partial\alpha_\nu}\frac{\partial T}{\partial x^i}\circ F_\alpha, \qquad F_\alpha = (F^1_\alpha, F^2_\alpha, \cdots, F^d_\alpha)$$

which is easily verified by a limiting argument as $\phi_n \to T$, $\phi_n \in \mathscr{S}(\boldsymbol{R}^d)$.

Finally, as an example of dependence on parameters, we discuss on the *asymptotic expansion of Wiener functionals.* Consider a family $\{F(\varepsilon, w)\}$ of E-valued Wiener functional depending on a parameter $\varepsilon \in (0, 1]$ where E is a separable Hilbert space. We assume

$$F(\varepsilon, w) \in \boldsymbol{D}_\infty(E) \qquad \text{for every } \varepsilon.$$

We define $F(\varepsilon, w) = O(\varepsilon^m)$ in $\boldsymbol{D}_\infty(E)$ as $\varepsilon \downarrow 0$ if $F(\varepsilon, w) = O(\varepsilon^m)$ in $\boldsymbol{D}_{p,k}(E)$ for every $1 < p < \infty$ and $k \in N$ as $\varepsilon \downarrow 0$, i.e.

$$\overline{\lim_{\varepsilon\downarrow 0}}\, \varepsilon^{-m}\|F(\varepsilon, w)\|_{p,k} < \infty.$$

Let $f_0, f_1, \cdots \in \boldsymbol{D}_\infty(E)$. We define

$$(9.17) \qquad F(\varepsilon, w) \sim f_0 + \varepsilon f_1 + \varepsilon^2 f_2 + \cdots \quad \text{in } \boldsymbol{D}_\infty(E) \text{ as } \varepsilon \downarrow 0$$
if, for every $n = 0, 1, 2, \cdots$,

$$F(\varepsilon, w) - (f_0 + \varepsilon f_1 + \cdots + \varepsilon^n f_n) = O(\varepsilon^{n+1}) \quad \text{in } \boldsymbol{D}_\infty(E) \quad \text{as } \varepsilon \downarrow 0.$$

Also, if $\{\Phi(\varepsilon, w)\}$ is a family of elements in $\boldsymbol{D}_{-\infty}$ depending on $\varepsilon \in (0, 1]$ and $\psi_0, \psi_1, \cdots \in \boldsymbol{D}_{-\infty}(E)$, we define

$$(9.18) \qquad \Phi(\varepsilon, w) \sim \psi_0 + \varepsilon\psi_1 + \varepsilon^2\psi_2 + \cdots \quad \text{in } \boldsymbol{D}_{-\infty}(E) \quad \text{as } \varepsilon \downarrow 0$$

if, for every $n = 0, 1, 2, \cdots$, we can find $k \in \boldsymbol{N}$ and $1 < p < \infty$ such that

$$\Phi(\varepsilon, w),\ \varepsilon \in (0, 1] \quad \text{and} \quad \psi_0, \psi_1, \cdots, \psi_n \text{ are all in } \boldsymbol{D}_{p,-k}(E)$$

and

$$\Phi(\varepsilon, w) - (\psi_0 + \varepsilon\psi_1 + \cdots + \varepsilon^n\psi_n) = O(\varepsilon^{n+1}) \text{ in } \boldsymbol{D}_{p,-k}(E) \quad \text{as } \varepsilon \downarrow 0.$$

We define $\tilde{\boldsymbol{D}}_\infty(E)$ and $\tilde{\boldsymbol{D}}_{-\infty}(E)$ for general E in the same way as (9.13) and (9.14), i.e.

$$\tilde{\boldsymbol{D}}_\infty(E) = \bigcap_{k=1}^{\infty} \bigcup_{1<p<\infty} \boldsymbol{D}_{p,k}(E)$$

and

$$\tilde{\boldsymbol{D}}_{-\infty}(E) = \bigcup_{k=1}^{\infty} \bigcap_{1<p<\infty} \boldsymbol{D}_{p,-k}(E).$$

If $\{G(\varepsilon, w)\}$ is a family of elements in $\tilde{\boldsymbol{D}}_\infty(E)$ depending on $\varepsilon \in (0.1]$ and $g_0, g_1, \cdots \in \tilde{\boldsymbol{D}}_\infty(E)$, we define

$$(9.19) \qquad G(\varepsilon, w) \sim g_0 + \varepsilon g_1 + \varepsilon^2 g_2 + \cdots \quad \text{in } \tilde{\boldsymbol{D}}_\infty(E) \quad \text{as } \varepsilon \downarrow 0$$

if, for every $n = 0, 1, \cdots$ and $k \in \boldsymbol{N}$, we can find $1 < p < \infty$ such that

$$G(\varepsilon, w),\ \varepsilon \in (0, 1] \quad \text{and} \quad g_0, g_1, \cdots, g_n \text{ are all in } \boldsymbol{D}_{p,k}(E)$$

and

$$G(\varepsilon, w) - (g_0 + \varepsilon g_1 + \cdots + \varepsilon^n g_n) = O(\varepsilon^{n+1}) \quad \text{in } \boldsymbol{D}_{p,k}(E) \quad \text{as } \varepsilon \downarrow 0.$$

If $\{\Psi(\varepsilon, w)\}$ is a family of elements in $\tilde{\boldsymbol{D}}_{-\infty}(E)$ depending on $\varepsilon \in (0,1]$ and if $\psi_0, \psi_1, \cdots \in \tilde{\boldsymbol{D}}_{-\infty}(E)$, we define

$$\Psi(\varepsilon, w) \sim \psi_0 + \varepsilon\psi_1 + \cdots \quad \text{in } \tilde{\boldsymbol{D}}_{-\infty}(E) \quad \text{as } \varepsilon \downarrow 0 \tag{9.20}$$

if, for every $n = 0, 1, 2, \cdots$, we can find $k \in \boldsymbol{N}$ such that

$$\Psi(\varepsilon, w),\ \varepsilon \in (0,1] \quad \text{and} \quad \psi_0, \psi_1, \cdots, \psi_n \text{ are all in } \bigcap_{1<p<\infty} \boldsymbol{D}_{p,-k}(E)$$

and, for every $1 < p < \infty$,

$$\Psi(\varepsilon, w) - (\psi_0 + \varepsilon\psi_1 + \cdots + \varepsilon^n\psi_n) = O(\varepsilon^{n+1}) \quad \text{in } \boldsymbol{D}_{p,-k}(E) \text{ as } \varepsilon \downarrow 0.$$

Noting the continuity of multiplications in Sobolev spaces as given by inequalities (8.46) and (8.47), it is easy to prove the following:

Proposition 9.3. (i) If $F(\varepsilon, w) \in \boldsymbol{D}_\infty(E)$ and $G(\varepsilon, w) \in \boldsymbol{D}_\infty$ $(\tilde{\boldsymbol{D}}_\infty)$, $\varepsilon \in (0,1]$ such that

$$F(\varepsilon, w) \sim f_0 + \varepsilon f_1 + \cdots \quad \text{in } \boldsymbol{D}_\infty(E) \quad \text{as } \varepsilon \downarrow 0 \tag{9.21}$$

with $f_i \in \boldsymbol{D}_\infty(E)$, $i = 0, 1, \cdots$ and

$$G(\varepsilon, w) \sim g_0 + \varepsilon g_1 + \cdots \quad \text{in } \boldsymbol{D}_\infty \text{ (resp. } \tilde{\boldsymbol{D}}_\infty) \quad \text{as } \varepsilon \downarrow 0 \tag{9.22}$$

with $g_i \in \boldsymbol{D}_\infty$ (resp. $\tilde{\boldsymbol{D}}_\infty$) $i = 0, 1, \cdots$, then $H(\varepsilon, w) = G(\varepsilon, w)F(\varepsilon, w)$ satisfies

$$H(\varepsilon, w) \sim h_0 + \varepsilon h_1 + \cdots \quad \text{in } \boldsymbol{D}_\infty(E) \text{ (resp. } \tilde{\boldsymbol{D}}_\infty(E)) \quad \text{as } \varepsilon \downarrow 0 \tag{9.23}$$

and $h_0, h_1, \cdots \in \boldsymbol{D}_\infty(E)$ (resp. $\tilde{\boldsymbol{D}}_\infty(E)$) are obtained by the formal multiplication:

$$h_0 = g_0 f_0,\ h_1 = g_0 f_1 + g_1 f_0,\ h_2 = g_0 f_2 + g_1 f_1 + g_2 f_0, \cdots. \tag{9.24}$$

(ii) If $G(\varepsilon, w) \in \boldsymbol{D}_\infty(\tilde{\boldsymbol{D}}_\infty)$ and $\Phi(\varepsilon, w) \in \tilde{\boldsymbol{D}}_{-\infty}(E)$, $\varepsilon \in (0, 1]$ such that

$$G(\varepsilon, w) \sim g_0 + \varepsilon g_1 + \cdots \quad \text{in } \boldsymbol{D}_\infty \text{ (resp. } \tilde{\boldsymbol{D}}_\infty) \quad \text{as } \varepsilon \downarrow 0 \tag{9.25}$$

with $g_i \in \boldsymbol{D}_\infty$ (resp. $\tilde{\boldsymbol{D}}_\infty$), $i = 0, 1, \cdots$ and

(9.26) $\Phi(\varepsilon, w) \sim \phi_0 + \varepsilon\phi_1 + \cdots$ in $\tilde{\boldsymbol{D}}_{-\infty}(E)$ as $\varepsilon \downarrow 0$

with $\phi_i \in \tilde{\boldsymbol{D}}_{-\infty}(E)$, $i = 0, 1, \cdots$, then $\Psi(\varepsilon, w) = G(\varepsilon, w)\,\Phi(\varepsilon, w)$ satisfies

(9.27) $\Psi(\varepsilon, w) \sim \psi_0 + \varepsilon\psi_1 + \cdots$ in $\tilde{\boldsymbol{D}}_{-\infty}(E)$ (resp. $\boldsymbol{D}_{-\infty}(E)$) as $\varepsilon \downarrow 0$

and $\psi_0, \psi_1, \cdots \in \tilde{\boldsymbol{D}}_{-\infty}(E)$ (resp. $\boldsymbol{D}_{-\infty}(E)$) are obtained by the formal multiplication:

(9.28) $\psi_0 = g_0\phi_0,\ \psi_1 = g_0\phi_1 + g_1\phi_0,\ \psi_2 = g_0\phi_2 + g_1\phi_1 + g_2\phi_0, \cdots.$

(iii) If $G(\varepsilon, w) \in \boldsymbol{D}_\infty$ and $\Phi(\varepsilon, w) \in \boldsymbol{D}_{-\infty}(E)$, $\varepsilon \in (0, 1]$ such that

(9.29) $G(\varepsilon, w) \sim g_0 + \varepsilon g_1 + \cdots$ in $\boldsymbol{D}_\infty$ as $\varepsilon \downarrow 0$

with $g_i \in \boldsymbol{D}_\infty$, $i = 0, 1, \cdots$ and

(9.30) $\Phi(\varepsilon, w) \sim \phi_0 + \varepsilon\phi_1 + \cdots$ in $\boldsymbol{D}_{-\infty}(E)$ as $\varepsilon \downarrow 0$

with $\phi_i \in \boldsymbol{D}_{-\infty}(E)$, $i = 0, 1, \cdots$, then (9.27) holds in $\boldsymbol{D}_{-\infty}(E)$ with $\psi_i \in \boldsymbol{D}_{-\infty}(E)$, $i = 0, 1, \cdots$ given by (9.28).

Theorem 9.4. Let $\{F(\varepsilon, w),\ \varepsilon \in (0,1]\}$ be a family of elements in $\boldsymbol{D}_\infty(\boldsymbol{R}^d)$ such that it has the asymptotic expansion:

(9.31) $F(\varepsilon, w) \sim f_0 + \varepsilon f_1 + \cdots$ in $\boldsymbol{D}_\infty(\boldsymbol{R}^d)$ as $\varepsilon \downarrow 0$

with $f_i \in \boldsymbol{D}_\infty(\boldsymbol{R}^d)$, $i = 0, 1, \cdots$ and satisfies

(9.32) $\overline{\lim_{\varepsilon \downarrow 0}} \|(\det \sigma(\varepsilon))^{-1}\|_p < \infty$ for all $1 < p < \infty$

where $\sigma(\varepsilon) = (\sigma^{ij}(\varepsilon))$ is the Malliavin covariance of $F(\varepsilon, w)$: $\sigma^{ij}(\varepsilon) = \langle DF^i(\varepsilon, w), DF^j(\varepsilon, w)\rangle_H$. Let $T \in \mathscr{S}'(\boldsymbol{R}^d)$. Then $\Phi(\varepsilon, w) = T \circ F(\varepsilon, w)$ has the asymptotic expansion in $\tilde{\boldsymbol{D}}_{-\infty}$ (and a fortiori in $\boldsymbol{D}_{-\infty}$):

(9.33) $\Phi(\varepsilon, w) \sim \phi_0 + \varepsilon\phi_1 + \cdots$ in $\tilde{\boldsymbol{D}}_{-\infty}$ as $\varepsilon \downarrow 0$

and $\phi_i \in \tilde{\boldsymbol{D}}_{-\infty}$, $i = 0, 1, \cdots$ are determined by the formal Taylor expansion:

$$\begin{aligned} \Phi(\varepsilon, w) &= T(f_0 + [\varepsilon f_1 + \varepsilon^2 f_2 + \cdots]) \\ &= \sum_\alpha \frac{1}{\alpha!}(D_\alpha T) \circ f_0[\varepsilon f_1 + \varepsilon^2 f_2 + \cdots]^\alpha \\ &= \phi_0 + \varepsilon\phi_1 + \cdots \end{aligned} \tag{9.34}$$

where (i) the summation is taken over all multi-indices and (ii) for every multi-index $\alpha = (\alpha_1, \alpha_2, \cdots, \alpha_d)$ and $a = (a_1, a_2, \cdots, a_d) \in \boldsymbol{R}^d$, we set as usual

$$\alpha! = \alpha_1!\alpha_2! \cdots \alpha_d! \quad \text{and} \quad a^\alpha = a_1^{\alpha_1}a_2^{\alpha_2} \cdots a_d^{\alpha_d}.$$

In particular, denoting $\partial_i = \partial/\partial x^i$, $i = 1, 2, \cdots, d$,

$$\begin{aligned} &\phi_0 = T(f_0), \quad \phi_1 = \sum_{i=1}^d f_1^i(\partial_i T)(f_0) \\ &\phi_2 = \sum_{i=1}^d f_2^i(\partial_i T)(f_0) + \frac{1}{2!}\sum_{i,j=1}^d f_1^i f_1^j(\partial_i\partial_j T)(f_0) \\ &\phi_3 = \sum_{i=1}^d f_3^i(\partial_i T)(f_0) + \frac{2}{2!}\sum_{i,j=1}^d f_1^i f_2^j(\partial_i\partial_j T)(f_0) \\ &\qquad + \frac{1}{3!}\sum_{i,j,k=1}^d f_1^i f_1^j f_1^k(\partial_i\partial_j\partial_k T)(f_0) \end{aligned} \tag{9.35}$$

and etc. Here

$$f_k = (f_k^1, f_k^2, \cdots, f_k^d) \ (\in \boldsymbol{D}_\infty(\boldsymbol{R}^d)), \quad k = 0, 1, 2, \cdots.$$

Remark. $f_0 \in \boldsymbol{D}_\infty(\boldsymbol{R}^d)$ satisfies the condition (A.2) by virtue of (9.32).

Proof. We set $F(0, w) = f_0$, $\sigma^{ij}(0) = \langle Df_0^i, Df_0^j\rangle_H$, $\gamma(\varepsilon) = (\gamma_{ij}(\varepsilon))$: $= \sigma^{-1}(\varepsilon)$, $\varepsilon \in [0,1]$. Then it is easy to see from (9.32) that $\gamma_{ij}(\varepsilon)$ has the asymptotic expansion:

$$\gamma_{ij}(\varepsilon) \sim \gamma_{ij}(0) + \varepsilon\gamma_{ij}^{(1)} + \varepsilon^2\gamma_{ij}^{(2)} + \cdots.$$

with $\gamma_{ij}^{(k)} \in \boldsymbol{D}^\infty$, $i, j = 1, 2, \cdots, d$, $k = 1, 2, \cdots$.

First, we show that, for every $k \in \boldsymbol{N}$, we can find $s \in \boldsymbol{R}$ and ϕ_0, $\phi_1, \cdots, \phi_{k-1} \in \bigcap_{1<p<\infty} \boldsymbol{D}_{p,s}$ such that

$$T \circ F(\varepsilon, w) \in \bigcap_{1<p<\infty} \boldsymbol{D}_{p,s} \qquad \text{for all } \varepsilon \in [0, 1]$$

and

$$T\circ F(\varepsilon, w)=\phi_0+\varepsilon\phi_1+\cdots+\varepsilon^{k-1}\phi_{k-1}+O(\varepsilon^k)\quad\text{in } \boldsymbol{D}_{p,s}\quad\text{as } \varepsilon\downarrow 0 \tag{9.36}$$

for all $1<p<\infty$. To prove this, we take $m\in\boldsymbol{N}$ so large that $\phi=(1+|x|^2-\Delta/2)^{-m}T$ is a bounded function on $\boldsymbol{R}^d$, k-times continuously differentiable with bounded derivatives up to k-th order. Then, for every $J\in\boldsymbol{D}_\infty$, we have, by the same integration by parts as (9.8),

$$E[T\circ F(\varepsilon, w)\cdot J]=E[\phi\circ F(\varepsilon, w)\cdot l_\varepsilon(J)]$$

where $l_\varepsilon(J)\in\boldsymbol{D}_\infty$ has a similar expression as (9.7):

$$l_\varepsilon(J)=\sum_{i=0}^{2m}<P_i(\varepsilon, w), D^iJ>_{H^{\otimes i}}$$

where $P_i(\varepsilon, w)\in\boldsymbol{D}^\infty(H^{\otimes i})$, $i=0, 1, \cdots, 2m$, which are polynomials in $F(\varepsilon, w)$, its derivatives and $\gamma(\varepsilon)$. By the assumption on ϕ, we have

$$\phi\circ F(\varepsilon, w)=\sum_{|\alpha|\leqq k-1}\frac{1}{\alpha!}(D_\alpha\phi)\circ f_0\cdot[F(\varepsilon, w)-f_0]^\alpha+V_k(\varepsilon, w)$$

where for some constant $M>0$,

$$|V_k(\varepsilon, w)|\leqq M\sum_{|\alpha|=k}|[F(\varepsilon, w)-f_0]^\alpha|.$$

Hence, for every $p'\in(1,\infty)$, we can find $C_1>0$ such that

$$\|V_k(\varepsilon, w)\|_{p'}\leqq C_1\varepsilon^k\qquad\text{for all } \varepsilon\in[0, 1].$$

Let $q'\in(1,\infty)$ such that $1/p'+1/q'=1$ and choose q such that $q>q'$. Then a positive constant C_2 exists such that

$$\|l_\varepsilon(J)\|_{q'}\leqq C_2\|J\|_{q,2m}\qquad\text{for all } \varepsilon\in[0, 1]\quad\text{and}\quad J\in\boldsymbol{D}_\infty.$$

Hence, for all $\varepsilon\in[0, 1]$ and $J\in\boldsymbol{D}_\infty$,

$$|E[V_k(\varepsilon, w)l_\varepsilon(J)]|\leqq\|V_k(\varepsilon, w)\|_{p'}\|l_\varepsilon(J)\|_{q'}\leqq C_1C_2\|J\|_{q,2m}\varepsilon^k.$$

Also

$$\begin{aligned}&\sum_{|\alpha|\leqq k-1}\frac{1}{\alpha!}(D_\alpha\phi)\circ f_0\cdot[F(\varepsilon, w)-f_0]^\alpha l_\varepsilon(J)\\ &=\sum_{i=0}^{2m}\sum_{|\alpha|\leqq k-1}<\frac{1}{\alpha!}(D_\alpha\phi)\circ f_0\cdot[F(\varepsilon, w)-f_0]^\alpha P_i(\varepsilon, w), D^iJ>_{H^{\otimes i}}\end{aligned} \tag{9.37}$$

and since $[F(\varepsilon, w) - f_0]^\alpha P_\iota(\varepsilon, w)$ has the asymptotic expansion

$$[F(\varepsilon, w) - f_0]^\alpha P_\iota(\varepsilon, w) \sim \varepsilon^{|\alpha|} e_{\alpha,\iota,0}(w) + \varepsilon^{|\alpha|+1} e_{\alpha,\iota,1}(w) + \cdots \quad \text{in } \boldsymbol{D}^{\infty}(H^{\otimes \iota}) \quad \text{as } \varepsilon \downarrow 0,$$

we have

$$(9.37) = Z_0(w) + \varepsilon Z_1(w) + \cdots + \varepsilon^{k-1} Z_{k-1}(w) + U_k(\varepsilon, w)$$

where

$$Z_l(w) = \sum_{\iota=0}^{2m} \sum_{|\alpha|+\nu=l} <\frac{1}{\alpha!}(D_\alpha \phi) \circ f_0 \cdot e_{\alpha,\iota,\nu},\ D^\iota J>_{H^{\otimes \iota}}.$$

We can easily find $C_3 > 0$ such that

$$|E[U_k(\varepsilon, w)]| \leqq C_3 \varepsilon^k \|J\|_{q,2m}$$

for all $\varepsilon \in [0, 1]$ and $J \in \boldsymbol{D}_\infty$. Set

$$\phi_l = \sum_{\iota=0}^{2m} \sum_{|\alpha|+\nu=l} (D^*)^\iota \left(\frac{1}{\alpha!}(D_\alpha \phi) \circ f_0 \cdot e_{\alpha,\iota,\nu}\right), \quad l = 0, 1, \cdots, k-1.$$

It is easy to see that

$$\phi_l \in \bigcap_{1<p<\infty} \boldsymbol{D}_{p,-2m}$$

and this, combined with the above, yields

$$\left|E[T \circ F(\varepsilon, w) \cdot J] - \sum_{l=0}^{k-1} \varepsilon^l E[\phi_l J]\right| \leqq (C_1 C_2 + C_3) \varepsilon^k \|J\|_{q,2m}$$

for all $\varepsilon \in [0,1]$ and $J \in \boldsymbol{D}_\infty$. Hence we have

$$\left\|T \circ F(\varepsilon, w) - \sum_{l=0}^{k-1} \varepsilon^l \phi_l\right\|_{p,-2m} \leqq (C_1 C_2 + C_3) \varepsilon^k$$

where $1/p + 1/q = 1$. It is clear in the above argument that $p \in (1, \infty)$ can be chosen arbitrarily and thus (9.36) is obtained.

It remains only to show that ϕ_l can be determined through the expression (9.34). Since

$$\phi_l = \sum_{i=0}^{2m} \sum_{|\alpha|+\nu=l} (D^*)^i\Big(\frac{1}{\alpha!}(D_\alpha\{(1 + |x|^2 - \Delta/2)^{-m}T\}) \circ f_0 \cdot e_{\alpha,i,\nu}\Big)$$

we see that the mapping $\mathscr{S}'(\boldsymbol{R}^d) \ni T \to \phi_l \in \tilde{\boldsymbol{D}}_{-\infty}$ is continuous in the sense that for every $n \in \boldsymbol{N}$, there exist $s > 0$, $p \in (1, \infty)$ and C such that

$$\|\phi_l\|_{p,-s} \leqq C\|T\|_{-2n}.$$

On the other hand, it is clear that ϕ_l is uniquely determined from T and must be calculated from the expression (9.34) in the case $T \in \mathscr{S}(\boldsymbol{R}^d)$. Hence, it must also be given by (9.34) for general $T \in \mathscr{S}'(\boldsymbol{R}^d)$. This completes the proof.

10. The case of stochastic differential equations: Applications to heat kernels

Let $X = (X(t, x, w))$ be the solution of a stochastic differential equation on $\boldsymbol{R}^d$ given in the same way as in Section 2:

$$\text{(2.1)} \qquad \begin{cases} dX(t) = \sigma_\alpha(X(t))dw^\alpha(t) + b(X(t))dt \\ X(0) = x \end{cases}$$

which is realized on the r-dimensional Wiener space $(\boldsymbol{W}, P)$. We assume, as before, that all coefficients $\sigma_\alpha(x) = (\sigma_\alpha^i(x))$ and $b(x) = (b^i(x))$ are $\boldsymbol{C}^\infty$ with bounded derivatives of all orders $\geqq 1$. We know from the results of Section 2 that, for almost all $w \in \boldsymbol{W}(P)$,

(i) $(t, x) \to X(t, x, w)$ is continuous,

(ii) $x \to X(t, x, w)$ is a diffeomorphism of $\boldsymbol{R}^d$.

If $Y(t) = (Y_j^i(t))$ is defined by

$$Y_j^i(t) = \frac{\partial}{\partial x^j} X^i(t, x, w),$$

then $Y(t)$ satisfies

$$\text{(2.16)} \qquad \begin{cases} dY(t) = \sigma'_\alpha(X(t))Y(t)dw^\alpha(t) + b'(X(t))Y(t)dt \\ Y(0) = I \end{cases}$$

where

$$\sigma'_\alpha(x)^i_j = \frac{\partial}{\partial x^j}\sigma^i_\alpha(x) \quad \text{and} \quad b'(x)^i_j = \frac{\partial}{\partial x^j} b^i(x)$$

and $I = (\delta^i_j)$ is the $d \times d$ identity matrix. The equations (2.1) and (2.16), put together, define a stochastic differential equation for the process $r(t) = (X(t), Y(t))$ and $r(t)$ takes values in $\boldsymbol{R}^d \times GL(d, \boldsymbol{R})$ which can be identified with *the bundle of linear frames* $GL(\boldsymbol{R}^d)$ over $\boldsymbol{R}^d$. If we introduce a system of vector fields V_α, $\alpha = 0, 1, 2, \cdots, r$ by

$$V_\alpha(x) = \sigma^i_\alpha(x)\frac{\partial}{\partial x^i}, \qquad \alpha = 1, 2, \cdots, r$$

and

$$V_0(x) = [b^i(x) - \frac{1}{2}\sum_{\alpha=1}^r \sigma^j_\alpha(x)^i_j \sigma^j_\alpha(x)]\frac{\partial}{\partial x^i}$$

then (2.1) and (2.16) can be rewritten as

$$(10.1) \qquad \begin{cases} dX(t) = \sum_{\alpha=1}^r V_\alpha(X(t)) \circ dw^\alpha(t) + V_0(X(t))dt \\ X(0) = x \end{cases}$$

and

$$(10.2) \qquad \begin{cases} dY(t) = \sum_{\alpha=1}^r \partial V_\alpha(X(t))Y(t) \circ dw^\alpha(t) + \partial V_0(X(t))Y(t)dt \\ Y(0) = I \end{cases}$$

where

$$(\partial V_\alpha(x))^i_j = \frac{\partial}{\partial x^j} V^i_\alpha(x).$$

Proposition 10.1. Let $t > 0$ and $x \in \boldsymbol{R}^d$ be fixed. Then
(i) $X(t, x, w) \in \boldsymbol{D}_\infty(\boldsymbol{R}^d)$, i.e., $X^i(t, x, w) \in \boldsymbol{D}_\infty$, $i = 1, 2, \cdots, d$,
(ii) the Malliavin covariance $\sigma(t) = (\sigma^{ij}(t))$, $\sigma^{ij}(t) = \langle DX^i(t), DX^j(t)\rangle_H$, is given by

$$(10.3) \qquad \sigma^{ij}(t) = \sum_{\alpha=1}^r \int_0^t [Y(t)Y(s)^{-1}V_\alpha(X(s))]^i [Y(t)Y(s)^{-1}V_\alpha(X(s))]^j ds. \, *$$

Proof. For $X(t) = X(t, x, w)$, we define $X^{(n)}(t) = X^{(n)}(t, x, w)$ as

* If $A = (A^i_j)$ is a $d \times d$ matrix and $x = (x^i) \in \boldsymbol{R}^d$, we denote $(Ax)^i = A^i_j x^j$, $i = 1, 2, \cdots, d$.

in Section 2:

$$X^{(n)}(t) = x + \int_0^t \sigma_\alpha(X^{(n)}(\phi_n(s))dw^\alpha(s) + \int_0^t b(X^{(n)}(\phi_n(s)))ds \tag{10.4}$$

where

$$\phi_n(s) = k/2^n \qquad \text{if } s \in [k/2^n, (k+1)/2^n), k = 0, 1, \cdots.$$

We know that $X^{(n)}(t) \in \boldsymbol{S}$ and $\|X^{(n)}(t) - X(t)\|_p \to 0$ as $n \to \infty$ for every $1 < p < \infty$. It is easy to see that $D_h X^{(n)}(t) = (D_h X^{(n),i}(t))_{i=1}^d$ defined by

$$D_h X^{(n),j}(t) = \langle DX^{(n),j}(t), h\rangle_H, \quad h \in H$$

satisfies

$$D_h X^{(n)}(t) = \int_0^t \sigma'_\alpha(X^{(n)}(\phi_n(s)))D_h X^{(n)}(\phi_n(s))dw^\alpha(s)$$
$$+ \int_0^t b'(X^{(n)}(\phi_n(s)))D_h X^{(n)}(\phi_n(s))ds + \int_0^t \sigma_\alpha(X^{(n)}(\phi_n(s)))\, \dot{h}^\alpha(s)ds$$

and hence, if we denote

$$D_h X^{(n)}(t) = \int_0^t \xi_{t,\beta}^{(n)}(v)\dot{h}^\beta(v)dv, \qquad h = (h^\beta(t)) \in H,$$

then $\xi_{t,\beta}^{(n)}(v)$, $t \geqq v$, $\beta = 1, 2, \cdots, r$, satisfies

$$\xi_{t,\beta}^{(n)}(v) = \int_{\psi_n(v)\wedge t}^t \sigma'_\alpha(X^{(n)}(\phi_n(s)))\xi_{\phi_n(s),\beta}^{(n)}(v)dw^\alpha(s)$$
$$+ \int_{\psi_n(v)\wedge t}^t b'(X^{(n)}(\phi_n(s)))\xi_{\phi_n(s),\beta}^{(n)}(v)ds + \sigma_\beta(X^{(n)}(\phi_n(v)), \tag{10.5}$$

where

$$\psi_n(v) = k/2^n \qquad \text{if } v \in ((k-1)/2^n, k/2^n], k = 0, 1, \cdots.$$

By applying a slight modification of Lemma 2.1 to (10.4) and (10.5), we have, for every $1 < p < \infty$,

$$E[\sup_{s\in[0,t]} |X^{(n)}(s) - X(s)|^p] \to 0 \tag{10.6}$$

and

$$\sup_{0 \leqq v \leqq t} E[\sup_{s \in [v,t]} |\xi^{(n)}_{s,\beta}(v) - \xi_{s,\beta}(v)|^p] \to 0 \tag{10.7}$$

as $n \to \infty$ where $\xi_{t,\beta}(v)$, $t \geqq v \geqq 0$, satisfies

$$\begin{aligned} \xi_{t,\beta}(v) &= \int_v^t \sigma'_\alpha(X(s))\xi_{s,\beta}(v)dw^\alpha(s) \\ &\quad + \int_v^t b'(X(s))\xi_{s,\beta}(v)ds + \sigma_\beta(X(v)). \end{aligned} \tag{10.8}$$

Note that $\xi_{t,\beta}(v)$ is uniquely determined by (10.8) and is given by

$$\xi_{t,\beta}(v) = Y(t)Y(v)^{-1}\sigma_\beta(X(v)), \qquad \beta = 1, 2, \cdots, r. \tag{10.9}$$

Indeed, it is immediately seen that $\xi_{t,\beta}(v)$ given by (10.9) is a solution of (10.8) and the uniqueness is obvious. This proves

$$X(t) \in \bigcap_{1<p<\infty} \boldsymbol{D}_{p,1}(\boldsymbol{R}^d) \qquad \text{for each } t \geqq 0$$

and

$$<DX^i(t), h> = \int_0^t \xi^i_{t,\beta}(v)\dot{h}^\beta(v)dv, \qquad h \in H \tag{10.10}$$

where $\xi_{t,\beta}(v) = (\xi^i_{t,\beta}(v))^d_{i=1}$. From this, (10.3) can be concluded. Repeating this argument as in the proof of Proposition 2.1, we see easily $X(t) \in \boldsymbol{D}_\infty(\boldsymbol{R}^d)$.

Next we shall investigate the assumption (A.2), namely the condition

$$(\det \sigma(t))^{-1} \in \bigcap_{1<p<\infty} \boldsymbol{L}_p \tag{10.11}$$

where $\sigma(t) = (\sigma^{ij}(t))$ is the Malliavin covariance of $X(t, x, w)$. It is easy to see that $\sigma(t)$ can be written as

$$\sigma^{ij}(t) = \sum_{\alpha=1}^r \int_0^t [Y(t)f_{V_\alpha}(r(s))]^i[Y(t)f_{V_\alpha}(r(s))]^j ds \tag{10.12}$$

where $f_{V_\alpha}(r)$, $r \in GL(\boldsymbol{R}^d)$, is defined by (2.34). Hence

$$\sigma(t) = Y(t)\hat{\sigma}(t)Y(t)^{*} \; * \tag{10.13}$$

where

* $Y(t)^*$ is the transpose of $Y(t)$.

(10.14) $\qquad \hat{\sigma}^{ij}(t) = \sum_{\alpha=1}^{r} \int_0^t f^i_{V_\alpha}(r(s)) f^j_{V_\alpha}(r(s)) ds.$

We have seen in Section 2 that

$$E[\|Y(t)\|^p + \|Y(t)^{-1}\|^p] < \infty \qquad \text{for all } 1 < p < \infty$$

and hence (10.11) is equivalent to

(10.15) $\qquad (\det \hat{\sigma}(t))^{-1} \in \bigcap_{1<p<\infty} \boldsymbol{L}_p.$

Note that $f_V(r(t))$, $V \in X(\boldsymbol{R}^d)$, is a d-dimensional Itô process such that

(10.16)
$$f_V(r(t)) - f_V(r(0)) = \sum_{\alpha=1}^{r} \int_0^t f_{[V_\alpha, V]}(r(s)) \circ dw^\alpha(s) + \int_0^t f_{[V_0, V]}(r(s)) ds$$

as we know from (2.36) of Section 2.

Now we obtain a general result on Itô processes by which we can discuss the condition (10.15). First we state several simple lemmas. In the following, $c_0, c_1, c_2, \cdots$ and $a_0, a_1, a_2, \cdots$ are positive constants independent of $n = 1, 2, \cdots$ and $w \in \boldsymbol{W}$.

Lemma 10.1. Let η: $\boldsymbol{W} \to \boldsymbol{R}$ be a real Wiener functional. Suppose that c_i, $i = 0, 1, 2, 3$ exist such that

(10.17) $\qquad P[|\eta| < n^{-c_0}] \leq c_1 \exp[-c_2 n^{c_3}], \; n = 1, 2, \cdots.$

Then $\quad E[|\eta|^{-p}] < \infty \qquad$ for all $p \geq 1$.

The proof is obvious and is ommitted.

Let $\mathscr{S}^d$ be the totality of $d \times d$ symmetric, non-negative definite matrices. For $g = (g^{ij}) \in \mathscr{S}^d$ and $l = (l^i) \in \boldsymbol{S}^{d-1} = \{x \in \boldsymbol{R}^d;\ |x| = 1\}$, we set

$$\|g\| = \{\sum_{i,j=1}^{d} (g^{ij})^2\}^{1/2} \quad \text{and} \quad g(l) = \sum_{i,j=1}^{d} g^{ij} l^i l^j.$$

We write $g_1 \geq g_2$, $g_1, g_2 \in \mathscr{S}^d$ if $g_1 - g_2$ is non-negative definite.

Lemma 10.2. Let η and $\tilde{\eta}$ be $\mathscr{S}^d$-valued Wiener functionals such that $\eta \geqq \tilde{\eta}$ a.s. Suppose that c_i, $i = 0, 1, 2, 3, 4$ exist such that $\|\tilde{\eta}\| \leqq c_0$ a.a.w. and

(10.18) $\sup_{l \in S^{d-1}} P[\tilde{\eta}(l) < n^{-c_1}] \leq c_2 \exp[-c_3 n^{c_4}]$, for $n = 1, 2, \cdots$.

Then

$$E[(\det \eta)^{-p}] < \infty \qquad \text{for all} \quad 1 < p < \infty.$$

Proof. Let $\bar{W} = \{w;\ \eta \geq \tilde{\eta} \text{ and } \|\tilde{\eta}\| \leq c_0\}$. Then $P(\bar{W}) = 1$. Set

$$W_n(l) = \{w \in \bar{W};\ \tilde{\eta}(l) \geq n^{-c_1}\} \quad \text{for } l \in S^{d-1} \text{ and } n = 1, 2, \cdots.$$

Then (10.18) implies that $P[W_n(l)^c] \leq c_2 \exp[-c_3 n^{c_4}]$. If $w \in \bar{W}$,

$$|\tilde{\eta}(l) - \tilde{\eta}(l')| \leq 2c_0 |l - l'| \qquad \text{for all} \quad l, l' \in S^{d-1}.$$

Clearly there exist $l_i \in S^{d-1}$, $i = 1, 2, \cdots, m$ such that

$$\bigcup_{k=1}^{m} \{l \in S^{d-1};\ |l - l_k| < 1/(4c_0 n^{c_1})\} = S^{d-1}$$

where $m \leq c_5 n^{(d-1)c_1}$ for some c_5. Then, $w \in \bigcap_{k=1}^{m} W_n(l_k)$ implies that

$$\inf_{|l|=1} \tilde{\eta}(l) \geq 1/(2n^{c_1})$$

and hence $\det \tilde{\eta} \geq 1/(2n^{c_1})^{d/2}$. Since $\det \eta \geq \det \tilde{\eta}$ if $w \in \bar{W}$, we have

$$P[\det \eta < 1/(2n^{c_1})^{\frac{d}{2}}] \leq P[\{\bigcap_{k=1}^{m} W_n(l_k)\}^c]$$
$$\leq c_5 n^{(d-1)c_1} c_2 \exp[-c_3 n^{c_4}] \leq a_0 \exp[-a_1 n^{a_2}].$$

By Lemma 10.1, we have

$$E[(\det \eta)^{-p}] < \infty \qquad \text{for all } 1 < p < \infty$$

which completes the proof.

In the following, we consider the case when $\eta = (\eta^{ij}) \in \mathscr{S}^d$ is given by

$$\eta^{ij} = \int_0^1 \sum_{\alpha=1}^{r} f_\alpha^i(s) f_\alpha^j(s) ds$$

and $s \to f_\alpha(s) \in R^d$, $\alpha = 1, 2, \cdots, r$, are $\{\mathscr{B}_t\}$-adapted continuous processes where $\{\mathscr{B}_t\}$ is the proper reference family on W for the r-dimensional Wiener process $w(t) = (w^\alpha(t))$ canonically realized on

(W, P). By Lemma 10.2, we can easily conclude the following:

Lemma 10.3. Suppose that there exist constants c_i, $i = 0, 1, 2, 3, 4$ and $\{\mathscr{B}_t\}$-stopping times $0 \leq \sigma_1 \leq \sigma_2 \leq 1$ such that $|f_\alpha(s)| \leq c_0$ for $s \in (\sigma_1, \sigma_2]$, $\sigma = 1, 2, \cdots, r$ a.s. and

$$(10.19) \quad \sup_{l \in S^{d-1}} P[\sum_{\alpha=1}^{r} \int_{\sigma_1}^{\sigma_2} [l \cdot f_\alpha(s)]^2 ds < n^{-c_1}] \leq c_2 \exp[-c_3 n^{c_4}], \quad \text{for every } n = 1, 2, \cdots,$$

where

$$l \cdot f_\alpha(s) = \sum_{i=1}^{d} l^i f_\alpha^i(s).$$

Then

$$E[(\det \eta)^{-p}] < \infty \qquad \text{for all } 1 < p < \infty.$$

Proof is immediate if we notice

$$\eta(l) = \sum_{\alpha=1}^{r} \int_0^1 [l \cdot f_\alpha(s)]^2 ds$$

and choose

$$\tilde{\eta}^{ij} = \sum_{\alpha=1}^{r} \int_{\sigma_1}^{\sigma_2} f_\alpha^i(s) f_\alpha^j(s) ds, \qquad i, j = 1, 2, \cdots, d,$$

in Lemma 10.2.

As usual, we call a continuous $\{\mathscr{B}_t\}$-adapted process $\xi(t)$: $[0, \infty) \to \boldsymbol{R}$ an Itô process with the characteristic $\xi_\alpha(t)$, $\alpha = 0, 1, \cdots, r$, if

$$\xi(t) = \xi(0) + \sum_{\alpha=1}^{r} \int_0^t \xi_\alpha(s) dw^\alpha(s) + \int_0^t \xi_0(s) ds$$

where $\xi_\alpha(s)$, $\alpha = 0, 1, \cdots, r$, are $\{\mathscr{B}_s\}$-adapted measurable processes such that

$$\sum_{\alpha=1}^{r} \int_0^t |\xi_\alpha(s)|^2 ds + \int_0^t |\xi_0(s)| ds < \infty \qquad \text{for every } t > 0, \quad \text{a.s.}$$

We are now in position to state a key result for Itô processes which plays a fundamental role in obtaining sufficient conditions for (10.15). This result is first obtained in a weaker form by Malliavin [107] and then improved by ideas of Kusuoka and Stroock (cf. [211] and [212]).

Lemma 10.4. (Key lemma). Let $\xi(t)$ be an Itô process with characteristics $\xi_\alpha(t)$, $\alpha = 0, 1, \cdots, r$. Assume furthermore, that $\xi_0(t)$ is also an Itô process with characteristics $\xi_{0,\alpha}(t)$, $\alpha = 0, 1, \cdots, r$. Suppose that there exist c_i, $i = 0, 1, 2, 3$ and $\{\mathscr{B}_t\}$-stopping times $0 \leq \sigma_1 \leq \sigma_2 \leqq 1$ such that

(i) for almost all w (P)

$$|\xi(s)| + \sum_{\alpha=1}^{r} |\xi_\alpha(s)|^2 + |\xi_0(s)| + \sum_{\beta=0}^{r} |\xi_{0,\beta}(s)| \leq c_0 \qquad \text{for all } s \in [\sigma_1, \sigma_2]$$

and

(ii) for every $n = 1, 2, \cdots$

$$P[\sigma_2 - \sigma_1 < \frac{1}{n}] \leq c_1 \exp[-c_2 n^{c_3}].$$

Then for every given c_4, there exist c_i, $i = 5, 6, 7, 8$ (which depend only on c_0, c_1, c_2, c_3 and c_4) such that

$$(10.20) \qquad P[\int_{\sigma_1}^{\sigma_2 \wedge (\sigma_1 + 1/n)} |\xi(s)|^2 ds \leq 1/n^{c_5}, \sum_{\alpha=0}^{r} \int_{\sigma_1}^{\sigma_2 \wedge (\sigma_1 + 1/n)} |\xi_\alpha(s)|^2 ds \geq 1/n^{c_4}] \leq c^6 \exp[-c_7 n^{c_8}] \qquad \text{for every } n = 1, 2, \cdots.$$

Here we present a proof of this lemma along the lines of [203] and [232]. It should be remarked that a much simplified different proof was given recently by Norris [222]. First we give a series of probabilistic lemmas.

Lemma 10.5. Let $K > 0$ and $X(t)$ be a one-dimensional continuous semimartingale

$$X(t) = X(0) + m(t) + A(t)$$

such that $\langle m \rangle(t) = \int_0^t \alpha(s)ds$, $A(t) = \int_0^t \beta(s)ds$ and $|\alpha(s)| \leq K$, $|\beta(s)| \leq K$. Then

$$P(\sigma_a < \lambda) \leq \frac{4}{\sqrt{\pi a}} \exp[-a^2/(8K\lambda)] \qquad \text{for all} \quad a > 0 \quad \text{and} \quad \lambda \in (0, a/2K]$$

where $\sigma_a = \inf\{t; |X(t) - X(0)| > a\}$.

*Proof.** By Theorem II-7.2′, there exists a one-dimensional Brownian motion $b(t)$ ($b(0) = 0$) such that

$$X(t) - X(0) = b(\langle m\rangle(t)) + A(t).$$

Since $\{|X(t) - X(0)| > a\} \subset \{|b(\langle m\rangle(t))| > a/2\} \cup \{|A(t)| > a/2\}$, we have $\sigma_a \geq (a/2K) \wedge (\hat{\sigma}_{a/2}/K)$ where $\hat{\sigma}_{a/2} = \inf\{t;\ |b(t)| > a/2\}$. Hence, if $0 < \lambda < a/2K$

$$\begin{aligned} P(\sigma_a < \lambda) &\leq P(\hat{\sigma}_{a/2} < K\lambda) \\ &\leq \frac{4}{\sqrt{2\pi K\lambda}} \int_{a/2}^{\infty} e^{-x^2/2K\lambda}\, dx \leq \frac{8}{\sqrt{2\pi K\lambda}a}(K\lambda\ \exp[-a^2/(8K\lambda)]) \\ &\leq 4\sqrt{1/(\pi a)}\exp[-a^2/(8K\lambda)] \end{aligned}$$

This completes the proof.

Before stating the next lemma, we shall prepare some notation. For a finite interval I and a square-integrable function $a(s)$ defined on I, we set

$$\sigma_I^2(a) = \frac{1}{|I|}\int_I [a(s)^2 - \bar{a}^2]ds$$

where

$$\bar{a} = \frac{1}{|I|}\int_I a(s)ds.$$

It is easy to see that

$$\sigma_I(a_1 + a_2) \leq \sigma_1(a_1) + \sigma_I(a_2).$$

Lemma 10.6. Let $b = (b(t))$ be a one-dimensional Brownian motion. Then for every $a > 0$ and $\varepsilon > 0$,

$$P(\sigma_{[0,a]}(b) < \varepsilon) \leq \sqrt{2}\exp\left[-\frac{a}{2^7\varepsilon^2}\right]. \tag{10.21}$$

Proof. Without loss of generality, we may assume that $b(0) = 0$. We use the following well known expansion formula of $b(t)$ in Fourier series:

* The proof was essentially given in Theoren IV-2.1. We shall repeat it however because of its importance.

$$b(t) = t\xi_0 + \sqrt{2}\sum_{k=1}^{\infty}\left[\xi_k\frac{(\cos 2\pi kt - 1)}{2\pi k} + \eta_k\frac{\sin 2\pi kt}{2\pi k}\right]$$

where $\{\xi_k, \eta_k\}$ are independent random variables with common distribution $N(0, 1)$ ([75], [136]).* Then

$$b(t) - \int_0^1 b(s)ds = (t - 1/2)\xi_0 + \sqrt{2}\sum_{k=1}^{\infty}\left[\xi_k\frac{\cos 2\pi kt}{2\pi k} + \eta_k\frac{\sin 2\pi kt}{2\pi k}\right]$$

and since $\{\cos 2\pi kt\}$ and $\{(t - 1/2), \sin 2\pi kt\}$ are mutually orthogonal in $\mathscr{L}_2([0, 1])$, we have

$$\sigma^2 := \sigma^2_{[0,1]}(b) \geq \sum_{k=1}^{\infty}\xi_k^2/(2\pi k)^2.$$

Hence for $z > 0$,

$$\begin{aligned} E(e^{-2z^2\sigma^2}) &\leq E(\exp\{-2z^2\sum_{k=1}^{\infty}\xi_k^2/(2\pi k)^2\}) \\ &= \prod_{k=1}^{\infty} E(\exp\{-z^2\xi_k^2/2\pi^2k^2\}) = \prod_{k=1}^{\infty}(1 + z^2/\pi^2k^2)^{-1/2} \\ &= \sqrt{z/\sinh z} \leq \sqrt{2e^{-z/2}}. \end{aligned}$$

Consequently

$$P(\sigma < \varepsilon) \leq \exp[2z^2\varepsilon^2]\, E(\exp[-2z^2\sigma^2]) \leq \sqrt{2}\exp[2z^2\varepsilon^2 - z/4]$$

and by setting $z = 1/16\varepsilon^2$ we have

$$P(\sigma < \varepsilon) \leq \sqrt{2}\exp\left[-\frac{1}{2^7\varepsilon^2}\right] \tag{10.22}$$

Since $\left\{\frac{1}{\sqrt{a}}b(at)\right\} \overset{\mathscr{L}}{\sim} \{b(t)\}$, we have $a\sigma^2_{[0,1]}(b) \overset{\mathscr{L}}{\sim} \sigma^2_{[0,a]}(b)$. Then(10.21) follows from (10.22)

Lemma 10.7. Let $b(t)$ be a one-dimensional Brownian motion on $[0, A]$ where A is a positive constant. Then, for every $0 < \gamma < 1/2$, there exist positive constants a_1, a_2, a_3 such that

$$P[\sup_{\substack{t,s\in[0,A]\\ t\neq s}}\frac{|b(t) - b(s)|}{|t - s|^{\gamma}} > n] \leq a_1\exp[-a_2n^{a_3}] \tag{10.23}$$

* $N(0, 1)$ stands for the normal distribution with mean 0 and variance 1.

for every $n = 1, 2, \cdots$.

Proof. First we note

$$(10.24)\quad P[\sup_{\substack{t,s\in[0,A]\\ t\neq s}} \frac{|b(t)-b(s)|}{|t-s|^{\gamma}} < \infty] = 1 \quad \text{for every } 0<\gamma<1/2.$$

Indeed it is proved in the proof of Corollary to Theorem I-4.3 that

$$P[\sup_{\substack{t,s\in[0,A]\\ t\neq s}} \frac{|X(t)-X(s)|}{|t-s|^{a}} < \infty] = 1$$

for any continuous stochastic process $X(t)$, $t \in [0, A]$, satisfying

$$E[|X(t)-X(s)|^{\alpha}] \leq M|t-s|^{1+\beta}, \qquad t, s \in [0, A],$$

where α, β, M are positive constants and a is any constant such that $0 < a < \beta/\alpha$. Since

$$E[|b(t)-b(s)|^{2m}] = (2m-1)(2m-3)\cdots 3\times 1\times |t-s|^{m}, \qquad m = 1, 2, \cdots,$$

(10.24) is now easily concluded.

For one-dimensional Wiener space (W_0^1, P) with time interval restricted to $[0, A]$, set

$$W_{0,\gamma}^1 = \{w \in W_0^1;\ \|w\|_{\gamma} < \infty\}$$

where

$$\|w\|_{\gamma} = \sup_{0\leq s\leq A} |w(s)| + \sup_{0\leq s<t\leq A} \frac{|w(t)-w(s)|}{|t-s|^{\gamma}}.$$

Then $W_{0,\gamma}^1$ is a Banach space which is however not separable and, if $0<\gamma<1/2$, (10.24) implies that $P(W_{0,\gamma}^1) = 1$. If W_{γ} is the closure of the Cameron-Martin Hilbert space H with respect to $\|\ \ \|_{\gamma}$-norm, then W_{γ} is a separable Banach space with respect to $\|\ \ \|_{\gamma}$-norm and it is easy to see that

$$W_{0,\gamma'}^1 \subset W_{\gamma} \qquad \text{if } 0<\gamma<\gamma'<1/2.$$

In particular, $P(W_{\gamma}) = 1$ for every $0<\gamma<1/2$. Then applying Fernique's theorem to the separable Banach space $B = W_{\gamma}$ for $0<\gamma<1/2$,

we have

$$K_\varepsilon := E[\exp\{\varepsilon \|w\|_\gamma^2\}] < \infty \qquad \text{for some } \varepsilon > 0.$$

Hence

$$P[\sup_{0 \le s < t \le A} \frac{|b(t) - b(s)|}{|t - s|^\gamma} > n]$$
$$\le P[\|w\|_\gamma > n] \le K_\varepsilon \exp[-\varepsilon n^2], \qquad n = 1, 2, \cdots .$$

For completeness, we give a proof of Fernique's theorem following Kuo [210].

Fernique's theorem: If B is a separable real Banach space with norm $\| \ \|$ and μ is a mean zero Gaussian measure (i.e. $\{l(x), l \in B'\} \subset L_2(B, \mu)$ and it is a mean zero Gaussian system with respect to μ). Then $\varepsilon > 0$ exists such that

$$\int_B \exp[\varepsilon \|x\|^2] \mu(dx) < \infty.$$

Proof. Consider the product probability space $(B \times B, \mu \otimes \mu)$. Then

$$\frac{x + y}{\sqrt{2}} \quad \text{and} \quad \frac{x - y}{\sqrt{2}}, \qquad \text{for } (x, y) \in B \times B$$

are B-valued random variables over this probability space such that they are mutually independent and both of them are μ-distributed. Hence, for $t < s < 0$,

$$\begin{aligned}
&\mu(\|x\| \le s)\, \mu(\|x\| > t) \\
&= \mu \otimes \mu\Big(\frac{\|x - y\|}{\sqrt{2}} \le s\Big) \mu \otimes \mu\Big(\frac{\|x + y\|}{\sqrt{2}} > t\Big) \\
&= \mu \otimes \mu\Big(\frac{\|x - y\|}{\sqrt{2}} \le s \text{ and } \frac{\|x + y\|}{\sqrt{2}} > t\Big) \\
&\leqq \mu \otimes \mu(|\|x\| - \|y\|| \le \sqrt{2} s \text{ and } \|x\| + \|y\| > \sqrt{2} t) \\
&\leqq \mu \otimes \mu\Big(\|x\| > \frac{t - s}{\sqrt{2}} \text{ and } \|y\| > \frac{t - s}{\sqrt{2}}\Big) \\
&= \mu\Big(\|x\| > \frac{t - s}{\sqrt{2}}\Big)^2.
\end{aligned}$$

For given $\varepsilon > 0$, define t_n, $n = 0, 1, \cdots$ by $t_0 = s$ and $t_{n+1} = s + \sqrt{2}\, t_n$. Then

$$t_n = \frac{(\sqrt{2})^{n+1} - 1}{\sqrt{2} - 1} s = ((\sqrt{2})^{n+1} - 1)(\sqrt{2} + 1)s.$$

Set

$$\alpha_n = \mu(\|x\| > t_n)/\mu(\|x\| \leq s).$$

Then

$$\begin{aligned}\alpha_{n+1} &= \mu(\|x\| > s + \sqrt{2}\, t_n)/\mu(\|x\| \leq s)\\ &= \mu(\|x\| > s + \sqrt{2}\, t_n)\mu(\|x\| \leq s)/(\mu(\|x\| \leq s))^2\end{aligned}$$

and applying the above obtained inequality to the numerator, we obtain

$$\alpha_{n+1} \leq \mu(\|x\| > t_n)^2/(\mu(\|x\| \leq s))^2 = \alpha_n^2.$$

Hence

$$\alpha_n \leq \alpha_0^{2^n} = \exp[2^n \log \alpha_0],$$

that is

$$\mu(\|x\| > (2^{\frac{n+1}{2}} - 1)(2^{\frac{1}{2}} + 1)s) \leq \mu(\|x\| \leq s)\exp[2^n \log \alpha_0], \qquad n = 1, 2, \cdots.$$

Choose s so large that

$$\mu(\|x\| \leq s) > 0 \quad \text{and} \quad \alpha_0 = \mu(\|x\| > s)/\mu(\|x\| \leq s) < 1.$$

Then the estimate obtained above can be written in the form

$$\mu(\|x\| > u) \leq b \exp[-au^2] \qquad \text{for all } u \geq c$$

with some positive constants a, b, c and the Fernique theorem follows at once from this.

Let $\eta(t)$, $t \in [0, 1]$, be an Itô process with bounded characteristic $\eta_\alpha(s)$. By Theorem II-7.2′, we can find a Brownian motion $b(t)$ such that

$$\eta(t) = \eta(0) + b(a(t)) + \int_0^t \eta_0(s)ds$$

where

$$a(t) = \sum_{\alpha=1}^{r} \int_0^t \eta_\alpha(s)^2 ds.$$

Hence Lemma 10.7 implies immediately the following:

Lemma 10.8. Let $\eta(t)$, $t \in [0, 1]$, be an Itô process with bounded characteristics.

Then c_i, $i = 0, 1, 2$ exist such that

$$(10.25) \qquad P[\sup_{0 \leq s < t \leq 1} \frac{|\eta(t) - \eta(s)|}{|t - s|^{1/3}} > n] \leqq c_0 \exp[-c_1 n^{c_2}],$$
$$\text{for } n = 1, 2, \cdots.$$

Finally we need the following real variable lemma due to Kusuoka and Stroock.

Lemma 10.9. Let $f(t)$ be a continuous function on $[a, b]$ and set

$$g(t) = g(a) + \int_a^t f(s)ds, \quad t \in [a, b].$$

If

$$\sup_{a \leq s < t \leq b} \frac{|f(t) - f(s)|}{|t - s|^{1/3}} \leq K$$

and

$$\int_a^b |f(s)|^2 ds > \varepsilon^2 \quad \text{for some } \varepsilon > 0 \text{ such that } \varepsilon^3 \leq 2^2 K^3 (b - a)^{5/2},$$

then

$$(10.26) \qquad (b - a)\sigma^2_{[a,b]}(g) \geq 2^{-13} 3^{-1} K^{-9} \varepsilon^{11} (b - a)^{-11/2}.$$

Proof is easily provided if we note that, first $t_0 \in [a, b]$ exists such that

$$|f(t_0)| > \varepsilon(b - a)^{-1/2}$$

and then an interval I, $I \subset [a, b]$, exists with length $\varepsilon^3 K^{-3} 2^{-3} (b - a)^{-3/2}$ on which f is of constant sign and $|f| \geq \varepsilon 2^{-1} (b - a)^{-1/2}$. We now note that there exists $t_1 \in I$ such that

$$g(t_1) = \overline{(g|_I)}, \quad \overline{(g|_I)} = \frac{1}{|I|}\int_{|I|} g(t)dt.$$

Then

$$\begin{aligned}(b-a)\sigma^2{}_{[a,b]}(g) &\geq \int_I (g(s)-\bar{g})^2 ds \\ &\geq \int_I \{\int_{t_1}^t f(s)ds\}^2 dt \\ &\geq \varepsilon^2 4^{-1}(b-a)^{-1}\int_I (t-t_1)^2 dt \\ &\geq (48)^{-1}\varepsilon^2(b-a)^{-1}|I|^3\end{aligned}$$

which completes the proof.

Proof of the key lemma. In proving (10.20), we may clearly assume $n \geq 2$ and $c_0 \geq 1$. In the following a_i and d_i, $i = 1, 2, \cdots$ are positive constants independent of n and w. We know(Theorem II-7.2′) that a one-dimensional Brownian motion $B(t)$ with $B(0) = 0$ exists such that

$$\xi(t) = \xi(\sigma_1) + B(A(t)) + g(t), \ \sigma_1 \leq t \leq 1$$

where

$$A(t) = \sum_{\alpha=1}^r \int_{\sigma_1}^t |\xi_\alpha(s)|^2 ds \quad \text{and} \quad g(t) = \int_{\sigma_1}^t \xi_0(s)ds.$$

Set

$$W_1 = [\sum_{\alpha=0}^r \int_{\sigma}^{\sigma_2\wedge(\sigma_1+1/n)} |\xi_\alpha(t)|^2 dt \geq 1/n^{c_4}]$$

and

$$W_2 = [\int_{\sigma_1}^{\sigma_2\wedge(\sigma_1+1/n)} |\xi(t)|^2 dt < 1/n^{c_5}].$$

Here c_4 is given and c_5 will be determinded later. Set

$$W_{1,1} = [\int_{\sigma_1}^{\sigma_2\wedge(\sigma_1+1/n)} |\xi_0(t)|^2 dt \geq 1/(2n^{c_4}), \ A(\sigma_2\wedge(\sigma_1+1/n)) < 1/n^{a_1}]$$

and

$$W_{1,2} = [A(\sigma_2 \wedge (\sigma_1 + 1/n)) \geq 1/n^{a_1}].$$

Then

$$W_1 \subset W_{1,1} \cup W_{1,2} \quad \text{if } a_1 \geq c_4 + 1 \tag{10.27}$$

since

$$1/n^{a_1} \leq 1/n^{c_4+1} \leq 1/(2n^{c_4}) \quad \text{for } n \geq 2.$$

Set

$$W_3 = [\sigma_2 - \sigma_1 \geq 1/n],$$
$$W_4 = [\sup_{\sigma_1 \leq s < t \leq \sigma_2} \frac{|\xi_0(t) - \xi_0(s)|}{|t - s|^{1/3}} < n]$$

and

$$W_5 = [\sup_{0 \leq t \leq 1/n^{a_1}} |B(t)| < 1/n^{a_2}].$$

By Lemma 10.9, we can find a_3 such that

$$\begin{aligned} &W_3 \cap W_4 \cap [\int_{I_n} |\xi_0(t)|^2 dt \geq 1/(2n^{c_4})] \\ &\subset W_3 \cap [\sigma^2_{I_n}(g) \geq 1/n^{a_3}] \end{aligned} \tag{10.28}$$

where $I_n = [\sigma_1, \sigma_1 + 1/n]$. Hence

$$\begin{aligned} &W_3 \cap [\sigma^2_{I_n}(g) < 1/n^{a_3}, \int_{I_n} |\xi_0(t)|^2 dt \geq 1/(2n^{c_4})] \\ &\subset W_3 \cap W_4^c. \end{aligned} \tag{10.29}$$

We have

$$\begin{aligned} W_{1,1} &\cap W_5 \cap W_3 \\ &\subset [\sup_{t \in I_n} |B(A(t))| < 1/n^{a_2}] \cap W_3 \\ &\subset [\sigma^2_{I_n}(B(A(\cdot))) \leq 1/n^{2a_2}] \cap W_3 \\ &\subset [\sigma^2_{I_n}(B(A(\cdot))) \leq 1/(4n^{a_3})] \cap W_3 \end{aligned}$$

if $a_2 > a_3/2 + 1$. Also

$$
\begin{aligned}
W_2 \cap W_3 &\subset [\int_{I_n} |\xi(t)|^2 ds/|I_n| < 1/n^{c_5-1}] \cap W_3 \\
&\subset [\sigma_{I_n}^2(\xi) < 1/n^{c_5-1}] \cap W_3 \\
&\subset [\sigma_{I_n}^2(\xi) < 1/(4n^{a_3})] \cap W_3
\end{aligned}
$$

if $c_5 > a_3 + 3$. Since

$$\sigma_{I_n}(g) \leq \sigma_{I_n}(B(A(\cdot))) + \sigma_{I_n}(\xi),$$

if $a_2 > a_3/2 + 1$ and $c_5 \geq a_3 + 3$,

$$
\begin{aligned}
&W_{1,1} \cap W_2 \cap W_3 \cap W_5 \\
&\subset [\sigma_{I_n}^2(g) < 1/n^{a_3}] \cap W_{1,1} \cap W_3 \qquad (10.30)\\
&\subset W_4^c
\end{aligned}
$$

by (10.29). Hence, if $a_1 \geq c_4 + 1$, $a_2 > a_3/2 + 1$ and $c_5 > a_3 + 3$,

$$W_{1,1} \cap W_2 \subset W_3^c \cup W_4^c \cup W_5^c.$$

By Lemma 10.5,

$$P[W_5^c] \leq d_1 \exp[-d_2 n^{a_1 - 2a_2}] \qquad \text{if } a_1 > 2a_2. \tag{10.31}$$

Hence we can conclude, by the assumption (ii) in Lemma 10.4, Lemma 10.8 and (10.31), that if $c_5 \geq a_3 + 3$ and $a_1 > (a_3 + 2) \vee (c_4 + 1)$,

$$P[W_{1,1} \cap W_2] \leq d_3 \exp[-d_4 n^{d_5}]. \tag{10.32}$$

Next we estimate $P(W_{1,2} \cap W_2)$. The idea is to use the fact $\sigma_I(B(A(\cdot)))$ is comparatively larger than $\sigma_I(g)$ if the interval I is sufficiently small. If $w \in W_3$, $I_n \subset [\sigma_1, \sigma_2]$. We divide I_n into n^m equal intervals

$$I_{n,k} = [\sigma_1 + k/n^{1+m}, \sigma_1 + (k+1)/n^{1+m}],\ k = 0, 1, \cdots, n^m - 1$$

where we choose $m \in N$ such that

$$m \geqq a_1. \tag{10.33}$$

Then if $w \in W_3$

$$
\begin{aligned}
\int_{I_{n,k}} |\xi(t)|^2 dt &= \int_{I_{n,k}} |\xi(\sigma_1) + B(A(t)) + g(t)|^2 dt \\
&\geq \int_{A(I_{n,k})} |\xi(\sigma_1) + B(s) + g(A^{-1}(s))|^2 dA^{-1}(s)
\end{aligned}
$$

$$\geq \frac{1}{c_0}\int_{A(I_{n,k})} |\xi(\sigma_1) + B(s) + g(A^{-1}(s))|^2 ds.$$

Here we note that

$$\sum_{\alpha=1}^{r} |\xi_\alpha(s)|^2 \leq c_0$$

by the assumption (i) in Lemma 10.4 and we set

$$A(I_{n,k}) = \{A(t);\ t \in I_{n,k}\}.$$

Set

$$J_k = [A(\sigma_1 + k/n^{1+m}), A(\sigma_1 + k/n^{1+m}) + 1/n^{a_1+m}],$$
$$k = 0, 1, \cdots, n^m - 1.$$

Then, setting $\bar{g}(s) = g(A^{-1}(s))$,

$$\begin{aligned}
&W_3 \cap \{|A(I_{n,k})| \geq 1/n^{a_1+m}\}\\
&= W_3 \cap \{A(I_{n,k}) \supset J_k\}\\
&\subset W_3 \cap \{\int_{I_{n,k}} |\xi(t)|^2 dt \geq \frac{1}{c_0}\int_{J_k} |\xi(\sigma_1) + B(s) + \bar{g}(s)|^2 ds\}\\
&\subset W_3 \cap \{\int_{I_{n,k}} |\xi(t)|^2 dt \geq \frac{|J_k|}{c_0}\sigma_{J_k}^2(B(\cdot) + \bar{g}(\cdot))\}\\
&\subset W_3 \cap \{\int_{I_{n,k}} |\xi(t)|^2 dt \geq \frac{|J_k|}{c_0}[\sigma_{J_k}(B) - \sigma_{J_k}(\bar{g})]^2\}.
\end{aligned}$$

Since

$$g(s) = \int_{\sigma_1}^{s} \xi_0(u)du \quad \text{and} \quad |\xi_0(u)| \leq c_0 \text{ on } [\sigma_1, \sigma_2],$$

we have

$$|\bar{g}(t) - \bar{g}(s)| \leq c_0|A^{-1}(t) - A^{-1}(s)|.$$

Therefore, setting $t_0 = A(\sigma_1 + k/n^{1+m})$,

$$\begin{aligned}
\sigma_{J_k}^2(\bar{g}) &\leq \frac{1}{|J_k|}\int_{J_k} |\bar{g}(t) - \bar{g}(t_0)|^2 dt\\
&\leq \frac{{c_0}^2}{|J_k|}\int_{J_k} (A^{-1}(t) - A^{-1}(t_0))^2 dt
\end{aligned}$$

$$\leq c_0^2[A^{-1}(A(\sigma_1 + (k+1)/n^{1+m})) - A^{-1}(A(\sigma_1 + k/n^{1+m}))]^2$$
$$= c_0^2/n^{2+2m}$$

if $w \in W_3 \cap \{ |A(I_{n,k})| \geq 1/n^{a_1+m}\}$ and hence $J_k \subset A(I_{n,k})$. Therefore

$$W_3 \cap [\sigma_{J_k}(B) \geq 2c_0/n^{1+m},\ |A(I_{n,k})| \geq 1/n^{a_1+m}\}$$
$$\subset W_3 \cap [\int_{I_{n,k}} |\xi(s)|^2 ds \geqq \frac{|J_k|}{c_0}\left(\frac{c_0}{n^{1+m}}\right)^2 = \frac{c_0}{n^{2+3m+a_1}}].$$

Set

$$W_6 = \bigcap_{k=0}^{n^m-1} \{\sigma_{J_k}(B) > 2c_0/n^{1+m}\}.$$

Since

$$A(\sigma_2 \wedge (\sigma_1 + 1/n)) = A(\sigma_1 + 1/n) = \sum_{k=0}^{n^m-1} |A(I_{n,k})| \geq 1/n^{a_1}$$
$$\text{on } W_3 \cap W_{1,2},$$

there exists k such that $|A(I_{n,k})| \geq 1/n^{a_1+m}$. Hence

$$W_{1,2} \cap W_3 \cap W_6$$
$$\subset \bigcup_{k=0}^{n^m-1} \{ |A(I_{n,k})| \geq 1/n^{a_1+m},\ \sigma_{J_k}(B) > 2c_0/n^{1+m}\} \cap W_3$$
$$\subset \bigcup_{k=0}^{n^m-1} \int_{I_{n,k}} |\xi(s)|^2 ds \geq c_0/n^{2+3m+a_1}\} \cap W_3$$
$$\subset \{\int_{\sigma_1}^{\sigma_2\wedge(\sigma_1+1/n)} |\xi(s)|^2 ds \geq 1/n^{2+3m+a_1}\}$$

since we assumed $c_0 \geqq 1$. Therefore if $c_5 \geq 2 + 3m + a_1$,

$$W_{1,2} \cap W_2 \subset W_3^c \cup W_6^c$$

and by Lemma 10.6 (note that $A(\sigma_1 + k/2^{1+m})$ is a stopping time for $B(t)$),

$$P[W_6^c] \leq \sum_{k=0}^{n^m-1} P[\sigma_{J_k}(B) \leq 2c_0/n^{1+m}]$$
$$\leq \sum_{k=0}^{n^m-1} d_6 \exp[-d_7|J_k|(1/n^{1+m})^{-2}]$$
$$\leq d_8 \exp[-d_9 n^{m+2-a_1}]$$

$$\leq d_8 \exp[-d_9 n^2]$$

because of (10.33). Noting the assumption (ii) in Lemma 10.4, we have thus obtained

$$(10.34) \qquad P[W_{1,2} \cap W_2] \leq d_{10} \exp[-d_{11} n^{d_{12}}].$$

Therefore, if c_5 is chosen sufficiently large, (10.32) and (10.34) imply the estimate (10.20) because $W_1 \cap W_2 \subset (W_{1,1} \cap W_2) \cup (W_{1,2} \cap W_2)$. This completes the proof of Lemma 10.4.

Remark 10.1. In the key lemma, the variable $n \in \boldsymbol{N}$ can be replaced obviously by a continuous variable $T \in [1, \infty)$.

Remark 10.2. By examining the above proof carefully, we can deduce the following: Let $\theta \in (0, 1]$ and we assume in Lemma 10.4 that $\xi(t)$, $\xi_\alpha(t)$, σ_1 and σ_2 may depend on the parameter θ but they satisfy the same assumptions (i) and (ii) where the constants c_0, c_1, c_2, c_3 are independent of θ. Then, for any given c_4 there exist c_i, $i = 5, 6, 7, 8, 9, 10$ (which depend only on c_0, c_1, c_2, c_3, c_4 and hence are independent of θ, in particular) such that

$$P\Big[\int_{\sigma_1}^{\sigma_2 \wedge (\sigma_1 + 1/n)} |\xi(s)|^2 ds \leq \theta/n^{c_5}, \sum_{\alpha=0}^{r} \int_{\sigma_1}^{\sigma_2 \wedge (\sigma_1 + 1/n)} |\xi_\alpha(s)|^2 ds \geq \theta/n^{c_4}\Big]$$
$$\leq c_6 n^{c_7} \exp[-c_8 \theta^{c_9} n^{c_{10}}] \quad \text{for all } n = 1, 2, \cdots \text{ and } \theta \in (0, 1].$$

Now we return to the stochastic differential equation (10.1). We introduce the following notation: for $V \in \mathfrak{X}(M)$,

$$(10.35) \qquad \begin{aligned} (V_\alpha, V) &= [V_\alpha, V] \qquad \alpha = 1, 2, \cdots, r \\ (V_0, V) &= [V_0, V] + \frac{1}{2}\sum_{\beta=1}^{r} [V_\beta, [V_\beta, V]]. \end{aligned}$$

Then, by (10.16), we have

$$(10.36) \qquad \begin{aligned} &l \cdot f_V(r(t)) - l \cdot f_V(r(0)) \\ &= \sum_{\alpha=1}^{r} \int_0^t l \cdot f_{[V_\alpha, V]}(r(s)) \circ dw^\alpha(s) + \int_0^t l \cdot f_{[V_0, V]}(r(s)) ds \\ &= \sum_{\alpha=1}^{r} \int_0^t l \cdot f_{(V_\alpha, V)}(r(s)) dw^\alpha(s) + \int_0^t l \cdot f_{(V_0, V)}(r(s)) ds \end{aligned}$$

for every $l \in S^{d-1}$ and $V \in \mathfrak{X}(R^d)$. For each $n = 0, 1, \cdots$, define a set $\sum_n \subset \mathfrak{X}(R^d)$ successively by

$$\textstyle\sum_0 = \{V_1, V_2, \cdots, V_r\}$$

(10.37)

$$\textstyle\sum_n = \{(V_\alpha, V);\ V \in \sum_{n-1}, \alpha = 0, 1, \cdots, r\} \quad \text{for } n = 1, 2, \cdots$$

and set

$$\hat{\Sigma}_n = \textstyle\sum_0 \cup \sum_1 \cup \cdots \cup \sum_n, \qquad \text{for } n = 0, 1, \cdots. \tag{10.38}$$

Now we introduce the following hypoellipticity condition of Hörmander type for the vector fields V_α, $\alpha = 0, 1, \cdots, r$.

Definition 10.1. We say that $V_\alpha, \alpha = 0, 1, \cdots, r$ satisfy the assumption (H) at $x \in R^d$ if there exists $M \in N$ and $A_1, A_2, \cdots A_d \in \hat{\Sigma}_M$ such that $A_1(x), A_2(x), \cdots, A_d(x)$ are linearly independent.

Cleary (H) is satisfied at x if and only if there exists $M \in N$ such that

$$\inf_{l \in S^{d-1}} \sum_{A \in \hat{\Sigma}_M} [l \cdot A(x)]^2 > 0. \tag{10.39}$$

If (10.39) is satisfied, then we can find $\delta > 0$ and bounded neighborhood $U(x)$ in R_d and $U(I)$ in $GL(d, R)$ of x and $I = (\delta^i_j)$ respectively such that

$$\inf_{l \in S^{d-1}} \sum_{A \in \hat{\Sigma}_M} [l \cdot f_A(x)]^2 \geq \delta \qquad \text{if } r \in U(x) \times U(I). \tag{10.40}$$

Set $\sigma = \inf\{s;\ r(s) \notin U(x) \times U(I)\}$ where $r(s)$ is the solution of (10.1) and (10.2) put together. Set $\sigma_1 \equiv 0$ and $\sigma_2 = \sigma \wedge 1$. Then σ_1 and σ_2 satisfy the condition (ii) of Lemma 10.4 by Lemma 10.5. Cleary

$$\inf_{l \in S^{d-1}} \int_{\sigma_1}^{\sigma_2 + (\sigma_1 + 1/n)} \{\sum_{A \in \hat{\Sigma}_M} [l \cdot f_A(r(s))]^2\} ds \geq \delta/n$$

on the set

$$W_n = [\sigma_2 - \sigma_1 \geq 1/n]$$

and $P[W_n^c] \leq a_1 \exp[-a_2 n]$, $n = 1, 2, \cdots$. Let $N = \#(\hat{\Sigma}_M)$. Then for every $l \in S^{d-1}$, we can find $\alpha_0, \alpha_1, \cdots, \alpha_k$, $0 \leqq k \leqq M$ such that $1 \leq \alpha_0 \leq r, 0 \leq \alpha_1, \alpha_2, \cdots, \alpha_k \leq r$ and

$$(10.41) \qquad \int_{\sigma_1}^{\sigma_1 \wedge (\sigma_2 + 1/n)} [l \cdot f_{V_{\alpha_0, \alpha_1, \cdots, \alpha_k}}(r(s))]^2 ds \geq \delta/(nN) \qquad \text{on } \boldsymbol{W}_n$$

where

$$V_{\alpha_0, \alpha_1, \cdots, \alpha_k} = (V_{\alpha_k}, (V_{\alpha_{k-1}}, (\cdots (V_{\alpha_1}, V_0)) \cdots).$$

Noting (10.36) and applying the key lemma successively, we can easily deduce the following: For each $j = 0, 1, \cdots, k$, there exist positive constant $c_{i,j}$, $i = 1, 2, 3, 4$, independent of $n = 1, 2, \cdots$ and $l \in S^{d-1}$ such that

$$P[\int_{\sigma_1}^{\sigma_2 \wedge (\sigma_1 + 1/n)} [l \cdot f_{V_{\alpha_0, \alpha_1, \cdots, \alpha_j}} (r(s))]^2 ds \leq n^{-c_{1,j}}]$$
$$\leq c_{2,j} \exp[-c_{3,j} n^{c_{4,j}}] \qquad \text{for all } n = 1, 2, \cdots .$$

In particular, we can conclude that positive constants c_i, $i = 1, 2, 3, 4$ exist independent of n such that

$$(10.42) \qquad \begin{aligned} &\sup_{l \in S^{d-1}} P[\sum_{\alpha=1}^{r} \int_{\sigma_1}^{\sigma_2} [l \cdot f_{V_\alpha}(r(s))]^2 ds \leq n^{-c_1}] \\ &\leq c_2 \exp[-c_3 n^{c_4}] \qquad \text{for every } n = 1, 2, \cdots . \end{aligned}$$

By Lemma 10.3 and (10.14), we obtain

$$(10.43) \qquad \|(\det \hat{\sigma}(1))^{-1}\|_p < \infty \qquad \text{for all } 1 < p < \infty.$$

Now introduce a parameter $1 \geq \varepsilon > 0$ and consider the following stochastic differential equations

$$(10.1)^\varepsilon \qquad \begin{cases} dX(t) = \varepsilon \sum_{\alpha=1}^{r} V_\alpha(X(t)) \circ dw^\alpha(t) + \varepsilon^2 V_0(X(t))dt \\ X(0) = x \end{cases}$$

and

$$(10.2)^\varepsilon \qquad \begin{cases} dY(t) = \varepsilon^2 \sum_{\alpha-1}^{r} \partial V_\alpha(X(t)) Y(t) \circ dw^\alpha(t) + \varepsilon^2 \partial V_0(X(t))dt \\ Y(0) = \boldsymbol{I}. \end{cases}$$

Denote the solution by $r^\varepsilon(t) = (X^\varepsilon(t), Y^\varepsilon(t))$. It is immediately seen that

$\{r(\varepsilon^2 t),\ t \geq 0\}$ is equivalent in law to $\{r^\varepsilon(t),\ t \geq 0\}$. In particular, $r(\varepsilon^2)$ is equivalent in law to $r^\varepsilon(1)$ and $\hat{\sigma}^\varepsilon(1)$ defined similarly for $r^\varepsilon(t)$ is equivalent in law to $\hat{\sigma}(\varepsilon^2)$. Suppose that $\{V_\alpha,\ \alpha = 0, 1, \cdots, r\}$ satisfy the condition (H) at x. We can give the same arguments as above for the vector field V_α^ε, $\varepsilon \in (0, 1]$, where $V_\alpha^\varepsilon = \varepsilon V_\alpha$, $\alpha = 1, 2, \cdots, r$ and $V_0^\varepsilon = \varepsilon^2 V_0$. In this case, however, $\delta > 0$ above is replaced by $\delta\varepsilon^k$ for some $k \in \boldsymbol{N}$ and $\delta > 0$ which is independent of $\varepsilon \in (0, 1]$ and we apply the key lemma in a generalized form as stated in Remark 10.2. Finally we obtain, instead of (10.42),

$$\sup_{l \in S^{d-1}} P\Big[\sum_{\alpha=1}^{r} \int_{\sigma_1^\varepsilon}^{\sigma_2^\varepsilon} [l \cdot f_{V_\alpha^\varepsilon}(r^\varepsilon(s))]^2 ds \leq \varepsilon^k n^{-c_1}\Big]$$
$$\leq c_2 n^{c_3} \exp[-\, c_4 \varepsilon^{c_5} n^{c_6}]$$

for $n = 1, 2, \cdots$ and $\varepsilon \in (0,1]$ where c_i, $i = 1, 2, 3, 4, 5, 6$ are independent of n and ε. Here σ_1^ε and σ_2^ε are also defined in the same way for the process $r^\varepsilon(t)$. From this we can deduce, as above, the following estimate:

$$P[\det \hat{\sigma}^\varepsilon(1) \leq \frac{1}{T}] \leq a_1 T^{a_2} \exp[-\, a_3 \varepsilon^{a_4} T^{a_5}]$$

for all $T \in [1, \infty)$ and $\varepsilon \in (0, 1]$ where a_i, $i = 1, 2, 3, 4, 5$ are positive constants independent of ε and T. We can now conclude that

$$\|(\det \hat{\sigma}^\varepsilon(1))^{-1}\|_p = \|(\det \hat{\sigma}(\varepsilon^2))^{-1}\|_p \leq K_1(p)\varepsilon^{-K_2} \quad \text{for all } p \in (1, \infty) \text{ and } \varepsilon \in (0, 1] \tag{10.44}$$

where $K_1(p) > 0$ may depend on p but not on ε and $K_2 > 0$ does not depend on p and ε.

Thus we obtain the following result due to Kusuoka and Stroock.

Theorem 10.2. Let $V_0, V_1, \cdots, V_r$ satisfy the assumption (H) at $x \in R^d$. Then, for every $t > 0$, $X(t, x, w) \in \boldsymbol{D}_\infty(\boldsymbol{R}^d)$ and it satisfies (A. 2): More precisely, there exists a positive constant K_2 and for every $1 < p < \infty$, there exists a positive constant $K_1(p)$ such that

$$\|(\det \sigma(t))^{-1}\|_p \leq K_1(p) t^{-K_2} \tag{10.45}$$

for all $t \in (0, 1]$ and $1 < p < \infty$.

Remark 10.3. If (H) is satisfied everywhere in a domain $D \subset \boldsymbol{R}^d$, the estimate (10.44) holds uniformly in $x \in \tilde{D}$ for every $\tilde{D} \Subset D$. This is clear from the above proof.

Thus if (H) is satisfied at $x \in \boldsymbol{R}^d$, then for any Schwartz distribution $T \in \mathscr{S}'(\boldsymbol{R}^d)$, $T(X(t, x, w)) \in \tilde{\boldsymbol{D}}_{-\infty}$ is defined for every $t > 0$. In particular, $\delta_y(X(t, x, w))$ is defined for every $y \in \boldsymbol{R}^d$ and

$$p(t, x, y) = E[\delta_y(X(t, x, w))]$$

which is $\boldsymbol{C}^\infty$ in y, coincides with the fundamental solution of

$$\frac{\partial u}{\partial t} = Au$$

where

$$A = \frac{1}{2}\sum_{\alpha=1}^{r} V_\alpha^2 + V_0 \quad *$$

i.e.

$$u(t, x) = \int_{\boldsymbol{R}^d} p(t, x, y)f(y)dy$$

solves the initial value problem

$$\frac{\partial u}{\partial t} = Au, \quad u|_{t=0} = f$$

uniquely among tempered (i.e. of polynomial growth) solutions for a tempered function f on $\boldsymbol{R}^d$. More generally, for $g(w) \in \tilde{\boldsymbol{D}}_\infty$,

$$E[g(w)\delta_y(X(t, x, w))]$$

is well-defined. In particular, if $C(x)$ is a tempered $\boldsymbol{C}^\infty$-function such that $C(x) \leqq K$ for some constant $K > 0$, then

$$g(w) = \exp[\int_0^t C(X(s, x, w))ds] \in \boldsymbol{D}_\infty$$

and

$$E[g(w)\delta_y(X(t, x, w))] = p^C(t, x, y)$$

* The vector field V_α is regarded as a differential operator.

is the fundamental solution of

$$\frac{\partial u}{\partial t} = (A + C)u.$$

Now assume that V_α, $\alpha = 0, 1, \cdots, r$, are bounded on $\boldsymbol{R}^d$ and satisfy (H) at $x \in \boldsymbol{R}^d$. Let $\phi(\xi)$ be a $\boldsymbol{C}^\infty$-function on $\boldsymbol{R}^d$ such that $\phi(\xi) = 1$ if $|\xi| \leq 1/3$ and $\phi(\xi) = 0$ if $|\xi| \geq 2/3$. Let $y \in \boldsymbol{R}^d$ and define

$$\psi(z) = \begin{cases} \phi\left(\dfrac{z - y}{|x - y|}\right) & \text{if} \quad x \neq y \\ 1 & \text{if} \quad x = y. \end{cases}$$

Then $\psi\delta_y = \delta_y$ and hence

$$p(t, x, y) = E[\delta_y(X(t, x, w))] = E[\psi(X(t, x, w))\delta_y(X(t, x, w))].$$

By the integration by parts as discussed in Section 9, this is easily seen to be a finite sum

$$\sum_i E[a_i(X(t, x, w))F_i(w)]$$

where $a_i(x)$ is of the form of a finite sum

$$\sum_j b_j(x)D_{\beta_j}\psi(x)$$

with bounded continuous functions $b_j(x)$ and $F_i(w) \in \boldsymbol{D}_\infty$ is a polynomial in the components of $X(t, x, w)$, their derivatives and $[\det \sigma(t)]^{-1}$. By Lemma 10.5, we can deduce that

$$P[|X(t, x, w) - x| \geq |x - y|/3] \leq a_1 \exp[- a_2|x - y|^2/t]$$

where a_1 and a_2 are positive constants independent of $x, y \in \boldsymbol{R}^d$ and $t \in [0, T]$. Combining this with (10.44), we obtain the following estimate:

Theorem 10.3. Suppose that V_α, $\alpha = 0, 1, \cdots, r$, are bounded and satisfy (H) everywhere in a domain D of $\boldsymbol{R}^d$. Then, $\nu > 0$ exists and for every compact set $K \Subset D$ and $T > 0$, positive constans c_1 and c_2 exist such that

$$p(t, x, y) \leq \frac{c_1}{t^\nu} \exp[- c_2|x - y|^2/t] \tag{10.46}$$

$$\text{for all} \quad t \in (0, T],\ x \in K \text{ and } y \in \boldsymbol{R}^d.$$

Next, we shall study the short time asymptotics of the fundamental solution $p(t, x, y)$. For this we introduce, as before, a parameter $\varepsilon \in (0, 1]$ and consider the equations $(10.1)^\varepsilon$ and $(10.2)^\varepsilon$. We denote the solutions by $X^\varepsilon(t, x, w)$ and $Y^\varepsilon(t, x, w)$, sometimes more simply by $X^\varepsilon(t)$ and $Y^\varepsilon(t)$, respectively. Then we have

$$(10.46) \qquad p(\varepsilon^2, x, y) = E[\delta_y(X^\varepsilon(1, x, w))], \qquad x, y \in \boldsymbol{R}^d.$$

We introduce the following notation: For $\alpha = (\alpha_1, \alpha_2, \cdots, \alpha_m) \in \{0, 1, 2, \cdots, r\}^m$, $m = 1, 2, \cdots$, we set

$$\|\alpha\| = m + \#\{\nu;\ \alpha_\nu = 0\}.$$

Also, let

$$S^\alpha(t, w) = \int_0^t \circ\, dw^{\alpha_1}(t_1)\int_0^{t_1} \circ\, dw^{\alpha_2}(t_2) \cdots \int_0^{t_{m-1}} \circ\, dw^{\alpha_m}(t_m)$$

be a multiple stochastic integral in the Stratonovich sense for $\alpha = (\alpha_1, \alpha_2, \cdots, \alpha_m) \in \{0, 1, 2, \cdots, r\}^m$ where we set $w^0(t) = t$.

Theorem 10.4. Let $x \in \boldsymbol{R}^d$ be fixed. Then $X^\varepsilon(1, x, w) \in \boldsymbol{D}_\infty(\boldsymbol{R}^d)$ has the asymptotic expansion

$$(10.47) \qquad X^\varepsilon(1, x, w) \sim f_0 + \varepsilon f_1 + \varepsilon^2 f_2 + \cdots \qquad \text{in } \boldsymbol{D}_\infty(\boldsymbol{R}^d) \text{ as } \varepsilon \downarrow 0$$

and $f_n \in \boldsymbol{D}_\infty(\boldsymbol{R}^d)$, $n = 0, 1, \cdots$, are given by

$$(10.48) \qquad f_0 = x$$

and

$$(10.49) \qquad f_n = \sum_{\alpha:\|\alpha\|=n} \tilde{V}_{\alpha_m} \cdot \tilde{V}_{\alpha_{m-1}} \cdots \tilde{V}_{\alpha_2}(V_{\alpha_1})(x) S^\alpha(1, w), \qquad n = 1, 2, \cdots.$$

Here the vector field V_α, regarded as a differential operator, is denoted by $\tilde{V}_\alpha$, i.e.

$$\tilde{V}_\alpha(f)(x) = V_\alpha^i(x)\frac{\partial f}{\partial x^i}(x), \qquad f \in C^\infty(\boldsymbol{R}^d \to \boldsymbol{R})$$

and

$$V_\alpha(y) = (V_\alpha^i(y)) \in C^\infty(\boldsymbol{R}^d \to \boldsymbol{R}^d), \qquad \alpha = 0, 1, \cdots, r.$$

In particular,

$$(10.50)\qquad f_1(w) = \sum_{\alpha=1}^{r} V_\alpha(x) w^\alpha(1).$$

The expression in (10.47) is uniform in $x \in K$ for any bounded set K in $\boldsymbol{R}^d$.

Proof. The proof is easily provided by successive applications of the Itô formula:

$$\begin{aligned}
&X^\varepsilon(1, x, w) - x \\
&= \varepsilon \sum_{\alpha=1}^{r} \int_0^1 V_\alpha(X^\varepsilon(s)) \circ dw_s^\alpha + \varepsilon^2 \int_0^1 V_0(X^\varepsilon(s))ds \\
&= \varepsilon \sum_{\alpha=1}^{r} V_\alpha(x) w^\alpha(1) + \varepsilon \sum_{\alpha=1}^{r} \int_0^1 [V_\alpha(X_0^\varepsilon(s)) - V_\alpha(x)] \circ dw_s^\alpha \\
&\quad + \varepsilon^2 \int_0^1 V_0(X^\varepsilon(s))ds \\
&= \varepsilon f_1 + \varepsilon^2 [\sum_{\alpha_1=1}^{r} \sum_{\alpha_2=1}^{r} \int_0^1 \{\int_0^s \tilde{V}_{\alpha_2}(V_{\alpha_1})(X^\varepsilon(u)) \circ dw^{\alpha_2}(u)\} \circ dw^{\alpha_1}(s) \\
&\quad + \int_0^1 V_0(X^\varepsilon(s))ds + \varepsilon \sum_{\alpha=1}^{r} \int_0^1 \{\int_0^s \tilde{V}_0(V_\alpha)(X^\varepsilon(u))du\} \circ dw^\alpha(s)] \\
&= \varepsilon f_1 + \varepsilon^2 [V_0(x) + \sum_{\alpha_1=1}^{r} \sum_{\alpha_2=1}^{r} \tilde{V}_{\alpha_2}(V_{\alpha_1})(x) S^{(\alpha_1, \alpha_2)}(1)] \\
&\quad + \varepsilon^2 [\sum_{\alpha_1=1}^{r} \sum_{\alpha_2=1}^{r} \int_0^1 \{\int_0^s (\tilde{V}_{\alpha_2}(V_{\alpha_1})(X^\varepsilon(u)) - \tilde{V}_{\alpha_2}(V_{\alpha_1})(x)) \circ dw^{\alpha_2}(u)\} \\
&\quad \circ dw^{\alpha}{}_1(s) + \int_0^1 [V_0(X^\varepsilon(s)) - V_0(x)]ds \\
&\quad + \varepsilon \sum_{\alpha=1}^{r} \int_0^1 \{\int_0^s \tilde{V}_0(V_\alpha)\,(X^\varepsilon(u))du\} \circ dw^\alpha(s)]
\end{aligned}$$

and so on. Continuing this process and applying Theorem III-3.1, it is easy to see that

$$X^\varepsilon(1, x, w) = f_0 + \varepsilon f_1 + \cdots + \varepsilon^n f_n + O(\varepsilon^{n+1})$$

in $\boldsymbol{L}_p(\boldsymbol{R}^d)$ for every $p \in (1, \infty)$. Since $D^k X^\varepsilon(t, x, w)\ [h_1, h_2, \cdots, h_k]$ are determined successively by stochastic differential equations as we saw in the proof of Proposition 10.1, it is easy to conclude that

$$D^k X^\varepsilon(1, x, w) = f_0^{(k)} + \varepsilon f_1^{(k)} + \cdots \varepsilon^n f_n^{(k)} + O(\varepsilon^{n+1})$$

in $L_p(H^{\otimes k} \otimes \boldsymbol{R}^d)$ as $\varepsilon \downarrow 0$. This completes the proof.

$X^\varepsilon(1, x, w)$ is not uniformly non-degenerate in the sense of (9.32) because $f_0 = x$ which is completely degenerate. However, if we consider

$$F(\varepsilon, w) = \frac{X^\varepsilon(1, x, w) - x}{\varepsilon}, \qquad \varepsilon \in (0, 1].$$

then we have the following:

Theorem 10.5. The family $F(\varepsilon, w)$, $\varepsilon \in (0, 1]$ defined above, satisfies (9.32) if and only if $(A^{ij}(x))$ defined by

$$A^{ij}(x) = \sum_{\alpha=1}^{r} V_\alpha^i(x) V_\alpha^j(x)$$

is non-degenerate, i.e. $\det(A^{ij}(x)) > 0$.

Proof. Since

$$F(\varepsilon, w) \sim f_1 + \varepsilon f_2 + \varepsilon^2 f_3 + \cdots \qquad \text{in } \boldsymbol{D}_\infty(\boldsymbol{R}^d)$$

where f_i are given in Theorem 10.4 and the Malliavin covariance of f_1 coincides with $A(x) = (A^{ij}(x))$, $A(x)$ must be non-degenerate in order that $F(\varepsilon, w)$ satisfies (9.32). Conversely, suppose that $\det(A^{ij}(x)) > 0$. Denoting by $\sigma(\varepsilon) = (\sigma^{ij}(\varepsilon))$ the Malliavin covariance of $F(\varepsilon, w)$, we have by Proposition 10.1 that

$$\sigma(\varepsilon) = \int_0^1 Y^\varepsilon(1)(Y^\varepsilon(s))^{-1} A(X^\varepsilon(s))[Y^\varepsilon(1)(Y^\varepsilon(s))^{-1}]^* ds.$$

Set

$$\tau = \inf\{s;\ (Y^\varepsilon(s))^{-1} A(X^\varepsilon(s))((Y^\varepsilon(s))^{-1})^* \leq A(x)/2\}.$$

Applying Lemma 10.5, it is easy to see that

$$P[\tau < 1/n] \leq c_1 \exp[-c_2 n^{c_3}], \qquad n = 1, 2, \cdots,$$

where c_i, $i = 1, 2, 3$, are positive constants independent of $\varepsilon \in (0, 1]$ and n. Note also

$$\sup_{\varepsilon \in (0,1)} \|\det(Y^\varepsilon(1))^{-1}\|_p < \infty \qquad \text{for all} \quad p \in (1, \infty)$$

because $(Y^\varepsilon(t))^{-1}$ is the solution of the equation (2.27) in which σ'_α and b' are replaced by $\varepsilon\sigma'_\alpha$ and $\varepsilon^2 b'$ respectively. Then

$$\det \sigma(\varepsilon) \geq (\det Y^\varepsilon(1))^2 \det\int_0^{1\wedge\tau}(Y^\varepsilon(s))^{-1}A(X^\varepsilon(s))((Y^\varepsilon(s))^{-1})^* ds$$

$$\geq \frac{1}{2^d}(\det Y^\varepsilon(1))^2\det(A(x))(1\wedge\tau)^d$$

and we can now easily conclude that

$$\sup_{\varepsilon\in(0,1)} \|\det(\sigma(\varepsilon))^{-1}\|_p < \infty \qquad \text{for all } p \in (1,\infty),$$

which complete the proof.

Suppose that $\det(A(x)) > 0$. Then by Theorem 9.4, $T(F(\varepsilon, x, w))$, $T \in \mathscr{S}(\boldsymbol{R}^d)$, has the asymptotic expansion in $\tilde{\boldsymbol{D}}_{-\infty}$. Since

$$\delta_x(X^\varepsilon(1, x, w)) = \delta_x(x + \varepsilon F(\varepsilon, w)) = \varepsilon^{-d}\delta_0(F(\varepsilon, w)),$$

we have the following:

Theorem 10.6. Suppose that $\det(A(x)) > 0$. Then $\delta_x(X^\varepsilon(1,x,w))$ has the following asymptotic expansion in $\tilde{\boldsymbol{D}}_{-\infty}$ as $\varepsilon \downarrow 0$:

$$\delta_x(X^\varepsilon(1, x, w)) \sim \varepsilon^{-d}(\Phi_0 + \varepsilon\Phi_1 + \varepsilon^2\Phi_2 + \cdots) \tag{10.51}$$

and $\Phi_k \in \tilde{\boldsymbol{D}}_{-\infty}$ can be obtained explicitly by Theorem 9.4:

$$\Phi_0 = \delta_0(f_1) \tag{10.52}$$

and

$$\Phi_k(w) = \sum_\alpha \sum_n \frac{1}{|\alpha|!} D_\alpha\delta_0(f_1) f_{n_1}^{\alpha_1} f_{n_2}^{\alpha_2} \cdots f_{n_l}^{\alpha_l} \qquad \text{for } k = 1, 2, \cdots \tag{10.53}$$

where $f_n = (f_n^i)$ is given by (10.49), f_1 by (10.50) in particular. In (10.53), the summation extends over all $\alpha = (\alpha_1, \alpha_2, \cdots, \alpha_l) \in \{1, 2, \cdots, d\}^l$ and $n = (n_1, n_2, \cdots, n_l)$, $n_i \geq 2$, $l = 1, 2, \cdots$ such that $n_1 + n_2 + \cdots + n_l - l = k$. Also $D_\alpha = \partial_{\alpha_1}\partial_{\alpha_2}\cdots\partial_{\alpha_n}$, $\partial_k = \dfrac{\partial}{\partial x^k}$ and $|\alpha| = l$ if $\alpha = (\alpha_1, \alpha_2, \cdots, \alpha_l)$.

From this we see that $E[\Phi_k(w)] = 0$ if k is odd since, then, $\Phi_k(-w)$

$= -\Phi_k(w)$ and the mapping: $\boldsymbol{W} \ni w \to -w \in \boldsymbol{W}$ preserves the measure P.

Corollary. Suppose that $\det A(x) > 0$ where $A(x) = (A^{ij}(x))$. Then $p(t, x, x)$, $t > 0$, $x \in \boldsymbol{R}^d$, has the asymptotic expansion

$$(10.55) \qquad p(t, x, x) \sim t^{-d/2}(c_0(x) + c_1(x)t + \cdots)$$

and $c_i(x)$ is given by

$$(10.56) \qquad c_i(x) = E[\Phi_{2i}], \qquad i = 0, 1, \cdots$$

where Φ_i is given by (10.52) and (10.53).

In particular

$$c_0 = \{(2\pi)^d \det A(x)\}^{-1/2}.$$

Remark 10.4. We can discuss by a similar method the asymptotic expansion of $p(t, x, x)$ in the degenerate case $\det A(x) = 0$ under some hypoellipticity condition and also the expansion of $p(t, x, y)$, $x \neq y$, as $t \downarrow 0$, cf. [192], [212], [214], [215] and [228].

We now consider an application of our results to heat kernels on manifolds. Let M be a compact Riemannian manifold and $b \in \mathfrak{X}(M)$. Let $e(t, x, y)$ be the fundamental solution of the heat equation

$$\frac{\partial u}{\partial t} = \frac{1}{2} \Delta_M u + b(u)$$

$$u_{|t=0} = f$$

where Δ_M is the Laplce-Beltrami operator for the Riemannian metric g on M i.e.

$$u(t, x) = \int_M e(t, x, y) f(y) m(dy) \quad (m(dy)\text{: the Riemannian volume})$$

solves the above initial value problem. Let $r(t, r, w) = (X(t, r, w), e(t, r, w))$ be the stochastic moving frame realized on the d-dimensional Wiener space $(\boldsymbol{W}, P)$ by the solution of stochastic differential equation (4.43) with the initial value $r_0 = r \in O(M)$. Let δ_y be the Dirac δ-function at $y \in M$, i.e. the Schwartz distribution on M such that

$$\delta_y(\phi) = \phi(y) \qquad \text{for every } \phi \in F(M).$$

Then we can define $\delta_y(X(t, r, w)) \in \tilde{\boldsymbol{D}}_{-\infty}$ and $e(t, x, y)$ can be expressed as

$$e(t, x, y) = E[\delta_y(X(t, r, w))]$$

where $\pi(r) = x$. Actually, we can extend our results obtained so far to the case of manifolds: We can develop a general theory of Wiener mappings with values in manifolds and pull-back of Schwartz distributions on M by these mappings. We do not go into details on these topics, however, only referring interested readers to [230] and [190].

In order to apply our results above to the asymptotic expansion of $e(t, x, x)$, we take two bounded coordinate neighborhoods U_1 and U_2 of x such that $U_1 \subset U_2$, view U_2 as a part in $\boldsymbol{R}^d$ and extend the Riemannian metric $(g_{ij}(y))$ and the vector field $b = (b^i(y))$ outside U_2 to a metric $g' = (g'_{ij}(x))$ and a vector field $b' = (b'^i(x))$ on $\boldsymbol{R}^d$ respectively, so that $g'_{ij}(y) = \delta_{ij}$ and $b'^i(y) = 0$ near ∞. Let $e'(t, x, y)$ be the fundamental solution of the heat equation

$$\frac{\partial u}{\partial t} = \frac{1}{2}\Delta' u + b'(u) \qquad \text{on } \boldsymbol{R}^d$$

where Δ' is the Laplace-Beltrami operator for g'. Then we have

$$\sup_{y, z \in U_1} |e(t, z, y) - e'(t, z, y)| \leq c_1 \exp[-c_2/t] \qquad \text{as } t \downarrow 0 \tag{10.57}$$

for some positive constants c_1 and c_2. For the proof, let $v \in \boldsymbol{C}^\infty(\boldsymbol{R}^d)$ with support contained in U_1 and set

$$u(t, z) = \int_{U_1} [e(t, z, y) - e'(t, z, y)]v(y)\sqrt{\det g(y)}dy, \qquad z \in U_2.$$

Then u is a solution of

$$\frac{\partial u}{\partial t} = \frac{1}{2}\Delta_M u + b(u) \qquad \text{on } U_2$$

and

$$\lim_{t \downarrow 0} u(t, z) = 0 \qquad \text{on } U_2.$$

By Itô's formula applied to $u(t\text{–}s, X_s)$ where $X_s = X(s, r, w)$ with $\pi(r) = z$, we deduce that

$$u(t, z) = E[u(t-\sigma, X_\sigma); \sigma \leq t], \qquad z \in U_2$$

where

$$\sigma = \inf\{t; X_t \in \partial U_2\}.$$

Hence

$$(10.58) \qquad |u(t, z)| \leq \sup_{s \in [0,t], \xi \in \partial U_2} |u(s, \xi)|, \qquad z \in U_2$$

Noting that $e(t, z, y)$ and $e'(t, z, y)$ have estimates similar to (10.46), in particular,

$$\sup_{\xi \in \partial U_2, y \in U_1} |e(s, \xi, y) - e'(s, \xi, y)| \leq c_1 \exp[-c_2/s], \qquad s \in [0, 1]$$

we obtain from (10.58) that

$$|u(t, z)| \leq c_1(\exp[-c_2 t])\|v\|_{L_1}, \; z \in U_1, \; t \in (0, 1]$$

and (10.57) follows from this.

Hence, in order to obtain the asymptotic expansion of $e(t, x, x)$ in powers of small t, we can instead take $e'(t, x, x)$. This e' can be given by a generalized Wiener functional expectation as

$$e'(\varepsilon^2, x, x) = E[\delta_x(X^\varepsilon(1, x, w)]$$

where $X^\varepsilon(t, x, w)$ is the solution of the following stochastic differential equation on $\boldsymbol{R}^d$:

$$\begin{cases} dX^i(t) = \varepsilon \sum_{k=1}^d \sigma_k^i(X(t))dw^k(t) - \varepsilon^2[\frac{1}{2} \sum_{j,k=1}^d g^{jk}(X(t))\Gamma_{jk}^i(X(t)) \\ \qquad\qquad + b^i(X(t))]dt \\ X(0) = x. \end{cases}$$

Here, we denote the components of the above extended g' and b' by g_{ij} and b^i; Γ_{jk}^i is the Christoffel symbol for g_{ij}. Also, (g^{ij}) is the inverse of (g_{ij}) and (σ_j^i) is the square-root of (g^{ij}), so that $g^{ij} = \sum_{k=1}^d \sigma_k^i \sigma_k^j$. Then, in the same way as above, we can obtain

$$e(t, x, x) = t^{-d/2}(c_0(x) + tc_1(x) + \cdots) \qquad \text{as } t \downarrow 0$$

and each $c_i(x)$ can be expressed explicitly by a generalized Wiener functional expectation. Usually, $c_i(x)$ is a very complicated polynomial in components of curvature tensor, their covariant derivatives and the vector fields b. By choosing our coordinates to be normal for the metric g, we can compute these expectations to obtain

$$c_0(x) = (2\pi)^{-d/2}$$

and

$$c_1(x) = (2\pi)^{-d/2}\left(\frac{1}{12}R(x) - \frac{1}{2}\operatorname{div}(b)(x) - \frac{1}{2}\|b\|^2(x)\right)$$

where $R(x) = R^i_i(x)$ is the scalar curvature. However the details are left to the reader. For related topics, cf. [217] and [231].

Finally we apply our method to prove the Gauss-Bonnet-Chern theorem. Let M be a compact oriented manifold of even dimension $d = 2l$. As Section 4, let $\Lambda_p(M)$, $p = 0, 1, \cdots, d$ be the spaces of differential p-forms and

$$\Lambda(M) = \sum_{p=0}^{d} \oplus \Lambda_p(M).$$

Consider the heat equation on $\Lambda(M)$

$$\text{(5.4)} \qquad \begin{cases} \dfrac{\partial u}{\partial t} = \dfrac{1}{2}\square u \\ u_{|t=0} = \alpha \end{cases}$$

where $\square = -(d+\delta)(d+\delta) = -(d\delta + \delta d)$ is the Laplacian of de Rham-Kodaira. In Section 5, we solved this initial value problem by a probabilistic way in the form (5.32) where $U(t, r)$ is the scalarization of $u(t, x)$. For given $r = (x, e) \in O(M)$, there is a *canonical isomorphism*

$$\tilde{r}: \overset{p}{\Lambda}\boldsymbol{R}^d \to \overset{p}{\Lambda}T^*_x(M) \qquad \text{for each} \qquad p = 0, 1, \cdots, d,$$

and hence

$$\tilde{r}: \Lambda\boldsymbol{R}^d = \sum_{p=0}^{d} \oplus \overset{p}{\Lambda}\boldsymbol{R}^d \to \Lambda T^*_x(M) = \sum_{p=0}^{d} \oplus \overset{p}{\Lambda}T^*_x(M),$$

defined by

$$\tilde{r}(\delta^{\alpha_1} \wedge \delta^{\alpha_2} \wedge \cdots \wedge \delta^{\alpha_p}) = f^{\alpha_1} \wedge f^{\alpha_2} \wedge \cdots f^{\alpha_p}$$

where $\{f^1, f^2, \cdots, f^d\}$ is the ONB in $T_x^*(M)$ which is dual to the ONB $e = \{e_1, e_2, \cdots, e_d\}$ in $T_x(M)$ and δ^α is defined by $\delta^\alpha = (0, \cdots, 0, \overset{\alpha-th}{1}, 0, \cdots, 0)$, (cf. Section 5). Using this isomorphism, the scalarization $F_\alpha \in C^\infty(O(M) \to \Lambda \boldsymbol{R}^d)$ of $\alpha \in \Lambda(M)$ can be written as

$$F_\alpha(r) = \tilde{r}^{-1}(\alpha(x)), \qquad r = (x, e) \in O(M).$$

Then by (5.32), the solution $u(t, x)$ of (5.4) is expressed as

$$u(t, r) = \tilde{r}E[M(t, r, w)\, \tilde{r}(t, r, w)^{-1}\alpha(X(t, r, w))] \tag{10.59}$$

where $r(t, r, w) = (X(t, r, w), e(t, r, w))$ is the stochastic moving frame, i.e. the solution of (5.5). Note that the right-hand side of (10.59) is independent of a particular choice of a frame $r = (x, e) \in O(M)$ over x: This just corresponds to the fact that $U(t, r) = \tilde{r}^{-1}u(t, x)$ is $O(d)$-equivariant, (cf. Section 5). Hence

$$u(t, x) = \int_M e(t, x, y)\alpha(y)m(dy)$$

and the fundamental solution

$$e(t, x, y) \in \mathrm{Hom}(\Lambda T_y^*(M), \Lambda T_x^*(M))$$

can be expressed by a generalized Wiener functional expectation as

$$e(t, x, y) = E[\tilde{r}M(t, r, w)\tilde{r}(t, r, w)^{-1}\delta_y(X(t, r, w))]. \tag{10.60}$$

We now introduce the decomposition

$$\Lambda(M) = \Lambda_+(M) \oplus \Lambda_-(M)$$

where

$$\Lambda_+(M) = \sum_{p:\text{even}} \oplus \Lambda_p(M) \quad \text{and} \quad \Lambda_-(M) = \sum_{p:\text{odd}} \oplus \Lambda_p(M).$$

Define a linear operator $(-1)^F$ on $\Lambda(M)$ by

$$(-1)^F\omega = \begin{cases} \omega & \text{if } \omega \in \Lambda_+(M) \\ -\omega & \text{if } \omega \in \Lambda_-(M). \end{cases}$$

Then the operator $Q = (d + \delta)$ satisfies

$$Q(-1)^F = -(-1)^F Q,$$

i.e., $\{(-1)^F, Q\} = (-1)^F Q + Q(-1)^F \doteq 0$. In particular, Q sends $\Lambda_{\pm}(M)$ into $\Lambda_{\mp}(M)$. For $\lambda \geqq 0$, let $H_\lambda = \{\omega \in \Lambda(M); \square\omega + \lambda\omega = 0\}$ and $H_\lambda^{\pm} = \{\omega \in \Lambda_{\pm}(M); \square\omega + \lambda\omega = 0\}$. Then $H_\lambda = H_\lambda^+ \oplus H_\lambda^-$ and $0 \leqq \dim H_\lambda < \infty$ (cf. [145]). We have

$$\text{(10.61)} \qquad \dim H_\lambda^+ = \dim H_\lambda^- \qquad \text{if } \lambda > 0,$$

because the mapping Q: $H_\lambda^+ \to H_\lambda^-$ is an isomorphism. Indeed, if $\omega \in H_\lambda^+$, then

$$\square Q\omega = -Q^3\omega = Q\square\omega = -\lambda Q\omega$$

showing that $Q\omega \in H_\lambda^-$. If $\omega \in H_\lambda^+$ and $Q\omega = 0$, then $\omega = (Q^2\omega)/\lambda = 0$ showing that the mapping Q: $H_\lambda^+ \to H_\lambda^-$ is one to one. Finally if $\omega \in H_\lambda^-$, then $\omega = Q[(Q\omega)/\lambda]$ and $(Q\omega)/\lambda \in H_\lambda^+$ showing that the mapping Q: $H_\lambda^+ \to H_\lambda^-$ is onto.

In the same way, we define the decomposition

$$\Lambda T_x^*(M) = \Lambda_+ T_x^*(M) \oplus \Lambda_- T_x^*(M), \qquad x \in M,$$

and

$$\Lambda \boldsymbol{R}^d = \Lambda_+ \boldsymbol{R}^d \oplus \Lambda_- \boldsymbol{R}^d.$$

Then $(-1)^F \in \operatorname{End}(\Lambda T_x^*(M))$ and $(-1)^F \in \operatorname{End}(\Lambda \boldsymbol{R}^d)$ are defined similary as above. For $A \in \operatorname{End}(\Lambda T_x^*(M))$ or $A \in \operatorname{End}(\Lambda \boldsymbol{R}^d)$, define the *supertrace* $\operatorname{Str}(A)$ of A by

$$\operatorname{Str}(A) = \operatorname{Tr}((-1)^F A).$$

If $A \in \operatorname{End}(\Lambda T_x^*(M))$ leaves $\Lambda_{\pm} T_x^*(M)$ invariant, then clearly

$$\operatorname{Str}(A) = \operatorname{Tr}(A|_{\Lambda_+(T_x^*(M))}) - \operatorname{Tr}(A|_{\Lambda_-(T_x^*(M))}).$$

Now we claim that

$$\text{(10.62)} \qquad \int_M \operatorname{Str}[e(t, x, x)]m(dx) = \chi(M), \qquad t > 0$$

$\chi(M)$ being the Euler characteristic of M, (cf. [145]). Indeed, by the eigenfunction expansion of $e(t, x, y)$,

$$e(t, x, y) = \sum_{n=0}^{\infty} e^{-\lambda_n t} \sum_{\phi:ONB\ in\ H_{\lambda_n}} \phi(x) \otimes \phi(y)$$

for some $0 \leq \lambda_0 < \lambda_1 < \cdots$ and hence

$$\text{(10.63)} \quad \begin{aligned} &\mathrm{Str}[e(t, x, x)] \\ &= \sum_{n=0}^{\infty} e^{-\lambda_n t} \sum_{\phi:ONB\ in\ H_{\lambda_n}^{+}} \|\phi(x)\|^2 - \sum_{n=0}^{\infty} e^{-\lambda_n t} \sum_{\phi:ONB\ in\ H_{\lambda_n}^{-}} \|\phi(x)\|^2 \end{aligned}$$

where $\| \quad \|$ is the Riemannian norm of $\Lambda T_x^*(M)$. Hence

$$\int_M \mathrm{Str}[e(t, x, x)]m(dx) = \sum_{n=0}^{\infty} e^{-\lambda_n t}(\dim H_{\lambda_n}^{+} - \dim H_{\lambda_n}^{-})$$

and, noting (10.61), this is equal to

$$\dim H_0^{+} - \dim H_0^{-} = \sum_{p=0}^{d} (-1)^p \dim H_p(M)$$

where $H_p(M) = \{\omega \in \Lambda_p(M);\ \square\omega = 0\}$ is the space of harmonic p-forms. Since

$$\dim H_p(M) = p\text{-th Betti number of } M$$

by the de Rham theorem, (cf. [145]) (10.62) is proved. Hence if we can show that

$$\text{(10.64)} \qquad \mathrm{Str}[e(t, x, x)] = C(x) + o(1)$$

as $t \downarrow 0$ uniformly on M, then

$$\chi(M) = \int_M C(x)m(dx).$$

Now we show that (10.64) actually holds and we identity the limit $C(x)$ with an explicit polynomial in terms of curvature tensor as given by (10.89) below. This result was first obtained by Patodi [223]: In the same way as above, $e(t, x, x) \in \mathrm{End}(\Lambda T_x^* M)$ has the asymptotic expansion

$$e(t, x, x) \sim t^{-l}(c_0(x) + c_1(x)t + \cdots + c_l(x)t^l + c_{l+1}(x)t^{l+1} + \cdots) \quad \text{as } t \downarrow 0,$$

where $l = d/2$ and $c_i(x) \in \mathrm{End}(\Lambda T_x^* M)$, $i = 0, 1, \cdots$, $c_i(x)$ being a very complicated polynomial in components of curvature tensor and its covariant derivatives. However, Patodi showed that a remarkable cancellation takes place for the supertrace, namely,

$$\mathrm{Str}[c_i(x)] = 0, \qquad i = 0, 1, \cdots, l-1$$

and furthermore, Str $[c_l(x)]$ is a polynomial of curvature tensor only, the terms involving covariant derivatives being completely cancelled. After Patodi, many simpler proofs were invented (cf. Gilkey [202], Getzler [201] etc.). A probabilistic method was first given by Bismut [193]. Our method, like Bismut, consists in obtaining cancellation at the level of functionals before taking expectation which simplifies much the proof of cancellation.

In the study of short time asymptotics of $e(t, x, x)$, we may assume as before, by choosing a coordinate neighborhood and extending the components of the metric to whole Euclidean space, that the metric tensor $g_{ij}(y)$ are defined globally on $\boldsymbol{R}^d$ and furthermore, we may choose a normal coordinate around x so that x is the origin of $\boldsymbol{R}^d$ and near the origin

$$g_{ij}(y) = \delta_{ij} + \frac{1}{3} R_{imkj}(0) y^m y^k + O(|y|^3) \tag{10.65}$$

and

$$\Gamma^i_{jk}(y) = \frac{1}{3} R_{ijkm}(0) y^m - \frac{1}{3} R_{ikjm}(0) y^m + O(|y|^2). \tag{10.66}$$

Let $(g^{ij}(y))$ and $(\sigma^i_j(y))$ be the inverse of $(g_{ij}(y))$ and the square root of $(g^{ij}(x))$ respectively. Consider the following stochastic differential equation on $\boldsymbol{R}^d \times GL(d, \boldsymbol{R})$ with a parameter $\varepsilon \in (0, 1]$:

$$\begin{cases} dX^i(t) = \varepsilon\sigma^i_k(X(t))dw^k(t) - \dfrac{\varepsilon^2}{2} g^{jk}(X(t))\Gamma^i_{jk}(X(t))dt \\ de^i_j(t) = -\Gamma^i_{mk}(X(t))e^k_j(t) \circ dX^m(t) \\ \qquad\qquad\qquad\qquad\qquad\qquad i, j = 1, 2, \cdots, d. \\ (X(0), e(0)) = (0, \boldsymbol{I}) \end{cases} \tag{10.67}$$

Denote this solution by $r^\varepsilon(t) = (X^\varepsilon(t), e^\varepsilon(t))$ and their components by $X^\varepsilon(t)^i$ and $e^\varepsilon(t)^i_j$ respectively. We know that $r^\varepsilon(t) \in O(M) = \{(y, e) \in \boldsymbol{R}^d \times GL(d, \boldsymbol{R});\ g_{ij}(y)e^i_\alpha e^j_\beta = \delta_{\alpha\beta}\}$ for all $t \geq 0$ a.s. and that $\{r^\varepsilon(t)\}$ is equivalent in law to $\{r(\varepsilon^2 t, r, w)\}$ where $r(t, r, w)$ is the solution of (5.5). Let $\Pi^\varepsilon(t) \in \mathrm{End}(\Lambda\boldsymbol{R}^d)$ be defined by

$$\Pi^\varepsilon(t)\colon \delta^{i_1} \wedge \delta^{i_2} \wedge \cdots \wedge \delta^{i_p} \to e^\varepsilon(t)^{i_1} \wedge e^\varepsilon(t)^{i_2} \wedge \cdots \wedge e^\varepsilon(t)^{i_p} \tag{10.68}$$

where $e^\varepsilon(t)^i = (e^\varepsilon(t)^i_1, e^\varepsilon(t)^i_2, \cdots, e^\varepsilon(t)^i_d) \in \boldsymbol{R}^d$, $i = 1, 2, \cdots, d$. Define $M^\varepsilon(t) \in \mathrm{End}(\Lambda \boldsymbol{R}^d)$ by the unique solution of

$$\text{(10.69)} \qquad \begin{cases} \dfrac{dM(t)}{dt} = \varepsilon^2 M(t)\tilde{D}_2[\dfrac{1}{2}J_{ijkm}(r^\varepsilon(t))] \\ M(0) = \boldsymbol{I}, \end{cases}$$

(cf.(5.30)) of Section 5 for notations). Then, by (10.60), we can conclude that

$$\text{(10.70)} \qquad \begin{aligned} e(\varepsilon^2, x, x) &= e(\varepsilon^2, 0, 0) \\ &= E[M^\varepsilon(1)\Pi^\varepsilon(1)\delta_0(X^\varepsilon(1))], \qquad \varepsilon \in (0, 1]. \end{aligned}$$

By (10.65), (10.66) and (10.67), we can easily deduce as in Theorem 10.4 that

$$\text{(10.71)} \qquad X^\varepsilon(1) = \varepsilon w(1) + O(\varepsilon^2) \quad \text{in } \boldsymbol{D}_\infty(\boldsymbol{R}^d) \quad \text{as } \varepsilon \downarrow 0$$

and, by Theorem 10.5, we see that $X^\varepsilon(1)/\varepsilon$ is uniformly non-degenerate in the sense of (9.32). Hence by Theorem 9.4, we have

$$\text{(10.72)} \qquad \delta_0(X^\varepsilon(1)) = \varepsilon^{-d}\delta_0(w(1)) + O(\varepsilon^{-d+1}) \quad \text{in} \quad \tilde{\boldsymbol{D}}^{-\infty} \quad \text{as} \quad \varepsilon \downarrow 0.$$

Similarly by (10.65), (10.66) and (10.67), we can deduce

$$\text{(10.73)} \qquad \begin{aligned} e^\varepsilon(1)^i_j = \delta^i_j &+ \frac{\varepsilon^2}{3}[R_{imjk}(0) + R_{ijmk}(0)]\int_0^1 w^k(s) \circ dw^m(s) \\ &+ O(\varepsilon^3) \quad \text{in } \tilde{\boldsymbol{D}}_\infty \quad \text{as } \varepsilon \downarrow 0,\ i, j = 1, 2, \cdots, d. \end{aligned}$$

From (10.67) and (10.68), we see that $\Pi^\varepsilon(t)$ is the unique solution of

$$\text{(10.74)} \qquad \Pi^\varepsilon(t) = I + \int_0^t \Pi^\varepsilon(s) \circ d\Theta^\varepsilon(s)$$

where $\Theta^\varepsilon(t)$ is an $\mathrm{End}(\Lambda \boldsymbol{R}^d)$-valued semimartingale defined by

$$\text{(10.75)} \qquad \Theta^\varepsilon(t) = D_1(\theta^\varepsilon(t)) = \sum_{i,j=1}^d \theta^\varepsilon_{ij}(t)a_i^* a_j$$

(cf. (5.23) for notations) where $\theta^\varepsilon(t)$ is $\boldsymbol{R}^d \times \boldsymbol{R}^d$-valued semimartingale given by

$$(10.76)\qquad \theta^{\varepsilon}_{ij}(t) = -\int_0^t \Gamma^i_{mj}(X^{\varepsilon}(s)) \circ dX^{\varepsilon}(s)^m.$$

By (10.66) we can deduce

$$(10.77)\qquad \theta^{\varepsilon}_{ij}(t) = \varepsilon^2 C_{ij}(t) + O(\varepsilon^3) \quad \text{in } \boldsymbol{D}_{\infty} \quad \text{as } \varepsilon \downarrow 0,$$

where

$$C_{ij}(t) = \frac{1}{3}[R_{imjk}(0) + R_{ijmk}(0)]\int_0^t w^k(s) \circ dw^m(s),$$
$$i, j = 1, 2, \cdots, d.$$

By (10.74), we have

$$\begin{aligned}(10.78)\qquad \Pi^{\varepsilon}(1) &= \boldsymbol{I} + \int_0^1 \Pi^{\varepsilon}(s) \circ d\Theta^{\varepsilon}(s) \\ &= \boldsymbol{I} + \Theta^{\varepsilon}(1) + \int_0^1\int_0^{t_1} \Pi^{\varepsilon}(t_2) \circ d\Theta^{\varepsilon}(t_2) \circ d\Theta^{\varepsilon}(t_1) \\ &= \boldsymbol{I} + \Theta^{\varepsilon}(1) + \int_0^1 \Theta^{\varepsilon}(t_1) \circ d\Theta^{\varepsilon}(t_1) \\ &\quad + \int_0^1\int_0^{t_1}\int_0^{t_2} \Pi^{\varepsilon}(t_3) \circ d\Theta^{\varepsilon}(t_3) \circ d\Theta^{\varepsilon}(t_2) \circ d\Theta^{\varepsilon}(t_1)\end{aligned}$$

and so on. Hence, setting

$$(10.79)\qquad A_m = \int_0^1\int_0^{t_1} \cdots \int_0^{t_{m-1}} \circ\, d\Theta^{\varepsilon}(t_m) \circ d\Theta^{\varepsilon}(t_{m-1}) \circ \cdots \circ d\Theta^{\varepsilon}(t_1),$$

it is easy to conclude

$$(10.80)\qquad \Pi^{\varepsilon}(1) = I + A_1 + \cdots + A_l + O(\varepsilon^{2l+2})$$
$$\text{in} \quad \boldsymbol{D}_{\infty}(\mathrm{End}(\Lambda \boldsymbol{R}^d)) \quad \text{as } \varepsilon \downarrow 0.$$

Furthermore, by (10.77), we can deduce for each $m = 1, 2, \cdots$ that

$$(10.81)\qquad \begin{aligned} A_m &= \varepsilon^{2m}\int_0^1\int_0^{t_1} \cdots \int_0^{t_{m-1}} \circ\, dD_1[C(t_m)] \circ dD_1[C(t_{m-1})] \circ \\ &\quad \cdots \circ dD_1[C(t_1)] + O(\varepsilon^{2m+1}) \quad \text{in} \quad \boldsymbol{D}_{\infty}(\mathrm{End}(\Lambda \boldsymbol{R}^d)) \quad \text{as } \varepsilon \downarrow 0.\end{aligned}$$

On the other hand, denoting by $J^{\varepsilon}(t)$ the $\boldsymbol{R}^d \otimes \boldsymbol{R}^d \otimes \boldsymbol{R}^d \otimes \boldsymbol{R}^d$-valued process $J_{ijkm}(r^{\varepsilon}(t))/2$, we see from (10.69) that

$$\begin{aligned}M^\varepsilon(1) &= I + \varepsilon^2\int_0^1 M^\varepsilon(t)\tilde{D}_2[J^\varepsilon(t)]dt \\ &= I + \varepsilon^2\int_0^1 \tilde{D}_2[J^\varepsilon(t)]dt \\ &\quad + \varepsilon^4\int_0^1\int_0^{t_1} M^\varepsilon(t_2)\tilde{D}_2[J^\varepsilon(t_2)]\tilde{D}_2[J^\varepsilon(t_1)]dt_2dt_1\end{aligned}$$

and so on. Hence, setting

$$B_n = \varepsilon^{2n}\int_0^1\int_0^{t_1}\cdots\int_0^{t_{n-1}}\tilde{D}_2[J^\varepsilon(t_n)]\tilde{D}_2[J^\varepsilon(t_{n-1})]\cdots\tilde{D}_2[J^\varepsilon(t_1)]\, dt_ndt_{n-1}\cdots dt_1, \tag{10.82}$$

it is easy to conclude that

$$M^\varepsilon(1) = I + B_1 + \cdots + B_l + O(\varepsilon^{2l+2}) \quad \text{in} \quad \boldsymbol{D}_\infty(\mathrm{End}(\varLambda\boldsymbol{R}^d)) \quad \text{as } \varepsilon\downarrow 0. \tag{10.83}$$

Furthermore, it is easy to deduce of each $n = 1, 2, \cdots$, that

$$B_n = \frac{\varepsilon^{2n}}{n!}(\tilde{D}_2[J])^n + O(\varepsilon^{2n+1}) \quad \text{in} \quad \boldsymbol{D}_\infty(\mathrm{End}(\varLambda\boldsymbol{R}^d)) \quad \text{as } \varepsilon\downarrow 0 \tag{10.84}$$

where $J = J^\varepsilon(0) = (R_{ijkm}(0)/2) \in \boldsymbol{R}^d\otimes\boldsymbol{R}^d\otimes\boldsymbol{R}^d\otimes\boldsymbol{R}^d$.

We now need the following algebraic lemma:

Lemma 10.10. Assume that $d = 2l$ is even.

(i) Let $a_1, a_2, \cdots, a_m \in \boldsymbol{R}^d\otimes\boldsymbol{R}^d$ and $b_1, b_2, \cdots, b_n \in \boldsymbol{R}^d\otimes\boldsymbol{R}^d\otimes\boldsymbol{R}^d\otimes\boldsymbol{R}^d$. Let $A \in \mathrm{End}(\varLambda\boldsymbol{R}^d)$ be a product of $D_1[a_1], D_1[a_2], \cdots, D_1[a_m]$ and $\tilde{D}_2[b_1], \tilde{D}_2[b_2], \cdots, \tilde{D}_2[b_n]$ in some order. Then

$$\mathrm{Str}(A) = 0 \quad \text{if } m + 2n < d = 2l. \tag{10.85}$$

(ii) Let $b = (b_{ijkm}) \in \boldsymbol{R}^d\otimes\boldsymbol{R}^d\otimes\boldsymbol{R}^d\otimes\boldsymbol{R}^d$ and suppose it satisfies

$$\begin{cases} b_{ijkm} = -b_{jikm} = -b_{ijmk} = b_{kmij} \quad \text{and} \\ b_{ijkm} + b_{ikmj} + b_{imjk} = 0. \end{cases} \tag{10.86}$$

Then

$$\begin{aligned}&\mathrm{Str}((\tilde{D}_2([b])^l)\\ (10.87)\qquad &= \frac{1}{2^l}\sum_{\nu,\mu\in\mathfrak{S}(2l)}\mathrm{sgn}(\nu)\mathrm{sgn}(\mu)b_{\nu(1)\nu(2)\mu(1)\mu(2)}\\ &\quad\times b_{\nu(3)\nu(4)\mu(3)\mu(4)}\cdots b_{\nu(2l-1)\nu(2l)\mu(2l-1)\mu(2l)}\end{aligned}$$

where $\mathfrak{S}(2l)$ is the permutation group of order $2l$.

Proof will be given at the end of this section.

Remark 10.5. The curvature tensor R_{ijkm} satisfies the relation (10.86).

By (i) of Lemma 10.10, we have

$$\mathrm{Str}[B_nA_m] = 0 \quad \text{if } 2n+m<2l=d.$$

If $2n+m=2l$ and $m>0$ or if $2n+m>2l$, then by (10.81) and (10.84),

$$B_nA_m = O(\varepsilon^{2m+2n}) = O(\varepsilon^{2l+1}) \quad \text{in } \boldsymbol{D}_\infty(\mathrm{End}(\Lambda \boldsymbol{R}^d)) \quad \text{as } \varepsilon\downarrow 0.$$

Hence we obtain by (10.80) and (10.83) that

$$\mathrm{Str}[M^\varepsilon(1)\Pi^\varepsilon(1)] = \varepsilon^{2l}\mathrm{Str}[B_l] + O(\varepsilon^{2l+1}) \quad \text{in } \boldsymbol{D}_\infty \quad \text{as } \varepsilon\downarrow 0.$$

Therefore, it follows from (10.84), (10.87) and Remark 10.5 that

$$\begin{aligned}&\mathrm{Str}\,[M^\varepsilon(1)\Pi^\varepsilon(1)] = \frac{\varepsilon^{2l}}{l!}\mathrm{Str}[(\tilde{D}_2[R_{ijkm}(0)/2])^l] + O(\varepsilon^{2l+1})\\ (10.88)\qquad &= \frac{\varepsilon^{2l}}{2^{2l}l!}\sum_{\nu,\mu\in\mathfrak{S}(2l)}\mathrm{sgn}(\nu)\mathrm{sgn}(\mu)R_{\nu(1)\nu(2)\mu(1)\mu(2)}(0)\\ &\quad\times R_{\nu(3)\nu(4)\mu(3)\mu(4)}(0)\cdots R_{\nu(2l-1)\nu(2l)\mu(2l-1)\mu(2l)}(0)\\ &\quad + O(\varepsilon^{2l+1}) \quad \text{in } \boldsymbol{D}_\infty \quad \text{as } \varepsilon\downarrow 0.\end{aligned}$$

Combining this with (10.70), (10.71) and (10.72) and noting

$$E[\delta_0(w(1))] = (2\pi)^{-d/2} = (2\pi)^{-l}$$

we finally obtained the following:

$$\begin{aligned}&\mathrm{Str}[e(\varepsilon^2,0,0)] = E[\mathrm{Str}[M^\varepsilon(1)\Pi^\varepsilon(1)]\delta_0(X^\varepsilon(1))]\\ (10.89)\qquad &= \frac{1}{2^{3l}\pi^l l!}\sum_{\nu,\mu\in\mathfrak{S}(2l)}\mathrm{sgn}(\nu)\mathrm{sgn}(\mu)R_{\nu(1)\nu(2)\mu(1)\mu(2)}(0)\end{aligned}$$

$$\times R_{\nu(3)\nu(4)\mu(3)\mu(4)}(0) \cdots R_{\nu(2l-1)\nu(2l)\mu(2l-1)\mu(2l)}(0)$$
$$+ O(\varepsilon)$$
$$\text{as } \varepsilon \downarrow 0.$$

It is clear that $O(\varepsilon)$ in (10.89) can be estimated uniformly with respect to $x \in M$ which we take to be the origin of the normal coordinate.

Generally, denoting by $J_{ijkm}(r)$ the scalarization of the curvature tensor $R_{ijkm}(x)$ as before, define a function $C(r)$ on $O(M)$ by

$$C(r) = \frac{1}{2^{3l}\pi^l l!} \sum_{\nu,\mu \in \mathfrak{S}(2l)} \operatorname{sgn}(\nu)\operatorname{sgn}(\mu) J_{\nu(1)\nu(2)\mu(1)\mu(2)}(r)$$
$$\times J_{\nu(3)\nu(4)\mu(3)\mu(4)}(r) \cdots J_{\nu(2l-1)\nu(2l)\mu(2l-1)\mu(2l)}(r), \quad r \in O(M). \tag{10.90}$$

It is easy to see that

$$C(ra) = C(r) \qquad \text{for every } a \in O(d)$$

and hence $C(r)$ depends only on $\pi(r) = x \in O(M)$. Thus we may write $C(r)$ by $C(\pi(r))$. This function $C(x)$ on M is known as the *Chern polynomial.* We have now established (10.64) and therefore

$$\chi(M) = \int_M C(x)m(dx).$$

This formula is known as the Gauss-Bonnet-Chern theorem. In the case $d = 2$, $C(x) = K(x)/2\pi$ where

$$K(x) = \frac{R_{1212}(x)}{\det(g_{ij}(x))}$$

is the Gauss total curvature and hence

$$\frac{1}{2\pi}\int_M K(x)m(dx) = \chi(M) = 2(1-g)$$

where g is the genus of the surface M. This is the classical Gauss-Bonnet theorem.

Proof of Lemma 10.10. A direct proof may be found in Patodi [223]. Following [194], we now give a proof based on a unifying idea of supersymmetry. Let
$\Lambda \boldsymbol{R}^d = \sum_{p=0}^{d} \oplus \overset{p}{\Lambda} \boldsymbol{R}^d$ be the exterior algebra over $\boldsymbol{R}^d$ and $a_i^*, a_i \in \operatorname{End}(\Lambda \boldsymbol{R}^d)$

be defined by (5.19) and its dual. Let $\Lambda_C \boldsymbol{R}^d$ be the complexification of $\Lambda \boldsymbol{R}^d$ and let $\gamma_i \in \text{End}(\Lambda_C \boldsymbol{R}^d)$, $i = 1, 2, \cdots, 2d$, be defined by

$$\begin{aligned} \gamma_{2i-1} &= a_i + a_i^* \\ \gamma_{2i} &= \sqrt{-1}(a_i^* - a_i) \end{aligned} \qquad i = 1, 2, \cdots, d.$$

Then from (5.20) we have $\{\gamma_i, \gamma_j\} = 2\delta_{ij}\boldsymbol{I}$ and $\gamma_i^* = \gamma_i$. For every subset $\boldsymbol{K} = \{\mu_1, \mu_2, \cdots, \mu_k\}$, $\mu_1 < \mu_2 < \cdots < \mu_k$ of $\{1, 2, \cdots, 2d\}$ let

$$\gamma_{\boldsymbol{K}} = (\sqrt{-1})^{k(k-1)/2}\gamma_{\mu_1}\gamma_{\mu_2}\cdots\gamma_{\mu_k} \quad \text{and}$$
$$\gamma_\phi = \boldsymbol{I}.$$

Then $\gamma_{\boldsymbol{K}}^2 = \boldsymbol{I}$ and $\gamma_{\boldsymbol{K}}^* = \gamma_{\boldsymbol{K}}$. Also, it is easy to see that

$$\text{Tr}(\gamma_A) = \begin{cases} 0 & \text{if } A \neq \phi \\ 2^d & \text{if } A = \phi. \end{cases}$$

Indeed if $A \neq \phi$, let $\mu_1 = \min\{i \mid i \in A\}$ and let $\tilde{A} = A \backslash \{\mu_1\}$. Then, if $\# A = k = $ even,

$$\begin{aligned} \text{Tr}(\gamma_A) &= (\sqrt{-1})^{k-1}\text{Tr}(\gamma_{\mu_1}\cdot\gamma_{\tilde{A}}) \\ &= -(\sqrt{-1})^{k-1}\text{Tr}(\gamma_{\tilde{A}}\cdot\gamma_{\mu_1}) \\ &= -(\sqrt{-1})^{k-1}\text{Tr}(\gamma_{\mu_1}\cdot\gamma_{\tilde{A}}) \\ &= -\text{Tr}(\gamma_A) \end{aligned}$$

and hence, $\text{Tr}(\gamma_A) = 0$. If $\# A = $ odd, choose $\mu \notin A$ and write

$$\begin{aligned} \text{Tr}(\gamma_A) &= \text{Tr}(\gamma_\mu \cdot \gamma_\mu\gamma_A) = -\text{Tr}(\gamma_\mu\cdot\gamma_A\cdot\gamma_\mu) \\ &= -\text{Tr}(\gamma_\mu\cdot\gamma_A\cdot\gamma_\mu^{-1}) = -\text{Tr}(\gamma_A). \end{aligned}$$

Hence $\text{Tr}(\gamma_A) = 0$.

It is easy to see that the system $\{\gamma_{\boldsymbol{K}}\}$, where $\boldsymbol{K}$ is a subset of $\{1, 2, \cdots, 2d\}$, forms a basis of $\text{End}(\Lambda_C \boldsymbol{R}^d)$: Independence of this system is easily seen from

$$\text{Tr}(\gamma_{\boldsymbol{K}}\gamma_{\boldsymbol{K}'}) = \begin{cases} 0 & \text{if } \boldsymbol{K} \neq \boldsymbol{K}' \\ 2^d & \text{if } \boldsymbol{K} = \boldsymbol{K}'. \end{cases}$$

Also, $\dim \text{End}(\Lambda_C \boldsymbol{R}^d) = (2^d)^2 = 2^{2d} = $ the number of the system $\{\gamma_{\boldsymbol{K}}\}$

and the assertion follows. Thus every $A \in \mathrm{End}\Lambda(\boldsymbol{R}^d)$ is expressed uniquely as

$$A = \sum_{K} C_K(A)\gamma_K \qquad C_K(A) \in \boldsymbol{C} \tag{10.91}$$

and

$$\mathrm{Tr}(A) = 2^d C_\phi(A). \tag{10.92}$$

(10.92) is known as the *Berezin formula*. Next we claim that

$$(-1)^F = (-1)^d \gamma_{(1,2,\cdots,2d)}. \tag{10.93}$$

Indeed, if we denote the right-hand side by α, then $\{\gamma_\mu, \alpha\} = 0$ and hence $\{a_i^*, \alpha\} = 0$, $i = 1, 2, \cdots, d$. From this, it is easy to conclude $\alpha = (-1)^F$ if we can show that $\alpha \cdot \omega = \omega$ for $\omega \in \mathring{\Lambda}_C(\boldsymbol{R}^d) = \boldsymbol{C}$. But

$$\gamma_{2k-1}\gamma_{2k} = \sqrt{-1}(a_k a_k^* - a_k^* a_k)$$

and hence

$$\gamma_{2k-1}\gamma_{2k}\omega = \sqrt{-1}\omega \qquad \text{if } \omega \in \boldsymbol{C}.$$

Thus

$$\alpha\omega = (-1)^d(\sqrt{-1})^{2d(2d-1)/2}(\sqrt{-1})^d\omega = (-1)^d(\sqrt{-1})^{2d^2}\omega = \omega.$$

Hence, combining this with (10.92), we have

$$\mathrm{Str}(A) = \mathrm{Tr}((-1)^F A) = (-1)^d 2^d C_{(1,2,\cdots,2d)}(A). \tag{10.94}$$

Now the proof of (i) is easy: If we express A in the form (10.91), then, in each term, γ's appear at most $2m + 4n$ times. Hence if $m + 2n < d$,

$$C_{(1,2,\cdots,2d)}(A) = 0.$$

For the proof of (ii), we first note that

$$\tilde{D}_2[b] = \frac{1}{2^3}\sum_{i,j,k,m=1}^{d} b_{ijkm}\gamma_{2i}\gamma_{2j}\gamma_{2k-1}\gamma_{2m-1} \tag{10.95}$$
$$+ \text{ a polynomial in } \gamma\text{'s of degree } 2$$

if (b_{ijkm}) satisfies (10.86). Indeed, omitting the summation sign for repeated indices, we have

$$\begin{aligned}\tilde{D}_2[b] &= b_{ijkm}a_i^* a_j a_k^* a_m \\ &= \frac{1}{2^4}\sum \pm(\sqrt{-1})^{\varepsilon} b_{ijkm}\tilde{\gamma}_i\tilde{\gamma}_j\tilde{\gamma}_k\tilde{\gamma}_m\end{aligned}$$

where $\tilde{\gamma}_i = \gamma_{2i}$ or γ_{2i-1} and the exponent ε depends only on the way of this choice. Noting the well-known property of (b_{ijkm}) satisfying (10.86) that *the alternation over any three indices vanishes*, it is easy to deduce that

$$\begin{aligned}&b_{ijkm}a_i^* a_j a_k^* a_m \\ &= \frac{1}{2^4}\{b_{ijkm}\gamma_{2i}\gamma_{2j}\gamma_{2k-1}\gamma_{2m-1} + b_{ijkm}\gamma_{2i-1}\gamma_{2j-1}\gamma_{2k}\gamma_{2m} \\ &\quad + b_{ijkm}\gamma_{2i}\gamma_{2j-1}\gamma_{2k-1}\gamma_{2m} + b_{ijkm}\gamma_{2i-1}\gamma_{2j}\gamma_{2k}\gamma_{2m-1} \\ &\quad - b_{ijkm}\gamma_{2i}\gamma_{2j-1}\gamma_{2k}\gamma_{2m-1} - b_{ijkm}\gamma_{2i-1}\gamma_{2j}\gamma_{2k-1}\gamma_{2m}\} \\ &\quad + \text{a polynomial in } \gamma\text{'s of degree 2.}\end{aligned}$$

It is easy to deduce from (10.86) that the first two terms are equal and the remaining four terms cancel. Hence the proof of (10.95) is finished. From this and (10.94), we can easily conclude (10.87).

Remark 10.5. If we take another supertrace

$$\mathrm{Str}^{(2)}(A) = \mathrm{Tr}[\gamma_{(2,4,\ldots,2n)}A\}$$

and do the same calculation as above, we can obtain the Hirzebruch signature theorem. Main difference in this case is that A_m in (10.83) does not disappear in the process of cancellation, that is, an effect of the parallel translation operator $\Pi^{\varepsilon}(1)$ remains in the final form of signature theorem. For details, cf. [204]. Indeed, we can give a similar probabilistic proof of Atiyah-Singer index theorem for every classical complex, cf. [193] and [224].

CHAPTER VI

Theorems on Comparison and Approximation and their Applications

1. A comparison theorem for one-dimensional Itô processes

Suppose now that we are given the following:

(i) a strictly increasing function defined on $[0, \infty)$ such that $\rho(0) = 0$ and

$$(1.1) \qquad \int_{0+} \rho(\xi)^{-2}\, d\xi = \infty;$$

(ii) a real continuous function $\sigma(t, x)$ defined on $[0, \infty)\times \boldsymbol{R}^1$ such that

$$(1.2) \qquad |\sigma(t,x) - \sigma(t,y)| \leq \rho(|x - y|), \qquad x, y \in \boldsymbol{R}^1, \quad t \geq 0;$$

(iii) two real continuous functions $b_1(t, x)$ and $b_2(t, x)$ defined on $[0, \infty)\times \boldsymbol{R}^1$ such that

$$(1.3) \qquad b_1(t, x) < b_2(t, x) \qquad t \geq 0, \quad x \in \boldsymbol{R}^1.$$

Let $(\Omega,\mathscr{F},P)$ be a probability space with a reference family $(\mathscr{F}_t)_{t\geq 0}$.

Theorem 1.1. * Suppose that we are given the following stochastic processes:

(1) two real $(\mathscr{F}_t)$-adapted continuous processes $x_1(t,\omega)$ and $x_2(t,\omega)$;

(2) a one-dimensional $(\mathscr{F}_t)$-Brownian motion $B(t,\omega)$ such that $B(0) = 0$, a.s.;

* This theorem covers the previous results obtained by e.g. Skorohod [150], Yamada [182] and Malliavin [108].

(3) two real $(\mathscr{F}_t)$-adapted well-measurable processes $\beta_1(t,\omega)$ and $\beta_2(t,\omega)$.
We assume that they satisfy the following conditions with probability one:

$$x_i(t) - x_i(0) = \int_0^t \sigma(s,x_i(s))dB(s) + \int_0^t \beta_i(s)ds, \qquad i = 1, 2, \tag{1.4}$$

$$x_1(0) \leq x_2(0), \tag{1.5}$$

$$\beta_1(t) \leq b_1(t,x_1(t)) \qquad \text{for every} \quad t \geq 0, \tag{1.6}$$

$$\beta_2(t) \geq b_2(t,x_2(t)) \qquad \text{for every} \quad t \geq 0. \tag{1.7}$$

Then, with probability one, we have

$$x_1(t) \leq x_2(t) \qquad \text{for every} \quad t \geq 0. \tag{1.8}$$

Furthermore, if the pathwise uniqueness of solutions holds for at least one of the following stochastic differential equations

$$dX(t) = \sigma(t,X(t))dB(t) + b_i(t,X(t))dt, \qquad i = 1, 2, \tag{1.9}$$

then the same conclusion (1.8) holds under the weakened condition

$$b_1(t,x) \leq b_2(t,x) \qquad \text{for} \quad t \geq 0, \quad x \in \boldsymbol{R}^1. \tag{1.3$'$}$$

Proof. * By a usual localization argument we may assume that $\sigma(t,x)$ and $b_i(t,x)$ are bounded.

Step 1. We assume that $b_1(t,x)$ is Lipschitz continuous, i.e., there exists a constant $K > 0$ such that

$$|b_1(t,x) - b_1(t,y)| \leq K|x - y|, \qquad x, y \in \boldsymbol{R}^1.$$

Choose a sequence $\psi_n(u)$, $n = 1, 2, \ldots$ of continuous functions as in the proof of Theorem IV-3.2 and set

$$\phi_n(x) = \begin{cases} 0, & x \leq 0, \\ \int_0^x dy \int_0^y \psi_n(u)du, & x > 0. \end{cases}$$

* The following proof is due to T. Shiga [146].

It is easy to see that $\phi_n \in C^2(\boldsymbol{R}^1)$, $\phi_n(x) = 0$ for $x \leq 0$, $0 \leq \phi_n'(x) \leq 1$ and $\phi_n(x) \uparrow x^+$ as $n \longrightarrow \infty$. An application of Itô's formula yields

$$\phi_n(x_1(t)-x_2(t)) = I_1(n) + I_2(n) + I_3(n),$$

where

$$I_1(n) = \int_0^t \phi_n'(x_1(s)-x_2(s))\{\sigma(s,x_1(s)) - \sigma(s,x_2(s))\}dB(s),$$

$$I_2(n) = \int_0^t \phi_n'(x_1(s)-x_2(s))\{\beta_1(s) - \beta_2(s)\}ds$$

and

$$I_3(n) = \frac{1}{2}\int_0^t \phi_n''(x_1(s)-x_2(s))\{\sigma(s,x_1(s)) - \sigma(s,x_2(s))\}^2 ds.$$

It is clear that $E(I_1(n)) = 0$ and

$$E(I_3(n)) \leq \frac{1}{2} E(\int_0^t \phi_n''(x_1(s)-x_2(s))\rho(|x_1(s)-x_2(s)|)^2 ds) \leq \frac{t}{n}.$$

Also,

$$\begin{aligned}
I_2(n) &\leq \int_0^t \phi_n'(x_1(s)-x_2(s))\{b_1(s,x_1(s)) - b_2(s,x_2(s))\}\,ds \\
&= \int_0^t \phi_n'(x_1(s)-x_2(s))\{b_1(s,x_1(s)) - b_1(s,x_2(s))\}\,ds \\
&\quad + \int_0^t \phi_n'(x_1(s)-x_2(s))\{b_1(s,x_2(s)) - b_2(s,x_2(s))\}\,ds \\
&\leq \int_0^t \phi_n'(x_1(s)-x_2(s))\{b_1(s,x_1(s)) - b_1(s,x_2(s))\}\,ds \\
&\leq K\int_0^t I_{[x_1(s)>x_2(s)]}\,|x_1(s) - x_2(s)|\,ds \\
&\leq K\int_0^t (x_1(s) - x_2(s))^+ ds.
\end{aligned}$$

Hence, by letting $n \longrightarrow \infty$, we have

$$\begin{aligned}
E[(x_1(t) - x_2(t))^+] &\leq KE[\int_0^t (x_1(s) - x_2(s))^+ ds] \\
&= K\int_0^t E((x_1(s) - x_2(s))^+)ds.
\end{aligned}$$

We can deduce from this that

$$E[(x_1(t) - x_2(t))^+] = 0 \qquad \text{for all} \quad t \geq 0,$$

that is

$$P(x_1(t) \leq x_2(t)) = 1 \qquad \text{for all} \quad t \geq 0,$$

and so by the continuity of paths we conclude that (1.8) holds.

If instead we assume that $b_2(t,x)$ is Lipschitz continuous, then a similar argument leads to the same conclusion (1.8).

Step 2. In the general case we choose $b(t,x)$ such that

$$b_1(t,x) < b(t,x) < b_2(t,x)$$

and $b(t,x)$ is Lipschitz continuous. Let $X(t)$ be the unique solution of the following stochastic differential equation

$$\begin{cases} dX(t) = \sigma(t,X(t))dB(t) + b(t,X(t))dt \\ X(0) = x_2(0). \end{cases}$$

Then the result from step 1 gives that $X(t) \leq x_2(t)$ and $x_1(t) \leq X(t)$ for all $t \geq 0$ a.s. Consequently we can conclude that (1.8) holds.

Step 3. We now turn to the proof of the second assertion. We assume that the pathwise uniqueness of solutions holds for one of the stochastic differential equations (1.9), say for $i = 1$. Let $X(t)$ be the solution of the equation

$$\begin{cases} dX(t) = \sigma(t,X(t))dB(t) + b_1(t,X(t))dt \\ X(0) = x_1(0). \end{cases} \tag{1.10}$$

For $\varepsilon > 0$, let $X^{(\pm\varepsilon)}(t)$ be the solutions of

$$\begin{cases} dX(t) = \sigma(t,X(t))dB(t) + (b_1(t,X(t)) \pm \varepsilon)dt \\ X(0) = x_1(0) \end{cases}$$

respectively. By what we have already proven,

$$X^{(-\varepsilon)}(t) \leq X(t) \leq X^{(+\varepsilon)}(t) \qquad \text{for every} \quad t \geq 0,$$

and if $0 < \varepsilon_2 < \varepsilon_1$, then

$$X^{(-\varepsilon_1)}(t) \leq X^{(-\varepsilon_2)}(t) \quad \text{and} \quad X^{(+\varepsilon_2)}(t) \leq X^{(+\varepsilon_1)}(t) \quad \text{for every} \quad t \geq 0.$$

Hence by the continuity of $b_1(t,x)$ and the pathwise uniqueness of solutions for (1.10), we have

$$\lim_{\varepsilon \downarrow 0} X^{(-\varepsilon)}(t) = \lim_{\varepsilon \downarrow 0} X^{(+\varepsilon)}(t) = X(t) \qquad \text{for every} \quad t \geq 0.$$

Again applying what we have already proven to $x_1(t)$ and $X^{(+\varepsilon)}(t)$, we have

$$\text{(1.11)} \qquad x_1(t) \leq X^{(+\varepsilon)}(t) \qquad \text{for every} \quad t \geq 0$$

since $\beta_1(t) \leq b_1(t,x_1(t))$ a.s. and $b_1(t,x) < b_1(t,x) + \varepsilon$. Hence, by letting $\varepsilon \downarrow 0$ in (1.11),

$$\text{(1.12)} \qquad x_1(t) \leq X(t), \qquad \text{for every} \quad t \geq 0.$$

Since $\beta_2(t) \geq b_2(t,x_2(t))$ a.s. and $b_2(t,x) > b_1(t,x) - \varepsilon$, we have $X^{(-\varepsilon)}(t) \leq x_2(t)$; letting $\varepsilon \downarrow 0$ gives

$$X(t) \leq x_2(t) \qquad \text{for every} \quad t \geq 0.$$

Combining this with (1.12) we obtain the inequality

$$x_1(t) \leq X(t) \leq x_2(t), \qquad \text{for every} \quad t \geq 0,$$

which completes the proof of the second assertion.

2. An application to an optimal control problem

As an example of an application of the comparison theorem in the previous section we shall consider the following stochastic optimization problem.

Let $k(r)$ be a non-decreasing and non-negative function defined on $[0, \infty)$. Let (B_t, u_t) be a system of stochastic processes defined on a probability space $(\Omega,\mathscr{F},P)$ with a reference family $(\mathscr{F}_t)$ such that

(i) B_t $(B_0 = 0)$ is a d-dimensional $(\mathscr{F}_t)$-Brownian motion and

(ii) u_t is a d-dimensional $(\mathscr{F}_t)$-well measurable process such that $|u_t| \leq 1$ for all $t \geq 0$ a.s.

Such a system (B_t, u_t) is called an *admissible system* or an *admissible control*. Let $x \in \boldsymbol{R}^d$ be given and fixed. For a given admissible system, the *response* X_t^u is defined by

$$(2.1) \qquad X_t^u = x + B_t + \int_0^t u_s ds.$$

The optimization problem then is to *minimize the expectation* $E(k(|X_1^u|))$ among all possible admissible systems. The solution is given as follows.

Let $U(y)$ be defined by

$$(2.2) \qquad U(y) = \begin{cases} -y/|y|, & y \in \boldsymbol{R}^d \setminus \{0\} \\ 0, & y = 0 \in \boldsymbol{R}^d. \end{cases}$$

Consider the following stochastic differential equation

$$(2.3) \qquad \begin{cases} dX_t = dB_t + U(X_t)dt \\ X_0 = x. \end{cases}$$

By the corollary to Theorem IV-4.2 we know that a solution (X_t°, B_t°) exists uniquely. Set

$$(2.4) \qquad u_t^\circ = U(X_t^\circ).$$

Then the admissible system (B_t°, u_t°) gives an optimal control; that is, for any admissible system (B_t, u_t), we have

$$(2.5) \qquad E(k(|X_1^\circ|)) \leq E(k(|X_1^u|)).$$

We shall now prove this fact as a simple consequence of Theorem 1.1. It was originally obtained by Beneš [3] by different techniques.

Theorem 2.1. Let (B_t, u_t) be any given admissible system and, for a fixed $x \in \boldsymbol{R}^d$, let the response $\{X_t^u\}$ be defined by (2.1). Then on an appropriate probability space we can construct $\boldsymbol{R}^d$-valued processes $\{\hat{X}_t^u\}$ and $\{\hat{X}_t^\circ\}$ such that

(i) $\{X_t^u\} \overset{\mathscr{L}}{\approx} \{\hat{X}_t^u\}$,

(ii) $\{X_t^\circ\} \overset{\mathscr{L}}{\approx} \{\hat{X}_t^\circ\}$

and

(iii) $|\hat{X}_t^\circ| \leq |\hat{X}_t^u|$ for every $t \geq 0$, a.s.

Corollary. Let $W^d = C([0, \infty) \longrightarrow R^d)$ and $F(w)$ be a non-negative Borel function on W^d with the following property:

$$\text{if } w_1, w_2 \in W^d \text{ and } |w_1(t)| \leq |w_2(t)| \text{ for every } t \geq 0, \text{ then } F(w_1) \leq F(w_2). \tag{2.6}$$

Then for any admissible system (B_t, u_t) we have

$$E(F(X^\circ_\cdot)) \leq E(F(X^u_\cdot)). \tag{2.7}$$

That is, the solution $\{X^\circ_t\}$ of (2.3) is optimal in the sense of minimizing the expectation of $F(X^u_\cdot)$.

Note that the particular case of $F(w) = k(|w(1)|)$ clearly satisfies (2.6).

In order to prove the theorem we first state the following lemma.

Lemma 2.1. Let (X_t, B_t) be a pair of d-dimensional continuous $(\mathscr{F}_t)$-adapted processes defined on a probability space $(\Omega, \mathscr{F}, P)$ with a reference family $(\mathscr{F}_t)$ such that $\{B_t\}$ is a d-dimensional Brownian motion with $B_0 = 0$. Let (Y_t, B'_t) be a similar pair defined on another space $(\Omega', \mathscr{F}' P')$ with $(\mathscr{F}_t')$. Then we can construct a probability space $(\hat{\Omega}, \hat{\mathscr{F}}, \hat{P})$ with a reference family $(\hat{\mathscr{F}}_t)$ and a triple $(\hat{X}_t, \hat{Y}_t, \hat{B}_t)$ of d-dimensional $(\hat{\mathscr{F}}_t)$-adapted processes such that

(i) $(X_t, B_t) \overset{\mathscr{L}}{\approx} (\hat{X}_t, \hat{B}_t)$,

(ii) $(Y_t, B'_t) \overset{\mathscr{L}}{\approx} (\hat{Y}_t, \hat{B}_t)$

and

(iii) $(\hat{B}_t)$ is a d-dimensional $(\hat{\mathscr{F}}_t)$-Brownian motion.

The proof follows in exactly the same way as in Theorem IV-1.1.

Proof of the theorem. Let (B_t, u_t) be a given admissible system and $X^u_t = x + B_t + \int_0^t u_s ds$ be the response. Let (X°_t, B°_t) be a solution of (2.3). Choose an $O(d)$-valued Borel function $(p_{ij}(x))$ such that

$$p_{1j}(x) = \begin{cases} x^j/|x| & x = (x^1, x^2, \ldots, x^d) \neq 0 \\ \delta_{1j} & x = 0. \end{cases} \tag{2.8}$$

Set $\bar{B}_t = \int_0^t p(X^u_s) dB_s$ and $\bar{B}^\circ_t = \int_0^t p(X^\circ_s) dB^\circ_s$. Then we have

$$X^u_t = x + \int_0^t p^{-1}(X^u_s) d\bar{B}_s + \int_0^t u_s ds \tag{2.9}$$

and

$$X_t^o = x + \int_0^t p^{-1}(X_s^o)d\bar{B}_s^o + \int_0^t U(X_s^o)ds. \tag{2.10}$$

Now we apply Lemma 2.1 to $(X_t^u, \bar{B}_t)$ and $(X_t^o, \bar{B}_t^o)$. We then have a triple $(\hat{X}_t^u, \hat{X}_t^o, \hat{B}_t)$ of $(\mathscr{F}_t)$-adapted processes on a probability space with a reference family $(\mathscr{F}_t)$ such that $\{\hat{B}_t\}$ is a d-dimensional $(\mathscr{F}_t)$-Brownian motion and $(X_t^u, \bar{B}_t) \approx (\hat{X}_t^u, \hat{B}_t)$, $(X_t^o, \bar{B}_t^o) \approx (\hat{X}_t^o, \hat{B}_t)$. Clearly there exists an $(\mathscr{F}_t)$-well measurable d-dimensional process $\{\hat{u}_t\}$ such that $|\hat{u}_t| \leq 1$ for every $t \geq 0$ and

$$\hat{X}_t^u = x + \int_0^t p^{-1}(\hat{X}_s^u)d\hat{B}_s + \int_0^t \hat{u}_s ds.$$

Applying Itô's formula to $x_1(t) = |\hat{X}_t^o|^2$ and $x_2(t) = |\hat{X}_t^u|^2$, we have

$$\begin{aligned} dx_2(t) &= 2\hat{X}_t^u \cdot p^{-1}(\hat{X}_t^u)d\hat{B}_t + 2\hat{X}_t^u \cdot \hat{u}_t dt + d\,dt \\ &= 2|\hat{X}_t^u|\,d\hat{B}_t^1 + [2\hat{X}_t^u \cdot \hat{u}_t + d]dt \\ &= 2\sqrt{x_2(t)}\,d\hat{B}_t^1 + [2\hat{X}_t^u \cdot \hat{u}_t + d]\,dt \end{aligned}$$

and

$$\begin{aligned} dx_1(t) &= 2\hat{X}_t^o \cdot p^{-1}(\hat{X}_t^o)d\hat{B}_t + 2\hat{X}_t^o \cdot U(\hat{X}_t^o)dt + d\,dt \\ &= 2|\hat{X}_t^o|\,d\hat{B}_t^1 + [-2|\hat{X}_t^o| + d]dt \\ &= 2\sqrt{x_1(t)}\,d\hat{B}_t^1 + [-2\sqrt{x_1(t)} + d]dt, \end{aligned}$$

where $\hat{B}_t = (\hat{B}_t^1, \hat{B}_t^2, \ldots, \hat{B}_t^d)$. (Note that $[x \cdot p^{-1}(x)]_i = \sum_j x_j(p^{-1}(x))_{ji} = \sum_j x_j p_{ij}(x) = \delta_{i1}|x|$). Set $\sigma(t,x) = 2\sqrt{x \vee 0}$, $b_1(t,x) = b_2(t,x) = -2\sqrt{x \vee 0} + d$, $\beta_1(t) = -2\sqrt{x_1(t)} + d$ and $\beta_2(t) = 2\hat{X}_t^u \cdot \hat{u}_t + d$. Then clearly $\beta_1(t) = b_1(t, x_1(t))$ and $\beta_2(t) \geq -2|\hat{X}_t^u| + d = b_2(t, x_2(t))$. As we shall see in the next lemma, the pathwise uniqueness of solutions holds for the stochastic differential equation

$$\begin{aligned} dX(t) &= \sigma(t, X(t))dB(t) + b_1(t, X(t))dt \\ &= 2(X(t) \vee 0)^{1/2}dB(t) + [-2(X(t) \vee 0)^{1/2} + d]dt. \end{aligned} \tag{2.11}$$

Therefore we can apply the second assertion of Theorem 2.1 and obtain the result that $x_1(t) \leq x_2(t)$ for all $t \geq 0$, a.s., that is $|\hat{X}_t^o| \leq |\hat{X}_t^u|$ for all $t \geq 0$, a.s.

Lemma 2.2. The pathwise uniqueness of solutions holds for the equation (2.11).

Proof. First we remark that the maximal and minimal solutions of (2.11) exist; that is, there exist strong solutions $X_1=F_1(X_1(0),B)$ and $X_2=F_2(X_2(0), B)$ of (2.11) such that if $(X(t), B(t))$ is any solution of (2.11) and $X_1(0)=X(0)=X_2(0)$, then $X_1(t) \leq X(t) \leq X_2(t)$ for every $t \geq 0$, a.s. Indeed, choose smooth functions $b_n^{(1)}(x)$ and $b_n^{(2)}(x)$ such that

$$b_1^{(1)}(x) \leq b_2^{(1)}(x) < \cdots \leq b_n^{(1)}(x) \leq \cdots \leq b(x) \leq \\ \cdots \leq b_n^{(2)}(x) \leq \cdots \leq b_2^{(2)}(x) \leq b_1^{(2)}(x)$$

and

$$\lim_{n\to\infty} b_n^{(1)}(x) = \lim_{n\to\infty} b_n^{(2)}(x) = b(x).$$

Here $b(x) = b_1(t, x) = -2(x\vee 0)^{1/2}+d$. Then the solutions $X_n(t)$ of

$$\begin{cases} dX(t) = \sigma(t,X(t))dB(t) + b_n^{(1)}(X(t))dt \\ X(0) = x \end{cases}$$

satisfy that $X_n(t) \leq X_{n+1}(t), n = 1, 2, \ldots,$ by Theorem 1.1 and we set $F_1(x, B) = \lim_{n\to\infty} X_n(\cdot)$. $F_2(x, B)$ is similarly defined. Clearly they possess the above property. Set

$$s(x) = \int_1^x \exp\left[-\int_1^y \frac{b(z)}{2z}\,dz\right] dy, \qquad x > 0.$$

Then $s(0+) > -\infty$ if $d = 1$ and $s(0+) = -\infty$ if $d \geq 2$. First we consider the case $d = 1$. By Itô's formula

$$s(X(t)) - s(X(0)) = \int_0^t s'(X(s))\sigma(s,X(s))dB(s)$$

for any solution $X(t)$ of (2.11) with $X(0) \geq 0$. Hence we have

$$E(s(X(t))) = E(s(X(0))).$$

From this we can conclude that $X_1(t) = X_2(t)$ a.s. for all $t \geq 0$ if $X_1(0) = X_2(0)$ a.s. This implies the pathwise uniqueness of solutions for (2.11). If $d \geq 2$, we see by Theorem 3.1 given below that $\inf_{0\leq t\leq T} X(t) > 0$ a.s. for every

$T > 0$ if $X(0) > 0$ a.s. The uniqueness of solutions for (2.11) can be easily deduced from this; we leave the details to the reader.

The optimization problem discussed above is a special example of stochastic control problems. For the recent development of the stochastic control theory, we refer the reader to Krylov [91].*[1]

3. Some results on one-dimensional diffusion processes*[2]

Let $I = (l, r)$ be an open interval in $\boldsymbol{R}^1$ $(-\infty \leq l < r \leq \infty)$. Let $\sigma(x)$ and $b(x)$ be sufficiently smooth real valued functions*[3] on I such that $\sigma^2(x) > 0$ for all $x \in I$. For $x \in I$, the stochastic differential equation

$$(3.1) \qquad \begin{cases} dX(t) = \sigma(X(t))dB(t) + b(X(t))dt \\ X(0) = x \end{cases}$$

has a unique solution $X^x(t)$ up to the explosion time $e = \lim_{n \uparrow \infty} \tau_n$, where $\tau_n = \inf\{t;\, X^x(t) \notin [a_n, b_n]\}$ (a_n and b_n $(n = 1, 2, \ldots)$ are chosen such that $l < a_n < b_n < r$ and $a_n \downarrow l$ and $b_n \uparrow r$). By the same proof as in Lemma IV-2.1, we can show that $\lim_{t \uparrow e} X^x(t)$ exists and is equal to l or r a.s. on the set $\{e < \infty\}$. We define $X^x(t)$ to be this limit for $t \geq e$ on the set $\{e < \infty\}$. Let $\hat{W}_I$ be the set of all continuous paths w: $[0, \infty) \longrightarrow [l, r]$ such that $w(0) \in I$ and $w(t) = w(e(w))$ for all $t \geq e(w) := \inf\{t;\, w(t) = l$ or $w(t) = r\}$. $e(w)$ is called the explosion time of the path w. Then $X^x = \{X^x(t)\}$ defines a $\hat{W}_I$-valued random variable with $e[X^x] = e$. Let P_x be the probability law on $\hat{W}_I$ of X^x. As we saw in Chapter IV, $\{P_x\}$ defines a diffusion process on I. It is called the *minimal L-diffusion*, where the differential operator L is defined by

$$(3.2) \qquad L = \frac{1}{2}\sigma^2(x)\frac{d^2}{dx^2} + b(x)\frac{d}{dx}.$$

Let $c \in I$ be fixed and set

$$(3.3) \qquad s(x) = \int_c^x \exp\left\{-\int_c^y \frac{2b(z)}{\sigma^2(z)}\,dz\right\} dy.$$

*[1] In particular, this book contains important results by Krylov on the estimates for the distributions of stochastic integrals, an extension of Itô's formula and stochastic differential equations with measurable coefficients.

*[2] Cf. [73] for more complete information.

*[3] In the following discussion it is sufficient to assume that σ and b are of class $\boldsymbol{C}^1$.

Then $s(x)$ is a strictly increasing smooth function on I satisfying

(3.4) $\qquad Ls(x) \equiv 0 \qquad$ on I.

Theorem 3.1. (1) If $s(l+) = -\infty$ and $s(r-) = \infty$, then

(3.5) $$P_x(e = \infty) = P_x(\overline{\lim_{t \uparrow \infty}} X(t) = r) = P_x(\underline{\lim_{t \uparrow \infty}} X(t) = l) = 1$$

for every x.* In particular, the process is recurrent, i.e. $P_x(\sigma_y < \infty) = 1$ for every $x, y \in I$ where $\sigma_y = \inf\{t; X(t) = y\}$.

(2) If $s(l+) > -\infty$ and $s(r-) = \infty$, then $\lim_{t \uparrow e} X(t)$ exists a.s. (P_x) and

(3.6) $$P_x(\lim_{t \uparrow e} X(t) = l) = P_x(\sup_{t<e} X(t) < r) = 1$$

for every x. A similar assertion holds if the roles of l and r are interchanged.

(3) If $s(l+) > -\infty$ and $s(r-) < \infty$, then $\lim_{t \uparrow e} X(t)$ exists a.s. and

(3.7) $$P_x(\lim_{t \uparrow e} X(t) = l) = 1 - P_x(\lim_{t \uparrow e} X(t) = r) = \frac{s(r-) - s(x)}{s(r-) - s(l+)}.$$

Thus the process is not recurrent in cases (2) and (3).

Proof. Let $l < a < x < b < r$ and $\tau = \inf\{t;\ X_t \notin [a, b]\}$. By Itô's formula and (3.4),

$$s(X^x(t \wedge \tau)) - s(x) = \int_0^{t \wedge \tau} s'(X^x(s))\sigma(X^x(s))dB(s)$$

and hence

$$E_x[s(X(t \wedge \tau))] = s(x).$$

By letting $t \uparrow \infty$, we have

$$E_x[s(X(\tau))] = s(a)P_x(X(\tau) = a) + s(b)P_x(X(\tau) = b) = s(x).$$

Combining this with

$$1 = P_x(X(\tau) = a) + P_x(X(\tau) = b)$$

* $X(t) = X(t, w) = w(t)$, $w \in \hat{W}_I$. Also $e = e(w)$.

we have

$$(3.8)\qquad P_x(X(\tau)=a)=\frac{s(b)-s(x)}{s(b)-s(a)},\qquad P_x(X(\tau)=b)=\frac{s(x)-s(a)}{s(b)-s(a)}.$$

Suppose $s(r-)=-s(l+)=\infty$. Then

$$\lim_{a\downarrow l} P_x(X(\tau)=b)=1.$$

Hence

$$P_x(\sup_{t<e} X(t)\geq b)\geq \lim_{a\downarrow l} P_x(X(\tau)=b)=1$$

for every $b<r$ and consequently $P_x(\sup_{t<e} X(t)=r)=1$. Similarly $P_x(\inf_{t<e} X(t)=l)=1$. Now it is easy to conclude the assertion of (1). Next, assume that $s(l+)>-\infty$ and $s(r-)=\infty$. We can deduce as above that

$$(3.9)\qquad P_x(\inf_{t<e} X(t)=l)=1.$$

Now $Y_t^{a,b}=s(X(t\wedge\tau))-s(l+)$ is a non-negative martingale and, letting $a\downarrow l$ and $b\uparrow r$, we can easily see that $Y_t=s(X(t\wedge e))-s(l+)$ is a non-negative supermartingale. Consequently $\lim_{t\uparrow\infty} Y_t=\lim_{t\uparrow e} s(X_t)-s(l+)$ exists as a finite limit a.s. (Theorem I-6.4). Therefore $\lim_{t\uparrow e} X_t$ exists a.s. and in view of (3.9) we must have (3.6).

The proof of (3) is similar and thus omitted.

Let $l<c<r$ be fixed and set

$$(3.10)\qquad \kappa(x)=2\int_c^x \exp\left[-\int_c^\xi \frac{2b(z)}{\sigma^2(z)}dz\right]\left\{\int_c^\xi \exp\left[\int_c^\eta \frac{2b(z)}{\sigma^2(z)}dz\right]\frac{d\eta}{\sigma^2(\eta)}\right\}d\xi,\quad x\in I.$$

Lemma 3.1. Let $u(x)$ be the unique solution of

$$(3.11)\qquad \begin{cases} Lu(x)=u(x)\\ u(c)=1\\ u'(c)=0.\end{cases}$$

Then

$$(3.12) \qquad 1+\kappa(x) \le u(x) \le \exp\{\kappa(x)\}, \qquad x \in I.$$

Proof. $u(x)$ is the solution of (3.11) if and only if it satisfies

$$u(x) = 1 + \int_c^x ds(y) \int_c^y u(z)dm(z),$$

where

$$ds(y) = \exp\left[-\int_c^y \frac{2b(z)}{\sigma^2(z)}dz\right]dy$$

and

$$dm(z) = 2\exp\left[\int_c^z \frac{2b(\eta)}{\sigma^2(\eta)}d\eta\right]\frac{1}{\sigma^2(z)}dz.$$

Hence

$$u(x) = \sum_{k=0}^{\infty} u_k(x)$$

with $u_0(x) \equiv 1$ and $u_n(x) = \int_c^x ds(y) \int_c^y u_{n-1}(z)dm(z)$. Clearly $u_n(x) \ge 0$ and $u_1(x) = \kappa(x)$. Consequently

$$u(x) \ge 1 + u_1(x) = 1 + \kappa(x).$$

On the other hand, if we suppose that

$$u_n(z) \le \kappa(z)^n/n!,$$

then

$$\begin{aligned} u_{n+1}(x) &= \int_c^x ds(y) \int_c^y u_n(z)dm(z) \le \frac{1}{n!}\int_c^x ds(y)\int_c^y \kappa(z)^n dm(z) \\ &\le \frac{1}{n!}\int_c^x ds(y)\kappa(y)^n \int_c^y dm(z) = \frac{1}{n!}\int_c^x \kappa(y)^n d[\kappa(y)] \\ &= \frac{\kappa(x)^{n+1}}{(n+1)!}. \end{aligned}$$

Consequently

$$u(x) = \sum_{n=0}^{\infty} u_n(x) \leq \sum_{n=0}^{\infty} \frac{\kappa(x)^n}{n!} = \exp\{\kappa(x)\}.$$

Remark 3.1. It is obvious that

(i) $\kappa(r-) < \infty \Rightarrow s(r-) < \infty$

and

(ii) $\kappa(l+) < \infty \Rightarrow s(l+) > -\infty$.

Theorem 3.2. (1) If $\kappa(r-) = \kappa(l+) = \infty$, then

(3.13) $$P_x(e = \infty) = 1 \quad \text{for all} \quad x \in I.$$

(2) If $\kappa(r-) < \infty$ or $\kappa(l+) < \infty$, then

(3.14) $$P_x(e < \infty) > 0 \quad \text{for all} \quad x \in I.$$

(3)
$$P_x(e < \infty) = 1 \quad \text{for all} \quad x \in I$$

if and only if one of the following cases occurs

(i) $\kappa(r-) < \infty$ and $\kappa(l+) < \infty$.
(ii) $\kappa(r-) < \infty$ and $s(l+) = -\infty$.
(iii) $\kappa(l+) < \infty$ and $s(r-) = \infty$.

Proof. Let $u(x)$ be defined by (3.11). Let $l < a < x < b < r$ and $\tau = \inf\{t;\, X(t) \notin (a, b)\}$. By Itô's formula

$$\begin{aligned} de^{-t}u(X(t)) &= e^{-t}u'(X(t))\sigma(X(t))dB(t) + e^{-t}(-u(X(t)) + (Lu)(X(t)))dt \\ &= e^{-t}u'(X(t))\sigma(X(t))dB(t) \end{aligned}$$

and hence $e^{-t\wedge\tau}u(X(t\wedge\tau))$ is a martingale. Letting $a \downarrow l$ and $b \uparrow r$ we immediately see that $\exp(-t\wedge e)u(X(t\wedge e))$ is a non-negative supermartingale. If $\kappa(r-) = \kappa(l+) = \infty$, then by Lemma 3.1, $\lim_{x\downarrow l} u(x) = \lim_{x\uparrow r} u(x) = \infty$. Consequently $P_x(e < \infty) > 0$ is impossible because $t \longmapsto \exp(-(t\wedge e))u(X(t\wedge e))$ is bounded a.s. as a non-negative supermartingale. This proves (1).

Next suppose that $\kappa(r-) < \infty$. Without loss of generality, we may assume that $c < x$. Let $\tau' = \inf\{t;\, X(t) = c\}$. By Lemma 3.1, $u(r-) < \infty$ and hence

(3.15) $$\exp(-t\wedge\tau')u(X(t\wedge\tau')) \quad \text{is a bounded } P_x\text{-martingale.}$$

Consequently

$$u(x) = E_x[\exp(-t \wedge \tau')u(X(t \wedge \tau'))]. \tag{3.16}$$

Letting $t \uparrow \infty$ in (3.16), we have

$$\begin{aligned} u(x) = E_x[\exp(-e)u(r-);\ \lim_{t \uparrow e \wedge \tau'} X(t) = r] \\ + E_x[\exp(-\tau')u(c);\ \lim_{t \uparrow e \wedge \tau'} X(t) = c]. \end{aligned}$$

If $E_x[\exp(-e);\ \lim_{t \uparrow e \wedge \tau'} X(t) = r] = 0$ then we have

$$u(x) = u(c)E_x[\exp(-\tau');\ \lim_{t \uparrow e \wedge \tau'} X(t) = c] \leq u(c).$$

This is clearly a contradiction. Therefore $E_x[\exp(-e);\ \lim_{t \uparrow e \wedge \tau'} X(t) = r] > 0$ which implies that $P_x(e < \infty) > 0$.

Finally we shall prove (3). If $\kappa(r-) < \infty$ and if none of (i), (ii) or (iii) hold then we must have $s(l+) > -\infty$ and $\kappa(l+) = \infty$. Then $P_x(\lim_{t \uparrow e} X(t) = l) > 0$. Since $\exp(-t \wedge e)u(X(t \wedge e))$ is a non-negative supermartingale and $\lim_{x \downarrow l} u(x) = \infty$, we must have $e = \infty$ a.s. on the set $\{\lim_{t \uparrow e} X(t) = l\}$. Consequently $P_x(e = \infty) > 0$. Therefore if $P_x(e < \infty) = 1$, then at least one of the conditions (i), (ii) and (iii) must hold.

It is now sufficient to show that each of conditions (i), (ii) and (iii) implies that $P_x(e < \infty) = 1$ for all x. First we shall assume (i). Define $G(x, y)$, $x, y \in I$, by

$$G(x, y) = \begin{cases} \dfrac{(s(x) - s(l-))(s(r-) - s(y))}{s(r-) - s(l+)} & x < y, \\[2ex] \dfrac{(s(y) - s(l-))(s(r-) - s(x))}{s(r-) - s(l+)} & y \leq x. \end{cases}$$

If $f(x)$ is a bounded continuous function on I, then the condition (i) implies that

$$u(x) = \int_I G(x, y)f(y)m(dy)$$

is a bounded function. In particular

$$u_1(x) = \int_I G(x, y)m(dy)$$

is a bounded function. It is easy to prove that $u \in C^2(I)$, $u(l+) = u(r-) = 0$ and $Lu = -f$. In particular, $Lu_1 = -1$. Hence by Itô's formula,

$$u_1(X(t\wedge e)) - u_1(x) + t\wedge e = \int_0^{t\wedge e} u_1'(X(s))dB(s).$$

Consequently

$$-E_x(u_1(X(t\wedge e))) + u_1(x) = E_x(t\wedge e).$$

Letting $t \uparrow \infty$ yields $u_1(x) = E_x(e) < \infty$. This proves that $P_x(e < \infty) = 1$. Next we shall assume (ii). For each $n = 1, 2, \ldots$, set $\bar{\sigma}_n = \inf\{t;\, X(t) = l+1/n\}$ and $\sigma_r = \inf\{t; X(t) = r\}$. Then $\lim_{n\uparrow\infty} \bar{\sigma}_n \wedge \sigma_r = e$. By the result of case (i), $P_x(\bar{\sigma}_n \wedge \sigma_r < \infty) = 1$ for all x. By Theorem 3.1, $\lim_{n\uparrow\infty} P_x(\bar{\sigma}_n > \sigma_r) = 1$ since $s(r-) < \infty$ and $s(l+) = -\infty$. Clearly $\{\sigma_r < \infty\} = \bigcup_n \{\sigma_r < \bar{\sigma}_n\}$ and hence $P_x(\sigma_r < \infty) = 1$. Consequently

$$P_x(e < \infty) = P_x(\sigma_r = e < \infty) = 1.$$

(iii) is obtained by interchanging the roles of l and r.

4. Comparison theorem for one-dimensional projection of diffusion processes

Let $\sigma = (\sigma_k^i(x))$ be a sufficiently smooth function: $\boldsymbol{R}^d \ni x \longmapsto \sigma(x) \in \boldsymbol{R}^d \otimes \boldsymbol{R}^d$ and $b = (b^i(x))$ be a sufficiently smooth function: $\boldsymbol{R}^d \ni x \longmapsto b(x) \in \boldsymbol{R}^d$. We consider a diffusion process $X = (X(t))$ on $\boldsymbol{R}^d$ defined by the solutions of the following stochastic differential equation

$$(4.1)\qquad dX_t^i = \sum_{k=1}^d \sigma_k^i(X_t)dB_t^k + b^i(X_t)dt, \qquad i = 1, 2, \ldots, d.$$

The diffusion X is defined up to the explosion time e (cf. Chapter IV, Section 2). As explained in Section 6 of Chapter IV, this is the diffusion process generated by the differential operator

$$(4.2)\qquad L = \frac{1}{2}\sum_{i,j=1}^d a^{ij}(x)\frac{\partial^2}{\partial x^i \partial x^j} + \sum_{i=1}^d b^i(x)\frac{\partial}{\partial x^i}$$

where

$$a^{ij}(x) = \sum_{k=1}^{d} \sigma_k^i(x)\sigma_k^j(x).$$

Let $p(x)$ be a smooth real function on $\boldsymbol{R}^d$ and let $I = \{\xi = p(x);\, x \in \boldsymbol{R}^d\}$. Then I is an interval in $\boldsymbol{R}^1$. Let S be the set (possibly empty) of all $x \in \boldsymbol{R}^d$ such that $p(x)$ is an end point of I. Let I^0 be the maximal open interval contained in I and $\bar{I}$ be the minimal closed interval in $[-\infty, \infty]$ which contains I. We assume that

$$|\nabla p(x)| = \left(\sum_{i,j=1}^{d} a^{ij}(x)\frac{\partial p}{\partial x^i}(x)\frac{\partial p}{\partial x^j}(x)\right)^{1/2} > 0 \qquad \text{for all} \quad x \in \boldsymbol{R}^d \setminus S.$$

Set

$$a(x) = \sum_{i,j=1}^{d} a^{ij}(x)\frac{\partial p}{\partial x^i}(x)\frac{\partial p}{\partial x^j}(x), \tag{4.3}$$

$$b(x) = (Lp)(x)/a(x), \qquad x \in \boldsymbol{R}^d \setminus S \tag{4.4}$$

and

$$\begin{cases} a^+(\xi) = \sup\limits_{x \in D(\xi;p)} a(x), & a^-(\xi) = \inf\limits_{x \in D(\xi;p)} a(x) \\ b^+(\xi) = \sup\limits_{x \in D(\xi;p)} b(x), & b^-(\xi) = \inf\limits_{x \in D(\xi;p)} b(x), \quad \xi \in I^0, \end{cases} \tag{4.5}$$

where $D(\xi;p) = \{x;\, p(x) = \xi\}$ for $\xi \in I^0$. We assume that $a^{\pm}(\xi)$ and $b^{\pm}(\xi)$ are locally Lipschitz continuous functions on I^0 and $a^{\pm}(\xi) > 0$.

On the interval I^0 we consider the following four minimal diffusion processes $\xi^{\pm\pm} = (\xi^{\pm\pm}(t))$ which are generated by the operators $L^{\pm\pm}$ respectively;

$$\begin{aligned} L^{++} &= a^+(\xi)\left(\frac{1}{2}\frac{d^2}{d\xi^2} + b^+(\xi)\frac{d}{d\xi}\right), \\ L^{+-} &= a^+(\xi)\left(\frac{1}{2}\frac{d^2}{d\xi^2} + b^-(\xi)\frac{d}{d\xi}\right), \\ L^{-+} &= a^-(\xi)\left(\frac{1}{2}\frac{d^2}{d\xi^2} + b^+(\xi)\frac{d}{d\xi}\right), \\ L^{--} &= a^-(\xi)\left(\frac{1}{2}\frac{d^2}{d\xi^2} + b^-(\xi)\frac{d}{d\xi}\right). \end{aligned} \tag{4.6}$$

Each of the diffusions is given as in Section 3; thus, the sample paths $\xi^{\pm\pm}(t)$ are defined for all $t \geq 0$ and are $\bar{I}$-valued continuous paths with $\bar{I}\setminus I^0$ as traps.

Let X_t be a sample path of the above diffusion (4.1) starting at $x_0 \in \boldsymbol{R}^d\setminus S$ and set

$$\zeta = \inf\{t;\ t \leq e,\ X_t \in S\}.$$

We assume that if $\zeta < \infty$ then $\lim_{s\uparrow\zeta} p(X_s)$ exists in $\bar{I}$. We set $p(X_t) = \lim_{s\uparrow\zeta} p(X_s)$ for $t \geq \zeta$. Thus $p(X_t)$ is defined for all $t \geq 0$ as an $\bar{I}$-valued continuous path.

Theorem 4.1 ([54]). Let $x_0 \in \boldsymbol{R}^d\setminus S$ be fixed and $\xi_0 = p(x_0) \in I^0$. Let $X = (X_t)$ be the above diffusion starting at x_0. Then we can construct $\bar{I}$-valued continuous stochastic processes η_t, η_t^{++}, η_t^{+-}, η_t^{-+}, η_t^{--} on the same probability space such that the following properties (i), (ii) and (iii) hold.

(i) (η_t) has the same law as $(p(X_t))$.

(ii) $(\eta_t^{\pm\pm})$ has the same law as $\xi^{\pm\pm} = (\xi^{\pm\pm}(t))$ starting at ξ_0 for each of four combinations of $\pm\pm$.

(iii) If we set

$$\bar{\eta}_t = \max_{0\leq s\leq t} \eta_s \qquad \underline{\eta}_t = \min_{0\leq s\leq t} \eta_s$$

and

$$\bar{\eta}_t^{\pm\pm} = \max_{0\leq s\leq t} \eta_s^{\pm\pm} \qquad \underline{\eta}_t^{\pm\pm} = \min_{0\leq s\leq t} \eta_s^{\pm\pm},$$

then with probability one we have

$$\bar{\eta}_t^{--} \leq \bar{\eta}_t \leq \bar{\eta}_t^{++} \qquad \text{for all} \quad t \geq 0 \tag{4.7}$$

and

$$\underline{\eta}_t^{+-} \leq \underline{\eta}_t \leq \underline{\eta}_t^{-+} \qquad \text{for all} \quad t \geq 0. \tag{4.8}$$

Proof. For simplicity we assume that $\zeta = \infty$ a.s. and that $(\xi_t^{\pm\pm})$ are all conservative diffusion processes on I^0; the general case can be proved with a slight modification.

Set

$$\phi^+(t) = \int_0^t [a(X_s)/a^+(p(X_s))]ds \quad \text{and} \quad \phi^-(t) = \int_0^t [a(X_s)/a^-(p(X_s))]ds.$$

Then clearly

$$\phi^+(t) \leq t \leq \phi^-(t) \qquad \text{for all} \quad t \geq 0.$$

Let $\psi^+(t)$ and $\psi^-(t)$ be the inverse functions of $t \longmapsto \phi^+(t)$ and $t \longmapsto \phi^-(t)$ respectively and set

$$X_t^{(+)} = X(\psi^+(t)) \qquad \text{and} \quad X_t^{(-)} = X(\psi^-(t)).$$

As in Chapter IV, Section 4, we see that $X_t^{(+)} = (X_t^{(+)1}, X^{(+)2}, \ldots, X_t^{(+)d})$ and $X_t^{(-)} = (X_t^{(-)1}, X_t^{(-)2}, \ldots, X_t^{(-)d})$ satisfy the following stochastic differential equations with appropriate d-dimensional Brownian motions $B_t^{(+)} = (B_t^{(+)1}, B_t^{(+)2}, \ldots, B_t^{(+)d})$ with $B_0^{(+)} = 0$ and $B_t^{(-)} = (B_t^{(-)1}, B_t^{(-)2}, \ldots, B_t^{(-)d})$ with $B_0^{(-)} = 0$ respectively:

$$(4.9) \quad \begin{cases} dX_t^{(+)i} = (a^+(p(X_t^{(+)}))/a(X_t^{(+)}))^{1/2} \sum_{k=1}^{d} \sigma_k^i(X_t^{(+)})dB_t^{(+)k} \\ \qquad + [a^+(p(X_t^{(+)}))/a(X_t^{(+)})]b^i(X_t^{(+)})dt, \quad i = 1,2, \ldots, d \\ X_0^{(+)} = x_0, \end{cases}$$

$$(4.10) \quad \begin{cases} dX_t^{(-)i} = (a^-(p(X_t^{(-)}))/a(X_t^{(-)}))^{1/2} \sum_{k=1}^{d} \sigma_k^i(X_t^{(-)})dB_t^{(-)k} \\ \qquad + [a^-(p(X_t^{(-)}))/a(X_t^{(-)})]b^i(X_t^{(-)})dt, \quad i = 1,2, \ldots, d \\ X_0^{(-)} = x_0. \end{cases}$$

By Itô's formula,

$$(4.11) \quad \begin{cases} dp(X_t^{(+)}) = (a^+(p(X_t^{(+)}))/a(X_t^{(+)}))^{1/2} \sum_{i,j=1}^{d} \sigma_j^i(X_t^{(+)})\frac{\partial p}{\partial x^i}(X_t^{(+)})dB_t^{(+)j} \\ \qquad + [a^+(p(X_t^{(+)}))/a(X_t^{(+)})]\,(Lp)(X_t^{(+)})dt \\ p(X_0^{(+)}) = \xi_0 \end{cases}$$

and

$$(4.12) \quad \begin{cases} dp(X_t^{(-)}) = (a^-(p(X_t^{(-)}))/a(X_t^{(-)}))^{1/2} \sum_{i,j=1}^{d} \sigma_j^i(X_t^{(-)})\frac{\partial p}{\partial x^i}(X_t^{(-)})dB_t^{(-)j} \\ \qquad + [a^-(p(X_t^{(-)}))/a(X_t^{(-)})](Lp)(X_t^{(-)})dt \\ p(X_t^{(-)}) = \xi_0. \end{cases}$$

Hence, if we set

$$\tilde{B}_t^{\pm} = \sum_{i,j=1}^{d} \int_0^t (1/a(X_s^{(\pm)}))^{1/2} \sigma_j^i(X_s^{(\pm)}) \frac{\partial p}{\partial x^i}(X_s^{(\pm)}) dB_s^{(\pm)j}$$

then $(\tilde{B}_t^+)$ and $(\tilde{B}_t^-)$ are 1-dimensional Brownian motions and we have

$$(4.13) \qquad \begin{cases} dp(X_t^{(+)}) = (a^+(p(X_t^{(+)})))^{1/2} d\tilde{B}_t^+ + a^+(p(X_t^{(+)}))b(X_t^{(+)})dt \\ p(X_0^{(+)}) = \xi_0 \end{cases}$$

and

$$(4.14) \qquad \begin{cases} dp(X_t^{(-)}) = (a^-(p(X_t^{(-)})))^{1/2} d\tilde{B}_t^- + a^-(p(X_t^{(-)}))b(X_t^{(-)})dt \\ p(X_0^{(-)}) = \xi_0. \end{cases}$$

Let $\eta_t^+ = p(X_t^{(+)})$ and $\eta_t^- = p(X_t^{(-)})$. Then by (4.13) and (4.14) we have

$$(4.15) \qquad \begin{cases} d\eta_t^+ = (a^+(\eta_t^+))^{1/2} d\tilde{B}_t^+ + a^+(\eta_t^+)b(X_t^{(+)})dt \\ \eta_0^+ = \xi_0 \end{cases}$$

and

$$(4.16) \qquad \begin{cases} d\eta_t^- = (a^-(\eta_t^-))^{1/2} d\tilde{B}_t^- + a^-(\eta_t^-)b(X_t^{(-)})dt \\ \eta_0^- = \xi_0. \end{cases}$$

Consider the following stochastic differential equation

$$(4.17) \qquad \begin{cases} d\eta_t^{++} = (a^+(\eta_t^{++}))^{1/2} d\tilde{B}_t^+ + a^+(\eta_t^{++})b^+(\eta_t^{++})dt, \\ \eta_0^{++} = \xi_0. \end{cases}$$

In Theorem 1.1, take

$$X_1(t) = \eta_t^+, \qquad X_2(t) = \eta_t^{++},$$
$$\sigma(t, \xi) = (a^+(\xi))^{1/2}, \ \beta_1(t) = b(X_t^{(+)})a^+(\eta_t^+), \ \beta_2(t) = b^+(\eta_t^{++})a^+(\eta_t^{++})$$

and

$$b_1(t, \xi) = b_2(t, \xi) = b^+(\xi)a^+(\xi).$$

Since we assumed that $a^+(\xi)$ and $b^+(\xi)$ are locally Lipschitz continuous, the pathwise uniqueness of solutions holds for the equation (4.17) and hence, by Theorem 1.1, we have

(4.18) $\eta_t^+ \leq \eta_t^{++}$ for all $t \geq 0$, a.s.

Similarly, if η_t^{+-} is the solution of the stochastic differential equation

$$\text{(4.19)}\qquad \begin{cases} d\eta_t^{+-} = (a^+(\eta_t^{+-}))^{1/2} d\tilde{B}_t^+ + a^+(\eta_t^{+-}) b^-(\eta_t^{+-}) dt \\ \eta_0^{+-} = \xi_0, \end{cases}$$

then we have

(4.20) $\eta_t^{+-} \leq \eta_t^+$ for all $t \geq 0$, a.s.

Also, if η_t^{-+} is the solution of the stochastic differential equation

$$\text{(4.21)}\qquad \begin{cases} d\eta_t^{-+} = (a^-(\eta_t^{-+}))^{1/2} d\tilde{B}_t^- + a^-(\eta_t^{-+}) b^+(\eta_t^{-+}) dt \\ \eta_0^{-+} = \xi_0, \end{cases}$$

then we have

(4.22) $\eta_t^- \leq \eta_t^{-+}$ for all $t \geq 0$, a.s.

and if η_t^{--} is the solution of

$$\text{(4.23)}\qquad \begin{cases} d\eta_t^{--} = (a^-(\eta_t^{--}))^{1/2} d\tilde{B}_t^- + a^-(\eta_t^{--}) b^-(\eta_t^{--}) dt \\ \eta_0^{--} = \xi_0, \end{cases}$$

then

(4.24) $\eta_t^{--} \leq \eta_t^-$ for all $t \geq 0$, a.s.

Finally, set $\eta_t = p(X_t)$. Then, since

$$\max_{0\leq s\leq t} \eta_s^+ = \max_{0\leq s\leq \psi^+(t)} \eta_s, \qquad \min_{0\leq s\leq t} \eta_s^+ = \min_{0\leq s\leq \psi^+(t)} \eta_s$$

and $t \leq \psi^+(t)$, we have

(4.25) $\max\limits_{0\leq s\leq t} \eta_s^+ \geq \max\limits_{0\leq s\leq t} \eta_s$ and $\min\limits_{0\leq s\leq t} \eta_s^+ \leq \min\limits_{0\leq s\leq t} \eta_s$.

Similarly, using the fact that $t \geq \psi^-(t)$, we have

(4.26) $\max\limits_{0\leq s\leq t} \eta_s^- \leq \max\limits_{0\leq s\leq t} \eta_s$ and $\min\limits_{0\leq s\leq t} \eta_s^- \geq \min\limits_{0\leq s\leq t} \eta_s$.

Combining (4.18), (4.24), (4.25) and (4.26), we have

$$\bar{\eta}_t^{--} = \max_{0\le s\le t} \eta_s^{--} \le \max_{0\le s\le t} \eta_s^{-} \le \max_{0\le s\le t} \eta_s = \bar{\eta}_t$$

and

$$\bar{\eta}_t = \max_{0\le s\le t} \eta_s \le \max_{0\le s\le t} \eta_s^{+} \le \max_{0\le s\le t} \eta_s^{++} = \bar{\eta}_t^{++}.$$

Similarly we can obtain

$$\underline{\eta}_t^{+-} \le \underline{\eta}_t \le \underline{\eta}_t^{-+}.$$

This proves the theorem.

Remark 4.1. If $a(x)$ in Theorem 4.1 depends only on $p(x)$, i.e., if there exists a function $\tilde{a}(\xi)$ defined on I such that $a(x) = \tilde{a}(p(x))$, then $a^+(\xi) = a^-(\xi) = \tilde{a}(\xi)$ and therefore we may assume that $\eta_t^{++} = \eta_t^{-+}$ and $\eta_t^{+-} = \eta_t^{--}$. In this case we have

$$\eta_t^{--} \le \eta_t \le \eta_t^{++} \qquad \text{for all} \quad t \ge 0, \qquad \text{a.s.}$$

As an application of Theorem 4.1 we shall now investigate the possibility of explosions for non-singular diffusions. Let $X=(X(t))$ be the d-dimensional diffusion process determined by the solution of (4.1). We assume $d \ge 2$ and $\det(a^{ij}(x)) > 0$ for all $x \in \boldsymbol{R}^d$. Our problem then is to find a criterion which determines whether or not explosion happens in finite time, i.e., whether e is finite or infinite. It is easy to see that $X(t)$ leaves any bounded set containing $X(0)$ in a finite time a.s., and hence we may assume for this problem that $a^{ij}(x) = \delta_{ij}$ and $b^i(x) = 0$ for $|x| \le 1$ and $i,j = 1, 2, \ldots, d$. Let $p(x) = |x|^2/2 = \sum_{i=1}^{d} (x^i)^2/2$, $x \in \boldsymbol{R}^d$. Then $I = [0, \infty)$ and $S = \{0\}$. In this case,

$$a(x) = \sum_{i,j=1}^{d} a^{ij}(x) x^i x^j$$

and

$$b(x) = \frac{1}{a(x)} \{\sum_{i=1}^{d} a^{ii}(x)/2 + \sum_{i=1}^{d} b^i(x) x^i\} \qquad \text{for} \quad x \in \boldsymbol{R}^d \setminus \{0\}.$$

Following (4.5), we set

$$a^+(r) = \max_{|x|=\sqrt{2r}} a(x), \qquad a^-(r) = \min_{|x|=\sqrt{2r}} a(x)$$

and

$$b^+(r) = \max_{|x|=\sqrt{2r}} b(x), \qquad b^-(r) = \min_{|x|=\sqrt{2r}} b(x)$$

for $r \in (0, \infty)$. It is easy to see that these functions are locally Lipschitz continuous and $a^\pm(r) > 0$ for $r \in (0, \infty)$. Let $r^+ = (r_t^+)$ and $r^- = (r_t^-)$ be the minimal diffusion processes generated by the operators

$$a^+(r)\left(\frac{1}{2}\frac{d^2}{dr^2} + b^+(r)\frac{d}{dr}\right)$$

and

$$a^-(r)\left(\frac{1}{2}\frac{d^2}{dr^2} + b^-(r)\frac{d}{dr}\right)$$

respectively. Set

$$c^+(r) = \exp\int_1^r 2b^+(u)du, \qquad c^-(r) = \exp\int_1^r 2b^-(u)du$$

and

$$s^+(r) = \int_1^r \frac{1}{c^+(u)}du, \qquad s^-(r) = \int_1^r \frac{1}{c^-(u)}du$$

for $r \in (0, \infty)$. By the above assumption,

$$s^+(r) = s^-(r) = \begin{cases} -\left(\frac{d}{2} - 1\right)(1/r^{\frac{d}{2}-1} - 1) & d > 2 \\ -\log\frac{1}{r} & d = 2 \end{cases}$$

for $r \in (0,1)$. Hence $s^+(0-) = s^-(0-) = -\infty$. By Theorem 3.2, if e^+ and e^- are the explosion times for $(r^+(t))$ and $(r^-(t))$ respectively, then

$$P_r^+\{e^+ = \infty\} = 1 \quad \text{or} \quad P_r^+\{e^+ < \infty\} = 1 \qquad \text{for} \quad r \in (0, \infty)$$

accordingly as

$$\int_1^\infty [1/c^+(r)] \int_1^r [c^+(u)/a^+(u)]dudr = \infty \quad \text{or} \quad < \infty;$$

also

$$P_r^-\{e^- = \infty\} = 1 \quad \text{or} \quad P_r^-\{e^- < \infty\} = 1 \qquad \text{for} \quad r \in (0, \infty)$$

accordingly as

$$\int_1^\infty [1/c^-(r)] \int_1^r [c^-(u)/a^-(u)]\, dudr = \infty \quad \text{or} \quad < \infty.$$

Here P_r^+ and P_r^- are the probability laws of $r^+ = (r^+(t))$ with $r^+(0)=r$ and $r^- = (r^-(t))$ with $r^-(0)=r$ respectively. By Theorem 4.1, we may assume that

$$\max_{0 \le s \le t} r^-(s) \le \max_{0 \le s \le t} |X(s)|^2/2 \le \max_{0 \le s \le t} r^+(s), \quad \text{a.s.}$$

and hence

$$e^+ \le e \le e^-, \qquad \text{a.s.}$$

Consequently we have the following result.

Theorem 4.2. (Hashiminsky [46]). Let P_x be the probability law of the solution $X = (X(t))$ with $X(0) = x$ of (4.1).

(i) If

$$\int_1^\infty 1/c^+(r) \int_1^r [c^+(u)/a^+(u)]dudr = \infty,$$

then $P_x(e = \infty) = 1$ for all $x \in \boldsymbol{R}^d$.

(ii) If

$$\int_1^\infty 1/c^-(r) \int_1^r [c^-(u)/a^-(u)]dudr < \infty,$$

then $P_x(e < \infty) = 1$ for all $x \in \boldsymbol{R}^d$.

5. Applications to diffusions on Riemannian manifolds

First we shall introduce some necessary notions in differential geometry.

Let M be a d-dimensional complete Riemannian manifold and ∇ be the Riemannian connection (cf. Chapter V, Section 4). To every pair $X, Y \in \mathfrak{X}(M)$ we associate a mapping R_{XY}: $\mathfrak{X}(M) \longrightarrow \mathfrak{X}(M)$, called the *curvature transform*, by

$$(5.1) \qquad R_{XY} = \nabla_{[X,Y]} - (\nabla_X \nabla_Y - \nabla_Y \nabla_X).$$

In local coordinates $(x^1, x^2, \dots, x^d)$,

$$(5.2) \qquad R_{XY}Z = X^i Y^j Z^k R_{\partial_i \partial_j} \partial_k$$

where $X = X^i \partial_i$, $Y = Y^i \partial_i$ and $Z = Z^i \partial_i$ $\left(\partial_i = \dfrac{\partial}{\partial x^i}\right)$; furthermore,

$$(5.3) \qquad \langle R_{\partial_i \partial_j} \partial_k, \partial_l \rangle = R_{ijkl} \ * \quad \text{and} \quad R_{ijkl} = g_{ih} R^h{}_{jkl}$$

where $R^i{}_{jkl}$ was defined in Chapter V, Section 5 by

$$R^i{}_{jkl} = \partial_k \left\{ {}_l{}^i{}_j \right\} - \partial_l \left\{ {}_k{}^i{}_j \right\} + \left(\left\{ {}_l{}^a{}_j \right\} \left\{ {}_k{}^i{}_a \right\} - \left\{ {}_k{}^a{}_j \right\} \left\{ {}_l{}^i{}_a \right\} \right).$$

A *plane section* ξ *at* $x \in M$ is a 2-dimensional subspace of $T_x(M)$. Let ξ be a plane section at x and let $X, Y \in T_x(M)$ be an orthonormal basis for ξ. The *sectional curvature* $K(\xi)$ of ξ is defined by

$$K(\xi) = \langle R_{XY} Y, X \rangle.$$

It can be shown that $K(\xi)$ depends on ξ alone and is independent of the particular choice of X and Y. In local coordinates,

$$K(\xi) = X^i Y^j Y^k X^l R_{ijkl}$$

if $X = X^i \partial_i$, $Y = Y^i \partial_i$ and $\{X, Y\}$ are orthonormal. For $X \in \mathfrak{X}(M)$ such that $\|X\| = \sqrt{\langle X, X \rangle} = 1$, the *Ricci curvature* $\rho(X)$ *of the direction* X is defined by

$$\rho(X) = \sum_{i=2}^{d} K(\{\{X, Y_i\}\}),$$

where $\{X, Y_2, \dots, Y_d\}$ is an orthonormal basis at $T_x(M)$ for each x and $\{\{X, Y_i\}\}$ is the plane section generated by X and Y_i. $\rho(X)$ is independent of the choice of $\{Y_2, Y_3, \dots, Y_d\}$. In local coordinates,

* $\langle \cdot, \cdot \rangle$ is the inner product in each tangent space.

$$\rho(X) = X^i X^j R_{ij} \quad \text{and} \quad R_{ij} = R^k{}_{ijk}$$

where $X = X^i \partial_i$ with $\|X\| = 1$.

Definition 5.1. M is called a space of constant curvature if $K(\xi)$ is independent of both the choice of the plane section ξ at each $x \in M$ and of the point $x \in M$.*[1]

Definition 5.2. Let γ: $I \longrightarrow M$ be a smooth curve. γ is called a *geodesic* if the tangent vectors $\gamma_* = \dfrac{d\gamma}{dt}$ are parallel along γ; in local coordinates, $\gamma(t) = (\gamma^1(t), \gamma^2(t), \ldots, \gamma^d(t))$ is a geodesic if and only if

$$\frac{d^2\gamma^k}{dt^2}(t) + \left\{{}_{i}{}^{k}{}_{j}\right\} \frac{d\gamma^i}{dt}\frac{d\gamma^j}{dt} = 0, \qquad t \in I. \tag{5.4}$$

The following are a consequence of the existence and uniqueness theorems for the differential equation (5.4):*[2] for every $x \in M$ and $X \in T_x(M)$ there exists a unique open interval $J(X)$ in $\boldsymbol{R}^1$ which contains 0 such that

(i) there exists a geodesic γ: $J(X) \longrightarrow M$ such that $\gamma(0) = x$ and $\gamma_*(0) = X$;

(ii) if $a < 0 < b$ and c: $(a,b) \longrightarrow M$ is geodesic such that $c(0) = x$ and $c_*(0) = X$, then $(a,b) \subset J(X)$ and $c(t) = \gamma(t)$ for $t \in (a,b)$.

Definition 5.3. The above geodesic γ is denoted by γ_X. We set

$$\widehat{T_x(M)} = \{X \in T_x(M);\ 1 \in J(X)\}$$

and define the *exponential mapping* $\exp$: $\widehat{T_x(M)} \longrightarrow M$ by

$$\exp X = \gamma_X(1). \tag{5.5}$$

It is also denoted by $\exp_x X$. Clearly $\exp tX = \gamma_X(t)$.

Set

$$r_0 = r_{x_0}(M) = \max\{r;\ \text{there exists a ball in } T_{x_0}(M) \text{ of radius } r \text{ about } 0 \text{ on which } \exp_{x_0} \text{ is a diffeomorphism}\}.$$

---

*[1] By Schur's theorem, the first property implies the second property if $d \geq 3$ ([7]).

*[2] e.g. [47].

If M is a simply connected space of constant curvature c then r_0 is given as

$$(5.6)\qquad r_0(c) = \begin{cases} \pi/\sqrt{c}, & c > 0 \\ \infty, & c \leq 0. \end{cases}$$

We also state the following fact known as Meyer's theorem. If the lower bound of the Ricci curvature is not less than $(d-1)a^2 > 0$ * then M is compact and the diameter of M is not greater than π/a. Furthermore, the fundamental group of M is finite ([7] and [122]).

Let $\xi = (\theta^1, \theta^2, \ldots, \theta^{d-1})$ be the spherical polar coordinate on the sphere in $T_{x_0}(M)$. $(r = d(x_0, x), \theta^1, \theta^2, \ldots, \theta^{d-1})$ induces a local coordinate in a neighbourhood U of x_0, called the *geodesic polar coordinate*, by the exponential map

$$(r, \xi) \longmapsto \exp_{x_0} r\xi.$$

Under this local coordinate, the Riemannian metric has the form

$$(g_{ij}) = \begin{pmatrix} 1 & 0 \\ 0 & (\bar{g}_{ij}) \end{pmatrix},$$

where $(\bar{g}_{ij}(r, \theta^1, \theta^2, \ldots, \theta^{d-1}))$ is a Riemannian metric on the geodesic sphere $S(x_0; r) = \{x; d(x_0, x) = r, x \in M\}$. The Laplace-Beltrami operator has the following expression

$$(5.7)\qquad \begin{aligned} \Delta = \frac{\partial^2}{\partial r^2} &+ \frac{1}{\sqrt{\det G}} \left(\frac{\partial}{\partial r} \sqrt{\det G}\right) \frac{\partial}{\partial r} \\ &+ \frac{1}{\sqrt{\det G}} \sum_{i,j=1}^{d-1} \frac{\partial}{\partial \theta^i} \left(\bar{g}^{ij} \sqrt{\det G} \frac{\partial}{\partial \theta^j}\right) \end{aligned}$$

where $G = (g_{ij})$ ([47]).

Let h be the Jacobian determinant of $\exp_{x_0}$. Then we have

$$(5.8)\qquad \frac{\partial}{\partial r} \log h + \frac{d-1}{r} = \frac{\partial}{\partial r} \log \sqrt{\det G}\,.$$

Also, if M is a space of constant curvature c, then

* We assume $a > 0$.

$$(5.9) \qquad \frac{\partial}{\partial r}\log\sqrt{\det G} = \frac{\partial}{\partial r}\log A, \qquad 0 < r < r_0(c),$$

where

$$A = A(r, c) = \begin{cases} \left(\frac{1}{\sqrt{c}}\sin\sqrt{c}\,r\right)^{d-1}, & c > 0 \\ r^{d-1}, & c = 0 \\ \left(\frac{1}{\sqrt{-c}}\sinh\sqrt{-c}\,r\right)^{d-1}, & c < 0. \end{cases}$$

The following lemma plays an important role in our further discussions.

Lemma 5.1. Let the Ricci curvature $\rho(X)$ satisfy that

$$(d-1)b \leq \rho(X), \qquad X \in T_x(M), \qquad \|X\| = 1,$$

and the sectional curvature $K(\xi)$ satisfy that $K(\xi) \leq a$ for every plane section ξ. Then we have

$$\text{(i)} \qquad \frac{\partial}{\partial r}\log h \leq \frac{\partial}{\partial r}\log A(r,b) - \frac{d-1}{r}, \qquad r \in (0, r_{x_0}(M))$$

$$\frac{\partial}{\partial r}\log h \geq \frac{\partial}{\partial r}\log A(r,a) - \frac{d-1}{r}, \qquad r \in (0, r_0(a))$$

and

$$\text{(ii)} \qquad h(r) \leq A(r, b)/r^{d-1}, \qquad r \in (0, r_{x_0}(M))$$

$$h(r) \geq A(r, a)/r^{d-1}, \qquad r \in (0, r_0(a)).$$

For a proof, see [7] (pp. 253 ~ 256) and [42].

Generally, a coordinate mapping $\phi: U \longrightarrow \boldsymbol{R}^d$, $U \subset M$ is called a *normal coordinate mapping* at $x = \phi^{-1}(0)$ if the inverse images of rays through $0 \in \boldsymbol{R}^d$ are geodesics. A normal coordinate neighborhood N, the domain of a normal coordinate mapping ϕ, has the property that every $y \in N$ can be joined to $\phi^{-1}(0)$ by a unique geodesic in N. As a corollary to Lemma 5.1 we have the following.

Corollary. If the Ricci curvature ρ satisfies $\rho(X) \geq (d-1)c$, $X \in T_x(M)$, $\|X\| = 1$, $x \in M$, then the volume of a normal coordinate ball

$B(x_0;r)$ is smaller than or equal to the volume of a normal coordinate ball of the same size in the simple space form of constant curvature c.

By the simple space form we mean the sphere if $c > 0$, the Euclidean space if $c = 0$ and the hyperbolic space if $c < 0$, (cf. [7], p 256).

Now we shall apply the comparison theorem to the Brownian motions on M. Let $x_0 \in M$ be fixed and let $B(x_0,r) = \{x \in M; d(x_0,x) < r\}$. For a fixed $r^* \leq r_{x_0}(M)$, let $X = (X_t)$ be the minimal Brownian motion on $B(x_0,r^*)$. As we saw in Chapter V, Section 4 and Section 8 the transition probability density $p(t,x,y)$ of X exists with respect to the Riemannian volume $m(dx)$:

$$P_x(X_t \in dy) = p(t, x, y)m(dy);$$

moreover $p(t, x, y)$ is C^∞ in $(0, \infty) \times B(x_0,r^*) \times B(x_0,r^*)$ and $p(t, x, y) = p(t, y, x)$. Let $r^* \leq r_0(c)$ and let $\xi^c = (\xi^c(t))$ be the minimal $L^{(c)}$-diffusion on $(0, r^*)$, where the operator $L^{(c)}$ is given by

$$L^{(c)} = \frac{1}{2}\left(\frac{d^2}{dr^2} + \left(\frac{d}{dr}\log A(r, c)\right)\frac{d}{dr}\right) = \frac{1}{2}\,\frac{1}{A(r,c)}\,\frac{d}{dr}\left(A(r, c)\frac{d}{dr}\right).$$

Let us assume that $d \geq 2$. Then

$$s(r) = \int_{r_0}^{r} \exp\left(-\int_{r_0}^{y} \frac{A'(z, c)}{A(z, c)}\,dz\right) dy = \int_{r_0}^{r} \frac{A(r_0,c)}{A(y,c)}\,dy \quad {}^{*1}$$

satisfies $s(0+) = -\infty$. Hence the results of Section 3 imply that 0 cannot be reached in a finite time for ξ^c. Also, the transition probability density $p^{(c)}(t, \xi, \eta)$ exists with respect to the measure $A(\eta, c)d\eta$, is C^∞ in $(0, \infty) \times (0, r^*) \times (0, r^*)$ and satisfies $p^{(c)}(t, \xi, \eta) = p^{(c)}(t, \eta, \xi)$. Noting now (5.7), (5.8) (5.9) and Lemma 5.1, the comparison theorem immediately implies the following results.

Theorem 5.1. Let the Ricci curvature $\rho(X)$ satisfy that $(d-1)b \leq \rho(X)$ for all $X \in T_x(M)$, $\|X\| = 1$ and $x \in M$. (Let the sectional curvature $K(\xi)$ satisfy that $K(\xi) \leq a$ for all plane sections ξ at every point x.) Let us fix r^* such that $0 < r^* \leq r_{x_0}(M)$ (respectively $0 < r^* \leq r_0(a)$).*2 Then we can construct, on a suitable probability space, the above diffusions (X_t) and $(\xi_t^{(b)})$ (respectively (X_t) and $(\xi_t^{(a)})$) such that $\xi_0^{(b)} = d(x_0,X_0)$

*1 $0 < r_0 < r^*$.

*2 We have automatically that $r^* \leq r_0(b)$ (respectively $r^* \leq r_{x_0}(M)$).

(respectively $\xi_0^{(a)} = d(x_0, X_0)$) and the following is satisfied: if $r_t = d(x_0, X_t)$, then

$$\text{(5.10)} \qquad \xi_t^{(b)} \geq r_t \qquad \text{(respectively} \quad \xi_t^{(a)} \leq r_t)$$

for $t \geq 0$, a.s.

Corollary 1. (A.Debiard-B.Gaveau-E.Mazet [14]).
(i)

$$\lim_{r_1 \downarrow 0} p^{(b)}(t, r, r_1) \leq p(t, y, x_0), \quad y = (r, \theta), \quad 0 < r < r^*$$
$$\text{(respectively} \quad \lim_{r_1 \downarrow 0} p^{(a)}(t, r, r_1) \geq p(t, y, x_0)).$$

(ii) If ζ is the explosion time (i.e. the hitting time to r^* for $(\xi_t^{(b)})$ (respectively $(\xi_t^{(a)})$) and the hitting time to $\partial B(x_0, r^*)$ for (X_t)), then $E_r^{(b)}(\zeta) \leq E_y(\zeta)$, $y = (r, \theta)$, $0 < r < r^*$ (respectively $E_r^{(a)}(\zeta) \geq E_y(\zeta)$) where $E_r^{(c)}$ and E_x stand for the expectations with respect to the probability measures of $(\xi_t^{(c)})$ with $\xi_0^{(c)} = r$ and (X_t) with $X_0 = x$ respectively.

Remark 5.1. 0 is an entrance boundary for $(\xi_t^{(a)})$ and $(\xi_t^{(b)})$ in the sense of Itô-McKean [73]. It is well known then that $\lim_{r_1 \downarrow 0} p^{(a)}(t, r, r_1)$ and $\lim_{r_1 \downarrow 0} p^{(b)}(t, r, r_1)$ exist.

The proof is immediate from the above theorem and the corollary to Lemma 5.1.

Corollary 2. Let M be a connected, complete d-dimensional Riemannian manifold ($d \geqq 2$) with non-positive sectional curvature for all plane sections. Furthermore, we assume that the Ricci curvature ρ satisfies the condition

$$0 \geq \rho(X) \geq (d-1)c > -\infty, \; X \in T_x(M), \; \|X\| = 1, \; x \in M,$$

then (X_t) is conservative i.e.,

$$P(t, x, M) = 1 \qquad \text{for all } x \in M.$$

Proof. Without loss of generality, we may assume that M is simply connected because the Brownian motion on M is obtained as the projection of the Brownian motion on the universal covering space $\tilde{M}$ of M. Then $r^* = \infty$ by the Cartan-Hadamard theorem ([7]) and the assertion follows immediately from the theorem.

In this result, the condition on sectional curvature can be dropped. For details, cf. [186] and [206]. Similarly, we see that if M is a complete simply connected Riemannian manifold and satisfies the condition

$$0 > a \geq K(\xi) > -\infty \qquad \text{for all plane sections} \quad \xi$$

then (X_t) is non-recurrent.

Further applications of comparison theorems, especially, applications to the smallest eigenvalue of the Laplacian, can be found in Azencott [191] Malliavin [108], A. Debiard-B. Gaveau-E. Mazet [14] and M. Pinsky [140].

6. Stochastic line integrals along the paths of diffusion processes

Let M be a manifold and γ be a smooth bounded curve in M. Then the line integral $\int_\gamma \alpha$ is defined for any differential 1-form α. Now let $X = (X(t))$ be an M-valued quasimartingale, i.e., $t \longmapsto f(X(t))$ is a quasimartingale in the sense of Chapter III, Section 1, for every $f \in F_0(M)$. We can then also define the "*stochastic*" line integral $\int_{X[0,t]} \alpha$ along the curve $\{X(s),\ s \in [0,t]\}$ for any differential 1-form α by making use of stochastic calculus; moreover the resulting process $t \longmapsto \int_{X[0,t]} \alpha$ is a quasimartingale. One standard way of defining this is as follows.

We choose a locally finite covering $\{W_n\}$ of M consisting of coordinate neighborhoods and choose coverings $\{U_n\}$ and $\{V_n\}$ of M such that

$$\bar{U}_n \subset V_n \subset \bar{V}_n \subset W_n.$$

Define sequences $\{\sigma_k^{(n)}\}$ and $\{\tau_k^{(n)}\}$ of stopping times by

$$\begin{aligned}
\sigma_{-1}^{(n)} &= 0, \\
\tau_k^{(n)} &= \inf\{t\,;\, t > \sigma_{k-1}^{(n)},\ X_t \in U_n\} \\
\sigma_k^{(n)} &= \inf\{t\,;\, t > \tau_k^{(n)},\ X_t \notin V_n\}.
\end{aligned}$$

Let $\alpha = \alpha_i(x)dx^i$ and $X(t) = (X^i(t)),\ \tau_k^{(n)} \leq t \leq \sigma_k^{(n)}$, under the local coordinate in W_n. We define $\int_{X[0,t]} \alpha$ by setting

$$\int_{X[0,t]} \alpha = \sum_{n=1}^{\infty} \sum_{k=1}^{\infty} \sum_{i=1}^{d} \int_{\tau_k^{(n)} \wedge t}^{\sigma_k^{(n)} \wedge t} (\psi_n \alpha_i)\,(X(s)) \circ dX^i(s)$$

where $\{\psi_n\}$ is a partition of unity subordinate to $\{U_n\}$. It is easy to see that

$\int_{X[0,t]} \alpha$ is well defined and is independent of the particular choice of coordinate neighbourhoods (cf. [50] and [51]). In the following, we shall restrict ourselves to the case when $X(t)$ is a non-singular diffusion on M. As we shall see, $\int_{X[0,t]} \alpha$ can be defined more directly in this case.

Let M be a d-dimensional manifold and $X = (X(t))$ be a non-singular diffusion process on M. As we saw in Remark V-4.2, we may assume that M is a Riemannian manifold and $X(t)$ is obtained as the projection onto M of the process $r(t)$ on $O(M)$ which is the solution of

$$(6.1) \qquad \begin{cases} dr(t) = \tilde{L}_k(r(t))\circ dw^k(t) + \tilde{L}_0(r(t))dt \\ r(0) = r. \text{ *} \end{cases}$$

Here $\{\tilde{L}_1, \tilde{L}_2, \ldots, \tilde{L}_d\}$ is the system of canonical vector fields corresponding to the Riemannian connection and $\tilde{L}_0$ is the horizontal lift of a vector field b on M. The projection $X(t) = \pi[r(t)]$ of $r(t)$ onto M is the diffusion process on M generated by the operator $\frac{1}{2}\Delta + b$. For simplicity, we shall assume that $X(t)$ is conservative. Let $\alpha \in \Lambda_1(M)$ be a differential 1-form and $(\bar{\alpha}_1(r), \bar{\alpha}_2(r), \ldots, \bar{\alpha}_d(r))$ be its scalarization. Recall that $\bar{\alpha}_i(r) = \alpha_j(x)e^j_i$, $i = 1, 2, \ldots, d$, if $\alpha(x) = \alpha_j(x)dx^j$ and $r = (x^i, e^j_i)$ in local coordinates. Let $\alpha(b) \in F(M)$ be defined by $\alpha(b)(x) = \alpha_i(x)b^i(x)$ if $\alpha = \alpha_i(x)dx^i$ and $b = b^i(x)\frac{\partial}{\partial x^i}$ in local coordinates.

Definition 6.1. We set

$$(6.2) \qquad \int_{X[0,t]} \alpha = \int_0^t \bar{\alpha}_k(r(s))\circ dw^k(s) + \int_0^t \alpha(b)(X(s))ds.$$

$\int_{X[0,t]} \alpha$ is called the *stochastic line integral of $\alpha \in \Lambda_1(M)$ along the curve* $\{X(s), s \in [0, t]\}$.

This definition coincides with the above mentioned one. Indeed, if U is a coordinate neighbourhood and $\sigma \leq \tau$ are any stopping times such that $X(t) \in U$ for all $t \in [\sigma, \tau]$, then

$$\int_\sigma^t \bar{\alpha}_k(r(s))\circ dw^k(s) + \int_\sigma^t \alpha(b)(X(s))ds$$
$$= \int_\sigma^t \alpha_j(X(s))[e^j_k(s)\circ dw^k(s) + b^j(X(s))ds]$$

* As in Chapter V, Section 4, $(w^k(t))$ is the canonical realization of the d-dimensional Wiener process.

$$= \int_\sigma^t \alpha_j(X(s)) \circ dX^j(s) \qquad \text{for any} \quad t \in [\sigma, \tau].$$

It is obvious that $t \longmapsto \int_{X[0,t]} \alpha$ is a quasimartingale. Its canonical decomposition is given in the following theorem.

Theorem 6.1.

$$\int_{X[0,t]} \alpha = \int_0^t \bar{\alpha}_k(r(s))dw^k(s) + \int_0^t \left(\alpha(b) - \frac{1}{2}\delta\alpha\right)(X(s))ds \tag{6.3}$$

where δ: $\Lambda_1(M) \longrightarrow F(M)$ is defined as in Chapter V, Section 4.

Proof. We have

$$\begin{aligned}\int_{X[0,t]} \alpha &= \int_0^t \bar{\alpha}_k(r(s)) \circ dw^k(s) + \int_0^t \alpha(b)(X(s))ds \\ &= \int_0^t \bar{\alpha}_k(r(s))dw^k(s) + \frac{1}{2}\int_0^t d[\bar{\alpha}_k(r(s))] \cdot dw^k(s) \\ &\quad + \int_0^t \alpha(b)(X(s))ds.\end{aligned}$$

Since $r(t)$ is the solution of (6.1),

$$d[\bar{\alpha}_k(r(s))] = (\tilde{L}_j\bar{\alpha}_k)(r(s)) \circ dw^j(s) + (\tilde{L}_0\bar{\alpha}_k)(r(s))ds$$

and hence

$$\begin{aligned}d[\bar{\alpha}_k(r(s))] \cdot dw^k(s) &= (\tilde{L}_j\bar{\alpha}_k)(r(s))dw^j(s) \cdot dw^k(s) \\ &= \sum_{k=1}^d (\tilde{L}_k\bar{\alpha}_k)(r(s))ds.\end{aligned}$$

By Proposition V-4.1, we have

$$\tilde{L}_k\bar{\alpha}_k = \overline{(\nabla\alpha)}_{kk}$$

where $\{\overline{(\nabla\alpha)}_{ij}\}$ is the scalarization of the (0,2)-tensor $\nabla\alpha$. Consequently

$$\tilde{L}_k\bar{\alpha}_k = (\nabla\alpha)_{ij}e^i_ke^j_k,$$

and by (4.22) and (5.14) of Chapter V,

$$\sum_{k=1}^{d} \tilde{L}_k \tilde{\alpha}_k = \sum_{k=1}^{d} (\nabla \alpha)_{ij} e_k^i e_k^j = g^{ij}(\nabla \alpha)_{ij} = -\delta \alpha.$$

Example 6.1. Let $M = \mathbf{R}^2$ and $X = (X_1(t), X_2(t))$ be the two dimensional Brownian motion. If

$$\alpha = \frac{1}{2}(x^1 dx^2 - x^2 dx^1), \qquad x = (x^1, x^2),$$

then

$$\int_{X[0,t]} \alpha = \frac{1}{2}\left\{\int_0^t X_1(s)dX_2(s) - \int_0^t X_2(s)dX_1(s)\right\}.$$

This stochastic integral in the case $X(0) = 0$ was introduced by P. Lévy ([100] and [101]) as the *stochastic area* enclosed by a Brownian curve and its chord. We set

$$S(t) = \int_{X[0,t]} \alpha = \frac{1}{2}\left\{\int_0^t X_1(s)dX_2(s) - \int_0^t X_2(s)dX_1(s)\right\} \tag{6.4}$$

and

$$r(t) = \sqrt{X_1(t)^2 + X_2(t)^2}.$$

Then by Itô's formula,

$$\frac{r(t)^2}{2} = \int_0^t r(s)dB_1(s) + t \tag{6.5}$$

where

$$B_1(t) = \int_0^t \frac{X_1(s)}{r(s)} dX_1(s) + \int_0^t \frac{X_2(s)}{r(s)} dX_2(s).$$

Clearly the system of martingales $(S(t), B_1(t))$ satisfies the relations $\langle B_1, B_1 \rangle_t = t$, $\langle B_1, S \rangle_t = 0$ and $\langle S, S \rangle_t = \frac{1}{4}\int_0^t r(s)^2 ds$. Let $B_2(t) = S(\phi_t)$ where ϕ_t is the inverse function of

$$t \longmapsto \frac{1}{4}\int_0^t r(s)^2 ds.$$

By Theorem II-7.3, the Brownian motions $B_1(t)$ and $B_2(t)$ are mutually independent. We suppose from now on that $X(0) = x$ is a fixed point in $\boldsymbol{R}^2$. Then we know that $r(t)$ is the pathwise unique solution of (6.5) with $r(0) = |x|$ (Chapter IV, Example 8.3). In particular, this implies that $\sigma[r(s), s \leq t] \subset \sigma[B_1(s), s \leq t]$, and consequently the processes $\{B_2(t)\}$ and $\{r(t)\}$ are mutually independent. Thus we have obtained the following result. *The stochastic area* $S(t)$ *has the expression*

$$S(t) = B_2\left(\frac{1}{4}\int_0^t r(s)^2 ds\right) \tag{6.6}$$

where $B_2(t)$ *is a one-dimensional Brownian motion independent of the process* $\{r(t)\}$. Suppose that $X(0)=x \neq 0$. Then $X(t) \neq 0$ for all $t \geq 0$ and hence we can introduce the polar coordinate representation: $X_1(t) = r(t)\cos\theta(t)$ and $X_2(t) = r(t)\sin\theta(t)$. An application of Itô's formula yields

$$\begin{aligned} dS(t) &= \frac{1}{2}\{r(t)\cos\theta(t)\sin\theta(t)dr(t) + r(t)^2(\cos\theta(t))^2 d\theta(t) \\ &\quad - r(t)\cos\theta(t)\sin\theta(t)dr(t) + r(t)^2(\sin\theta(t))^2 d\theta(t)\} \\ &= \frac{1}{2} r(t)^2 d\theta(t), \end{aligned}$$

and hence

$$S(t) = \frac{1}{2}\int_0^t r(s)^2 d\theta(s).$$

Then

$$\tilde{\theta}(t) := \theta(t) - \theta(0) = \int_0^t \frac{2}{r(s)^2} dS(r),$$

and consequently $d\langle B_1, \tilde{\theta}\rangle_t = \dfrac{2}{r(t)^2} d\langle B_1, S\rangle_t = 0$ and

$$d\langle\tilde{\theta}, \tilde{\theta}\rangle_t = \frac{4}{r(t)^4} d\langle S, S\rangle_t = \frac{1}{r(t)^2} dt.$$

Again by Theorem II-7.3, there exists a Brownian motion $\{B_3(t)\}$ ($B_3(0) = 0$) independent of $\{B_1(t)\}$ (and hence independent of the process $\{r(t)\}$) such that

$$\tilde{\theta}(t) = B_3\left(\int_0^t \frac{1}{r(s)^2}\,ds\right).$$

The formula

$$(6.7)\qquad \begin{cases} X_1(t) = r(t)\cos\left[\theta(0) + B_3\left(\int_0^t \frac{1}{r(s)^2}\,ds\right)\right] \\ X_2(t) = r(t)\sin\left[\theta(0) + B_3\left(\int_0^t \frac{1}{r(s)^2}\,ds\right)\right] \end{cases}$$

is known as the *skew product representation* of two dimensional Brownian motion.

As an application of (6.6) we can obtain the following formula in the case $X(0) = 0$:

$$(6.8)\qquad E(e^{i\xi S(t)}) = \left(\cosh\frac{\xi t}{2}\right)^{-1} \qquad \text{for}\quad \xi \in \boldsymbol{R}.$$

Indeed, noting the independence of $\{B_2(t)\}$ and $\{r(t)\}$, we have

$$\begin{aligned} E(e^{i\xi S(t)}) &= E\left(\exp\left[i\xi\, B_2\left(\frac{1}{4}\int_0^t r(s)^2 ds\right)\right]\right) \\ &= E\left(\exp\left[-\frac{\xi^2}{8}\int_0^t r(s)^2 ds\right]\right) \\ &= \left\{E\left(\exp\left[-\frac{\xi^2}{8}\int_0^t \eta(s)^2 ds\right]\right)\right\}^2 \end{aligned}$$

where $\eta(t)$ is a one-dimensional Brownian motion with $\eta(0) = 0$. Since $\left\{\frac{1}{\sqrt{c}}\eta(ct)\right\} \overset{\mathscr{L}}{\approx} \{\eta(t)\}$ for every $c > 0$,

$$E(e^{i\xi S(t)}) = \left\{E\left(\exp\left[-\frac{\xi^2}{8}\,t^2\int_0^1 \eta(s)^2 ds\right]\right)\right\}^2.$$

Therefore it is sufficient to prove the formula (Cameron-Martin [10] and Kac [80])

$$(6.9)\qquad E(\exp[-\lambda\int_0^1 \eta(s)^2 ds]) = (\cosh\sqrt{2\lambda})^{-1/2} \qquad (\lambda > 0).$$

Let $K(t, s) = t \wedge s$ and consider the eigenvalue problem in $\mathscr{L}_2([0, 1])$:

$$\lambda \int_0^1 K(t,s)\phi(s)ds = \phi(t).$$

The eigenvalues and eigenfunctions are given by

$$\begin{cases} \lambda_n = \left[\left(n + \frac{1}{2}\right)\pi\right]^2, \\ \phi_n = \sqrt{2}\,\sin\left(n + \frac{1}{2}\right)\pi x, \end{cases} \qquad n = 0, 1, \ldots .$$

Now it is easy to see that

$$\eta(t) = \sum_{n=0}^{\infty} \frac{1}{\sqrt{\lambda_n}} \xi_n \phi_n(t)$$

where the ξ_n are independent random variables with common distribution $N(0,1)$. Then

$$\int_0^1 \eta^2(s)ds = \sum_{n=0}^{\infty} \frac{\xi_n^2}{\lambda_n},$$

and hence

$$E(\exp[-\lambda \int_0^1 \eta^2(s)ds]) = \prod_{n=0}^{\infty} E\left(\exp\left[-\frac{\lambda}{\lambda_n}\xi_n^2\right]\right) = \prod_{n=0}^{\infty} \sqrt{\frac{1}{1 + 2\frac{\lambda}{\lambda_n}}}$$

$$= \prod_{n=0}^{\infty} \left(1 + \frac{8\lambda}{(2n+1)^2\pi^2}\right)^{-1/2} = (\cosh \sqrt{2\lambda})^{-1/2}.$$

We can also obtain a formula which is more detailed than (6.8) (Gaveau [33]). Let $X(t)$ be the two dimensional Brownian motion with $X(0) = 0$. Then for every $x \in \boldsymbol{R}^2$, $t > 0$ and $\xi \in \boldsymbol{R}$,

$$E(e^{i\xi S(t)} \mid X(t) = x) = \frac{\xi t}{2\sinh \frac{\xi t}{2}} \exp\left[\left(1 - \frac{\xi t}{2}\coth\frac{\xi t}{2}\right)\frac{|x|^2}{2t}\right]. \tag{6.10}$$

(6.8) is an easy consequence of (6.10) and hence the following proof also provides us with another proof of (6.8) (or (6.9)). *

* Still another proof of (6.9) may be found in [114].

Proof. By the rotation invariance of the Brownian motion it is easy to see that

$$E(e^{i\xi S(t)} \mid X(t) = x) = E(e^{i\xi S(t)} \mid r(t) = r)$$

where $r = |x|$. By (6.6), we have

$$E(e^{i\xi S(t)} \mid X(t) = x) = E\left(\exp\left[-\frac{\xi^2}{8}\int_0^t r(s)^2 ds\right] \mid r(t) = r\right).$$

But we know (Chapter V, Section 3) that

$$u(t, x) = E((\exp[-\alpha\int_0^t |x + X(s)|^2 ds])f(x+X(t))), \quad t > 0, \quad x \in \boldsymbol{R}^2,$$

where $\alpha > 0$ is a given constant, is the solution of the initial value problem

$$\text{(6.11)} \qquad \begin{cases} \dfrac{\partial u}{\partial t} = \dfrac{1}{2}\Delta u - \alpha |x|^2 u \\ u|_{t=0} = f. \end{cases}$$

Let $\{H_n(x); n = 0, 1, \ldots\}$ be the Hermite polynomials and set $\phi_{n,m}(x) = (\exp[-|x|^2])H_n(x_1)H_m(x_2)$ for $x = (x_1, x_2)$ and $n, m = 0, 1, \ldots$. Since $\phi_{n,m}$ satisfies

$$(\Delta - |x|^2)\phi_{n,m} + 2(n + m + 1)\phi_{n,m} = 0,$$

we can solve (6.11) by the method of eigenfunction expansion:

$$u(t, x) = \int_{\boldsymbol{R}^2} p_\alpha(t, x, y) f(y) dy$$

where

$$p_\alpha(t, x, y) = \sum_{n,m=0}^{\infty} e^{-\sqrt{2\alpha}(n+m+1)t} e_{n,m}(x) e_{n,m}(y)$$

and

$$e_{n,m}(x) = (\pi n! m! 2^{n+m}(2\alpha)^{-1/2})^{-1/2}\phi_{n,m}((2\alpha)^{1/4}x).$$

A formula on Hermite polynomials yields

$$p_\alpha(t,x,y)=\prod_{i=1}^{2}\sum_{n=0}^{\infty} e^{-\sqrt{2\alpha}(n+\frac{1}{2})t}\, e^{-\sqrt{2\alpha}(x_i^2+y_i^2)/2}\,\frac{H_n((2\alpha)^{1/4}x_i)H_n((2\alpha)^{1/4}y_i)}{\sqrt{\pi}\,2^n n!(2\alpha)^{-1/4}}$$

$$(6.12)\qquad =\prod_{i=1}^{2}\frac{\exp\left[-\frac{\sqrt{2\alpha}}{2}\coth\sqrt{2\alpha}t\,[x_i^2-2x_iy_i\operatorname{sech}\sqrt{2\alpha}t+y_i^2]\right]}{\sqrt{2\pi}\sinh\sqrt{2\alpha}t\,(2\alpha)^{-1/2}}$$

$$x=(x_1,x_2),\qquad y=(y_1,y_2).$$

Therefore,

$$E(e^{i\xi S(t)}\mid X(t)=x)=p_{\xi^2/8}(t,0,x)\left(\frac{1}{2\pi t}\exp\left[-\frac{|x|^2}{2t}\right]\right)^{-1}$$

$$=\frac{\xi t\exp\left[\left(1-\frac{\xi t}{2}\coth\frac{\xi t}{2}\right)\frac{|x|^2}{2t}\right]}{2\sinh\frac{\xi t}{2}}.$$

Finally we consider the three dimensional process

$$X(t)=(X_1(t),X_2(t),X_3(0)+S(t)),$$

where $(X_1(t), X_2(t))$ is a two-dimensional Brownian motion and $S(t)$ is defined by (6.4). It is a diffusion process on $\boldsymbol{R}^3$ with generator

$$A=\frac{1}{2}(L_1^2+L_2^2),$$

where

$$L_1=\frac{\partial}{\partial x_1}-\frac{x_2}{2}\frac{\partial}{\partial x_3}\quad\text{and}\quad L_2=\frac{\partial}{\partial x_2}+\frac{x_1}{2}\frac{\partial}{\partial x_3}.$$

It is a degenerate diffusion, but since $[L_1,L_2]=L_1L_2-L_2L_1=\frac{\partial}{\partial x_3}$, $\dim\mathfrak{L}\{L_1,L_2\}_x=3$ for every x. Thus the smooth transition probability density $p(t,x,y)$ exists by Theorem V-8.1. It is easy to obtain from (6.10) that

$$p(t,0,x)=\left(\frac{1}{2\pi t}\right)^2\int_{-\infty}^{\infty}\frac{\xi/2}{\sinh\xi/2}\exp\left\{i\frac{\xi}{t}x_3-\frac{x_1^2+x_2^2}{2t}\frac{\xi/2}{\tanh\xi/2}\right\}d\xi,\qquad x=(x_1,x_2,x_3)$$

(cf. Gaveau [33]). Also, by Example 8.1 given below, we see that the topological support $\mathscr{S}(P_x)$ of the law P_x of $\{X(t)\}$ with $X(0) = x$ is the whole space $W_x^3 = \{w; C([0, \infty) \longrightarrow R^3), w(0) = x\}$. For a further detailed study of this diffusion, we refer the reader to Gaveau [33].

Remarks 6.1. The formulas (6.8) and (6.10) can be obtained more directly by the following Fourier series expansion, (cf. [33], [216]): Let $X(t) = (X^1(t), X^2(t))$ be a two-dimensional Brownian motion such that $X(0) = 0$ and set, for $i = 1$ and 2,

$$\xi_k^{(i)} = \sqrt{2}\int_0^1 \sin 2\pi kt\, dX^i(t) \quad \text{and}$$

$$\eta_k^{(i)} = \sqrt{2}\int_0^1 \cos 2\pi kt\, dX^i(t), \qquad k = 1, 2, \cdots.$$

Note that $\{X^1(1), X^2(1), \xi_k^{(1)}, \xi_k^{(2)}, \eta_k^{(1)}, \eta_k^{(2)}, k = 1, 2, \cdots\}$ is a system of independent Gaussian random variables with mean 0 and variance 1. Let $S(1)$ be the stochastic area:

$$S(1) = \frac{1}{2}\int_0^1 \{X^1(s)dX^2(s) - X^2(s)dX^1(s)\}.$$

Then it admits the following expansion in the sense of $\mathscr{L}_2(P)$ and almost surely:

$$(*) \quad S(1) = \sum_{k=1}^{\infty} \frac{1}{2\pi k}\{(\eta_k^{(1)} - \sqrt{2}X^1(1))\xi_k^{(2)} - (\eta_k^{(2)} - \sqrt{2}X^2(1))\xi_k^{(1)}\}.$$

This can be deduced, at least heuristically, from the Fourier series expansion of $X^i(t)$:

$$X^i(t) = tX^i(1) + \sqrt{2}\sum_{k=1}^{\infty}(\xi_k^{(i)}\int_0^t \sin 2\pi ks\, ds + \eta_k^{(i)}\int_0^t \cos 2\pi ks\, ds) \qquad i = 1,2.$$

It can be proved rigorously as follows: For $F(s, t) \in \mathscr{L}_2([0, 1]^2)$, the multiple stochastic integral

$$I(F) = \int_0^1\int_0^1 F(s, t)dX^1(s)dX^2(t)$$

is defined to be the iterated stochastic integral

$$\int_0^1 \{ \int_0^1 F(s, t) dX^1(s) \} dX^2(s).$$

This is well-defined because, setting

$$\mathscr{F}_t = \sigma[X^1(u), X^2(v);\ 0 \leqq u \leqq 1, 0 \leqq v \leqq t\} \qquad \text{for } 0 \leqq t \leqq 1,$$

$\Phi(t) := \int_0^1 F(s, t) dX^1(s)$ is $(\mathscr{F}_t)$-adapted measurable process and $X^2(t)$ is an $(\mathscr{F}_t)$-Brownian motion. It is easy to see that $I(F)$ is also given by

$$\int_0^1 \{ \int_0^1 F(s, t) dX^2(t) \} dX^1(s)$$

and that, if $F, G \in \mathscr{L}_2([0, 1]^2)$, then

$$E[\{ I(F) - I(G) \}^2] = \int_0^1 \int_0^1 [F(s, t) - G(s, t)]^2 \, dsdt.$$

In particular, if F_n, $F \in \mathscr{L}_2([0, T]^2)$ and $F_n \to F$ in $\mathscr{L}_2([0, 1]^2)$, then $I(F_n) \to I(F)$ in $\mathscr{L}_2(P)$. Also it is clear that, if $F(s, t) = f(s)g(t)$, $f, g \in \mathscr{L}_2([0, 1])$, then

$$I(F) = \int_0^1 f(s) dX^1(s) \int_0^1 g(s) dX^2(s).$$

Now define $F(s, t) \in \mathscr{L}_2([0, 1]^2)$ by

$$F(s, t) = \begin{cases} 1/2, & 0 \leq s \leq t \leq 1 \\ -1/2, & 0 \leq t < s \leq 1. \end{cases}$$

Then $I(F) = S(1)$. Also $F(s, t)$ can be expanded into Fourier series in $\mathscr{L}_2([0, 1]^2)$:

$$F(s, t) = \sum_{k=1}^{\infty} \left\{ \frac{(\cos 2\pi ks - \sqrt{2}) \sin 2\pi kt}{2\pi k} - \frac{\sin 2\pi ks (\cos 2\pi kt - \sqrt{2})}{2\pi k} \right\}.$$

Now, (*) can be easily calculated in $\mathscr{L}_2(P)$-sense and, since this is a sum of independent random variables, a.s. covergence also holds.

Then, if $|\sigma| < 2\pi$,

$$I: = E[e^{\sigma S(1)} \mid X^1(1) = x^1, X^2(1) = x^2\}$$
$$= E\left[\exp\left[\sum_{k=1}^{\infty} \frac{\sigma}{2\pi k}\{(\eta_k^{(1)} - \sqrt{2}\,x^1)\xi_k^{(2)} - (\eta_k^{(2)} - \sqrt{2}\,x^2)\xi_k^{(1)}\}\right]\right]$$
$$= \prod_{k=1}^{\infty} E\left[\exp\left[\frac{\sigma}{2\pi k}(\eta_k^{(1)} - \sqrt{2}\,x^1)\xi_k^{(2)}\right]\right] E\left[\exp\left[-\frac{\sigma}{2\pi k}(\eta_k^{(2)} - \sqrt{2}\,x^2)\xi_k^{(1)}\right]\right]$$

for $(x^1, x^2) \in R^2$. We see easily that, if $a \in R$,

$$E\left[\exp\left[\frac{\sigma}{2\pi k}(\eta_k^{(1)} - \sqrt{2}\,a)\xi_k^{(2)}\right]\right]$$
$$= E\left[\exp\left[-\frac{\sigma}{2\pi k}(\eta_k^{(2)} - \sqrt{2}\,a)\xi_k^{(1)}\right]\right]$$
$$= (2\pi)^{-1}\int_{-\infty}^{\infty}\int_{-\infty}^{\infty} \exp\left[\frac{\sigma}{2\pi k}(x - \sqrt{2}\,a)y - (x^2 + y^2)/2\right]dxdy$$
$$= (1 - \left(\frac{\sigma}{2\pi k}\right)^2)^{-1/2} \exp[(1 - \left(\frac{\sigma}{2\pi k}\right)^2)^{-1}\left(\frac{\sigma}{2\pi k}\right)^2 a^2].$$

Hence

$$I = \prod_{k=1}^{\infty}(1 - \left(\frac{\sigma}{2\pi k}\right)^2)^{-1} \exp\left[\sum_{k=1}^{\infty}(1 - \left(\frac{\sigma}{2\pi k}\right)^2)^{-1}\left(\frac{\sigma}{2\pi k}\right)^2 |x|^2\right]$$
$$= \frac{\sigma}{2\sin\sigma/2}\exp\left[\left(1 - \frac{\sigma}{2}\cot\frac{\sigma}{2}\right)|x|^2\right]$$

by using well-known formulas

$$\frac{\sin x}{x} = \prod_{n=1}^{\infty}\left(1 - \frac{x^2}{n^2\pi^2}\right), \quad \cot x = \frac{1}{x} + 2x\sum_{n=1}^{\infty}\frac{1}{x^2 - n^2\pi^2}.$$

By an analytic continuation and the scaling property of the Wiener process, we obtain (6.10). The proof of (6.8) is similar.

For another interesting proof of (6.10), Cf. Yor [236].

Example 6.2. Let M be a Riemannian manifold. The horizontal Brownian motion $r = \{r(t)\}$ is a diffusion process on $O(M)$ determined by the equation (6.1) with $\tilde{L}_0 = 0$:

$$(6.13) \qquad \begin{cases} dr(t) = \tilde{L}_k(r(t))\circ dw^k(t) \\ r(0) = r. \end{cases}$$

Let b be a vector field on M. The differential 1-form $\omega(b)$ is defined as usual by $\omega(b) = b_i(x)dx^i$ where $b_i(x) = g_{ij}(x)b^j(x)$ and $b = b^i(x)\dfrac{\partial}{\partial x^i}$. It is easy to see that the scalarizations of b and $\omega(b)$ coincide:

$$\overline{\omega(b)}_i(r) = b_j(x)e_i^j = b^j(x)f_j^i = \bar{b}^i(r), \qquad i = 1, 2, \ldots, d.$$

Suppose that $\bar{b}^i(r) = \overline{\omega(b)}_i(r)$ is bounded on $O(M)$ for $i = 1, 2, \ldots, d$. Then

$$M(t) = \exp\{\sum_{i=1}^{d}\int_0^t \bar{b}^i(r(s))dw^i(s) - \frac{1}{2}\int_0^t \sum_{i=1}^{d}[\bar{b}^i(r(s))]^2 ds\} \tag{6.14}$$

is an exponential martingale. By the transformation of the drift determined by $M(t)$ (cf. Chapter IV, Section 4.1), we obtain a d-dimensional Wiener process $\tilde{w}(t) = (\tilde{w}^i(t))$ where $\tilde{w}^i(t) = w^i(t) - \int_0^t \bar{b}^i(r(s))ds$. The equation (6.1) now becomes

$$\begin{cases} dr(t) = \tilde{L}_k(r(t))\circ d\tilde{w}^k(t) + \tilde{L}_0(r(t))dt \\ r(0) = r \end{cases} \tag{6.15}$$

where $\tilde{L}_0$ is a vector field on $O(M)$ given by $\tilde{L}_0(r) = \bar{b}^k(r)\tilde{L}_k(r)$. It is immediately seen that $\tilde{L}_0$ coincides with the horizontal lift of b. Thus the solution of (6.1) is obtained from the solution of (6.13) by the transformation of drift determined by $M(t)$. In particular, the diffusion generated by the operator $\frac{1}{2}\Delta + b$ is obtained from the Brownian motion by the same transformation. Note that $M(t)$ can also be expressed in the form

$$\begin{aligned} M(t) &= \exp\Big[\int_{X[0,t]} \omega(b) + \frac{1}{2}\int_0^t \delta[\omega(b)](X(s))ds \\ &\qquad - \frac{1}{2}\int_0^t \|\omega(b)\|^2(X(s))ds\Big] \\ &= \exp\Big[\int_{X[0,t]} \omega(b) - \frac{1}{2}\int_0^t \operatorname{div}(b)(X(s))ds \\ &\qquad - \frac{1}{2}\int_0^t \|b\|^2(X(s))ds\Big]. \end{aligned}$$

This follows immediately from (6.3) and

$$\sum_{i=1}^{d} \bar{b}^i(r)^2 = \sum_{i=1}^{d} e_i^j e_i^l b_j(x) b_l(x) = g^{jl}(x) b_j(x) b_l(x)$$
$$= g_{jl}(x) b^j(x) b^l(x) = \|b\|^2(x).$$

7. Approximation theorems for stochastic integrals and stochastic differential equations

As we have seen in this book, stochastic integrals and solutions of stochastic differential equations are most important and typical examples of Wiener functionals. If, in these definitions of integrals or equations, a Brownian path is replaced by a smooth path, then they are defined by using ordinary calculus and we are thus provided with functionals defined on smooth paths. A question naturally arises: if we substitute in these functionals defined on smooth paths an approximation of Wiener process (i.e., a process consisting of smooth sample paths which converge to Brownian paths uniformly a.s.), do they converge to original Wiener functionals? Since these functionals are usually not continuous in the uniform topology of the path space, the answer is negative in general and the limiting Wiener functionals, if they exist, need to be modified according to ways of approximations. For such familiar approximations as piecewise linear approximations or regularizations by convolution (mollifiers), however, the answer is affirmative if we adopt the symmetric multiplication in the definition of stochastic integrals or stochastic differential equations.

Let $(\boldsymbol{W}_0^r, P^W)$ be the r-dimensional Wiener space * with the usual reference family $\{\mathscr{B}_t\}$. P^W is simply denoted by P below. The shift operator θ_t $(t \geq 0)$: $\boldsymbol{W}_0^r \longrightarrow \boldsymbol{W}_0^r$ is defined by $(\theta_t w)(s) = w(t+s) - w(t)$. We shall consider the following class of approximations for a Wiener process.

Definition 7.1. By an *approximation of a Wiener process*, we mean a family $\{B_\delta(t,w) = (B_\delta^1(t,w), B_\delta^2(t,w), \ldots, B_\delta^r(t,w))\}_{\delta>0}$ of r-dimensional continuous processes defined over the Wiener space $(\boldsymbol{W}_0^r, P)$ such that

(i) for every w, $t \longmapsto B_\delta(t, w)$ is piecewise continuously differentiable,

(ii) $B_\delta(0,w)$ is $\mathscr{B}_\delta$-measurable,

(iii) $B_\delta(t+k\delta,w) = B_\delta(t, \theta_{k\delta}w) + w(k\delta)$ for every $k = 1, 2, \ldots, t \geq 0$ and w,

(iv)

$$E[B_\delta^i(0,w)] = 0 \qquad \text{for} \quad i = 1, 2, \ldots, r,$$

* $\boldsymbol{W}_0^r = \{w \in \boldsymbol{C}([0, \infty) \longrightarrow \boldsymbol{R}^r);\ w(0) = 0\}$ and P^W is the Wiener measure on $\boldsymbol{W}_0^r$.

(v)

$$E[|B_\delta^i(0,w)|^6] \le c\delta^3 \qquad \text{for} \quad i=1,2,\dots,r,\text{*}$$

(vi)

$$E[(\int_0^\delta |\dot{B}_\delta^i(s,w)|\,ds)^6] \le c\delta^3 \qquad \text{for} \quad i=1,2,\dots,r,$$

where

$$\dot{B}_\delta^i(s,w) = \frac{d}{ds} B_\delta^i(s,w).$$

If $\{B_\delta(t,w)\}_{\delta>0}$ satisfies the above conditions, then for every $T>0$ we have that

$$(7.1) \qquad E\{\max_{0\le t\le T} |w(t) - B_\delta(t,w)|^2\} \longrightarrow 0 \qquad \text{as} \quad \delta \downarrow 0.$$

Thus $\{B_\delta(t,w)\}_{\delta>0}$ actually approximates the Wiener process $\{w(t)\}$. (7.1) is an obvious consequence of Theorem 7.1 below. A simple application of Hölder's inequality yields that

$$E[(\int_0^\delta |\dot{B}_\delta^{i_1}(s,w)|\,ds)^{p_1}(\int_0^\delta |\dot{B}_\delta^{i_2}(s,w)|\,ds)^{p_2}\cdots(\int_0^\delta |\dot{B}_\delta^{i_m}(s,w)|\,ds)^{p_m}]$$
$$\le c_1\delta^{\frac{1}{2}(p_1+p_2+\cdots+p_m)} \quad \text{if} \quad p_i \ge 1 \text{ and } \quad p_1+p_2+\cdots+p_m \le 6.$$

By (iii), this estimate also holds if $\int_0^\delta$ is replaced by $\int_{k\delta}^{(k+1)\delta}$. Again by Hölder's inequality, it follows that

$$(7.2) \qquad \begin{aligned} &E[(\int_0^{n_1\delta} |\dot{B}_\delta^{i_1}(s,w)|\,ds)^{p_1}(\int_0^{n_2\delta} |\dot{B}_\delta^{i_2}(s,w)|\,ds)^{p_2}\cdots(\int_0^{n_m\delta} |\dot{B}_\delta^{i_m}(s,w)|\,ds)^{p_m}] \\ &\le c_1 n_1^{p_1} n_2^{p_2}\cdots n_m^{p_m}\delta^{\frac{1}{2}(p_1+p_2+\cdots+p_m)} \end{aligned}$$

if $1 \le p_i$, $p_1+p_2+\cdots+p_m \le 6$ and $n_i \in \boldsymbol{Z}_+$.

Let us introduce the following notations:

$$(7.3) \qquad s_{ij}(t,\delta) := \frac{1}{t}E\left\{\frac{1}{2}\int_0^t [B_\delta^i(s,w)\dot{B}_\delta^j(s,w) - B_\delta^j(s,w)\dot{B}_\delta^i(s,w)]\,ds\right\}$$

* $c, c', c_1, c_2, \dots$ are positive constants independent of δ.

and

$$c_{ij}(t,\delta) := \frac{1}{t}E\left\{\int_0^t \dot{B}_\delta^i(s,w)[B_\delta^j(t,w) - B_\delta^j(s,w)]ds\right\} \tag{7.4}$$

for $i, j = 1, 2, \ldots, r$ and $t > 0$. Clearly $(s_{ij}(t,\delta))$ is a skew-symmetric $r\times r$-matrix for each t and δ. We shall make the following assumption.

Assumption 7.1. There exists a skew-symmetric $r\times r$-matrix (s_{ij}) such that

$$s_{ij}(\delta,\delta) \longrightarrow s_{ij} \quad \text{as} \quad \delta \downarrow 0.^{*} \tag{7.5}$$

Set

$$c_{ij} = s_{ij} + \frac{1}{2}\delta_{ij}, \qquad i, j = 1, 2, \ldots, r. \tag{7.6}$$

Lemma 7.1. Let $k(\delta)$: $(0, 1] \longrightarrow \boldsymbol{Z}_+$ such that $k(\delta) \uparrow \infty$ as $\delta \downarrow 0$. Then

$$\lim_{\delta\downarrow 0} c_{ij}(k(\delta)\delta, \delta) = c_{ij}. \tag{7.7}$$

Proof. We set $\delta^* = k(\delta)\delta$. Then for every $n = 1, 2, \ldots,$

$$\begin{aligned}
2n\delta\, s_{ij}(n\delta,\delta) &= \sum_{k=0}^{n-1} E\Big[\int_{k\delta}^{(k+1)\delta} \{B_\delta^i(s,w)\dot{B}_\delta^j(s,w) - B_\delta^j(s,w)\dot{B}_\delta^i(s,w)\}\,ds\Big] \\
&= \sum_{k=0}^{n-1} E\Big[\int_0^\delta \{(B_\delta^i(s,\theta_{k\delta}w) + w^i(k\delta))\dot{B}_\delta^j(s,\theta_{k\delta}w) \\
&\qquad - (B_\delta^j(s,\theta_{k\delta}w) + w^j(k\delta))\dot{B}_\delta^i(s,\theta_{k\delta}w)\}\,ds\Big] \\
&= 2n\delta s_{ij}(\delta,\delta) + \sum_{k=0}^{n-1} E[w^i(k\delta)(B_\delta^j(\delta,\theta_{k\delta}w) - B_\delta^j(0,\theta_{k\delta}w)) \\
&\qquad - w^j(k\delta)(B_\delta^i(\delta,\theta_{k\delta}w) - B_\delta^i(0,\theta_{k\delta}w))] \\
&= 2n\delta\, s_{ij}(\delta,\delta) + \sum_{k=0}^{n-1} \{E[w^i(k\delta)]E[B_\delta^j(\delta,w) - B_\delta^j(0,w)] \\
&\qquad - E[w^j(k\delta)]E[B_\delta^i(\delta,w) - B_\delta^i(0,w)]\}
\end{aligned}$$

* We may consider the case that $s_{ij}(\delta_n,\delta_n) \longrightarrow s_{ij}$ for some sequence $\delta_n \downarrow 0$. All results below also hold in this case if we replace $\{\delta\}$ by the sequence $\{\delta_n\}$.

$$= 2n\delta s_{ij}(\delta, \delta).$$

Hence

$$\begin{aligned}
&\delta^* s_{ij}(\delta, \delta)\\
&\quad = \delta^* s_{ij}(\delta^*, \delta)\\
&\quad = \frac{1}{2} E\Big[\int_0^{\delta^*} \frac{d}{ds}(B_\delta^i(s, w)B_\delta^j(s, w))ds\Big] - E\Big[\int_0^{\delta^*} B_\delta^j(s, w)\dot{B}_\delta^i(s, w)ds\Big]\\
&\quad = \frac{1}{2} E[B_\delta^i(\delta^*, w)B_\delta^j(\delta^*, w)] - \frac{1}{2} E[B_\delta^i(0, w)B_\delta^j(0, w)]\\
&\qquad + E\Big[\int_0^{\delta^*} \dot{B}_\delta^i(s, w)\{B_\delta^j(\delta^*, w) - B_\delta^j(s, w)\}\, ds\Big]\\
&\qquad - E[B_\delta^j(\delta^*, w)\{B_\delta^i(\delta^*, w) - B_\delta^i(0, w)\}]\\
&\quad = \frac{1}{2} E[(B_\delta^i(0, \theta_{\delta^*}w) + w^i(\delta^*))(B_\delta^j(0, \theta_{\delta^*}w) + w^j(\delta^*))]\\
&\qquad - \frac{1}{2} E[B_\delta^i(0, w)B_\delta^j(0, w)] + \delta^* c_{ij}(\delta^*, \delta)\\
&\qquad - E[(B_\delta^j(0, \theta_{\delta^*}w) + w^j(\delta^*))(B_\delta^i(0, \theta_{\delta^*}w) + w^i(\delta^*) - B_\delta^i(0, w))]\\
&\quad = \frac{1}{2} E[B_\delta^i(0, w)B_\delta^j(0, w)] + \frac{1}{2} E[w^i(\delta^*)w^j(\delta^*)]\\
&\qquad - \frac{1}{2} E[B_\delta^i(0, w)B_\delta^j(0, w)] + \delta^* c_{ij}(\delta^*, \delta) - E[B_\delta^i(0, w)B_\delta^j(0, w)]\\
&\qquad - E[w^i(\delta^*)w^j(\delta^*)] + E[w^j(\delta^*)B_\delta^i(0, w)]\\
&\quad = \delta^* c_{ij}(\delta^*, \delta) - \frac{1}{2} E[w^i(\delta^*)w^j(\delta^*)] - E[B_\delta^i(0, w)B_\delta^j(0, w)]\\
&\qquad + E[w^j(\delta^*)B_\delta^i(0, w)].
\end{aligned}$$

The assertion is now clear since

$$\begin{aligned}
&E[w^i(\delta^*)w^j(\delta^*)] = \delta_{ij}\delta^*,\\
&|E[B_\delta^i(0, w)B_\delta^j(0, w)]| \le (E[B_\delta^i(0, w)^2])^{1/2}(E[B_\delta^j(0, w)^2])^{1/2}\\
&\qquad\qquad \le c_2\delta = o(\delta^*)
\end{aligned}$$

and

$$\begin{aligned}
|E[w^j(\delta^*)B_\delta^i(0, w)]| &\le (E[w^j(\delta^*)^2])^{1/2}(E[B_\delta^i(0, w)^2])^{1/2}\\
&\le c_3\delta^{*1/2}\delta^{1/2} = o(\delta^*).
\end{aligned}$$

Before stating our main results, we shall first give several examples. Let Φ be the space of continuously differentiable functions ϕ on [0, 1] such that $\phi(0) = 0$, $\phi(1) = 1$. Let $\dot{\phi} = \frac{d}{dt}\phi$ and $\Delta_k w^i = w^i(k\delta+\delta) - w^i(k\delta)$.

Example 7.1. Choose $\phi^i \in \Phi$, $i = 1, 2, \ldots, r$, and set

$$B^i_\delta(t, w) = w^i(k\delta) + \phi^i((t - k\delta)/\delta)\Delta_k w^i, \tag{7.8}$$

$$\text{if } k\delta \leq t < (k+1)\delta,\ k = 0, 1, \ldots .$$

It is easy to see that $\{B_\delta(t, w)\}_{\delta>0}$ satisfies all conditions (i) $\sim$ (vi) of Definition 7.1 and hence it is an approximation of a Wiener process. For example,

$$\begin{aligned} E[\{\int_0^\delta |\dot{B}^i_\delta(s, w)|\,ds\}^6] &= E[|\Delta_0 w^i|^6](\int_0^1 |\dot{\phi}^i(s)|\,ds)^6 \\ &= 15(\int_0^1 |\dot{\phi}^i(s)|\,ds)^6\delta^3. \end{aligned}$$

Also, it is clear that $s_{ij}(\delta, \delta) = 0$ and hence Assumption 7.1 is satisfied with $s_{ij} = 0$. (Thus $c_{ij} = \frac{1}{2}\delta_{ij}$ in this example.) If, in particular, $\phi^i(t) = t$ for $i = 1, 2, \ldots, r$, then $\{B_\delta(t,w)\}_{\delta>0}$ is the familiar piecewise linear approximation.

Example 7.2. (McShane [115]). Let $r = 2$ and choose $\phi^i \in \Phi$, $i = 1, 2$. Set

$$B^i_\delta(t, w) = \begin{cases} w^i(k\delta) + \phi^i((t - k\delta)/\delta)\Delta_k w^i, & \Delta_k w^1 \Delta_k w^2 \geq 0, \\ w^i(k\delta) + \phi^{3-i}((t - k\delta)/\delta)\Delta_k w^i, & \Delta_k w^1 \Delta_k w^2 < 0, \end{cases} \tag{7.9}$$

$$\text{if } \quad k\delta \leq t < (k+1)\delta.$$

It is easy to see that $\{B_\delta(t, w)\}_{\delta>0}$ is an approximation of a Wiener process. In this case, however,

$$\begin{aligned} &\frac{1}{2}\int_0^\delta \{B^1_\delta(s, w)\dot{B}^2_\delta(s, w) - B^2_\delta(s, w)\dot{B}^1_\delta(s, w)\}\,ds \\ &= \frac{|\Delta_0 w^1 \Delta_0 w^2|}{2}(1 - 2\int_0^1 \dot{\phi}^1(s)\phi^2(s)ds), \end{aligned}$$

and hence Assumption 7.1 is satisfied with

$$s_{12} = \frac{1}{\pi}(1 - 2\int_0^1 \dot{\phi}^1(s)\phi^2(s)ds).$$

Example 7.3 (mollifiers). Let ρ be a non-negative C^∞-function whose support is contained in [0, 1] and $\int_0^1 \rho(s)ds = 1$. Set

$$\rho_\delta(s) = \frac{1}{\delta}\rho\left(\frac{s}{\delta}\right) \qquad \text{for} \quad \delta > 0$$

and

$$B_\delta^i(t, w) = \int_0^\infty w^i(s)\rho_\delta(s - t)ds = \int_0^\delta w^i(s + t)\rho_\delta(s)ds \tag{7.10}$$

$$\text{for} \quad i = 1, 2, \dots, r.$$

It is easy to verify that $\{B_\delta(t, w)\}_{\delta>0}$ is an approximation of a Wiener process. Indeed, (i) $\sim$ (iv) in Definition 7.1 are obvious and (v) and (vi) are verified as follows. We have

$$|B_\delta^i(0, w)|^{2m} = (\int_0^1 \frac{1}{\sqrt{\delta}} w^i(\delta s)\rho(s)ds)^{2m}\delta^m$$

and hence

$$E[|B_\delta^i(0,w)|^{2m}] = \delta^m E[(\int_0^1 w^i(s)\rho(s)ds)^{2m}].$$

Also,

$$\dot{B}_\delta^i(s, w) = -\frac{1}{\delta}\int_0^1 w^i(s+\delta\xi)\rho'(\xi)d\xi$$

and hence

$$\begin{aligned} &E[(\int_0^\delta |\dot{B}_\delta^i(s,w)|ds)^{2m}] \\ &= E[(\int_0^1 |\dot{B}_\delta^i(\delta s, w)|ds)^{2m}]\delta^{2m} \\ &= E[(\int_0^1 |\int_0^1 w^i(\delta(s + \xi))\rho'(\xi)d\xi|ds)^{2m}] \end{aligned}$$

$$= E[(\int_0^1 |\int_0^1 \frac{w^i(\delta(s+\xi))}{\sqrt{\delta}} \rho'(\xi)d\xi\,|\,ds)^{2m}]\delta^m$$

$$= E[(\int_0^1 |\int_0^1 w^i(s+\xi)\rho'(\xi)d\xi\,|\,ds)^{2m}]\delta^m.$$

Clearly $s_{ij}(\delta,\delta)= 0$ and consequently Assumption 7.1 is satisfied with $s_{ij} = 0$.

First, we shall consider the approximation of stochastic integrals. Let α be a differential 1-form on $\boldsymbol{R}^r$ given by

$$\alpha = \sum_{i=1}^{r} \alpha_i(x)dx^i.$$

We shall assume that all partial derivatives of $\alpha_i(x)$, $i = 1, 2, \ldots, r$, are bounded. In particular, we have

$$|\alpha_i(x)| \leq K(1 + |x|), \qquad i = 1, 2, \ldots, r,$$

for some positive constant K. Set

$$\text{(7.11)} \qquad A(t, \alpha; w) = \int_{w[0,t]} \alpha = \sum_{i=1}^{r} \int_0^t \alpha_i(w(s)) \circ dw^i(s)$$

and

$$\text{(7.12)} \qquad A(t, \alpha; B_\delta) = \int_{B_\delta[0,t]} \alpha = \sum_{i=1}^{r} \int_0^t \alpha_i(B_\delta(s))\dot{B}_\delta^i(s)ds.$$

Theorem 7.1. Let $\{B_\delta(t, w)\}_{\delta>0}$ be an approximation of a Wiener process satisfying Assumption 7.1. Then

$$\text{(7.13)} \qquad \lim_{\delta \downarrow 0} E[\sup_{0\leq t\leq T} |A(t, \alpha; B_\delta) - A(t, \alpha; w) - \int_0^t \sum_{i,j=1}^{r} s_{ij}\, \partial_i\alpha_j(w(s))ds\,|^2] = 0$$

for every $T > 0$. Here $\partial_i\alpha_j = \dfrac{\partial}{\partial x^i}\alpha_j$.

Proof. * It is sufficient to show that

* We are indebted to S. Nakao for the main idea of the proof.

$$\lim_{\delta\downarrow 0} E[\sup_{0\le t\le T} | \int_0^t u(B_\delta(s, w))\dot{B}_\delta^j(s, w)ds - \int_0^t u(w(s))\circ dw^j(s) - \int_0^t (\sum_{i=1}^r s_{ij}u_i(w(s)))ds|^2] = 0 \tag{7.14}$$

for every $T > 0$ and $j = 1, 2, \ldots, r$. Here u: $\boldsymbol{R}^r \longrightarrow \boldsymbol{R}$ is a C^2-function such that all partial derivatives are bounded and $u_i(x) = \frac{\partial}{\partial x^i} u(x)$.

First we shall introduce the following notation. Choose $n(\delta)$: $(0,1] \longrightarrow \boldsymbol{Z}_+$ such that

$$n(\delta)^4\delta \downarrow 0 \quad \text{and} \quad n(\delta) \uparrow \infty \quad \text{as} \quad \delta \downarrow 0. \tag{7.15}$$

We shall then write

$$\bar{\delta} = n(\delta)\delta, \tag{7.16}$$

$$\begin{aligned} [s]^+(\bar{\delta}) &= (k+1)\bar{\delta} \\ [s]^-(\bar{\delta}) &= k\bar{\delta} \end{aligned} \qquad \text{if} \quad k\bar{\delta} \le s < (k+1)\bar{\delta} \tag{7.17}$$

and

$$m(t) = m(t)(\delta) = [t]^-(\bar{\delta})/\bar{\delta}. \tag{7.18}$$

Integration by parts yields

$$\begin{aligned}
&\int_{k\bar{\delta}}^{(k+1)\bar{\delta}} u(B_\delta(s, w))dB_\delta^j(s, w) \\
&= -\int_{k\bar{\delta}}^{(k+1)\bar{\delta}} u(B_\delta(s, w))\frac{d}{ds}[B_\delta^j((k+1)\bar{\delta}, w) - B_\delta^j(s, w)]ds \\
&= u(B_\delta(k\bar{\delta}, w))[B_\delta^j(k\bar{\delta} + \bar{\delta}, w) - B_\delta^j(k\bar{\delta}, w)] \\
&\quad + \sum_{i=1}^r \int_{k\bar{\delta}}^{(k+1)\bar{\delta}} u_i(B_\delta(s, w))\dot{B}_\delta^i(s, w)[B_\delta^j(k\bar{\delta} + \bar{\delta}, w) - B_\delta^j(s, w)]ds \\
&= (u(B_\delta(k\bar{\delta}, w)) - u(w(k\bar{\delta})))[B_\delta^j((k+1)\bar{\delta}, w) - B_\delta^j(k\bar{\delta}, w)] \\
&\quad + u(w(k\bar{\delta}))[B_\delta^j((k+1)\bar{\delta}, w) - w^j((k+1)\bar{\delta})] \\
&\quad - u(w(k\bar{\delta}))[B_\delta^j(k\bar{\delta}, w) - w^j(k\bar{\delta})] \\
&\quad + u(w(k\bar{\delta}))[w^j((k+1)\bar{\delta}) - w^j(k\bar{\delta})] \\
&\quad + \sum_{i=1}^r \int_{k\bar{\delta}}^{(k+1)\bar{\delta}} u_i(B_\delta(s, w))\dot{B}_\delta^i(s, w)[B_\delta^j((k+1)\bar{\delta}, w) - B_\delta^j(s, w)]ds.
\end{aligned} \tag{7.19}$$

Also,

$$\begin{aligned}
&\int_{[t]^-(\delta)}^t u(B_\delta(s, w))dB_\delta^j(s, w) \\
&\quad = -u(B_\delta(t, w))[B_\delta^j([t]^+(\bar{\delta}), w) - B_\delta^j(t, w)] \\
&\qquad + u(B_\delta([t]^-(\bar{\delta}), w))[B_\delta^j([t]^+(\bar{\delta}), w) - B_\delta^j([t]^-(\bar{\delta}), w)] \\
&\qquad + \sum_{i=1}^r \int_{[t]^-(\delta)}^t u_i(B_\delta(s, w))\dot{B}_\delta^i(s, w)[B_\delta^j([t]^+(\bar{\delta}), w) - B_\delta^j(s, w)]ds \\
&\quad = -[u(B_\delta(t, w)) - u(B_\delta([t]^-(\bar{\delta}), w))][B_\delta^j([t]^+(\bar{\delta}), w) - B_\delta^j(t, w)] \\
&\qquad - [u(B_\delta([t]^-(\bar{\delta}), w)) - u(w([t]^-(\bar{\delta})))][B_\delta^j([t]^+(\bar{\delta}), w) - B_\delta^j(t, w)] \\
&\qquad - u(w([t]^-(\bar{\delta})))[B_\delta^j([t]^+(\bar{\delta}), w) - B_\delta^j(t, w)] \\
&\qquad + u(B_\delta([t]^-(\bar{\delta}), w))[B_\delta^j([t]^+(\bar{\delta}), w) - B_\delta^j([t]^-(\bar{\delta}), w)] \\
&\qquad + \sum_{i=1}^r \int_{[t]^-(\delta)}^t u_i(B_\delta(s, w))\dot{B}_\delta^i(s, w)[B_\delta^j([t]^+(\bar{\delta}), w) - B_\delta^j(s, w)]ds.
\end{aligned} \tag{7.20}$$

By (7.19) and (7.20) we have

$$\begin{aligned}
&\int_0^t u(B_\delta(s, w))dB_\delta^j(s, w) - \int_0^t u(w(s))\circ dw^j(s) - \int_0^t \sum_{i=1}^r s_{ij}u_i(w(s))ds \\
&\quad = \int_0^t u(B_\delta(s, w))dB_\delta^j(s, w) - \int_0^t u(w(s))dw^j(s) - \int_0^t \sum_{i=1}^r c_{ij}u_i(w(s))ds \\
&\quad = - [u(B_\delta(t, w)) - u(B_\delta([t]^-(\bar{\delta}), w))][B_\delta^j([t]^+(\bar{\delta}), w) - B_\delta^j(t, w)] \\
&\qquad - [u(B_\delta([t]^-(\bar{\delta}), w)) - u(w([t]^-(\bar{\delta})))][B_\delta^j([t]^+(\bar{\delta}), w) - B_\delta^j(t, w)] \\
&\qquad - u(w([t]^-(\bar{\delta})))[B_\delta^j([t]^+(\bar{\delta}), w) - B_\delta^j(t, w)] \\
&\qquad + [u(B_\delta([t]^-(\bar{\delta}), w)) - u(w([t]^-(\bar{\delta})))][B_\delta^j([t]^+(\bar{\delta}), w) - B_\delta^j([t]^-(\bar{\delta}), w)] \\
&\qquad + u(w([t]^-(\bar{\delta})))[B_\delta^j([t]^+(\bar{\delta}), w) - B_\delta^j([t]^-(\bar{\delta}), w)] \\
&\qquad + \sum_{i=1}^r \int_{[t]^-(\delta)}^t u_i(B_\delta(s, w))\dot{B}_\delta^i(s, w)[B_\delta^j([t]^+(\bar{\delta}), w) - B_\delta^j(s, w)]ds \\
&\qquad - u(w([t]^-(\bar{\delta})))[w^j(t) - w^j([t]^-(\bar{\delta}))] \\
&\qquad + [u(w([t]^-(\bar{\delta})))(w^j(t) - w^j([t]^-(\bar{\delta}))) - \int_{[t]^-(\delta)}^t u(w(s))dw^j(s)] \\
&\qquad - \int_{[t]^-(\delta)}^t \sum_{i=1}^r c_{ij}u_i(w(s))ds \\
&\qquad + \sum_{k=0}^{m(t)-1} [u(B_\delta(k\bar{\delta}, w)) - u(w(k\bar{\delta}))][B_\delta^j((k+1)\bar{\delta}, w) - B_\delta^j(k\bar{\delta}, w)]
\end{aligned} \tag{7.21}$$

$$+\sum_{k=0}^{m(t)-1} u(w(k\bar{\delta}))[B_\delta^j((k+1)\bar{\delta}, w) - w^j((k+1)\bar{\delta})]$$

$$-\sum_{k=0}^{m(t)-1} u(w(k\bar{\delta}))[B_\delta^j(k\bar{\delta}, w) - w^j(k\bar{\delta})]$$

$$+\Big[\sum_{k=0}^{m(t)-1} u(w(k\bar{\delta}))(w^j((k+1)\bar{\delta}) - w^j(k\bar{\delta})) - \int_0^{[t]^-(\bar{\delta})} u(w(s))dw^j(s)\Big]$$

$$+\sum_{i=1}^{r}\sum_{k=0}^{m(t)-1} \int_{k\bar{\delta}}^{(k+1)\bar{\delta}} [u_i(B_\delta(s, w)) - u_i(w(k\bar{\delta}))]\dot{B}_\delta^i(s, w)[B_\delta^j((k+1)\bar{\delta}, w) - B_\delta^j(s, w)]ds$$

$$+\sum_{i=1}^{r}\sum_{k=0}^{m(t)-1} u_i(w(k\bar{\delta})) \int_{k\bar{\delta}}^{(k+1)\bar{\delta}} [\dot{B}_\delta^i(s, w)(B_\delta^j((k+1)\bar{\delta}, w) - B_\delta^j(s, w)) - c_{ij}(\bar{\delta}, \delta)]ds$$

$$+\sum_{i=1}^{r}\sum_{k=0}^{m(t)-1} u_i(w(k\bar{\delta}))[c_{ij}(\bar{\delta}, \delta) - c_{ij}]\bar{\delta}$$

$$+\sum_{i=1}^{r}\sum_{k=0}^{m(t)-1} \int_{k\bar{\delta}}^{(k+1)\bar{\delta}} [u_i(w(k\bar{\delta})) - u_i(w(s))]ds\, c_{ij}$$

$$:= I_1(t) + I_2(t) + I_3(t) + I_4(t) + I_5(t) + \sum_{i=1}^{r} I_6^i(t) + I_7(t)$$
$$+ I_8(t) + I_9(t) + I_{10}(t) + I_{11}(t) + I_{12}(t) + I_{13}(t)$$
$$+ \sum_{i=1}^{r} I_{14}^i(t) + \sum_{i=1}^{r} I_{15}^i(t) + \sum_{i=1}^{r} I_{16}^i(t) + \sum_{i=1}^{r} I_{17}^i(t).$$

It is clear that

$$E[\sup_{0\le t\le T} |I_9(t)|^2] \longrightarrow 0 \quad \text{as} \quad \delta \downarrow 0 \tag{7.22}$$

and, by a martingale inequality,

$$\begin{aligned} E[\sup_{0\le t\le T} |I_{13}(t) + I_8(t)|^2] &\le c_4 E\Big[\int_0^T \{u(w([t]^-(\bar{\delta}))) - u(w(t))\}^2 dt\Big] \\ &\longrightarrow 0 \qquad \text{as} \quad \delta \downarrow 0. \end{aligned} \tag{7.23}$$

By Lemma 7.1, it is clear that

$$E[\sup_{0\le t\le T} |I_{16}^i(t)|^2] \longrightarrow 0 \quad \text{as} \quad \delta \downarrow 0. \tag{7.24}$$

We have

$$E[\sup_{0\le t\le T} |I_{17}^i(t)|^2]$$

$$\leq c_5 \sum_{i=1}^{r} E[(\int_0^T |w^i([s]^-(\tilde{\delta})) - w^i(s)|\,ds)^2]$$

$$\text{(7.25)} \qquad \leq c_5 T \sum_{i=1}^{r} \int_0^T E[|w^i([s]^-(\tilde{\delta})) - w^i(s)|^2]ds$$

$$\leq c_5 T^2 r\tilde{\delta} \longrightarrow 0 \qquad \text{as} \quad \delta \downarrow 0.$$

Also, by (7.2),

$$E[\sup_{0\leq t\leq T} |I_1(t)|^2]$$

$$\leq c_6 \sum_{i=1}^{r} E[\sup_{0\leq t\leq T} (B_\delta^i(t, w) - B_\delta^i([t]^-(\tilde{\delta}), w))^2 (B_\delta^j([t]^+(\tilde{\delta}), w) - B_\delta^j(t, w))^2]$$

$$\leq c_6 \sum_{i=1}^{r} E[\sup_{0\leq t\leq T} (\int_{[t]^-(\tilde{\delta})}^{[t]^+(\tilde{\delta})} |\dot{B}_\delta^i(s, w)|\,ds)^2 (\int_{[t]^-(\tilde{\delta})}^{[t]^+(\tilde{\delta})} |\dot{B}_\delta^j(s, w)|\,ds)^2]$$

$$\text{(7.26)} \quad = c_6 \sum_{i=1}^{r} E[\max_{0\leq k\leq m(T)} (\int_{k\tilde{\delta}}^{(k+1)\tilde{\delta}} |\dot{B}_\delta^i(s, w)|\,ds)^2 (\int_{k\tilde{\delta}}^{(k+1)\tilde{\delta}} |\dot{B}_\delta^j(s, w)|\,ds)^2]$$

$$\leq c_6 \sum_{i=1}^{r} \sum_{k=0}^{m(T)} E[(\int_{k\tilde{\delta}}^{(k+1)\tilde{\delta}} |\dot{B}_\delta^i(s, w)|\,ds)^2 (\int_{k\tilde{\delta}}^{(k+1)\tilde{\delta}} |\dot{B}_\delta^j(s, w)|\,ds)^2]$$

$$\leq c_6(m(T) + 1) \sum_{i=1}^{r} E[(\int_0^{\tilde{\delta}} |\dot{B}_\delta^i(s, w)|\,ds)^2 (\int_0^{\tilde{\delta}} |\dot{B}_\delta^j(s, w)|\,ds)^2]$$

$$\leq c_7(m(T) + 1)n(\delta)^4\delta^2$$

$$\leq c_8 n(\delta)^3\delta \longrightarrow 0 \qquad \text{as} \quad \delta \downarrow 0.$$

As for $I_2(t)$,

$$E[\sup_{0\leq t\leq T} |I_2(t)|^2]$$

$$\leq c_9 \sum_{i=1}^{r} E[\sup_{0\leq t\leq T} (B_\delta^i([t]^-(\tilde{\delta}), w) - w^i([t]^-(\tilde{\delta})))^2$$

$$\times (B_\delta^j([t]^+(\tilde{\delta}), w) - B_\delta^j(t, w))^2]$$

$$\leq c_9 \sum_{i=1}^{r} E[\sup_{0\leq t\leq T} (B_\delta^i([t]^-(\tilde{\delta}), w) - w^i([t]^-(\tilde{\delta})))^2 (\int_{[t]^-(\tilde{\delta})}^{[t]^+(\tilde{\delta})} |\dot{B}_\delta^j(s, w)|\,ds)^2]$$

$$= c_9 \sum_{i=1}^{r} E[\max_{0\leq k\leq m(T)} (B_\delta^i(k\tilde{\delta}, w) - w^i(k\tilde{\delta}))^2 (\int_{k\tilde{\delta}}^{(k+1)\tilde{\delta}} |\dot{B}_\delta^j(s, w)|\,ds)^2]$$

$$\text{(7.27)} \quad \leq c_9 \sum_{i=1}^{r} \sum_{k=0}^{m(T)} E[(B_\delta^i(k\tilde{\delta}, w) - w^i(k\tilde{\delta}))^2 (\int_{k\tilde{\delta}}^{(k+1)\tilde{\delta}} |\dot{B}_\delta^j(s, w)|\,ds)^2]$$

$$\leq c_9(m(T) + 1) \sum_{i=1}^{r} E[(B_\delta^i(0, w))^2 (\int_0^{\tilde{\delta}} |\dot{B}_\delta^j(s, w)|\,ds)^2]$$

$$\leq c_9(m(T) + 1) \sum_{i=1}^{r} \{E[(B_\delta^i(0, w))^4] E[(\int_0^{\tilde{\delta}} |\dot{B}_\delta^j(s, w)|\,ds)^4]\}^{1/2}$$

$$\leq c_{10}(m(T)+1)(\delta^2 n(\delta)^4\delta^2)^{1/2}$$
$$\leq c_{11}\delta n(\delta) \longrightarrow 0 \quad \text{as} \quad \delta \downarrow 0.$$

In the case of $I_3(t)$, we have

$$E[\sup_{0\leq t\leq T} |I_3(t)|^2]$$
$$\leq c_{12}\sum_{i=1}^{r} E[\sup_{0\leq t\leq T}(1+|w^i([t]^-(\delta))|^2)(\int_{[t]^-(\delta)}^{[t]^+(\delta)} |\dot{B}^j_\delta(s,w)|ds)^2]$$
$$\leq c_{12}\sum_{i=1}^{r} \{E[\sup_{0\leq t\leq T}(1+|w^i(t)|^2)^2]E[\max_{0\leq k\leq m(T)}(\int_{k\delta}^{(k+1)\delta} |\dot{B}^j_\delta(s,w)|ds)^4]\}^{1/2}$$
$$\leq c_{13}(E[\sum_{k=0}^{m(T)}(\int_{k\delta}^{(k+1)\delta} |\dot{B}^j_\delta(s,w)|ds)^4])^{1/2} \tag{7.28}$$
$$\leq c_{14}((m(T)+1)n(\delta)^4\delta^2)^{1/2}$$
$$\leq c_{15}(n(\delta)^3\delta)^{1/2} \longrightarrow 0 \quad \text{as} \quad \delta \downarrow 0.$$

Similarly, as in (7.27) and (7.28), we can prove that

$$E[\sup_{0\leq t\leq T} |I_4(t)|^2] \longrightarrow 0 \quad \text{as} \quad \delta \downarrow 0 \tag{7.29}$$

and

$$E[\sup_{0\leq t\leq T} |I_5(t)|^2] \longrightarrow 0 \quad \text{as} \quad \delta \downarrow 0. \tag{7.30}$$

For $I^i_6(t)$,

$$E[\sup_{0\leq t\leq T} |I^i_6(t)|^2]$$
$$\leq c_{16}E[\sup_{0\leq t\leq T}\{\int_{[t]^-(\delta)}^{[t]^+(\delta)} |\dot{B}^i_\delta(s,w)|ds\int_{[t]^-[\delta]}^{[t]^+(\delta)} |\dot{B}^j_\delta(s,w)|ds\}^2]$$
$$\leq c_{17}\sum_{k=0}^{m(T)} E[\{\int_{k\delta}^{(k+1)\delta} |\dot{B}^i_\delta(s,w)|ds\int_{k\delta}^{(k+1)\delta} |\dot{B}^j_\delta(s,w)|ds\}^2] \tag{7.31}$$
$$\leq c_{17}(m(T)+1)E[\{\int_0^\delta |\dot{B}^i_\delta(s,w)|ds\int_0^\delta |\dot{B}^j_\delta(s,w)|ds\}^2]$$
$$\leq c_{18}\frac{1}{n(\delta)\delta}n(\delta)^4\delta^2$$
$$= c_{18}n(\delta)^3\delta \longrightarrow 0 \quad \text{as} \quad \delta \downarrow 0.$$

As for $I_7(t)$,

$$E[\sup_{0\leq t\leq T} |I_7(t)|^2]$$

$$\leq c_{19}E[\sup_{0\leq t\leq T}(1 + |w([t]^-(\tilde{\delta}))|^2) \sup_{0\leq t\leq T}|w^j(t) - w^j([t]^-(\tilde{\delta}))|^2]$$

$$\leq c_{20}\{E[\max_{0\leq k\leq m(T)} \sup_{0\leq t\leq \delta}|w^j(t + k\tilde{\delta}) - w^j(k\tilde{\delta})|^4]\}^{1/2}$$

$$\leq c_{20}\{\sum_{k=0}^{m(T)} E[\sup_{0\leq t\leq \delta}|w^j(t + k\tilde{\delta}) - w^j(k\tilde{\delta})|^4]\}^{1/2} \tag{7.32}$$

$$\leq c_{20}\{(m(T) + 1)E[\sup_{0\leq t\leq \delta}|w^j(t)|^4]\}^{1/2}$$

$$\leq c_{21}\left[\frac{1}{n(\delta)\delta}(n(\delta)\delta)^2\right]^{1/2}$$

$$= c_{21}(n(\delta)\delta)^{1/2} \longrightarrow 0 \quad \text{as} \quad \delta \downarrow 0.$$

We estimate $I_{10}(t)$ by

$$E[\sup_{0\leq t\leq T}|I_{10}(t)|^2]$$

$$\leq E[\sup_{0\leq t\leq T} \sum_{k=0}^{m(t)-1}[u(B_\delta(k\tilde{\delta}, w)) - u(w(k\tilde{\delta}))]^2 \sum_{k=0}^{m(t)-1}[B_\delta^j((k + 1)\tilde{\delta}, w) - B_\delta^j(k\tilde{\delta}, w)]^2]$$

$$\leq E[\sum_{k=0}^{m(T)-1}[u(B_\delta(k\tilde{\delta}, w)) - u(w(k\tilde{\delta}))]^2 \sum_{k=0}^{m(T)-1}[B_\delta^j((k+ 1)\tilde{\delta}, w) - B_\delta^j(k\tilde{\delta}, w)]^2] \tag{7.33}$$

$$\leq \{E[m(T)\sum_{k=0}^{m(T)-1}[u(B_\delta(k\tilde{\delta}, w)) - u(w(k\tilde{\delta}))]^4] \times E[m(T)\sum_{k=0}^{m(T)-1}[B_\delta^j((k + 1)\tilde{\delta}, w) - B_\delta^j(k\tilde{\delta},w)]^4\}^{1/2}$$

$$\leq c_{22}m(T)\sum_{i=1}^{r}\{\sum_{k=0}^{m(T)-1}E[|B_\delta^i(k\tilde{\delta}, w) - w^i(k\tilde{\delta})|^4] \times \sum_{k=0}^{m(T)-1}E[|B_\delta^j((k + 1)\tilde{\delta}, w) - B_\delta^j(k\tilde{\delta}, w)|^4\}^{1/2}$$

$$\leq c_{23}m(T)^2\delta\{2E[|B_\delta^j(0, w)|^4] + E[|w(\tilde{\delta})|^4]\}^{1/2}$$

$$\leq c_{24}m(T)^2\delta\tilde{\delta}$$

$$\leq c_{25}n(\delta)^{-1} \longrightarrow 0 \quad \text{as} \quad \delta \downarrow 0.$$

In the case of $I_{11}(t)$, we first note that

$$\eta_n(w) = \sum_{k=0}^{n} u(w(k\tilde{\delta}))(B_\delta^j((k + 1)\tilde{\delta}, w) - w^j((k + 1)\tilde{\delta}))$$

$$= \sum_{k=0}^{n} u(w(k\tilde{\delta}))B_\delta^j(0, \theta_{(k+1)\tilde{\delta}}w)$$

is an $\{\mathscr{F}_n\}$-martingale where $\mathscr{F}_n = \mathscr{B}_{(n+2)\delta}$. Hence, by a martingale inequality,

$$
\begin{aligned}
&E[\sup_{0\le t\le T} |I_{11}(t)|^2] \\
&\quad = E[\max_{0\le n\le m(T)-1} |\eta_n(w)|^2] \\
&\quad \le c_{26} E[|\eta_{m(T)-1}(w)|^2] \\
&\quad = c_{26} \sum_{k=0}^{m(T)-1} E[u(w(k\bar{\delta}))^2 (B_\delta^j(0, \theta_{(k+1)\delta} w))^2] \\
&\quad = c_{26} \sum_{k=0}^{m(T)-1} E[u(w(k\bar{\delta}))^2] E[(B_\delta^j(0, w))^2] \\
&\quad \le c_{27} m(T)\delta \\
&\quad = c_{28} n(\delta)^{-1} \longrightarrow 0 \qquad \text{as} \quad \delta \downarrow 0.
\end{aligned} \tag{7.34}
$$

We can show in the same way that

$$
E[\sup_{0\le t\le T} |I_{12}(t)|^2] \longrightarrow 0 \qquad \text{as} \quad \delta \downarrow 0. \tag{7.35}
$$

As for $I_{14}^i(t)$,

$$
\begin{aligned}
&E[\sup_{0\le t\le T} |I_{14}^i(t)|^2] \\
&\le E[\sup_{0\le t\le T} \{ \sum_{k=0}^{m(t)-1} \int_{k\delta}^{(k+1)\delta} [|u_i(B_\delta(s, w)) - u_i(w(k\bar{\delta}))| \, |\dot{B}_\delta^i(s, w)| \\
&\qquad\qquad \times \int_s^{(k+1)\delta} |\dot{B}_\delta^j(\xi, w)| d\xi] ds\}^2] \\
&\le E[\{ \sum_{k=0}^{m(T)-1} \int_{k\delta}^{(k+1)\delta} [|u_i(B_\delta(s, w)) - u_i(w(k\bar{\delta}))| \, |\dot{B}_\delta^i(s, w)| \\
&\qquad\qquad \times \int_{k\delta}^{(k+1)\delta} |\dot{B}_\delta^j(\xi, w)| d\xi] ds\}^2] \\
&\le c_{29} \sum_{l=1}^{r} E[\{ \sum_{k=0}^{m(T)-1} \int_{k\delta}^{(k+1)\delta} [|B_\delta^l(0, \theta_{k\delta} w) + \int_0^{s-k\delta} \dot{B}_\delta^l(\eta, \theta_{k\delta} w) d\eta| \\
&\qquad\qquad \times |\dot{B}_\delta^i(s, w)| \int_{k\delta}^{(k+1)\delta} |\dot{B}_\delta^j(\xi, w)| d\xi] ds\}^2] \\
&\le c_{29} m(T) \sum_{l=1}^{r} E[\sum_{k=0}^{m(T)-1} \{ \int_{k\delta}^{(k+1)\delta} [(|B_\delta^l(0, \theta_{k\delta} w)| \\
&\quad + \int_0^{s-k\delta} |\dot{B}_\delta^l(\eta, \theta_{k\delta} w)| d\eta) |\dot{B}_\delta^i(s, w)| \int_{k\delta}^{(k+1)\delta} |\dot{B}_\delta^j(\xi, w)| d\xi] ds\}^2] \\
&\le c_{29} m(T)^2 \sum_{l=1}^{r} E[\{|B_\delta^l(0, w)| \int_0^\delta |\dot{B}_\delta^i(s, w)| ds \int_0^\delta |\dot{B}_\delta^j(s, w)| ds
\end{aligned} \tag{7.36}
$$

$$+ \int_0^{\delta} |\dot{B}_\delta^i(\eta, w)| d\eta \int_0^{\delta} |\dot{B}_\delta^i(s, w)| ds \int_0^{\delta} |\dot{B}_\delta^j(\xi, w)| d\xi\}^2]$$

$$\le c_{30} m(T)^2 \sum_{l=1}^{r} \{E[|B_\delta^l(0, w)|^2 (\int_0^{\delta} |\dot{B}_\delta^i(s, w)| ds)^2 (\int_0^{\delta} |\dot{B}_\delta^j(s, w)| ds)^2]$$

$$+ E[(\int_0^{\delta} |\dot{B}_\delta^l(\eta, w)| d\eta)^2 (\int_0^{\delta} |\dot{B}_\delta^i(s, w)|^2 ds)^2 (\int_0^{\delta} |\dot{B}_\delta^j(s, w)|^2 ds)^2]\}$$

$$\le c_{30} m(T)^2 \sum_{l=1}^{r} \{E[|B_\delta^l(0, w)|^6]^{1/3} E[(\int_0^{\delta} |\dot{B}_\delta^i(s, w)| ds)^3$$

$$\times (\int_0^{\delta} |\dot{B}_\delta^j(s, w)| ds)^3]^{2/3}$$

$$+ E[(\int_0^{\delta} |\dot{B}_\delta^l(\eta, w)| d\eta)^2 (\int_0^{\delta} |\dot{B}_\delta^i(s, w)| ds)^2 (\int_0^{\delta} |\dot{B}_\delta^j(s, w)| ds)^2]\}$$

$$\le c_{31} \frac{1}{n(\delta)^2 \delta^2} (\delta (n(\delta)^6 \delta^3)^{2/3} + n(\delta)^6 \delta^3)$$

$$\le c_{32} n(\delta)^4 \delta \longrightarrow 0 \quad \text{as} \quad \delta \downarrow 0.$$

Finally we shall prove that

$$E[\sup_{0 \le t \le T} |I_{15}^i(t)|^2] \longrightarrow 0 \quad \text{as} \quad \delta \downarrow 0. \tag{7.37}$$

For this, we first note that

$$\tilde{\delta} c_{ij}(\tilde{\delta}, \delta) = \delta c_{ij}^*(\tilde{\delta}, \delta) + (\tilde{\delta} - \delta) c_{ij}(\tilde{\delta} - \delta, \delta) \tag{7.38}$$

where

$$c_{ij}^*(\tilde{\delta}, \delta) = E[\int_0^{\delta} \dot{B}_\delta^i(s, w)[B_\delta^j(\tilde{\delta}, w) - B_\delta^j(s, w)] ds]/\delta.$$

Hence

$$\begin{aligned} I_{15}^i(t) = &\sum_{k=0}^{m(t)-1} u_i(w(k\tilde{\delta})) \int_{k\tilde{\delta}+\delta}^{(k+1)\tilde{\delta}} \{\dot{B}_\delta^i(s, w)(B_\delta^j((k+1)\tilde{\delta}, w) - B_\delta^j(s, w)) \\ &\qquad - c_{ij}(\tilde{\delta} - \delta, \delta)\} ds \\ &+ \sum_{k=0}^{m(t)-1} u_i(w(k\tilde{\delta})) \int_{k\tilde{\delta}}^{k\tilde{\delta}+\delta} \{\dot{B}_\delta^i(s, w)(B_\delta^j((k+1)\tilde{\delta}, w) - B_\delta^j(s, w)) \\ &\qquad - c_{ij}^*(\tilde{\delta}, \delta)\} ds \\ =: &\ J_1(t) + J_2(t). \end{aligned} \tag{7.39}$$

It is easy to see that

$$\eta_n(w) = \sum_{k=0}^{n} u_i(w(k\bar{\delta})) \int_{k\bar{\delta}+\delta}^{(k+1)\bar{\delta}} \{\dot{B}^i_\delta(s, w)(B^j_\delta((k+1)\bar{\delta}, w) - B^j_\delta(s, w)) - c_{ij}(\bar{\delta} - \delta, \delta)\} ds$$

$$= \sum_{k=0}^{n} u_i(w(k\bar{\delta})) \int_0^{\bar{\delta}-\delta} \{\dot{B}^i_\delta(s, \theta_{k\bar{\delta}+\delta}w)(B^j_\delta(\bar{\delta} - \delta, \theta_{k\bar{\delta}+\delta}w) - B^j_\delta(s, \theta_{k\bar{\delta}+\delta}w)) - c_{ij}(\bar{\delta} - \delta, \delta)\} ds$$

is an $\{\mathscr{F}_n\}$-martingale where $\mathscr{F}_n = \mathscr{B}_{(n+1)\bar{\delta}+\delta}$. Consequently

$$\begin{aligned}
E[\sup_{0\le t\le T} |J_1(t)|^2] &= E[\max_{0\le n\le m(T)-1} |\eta_n|^2] \\
&\le c_{33} E[|\eta_{m(T)-1}|^2] \\
&= c_{33} \sum_{k=0}^{m(T)-1} E[u_i(w(k\bar{\delta}))^2 (\int_0^{\bar{\delta}-\delta} \{\dot{B}^i_\delta(s, \theta_{k\bar{\delta}+\delta}w)(B^j_\delta(\bar{\delta} - \delta, \theta_{k\bar{\delta}+\delta}w) \\
&\qquad - B^j_\delta(s, \theta_{k\bar{\delta}+\delta}w)) - c_{ij}(\bar{\delta} - \delta, \delta)\} ds)^2] \\
&\le c_{34} m(T) E[(\int_0^{\bar{\delta}-\delta} \{\dot{B}^i_\delta(s, w)(B^j_\delta(\bar{\delta} - \delta, w) - B^j_\delta(s, w)) \\
&\qquad - c_{ij}(\bar{\delta} - \delta, \delta)\} ds)^2] \qquad (7.40)\\
&\le c_{34} m(T) E[(\int_0^{\bar{\delta}-\delta} \{\dot{B}^i_\delta(s, w)(B^j_\delta(\bar{\delta} - \delta, w) - B^j_\delta(s, w))\} ds)^2] \\
&\le c_{34} m(T) E[(\int_0^{\bar{\delta}} |\dot{B}^i_\delta(s, w)| ds)^2 (\int_0^{\bar{\delta}} |\dot{B}^j_\delta(s, w)| ds)^2] \\
&\le c_{35} \frac{1}{n(\delta)\delta} n(\delta)^4 \delta^2 \\
&= c_{35} n(\delta)^3 \delta \longrightarrow 0 \qquad \text{as} \quad \delta \downarrow 0.
\end{aligned}$$

Also

$$\begin{aligned}
E[\sup_{0\le t\le T} |J_2(t)|^2] \\
\le c_{36} E[\sup_{0\le t\le T} (\sum_{k=0}^{m(t)-1} |\int_{k\bar{\delta}}^{k\bar{\delta}+\delta} \{\dot{B}^i_\delta(s, w)(B^j_\delta((k+1)\bar{\delta}, w) \qquad (7.41)\\
- B^j_\delta(s, w))\} ds|)^2 + m(T)^2 \delta^2 c^*_{ij}(\bar{\delta}, \delta)^2].
\end{aligned}$$

By (7.38) and Lemma 7.1, we have

$$\frac{\delta}{\tilde{\delta}} c^*_{ij}(\tilde{\delta}, \delta) \longrightarrow 0 \quad \text{as} \quad \delta \downarrow 0$$

and consequently

$$m(T)^2\delta^2 c^*_{ij}(\tilde{\delta}, \delta)^2 \leq c_{37}\left(\frac{\delta}{\tilde{\delta}} c^*_{ij}(\tilde{\delta}, \delta)\right)^2 \longrightarrow 0 \quad \text{as} \quad \delta \downarrow 0. \tag{7.42}$$

On the other hand,

$$\begin{aligned}
&E[\sup_{0\leq t\leq T}(\sum_{k=0}^{m(t)-1} |\int_{k\tilde{\delta}}^{k\tilde{\delta}+\delta} \dot{B}^i_\delta(s, w)(B^j_\delta(k\tilde{\delta} + \tilde{\delta}, w) - B^j_\delta(s, w))ds|)^2] \\
&\leq E[\sup_{0\leq t\leq T} m(t) \sum_{k=0}^{m(t)-1} (\int_{k\tilde{\delta}}^{k\tilde{\delta}+\delta} \dot{B}^i_\delta(s, w)(B^j_\delta(k\tilde{\delta} + \tilde{\delta}, w) - B^j_\delta(s, w))ds)^2] \\
&\leq m(T) \sum_{k=0}^{m(T)-1} E[(\int_{k\tilde{\delta}}^{k\tilde{\delta}+\delta} \dot{B}^i_\delta(s, w)(B^j_\delta(k\tilde{\delta} + \tilde{\delta}, w) - B^j_\delta(s, w))ds)^2] \\
&= m(T)^2 E[(\int_0^\delta \dot{B}^i_\delta(s, w)(B^j_\delta(\tilde{\delta}, w) - B^j_\delta(s, w))ds)^2] \\
&\leq c_{38} m(T)^2 \{E[(\int_0^\delta \dot{B}^i_\delta(s, w)(B^j_\delta(\delta, w) - B^j_\delta(s ,w))ds)^2] \\
&\qquad + E[(B^i_\delta(\delta, w) - B^i_\delta(0, w))^2(B^j_\delta(\tilde{\delta}, w) - B^j_\delta(\delta, w))^2]\} \\
&\leq c_{38} m(T)^2 \{E[(\int_0^\delta |\dot{B}^i_\delta(s, w)| ds)^2(\int_0^\delta |\dot{B}^j_\delta(s, w)| ds)^2] \\
&\qquad + E[(\int_0^\delta |\dot{B}^i_\delta(s, w)| ds)^2 \{B^j_\delta(0, \theta_{\tilde{\delta}} w) - B^j_\delta(0, \theta_\delta w) \\
&\qquad\qquad + (w^j(\tilde{\delta}) - w^j(\delta))\}^2]\} \\
&\leq c_{39} \frac{1}{n(\delta)^2\delta^2}(\delta^2 + \delta^2 n(\delta)) \longrightarrow 0 \quad \text{as} \quad \delta \downarrow 0.
\end{aligned} \tag{7.43}$$

This completes the proof.

Next we obtain an approximation theorem for solutions of stochastic differential equations. Such theorems have been discussed by several authors: McShane [115], Wong-Zakai [181], Stroock-Varadhan [158], Kunita [93], Nakao-Yamato [126], Malliavin [106] and Ikeda-Nakao-Yamato [52]. Our result covers most of these results in a unified way. Main idea of the proof is due to S. Nakao.

Let

$$A_0 = \sum_{i=1}^d b^i(x) \frac{\partial}{\partial x^i} \quad \text{and} \quad A_n = \sum_{i=1}^d \sigma^i_n(x) \frac{\partial}{\partial x^i}, \qquad n = 1, 2, \ldots, r$$

be vector fields on $\boldsymbol{R}^d$ such that $\sigma_n^i \in C_b^2(\boldsymbol{R}^d)$ and $b^i \in C_b^1(\boldsymbol{R}^d)$. Let $\{B_\delta(t,w)\}_{\delta>0}$ be an approximation of Wiener process defined on the Wiener space (W_0^r, P). We assume that $\{B_\delta(t,w)\}_{\delta>0}$ satisfies the Assumption 7.1. Consider the following equations *[1]

$$(7.44)\qquad \begin{cases} dX_\delta(t,w) = \sum_{n=1}^{r} A_n(X_\delta(t,w))dB_\delta^n(t,w) + A_0(X_\delta(t,w))dt \\ X_\delta(0,w) = x \end{cases}$$

and

$$(7.45)\qquad \begin{cases} dX(t,w) = \sum_{n=1}^{r} A_n(X(t,w))\circ dw^n(t) + A_0(X(t,w))dt \\ \qquad\qquad + \sum_{1\le n\le m\le r} s_{nm}[A_n,A_m](X(t,w))dt \\ X(0,w) = x. \end{cases}$$

These are equivalent to the respective integral equations

$$(7.44)'\qquad X_\delta^i(t,w) - x^i = \sum_{n=1}^{r} \int_0^t \sigma_n^i(X_\delta(s,w))\dot{B}_\delta^n(s,w)ds + \int_0^t b^i(X_\delta(s,w))ds$$

and

$$(7.45)'\qquad \begin{aligned} X^i(t,w) - x^i &= \sum_{n=1}^{r} \int_0^t \sigma_n^i(X(s,w))dw^n(s) + \int_0^t b^i(X(s,w))ds \\ &\quad + \sum_{n,m=1}^{r} \sum_{\alpha=1}^{d} c_{nm} \int_0^t \sigma_n^\alpha(\partial_\alpha \sigma_m^i)(X(s,w))ds \quad i = 1, 2, \dots, d. \end{aligned}$$

For any given $x \in \boldsymbol{R}^d$, the solutions of (7.44) and (7.45) are unique; we shall denote them by $X_\delta = (X_\delta(t,x,w))$ and $X = (X(t,x,w))$. In the following discussion, the common initial value x is arbitrary but fixed and so we shall often suppress it.

Theorem 7.2. For every $T > 0$ we have

$$(7.46)\qquad \lim_{\delta\downarrow 0} E[\sup_{0\le t\le T} |X(t,w) - X_\delta(t,w)|^2] = 0.$$

*Proof.**[2] First we note by (7.44)′ that

---

*[1] We use the same notations as in Chapter V.

*[2] In the following, $K, K_1, K_2, \dots$ are positive constants. $\tilde{\delta}$ has the same meaning as in the proof of Theorem 7.1.

$$|X_\delta^i(t, w) - X_\delta^i(s, w)| \leq K(\sum_{n=1}^r \int_s^t |\dot{B}_\delta^n(u, w)| du + (t - s)) \tag{7.47}$$

for every $0 \leq s < t$. We have for each $i = 1, 2, \ldots, d$ that

$$X_\delta^i(t, w) - X^i(t, w) := H_1(t) + H_2(t) + H_3 + H_4(t) \tag{7.48}$$

where

$$\begin{aligned} H_1(t) = &\sum_{n=1}^r \int_{[t]^-(\delta)}^t \sigma_n^i(X_\delta(s, w))\dot{B}_\delta^n(s, w)ds \\ &- \sum_{n=1}^r \int_{[t]^-(\delta)}^t \sigma_n^i(X(s, w))dw^n(s) \\ &- \sum_{n,m=1}^r \sum_{\alpha=1}^d c_{nm} \int_{[t]^-(\delta)}^t (\sigma_n^\alpha \partial_\alpha \sigma_m^i)(X(s, w))ds, \end{aligned} \tag{7.49}$$

$$\begin{aligned} H_2(t) = &\sum_{n=1}^r \{ \int_\delta^{[t]^-(\delta)} \sigma_n^i(X_\delta(s, w))\dot{B}_\delta^n(s, w)ds \\ &- \int_\delta^{[t]^-(\delta)} \sigma_n^i(X(s, w))dw^n(s) \\ &- \sum_{m=1}^r \sum_{\alpha=1}^d c_{mn} \int_\delta^{[t]^-(\delta)} \sigma_m^\alpha \partial_\alpha \sigma_n^i(X(s, w))ds\}, \end{aligned} \tag{7.50}$$

$$\begin{aligned} H_3 = &\sum_{n=1}^r \int_0^\delta \sigma_n^i(X_\delta(s, w))\dot{B}_\delta^n(s, w)ds - \sum_{n=1}^r \int_0^\delta \sigma_n^i(X(s, w))dw^n(s) \\ &- \sum_{n,m=1}^r \sum_{\alpha=1}^d c_{nm} \int_0^\delta \sigma_n^\alpha \partial_\alpha \sigma_m^i(X(s, w))ds \end{aligned} \tag{7.51}$$

and

$$H_4(t) = \int_0^t b^i(X_\delta(s, w))ds - \int_0^t b^i(X(s, w))ds. \tag{7.52}$$

It is clear that

$$E[\sup_{0\leq t\leq t_1} |H_4(t)|^2] \leq K_1 \int_0^{t_1} E[|X_\delta(s, w) - X(s, w)|^2]ds \quad \text{for every} \quad 0 \leq t_1 \leq T. \tag{7.53}$$

We now write

$$H_1(t) = H_{11}(t) - \sum_{n=1}^r H_{12}^n(t) - H_{13}(t).$$

Then it is obvious that

$$(7.54)\qquad E[\sup_{0\le t\le T}|H_{13}(t)|^2]\le K_2\tilde{\delta}^2.$$

By a similar estimate as for (7.28), we have

$$(7.55)\qquad E[\sup_{0\le t\le T}|H_{11}(t)|^2]\le K_3(n(\delta)^3\delta)^{1/2}.$$

Also

$$\begin{aligned} H^n_{12}(t) &= \sigma^i_n(X([t]^-(\tilde{\delta}),w))(w^n(t)-w^n([t]^-(\tilde{\delta}))) \\ &\quad + \int_{[t]^-(\delta)}^t (\sigma^i_n(X(s,w))-\sigma^i_n(X([t]^-(\tilde{\delta}),w)))dw^n(s) \\ &:= H^n_{121}(t)+H^n_{122}(t). \end{aligned}$$

By a similar estimate as for (7.32), we have

$$(7.56)\qquad E[\sup_{0\le t\le T}|H^n_{121}(t)|^2]\le K_4(n(\delta)\delta)^{1/2}.$$

As for $H^n_{122}(t)$,

$$\begin{aligned} & E[\sup_{0\le t\le T}|H^n_{122}(t)|^2] \\ &\le E[\sup_{0\le k\le m(T)}\sup_{0\le t\le \delta}\{\int_{k\delta}^{k\delta+t}(\sigma^i_n(X(s,w))-\sigma^i_n(X([s]^-(\tilde{\delta}),w)))dw^n(s)\}^2] \\ &\le \sum_{k=0}^{m(T)} E'[E[\sup_{0\le t\le\delta}\{\int_0^t(\sigma^i_n(X(s,y,w))-\sigma^i_n(y))dw^n(s)\}^2]|_{y=X(k\delta,w')}]\ * \\ (7.57)\quad &\le K_5\sum_{k=0}^{m(T)} E'[E[\int_0^\delta(\sigma^i_n(X(s,y,w))-\sigma^i_n(y))^2ds]|_{y=X(k\delta,w')}] \\ &\le K_6\sum_{k=0}^{m(T)}\sum_{j=1}^{d} E'[E[\int_0^\delta|X^j(s,y,w)-y^j|^2ds]|_{y=X(k\delta,w')}] \\ &\le K_7(m(T)+1)\int_0^{\tilde{\delta}}[s+s^2]ds \\ &\le K_8\tilde{\delta}. \end{aligned}$$

Here we used the estimate

$$|X^j(s,y,w)-y^j|^2\le K_9[(\sum_{n=1}^{r}\int_0^s\sigma^j_n(X(\xi,y,w))dw^n(\xi))^2$$

* E' stands for the integration with respect to w'.

$$+ (\int_0^t \bar{b}^j(X(\xi, y, w))d\xi)^2], *$$

and hence

$$E[|X^j(s, y, w) - y^j|^2] \leq K_{10}(s + s^2).$$

Putting these estimate together, we conclude that

$$(7.58) \qquad E[\sup_{0 \leq t \leq T} |H_1(t)|^2] = o(1) \qquad \text{as} \quad \delta \downarrow 0.$$

It is obvious that $|H_3| \leq \sup_{0 \leq t \leq T} |H_1(t)|$ and hence

$$(7.59) \qquad E[|H_3|^2] = o(1) \qquad \text{as} \quad \delta \downarrow 0.$$

Finally we shall estimate $H_2(t)$. We have

$$\begin{aligned}
&\int_{k\delta}^{(k+1)\delta} \sigma_n^i(X_\delta(s,w))\dot{B}_\delta^n(s, w)ds \\
&= \sigma_n^i(X_\delta(k\bar{\delta}, w))[B_\delta^n((k+1)\bar{\delta}, w) - B_\delta^n(k\bar{\delta}, w)] \\
&+ \sum_{\beta=1}^d \int_{k\delta}^{(k+1)\delta} \partial_\beta \sigma_n^i(X_\delta(s, w))[\sum_{j=1}^r \sigma_j^\beta(X_\delta(s, w))\dot{B}_\delta^j(s, w) + b^\beta(X_\delta(s, w))] \\
&\qquad \times [B_\delta^n((k+1)\bar{\delta}, w) - B_\delta^n(s, w)]ds \\
&:= J_1(k) + \sum_{\beta=1}^d J_2^\beta(k).
\end{aligned}$$

Also

$$\begin{aligned}
J_1(k) &= \sigma_n^i(X_\delta(k\bar{\delta} - \delta, w))(w^n((k+1)\bar{\delta}) - w^n(k\bar{\delta})) \\
&+ [\sigma_n^i(X_\delta(k\bar{\delta},w)) - \sigma_n^i(X_\delta(k\bar{\delta} - \delta, w))](B_\delta^n((k+1)\bar{\delta}, w) - B_\delta^n(k\bar{\delta}, w)) \\
&+ \sigma_n^i(X_\delta(k\bar{\delta} - \delta, w))(B_\delta^n((k+1)\bar{\delta}, w) - w^n((k+1)\bar{\delta})) \\
&+ \sigma_n^i(X_\delta(k\bar{\delta} - \delta, w))(w^n(k\bar{\delta}) - B_\delta^n(k\bar{\delta}, w)) \\
&:= J_{11}(k) + J_{12}(k) + J_{13}(k) + J_{14}(k).
\end{aligned}$$

Hence, writing $H_2(t) = \sum_{n=1}^r H_2^n(t)$,

$$H_2^n(t) = I_1(t) + I_2(t) + I_3(t) + I_4(t) + I_5(t),$$

---

* $\{\bar{b}^j\}$ is the coefficient of the vector field $A_0 + \sum_{n<m} s_{nm}[A_n, A_m]$.

where

$$I_1(t) = \sum_{k=1}^{m(t)-1} J_{11}(k) - \int_{\delta}^{[t]^-(\delta)} \sigma_n^i(X(s, w))dw^n(s) \tag{7.60}$$

$$I_i(t) = \sum_{k=1}^{m(t)-1} J_{1i}(k), \qquad i = 2, 3, 4 \tag{7.61}$$

and

$$I_5(t) = \sum_{\beta=1}^{d} \Big\{ \sum_{k=1}^{m(t)-1} J_2^{\beta}(k) - \int_{\delta}^{[t]^-(\delta)} \sum_{j=1}^{r} c_{jn}(\sigma_j^{\beta}\partial_{\beta}\sigma_n^i)(X(s, w))ds\Big\}. \tag{7.62}$$

First we write $I_1(t)$ as

$$I_1(t) = \int_{\delta}^{[t]^-(\delta)} [\sigma_n^i(X_\delta([s]^-(\bar{\delta}) - \delta, w)) - \sigma_n^i(X(s, w))]dw^n(s)$$

and then use a martingale inequality to obtain

$$\begin{aligned} E[\sup_{0\le t\le t_1} |I_1(t)|^2] &\le K_{11}E[|I_1(t_1)|^2] \\ &\le K_{12}\int_{\delta}^{[t_1]^-(\delta)} E[|X_\delta([s]^-(\bar{\delta}) - \delta, w) - X(s, w)|^2]ds \\ &\le K_{13}\Big\{\int_0^{t_1} E[|X_\delta(s, w) - X(s, w)|^2]ds \\ &\qquad + \int_{\delta}^{[t]^-(\delta)} E[|X_\delta([s]^-(\bar{\delta}) - \delta, w) - X_\delta(s, w)|^2]ds\Big\} \\ &\le K_{14}\Big\{\int_0^{t_1} E[|X_\delta(s, w) - X(s, w)|^2]ds \\ &\qquad + \int_{\delta}^{T} E[\sum_{j=1}^{r}(\int_{[s]^-(\delta)-\delta}^{[s]^+(\delta)} |\dot{B}_\delta^j(\xi, w)|d\xi)^2 + (\bar{\delta} + \delta)^2]ds\Big\} \\ &\le K_{15}\Big\{\int_0^{t_1} E[|X_\delta(s, w) - X(s, w)|^2]ds + n(\delta)^2\delta\Big\} \\ &\qquad\qquad \text{for every} \quad 0 \le t_1 \le T. \end{aligned} \tag{7.63}$$

In the case of $I_2(t)$,

$$\begin{aligned} &E[\sup_{0\le t\le T} |I_2(t)|^2] \\ &\le E[\sum_{k=1}^{m(T)-1} (\sigma_n^i(X_\delta(k\bar{\delta}, w)) - \sigma_n^i(X_\delta(k\bar{\delta} - \delta, w)))^2 \end{aligned}$$

$$\times \sum_{k=1}^{m(T)-1} (B_\delta^n((k+1)\tilde{\delta}, w) - B_\delta^n(k\tilde{\delta}, w))^2]$$

$$\leq K_{16} \{m(T) \sum_{k=1}^{m(T)-1} E[|X_\delta(k\tilde{\delta}, w) - X_\delta(k\tilde{\delta} - \delta, w)|^4] \tag{7.64}$$

$$\times m(T) \sum_{k=1}^{m(T)-1} E[|B_\delta^n((k+1)\tilde{\delta}, w) - B_\delta^n(k\tilde{\delta}, w)|^4]\}^{1/2}$$

$$\leq K_{17}(m(T)^2\delta^2 m(T)^2\tilde{\delta}^2)^{1/2}$$

$$\leq K_{18}(n(\delta))^{-1} \longrightarrow 0 \qquad \text{as} \quad \delta \downarrow 0.$$

Similarly as for (7.34), we can show that

$$E[\sup_{0\leq t\leq T} |I_3(t)|^2] \leq K_{19}(n(\delta))^{-1} \longrightarrow 0 \qquad \text{as} \quad \delta \downarrow 0 \tag{7.65}$$

and

$$E[\sup_{0\leq t\leq T} |I_4(t)|^2] \leq K_{20}(n(\delta))^{-1} \longrightarrow 0 \qquad \text{as} \quad \delta \downarrow 0. \tag{7.66}$$

Finally we shall consider $I_5(t)$. Write $I_5(t)$ as

$$I_5(t) = \sum_{\beta=1}^{d} I_5^\beta(t),$$

where

$$I_5^\beta(t) = \sum_{k=1}^{m(t)-1} J_2^\beta(k) - \sum_{k=1}^{m(t)-1} \int_{k\delta}^{(k+1)\delta} \sum_{j=1}^{r} c_{jn}(\sigma_j^\beta \partial_\beta \sigma_n^i)(X(s, w))ds$$

$$= \sum_{j=1}^{r} \sum_{k=1}^{m(t)-1} \int_{k\delta}^{(k+1)\delta} [(\sigma_j^\beta \partial_\beta \sigma_n^i)(X_\delta(s, w))$$

$$- (\sigma_j^\beta \partial_\beta \sigma_n^i)(X_\delta(k\tilde{\delta}, w))]\dot{B}_\delta^j(s, w)[B_\delta^n((k+1)\tilde{\delta}, w) - B_\delta^n(s, w)]ds$$

$$+ \sum_{k=1}^{m(t)-1} \int_{k\delta}^{(k+1)\delta} (b^\beta \partial_\beta \sigma_n^i)(X_\delta(s, w))(B_\delta^n((k+1)\tilde{\delta}, w) - B_\delta^n(s, w))ds$$

$$+ \sum_{j=1}^{r} \sum_{k=1}^{m(t)-1} \int_{k\delta}^{(k+1)\delta} (\sigma_j^\beta \partial_\beta \sigma_n^i)(X_\delta(k\tilde{\delta}, w))[\dot{B}_\delta^j(s, w)(B_\delta^n((k+1)\tilde{\delta}, w)$$

$$- B_\delta^n(s, w)) - c_{jn}(\tilde{\delta}, \delta)]ds$$

$$+ \sum_{j=1}^{r} \sum_{k=1}^{m(t)-1} \int_{k\delta}^{(k+1)\delta} [(\sigma_j^\beta \partial_\beta \sigma_n^i)(X_\delta(k\tilde{\delta}, w)) - (\sigma_j^\beta \partial_\beta \sigma_n^i)(X(s, w))]ds\, c_{jn}$$

$$+ \sum_{j=1}^{r} \sum_{k=1}^{m(t)-1} \tilde{\delta}(\sigma_j^\beta \partial_\beta \sigma_n^i)(X_\delta(k\tilde{\delta}, w))(c_{jn}(\tilde{\delta}, \delta) - c_{jn})$$

$$:= \sum_{j=1}^{r} I_{51}^j(t) + I_{52}(t) + \sum_{j=1}^{r} I_{53}^j(t) + \sum_{j=1}^{r} I_{54}^j(t) + \sum_{j=1}^{r} I_{55}^j(t).$$

It is obvious that

$$E[\sup_{0\le t\le T} |I_{55}^j(t)|^2] \le K_{21}(c_{jn}(\tilde{\delta}, \delta) - c_{jn})^2 = o(1). \tag{7.67}$$

As for $I_{54}^j(t)$,

$$\begin{aligned}
&E[\sup_{0\le t\le t_1} |I_{54}^j(t)|^2] \\
&\le E[(\int_0^{t_1} |(\sigma_j^\beta \partial_\beta \sigma_n^i)(X_\delta([s]^-(\tilde{\delta}), w)) - (\sigma_j^\beta \partial_\beta \sigma_n^i)(X(s, w))| ds)^2] c_{jn}^2 \\
&\le TE[\int_0^{t_1} |(\sigma_j^\beta \partial_\beta \sigma_n^i)(X_\delta([s]^-(\tilde{\delta}), w)) - (\sigma_j^\beta \partial_\beta \sigma_n^i)(X(s, w))|^2 ds] c_{jn}^2 \\
&\le K_{22} \int_0^{t_1} E[|X(s, w) - X_\delta([s]^-(\tilde{\delta}), w)|^2] ds \qquad (7.68)\\
&\le K_{23}[\int_0^{t_1} E[|X(s, w) - X_\delta(s, w)|^2] ds \\
&\qquad + \int_0^T E[|X_\delta(s, w) - X_\delta([s]^-(\tilde{\delta}), w)|^2] ds] \\
&\le K_{24}[\int_0^{t_1} E[|X(s, w) - X_\delta(s, w)|^2] ds + n(\delta)^2\delta] \\
&\qquad\qquad \text{for} \quad 0 \le t_1 \le T.
\end{aligned}$$

In the case of I_{52}

$$\begin{aligned}
&E[\sup_{0\le t\le T} |I_{52}(t)|^2] \\
&\le E[\sup_{0\le t\le T} (\sum_{k=1}^{m(t)-1} \int_{k\tilde{\delta}}^{(k+1)\tilde{\delta}} |(b^\beta \partial_\beta \sigma_n^i)(X_\delta(s, w))| ds \int_{k\tilde{\delta}}^{(k+1)\tilde{\delta}} |\dot{B}_\delta^n(s, w)| ds)^2] \\
&\le K_{25} E[\{\sum_{k=1}^{m(T)-1} \tilde{\delta} \int_{k\tilde{\delta}}^{(k+1)\tilde{\delta}} |\dot{B}_\delta^n(s, w)| ds\}^2] \qquad (7.69)\\
&\le K_{26} \tilde{\delta}^2 m(T) \sum_{k=1}^{m(T)-1} E[(\int_{k\tilde{\delta}}^{(k+1)\tilde{\delta}} |\dot{B}_\delta^n(s, w)| ds)^2] \\
&\le K_{26} \tilde{\delta}^2 m(T)^2 n(\delta)^2 \delta \\
&\le K_{27} n(\delta)^2 \delta \longrightarrow 0 \quad \text{as} \quad \delta \downarrow 0.
\end{aligned}$$

We estimate $I_{51}^j(t)$ by

$$
\begin{aligned}
&E[\sup_{0\le t\le T} |I^j_{51}(t)|^2] \\
&\le K_{28} \sum_{l=1}^{d} E[\{\sum_{k=1}^{m(T)-1} \int_{k\delta}^{(k+1)\delta} |X^l_\delta(s, w) - X^l_\delta(k\tilde{\delta}, w)|\,|\dot{B}^j_\delta(s, w)|\,ds \\
&\qquad\qquad \times \int_{k\delta}^{(k+1)\delta} |\dot{B}^n_\delta(s, w)|\,ds\}^2] \\
&\le K_{29} E[\{\sum_{k=1}^{m(T)-1} (\sum_{l=1}^{d} \int_{k\delta}^{(k+1)\delta} |\dot{B}^l_\delta(s, w)|\,ds + \tilde{\delta}) \int_{k\delta}^{(k+1)\delta} |\dot{B}^j_\delta(s, w)|\,ds \\
&\qquad\qquad \times \int_{k\delta}^{(k+1)\delta} |\dot{B}^n_\delta(s, w)|\,ds\}^2] \\
&\le K_{30} \sum_{l=1}^{d} m(T) \sum_{k=1}^{m(T)-1} \{E[(\int_{k\delta}^{(k+1)\delta} |\dot{B}^l_\delta(s, w)|\,ds)^2 (\int_{k\delta}^{(k+1)\delta} |\dot{B}^j_\delta(s, w)|\,ds)^2 \\
&\qquad\qquad \times (\int_{k\delta}^{(k+1)\delta} |\dot{B}^n_\delta(s, w)|\,ds)^2] \\
&\qquad + \tilde{\delta}^2 E[(\int_{k\delta}^{(k+1)\delta} |\dot{B}^j_\delta(s, w)|\,ds)^2 (\int_{k\delta}^{(k+1)\delta} |\dot{B}^n_\delta(s, w)|\,ds)^2]\} \\
&\le K_{31} m(T)^2 (n(\delta)^6 \delta^3 + n(\delta)^6 \delta^4) \\
&\le K_{32} n(\delta)^4 \delta \longrightarrow 0 \qquad \text{as} \quad \delta \downarrow 0.
\end{aligned}
\tag{7.70}
$$

By repeating the same proof as for (7.37) we see that

$$
E[\sup_{0\le t\le T} |I^j_{53}(t)|^2] \le K_{33}\left(n(\delta)^3 \delta + \left(\frac{\delta}{\tilde{\delta}} c^*_{jn}(\tilde{\delta}, \delta)\right)^2 + n(\delta)^{-1}\right) \longrightarrow 0 \quad \text{as} \quad \delta \downarrow 0.
\tag{7.71}
$$

Now combining (7.67), (7.68), (7.69), (7.70) and (7.71), we obtain the estimate

$$
E[\sup_{0\le t\le t_1} |I_5(t)|^2] \le K_{34} \int_0^{t_1} E[|X(s, w) - X_\delta(s, w)|^2] ds + o(1) \quad \text{as} \quad \delta \downarrow 0,
\tag{7.72}
$$

where $o(1)$ is uniform in $t_1 \in [0, T]$. By (7.63), (7.64), (7.65), (7.66) and (7.72), we have

$$
E[\sup_{0\le t\le t_1} |H_2(t)|^2] \le K_{35} \int_0^{t_1} E[|X(s, w) - X_\delta(s, w)|^2] ds + o(1),
\tag{7.73}
$$

where again $o(1)$ is uniform in $t_1 \in [0, T]$. Finally, by (7.53), (7.58), (7.59) and (7.73), we have

$$E[\sup_{0\le t\le t_1}|X_\delta(t, w) - X(t, w)|^2] \le K_{36}\int_0^{t_1} E[|X_\delta(s, w) - X(s, w)|^2]ds + o(1). \tag{7.74}$$

As before $o(1)$ is uniform in $t_1\in[0, T]$. (7.46) then follows from (7.74) by a standard argument. This completes the proof.

Remark 7.1. If the vector fields $A_1, A_2, \dots, A_r$ are commutative, i.e., $[A_n, A_m] = 0$ for $n, m = 1, 2, \dots, r$, then the limiting process of $(X_\delta(t, w))$ does not depend on the choice of the approximation $\{B_\delta(t, w)\}$ as Theorem 7.2 shows. But this is seen more directly in this case if we notice that $X(\cdot, w)$ is a continuous functional on W_0^r by the result of Doss (Example 2.2 of Chapter III, Section 2). If $A_1, A_2, \dots, A_r$ are not commutative, however, $X(\cdot, w)$ is not continuous on W_0^r and the limiting processes of $X_\delta(\cdot, w)$ are controlled by $\{s_{ij}\}$. For piecewise linear approximations or approximations by mollifiers, we have $s_{ij} = 0$ and so the limiting process is the solution of the stochastic differential equation corresponding to vector fields $A_1, A_2, \dots, A_r$ and to A_0 as was defined in Chapter V, Section 1.

Remark 7.2. In the proof of Theorem 7.2, we have actually shown that

$$\lim_{\delta\downarrow 0}\sup_{x\in \mathbf{R}^d} E[\sup_{0\le t\le T}|X_\delta(t, x, w) - X(t, x, w)|^2] = 0. \tag{7.75}$$

In the following, we shall restrict ourselves to a class of piecewise linear approximations. It is an open question whether our results below can also be obtained for more general class of approximations. We shall assume that the coefficients σ_n^i and b^i of vector fields $A_0, A_1, \dots, A_r$ are C^∞ and bounded together with derivatives of all orders. In the following discussion, the terms involving the coefficients b^i do not cause any trouble and so we shall assume, just for simplicity of notations, that $A_0 \equiv 0$. Thus we consider the stochastic differential equation

$$dX(t) = \sigma(X(t))\circ dw(t). \, * \tag{7.76}$$

Also, setting for each $n = 1, 2, \dots,$

$$w_n(t) = n\left[\left(\frac{j+1}{n} - t\right)w\left(\frac{j}{n}\right) + \left(t - \frac{j}{n}\right)w\left(\frac{j+1}{n}\right)\right],$$

* We shall use the matrix notations below; in particular $\sigma = (\sigma_\alpha^i)$ and $w = (w^\alpha)$.

$$\text{if} \quad \frac{j}{n} \le t \le \frac{j+1}{n}, \quad j = 0, 1, \dots,$$

we consider the following ordinary differential equation

$$\frac{dX_n}{dt}(t) = \sigma(X_n(t)) \frac{dw_n}{dt}(t). \tag{7.77}$$

The solutions of (7.76) and (7.77) with the initial value $x \in \mathbf{R}^d$ are denoted by $X(t,x,w)$ and $X_n(t,x,w)$, respectively.

Lemma 7.2. Let T and N be arbitrary given positive constants. Then we have

$$\sup_{|x| \le N} \sup_n E[\sup_{0 \le s \le T} \|D^\alpha X_n(s, x, w)\|^p] < \infty \tag{7.78}$$

for every $p \ge 2$ and multi-index α.

Proof. First we consider the case $\alpha = (0, 0, \dots, 0)$, i.e., $D^\alpha X_n = X_n$. We denote

$$\Delta_k w = w\left(\frac{k+1}{n}\right) - w\left(\frac{k}{n}\right).$$

Then, if $\frac{k}{n} \le t \le \frac{k+1}{n}$,

$$\begin{aligned}
X_n(t, x, w) &- X_n\left(\frac{k}{n}, x, w\right) \\
&= \int_{k/n}^t \sigma(X_n(s, x, w)) \Delta_k w \, ds \, n \\
&= \sigma\left(X_n\left(\frac{k}{n}, x, w\right)\right) \Delta_k w \, n\left(t - \frac{k}{n}\right) \\
&\qquad + \int_{k/n}^t \left[\sigma(X_n(s, x, w)) - \sigma\left(X_n\left(\frac{k}{n}, x, w\right)\right)\right] ds \, \Delta_k w \, n
\end{aligned}$$

and hence

$$\begin{aligned}
X_n(t, x, w) &- x \\
&= \sum_{k=0}^{[nt]-1} \sigma\left(X_n\left(\frac{k}{n}, x, w\right)\right) \Delta_k w
\end{aligned}$$

$$+\sum_{k=0}^{[nt]-1}\int_{k/n}^{(k+1)/n}\left[\sigma(X_n(s,x,w))-\sigma\left(X_n\left(\frac{k}{n},x,w\right)\right)\right]ds\,\Delta_k w\,n$$
$$+\sigma\left(X_n\left(\frac{[nt]}{n},x,w\right)\right)\Delta_{[nt]}w\left(t-\frac{[nt]}{n}\right)n$$
$$+\int_{[nt]/n}^{t}\left[\sigma(X_n(s,x,w))-\sigma\left(X_n\left(\frac{[nt]}{n},x,w\right)\right)\right]ds\,\Delta_{[nt]}w\,n$$
$$:=I_1(t)+I_2(t)+I_3(t)+I_4(t).$$

By Theorem III-3.1,

$$\begin{aligned}
&E[\sup_{0\le t\le T}\|I_1(t)\|^p]\\
&\le K_1E\left[\left(\sum_{k=0}^{[nt]-1}\left\|\sigma\left(X_n\left(\frac{k}{n},x,w\right)\right)\right\|^2\right)^{p/2}\right]n^{-p/2} \text{ *1}\\
&\le K_2n^{p/2}n^{-p/2}\\
&=K_2<\infty .
\end{aligned}$$

Also

$$\begin{aligned}
&E[\sup_{0\le t\le T}|I_2(t)|^p]\\
&\le T^{p-1}n^{p-1}E\Big[\sup_{0\le t\le T}\sum_{k=0}^{[nt]-1}\Big\|\int_{k/n}^{(k+1)/n}\Big[\sigma(X_n(s,x,w))\\
&\qquad\qquad -\sigma\left(X_n\left(\frac{k}{n},x,w\right)\right)\Big]ds\,\Delta_k w\Big\|^p\Big]n^p \text{ *2}\\
&\le T^{p-1}n^{p-1}E\Big[\sum_{k=0}^{[nT]-1}\Big\|\int_{k/n}^{(k+1)/n}\Big[\sigma(X_n(s,x,w))\\
&\qquad\qquad -\sigma\left(X_n\left(\frac{k}{n},x,w\right)\right)\Big]ds\,\Delta_k w\Big\|^p\Big]n^p\\
&\le T^{p-1}n^{p-1}n^pn^{-(p-1)}\sum_{k=0}^{[nT]-1}\int_{k/n}^{(k+1)/n}E\Big[\Big\|\Big[\sigma(X_n(s,x,w))\\
&\qquad\qquad -\sigma\left(X_n\left(\frac{k}{n},x,w\right)\right)\Big]\Delta_k w\Big\|^p\Big]ds\\
&\le K_3n^p\sum_{k=0}^{[nT]-1}\int_{k/n}^{(k+1)/n}E\left[\left\|X_n(s,x,w)-X_n\left(\frac{k}{n},x,w\right)\right\|^p\|\Delta_k w\|^p\right]ds
\end{aligned}$$

*1 $K_1, K_2, \ldots$ are positive constants independent of n and x.

*2 Generally, if $A_1, A_2, \ldots, A_k$ are matrices of the same type $\|\sum_{i=1}^{k} A_i\|^p \le (\sum_{i=1}^{k}\|A_i\|)^p \le k^{p-1}\sum_{i=1}^{k}\|A_i\|^p$.

$$\leq K_3 n^{2p} \sum_{k=0}^{[nT]-1} \int_{k/n}^{(k+1)/n} E[\|\int_{k/n}^{s} \sigma(X_n(u, x, w))\Delta_k w du\|^p \|\Delta_k w\|^p] ds$$

$$\leq K_3 n^{2p} \sum_{k=0}^{[nT]-1} \int_{k/n}^{(k+1)/n} E[(\int_{k/n}^{s} \|\sigma(X_n(u, x, w))\| \, \|\Delta_k w\| du)^p \|\Delta_k w\|^p] ds$$

$$\leq K_4 n^{2p} \sum_{k=0}^{[nT]-1} \int_{k/n}^{(k+1)/n} \left(s - \frac{k}{n}\right)^p ds \, E(\|\Delta_k w\|^{2p})$$

$$\leq K_5 n^{2p} n \, n^{-(p+1)} \, n^{-p}$$

$$= K_5 < \infty .$$

As for $I_3(t)$,

$$E[\sup_{0\leq t\leq T} \|I_3(t)\|^p]$$

$$\leq E[\sup_{0\leq t\leq T} \|\sigma\left(X_n\left(\frac{[nt]}{n}, x, w\right)\right)\Delta_{[nt]} w\|^p]$$

$$\leq K_6 E[\sup_{0\leq t\leq T} \|\Delta_{[nt]} w\|^p]$$

$$\leq K_6 E[\max_{0\leq k\leq [nT]} \|\Delta_k w\|^p]$$

$$\leq K_6 \sum_{k=0}^{[nT]} E[\|\Delta_k w\|^p]$$

$$\leq K_7([nT] + 1) n^{-p/2}$$

$$\leq K_8 < \infty .$$

Finally as in the estimate for $I_2(t)$,

$$E[\sup_{0\leq t\leq T} \|I_4(t)\|^p]$$

$$\leq E[\sup_{0\leq t\leq T} \int_{[nt]/n}^{t} \|\Big(\sigma(X_n(s, x, w)) - \sigma\left(X_n\left(\frac{[nt]}{n}, x, w\right)\right)\Big)\Delta_{[nt]} w\|^p ds] n^p n^{-(p-1)}$$

$$\leq E[\sup_{0\leq t\leq T} \int_{[nt]/n}^{([nt]+1)/n} \|\sigma(X_n(s, x, w)) - \sigma\left(X_n\left(\frac{[nt]}{n}, x, w\right)\right)\|^p \|\Delta_{[nt]} w\|^p ds] \, n$$

$$\leq K_9 n \sum_{k=0}^{[nT]} E\left[\int_{k/n}^{(k+1)/n} \|X_n(s, x, w) - X_n\left(\frac{k}{n}, x, w\right)\|^p \|\Delta_k w\|^p ds\right]$$

$$\leq K_{10} n^{1-p}$$

$$\leq K_{10} < \infty .$$

Consequently, we have obtained the estimate

$$E[\sup_{0\leq t\leq T} \|X_n(t, x, w)\|^p] \leq K_{11}(1 + |x|^p). \tag{7.79}$$

Thus (7.78) is proved for $\alpha = (0, 0, \ldots, 0)$.

Next, we consider the case of the first order derivatives. Setting

$$Y_n(t, x, w) = \left(\frac{\partial}{\partial x^j} X_n^i(t, x, w)\right) \quad \text{and} \quad D\sigma = \left(\frac{\partial}{\partial x^j}\sigma_\alpha^i\right),$$

we have

$$Y_n(t, x, w) = I + \int_0^t D\sigma(X_n(s, x, w))Y_n(s, x, w)\dot{w}_n(s)ds. \,*$$

Hence

$$\begin{aligned}
&Y_n(t, x, w) - I \\
&= \sum_{k=0}^{[nt]-1} D\sigma\left(X_n\left(\frac{k}{n}, x, w\right)\right)Y_n\left(\frac{k}{n}, x, w\right)\Delta_k w \\
&+ \sum_{k=0}^{[nt]-1} \int_{k/n}^{(k+1)/n} \Big[D\sigma(X_n(s, x, w)) \\
&\qquad - D\sigma\left(X_n\left(\frac{k}{n}, x, w\right)\right)\Big]ds\, Y_n\left(\frac{k}{n}, x, w\right) \Delta_k w\, n \\
&+ \sum_{k=0}^{[nt]-1} \int_{k/n}^{(k+1)/n} D\sigma(X_n(s, x, w))\left(Y_n(s, x, w) - Y_n\left(\frac{k}{n}, x, w\right)\right)ds\,\Delta_k w\, n \\
&+ D\sigma\left(X_n\left(\frac{[nt]}{n}, x, w\right)\right) Y_n\left(\frac{[nt]}{n}, x, w\right)\Delta_{[nt]} w\, n\left(t - \frac{[nt]}{n}\right) \\
&+ \int_{[nt]/n}^t \Big[D\sigma(X_n(s, x, w)) \\
&\qquad - D\sigma\left(X_n\left(\frac{[nt]}{n}, x, w\right)\right)\Big]\, ds\, Y_n\left(\frac{[nt]}{n}, x, w\right)\Delta_k w\, n \\
&+ \int_{[nt]/n}^t D\sigma(X_n(s, x, w))\left(Y_n(s, x, w) - Y_n\left(\frac{[nt]}{n}, x, w\right)\right)ds\, \Delta_k w\, n \\
&:= J_1(t) + J_2(t) + J_3(t) + J_4(t) + J_5(t) + J_6(t).
\end{aligned}$$

* We use the following notations: for $A = (a_{j,\alpha}^i)$, $B = (b_j^i)$ and $C = (c^\alpha)$, $ABC = (d_j^i)$ where $d_j^i = \sum_{k,\alpha} a_{k,\alpha}^i b_j^k c^\alpha$. It is easy to see $\|ABC\| \leq \|A\|\|B\|\|C\|$ where $\|A\| = (\sum_{i,j,\alpha} |a_{j,\alpha}^i|^2)^{1/2}$.

Let $t_1 \in [0, T]$. By Theorem III-3.1,

$$
\begin{aligned}
&E[\sup_{0\le t\le t_1} \|J_1(t)\|^p] \\
&\le K_{12}E\left[\left(\sum_{k=0}^{[nt_1]-1} \left\|D\sigma\left(X_n\left(\frac{k}{n}, x, w\right)\right)\right\|^2 \left\|Y_n\left(\frac{k}{n}, x, w\right)\right\|^2\right)^{p/2}\right] n^{-p/2} \\
&\le K_{13}E\left[\left(\sum_{k=0}^{[nt_1]-1} \left\|Y_n\left(\frac{k}{n}, x, w\right)\right\|^2\right)^{p/2}\right] n^{-p/2} \\
&\le K_{14}n^{-1} \sum_{k=0}^{[nt_1]-1} E\left(\left\|Y_n\left(\frac{k}{n}, x, w\right)\right\|^p\right) \\
&\le K_{14} \int_0^{t_1} E(\sup_{0\le s\le t} \|Y_n(s, x, w)\|^p)dt.
\end{aligned}
\tag{7.80}
$$

As for $J_2(t)$,

$$
\begin{aligned}
&E[\sup_{0\le t\le t_1} \|J_2(t)\|^p] \\
&= E\Big[\sup_{0\le t\le t_1} \Big\|\sum_{k=0}^{[nt]-1} \int_{k/n}^{(k+1)/n} \Big[D\sigma(X_n(s, x, w)) \\
&\qquad - D\sigma\left(X_n\left(\frac{k}{n}, x, w\right)\right)\Big] ds\, Y_n\left(\frac{k}{n}, x, w\right)\Delta_k w\Big\|^p\Big] n^p \\
&\le T^{p-1}n^{2p-1}E\Big[\sup_{0\le t\le t_1} \sum_{k=0}^{[nt]-1} \Big\|\int_{k/n}^{(k+1)/n} \Big[D\sigma(X_n(s, x, w)) \\
&\qquad - D\sigma\left(X_n\left(\frac{k}{n}, x, w\right)\right)\Big] ds\, Y_n\left(\frac{k}{n}, x, w\right)\Delta_k w\Big\|^p\Big] \\
&\le T^{p-1}n^{2p-1} \sum_{k=0}^{[nt_1]-1} E\Big[\Big\|\int_{k/n}^{(k+1)/n} \Big[D\sigma(X_n(s, x, w)) \\
&\qquad - D\sigma\left(X_n\left(\frac{k}{n}, x, w\right)\right)\Big] ds\, Y_n\left(\frac{k}{n}, x, w\right)\Delta_k w\Big\|^p\Big] \\
&\le T^{p-1}n^{2p-1}n^{-(p-1)} \sum_{k=0}^{[nt_1]-1} \int_{k/n}^{(k+1)/n} E\Big[\Big\|\Big\{D\sigma(X_n(s, x, w)) \\
&\qquad - D\sigma\left(X_n\left(\frac{k}{n}, x, w\right)\right)\Big\} Y_n\left(\frac{k}{n}, x, w\right)\Delta_k w\Big\|^p\Big] ds \\
&\le K_{15}n^p \sum_{k=0}^{[nt_1]-1} \int_{k/n}^{(k+1)/n} E\Big[\Big\|X_n(s, x, w) - X_n\left(\frac{k}{n}, x, w\right)\Big\|^p \\
&\qquad \times \Big\|Y_n\left(\frac{k}{n}, x, w\right)\Big\|^p \|\Delta_k w\|^p\Big] ds \\
&= K_{15}n^{2p} \sum_{k=0}^{[nt_1]-1} \int_{k/n}^{(k+1)/n} E\Big[\Big\|\int_{k/n}^{s} \sigma(X_n(u, x, w))\, \Delta_k w\, du\Big\|^p
\end{aligned}
\tag{7.81}
$$

$$\times \|Y_n\Big(\frac{k}{n}, x, w\Big)\|^p \|\Delta_k w\|^p\Big] ds$$

$$\le K_{16} n^{2p} \sum_{k=0}^{[nt_1]-1} \int_{k/n}^{(k+1)/n} \Big(s - \frac{k}{n}\Big)^p ds\, E\Big[\|Y_n\Big(\frac{k}{n}, x, w\Big)\|^p \|\Delta_k w\|^{2p}\Big]$$

$$= K_{16} n^{2p} \sum_{k=0}^{[nt_1]-1} \int_{k/n}^{(k+1)/n} \Big(s - \frac{k}{n}\Big)^p ds\, E\Big[\|Y_n\Big(\frac{k}{n}, x, w\Big)\|^p\Big] E[\|\Delta_k w\|^{2p}]$$

$$\le K_{17} n^{2p} n^{-(p+1)} n^{-p} \sum_{k=0}^{[nt_1]-1} E\Big[\|Y_n\Big(\frac{k}{n}, x, w\Big)\|^p\Big]$$

$$= K_{17} n^{-1} \sum_{k=0}^{[nt_1]-1} E\Big[\|Y_n\Big(\frac{k}{n}, x, w\Big)\|^p\Big]$$

$$\le K_{17} \int_0^{t_1} E[\sup_{0\le s\le t} \|Y_n(s, x, w)\|^p] dt.$$

For $J_3(t)$

$$E[\sup_{0\le t\le t_1} \|J_3(t)\|^p]$$

$$= E\Big[\sup_{0\le t\le t_1} \|\sum_{k=0}^{[nt]-1} \int_{k/n}^{(k+1)/n} D\sigma(X_n(s, x, w))\Big(Y_n(s, x, w) - Y_n\Big(\frac{k}{n}, x, w\Big)\Big) ds\, \Delta_k w\, n\|^p\Big] \tag{7.82}$$

$$\le T^{p-1} n^p \sum_{k=0}^{[nt_1]-1} \int_{k/n}^{(k+1)/n} E\Big[\|D\sigma(X_n(s, x, w))\|^p \|Y_n(s, x, w) - Y_n\Big(\frac{k}{n}, x, w\Big)\|^p \|\Delta_k w\|^p\Big] ds$$

$$\le K_{18} n^p \sum_{k=0}^{[nt_1]-1} \int_{k/n}^{(k+1)/n} E\Big[\|Y_n(s, x, w) - Y_n\Big(\frac{k}{n}, x, w\Big)\|^p \|\Delta_k w\|^p\Big] ds.$$

If $\dfrac{k}{n} \le s \le \dfrac{k+1}{n}$,

$$\|Y_n(s, x, w) - Y_n\Big(\frac{k}{n}, x, w\Big)\|^2$$

$$= \|\int_{k/n}^{s} D\sigma(X_n(u, x, w)) Y_n(u, x, w) \Delta_k w\, du\|^2 n^2$$

$$\le n^2 \|\Delta_k w\|^2 \int_{k/n}^{s} \|D\sigma(X_n(u, x, w))\|^2 du \int_{k/n}^{s} \|Y_n(u, x, w)\|^2 du$$

$$\le K_{19} n^2 \|\Delta_k w\|^2 \Big(s - \frac{k}{n}\Big) \int_{k/n}^{s} \|Y_n(u, x, w)\|^2 du$$

$$\leq K_{20}\Big\{\|\Delta_k w\|^2 \Big\|Y_n\Big(\frac{k}{n}, x, w\Big)\Big\|^2 + n\|\Delta_k w\|^2 \int_{k/n}^{s} \Big\|Y_n(u, x, w) - Y_n\Big(\frac{k}{n}, x, w\Big)\Big\|^2 du\Big\}.$$

From this inequality, we obtain the following estimate:

$$\Big\|Y_n(s, x, w) - Y_n\Big(\frac{k}{n}, x, w\Big)\Big\|^2 \leq K_{20}\|\Delta_k w\|^2 \Big\|Y_n\Big(\frac{k}{n}, x, w\Big)\Big\|^2 \exp\Big\{K_{20} n\|\Delta_k w\|^2\Big(s - \frac{k}{n}\Big)\Big\}$$

and consequently,

$$\Big\|Y_n(s, x, w) - Y_n\Big(\frac{k}{n}, x, w\Big)\Big\|^p \leq K_{21}\|\Delta_k w\|^p \Big\|Y_n\Big(\frac{k}{n}, x, w\Big)\Big\|^p \exp\{K_{22}\|\Delta_k w\|^2\}. \tag{7.83}$$

Substituting (7.83) into (7.82) yields

$$\begin{aligned} E[\sup_{0\leq t\leq t_1} \|J_3(t)\|^p] &\leq K_{23} n^p n^{-1} \sum_{k=0}^{[nt_1]-1} E\Big[\Big\|Y_n\Big(\frac{k}{n}, x, w\Big)\Big\|^p\Big] \\ &\qquad \times E[\|\Delta_k w\|^{2p} \exp\{K_{22}\|\Delta_k w\|^2\}] \\ &\leq K_{24} \frac{1}{n} \sum_{k=0}^{[nt_1]-1} E\Big[\Big\|Y_n\Big(\frac{k}{n}, x, w\Big)\Big\|^p\Big] \\ &\leq K_{24} \int_0^{t_1} E[\sup_{0\leq s\leq t} \|Y_n(s, x, w)\|^p] dt \end{aligned} \tag{7.84}$$

since

$$\begin{aligned} E[\|\Delta_k w\|^{2p} \exp\{K_{22}\|\Delta_k w\|^2\}] &= \int_{R^r} \Big(2\pi \frac{1}{n}\Big)^{-r/2} |x|^{2p} \exp\Big\{K_{22}|x|^2 - \frac{n}{2}|x|^2\Big\} dx \\ &\leq K_{25} n^{-p} \qquad \text{for sufficiently large } n. \end{aligned}$$

$J_4(t)$ is estimated as follows:

$$E[\sup_{0\leq t\leq t_1} \|J_4(t)\|^p]$$

$$= E\Big[\sup_{0\le t\le t_1}\|D\sigma\Big(X_n\Big(\frac{[nt]}{n}, x, w\Big)\Big)\, Y_n\Big(\frac{[nt]}{n}, x, w\Big)n\Big(t-\frac{[nt]}{n}\Big)\Delta_{[nt]}w\|^p\Big]$$

$$\le K_{26}E\Big[\sup_{0\le t\le t_1}\|Y_n\Big(\frac{[nt]}{n}, x, w\Big)\|^p\,\|\Delta_{[nt]}w\|^p\Big]$$

$$\le K_{26}\sum_{k=0}^{[nt_1]}E\Big[\|Y_n\Big(\frac{k}{n}, x, w\Big)\|^p\|\Delta_k w\|^p\Big] \tag{7.85}$$

$$\le K_{27}n^{-p/2}\sum_{k=0}^{[nt_1]}E\Big[\|Y_n\Big(\frac{k}{n}, x, w\Big)\|^p\Big]$$

$$\le K_{28}\Big(n^{1-p/2}\int_0^{t_1}E[\sup_{0\le s\le t}\|Y_n(s, x, w)\|^p]dt + n^{-p/2}E\Big[\|Y_n\Big(\frac{[nt_1]}{n}, x, w\Big)\|^p\Big]\Big).$$

In the case of $J_5(t)$,

$$E[\sup_{0\le t\le t_1}\|J_5(t)\|^p]$$

$$\le K_{29}nE\Big[\sup_{0\le t\le t_1}\int_{[nt]/n}^{t}\|X_n(s, x, w) - X_n\Big(\frac{[nt]}{n}, x, w\Big)\|^p ds \times \|Y_n\Big(\frac{[nt]}{n}, x, w\Big)\|^p\|\Delta_k w\|^p\Big] \tag{7.86}$$

$$\le K_{30}n^{1-p}n^{-1}\sum_{k=0}^{[nt_1]}E\Big(\|Y_n\Big(\frac{k}{n}, x, w\Big)\|^p\Big)$$

$$\le K_{30}\Big(n^{1-p}\int_0^{t_1}E[\sup_{0\le s\le t}\|Y_n(s, x, w)\|^p]dt + n^{-p}E\Big[\|Y_n\Big(\frac{[nt_1]}{n}, x, w\Big)\|^p\Big]\Big)$$

by the same estimate as for $J_2(t)$. As for $J_6(t)$,

$$E[\sup_{0\le t\le t_1}\|J_6(t)\|^p]$$

$$\le K_{31}n\sum_{k=0}^{[nt_1]}\int_{k/n}^{(k+1)/n}E\Big[\|Y_n(s, x, w) - Y_n\Big(\frac{k}{n}, x, w\Big)\|^p\|\Delta_k w\|^p\Big]ds$$

$$\le K_{32}n^{-p}\sum_{k=0}^{[nt_1]}E\Big[\|Y_n\Big(\frac{k}{n}, x, w\Big)\|^p\Big] \tag{7.87}$$

$$\le K_{32}\Big(n^{1-p}\int_0^{t_1}E[\sup_{0\le s\le t}\|Y_n(s, x, w)\|^p]dt + n^{-p}E\Big[\|Y_n\Big(\frac{[nt_1]}{n}, x, w\Big)\|^p\Big]\Big)$$

by the same estimate as for $J_3(t)$. Combining (7.80), (7.81), (7.84), (7.85), (7.86) and (7.87) together, it is easy to conclude that

$$E[\sup_{0\le t\le t_1} \|Y_n(t, x, w)\|^p] \le K_{33}(1 + \int_0^{t_1} E[\sup_{0\le s\le t} \|Y_n(s, x, w)\|^p]dt).$$

By a standard truncation argument, we can easily deduce from this that

$$(7.88) \qquad E[\sup_{0\le t\le T} \|Y_n(t, x, w)\|^p] \le K_{33}e^{K_{33}T}.$$

Thus (7.78) is proved for every α such that $|\alpha| = 1$.*1

Now we proceed to the case of $|\alpha| = 2$. First we note that if $k/n \le t \le (k+1)/n$, we have

$$(7.89) \qquad E\left[\|X_n(t, x, w) - X_n\left(\frac{k}{n}, x, w\right)\|^p\right] \le K_{34}n^{-p/2}$$

and, also by (7.83) and (7.88),

$$(7.90) \qquad E\left[\|Y_n(t, x, w) - Y_n\left(\frac{k}{n}, x, w\right)\|^p\right] \le K_{35}n^{-p/2}.$$

Set

$$Y^{i,(n)}_{j_1,j_2}(t, x, w) = \frac{\partial^2}{\partial x^{j_1}\partial x^{j_2}} X^i_n(t, x, w).$$

Then

$$Y^{i,(n)}_{j_1,j_2}(t, x, w) = \sum_{k=1}^{d}\sum_{\alpha=1}^{r}\int_0^t \sigma'_\alpha(X_n(s, x, w))^i_k Y^{k,(n)}_{j_1,j_2}(s, x, w)\dot{w}^\alpha_n(s)ds$$
$$+ \sum_{k,l=1}^{d}\sum_{\alpha=1}^{r}\int_0^t \sigma''_\alpha(X_n(s, x, w))^i_{kl} Y_n(s, x, w)^k_{j_1} Y_n(s, x, w)^l_{j_2}\dot{w}^\alpha_n(s)ds.*2$$

We denote the second term on the right by $\alpha_n(t, x, w)$. Then, denoting as

$$\alpha_n(t, x, w) = \sum_{k,l=1}^{d}\sum_{\alpha=1}^{r}\Big[\sum_{m=0}^{[nt]-1}\int_{m/n}^{(m+1)/n} \cdot + \int_{[nt]/n}^{t} \cdot\Big]$$

*1 It was actually proved that $\sup_{|x|\le N}$ can be strengthened to $\sup_x$.

*2 $\sigma'_\alpha(x)^i_j = \frac{\partial}{\partial x^j}\sigma^i_\alpha(x)$ and $\sigma''_\alpha(x)^i_{kl} = \frac{\partial^2}{\partial x^k \partial x^l}\sigma^i_\alpha(x)$.

$$:= \sum_{k,l=1}^{d} \sum_{\alpha=1}^{r} (H_1(t;k,l,\alpha) + H_2(t;k,l,\alpha)),$$

$$H_1(t;k,l,\alpha) = \sum_{m=0}^{[nt]-1} \sigma''_\alpha\Big(X_n\Big(\frac{m}{n},x,w\Big)\Big)^i_{kl} Y_n\Big(\frac{m}{n},x,w\Big)^k_{j_1} Y_n\Big(\frac{m}{n},x,w\Big)^l_{j_2} \Delta_m w^\alpha$$

$$+ \sum_{m=0}^{[nt]-1} \int_{m/n}^{(m+1)/n} \Big[\sigma''_\alpha(X_n(s,x,w))^i_{kl} - \sigma''_\alpha\Big(X_n\Big(\frac{m}{n},x,w\Big)\Big)^i_{kl}\Big] ds$$

$$\times Y_n\Big(\frac{m}{n},x,w\Big)^k_{j_1} Y_n\Big(\frac{m}{n},x,w\Big)^l_{j_2} \Delta_m w^\alpha n$$

$$+ \sum_{m=0}^{[nt]-1} \int_{m/n}^{(m+1)/n} \sigma''_\alpha(X_n(s,x,w))^i_{kl} \Big[Y_n(s,x,w)^k_{j_1} - Y_n\Big(\frac{m}{n},x,w\Big)^k_{j_1}\Big] ds$$

$$\times Y_n\Big(\frac{m}{n},x,w\Big)^l_{j_2} \Delta_m w^\alpha n$$

$$+ \sum_{m=0}^{[nt]-1} \int_{m/n}^{(m+1)/n} \sigma''_\alpha(X_n(s,x,w))^i_{kl} Y_n(s,x,w)^k_{j_1} \Big[Y_n(s,x,w)^l_{j_2}$$

$$- Y_n\Big(\frac{m}{n},x,w\Big)^l_{j_2}\Big] ds\, \Delta_m w^\alpha n$$

$$:= K_1(t) + K_2(t) + K_3(t) + K_4(t).$$

Then, by Theorem III-3.1 and (7.88),

$$E[\sup_{0\le t\le T} |K_1(t)|^p] \le K_{36} E\Big[\Big(\frac{1}{n}\sum_{m=0}^{[nT]-1} \|Y_n\Big(\frac{m}{n},x,w\Big)\|^4\Big)^{p/2}\Big] \le K_{37} < \infty .$$

Also, by (7.89) with p replaced by $2p$,

$$E[\sup_{0\le t\le T} |K_2(t)|^p]$$

$$\le K_{38} n^{p-1} \sum_{m=0}^{[nT]-1} n^p n^{-(p-1)} \int_{m/n}^{(m+1)/n} E\Big[\|X_n(s,x,w)$$

$$- X_n\Big(\frac{m}{n},x,w\Big)\|^{2p}\Big]^{1/2} ds\, E\Big[\|Y_n\Big(\frac{m}{n},x,w\Big)\|^{4p}\Big]^{1/4} E[|\Delta_m w^\alpha|^{4p}]^{1/4}$$

$$\le K_{39} < \infty .$$

Using (7.90), similar estimates hold for $K_3(t)$ and $K_4(t)$. Thus we obtain

$$E[\sup_{0\le t\le T} |H_1(t;k,l,\alpha)|^p] \le K_{40} < \infty .$$

More easily we can obtain

$$E[\sup_{0\le t\le T} |H_2(t;k,l,\alpha)|^p] \le K_{41} < \infty .$$

Therefore

$$\sup_{n,x} E[\sup_{0\leq t\leq T} \|\alpha_n(t, x, w)\|^p] \leq K_{42} < \infty .$$

Using this,

$$\sup_{n,x} E[\sup_{0\leq t\leq T} | Y^{i,(n)}_{j_1,j_2}(t, x, w)|^p] < \infty$$

can be proved as in the case of $Y_n(t, x, w)$. By continuing this process step by step we can complete the proof of Lemma 7.2.

Now $(X(t, x, w), Y(t, x, w), \dots)$ is the solution of stochastic differential equation

$$\begin{cases} dX(t) = \sigma(X(t))\circ dw(t) \\ dY(t) = D\sigma(X(t))Y(t)\circ dw(t) \\ \quad\vdots \\ X(0) = x \\ Y(0) = I \\ \quad\vdots \end{cases}$$

The coefficients of this equation are not bounded and we cannot apply Theorem 7.2 directly. But Lemma 7.2 enables us to apply a standard truncation argument as in the proof of Lemma 2.1 of Chapter V and obtain the following: for every $T > 0$ and $N > 0$,

$$\lim_{n\to\infty} \sup_{|x|\leq N} E[\sup_{0\leq t\leq T} | D^\alpha X_n(t, x, w) - D^\alpha X(t, x, w)|^p] = 0 \tag{7.91}$$

for every $p \geq 2$ and multi-index α. Then, as in the proof of Proposition V-2.2, we can obtain the following result.

Theorem 7.3. For every $T > 0$ and $N > 0$,

$$\lim_{n\to\infty} E[\sup_{0\leq t\leq T} \sup_{|x|\leq N} |D^\alpha X_n(t, x, w) - D^\alpha X(t, x, w)|^p] = 0 \tag{7.92}$$

for every $p \geq 2$ and multi-index α.

We have assumed that coefficients are bounded together with all of their derivatives. Now we assume only that coefficients are C^∞ and that the stochastic differential equation and approximating ordinary differential equations possess global solutions for each fixed initial value. By a standard truncation argument, we can easily deduce from (7.92) the following:

Corollary. There exists a subsequence $\{n_k\}$ such that, with probability one,

$$D^\alpha X_{n_k}(t, x, w) \longrightarrow D^\alpha X(t, x, w)$$

as $k \longrightarrow \infty$ compact uniformly in (t, x) for every multi-index α. *[1]

8. The support of diffusion processes

Let $\sigma_k^i(x)$ and $b^i(x)$, $i = 1, 2, \ldots, d,\quad k = 1, 2, \ldots, r,$ be bounded smooth functions on $\boldsymbol{R}^d$ with bounded derivatives*[2] and consider the stochastic differential equation

$$(8.1) \qquad \begin{cases} dX^i(t) = \sum_{k=1}^r \sigma_k^i(X(t)) \circ dB^k(t) + b^i(X(t))dt \\ X(0) = x, \qquad\qquad i = 1, 2, \ldots, d. \end{cases}$$

Let P_x be the probability law of the solution $X = \{X(t)\}$. Then as we saw in Chapter IV, the system $\{P_x\}$ of probabilities on $\boldsymbol{W}^d$ constitutes the unique diffusion measure generated by the operator A, where

$$(8.2) \qquad Af(x) = \frac{1}{2} \sum_{i,j=1}^d a^{ij}(x) \frac{\partial^2 f}{\partial x^i \partial x^j} + \sum_{i=1}^d \tilde{b}^i(x) \frac{\partial f}{\partial x^i}(x), \qquad f \in \boldsymbol{C}_b^2(\boldsymbol{R}^d),$$

$$(8.3) \qquad a^{ij}(x) = \sum_{k=1}^r \sigma_k^i(x) \sigma_k^j(x)$$

and

$$(8.4) \qquad \tilde{b}^i(x) = b^i(x) + \frac{1}{2} \sum_{j=1}^d \sum_{k=1}^r \left(\frac{\partial}{\partial x^j} \sigma_k^i(x)\right) \sigma_k^j(x).$$

Now the path space $\boldsymbol{W}^d$ is a Fréchet space with the metric defined in Chapter IV, Section 1, or equivalently, the topology of $\boldsymbol{W}^d$ is defined by

---

*[1] The subsequence $\{n_k\}$ can be chosen from any given subsequence of $1, 2, \ldots, n, \ldots$.

*[2] To be precise, the following discussions will be valid if $\sigma \in \boldsymbol{C}_b^2(\boldsymbol{R}^d)$, $b \in \boldsymbol{C}_b^1(\boldsymbol{R}^d)$ and the second order derivatives of σ are uniformly continuous.

the system of seminorms $\{\|\cdot\|_T, T > 0\}$ where

$$\|w\|_T = \max_{0 \le t \le T} |w(t)| \qquad \text{for} \quad w \in \boldsymbol{W}^d. \tag{8.5}$$

The purpose of this section is to describe the topological support $\mathscr{S}(P_x)$ of the measure P_x, i.e., the smallest closed subset of $\boldsymbol{W}^d$ that carries probability 1. In order to do this we need to introduce the following subclasses of the space $\boldsymbol{W}_0^r = \{w \in \boldsymbol{W}^r;\ w(0) = 0\}$;

$$\mathscr{S} \subset \mathscr{S}_p \subset \boldsymbol{W}_0^r$$

where

$$\mathscr{S}_p = \{\phi \in \boldsymbol{W}_0^r;\ t \longmapsto \phi(t) \quad \text{is piecewise smooth}\}$$

and

$$\mathscr{S} = \{\phi \in \boldsymbol{W}_0^r;\ t \longmapsto \phi(t) \quad \text{is smooth}\}.$$

For $\phi \in \mathscr{S}_p$ and $x \in \boldsymbol{R}^d$, we obtain a d-dimensional curve $\xi = \xi(x, \phi) = (\xi_t(x, \phi))$ by solving of the ordinary differential equation

$$\begin{cases} d\xi_t^i = \sum_{k=1}^r \sigma_k^i(\xi_t)\dot{\phi}^k(t)dt + b^i(\xi_t)dt^* \\ \xi_0 = x, \qquad i = 1, 2, \ldots, d. \end{cases} \tag{8.6}$$

We define the subclasses $\mathscr{S}^x$ and $\mathscr{S}_p^x$ of $\boldsymbol{W}^d$ by

$$\mathscr{S}^x = \{\xi(x, \phi);\ \phi \in \mathscr{S}\}. \tag{8.7}$$

and

$$\mathscr{S}_p^x = \{\xi(x, \phi);\ \phi \in \mathscr{S}_p\}. \tag{8.8}$$

It is easy to see that the closures in $\boldsymbol{W}^d$ of $\mathscr{S}^x$ and $\mathscr{S}_p^x$ coincide; i.e., $\overline{\mathscr{S}^x} = \overline{\mathscr{S}_p^x}$. Now we can state the main theorem due to Stroock and Varadhan [158].

Theorem 8.1. $\mathscr{S}(P_x) = \overline{\mathscr{S}^x}$ for every $x \in \boldsymbol{R}^d$.

---

* $\dot{\phi} = \dfrac{d\phi}{dt}$.

Proof. First we shall prove the inclusion $\mathscr{S}(P_x) \subset \overline{\mathscr{S}^x}$. For each $n = 1, 2, \ldots$ and $t \geq 0$, we set $t_n = [2^n t]/2^n$ and $\bar{t}_n = ([2^n t] + 1)/2^n$. Let $\{B(t)\}$ be a given r-dimensional Brownian motion with $B(0) = 0$ and define $\{B_n(t)\}$ by

$$B_n(t) = \frac{(\bar{t}_n - t)}{(\bar{t}_n - t_n)} B(t_n) + \frac{(t - t_n)}{(\bar{t}_n - t_n)} B(\bar{t}_n).$$

Then $B_n \in \mathscr{S}_p$ and hence $\xi(x, B_n) \in \mathscr{S}_p^x$. By Theorem 7.2, we can conclude that $\xi(x, B_n) \longrightarrow X$ in W^d in probability. Hence $P_x^n \xrightarrow{w} P_x$ as $n \longrightarrow \infty$ where P_x^n is the probability law of $\xi(x, B_n)$. Thus

$$P_x(\overline{\mathscr{S}_p^x}) \geq \overline{\lim_{n\to\infty}}\, P_x^n(\overline{\mathscr{S}_p^x}) = 1$$

and consequently $\overline{\mathscr{S}^x} = \overline{\mathscr{S}_p^x} \supset \mathscr{S}(P_x)$.

The converse inclusion $\mathscr{S}(P_x) \supset \overline{\mathscr{S}^x}$ will follow immediately from the following theorem. We consider the equation (8.1) on the Wiener space (W_0^r, P^W) with respect to the canonical realization of the Wiener process; its solution is denoted by $X^x = (X(t, w))$.

Theorem 8.2. For every $\phi \in \mathscr{S}$, $T > 0$ and $\varepsilon > 0$,

$$P^W(\|X^x - \xi(x, \phi)\|_T < \varepsilon \,|\, \|w - \phi\|_T < \delta) \longrightarrow 1 \quad \text{as } \delta \downarrow 0. \tag{8.9}$$

Remark 8.1. It is well known that $P^W(\|w - \phi\|_T < \delta) > 0$ for every $\phi \in \mathscr{S}$, $T > 0$ and $\delta > 0$, i.e., $\mathscr{S}(P^W) = W_0^r$. Indeed, by the result of Chapter IV, Section 4,

$$P^W(\|w - \phi\|_T < \delta) = E^{P^W}(M(w): \|w\|_T < \delta) > 0$$

where

$$M(w) = \exp[-\sum_{k=1}^r \int_0^T \dot{\phi}^k(s) dw^k(s) - \frac{1}{2}\int_0^T |\dot{\phi}(s)|^2 ds].$$

To prove Theorem 8.2 we first introduce several lemmas.

Lemma 8.1. There exist positive constants c_1 and c_2 such that

$$P^W(\|w\|_T < \varepsilon) \sim c_1 \exp\left(-\frac{c_2}{\varepsilon^2}\right) \qquad \text{as } \varepsilon \downarrow 0. \tag{8.10}$$

Proof. Let P_x be the r-dimensional Wiener measure starting at $x \in \boldsymbol{R}^r$; so, in particular, $P_0 = P^W$. Let $D = \{x \in \boldsymbol{R}^r;\ |x| < 1\}$ and set

$$\sigma(w) = \inf\{t;\ w(t) \notin D\} \qquad \text{for} \quad w \in \boldsymbol{W}^r.$$

Then $u(t, x) = E_x[f(w(t))I_{\{\sigma(w)>t\}}]$, $x \in D$, $t > 0$, is the solution of the initial value problem

$$\begin{cases} \dfrac{\partial u}{\partial t} = \dfrac{1}{2}\Delta u & \text{in} \quad D, \\ u|_{\partial D} = 0 \\ u|_{t=0} = f. \end{cases}$$

Consequently

$$u(t, x) = \sum_{n=1}^{\infty} e^{-\lambda_n t}\, \phi_n(x) \int_D \phi_n(y) f(y) dy,$$

where $0 < \lambda_1 < \lambda_2 \leq \lambda_3 \leq \cdots$ are eigenvalues and $\{\phi_n(x)\}$ are corresponding eigenfunctions of the eigenvalue problem

$$\begin{cases} \dfrac{1}{2}\Delta\phi + \lambda\phi = 0 & \text{in} \quad D, \\ \phi|_{\partial D} = 0. \end{cases}$$

In particular,

$$\begin{aligned} P^W(\|w\|_T < \varepsilon) &= P^W(\max_{0\leq t\leq T} |w(t)| < \varepsilon) = P^W(\max_{0\leq t\leq T} |\varepsilon w(t/\varepsilon^2)| < \varepsilon) \\ &= P^W(\max_{0\leq t\leq T/\varepsilon^2} |w(t)| < 1) = P^W(\sigma(w) > T/\varepsilon^2) \\ &= \sum_{n=1}^{\infty} e^{-\lambda_n T/\varepsilon^2}\, \phi_n(0) \int_D \phi_n(y) dy, \end{aligned}$$

and consequently

$$P^W(\|w\|_T < \varepsilon) \sim e^{-\lambda_1 T/\varepsilon^2}\, \phi_1(0) \int_D \phi_1(y) dy.$$

Therefore (8.10) is proved with

$$c_1 = \phi_1(0)\int_D \phi_1(x)dx \qquad \text{and} \quad c_2 = \lambda_1 T.$$

Lemma 8.2. Set

$$\eta^{ij}(t) = \frac{1}{2}\int_0^t [w^i(s)dw^j(s) - w^j(s)dw^i(s)] \tag{8.11}$$

$$\text{for} \quad i, j = 1, 2, \ldots, r.$$

Then we have

$$\lim_{M\uparrow\infty} \sup_{0<\delta\leq 1} P^W[\, \|\eta^{ij}\|_T > M\delta \mid \|w\|_T < \delta] = 0. \tag{8.12}$$

Proof. Let $i \neq j$ be fixed and set

$$a(t) = \frac{1}{4}\int_0^t [(w^i)^2(s) + (w^j)^2(s)]ds.$$

We have remarked in Section 6 that $B(t) := \eta^{ij}(a^{-1}(t))$ is a Brownian motion independent of $\{(w^i)^2(t)+(w^j)^2(t)\}$ (and hence, independent of the radial process $\{|w(t)|\}$ and, in particular, of $\|w\|_T$). Then

$$P^W[\, \|\eta^{ij}\|_T > M\delta \mid \|w\|_T < \delta] = P^W[\, \|B(a(t))\|_T > M\delta \mid \|w\|_T < \delta]$$
$$\leq P^W[\max_{0\leq s\leq \delta^2 T/4} |B(s)| > M\delta] = P^W[\max_{0\leq s\leq T/4} |B(s)| > M],$$

and this proves (8.12).

Corollary. Let

$$\xi^{ij}(t) = \int_0^t w^i(s)\circ dw^j(s), \qquad i, j = 1, 2, \ldots, r. \tag{8.13}$$

Then for all $i, j = 1, 2, \ldots, r$,

$$\lim_{M\uparrow\infty} \sup_{0<\delta\leq 1} P^W(\, \|\xi^{ij}\|_T > M\delta \mid \|w\|_T < \delta) = 0. \tag{8.14}$$

In particular, for every $\varepsilon > 0$,

$$P^W(\, \|\xi^{ij}\|_T > \varepsilon \mid \|w\|_T < \delta) \longrightarrow 0 \qquad \text{as} \quad \delta \downarrow 0. \tag{8.15}$$

Proof. Since

$$\frac{1}{2} w^i(t)w^j(t) - \xi^{ji}(t) = \eta^{ij}(t),$$

(8.14) follows at once from (8.12).

The following is a key lemma for the proof of Theorem 8.2.

Lemma 8.3. Let $f(x): \boldsymbol{R}^d \longrightarrow \boldsymbol{R}$ be bounded and uniformly continuous. Then for all $\varepsilon > 0$ and $i, j = 1, 2, \ldots, r$,

$$P^W(\| \int_0^t f(X(s, w))d\xi^{ij}(s)\|_T > \varepsilon \mid \|w\|_T < \delta) \longrightarrow 0 \quad \text{as} \quad \delta \downarrow 0. \tag{8.16}$$

Proof. First we shall assume that $f \in C_b^2(\boldsymbol{R}^d)$. Then by Itô's formula,

$$\begin{aligned}
&\int_0^t f(X(s))d\xi^{ij}(s) \\
&= f(X(t))\xi^{ij}(t) - \int_0^t f_l(X(s))\sigma_k^l(X(s))\xi^{ij}(s)dw^k(s) \\
&\quad - \int_0^t (Af)(X(s))\xi^{ij}(s)ds - \int_0^t f_l(X(s))\sigma_j^l(X(s))w^i(s)ds \ ^{*1} \\
&:= I_1(t) + I_2(t) + I_3(t) + I_4(t)\,.
\end{aligned}$$

Clearly it is sufficient to show that

$$P^W(\|I_i(t)\|_T > \varepsilon \mid \|w\|_T < \delta) \longrightarrow 0$$

as $\delta \downarrow 0$ for every $\varepsilon > 0$ and $i = 1, 2, 3, 4$. This follows from (8.15) for $i = 1$ and $i = 3$ and it is obvious for $i = 4$. So we only consider $I_2(t)$. For simplicity we set $\alpha_k(x) = -f_l(x)\sigma_k^l(x)$. Then, by Itô's formula,

$$\begin{aligned}
I_2(t) &= \int_0^t \alpha_k(X(s))\xi^{ij}(s)dw^k(s) \\
&= \alpha_k(X(t))\xi^{ij}(t)w^k(t) - \int_0^t \alpha_{k,l}(X(s))\sigma_m^l(X(s))\xi^{ij}(s)w^k(s)dw^m(s) \ ^{*2}
\end{aligned}$$

---

*1 $f_l = \dfrac{\partial f}{\partial x^l}$. We also adopt the usual convention for summation.

*2 $\alpha_{k,l}(x) = \dfrac{\partial}{\partial x^l}\alpha_k(x)$.

$$
\begin{aligned}
&- \int_0^t (A\alpha_k)(X(s))\xi^{ij}(s)w^k(s)ds \\
&- \int_0^t \alpha_k(X(s))w^k(s)d\xi^{ij}(s) \\
&- \int_0^t \alpha_j(X(s))w^i(s)ds - \int_0^t \xi^{ij}(s)\alpha_{k,l}(X(s))\sigma^l_m(X(s))\delta^{km}ds \\
&- \int_0^t w^k(s)\alpha_{k,l}(X(s))\sigma^l_j(X(s))w^i(s)ds \\
:&= J_1(t) + J_2(t) + J_3(t) + J_4(t) + J_5(t) + J_6(t) + J_7(t).
\end{aligned}
$$

Again it is sufficient to show that

$$P^W(\,\|J_i(t)\|_T > \varepsilon \mid \|w\|_T < \delta) \longrightarrow 0$$

as $\delta \downarrow 0$ for all $\varepsilon > 0$ and $i = 1, 2, \ldots, 7$. This is no problem for $i = 1, 3, 5, 6, 7$. By Theorem II-7.2′, there exists a one-dimensional Brownian motion $B(t)$ such that

$$J_2(t) = B(a(t)),$$

where

$$a(t) = \langle J_2, J_2\rangle_t = \int_0^t [\xi^{ij}(s)]^2[\alpha_{k,l}\alpha_{k',l'}a^{ll'}](X(s))w^k(s)w^{k'}(s)ds.$$

Now

$$
\begin{aligned}
&P^W(\,\|J_2(t)\|_T > \varepsilon \mid \|w\|_T < \delta) \\
&\quad \leq P^W(\,\|\xi^{ij}\|_T > M\delta \mid \|w\|_T < \delta) \\
&\qquad\qquad + P^W(\|J_2(t)\|_T > \varepsilon,\ \|\xi^{ij}\|_T \leq M\delta \mid \|w\|_T < \delta).
\end{aligned}
$$

By (8.14), we can choose $M > 0$ for any given $\eta > 0$ such that the first term on the right is less than η for all $1 \geq \delta > 0$. Clearly $\|\xi^{ij}\|_T \leq M\delta$ and $\|w\|_T < \delta$ imply that $a(t) \leq a(T) \leq c_3\delta^4$,* and hence

$$
\begin{aligned}
&P^W(\|J_2(t)\|_T > \varepsilon,\quad \|\xi^{ij}\|_T \leq M\delta,\quad \|w\|_T < \delta) \\
&\quad \leq P^W(\max_{0\leq t\leq c_3\delta^4} |B(t)| > \varepsilon) = P^W(\max_{0\leq t\leq 1} |B(t)| > \varepsilon/\sqrt{c_3}\,\delta^2)
\end{aligned}
$$

* In the following, $c_3, c_4, \ldots$ are positive constants independent of δ.

$$= 2\int_{c_3^{-1/2}\varepsilon\,\delta^{-2}}^{\infty} \frac{1}{\sqrt{2\pi}} \exp\left[-\frac{x^2}{2}\right] dx \leq c_4 \exp\left[-c_5\frac{\varepsilon^2}{\delta^4}\right].$$

By Lemma 8.1,

$$P(\|w\|_T < \delta) \geq c_6 \exp\left[-c_7\frac{1}{\delta^2}\right].$$

Hence

$$\begin{aligned} &P^W(\ \|J_2(t)\|_T > \varepsilon,\ \|\xi^{ij}\|_T \leq M\delta \mid \|w\|_T < \delta) \\ &\leq c_8 \exp\left[-\frac{c_9\varepsilon^2 - c_{10}\delta^2}{\delta^4}\right] \longrightarrow 0 \qquad \text{as} \quad \delta \downarrow 0. \end{aligned}$$

Consequently,

$$\overline{\lim_{\delta\downarrow 0}}\, P^W(\ \|J_2(t)\|_T > \varepsilon \mid \|w\|_T < \delta) \leq \eta,$$

and since η is arbitrary, we obtain the desired conclusion. Next we consider $J_4(t)$.

$$\begin{aligned} J_4(t) &= -\int_0^t \alpha_k(X(s))w^k(s)w^i(s)dw^j(s) - \frac{\delta_{ij}}{2}\int_0^t \alpha_k(X(s))w^k(s)ds \\ &:= K_1(t) + K_2(t). \end{aligned}$$

Then it is obvious that for any $\varepsilon > 0$

$$P^W(\ \|K_2(t)\|_T > \varepsilon \mid \|w\|_T < \delta) \longrightarrow 0 \qquad \text{as} \quad \delta \downarrow 0.$$

$K_1(t)$ is a martingale with

$$\langle K_1, K_1\rangle_t = \int_0^t [\alpha_k(X(s))w^k(s)w^i(s)]^2 ds.$$

Therefore, if $\|w\|_T < \delta$ then $\langle K_1,K_1\rangle_T \leq c_{11}\delta^4$. By repeating the same argument as above we can conclude that

$$P^W(\|K_1(t)\|_T > \varepsilon \mid \|w\|_T < \delta) \leq c_{12} \exp\left[-\frac{c_{13}\varepsilon^2 - c_{14}\delta^2}{\delta^4}\right] \longrightarrow 0$$

as $\delta \downarrow 0$. Thus (8.16) is proved for $f \in C_b^2(\boldsymbol{R}^d)$. Now suppose that f is uniformly continuous and choose $f_n \in C_2^b(\boldsymbol{R}^d)$ such that $f_n \longrightarrow f$ uniformly. Set

$$Y_n(t) := \int_0^t f(X(s))d\xi^{ij}(s) - \int_0^t f_n(X(s))d\xi^{ij}(s)$$
$$= \int_0^t (f - f_n)(X(s))w^i(s)dw^j(s) + \frac{\delta_{ij}}{2}\int_0^t (f - f_n)(X(s))ds.$$

Then, for any given $\varepsilon > 0$, we have by the same argument as above that

$$P^W(\|Y_n\|_T > \varepsilon \mid \|w\|_T < \delta)$$
$$\leq P^W(\|\int_0^t (f - f_n)(X(s))w^i(s)dw^j(s)\|_T > \frac{\varepsilon}{2} \mid \|w\|_T < \delta)$$
$$\leq c_{15} \exp\left[-\frac{c_{16}\varepsilon^2}{\|f - f_n\|^2\delta^2}\right] \leq c_{15} \exp\left[-\frac{c_{16}\varepsilon^2}{\|f - f_n\|}\right]$$

for all large n and $0 < \delta \leq 1$. Hence

$$P^W(\|\int_0^t f(X(s))d\xi^{ij}(s)\|_T > \varepsilon \mid \|w\|_T < \delta)$$
$$\leq P^W(\|\int_0^t f_n(X(s))d\xi^{ij}(s)\|_T > \frac{\varepsilon}{2} \mid \|w\|_T < \delta)$$
$$+ P^W(\|Y_n\|_T > \frac{\varepsilon}{2} \mid \|w\|_T < \delta).$$

For given $\eta > 0$, we can choose n such that the second term is less than η for all $0 < \delta \leq 1$. Then letting $\delta \downarrow 0$ the first term tends to 0. Consequently

$$\overline{\lim_{\delta \downarrow 0}}\, P^W(\|\int_0^t f(X(s))d\xi^{ij}(s)\|_T > \varepsilon \mid \|w\|_T < \delta) \leq \eta,$$

and since η is arbitrary, the proof of (8.16) is now complete.

Proof of Theorem 8.2. First we shall prove (8.9) when $\phi(t) \equiv 0$. In this case, $\xi_t = (\xi_t(x, \phi))$ is the solution of

$$\begin{cases} d\xi_t^i = b^i(\xi_t)dt \\ \xi_0 = x. \end{cases}$$

Now

$$X^i(t) - \xi^i(t) = \sum_{k=1}^{r} \int_0^t \sigma_k^i(X(s)) \circ dw^k(s) + \int_0^t [b^i(X(s)) - b^i(\xi_s)]ds.$$

We shall prove that

$$P^W(\| \int_0^t \sigma_k^i(X(s)) \circ dw^k(s) \|_T > \varepsilon \mid \|w\|_T < \delta) \longrightarrow 0 \tag{8.17}$$

as $\delta \downarrow 0$ for every $\varepsilon > 0$. By Itô's formula

$$\begin{aligned}
&\int_0^t \sigma_k^i(X(s)) \circ dw^k(s) \\
&\quad = \sigma_k^i(X(t))w^k(t) - \int_0^t w^k(s) \circ d(\sigma_k^i(X(s))) \\
&\quad = \sigma_k^i(X(t))w^k(t) - \int_0^t \sigma_{k,l}^i(X(s))\sigma_m^l(X(s))w^k(s) \circ dw^m(s) \,^* \\
&\qquad\qquad - \int_0^t \sigma_{k,l}^i(X(s))b^l(X(s))w^k(s)ds \\
&\quad := I_1(t) + I_2(t) + I_3(t).
\end{aligned}$$

It is sufficient to show that for every $\varepsilon > 0$ and $i = 1, 2, 3$,

$$P^W(\|I_i(t)\| > \varepsilon \mid \|w\|_T < \delta) \longrightarrow 0 \quad \text{as} \quad \delta \downarrow 0.$$

Only the case of $i = 2$ needs to be examined. Then

$$\begin{aligned}
I_2(t) &= - \int_0^t \sigma_{k,l}^i(X(s))\sigma_m^l(X(s)) \circ d\xi^{km}(s) \\
&= - \int_0^t \sigma_{k,l}^i(X(s))\sigma_m^l(X(s))d\xi^{km}(s) \\
&\quad - \frac{1}{2} \int_0^t \frac{\partial}{\partial x^n} [\sigma_{k,l}^i \sigma_m^l](X(s))\sigma_\alpha^n(X(s))w^k(s)\delta^{\alpha m}ds \\
&= J_1(t) + J_2(t),
\end{aligned}$$

and we can conclude that for every $\varepsilon > 0$ and $i = 1, 2$

$$P^W(\|J_i(t)\|_T > \varepsilon \mid \|w\|_T < \delta) \longrightarrow 0 \quad \text{as} \quad \delta \downarrow 0.$$

---

* $\sigma_{k,l}^i = \dfrac{\partial}{\partial x^l} \sigma_k^i(x)$.

Indeed, it is obvious for $i=2$ and the case $i=1$ follows from Lemma 8.3.

On the set

$$\{w;\ \|\int_0^t \sigma_k^i(X(s))\circ dw^k(s)\|_T < \varepsilon\},$$

we have

$$|X(t) - \xi(t)| \leq \varepsilon + K\int_0^t |X(s) - \xi(s)|ds$$

where K is a positive constant. Hence $|X(t) - \xi(t)| \leq \varepsilon e^{Kt}$. This, combined with (8.17), yields

$$P^W(\ \|X(t) - \xi(t)\|_T > \varepsilon \mid \|w\|_T < \delta) \longrightarrow 0$$

as $\delta \downarrow 0$ for every $\varepsilon > 0$.

Now we consider the case of general $\phi \in \mathscr{S}$. Set

$$M(w) = \exp\{\sum_{k=1}^r \int_0^T \dot{\phi}^k(s)dw^k(s) - \frac{1}{2}\int_0^T |\dot{\phi}|(s)^2 ds\}$$

and define the measure $\bar{P}$ on $\boldsymbol{W}_0^r(T)$: = the restriction of $\boldsymbol{W}_0^r$ on the interval $[0, T]$ by

$$\frac{d\bar{P}}{dP^W} = M(w).$$

Then by Theorem IV-4.1, $\bar{w}(t) = w(t) - \phi(t)$ is an r-dimensional Brownian motion for $\bar{P}$, and $X(t)$ satisfies

$$X(t) = x + \int_0^t \sigma(X(s))\circ d\bar{w}(s) + \int_0^t \tilde{b}(s, X(s))ds$$

where $\tilde{b}(s, x) = b(x) + \sigma(x)\dot{\phi}(s)$. * Since $\xi = (\xi(x, \phi))$ satisfies

$$\begin{cases} d\xi_t = \tilde{b}(t, \xi_t)dt \\ \xi_0 = x, \end{cases}$$

we can conclude from the above that for every $\varepsilon > 0$,

---

* Thus we need to consider the case that the drift b dependent on t and $b \in \boldsymbol{C}^1([0, \infty) \times \boldsymbol{R}^d)$. All the above results remain valid in such a case with an obvious modification of the proof.

$$\bar{P}(\ \|X(t) - \xi_t\|_T > \varepsilon \mid \|\bar{w}\|_T < \delta) \longrightarrow 0 \qquad \text{as} \quad \delta \downarrow 0. \tag{8.18}$$

Noting that

$$M(w) = \exp\{\sum_{k=1}^{r} \dot{\phi}^k(T)w^k(T) - \sum_{k=1}^{r} \int_0^T w^k(s)\ddot{\phi}^k(s)ds - \frac{1}{2}\int_0^T |\dot{\phi}|(s)^2 ds\}$$

is continuous in w (i.e., if $\|w_n - w\|_T \longrightarrow 0$, then $M(w_n) \longrightarrow M(w)$), (8.18) implies that

$$\begin{aligned}
&\lim_{\delta\downarrow 0} P^W(\ \|X(t) - \xi_t\|_T < \varepsilon \mid \|w - \phi\|_T < \delta) \\
&\quad = \lim_{\delta\downarrow 0} \frac{P^W(\|X(t) - \xi_t\|_T < \varepsilon,\ \|w - \phi\|_T < \delta)}{E^W(M\colon \|X(t) - \xi_t\|_T < \varepsilon,\ \|w - \phi\|_T < \delta)} \\
&\qquad\qquad \times \frac{E^W(M\colon \|w - \phi\|_T < \delta)}{P^W(\|w - \phi\|_T < \delta)} \longrightarrow 1.
\end{aligned}$$

Example 8.1. Let $L_0, L_1, \ldots, L_r$ be vector fields on $\boldsymbol{R}^d$ whose coefficients (in the Euclidean coordinates) are bounded and smooth with bounded derivatives. Let $X^x = (X(t, w))$ be the solution of

$$\begin{cases} dX(t) = \sum_{i=1}^{r} L_i(X(t))\circ dw^i(t) + L_0(X(t))dt \\ X(0) = x. \end{cases}$$

Let P_x be the probability law of X^x. Theorem 8.1 implies that

$$\mathscr{S}(P_x) = \overline{\{\xi(x, \phi);\, \phi \in \mathscr{S}\}}.$$

$\xi(x, \phi)$ is the solution of the dynamical system

$$\begin{cases} \dfrac{d\xi_t}{dt} = \sum_{i=1}^{r} L_i(\xi_t)\dot{\phi}^i(t) + L_0(\xi_t). \\ \xi_0 = x. \end{cases}$$

It is known (cf. [95]) that $\overline{\{\xi(x, \phi);\, \phi \in \mathscr{S}\}}$ contains all curves η_t such that

$$\begin{cases} \dfrac{d\eta_t}{dt} = Z(\eta_t) + L_0(\eta_t) \\ \eta_0 = x, \end{cases}$$

where Z is an element of the Lie algebra $\mathfrak{L}(L_1, L_2, \ldots, L_r)$ generated by $L_1, L_2, \ldots, L_r$. In particular, if $\mathfrak{L}(L_1, L_2, \ldots, L_r)$ has rank d at every point, $\overline{\{\xi(x,\phi);\ \phi \in \mathscr{S}\}} = \boldsymbol{W}_x^d := \{w \in \boldsymbol{W}^d;\ w(0) = x\}$ and consequently

$$\mathscr{S}(P_x) = \boldsymbol{W}_x^d.$$

Example 8.2. (maximum principle). Let A be the differential operator defined by (8.2). A function defined in a domain $D \subset \boldsymbol{R}^d$ is said to be *A-subharmonic in D* if it is upper semicontinuous and $Au \geq 0$ (in a certain weak sense to be specified later). A classical maximum principle for the Laplacian asserts that any subharmonic function in D which attains its maximum in D must be a constant. If the operator A is degenerate, however, such a maximum principle no longer holds in general. For example, if $d = 2$,

$$A = \frac{\partial^2}{\partial x_1^2} + \frac{\partial}{\partial x_2}, \qquad x = (x_1, x_2)$$

and D is any domain intersecting the x_1-axis, the function $u(x)$ defined by

$$u(x) = \begin{cases} 0 & \text{if} \quad x_2 > 0 \\ -x_2^2 & \text{if} \quad x_2 \leq 0 \end{cases}$$

is a nonconstant A-subharmonic function which attains its maximum in D.

We are interested in the following problem. For a given domain D and a point $x \in D$, we want to determine a (relatively) closed subset $D(x)$ of D having the following properties:

(8.19) *for any A-subharmonic function $u(y)$ in D such that* $u(x) = \max_{y \in D(x)} u(y)$, *it holds that $u(y) = u(x)$ for every $y \in D(x)$*,

(8.20) *$D(x)$ is maximal with respect to the property* (8.19): i.e., *if $z \in D \setminus D(x)$, then there exists an A-subharmonic function $u(y)$ in D such that $u(x) = \max_{y \in D(x)} u(y)$ and $u(z) < u(x)$.*

It is clear that $D(x)$ is uniquely determined if it exists. The support theorem 8.1 enables us to describe the set $D(x)$. Before proceeding, however, we shall first make precise the notion of A-subharmonicity. For simplicity we shall restrict ourselves to locally bounded functions. Let P_x be the probability law of the solution of (8.1). Let D be a domain in $\boldsymbol{R}^d$ and $\{D_n\}$ be a compact exhaustion of D: D_n are bounded subdomains of

D such that $\bar{D}_n \subset D_{n+1}$ and $\bigcup_n D_n = D$. Let $\sigma_n = \sigma_n(w) = \inf\{t;\ w(t) \notin D_n\}$, $n = 1, 2, \ldots$, $w \in W^d$.

Definition 8.1. A function $u(x)$ defined in D is said to be *A-subharmonic in D* if
(i) it is locally bounded and upper semicontinuous and
(ii) for each $n = 1, 2, \ldots$ and $x \in D$, $t \longmapsto u(w(t \wedge \sigma_n)) - u(x)$ is a P_x-submartingale.

If u is in $C^2(D)$, then $u(w(t \wedge \sigma_n)) - u(x) = a$ *martingale* $+ \int_0^{t\wedge\sigma_n} (Au)(w(s))ds$. From this it is easy to see that u is A-subharmonic if and only if $Au \geq 0$ in D.

Now we shall describe the set $D(x)$. For simplicity, we shall assume that D has the property that

$$\sup_{y \in D} E_y[\int_0^\tau I_K(w(s))ds] < \infty$$

for every compact set $K \subset D$, where $\tau = \tau(w) = \inf\{t; w(t) \notin D\}$. Choose a compact exhaustion $\{D_n\}$ of D. Then the above assumption clearly implies that $\sup_{y \in D_n} E_y[\sigma_n] < \infty$ for every n. Set

$$D(x) = \overline{\{y;\ {}^\exists\phi \in \mathscr{S},\ {}^\exists t_0 > 0 \text{ such that } y = \xi(x, \phi)(t_0) \text{ and } \{\xi(x, \phi)(t);\ t \in [0, t_0]\} \subset D\}} \cap D. \tag{8.21}$$

Theorem 8.3.* $D(x)$ possesses the properties (8.19) and (8.20).

Proof. First we shall prove that the set $D(x)$ defined by (8.21) possesses the property (8.19). Let $y \in D(x)$. It follows from Theorem 8.1 that for every neighbourhood U of y there exists an n such that $P_x(\sigma_U < \sigma_n) > 0$, where $\sigma_U = \sigma_U(w) = \inf\{t;\ w(t) \in U\}$. Let $u(x)$ be an A-subharmonic function in D such that $u(x) = \max_{z \in D(x)} u(z)$. Then

$$u(x) \leq E_x(u(w(\sigma_U \wedge \sigma_n)));$$

moreover, Theorem 8.1 implies that $P_x(w(\sigma_U \wedge \sigma_n) \in D(x)) = 1$. Consequently, $P_x(u(w(\sigma_U \wedge \sigma_n)) = u(x)) = 1$. In particular, $u(w(\sigma_U)) = u(x)$ on the set $\{\sigma_U < \sigma_n\}$. From this it is easy to find a sequence $z_n \in D$ such that $z_n \longrightarrow y$ and $u(z_n) = u(x)$. Then

* This was first proven by Stroock-Varadhan [158], but they only discussed the case of operators of parabolic type.

$$u(x) \geq u(y) \geq \overline{\lim_{n\to\infty}}\, u(z_n) = u(x),$$

that is, $u(y) = u(x)$.

Next we shall prove that $D(x)$ defined by (8.21) possesses the property (8.20). For this we need the following lemma.

Lemma 8.4. If f is a nonpositive continuous function in D with compact support, then the function u defined by

$$u(y) = E_y[\int_0^\tau f(w(s))ds]$$

is a bounded A-subharmonic function in D.

Assuming this lemma for a moment, we shall now complete the proof of Theorem 8.3. Let $z \in D \setminus D(x)$. Then by Theorem 8.1 we can easily find a bounded neighbourhood U of z such that $P_x(\sigma_U < \tau) = 0$. We choose a nonpositive continuous function f in D such that $f(z) = -1$ and $f(y) = 0$ if $y \notin U$. Set

$$u(y) = E_y[\int_0^\tau f(w(s))ds].$$

Then u is a bounded nonpositive A-subharmonic function in D such that $u(x) = 0$ and $u(z) < 0$. This proves that $D(x)$ possesses the property (8.20).

Proof of Lemma 8.4. u is bounded because of the above assumption on D. First we shall prove that it is upper semicontinuous. We saw in the proof of Proposition V-2.1 that if $X(t, x)$ is the solution of (8.1), then for every $t > 0$, $\boldsymbol{R}^d \ni x \longmapsto X(t, x) \in \boldsymbol{W}^d$ is continuous a.s. Then we can easily see that

$$x \longmapsto I_{\{t<\tau[X(\cdot, x)]\}}$$

is lower semicontinuous a.s. and hence

$$x \longmapsto I_{\{t<\tau[X(\cdot,x)]\}}\, f(X(t, x))$$

is upper semicontinuous a.s. The upper semicontinuity of

$$x \longmapsto u(x) = E_x[\int_0^\tau f(w(s))ds] = \int_0^\infty E[f(X(t, x))I_{\{t<\tau[X(\cdot,x)]\}}]\, dt$$

follows at once from Fatou's lemma.

Next we shall prove that $t \longmapsto u(w(t \wedge \sigma_n))$ is a P_x-submartingale for every n. It suffices to prove that $E_x[u(w(\sigma \wedge \sigma_n))] \geq u(x)$ for every stopping time σ. But this is a consequence of the following formula (Dynkin's formula) which is immediately obtained from the strong Markov property:

$$E_x[u(w(\sigma \wedge \sigma_n))] - u(x) = -E_x\Big[\int_0^{\sigma\wedge\sigma_n} f(w(s))ds\Big].$$

9. Asymptotic evaluation of the diffusion measure for tubes around a smooth curve

Consider a non-singular diffusion process X on a manifold M. We are sometimes interested in the following questions: given two smooth curves starting at the same point, which one is more probable for the diffusion process, or, among all possible smooth curves connecting two given points, which one is most probable for the diffusion process? One way to answer these questions is to evaluate the measure of tubes around a smooth curve ([154]). As we know from Chapter V, Section 4, we may assume that M is a Riemannian manifold and the diffusion is generated by the operator A where

$$A = \frac{1}{2}\Delta + b.$$

Here Δ is the Laplace-Beltrami operator and b is a vector field. The Riemannian distance $\rho(x, y)$ is defined (if x and y is sufficiently near) as the minimum length of a geodesic curve from x to y. For a given smooth curve ϕ: $[0, T] \longrightarrow M$ such that $\phi(0) = x$, we want to evaluate

$$\mu_\varepsilon^x(\phi) = P_x\{w;\, \rho(X_t(w), \phi(t)) \leq \varepsilon \quad \text{for all} \quad t \in [0, T]\}.$$

If for any two such curves ϕ and ψ,

$$\lim_{\varepsilon \downarrow 0} \mu_\varepsilon^x(\phi)/\mu_\varepsilon^x(\psi)$$

exists and can be expressed as

$$\exp\Big[\int_0^T L(\dot\phi(s), \phi(s))ds - \int_0^T L(\dot\psi(s), \psi(s))ds\Big]$$

by some function $L(\dot x, x)$ on the tangent bundle TM, the above questions can be answered in terms of the function L Such a function is called the Onsager-Machlup function by physicists (cf. [20], [41], [56] and [134]. For simplicity, we discuss here the case of $M = \mathbf{R}^d$ and obtain the function L: we refer the reader to [199] and [227] for the results in general manifolds.

So let $M = \boldsymbol{R}^d$ and let $\{P_x\}$ be the diffusion measure (on $\boldsymbol{W}^d$) generated by the operator

$$A = \frac{1}{2}\Delta + \sum_{i=1}^{d} b^i(x)\frac{\partial}{\partial x^i} \qquad \left(\Delta = \sum_{i=1}^{d}\left(\frac{\partial}{\partial x^i}\right)^2\right).$$

We shall assume that $b^i(x) \in \boldsymbol{C}_b^2(\boldsymbol{R}^d)$, $i = 1, 2, \ldots, d$. Let P^W be the d-dimensional Wiener measure (on $\boldsymbol{W}^d$) starting at 0. By Lemma 8.1, we know that

$$P^W(\|w\|_T < \varepsilon) \sim c\exp\left[-\frac{\lambda_1 T}{\varepsilon^2}\right]^{*1} \tag{9.1}$$

where λ_1 is the first eigenvalue of the eigenvalue problem in $D = \{x \in \boldsymbol{R}^d; |x| < 1\}$ given by

$$\begin{cases} \dfrac{1}{2}\Delta\phi + \lambda\phi = 0 \\ \phi|_{\partial D} = 0, \end{cases} \tag{9.2}$$

and $c = \phi_1(0)\int_D \phi_1(x)dx$, $\phi_1(x)$ being the normalized eigenfunction for λ_1. In the following, $T > 0$ and $x \in \boldsymbol{R}^d$ are arbitrary but fixed.

Theorem 9.1. Let ϕ: $[0, T] \longrightarrow \boldsymbol{R}^d$ be a smooth curve*2 such that $\phi(0) = x$. Then we have (writing $b(x) = (b^1(x), b^2(x), \ldots, b^d(x))$

$$\begin{aligned} &P_x(w: \|w-\phi\|_T < \varepsilon) \\ &\sim \exp\left[-\frac{1}{2}\int_0^T |b(\phi(s)) - \dot{\phi}(s)|^2 ds - \frac{1}{2}\int_0^T (\operatorname{div} b)(\phi(s))ds\right] P^W(\|w\|_T < \varepsilon) \\ &\sim c\exp\left(-\frac{1}{2}\int_0^T |b(\phi(s)) - \dot{\phi}(s)|^2 ds - \frac{1}{2}\int_0^T (\operatorname{div} b)(\phi(s))ds\right)\exp\left[-\frac{\lambda_1 T}{\varepsilon^2}\right] \quad \text{as } \varepsilon \downarrow 0. \end{aligned} \tag{9.3}$$

*1 $\|w\|_T = \max_{0 \le t \le T} |w(t)|$, $w \in \boldsymbol{W}^d$.

*2 It is sufficient to assume that $\phi \in \boldsymbol{C}^2([0, T] \longrightarrow \boldsymbol{R}^d)$.

Here $(\mathrm{div}\ b)(x) = \sum_{i=1}^{d} \frac{\partial}{\partial x^i} b^i(x)$.

Corollary. The Onsager-Machlup function $L(\dot{x}, x)$ is given, up to an additive constant, by

$$(9.4)\qquad L(\dot{x}, x) = -\frac{1}{2}|\dot{x} - b(x)|^2 - \frac{1}{2}(\mathrm{div}\ b)(x).$$

Proof. By the transformation of drift discussed in Chapter IV, Section 4–1, we have

$$(9.5)\qquad P_x(B) = E^W\Big(\exp\Big[\int_0^T b(x + w(s))dw(s) - \frac{1}{2}\int_0^T |b(x + w(s))|^2 ds\Big]; B\Big)$$

for $B \in \mathscr{B}(W^d)$. Similarly,

$$(9.6)\qquad P^W(B) = E^W\Big[\exp\Big[-\int_0^T \dot{\phi}(s)dw(s) - \frac{1}{2}\int_0^T |\dot{\phi}(s)|^2 ds\Big] : (w; [w + \phi - x] \in B)\Big].^{*1}$$

By combining (9.5) and (9.6), we have

$$(9.7)\qquad \begin{aligned} &P_x(\|w - \phi\|_T < \varepsilon) \\ &= E^W\Big[\exp\Big[-\int_0^T \dot{\phi}(s)dw(s) - \frac{1}{2}\int_0^T |\dot{\phi}(s)|^2 ds \\ &\quad + \int_0^T b(w(s) + \phi(s))d[w(s) + \phi(s)] - \frac{1}{2}\int_0^T |b(w(s) + \phi(s))|^2 ds\Big] \\ &\quad : \|w\|_T < \varepsilon\Big]. \end{aligned}$$

If w satisfies $\|w\|_T < \varepsilon$, then

$$\Big|\int_0^T \dot{\phi}(s)dw(s)\Big| = \Big|\dot{\phi}(T)w(T) - \int_0^T w(s)\ddot{\phi}(s)ds\Big| \le A_1\varepsilon, ^{*2}$$

---

*1 For $w \in W^d$ and $x \in \boldsymbol{R}$, $w + x \in W^d$ is defined by $[w + x](t) = w(t) + x$.

*2 $A_1, A_2, \ldots$ and $K_1, K_2, \ldots$ are positive constants independent of ε.

$$\left|\int_0^T b(w(s)+\phi(s))\dot{\phi}(s)ds - \int_0^T b(\phi(s))\dot{\phi}(s)ds\right| \le A_2\varepsilon$$

and

$$\left|\int_0^T |b(w(s)+\phi(s))|^2 ds - \int_0^T |b(\phi(s))|^2 ds\right| \le A_3\varepsilon.$$

We immediately have from this that

$$\begin{aligned}
&\exp\left[-\frac{1}{2}\int_0^T |\dot{\phi}(s)-b(\phi(s))|^2 ds - (A_1+A_2+A_3)\varepsilon\right] \\
(9.8)\quad &\le P_x(\|w-\phi\|_T<\varepsilon)\{E^W[\exp[\int_0^T b(w(s)+\phi(s))dw(s)] : \|w\|_T<\varepsilon]\}^{-1} \\
&\le \exp\left[-\frac{1}{2}\int_0^T |\dot{\phi}(s)-b(\phi(s))|^2 ds + (A_1+A_2+A_3)\varepsilon\right].
\end{aligned}$$

Consequently it is sufficient to show that

$$(9.9)\quad E^W\left[\exp\left[\int_0^T b(w(s)+\phi(s))dw(s) + \frac{1}{2}\int_0^T (\operatorname{div} b)(\phi(s))ds\right] \,\Big|\, \|w\|_T<\varepsilon\right] \longrightarrow 1 \qquad \text{as } \varepsilon \downarrow 0.$$

Now

$$\begin{aligned}
&\int_0^T b(w(s)+\phi(s))dw(s) + \frac{1}{2}\int_0^T (\operatorname{div} b)(\phi(s))ds \\
&= \int_0^T b(\phi(s))dw(s) + \sum_{i,j=1}^d \int_0^T b^i{}_j(\phi(s))w^j(s)dw^i(s) \;* \\
&\quad + \frac{1}{2}\int_0^T (\operatorname{div} b)(\phi(s))ds + \sum_{i=1}^d \int_0^T \Phi_i(s,w)dw^i(s)
\end{aligned}$$

where

$$(9.10)\qquad \Phi_i(s,w) = b^i(w(s)+\phi(s)) - b^i(\phi(s)) - \sum_{j=1}^d b^i{}_j(\phi(s))w^j(s).$$

Since

---

* $b^i{}_j = \dfrac{\partial}{\partial x^j} b^i$.

$$\left| \int_0^T b(\phi(s))dw(s) \right| \leq A_4\varepsilon \qquad \text{if} \quad \|w\|_T < \varepsilon,$$

(9.9) is equivalent to

$$\begin{aligned} E^W\Big[\exp\Big\{\sum_{i,j=1}^d \int_0^T b^i{}_j(\phi(s))w^j(s)dw^i(s) \\ + \frac{1}{2}\int_0^T (\operatorname{div} b)(\phi(s))ds + \sum_{i=1}^d \int_0^T \Phi_i(s, w)dw^i(s)\Big\} \,\Big|\, \|w\|_T < \varepsilon\Big] \\ \longrightarrow 1 \qquad \text{as} \quad \varepsilon \downarrow 0. \end{aligned} \tag{9.11}$$

Note that

$$\begin{aligned} &\sum_{i,j=1}^d \int_0^T b^i{}_j(\phi(s))w^j(s)dw^i(s) + \frac{1}{2}\int_0^T (\operatorname{div} b)(\phi(s))ds \\ &= \sum_{i,j=1}^d \int_0^T b^i{}_j(\phi(s))\Big[w^j(s)dw^i(s) + \frac{1}{2}\delta_{ij}ds\Big] \\ &= \sum_{i,j=1}^d \int_0^T b^i{}_j(\phi(s))d\xi^{ji}(s) \end{aligned}$$

where

$$\xi^{ji}(t) = \int_0^t w^j(s)\circ dw^i(s).$$

Generally, if random variables Y_1, Y_2 satisfy

$$\overline{\lim_{\varepsilon\downarrow 0}}\, E^W[\, e^{cY_i} \mid \|w\|_T < \varepsilon] \leq 1, \qquad i = 1, 2$$

for every real constant c, then

$$\begin{aligned} &\overline{\lim_{\varepsilon\downarrow 0}}\, E^W[\exp[Y_1 + Y_2] \mid \|w\|_T < \varepsilon] \\ &\quad \leq \overline{\lim_{\varepsilon\downarrow 0}}\, \{E^W(\exp[2Y_1] \mid \|w\|_T < \varepsilon)\, E^W(\exp[2Y_2] \mid \|w\|_T < \varepsilon)\}^{1/2} \\ &\quad \leq 1, \end{aligned}$$

and similary

$$\overline{\lim_{\varepsilon\downarrow 0}}\, E^W(\exp(-Y_1 - Y_2) \mid \|w\|_T < \varepsilon) \leq 1.$$

But

$$E^W(\exp[Y_1 + Y_2] \mid \|w\|_T < \varepsilon)E^W(\exp[-Y_1 - Y_2] \mid \|w\|_T < \varepsilon)$$
$$\geq \left\{E^W\left(\exp\left[\frac{Y_1 + Y_2}{2}\right]\exp\left[\frac{-Y_1 - Y_2}{2}\right] \mid \|w\|_T < \varepsilon\right)\right\}^2 = 1,$$

and hence

$$\varliminf_{\varepsilon \downarrow 0} E^W(\exp[Y_1 + Y_2] \mid \|w\|_T < \varepsilon)$$
$$\geq \{\varlimsup_{\varepsilon \downarrow 0} E^W(\exp(-Y_1 - Y_2) \mid \|w\|_T < \varepsilon)\}^{-1} \geq 1.$$

In conclusion, we have

$$\lim_{\varepsilon \downarrow 0} E^W(\exp[Y_1 + Y_2] \mid \|w\|_T < \varepsilon) = 1.$$

This result can be extended easily to an arbitrary number of random variables: if $Y_1, Y_2, \ldots, Y_n$ satisfy

$$\varlimsup_{\varepsilon \downarrow 0} E^W(\exp[cY_i] \mid \|w\|_T < \varepsilon) \leq 1$$

for every real constant c and $i = 1, 2, \ldots, n$, then

$$\lim_{\varepsilon \downarrow 0} E^W(\exp[Y_1 + Y_2 + \cdots + Y_n] \mid \|w\|_T < \varepsilon) = 1.$$

Noting this fact, (9.11) will follow if we can show that for every real constant c and $i, j = 1, 2, \ldots, d$,

$$\varlimsup_{\varepsilon \downarrow 0} E^W(\exp[c\int_0^T b^i{}_j(\phi(s))d\xi^{ji}(s)] \mid \|w\|_T < \varepsilon) \leq 1 \tag{9.12}$$

and

$$\varlimsup_{\varepsilon \downarrow 0} E^W[\exp[c\int_0^T \Phi_i(s, w)dw^i(s)] \mid \|w\|_T < \varepsilon] \leq 1. \tag{9.13}$$

First we shall prove (9.13). Since

$$|c\Phi_i(s, w)| \leq A_5\varepsilon^2 \qquad \text{on the set} \quad \{w; \|w\|_T < \varepsilon\},$$

we have

$$P^W(\|\int_0^t c\Phi_i(s, w)dw^i(s)\|_T > \delta \mid \|w\|_T < \varepsilon)$$
$$\leq A_6\exp\left[-\frac{A_7\delta^2 - A_8\varepsilon^2}{\varepsilon^4}\right] \leq K_1\exp\left[-K_2\frac{\delta^2}{\varepsilon^4}\right] \qquad \text{if} \quad 0 < \varepsilon < K_3\delta$$

by a standard estimate as in the proof of Theorem 8.2. From this it is easy to conclude that

$$\text{(9.14)} \qquad \lim_{\varepsilon\downarrow 0} E^W(\exp[|c\int_0^T \Phi_i(s, w)dw^i(s)|] \mid \|w\|_T < \varepsilon) = 1.$$

Indeed, for each $\delta_0 > 0$ and $0 < \varepsilon < K_3\delta_0$,

$$E^W(\exp[|c\int_0^T \Phi_i(s, w)dw^i(s)|] \mid \|w\|_T < \varepsilon)$$
$$\leq e^{\delta_0} + K_1\int_{\delta_0}^\infty e^\xi\exp\left[-\frac{K_2\xi^2}{\varepsilon^4}\right]d\xi + K_1e^{\delta_0}\exp\left[-\frac{K_2\delta_0^2}{\varepsilon^4}\right].$$

(9.14) follows by first letting $\varepsilon \downarrow 0$ and then letting $\delta_0 \downarrow 0$.

Finally we shall prove (9.12). By writing $cb^i{}_j = c(x)$ it is sufficient to show that

$$\text{(9.15)} \qquad \overline{\lim_{\varepsilon\downarrow 0}}\, E^W(\exp[\int_0^T c(\phi(s))d\xi^{ji}(s)] \mid \|w\|_T < \varepsilon) \leq 1$$

for every $c(x) \in C_b^1(\boldsymbol{R}^d)$. Now

$$\int_0^T c(\phi(s))d\xi^{ji}(s) = c(\phi(T))\xi^{ji}(T) - \int_0^T \xi^{ji}(s)d\,[c(\phi(s))].$$

Also $\xi^{ji}(t) = \frac{1}{2}w^i(t)w^j(t) - \eta^{ij}(t)$ where $\eta^{ij}(t)$ is defined by (8.11). As we saw in the proof of Lemma 8.2,

$$\eta^{ij}(t) = B\left(\frac{1}{4}\int_0^t [(w^i)^2(s) + (w^j)^2(s)]ds\right)$$

where B is a Brownian motion independent of the radial process $\{|w(t)|\}$ (and hence independent of $\|w\|_T$). Then

$$E^W(\exp[\int_0^T c(\phi(s))d\xi^{ij}(s)] \mid \|w\|_T < \varepsilon)$$
$$\leq e^{K_4\varepsilon^2} E^W(\exp[K_5 \max_{0\leq t\leq \frac{1}{4}\varepsilon^2 T} |B(t)|])$$
$$= e^{K_4\varepsilon^2} \int_0^\infty \sqrt{\frac{2}{\pi}} \exp\left[-\frac{x^2}{2}\right] \exp\left[\frac{\varepsilon}{2}\sqrt{T}\, K_5 x\right] dx \longrightarrow 1 \quad \text{as } \varepsilon \downarrow 0.$$

This completes the proof.

In Chapter V, Section 4, we saw that the diffusion $\{P_x\}$ is symmetrizable if and only if the differential 1-form ω defined by $\omega = \sum_{i=1}^{d} b^i(x)dx^i$ is given as $\omega = dF$ for some $F \in C^\infty(\boldsymbol{R}^d \longrightarrow \boldsymbol{R})$, or equivalently the line integral $\int_\gamma \omega$ vanishes along every closed smooth curve. Using Theorem 9.1, this condition can be restated as follows.

Theorem 9.2. The diffusion $\{P_x\}$ is symmetrizable if and only if

(9.16) $$\lim_{\varepsilon\downarrow 0} \frac{P_x(\|w - \phi\|_T < \varepsilon)}{P_x(\|w - \phi_-\|_T < \varepsilon)} = 1$$

for every x and every smooth curve ϕ: $[0, T] \longrightarrow \boldsymbol{R}^d$ such that $\phi(0) = \phi(T) = x$. Here the curve ϕ_- is defined by

(9.17) $$\phi_-(t) = \phi(T - t), \qquad 0 \leq t \leq T.$$

Proof. By Theorem 9.1, the limit in (9.16) is equal to

$$\exp(2\int_0^T b(\phi(s))\dot{\phi}(s)ds) = \exp(2\int_\phi \omega).$$

Thus (9.16) holds if and only if $\int_\phi \omega = 0$.

A similar probabilistic characterization of the symmetry was obtained by Kolmogorov [87] in the case of a Markov chain.

Bibliography

1. S. Agmon, *Lectures on elliptic boundary value problems*, Van Nostrand, Princeton, New Jersey, 1965.
2. H. Airault, Perturbations singulières et solutions stochastiques de problèmes de D. Neumann-Spencer, *Jour. Math. Pures Appl.*, **55**(1976), 233–268.
3. V. E. Beneš, Composition and invariance methods for solving some stochastic control problems, *Adv. Appl. Prob.*, **7** (1975), 299–329.
4. S. Bernstein, Principes de la théorie des équations différentielles stochastiques, *Trudy Fiz-Mat. Stekrov Inst. Acad. Nauk*, **5** (1934), 95–124.
5. S. Bernstein, Équations différentielles stochastiques, *Act. Sci. et Ind.* **738**, *Conf. intern. Sci. Math. Univ. Genève*, 5–31, Herman, Paris, 1938.
6. P. Billingsley, *Convergence of probability measures*, John Wiley and Sons, New York, 1968.
7. R. Bishop and R. J. Crittenden, *Geometry of manifolds*, Academic Press, New York, 1964.
8. R. M. Blumenthal and R. K. Getoor, *Markov processes and potential theory*, Academic Press, New York, 1968.
9. R. H. Cameron and W. T. Martin, Transformation of Wiener integrals under translations, *Ann. Math.*, **45** (1944), 386–396.
10. R. H. Cameron and W. T. Martin, Evaluations of various Wiener integrals by use of certain Sturm-Liouville differential equations, *Bull. Amer. Math. Soc.*, **51** (1945), 73–90.
11. K. L. Chung and R. Durret, Downcrossing and local time, *Z. Wahr. verw. Geb.*, **35** (1976), 147–149.
12. B. S. Cirel'son, An example of a stochastic differential equation having no strong solution, *Theor. Prob. Appl.*, **20** (1975), 416–418.
13. P. E. Conner, *The Neumann's problem for differential forms on Riemannian manifolds, Mem. Amer. Math. Soc.*, **20** (1956).
14. A. Debiard, B. Gaveau and E. Mazet, Théorèmes de comparaison en géométrie riemannienne, *Publ. RIMS, Kyoto Univ.*, **12** (1976), 391–425.
15. C. Dellacherie, *Capacités et processus stochastiques*, Springer-Verlag, Berlin, 1972.
16. C. Dellacherie, Intégrals stochastiques par rapport aux processus de Wiener et de Poisson, *Séminaire de Prob. (Univ. de Strasbourg) VIII, Lecture Notes in Math.*, **381**, 25–26, Springer-Verlag, Berlin, 1974.
17. C. Doléans-Dade and P. A. Meyer, Intégrales stochastiques par rapport aux martingales locales, *Séminaire de Prob. (Univ. de Strasbourg) IV, Lecture Notes in Math.*, **124**, 77–107, Springer-Verlag, Berlin, 1970.
18. J. L. Doob, *Stochastic processes*, John Wiley and Sons, New York, 1953.
19. H. Doss, Liens entre équations différentielles stochastiques et ordinaires, *Ann. Inst. H. Poincaré*, **13** (1977), 99–125.
20. D. Dürr and A. Bach, The Onsager-Machlup function as Lagrangian for the most probable path of a diffusion process, *Comm. Math. Phys.*, **60** (1978), 153–170.
21. E. B. Dynkin, *Theory of Markov processes*, Pergamon Press, Oxford, 1960.
22. E. B. Dynkin, *Markov processes*, I, II, Springer-Verlag, Berlin, 1965.
23. J. Eells and K. D. Elworthy, Stochastic dynamical systems, *Control theory and*

topics in functional analysis, III, Intern. atomic energy agency, Vienna, 1976, 179–185.

24. A. Einstein, *Investigations on the theory of the Brownian movement*, Methuen, London, 1926.
25. K. D. Elworthy, Stochastic dynamical systems and their flows, *Stochastic Analysis* (ed. by A. Friedman and M. Pinsky), 79–95, Academic Press, New York, 1978.
26. W. Feller, Zur Theorie der Stochastischen Prozesse (Existenz- und Eindeutigkeitssätze), *Math. Ann.*, **113** (1936), 113–160.
27. D. L. Fisk, Quasi-martingales and stochastic integrals, *Tech. Rep.* **1**, Dept. Math. Michigan State Univ., 1963.
28. A. Friedman, *Stochastic differential equations and applications Vol. 1 and Vol. 2*, Academic Press, New York, 1975.
29. M. Fujisaki, G. Kallianpur and H. Kunita, Stochastic differential equations for the non linear filtering problem, *Osaka J. Math.*, **9** (1972), 19–40.
30. M. Fukushima, *Dirichlet forms and Markov processes*, Kodansha, Tokyo, 1980.
31. T. Funaki, Construction of a solution of random transport equation with boundary condition, *J. Math. Soc. Japan*, **31** (1979), 719–744.
32. A. M. Garsia, *Martingale inequalities, Seminar Notes on Recent Progress*, W. A. Benjamin, Reading, Massachusetts, 1973.
33. B. Gaveau, Principe de moindre action, propagation de la chaleur et estimées sous elliptiques sur certain groupes nilpotents, *Acta Math.*, **139** (1977), 95–153.
34. R. K. Getoor and M. J. Sharpe, Conformal martingales, *Invent. Math.*, **16** (1972), 271–308.
35. I. I. Gihman, A method of constructing random processes, *Dokl. Akad. Nauk SSSR*, **58** (1947), 961–964, (in Russian).
36. I. I. Gihman, Certain differential equations with random functions, *Ukr. Mat. Zh.*, **2** (1950), No. 3, 45–69, (in Russian).
37. I. I. Gihman, On the theory of differential equations of random processes, *Ukr. Mat. Zh.*, **2** (1950), No. 4, 37–63, (in Russian).
38. I. I. Gihman and A. V. Skorohod, *Stochastic differential equations*, Springer-Verlag, Berlin, 1972.
39. I. I. Gihman and A. V. Skorohod, *The theory of stochastic processes III*, Springer-Verlag, Berlin, 1979.
40. I. V. Girsanov, On transforming a certain class of stochastic processes by absolutely continous substitution of measures, *Theor. Prob. Appl.*, **5** (1960), 285–301.
41. R. Graham, Path integral formulation of general diffusion processes, *Z. Physik B*, **26** (1977), 281–290.
42. R. E. Greene and H. Wu, *Function theory on manifolds which possess a pole, Lecture Notes in Math.*, **699**, Springer-Verlag Berlin, 1979.
43. B. Grigelionis, On representation of integer-valued random measures by means of stochastic integrals with respect to the Poisson measure, *Lit. Mat. Sb.*, **11** (1971), 93–108, (in Russian).
44. B. Grigelionis, On the martingale characterization of stochastic processes with independent increments, *Lit. Mat. Sb.*, **17** (1977), 52–60.
45. P. R. Halmos, *Measure theory*, Van Nostrand, New York, 1950.
46. R. Z. Hashiminsky, Ergodic properties of recurrent diffusion processes and stabilization of the solution of the Cauchy problem for parabolic equations, *Theor. Prob. Appl.*, **5** (1960), 179–196.
47. S. Helgason, *Differential geometry and symmetric spaces*, Academic Press, New York, 1962.
48. L. Hörmander, Hypoelliptic second order differential equations, *Acta Math.*, **119** (1967), 147–171.
49. N. Ikeda, On the construction of two-dimensional diffusion processes satisfying Wentzell's boundary conditions and its application to boundary value problems, *Mem. Coll. Sci. Univ. Kyoto Math.*, **33** (1961), 367–427.
50. N. Ikeda and S. Manabe, Stochastic integral of differential forms and its appli-

cations, *Stochastic Analysis* (ed. by A. Friedman and M. Pinsky), 175–185, Academic Press, New York, 1978.
51. N. Ikeda and S. Manabe, Integral of differential forms along the path of diffusion processes, *Publ. RIMS, Kyoto Univ.*, **15** (1979), 827–852.
52. N. Ikeda, S. Nakao and Y. Yamato, A. class of approximations of Brownian motion, *Publ. RIMS, Kyoto Univ.*, **13** (1977), 285–300.
53. N. Ikeda and S. Watanabe, The local structure of a class of diffusions and related problems, *Proc. Second Japan-USSR Symp. Prob. Theor., Lecture Notes in Math.*, **330**, 124–169, Springer-Verlag, Berlin, 1973.
54. N. Ikeda and S. Watanabe, A comparison theorem for solutions of stochastic differential equations and its applications, *Osaka J. Math.*, **14** (1977), 619–633.
55. N. Ikeda and S. Watanabe, Heat equation and diffusion on Riemannian manifold with boundary, *Proc. Intern. Symp. SDE Kyoto 1976* (ed. by K. Itô), 75–94, Kinokuniya, Tokyo, 1978.
56. H. Ito, Probabilistic construction of Lagrangean of diffusion processes and its application, *Prog. Theoretical Phys.*, **59** (1978), 725–741.
57. K. Itô, Differential equations determining Markov processes, *Zenkoku Shijō Sūgaku Danwakai*, **244** (1942), No. 1077, 1352–1400, (in Japanese).
58. K. Itô, On stochastic processes (1) (Infinitely divisible laws of probability), *Japanese J. Math.*, **18** (1942), 261–301.
59. K. Itô, Stochastic integral, *Proc. Imp. Acad. Tokyo*, **20** (1944), 519–524.
60. K. Itô, On a stochastic integral equation, *Proc. Imp. Acad. Tokyo*, **22** (1946), 32–35.
61. K. Itô, Stochastic differential equations in a differentiable manifold, *Nagoya Math. J.*, **1** (1950), 35–47.
62. K. Itô, *On stochstic differential equations, Mem. Amer. Math. Soc.*, **4** (1951).
63. K. Itô, On a formula concerning stochastic differentials, *Nagoya Math. J.*, **3** (1951), 55–65.
64. K. Itô, Multiple Wiener integral, *J. Math. Soc. Japan.*, **3** (1951), 157–169.
65. K. Itô, Stochastic differential equations in a differentiable manifold (2), *Mem. Coll. Sci. Univ. Kyoto Math.*, **28** (1953), 81–85.
66. K. Itô, *Theory of probability*, Iwanami, Tokyo, 1953 (in Japanese).
67. K. Itô, *Lectures on stochastic processes.*, Tata Institute of Fundamental Research, Bombay, 1960.
68. K. Itô, The Brownian motion and tensor fields on Riemannian manifold, *Proc. Intern. Congr. Math.*, Stockholm, 536–539, 1963.
69. K. Itô, *Stochastic processes, Lecture Notes Series*, **16**, Aarhus Univ. 1969.
70. K. Itô, Poisson point processes attached to Markov processes, *Proc. Sixth Berkeley Symp. Math. Statist. Prob. III*, 225–239, Univ. California Press, Berkeley, 1972.
71. K. Itô, Stochastic differentials, *Appl. Math. Opt.*, **1** (1975), 374–381.
72. K. Itô, Stochastic parallel displacement, *Probabilistic methods in differential equations, Lecture Notes in Math.*, **451**, 1–7, Springer-Verlag, Berlin, 1975.
73. K. Itô and H.P. McKean, Jr., *Diffusion processes and their sample paths*, Springer-Verlag, Berlin, 1965.
74. K. Itô and M. Nisio, On stationary solutions of a stochastic differential equation, *J. Math. Kyoto Univ.*, **4** (1964), 1–75.
75. K. Itô and M. Nisio, On the convergence of sums of independent Banach space valued random variables, *Osaka J. Math.*, **5** (1968), 35–48.
76. K. Itô and S. Watanabe, Introduction to stochastic differential equations, *Proc. Intern. Symp. SDE Kyoto 1976* (ed. by K. Itô), i-xxx, Kinokuniya, Tokyo, 1978.
77. J. Jacod, *Calcul stochastique et problèmes de martingales, Lecture Notes in Math.*, **714**, Springer-Verlag, Berlin, 1979.
78. T. Jeulin, Grossissement d'une filtration et applications, *Séminaire de Prob. XIII (Univ. de Strasbourg), Lecture Notes in Math.*, **721**, 574–609, Springer-Verlag, Berlin, 1979.

79. M. Kac, On distributions of certain Wiener functionals, *Trans. Amer. Math. Soc.*, **65** (1949), 1–13.
80. M. Kac, On some connections between probability theory and differential and integral equations, *Proc. Second Berkeley Symp. Math. Statist. Prob.*, 189–215, Univ. California Press, Berkeley, 1951.
81. N.E. Karoui and J.P. Lepeltier, Représentation des processus ponctuels multivariés à l'aide d'un processus de Poisson, *Z. Wahr. verw. Geb.*, **39** (1977), 111–133.
82. Y. Kasahara and S. Kotani, On limit processes for a class of additive functionals of recurrent diffusion processes, *Z. Wahr. verw. Geb.*, **49** (1979), 133–159.
83. N. Kazamaki, The equivalence of two conditions on weighted norm inequalities for martingales, *Proc. Intern. Symp. SDE Kyoto 1976* (ed. by K. Itô), 141–152, Kinokuniya, Tokyo, 1978.
84. F.B. Knight, A reduction of continuous square-integrable martingales to Brownian motion, *Lecture Notes in Math.*, **190**, Springer-Verlag, Berlin, 1971.
85. A.N. Kolmogoroff, Über die analytischen Methoden in der Wahrscheinlichkeitsrechnung, *Math. Ann.*, **104** (1931), 415–458.
86. A.N. Kolmogoroff, Zufällige Bewegungen, *Ann. Math. II.*, **35** (1934), 116–117.
87. A.N. Kolmogoroff, Zur Theorie der Markoffschen Ketten, *Math. Ann.*, **112** (1936), 155–160.
88. A.N. Kolmogoroff, Zur Umkehrbarkeit der statistischen Naturgesetze, *Math. Ann.*, **113** (1937), 766–772.
89. M.G. Krein and M.A. Rutman, Linear operators leaving invariant a cone in a Banach space, *Amer. Math. Trans. Series 1*, **10** (1950), 199–325.
90. N.V. Krylov, On Ito's stochastic integral equations, *Theor. Prob. Appl.*, **14** (1969), 330–336.
91. N.V. Krylov, *Control of diffusion type processes*, Nauka, Moscow, 1977, (in Russian).
92. N.V. Krylov and M.V. Safonov, Estimate for the hitting probability of diffusion processes to a set of positive measure, *Dokl. Akad. Nauk SSSR.*, **245** (1979), 18–20, (in Russian).
93. H. Kunita, *Diffusion processes and control systems.*, Course at University of Paris VI, 1974.
94. H. Kunita, *Estimation of stochastic processes*, Sangyō Tosho, Tokyo, 1976 (in Japanese).
95. H. Kunita, Supports of diffusion processes and controllability problems, *Proc. Intern. Symp. SDE Kyoto 1976* (ed. by K. Itô), 163–185, Kinokuniya, Tokyo, 1978.
96. H. Kunita, On the representation of solutions of stochastic differential equations, *Seminaire de Prob. XIV, Lecture Notes in Math.*, **784**, 282–304, Springer-Verlag, Berlin, 1980.
97. H. Kunita and S. Watanabe, On square integrable martingales, *Nagoya Math. J.*, **30** (1967), 209–245.
98. K. Kuratowski, *Topology I.*, Academic Press, New York, 1966.
99. P. Lévy, *Théorie de l'addition des variables aléatoires*, Gauthier-Villars, Paris, 1937.
100. P. Lévy, Le mouvement brownien plan, *Amer. J. Math.*, **62** (1940), 487–550.
101. P. Lévy, *Processus stochastiques et mouvement brownien*, Gauthier-Villars, Paris, 1948.
102. R.S. Liptzer and A.N. Shiryaev, *Statistics of stochastic processes*, Nauka, Moscow, 1974, (in Russian).
103. B. Maisonneuve, Exit systems, *Ann. Prob.*, **3** (1975), 399–411.
104. P. Malliavin, Formule de la moyenne, calcul de perturbations et théorèmes d'annulation pour les formes harmoniques, *J. Funct. Analy.*, **17** (1974), 274–291.
105. P. Malliavin, Champs de Jacobi stochastiques, *C.R.A. Sc. Paris*, **285** (1977), 789–792.
106. P. Malliavin, Stochastic calculus of variation and hypoelliptic operators, *Proc. Intern. Symp. SDE Kyoto 1976* (ed. by K. Itô), 195–263, Kinokuniya, Tokyo, 1978.

107. P. Malliavin, C^k-hypoellipticity with degeneracy, *Stochastic Analysis* (ed. A. Friedman and M. Pinsky), 199–214, 327–340, Academic Press, New York, 1978.
108. P. Malliavin, *Géométrie différentielle stochastique*, Les Presses de l'Université de Montréal, Montréal, 1978.
109. S. Manabe, On the intersection number of the path of a diffusion and chains, *Proc. Japan Acad.*, **55** (1979), 23–26.
110, G. Maruyama, On the transition probability functions of the Markov process, *Nat. Sci. Rep. Ochanomizu Univ.*, **5** (1954), 10–20.
111. G. Maruyama, Continuous Markov processes and stochastic equations, *Rend. Circ. Mate. Palermo.*, **4** (1955), 48–90.
112. Y. Matsushima, *Differentiable manifolds*, Marcel Dekker, New York, 1972.
113. H.P. McKean Jr., *Stochastic integrals*, Academic Press, New York, 1969.
114. H.P. McKean, Brownian local times, *Adv. Math.*, **16** (1975), 91–111.
115. E.J. McShane, Stochastic differential equations and models of random processes, *Proc. Sixth Berkeley Symp. Math. Statist. Prob. III*, 263–294, Univ. California Press, Berkeley, 1972.
116. E.J. McShane, *Stochastic calculus and stochastic models.*, Academic Press, New York, 1974.
117. P.A. Meyer, *Probability and potentials*, Blaisdel, Waltham, Massachusetts, 1966.
118. P.A. Meyer, Intégrales stochastiques I-IV, *Séminaire de Prob. (Univ. de Strasbourg) I, Lecture Notes in Math.*, **39**, 72–162, Springer-Verlag, Berlin, 1967.
119. P.A. Meyer, Démonstration simplifiée d'un théorème de Knight, *Séminaire de Prob. (Univ. de Strasbourg) V, Lecture Notes in Math.*, **191**, 191–195, Springer-Verlag, Berlin, 1971.
120. P.A. Meyer, *Martingales and stochastic integrals I, Lecture Notes in Math.*, **284**, Springer-Verlag, Berlin, 1972.
121. P.A. Meyer, Un cours sur les intégrales stochastiques, *Séminaire de Prob. (Univ. de Strasbourg) X, Lecture Notes in Math.*, **511**, 245–400, Springer-Verlag, Berlin, 1976.
122. J. Milnor, *Morse theory*, Princeton Univ. Press, Princeton, New Jersey, 1963.
123. M. Motoo, Diffusion process corresponding to $1/2 \sum \partial^2/\partial x^{i2} + \sum b^i(x)\, \partial/\partial x^i$, *Ann. Inst. Statist. Math.*, **12** (1960–61), 37–61.
124. M. Motoo, Periodic boundary problems of the two dimensional Brownian motion on upperhalf plane, *Proc. Intern. Symp. SDE Kyoto 1976* (ed. by K. Itô), 265–281, Kinokuniya, Tokyo, 1978.
125. S. Nakao, On the pathwise uniqueness of solutions of one-dimensional stochastic differential equations, *Osaka J. Math.*, **9** (1972), 513–518.
126. S. Nakao and Y. Yamato, Approximation theorem on stochastic differential equations, *Proc. Intern. Symp. SDE Kyoto 1976* (ed. by K. Itô), 283–296, Kinokuniya, Tokyo, 1978.
127. E. Nelson, The adjoint Markoff process, *Duke Math. J.*, **25** (1958), 671–690.
128. E. Nelson, *Tensor analysis, Mathematical Notes*, Princeton Univ. Press, Princeton, New Jersey, 1967.
129. J. Neveu, *Bases mathématiques du calcul des probabilités*, Masson et Cie., Paris, 1964.
130. M. Nisio, On the existence of solutions of stochastic differential equations, *Osaka J. Math.*, **10** (1973), 185–208.
131. K. Nomizu, *Lie groups and differential geometry*, Publ. Math. Soc. Japan, Tokyo, 1956.
132. A.A. Novikov, On moment inequalities and identities for stochastic integrals, *Proc. Second Japan-USSR Symp. Prob. Theor., Lecture Notes in Math.*, **330**, 333–339, Springer-Verlag, Berlin, 1973.
133. S. Ogawa, On a Riemann definition of the stochastic integral, (I), (II), *Proc. Japan Acad.*, **46** (1970), 153–157, 158–161.
134. L. Onsager and S. Machlup, Fluctuations and irreversible processes, I, II, *Phys. Rev.*, **91** (1953), 1505–1512, 1512–1515.

135. S. Orey, Diffusion on the line and additive functionals of Brownian motion, *Proc. conf. on stochastic differential equations and applications* (ed. by J.D. Mason), 211–230, Academic Press, New York, 1977.
136. R.E.A.C. Paley and N. Wiener, *Fourier transforms in the complex domain, Amer. Math. Soc, Coll. Publ.*, **19**, 1934.
137. G.C. Papanicolaou, D.W. Stroock and S.R.S. Varadhan, Martingale approach to some limit theorems, *1976 Duke Turbulence Conf.*, Duke Univ. Math. Series III, 1977.
138. K.P. Parthasarathy, *Probability measures on metric spaces*, Academic Press, New York, 1967.
139. M. Pinsky, Multiplicative operator functionals and their asymptotic properties, *Advances in Probability* (ed. by P. Ney and S. Port), **3**, 1–100, Marcel Dekker, New York, 1974.
140. M. Pinsky, Stochastic Riemannian geometry, *Probabilistic analysis and related topics* (ed. by A.T. Bharucha-Reid), Academic Press, New York, 1978.
141. J.W. Pitman, One-dimensional Brownian motion and the three-dimensional Bessel process, *Adv. Appl. Prob.*, **7** (1975), 511–526.
142. Yu.V. Prohorov, Convergence of random processes and limit theorems in probability theory, *Theor. Prob. Appl.*, **1** (1956), 157–214.
143. K.M. Rao, On decomposition theorems of Meyer, *Math. Scand.*, **24** (1969), 66–78.
144. D.B. Ray and I.M. Singer, R-torsion and the Laplacian on Riemannian manifolds, *Adv. Math.*, **7** (1971), 145–210.
145. G.de Rham, *Variétés différentiables*, Hermann, Paris, 1960.
146. T. Shiga, Diffusion processes in population genetics, *J. Math. Kyoto Univ.*, Forthcoming.
147. T. Shiga and S. Watanabe, Bessel diffusions as a one-parameter family of diffusion processes, *Z. Wahr. verw. Geb.*, **27** (1973), 37–46.
148. I. Shigekawa, Derivatives of Wiener functionals and absolute continuity of induced measures, *J. Math. Kyoto Univ.*, **20** (1980), 263–289.
149. A.V. Skorohod, Stochastic equations for diffusion processes in a bounded region, *Theor. Prob. Appl.*, **6** (1961), 264–274.
150. A.V. Skorohod, *Studies in the theory of random processes*, Addison-Wesley, Reading, Massachusetts, 1965.
151. S.L. Sobolev, *Applications of functional analysis in mathematical physics*, Amer. Math. Soc., Providence, 1963.
152. E.M. Stein, *Singular integrals and differentiability properties of functions*, Princeton Univ. Press, Princeton, New Jersey, 1970.
153. R.L. Stratonovich, *Conditional Markov processes and their application to the theory of optimal control*, Amer. Elsevier, New York, 1968.
154. R.L. Stratonovich, On the probability functional of diffusion processes, *Selected Trans. in Math. Statist. Prob.*, **10** (1971), 273–286.
155. C. Stricker, Quasimartingales, martingales locales, semimartingales et filtration naturelle, *Z. Wahr. verw. Geb.*, **39** (1977), 55–63.
156. D.W. Stroock, On the growth of stochastic integrals, *Z. Wahr. verw. Geb.*, **18** (1971), 340–344.
157. D.W. Stroock and S.R.S. Varadhan, Diffusion processes with continuous coefficients, I, II, *Comm. Pure Appl. Math.*, **22** (1969), 345–400, 479–530.
158. D.W. Stroock and S.R.S. Varadhan, On the support of diffusion processes with applications to the strong maximum principle, *Proc. Sixth Berkeley Symp. Math. Statist. Prob. III.*, 333–359, Univ. California Press, Berkeley, 1972.
159. D.W. Stroock and S.R.S. Varadhan, Diffusion processes, *Proc. Sixth Berkeley Symp. Math. Statist. Prob. III.*, 361–368, Univ. California Press, Berkeley, 1972.
160. D.W. Stroock and S.R.S. Varadhan, *Multidimensional diffusion processes*, Springer-Verlag, Berlin, 1979.
161. H. Tanaka, On the uniqueness of Markov process associated with the Boltzmann

equation of Maxwellian molecules, *Proc. Intern. Symp. SDE Kyoto 1976* (ed. by K. Itô), 409–425, Kinokuniya, Tokyo, 1978.
162. H. Tanaka and M. Hasegawa, *Stochastic differential equations, Seminar on Prob.*, **19**, 1964, (in Japanese).
163. H.F. Trotter, A property of Brownian motion paths, *Illi. J. Math.*, **2** (1958), 425–433.
164. Y. Umemura (Y. Yamazaki), On the infinite dimensional Laplacian operator, *J. Math. Kyoto Univ.*, **4** (1965), 477–492.
165. S.R.S. Varadhan, *Stochastic processes, Lecture Notes*, Courant Institute of Math. Sci. New York Univ. (1967/68), 1968.
166. S. Watanabe, On stochastic differential equations for multi-dimensional diffusion processes with boundary conditions, I, II, *J. Math. Kyoto Univ.*, **11** (1971), 169–180, 545–551.
167. S. Watanabe, Solution of stochastic differential equations by random time change, *Appl. Math. Opt.*, **2** (1975), 90–96.
168. S. Watanabe, On time inversion of one-dimensional diffusion processes, *Z. Wahr. verw. Geb.*, **31** (1975), 115–124.
169. S. Watanabe, *Stochastic differential equations*, Sangyō Tosho, Tokyo, 1975 (in Japanese).
170. S. Watanabe, Construction of diffusion processes by means Poisson point process of Brownian excursions, *Proc. Third Japan-USSR Symp. Prob. Theor., Lecture Notes in Math.*, **550**, 650–654, Springer-Verlag, Berlin, 1976.
171. S. Watanabe, Poisson point process of Brownian excursions and its applications to diffusion processes, *Proc. Symp. Pure Math. Amer. Math. Soc.*, **31**, 153–164, 1977.
172. S. Watanabe, Excursion point process of diffusion and stochastic integral, *Proc. Intern. Symp. SDE Kyoto 1976* (ed. by K. Itô), 437–461, Kinokuniya, Tokyo, 1978.
173. S. Watanabe, Point processes and martingales, *Stochastic Analysis*, (ed. by A. Friedman and M. Pinsky), 315–326, Academic Press, New York, 1978.
174. S. Watanabe, Construction of diffusion processes with Wentzell's boundary conditions by means of Poisson point processes of Brownian excursions, *Probability theory, Banach Center Publications*, Vol. **5**, 255–271, Polish Scientific Publishers, Warsaw, 1979.
175. A.D. Wentzell, On boundary conditions for multidimensional diffusion processes, *Theor. Prob. Appl.*, **4** (1959), 164–177.
176. H. Whitney, *Geometric integration theory*, Princeton Univ. Press, Princeton, New Jersey, 1957.
177. N. Wiener, Differential space, *J. Math. Phys.*, **2** (1923), 131–174.
178. N. Wiener, *Nonlinear problems in random theory*, M.I.T. Press, Cambridge and John Wiley and Sons, New York, 1958.
179. D. Williams, Path decomposition and continuity of local time for one-dimensional diffusions, I, *Proc. London Math. Soc.* **(3)**, **28** (1974), 738–768.
180. D. Williams, Lévy's downcrossing theorem, *Z. Wahr. verw. Geb.*, **40** (1977), 157–158.
181. E. Wong and M. Zakai, On the relation between ordinary and stochastic differential equations, *Intern. J. Engng. Sci.*, **3** (1965), 213–229.
182. T. Yamada, On a comparison theorem for solutions of stochastic differential equations and its applications, *J. Math. Kyoto Univ.*, **13** (1973), 497–512.
183. T. Yamada and S. Watanabe, On the uniqueness of solutions of stochastic differential equations, *J. Math. Kyoto Univ.*, **11** (1971), 155–167.
184. Y. Yamato, Stochastic differential equations and nilpotent Lie algebras, *Z. Wahr. verw. Geb.*, **47** (1979), 213–229.
185. K. Yano and S. Bochner, *Curvature and Betti numbers, Ann. Math. Studies*, **32**, Princeton Univ. Press, Princeton, New Jersey, 1953.
186. S.T. Yau, On the heat kernel of a complete Riemannian manifold, *J. Math. Pures Appl.*, **57** (1978), 191–201.

187. M.P. Yershov, On stochastic equations, *Proc. Second Japan-USSR Symp. Prob. Theor., Lecture Notes in Math.*, **330**, 527–530, Springer-Verlag, Berlin, 1973.
188. M.P. Yershov, Localization of conditions on the coefficients of diffusion type equations in existence theorems, *Proc. Intern. Symp. SDE Kyoto 1976* (ed. by K. Itô), 493–507, Kinokuniya, Tokyo, 1978.
189. A.K. Zbonkin and N.V. Krylov, On strong solutions of stochastic differential equations, *Proc. School-Seminar on Theor. Random Processes (Druskininkai)*, Vilnius, Ac. Sci. Lit. SSR, part II, 9–88, 1975, (in Russian).

References Added in the Second Edition

190. H. Akiyama, Geometric aspects of Malliavin's calculus on vector bundles, *J. Math. Kyoto Univ.*, **26**, (1986), 673–696.
191. R. Azencott, Behavior of diffusion semi-groups at infinity, *Bull. Soc. Math. France*, **102** (1974), 193–240.
192. M. Bismut, *Large deviations and the Malliavin calculus,* Birkhäuse, Boston, 1984.
193. J-M. Bismut, The Atyah-Singer theorems: a probabilistic approach, I, the index theorem, II, the Lefschetz fixed point formulas, *J. Func. Anal.*, **57** (1984), 56–99 and 329–348.
194. H. J. Cycon, R. G. Froece, W. Kirsh and B. Simon, *Schrödinger operators with application to quantum mechanics and global geometry,* Texts and Monographs in Physics, Springer-Verlag, Berlin, 1987.
195. A. Debiard and B. Gaveau, Frontière de Shilov des domaines faiblement pseudo-convexes, *Bull. Sci. Math.*, **100** (1976), 17–31.
196. R. Durrett, *Brownian motion and matringales in analysis*, Wadsworth, Inc., 1984.
197. K.D. Elworthy, *Stochastic differential equations on manifolds,* London Math, Soc. Lect. Note Ser., **70**, Cambridge University Press, 1982.
198. M. Fukushima and M. Okada, On Dirichlet forms for plurisubharmonic functions, *Acta Math.*, **159** (1987), 171–213.
199. T. Fujita and S. Kotani, The Onsager-Machlup function for diffusion processes, *J. Math. Kyoto Univ.*, **22** (1982), 115–130.
200. B. Gaveau, Méthodes de contrôle optimal en analyse complexe. Résolution d'equation de Monge-Ampère complexe, *J. Funct. Analy.*, **24** (1977), 391–411.
201. E. Getzler, A short proof of the Atiyah-Singer index theorems, *Topology*, **25** (1986), 111–117.
202. P.B. Gilkey, *Invariance theory, the heat equation, and the Atiyah-Singer index theorem*, Math. Lecture Ser., **11**, Publish or Perish, Inc., 1984.
203. N. Ikeda and S. Watanabe, An introduction to Malliavin's calculus, *Stochastic Analysis, Proc. Taniguchi Intern. Symp. on Stochastic Analysis, Katata and Kyoto* 1982, (ed. by K. Itô), 1–52, Kinokuniya/North-Holland, Tokyo, 1984.
204. N. Ikeda and S. Watanabe, Malliavin calculus of Wiener functionals and its applications, *From local times to global geometry, control and physics,* (ed. by K. D. Elworthy), Pitman Research, *Notes in Math. Ser.*, **150**, 132–178, Longman Scientific and Technical, Horlow, 1987.
205. H. Kaneko and S. Taniguchi, A stochastic approach to the Šilov boundary, *J. Funct. Anal.*, **74** (1987), 415–429.
206. S.W. Kendall, Stochastic differential geometry, An introduction, *Acta Appl. Math.*, **9** (1987), 29–60.
207. S. Kobayashi and K. Nomizu, *Foundation of differential geometry*, Vol. I, II, Interscience, New York, 1963 and 1969.
208. M. Krée and P. Krée, Continute de la divergence dans les espaces de Sobolev relatifs à l'espace de Wiener, *C.R. Acad. Sci. Paris,* 296 (1983), 833–836.
209. H. Kunita, Stochastic differential equations and stochastic flows of diffeomor-

phisms, *Ecole d'Eté de Probabilités de Saint-Flour* XII-1982, *Lecture Notes in Math.*, **1097**, 143–303, Springer-Verlag, Berlin, 1984.
210. H-H. Kuo, Gaussian measures in Banach spaces, *Lecture Notes in Math.*, **463**, Springer-Verlag, Berlin, 1975.
211. S. Kusuoka and D.W. Stroock, Applications of the Malliavin calculus, Part I, *Stochastic Analysis, Proc. Taniguchi Intern. Symp. on Stochastic Analysis, Katata and Kyoto,* 1982, (ed. by K. Itô), 271–306, Kinokuniya/North-Holland, Tokyo, 1984.
212. S. Kusuoka and D.W. Stroock, Applications of the Malliavin calculus, Part II, *J. Fac. Sci. Univ. Tokyo Sect.* IA, Math., **32** (1985), 1–76.
213. S. Kusuoka and D.W. Stroock, Applications of the Malliavin calculus, Part III, *J. Fac. Sci., Univ. Tokyo Sect.*, IA, Math., **34** (1987), 391–442.
214. R. Léandre, Majoration en temps petit de la densité d'une diffusion dégénérée, *Probab. Th. Rel. Fields,* **74** (1987), 289–294.
215. R. Léandre, Dévelopement asymptotique de la densité d'une diffusion dégénérée, to appear in *J. Funct. Anal.*
216. P. Lévy, Random functions: General theory with special reference to Laplacian random functions, *Univ. of California Publ. in Statistics,* **I** (1953), 331–388.
217. H.P. McKean and I.M. Singer, Curvature and eigenvalues of the Laplacian, *J. Diff. Geom.*, **1** (1967), 43–69.
218. P.A. Meyer, Quelques resultats analytiques sur le semigroupe d'Ornstein-Uhlenbeck en dimension infinie, *Theory and applications of random fields, Proc.* IFIP-WG 7/1 *Working Conference,* (ed. by G. Kallianpur), *Lecture Notes in Contr. Inform.*, **49**, 201–214, Springer-Verlag, Berlin, 1983.
219. M. Motoo and S. Watanabe, On a class of additive functionals of Markov processes, *J. Math. Kyoto Univ.*, **4** (1964–65), 429–469.
220. E. Nelson, The free Markov field, *J. Funct. Anal.*, **12** (1973), 217–227.
221. J. Neveu, Sur l'esperance conditionnelle par rapport à un movement brownien, *Ann. Inst. Henri Poincaré,* **B XII**, (1976), 105–110.
222. J. Norris, Simplified Maliavin calculus, Seminaire de Prob. XX 1984/85 (ed. par J. Azema et M. Yor), *Lect Notes in Math.*, **1204**, 101–130, Springer-Verlag, Berlin, 1986.
223. V.K. Patodi, Curvature and eigenforms of the Laplace operator, *J. Diff. Geom.*, **5** (1971), 233–249.
224. I. Shigekawa and N. Ueki, A stochastic approach for the Riemann-Roch theorem, *Osaka J. Math.*, **25** (1988), 759–784.
225. H. Sugita, Sobolev spaces of Wiener functionals and Malliavin's calculus, *J. Math. Kyoto Univ.*, **25** (1985), 31–48.
226. H. Sugita, On a characterization of the Sobolev spaces over an abstract Wiener space, *J. Math. Kyoto, Univ.*, **25** (1985), 717–725.
227. Y. Takahashi and S. Watanabe, The probability functionals (Onsager-Machlup functions of diffusion processes, *Stochastic integrals* (ed. by D. Williams) *Lecture Notes in Math.*, **851**, 433–463, Springer-Verlag, Berlin, 1981.
228. S. Takanobu, Diagonal short time asymptotics of heat kernels for certain degenerate second order differential operators of Hörmander type, *Publ. RIMS, Kyoto Univ.*, **24** (1988), 169–203.
229. S. Takanobu and S. Watanabe, On the existence and uniqueness of diffusion processes with Wentzell's boundary conditions, *J. Math. Kyoto Univ.*, **28** (1988), 71–80.
230. S. Taniguchi, Malliavin's stochastic calculus of variations for manifold-valued Wiener functionals and its applications, *Z. Wahrsch, verw. Gebiete,* **65** (1983), 269–290.
231. H. Uemura, On a short time expansion of the fundamental solution of heat equations by the method of Wiener functionals, *J. Math. Kyoto Univ.*, **27** (1987), 417–431.

232. S. Watanabe, *Lectures on stochastic differential equations and Malliavin calculus*, Tata Institute of Fundamental Research/Springer, 1984.
233. S. Watanabe, Analysis of Wiener functions (Malliavin calculus) and its applications to heat kernels, *Ann. Probab.*, **15** (1987), 1–39.
234. S. Watanabe, Construction of semimartingales from pieces by the method of excursion point processes, *Ann. Inst. Henri Poincare*, Sup. au n° 2, **23** (1987), 297–320.
235. D. Williams, Markov properties of Brownian local time, *Bull. AMS*, **75** (1969), 1035–1036.
236. M. Yor, Remarques sur une formule de Paul Lévy, *Séminaire de Prob.* XIV, 1978/79, (éd. par J. Azéma et M. Yer), *Lect. Notes in Math.*, **784**, 343–346, Springer-Verlag, Berlin, 1980.

Index